T0181651

Graduate Texts in Physics

Graduate Texts in Physics

Graduate Texts in Physics publishes core learning/teaching material for graduate- and advanced-level undergraduate courses on topics of current and emerging fields within physics, both pure and applied. These textbooks serve students at the MS- or PhD-level and their instructors as comprehensive sources of principles, definitions, derivations, experiments and applications (as relevant) for their mastery and teaching, respectively. International in scope and relevance, the textbooks correspond to course syllabi sufficiently to serve as required reading. Their didactic style, comprehensiveness and coverage of fundamental material also make them suitable as introductions or references for scientists entering, or requiring timely knowledge of, a research field.

More information about this series at http://www.springer.com/series/8431

Karl F. Renk

Basics of Laser Physics

For Students of Science and Engineering

Second Edition

With 344 Figures

Karl F. Renk
Institut für Angewandte Physik
Universität Regensburg
Regensburg
Germany

ISSN 1868-4513 ISSN 1868-4521 (electronic)
Graduate Texts in Physics
ISBN 978-3-319-84453-4 ISBN 978-3-319-50651-7 (eBook)
DOI 10.1007/978-3-319-50651-7

Printed on acid-free paper

This Springer imprint is published by Springer Nature
The registered company is Springer International Publishing AG
The registered company address is: Gewerbestrasse 11, 6330 Cham, Switzerland

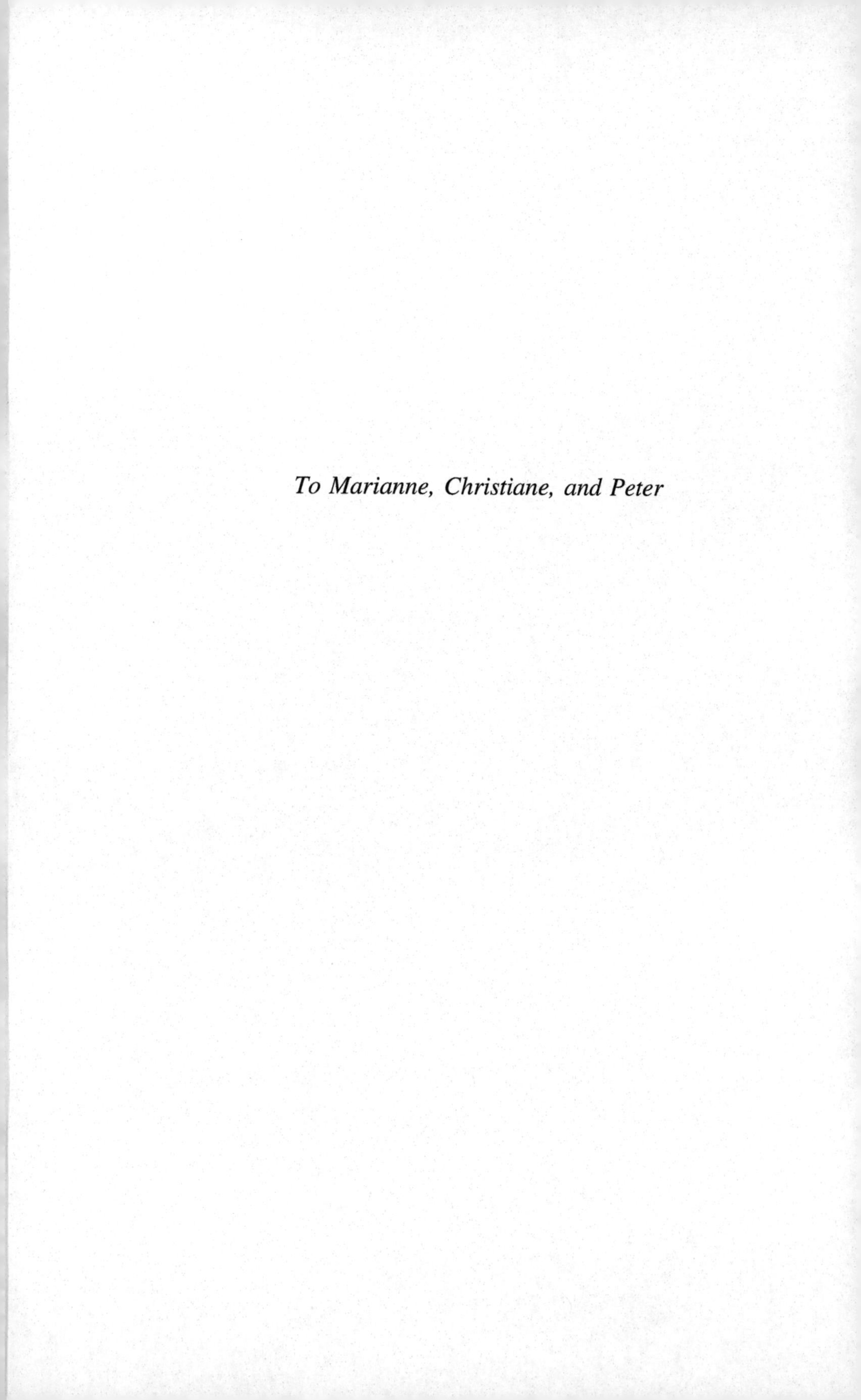

To Marianne, Christiane, and Peter

Preface to the Second Edition

Observation of a gravitational wave is the most spectacular recent application of a laser (published in Physical Review Letters, February 2016). Four Nobel Prizes in the last four years for achievements in physics and chemistry (see Sect. 1.9 of the book) demonstrate the significance of lasers for scientific research. There is a steady development of lasers and of their use in scientific research in physics, chemistry, engineering, biophysics, medicine, and technical applications. Important progress has been made in the last years in the development and application of infrared and far-infrared free-electron lasers, and of X-ray free-electron lasers. X-ray free-electron lasers are opening new possibilities in scientific research and in application.

The first edition of the textbook *Basics of Laser Physics* presented a *modulation model of a free-electron laser*, illustrating dynamical processes in a free-electron laser. The second edition gives a modified treatment of the model. The model provides analytical expressions for the gain and for the saturation field of radiation in a free-electron laser. The results drawn from the modulation model are consistent with the results of theory that is based on Maxwell's equations; main results of theory arise from numerical solutions of Maxwell's equations. In accord with the modulation model is a description of the active medium of a free-electron laser as a quantum system, already discussed in the first edition: an electron, which performs an oscillation in a spatially periodic magnetic field, may be describable as an electron occupying an energy level of an energy-ladder system; accordingly, electronic transitions between the energy levels are origin of spontaneous and stimulated emission of radiation.

In order to stress features that are common to a conventional laser and a free-electron laser or show differences, various points are clearly structured in the new edition, such as the role of dephasing between a radiation field and an oscillator or Lorentzian-like functions (denoted as "Lorentz functions") describing frequency dependences of gain near or outside resonances. The second edition contains additionally: classical oscillator model of a laser (van der Pol equation of a laser); onset of laser oscillation of a titanium–sapphire laser; discussion of differences between a conventional laser and a free-electron laser; and a modification of the description of the yet hypothetical Bloch laser.

Additional problems should provide a deepening of the understanding of lasers. Furthermore, errors are corrected. The principle of the overall representation remains unchanged: this book is designed in a way that a student can study many of the chapters without special knowledge of the preceding chapters. In most chapters, the content develops from a more general aspect to specific aspects. Let me mention a particular point-concerning notation. I am using, besides the letter N for the number of particles per unit volume, the letter Z for the number (=*Zahl*, German) of photons per unit volume, instead of common combinations of a Latin and a Greek letter, or of an upper- and a lower-case letter.

I am indebted to Manfred Helm for a number of very helpful comments to the first edition and Joachim Keller for discussions of basic questions concerning the free-electron laser. I would like to thank Sergey Ganichev, Rupert Huber, Alfons Penzkofer, Willli Prettl, and Stephan Winnerl for discussions. It is a pleasure to acknowledge encouragement by Claus Ascheron and the friendly collaboration with Adelheid Duhm and Elke Sauer at Springer Verlag. I thank Sameena Begum Khan and her production team at Springer Verlag for the commitment to the preparation of the book. Finally, I would like to thank my wife Marianne for her patience.

Regensburg, Germany Karl F. Renk

Contents

Part I General Description of a Laser and an Example

Part I
General Description of a Laser and an Example

Chapter 1
Introduction

We will ask and partly answer a few questions. What is the difference between a laser and a light bulb? In which frequency ranges are lasers available? Which are the sizes and the costs of lasers? Why is it necessary to have different types of lasers in the same frequency range? We will also mention some specific lasers and we will discuss the concept of the book.

1.1 Laser and Light Bulb

The *spatial and temporal* coherence makes the difference between a laser and a light bulb (Fig. 1.1). While a lamp emits uncorrelated wave trains into all spatial directions, a laser generates coherent waves and the waves can have a high directionality. Which are the possibilities of generation of spatially and temporally coherent waves? A laser can generate a coherent continuous wave or a coherent pulse train. Extreme cases of generation of visible radiation are as follows:

- The *continuous wave laser* (cw laser) emits a continuous electromagnetic wave. The field is spatially and temporally coherent.
- The *femtosecond laser* emits an electromagnetic wave consisting of a pulse train; the duration of a single pulse of a train can be as short as 5 fs (1 fs = 1 femtosecond $= 10^{-15}$ s). The field of a pulse train is spatially and temporally coherent too.

Besides continuous wave lasers and femtosecond lasers, there are pulsed lasers producing laser pulses with durations in the picosecond, nanosecond, microsecond, or millisecond ranges. We use the abbreviations:

- 1 ms = 1 millisecond $= 10^{-3}$ s
- 1 μs = 1 microsecond $= 10^{-6}$ s
- 1 ns = 1 nanosecond $= 10^{-9}$ s
- 1 ps = 1 picosecond $= 10^{-12}$ s

K.F. Renk, *Basics of Laser Physics*, Graduate Texts in Physics,
DOI 10.1007/978-3-319-50651-7_1

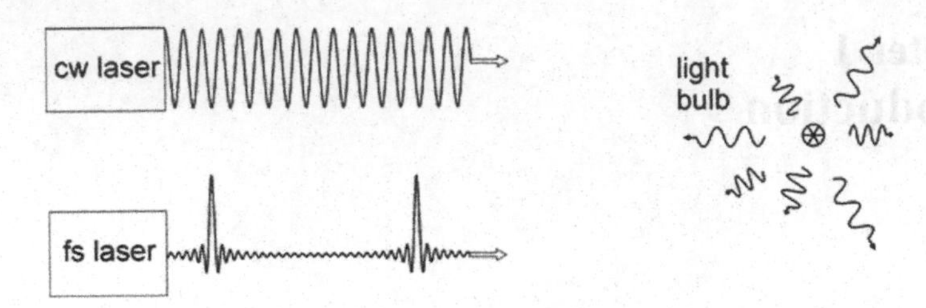

Fig. 1.1 Continuous wave (*cw*) laser, femtosecond (*fs*) laser and light bulb

- 1 fs = 1 femtosecond = 10^{-15} s
- 1 as = 1 attosecond = 10^{-18} s

The acronym LASER means: *L*ight *A*mplification by *S*timulated *E*mission of *R*adiation. It developed to *laser* = device for generation of coherent electromagnetic waves by stimulated emission of radiation. The *maser* (=*microwave laser*) makes use of microwave amplification by stimulated emission of radiation.

1.2 Spectral Ranges of Lasers and List of a Few Lasers

Figure 1.2 shows wavelengths and frequencies of spectral ranges of the electromagnetic spectrum—from X-rays over the ultraviolet (UV), the visible, the near infrared (NIR), the far infrared (FIR) spectral ranges to microwaves and radiowaves. The frequency ν of an electromagnetic wave in vacuum obeys the relation

$$\nu = c/\lambda, \tag{1.1}$$

where c (= $3 \times 10^8\,\mathrm{m\,s^{-1}}$) is the speed of light and λ the wavelength. Abbreviations of frequencies are as follows:

- 1 MHz = 1 megahertz = 10^6 Hz
- 1 GHz = 1 gigahertz = 10^9 Hz
- 1 THz = 1 terahertz = 10^{12} Hz
- 1 PHz = 1 petahertz = 10^{15} Hz

The visible spectral range corresponds to a frequency range of about 430–750 THz (wavelength range about 400–700 nm). Optics and light refer to electromagnetic waves with vacuum wavelengths smaller than about 1 mm, i.e., with frequencies above 300 GHz. Lasers are available in the ultraviolet, visible, near infrared, far infrared, and microwave regions. Lasers of the range of X-rays are being developed. The spectral ranges in which lasers are available extend from the GHz range over the THz range to the region above 1,000 THz.

The ancient Greeks understood $\mu\epsilon\gamma\alpha$ (mega) as something that was exceeding all measurable things, $\gamma\iota\gamma\alpha$ (giga) had to do with the giants, $\tau\epsilon\rho a$ (tera) included

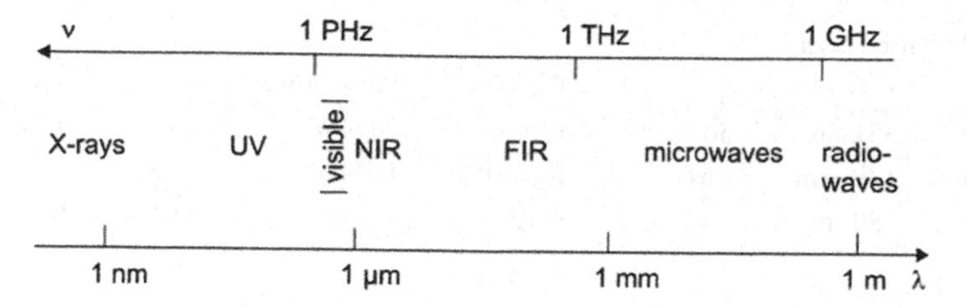

Fig. 1.2 Spectral ranges of lasers

Table 1.1 Laser wavelengths, frequencies, and quantum energies

Laser	λ	ν(THz)	$h\nu(10^{-19}$ J)	P_{out}
HeNe	633 nm	474	3.1	1–10 mW
CO_2	10.6 μm	28	0.18	1 W to 1 kW
Nd:YAG	1.06 μm	283	1.9	2 W
TiS	830 nm	360	2.4	100 mW to 5 W
Fiber	1.5 μm	200	1.3	1 W
Semiconductor	840 nm	357	2.4	10–100 mW
QCL	5 μm	60	0.25	10–100 mW

their gods, and πετα (peta) was the largest one could imagine—world, giants, gods, and all spheres together. The notation "terahertz" was introduced shortly after the discovery of the helium–neon laser, which emits coherent radiation at a frequency of 474 THz (wavelength 633 nm).

Table 1.1 shows data of a few continuous wave lasers. The data concern: λ = laser wavelength; ν = laser frequency; $h\nu$ = quantum energy of the photons of a laser field (= photon energy); $h = 6.6 \times 10^{-34}$ J s; P_{out} = output power.

- *Helium–neon laser* (HeNe laser). It generates red laser light of a power in the milliwatt range. Helium–neon lasers emitting radiation at other wavelengths are also available.
- *CO_2 laser* (carbon dioxide laser). It produces infrared radiation of high power at wavelengths around 9.6 and 10.6 μm.
- *Neodymium YAG laser* (Nd:YAG laser; YAG = yttrium aluminum garnet). The laser is a source of near infrared radiation (wavelength 1.06 μm).
- *Titanium–sapphire laser* (TiS laser). The laser operates as a continuous wave laser or as a femtosecond laser. The cw titanium–sapphire laser is tunable over a very wide spectral range (650–1080 nm).
- *Fiber laser.* Fiber lasers (= lasers with glass fibers doped with rare earth ions) operate in the wavelength range of about 0.7–3 μm.
- *Semiconductor laser*. Semiconductor lasers (more accurately: bipolar semiconductor lasers) are available in the entire visible, the near UV, and the near infrared. The

Table 1.2 Pulsed lasers

Laser	λ	t_p	W_p	Pulse power	ν_{rep}	P_{av}
Excimer	351 nm	50 ns	1 J	20 MW	10 Hz	10 W
Nd:YAG	1.06 μm	6 ns	100 mJ	16 MW	100 Hz	10 W
TiS	780 nm	10 fs	10 nJ	1 MW	50 MHz	0.5 W

wavelength and the power (from the nW range to the 100 mW range) of radiation generated by a semiconductor laser depend on its design. A stack of semiconductor lasers can produce radiation with a power up to the kW range.
- *Quantum cascade laser* (QCL). A QCL is a type of semiconductor laser that produces radiation in the infrared or in the far infrared. The laser wavelength of a quantum cascade laser depends on its design.

Table 1.2 shows data of a few pulsed lasers: t_p = pulse duration = halfwidth of a pulse on the time scale = FWHM = full width at half maximum; W_p = energy of radiation in a pulse = pulse energy; pulse power = W_p/t_p; ν_{rep} = repetition rate; P_{av} = average power.

- *Excimer laser.* It is able to produce UV radiation pulses of high pulse power; the wavelength given in the table is that of a laser operated with XeF excimers. Excimers with other materials generate radiation at other wavelengths (XeCl, λ = 308 nm; KrF, 248 nm; ArF, 193 nm).
- *Neodymium YAG laser.* Depending on the design of a pulsed neodymium YAG laser, the pulse duration can have a value between 5 ps or a value that is larger than that given in the table. The average power can be larger than 10 W.
- *Titanium–sapphire femtosecond laser.* The power is large during very short time intervals.

A laser system, consisting of a laser oscillator and a laser amplifier, can generate radiation pulses of much larger pulse power levels (Sect. 16.8).

1.3 Laser Safety

Laser safety has to be taken very seriously: a laser emitting visible radiation of 1 mW power leads to a power density in the focus (area λ^2) of a lens—for instance, in the focus of the lens of an eye—of the order of 10^9 W m^{-2} (10^5 W cm^{-2}). Such a power density can lead to damage of an eye. Dear reader, please take care of the corresponding safety rules when you experiment with a laser!

1.4 Sizes of Lasers, Cost of Lasers, and Laser Market

There are lasers of very different size.

- A gas laser or a solid state laser has a typical length of 1 m (down to 10 cm). The price of a laser is between 100 US dollars and 1 million dollars.
- A free-electron laser has a typical length of 10 m (not taking account of a much larger accelerator). The price of a free-electron laser lies, depending on its properties, between 10 million and billions of dollars.
- The smallest lasers are semiconductor lasers with sizes ranging from about 1 mm to smaller sizes. Microlasers with dimensions of the order of 10 μm can be fabricated; nanolasers—with extensions below 1 μm—may be suitable for special applications. Mass production (at a price of 10 dollars per laser or much less) resulted in a great variety of applications of semiconductor lasers.

The *laser market* (Fig. 1.3a) is strongly growing. The development may be similar as for the computer market. After the discovery of the transistor in 1946, it took about 50 years until the transistor became widely distributed—as the essential basis of a computer. The main breakthrough was due to miniaturization realized in the microelectronics and due to integration of transistors in large systems. The laser, with its first operation 14 years after the transistor, is beginning to be widely spread

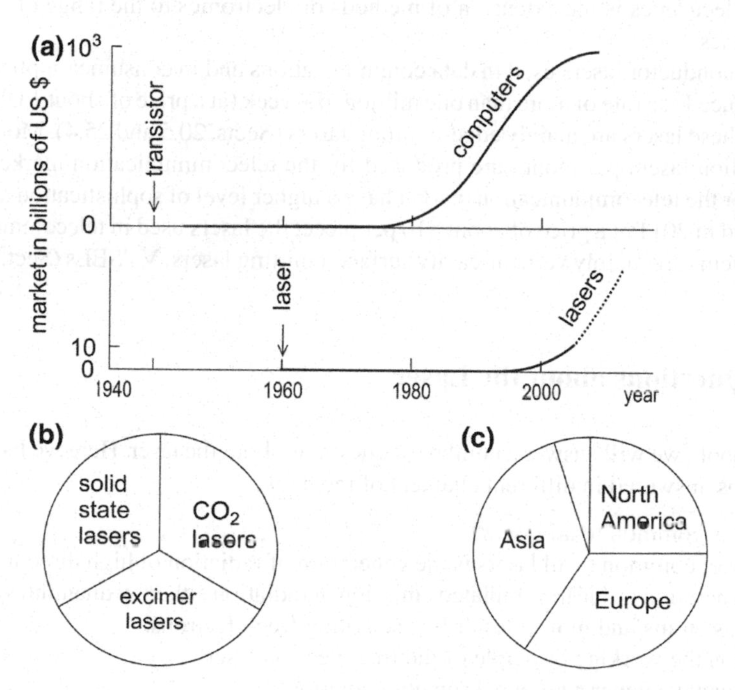

Fig. 1.3 Laser market. **a** General development. **b** Industrial lasers (in 2010). **c** Places of installation

as a part of devices of the daily life. The integration of the lasers in other devices and in large systems became possible by the development of the semiconductor lasers and their miniaturization.

The laser market offers a large variety of different lasers designed for particular applications. The laser field is in a rapid development; improvements of laser designs, new types of lasers, and new applications make the field strongly growing. We mention here the *industrial lasers*, machines suitable for various applications. In 2009, the main contributions to the turnover in the market of industrial lasers (Fig. 1.3b) came from the CO_2 lasers, the excimer lasers, and the solid state lasers (including a small portion of semiconductor lasers). Among the solid state lasers, there are different types, namely rod lasers, disk lasers, and fiber lasers. Industrial lasers find use in materials processing—cutting, welding, marking, engraving, and microprocessing; today (2016), disk lasers reach high output power levels and are developing as simple powerful industrial lasers. A main application of excimer lasers concerns structuring of semiconductors. The overall turnover of industrial lasers was about nine billion dollars in 2010. Most installations of industrial lasers (Fig. 1.3c) are in Asia, Europe, and North America.

Lasers are the basis of photonics (=photoelectronics) and optics. *Optoelectronics*—the counterpart at optical frequencies to electronics at radio and microwave frequencies—and *integrated optics* refer to optical systems used in optical communications, signal processing, sensing with radiation, and other fields. A characteristic of optoelectronics is the extension of methods of electronics to the range of optical frequencies.

Semiconductor lasers used in data communications and in consumer applications are produced at a rate of more than one million in a week (at a prize of about 1 US $ per piece); these lasers are mainly edge-emitting lasers (Sects. 20.5 and 25.4). More than one million lasers per month are produced for the telecommunication market. The lasers for the telecommunication market have a higher level of sophistication and are produced in 2011 at a price of about $10 per piece; the lasers used in telecommunication systems are mainly vertical-cavity surface-emitting lasers, VCSELs (Sect. 22.7).

1.5 Questions about the Laser

In this book, we will answer a number of questions about the laser. Here we list some questions answered in different chapters of the book.

- What is common to all lasers?
  Answer: common to all lasers is the generation of radiation of high directionality; the generation is due to stimulated emission of radiation either by quantum systems such as atoms and molecules or by oscillating free-electrons.
- What is the working principle of the free-electron laser?
- How can we generate monochromatic radiation?

- How can we generate femtosecond pulses?
- What is the role of diffraction in a laser? We will see that diffraction plays an important and favorable role: diffraction can suppress unwanted radiation.
- What is the angle of divergence of laser radiation? The angle of divergence is in general not determined by diffraction but by a kind of natural beams—Gaussian beams—that fit perfectly to resonators with two spherical mirrors. A laser is able to generate a Gaussian beam.
- How can we produce laser radiation in different ranges of the electromagnetic spectrum?
- What is the difference between a laser and a classical oscillator?

Laser physics connects optics with atomic physics, molecular physics, solid state physics (including semiconductor physics), and, of course, quantum mechanics, and furthermore with engineering, chemistry, biology, and medicine.

1.6 Different Types of Lasers in the Same Spectral Range

Why do we need different types of lasers for the same spectral range? Different types of lasers fulfill different tasks.

- If we need lasers of CD (compact disk) or blue ray players, semiconductor lasers, with small sizes and low power consumption, fulfill the task.
- To cut metal plates, a high power laser as the CO_2 laser is suitable. The efficiency of conversion of electric power to radiation power of a CO_2 laser is large (larger than 10%). The CO_2 laser emits radiation in the infrared spectral region. High power disk lasers (pumped with semiconductor lasers) emitting near infrared radiation have also high efficiencies for conversion of electric energy to radiation. These lasers may become able (2016) to compete with CO_2 lasers.
- To generate femtosecond optical pulses, with durations from 100 to 5 fs, only few of the many lasers have appropriate properties. The most prominent femtosecond laser is the titanium–sapphire femtosecond laser.

Large progress came with the miniaturized semiconductor lasers but also with the high-power semiconductor lasers—that can be applied, for example, as pump sources of other lasers.

1.7 Concept of the Book

Figure 1.4 gives a survey of the main topics treated in the book.

General description of a laser and an example. We will describe main properties of a laser and of the components, namely the active medium and the resonator. We will introduce the laser as an oscillator: an active medium drives the oscillation of

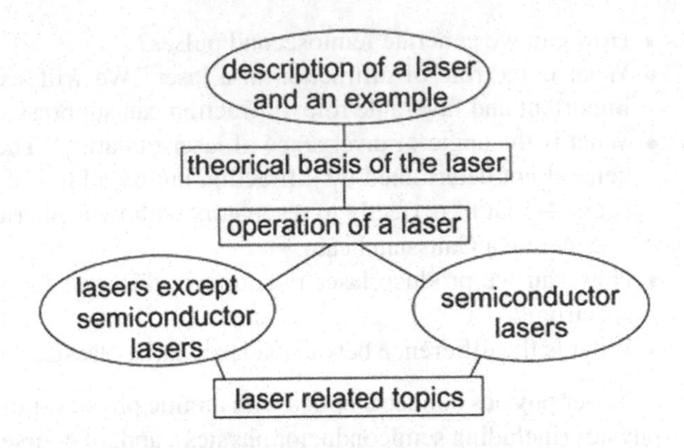

Fig. 1.4 Concept of the book

an electromagnetic field in a resonator. Early in the book we will discuss a particular laser (the titanium–sapphire laser) in some detail. This allows us to be specific, if necessary, during a treatment of the theory and thc discussion of the operation of a laser.

Theory of the laser. To describe the interaction of light with matter, we introduce the Einstein coefficients. A theoretical treatment of the laser oscillation yields the laser threshold condition and other important properties.

Operation of a laser. We will mention different techniques of operation of a laser as a continuous wave laser or as a pulsed laser. We will begin this part with a treatment of the properties of resonators and the description of Gaussian waves.

Lasers except semiconductor lasers

- Gas lasers. The active medium consists of atoms, ions, or molecules in gases. Gas lasers are available in the UV, visible, NIR, FIR, and microwave ranges. Two of the most important industrial lasers (the excimer and the CO_2 laser) are gas lasers.
- Solid state lasers (except semiconductor lasers). The active medium consists of ions in a dielectric solid; the solid is a host for ions. Solid state lasers, operated at room temperature, are available in the visible and the near infrared. Stimulated transitions between electronic states of ions give rise to generation of laser radiation. Besides crystals, other condensed matter materials—glasses, polymers, and liquids—are also suitable as host materials of ions, atoms, or molecules.
- Free-electron lasers. The basis is the emission of radiation by oscillating free-electrons. The electrons are passing at a velocity near the speed of light through a spatially periodic magnetic field. Free-electron lasers are available in the visible, infrared, and far infrared; free-electron lasers generating X-rays are being developed.

Semiconductor lasers (bipolar semiconductor lasers and quantum cascade lasers). Semiconductor lasers are solid state lasers that use conduction electrons in semiconductors. Semiconductor lasers are available in the visible, near UV, and NIR spectral ranges and are being developed for the FIR. Stimulated transitions are either due to electronic transitions between the conduction band and the valence band of a

semiconductor—in bipolar lasers—or between subbands of a conduction band—in quantum cascade lasers. Preparation of mixed semiconductor materials and of heterostructures makes it possible to realize new, artificial materials that are used in quantum well, quantum wire, quantum dot, and quantum cascade lasers. The laser wavelength is adjustable through an appropriate design of a heterostructure. We will, furthermore, present the idea of a *Bloch laser* (=superlattice Bloch laser=Bloch oscillator) that may become suitable for generation of FIR radiation.
Laser-related topics

- Optical communications. This is an important field of applications of semiconductor lasers.
- Light emitting diode (LED). The LED is the basis of many different kinds of illumination. The development of LEDs is going on in parallel to the development of semiconductor lasers. The organic LED (OLED) is suited to realize simple large area light sources.
- Nonlinear Optics. We will give a short introduction to the field of Nonlinear Optics. Our main aspect will be: how can we convert coherent laser radiation of one frequency to coherent radiation of other frequencies?

We will discuss various applications in connection with different topics.

1.8 References

References cited either at the end of a chapter or in the text include: textbooks on lasers; textbooks on optoelectronics and integrated optics; books on lasers and nonlinear optics; textbooks on other fields (optics, electromagnetism, atomic and molecular physics, quantum mechanics, solid state physics; microwave electronics, mathematical formulas; and to a small extent original literature). Original literature about lasers is well documented in different textbooks on lasers [1–11, 308]. Introductions to quantum optics are given, for instance, in [12, 13], or, to quantum electronics [309]. In connection with mathematical functions, *see*, for instance, [14–20].

1.9 A Remark About the History of the Laser

Data concerning the history of lasers

1865 James Clerk Maxwell (King's College, London): Maxwell's equations.
1888 Heinrich Hertz (University of Karlsruhe): generation and detection of electromagnetic waves.
1900 Max Planck (University of Berlin): quantization of radiation in a cavity.
1905 Albert Einstein (Patent office Bern): quantization of radiation.

1905 Niels Bohr (University of Copenhagen): quantization of the energy states of an atom.
1917 Einstein (then in Berlin): interaction of radiation with an atom, spontaneous and stimulated emission.
1923 Henryk A. Kramers: influence of stimulated emission on the refractive index of atomic gases containing excited atoms (a theoretical study).
1928 Rudolf Ladenburg (Kaiser Wilhelm Institute, Berlin): observation of an influence of stimulated emission on the refractive index of a gas of neon atoms excited by electron collisions in a gas discharge.
1951 Charles H. Townes (Columbia University): idea of a maser.
1954 Townes: ammonia maser (frequency 23.870 GHz, wavelength 1.25 cm); Nicolai Basov, Aleksandr Prokhorov (Lebedev Physical Institute, Moscow): idea of a maser in parallel to the development in the USA and realization of an ammonia laser.
1956 Nicolaas Bloembergen (Harvard University): proposal of the three-level maser (leading to solid state masers in various laboratories).
1958 Arthur L. Schawlow, Townes: proposal of infrared and optical masers (lasers) including the formulation of the threshold condition of laser oscillation; Prokhorov: general description of the principle of optical masers (lasers).
1959 Basov: proposal of the semiconductor laser.
1960 Theodore Maiman (Hughes Research Laboratories): ruby laser (694 nm).
1960 Ali Javan (Bell laboratories): helium-neon laser (1.15 μm, later 633 nm).
1961 L. F. Johnson, K. Nassau (Bell Laboratories): neodymium YAG laser.
1962 Robert N. Hall (General Electric Research Laboratories): semiconductor laser.
1963 Herbert Kroemer (University of California Santa Barbara): proposal of the heterostructure laser.
1964 C. Kumar N. Patel (Bell Laboratories): carbon dioxide laser; W. Bridges (Bell Laboratories): argon ion laser.
1966 Peter P. Sorokin (IBM Yorktown Heights) and Fritz P. Schäfer (Max-Planck-Institut für Biophysikalische Chemie, Göttingen): dye laser.
1968 William T. Silfvast (Bell Laboratories): metal vapor laser.
1975 Basov: excimer laser.
1977 John Madey, Luis Elias and coworkers (Stanford University): free-electron laser.
1979 J. C. Walling (Allied Chemical Corporation): alexandrite laser (first tunable solid state laser).
1982 P. Moulton (Schwartz Electro-Optics): titanium–sapphire laser.
1991 M. Haase and coworkers (3M Photonics): green diode laser (based on ZnSe).
1994 Federico Capasso, Jérome Faist, and coworkers (Bell laboratories): quantum cascade laser.
1997 Shuji Nakamura (Nichia Chemicals, Japan): blue diode laser (based on GaN).

For references concerning the history of lasers, *see* [20–24] and also Sects. 9.10 and 19.13. The acronym laser was introduced by Gordon Gould (Columbia University) at the Ann Arbor Conference on Optical Pumping in 1959. In his thesis (Moscow 1940—unknown until 1959) Vladimir Fabrikant discussed amplification of optical radiation by stimulated emission of radiation.

In 1951, Charles Townes, searching for an oscillator generating microwave radiation at higher frequencies than other microwave oscillators (magnetron, klystron) available at the time, had the idea of a maser [21]—based on three aspects (*see* Sect. 2.1): stimulated emission of radiation by an atomic system; creation of a population inversion (in a molecular beam); feedback of radiation by use of a resonator. The realization of the first maser (1954) stimulated the development of other types of masers, particularly of the solid state three-level maser. In 1958, Schawlow and Townes published an article on "infrared and optical masers." This paper described the conditions of operation of a laser (Chap. 8) and initiated the search for a concrete laser. Maiman was the first to operate a laser, the ruby laser, in May 1960. Later in the year, Javan reported operation of a helium–neon laser.

The application for a laser patent in mid-1958 by Bell Laboratories, with Schawlow (at Bell Laboratories) and Townes (Columbia University) as inventors, led to the first US patent on a laser, issued in 1960. A student, Gordon Gould, then working in a group at Columbia University on his PhD thesis, wrote in several notebooks (from 1958 on after the circulation of preprints of the 1958 paper of Schawlow and Townes) ideas about lasers, which later were the basis of patent applications. After many court cases, Gould succeeded to obtain patents on various aspects of lasers. Since it took a very long time to be issued (in 1976, 1978, 1988, and 1989), the patents allowed Gould and several companies he co-founded to get back the money (several tens of millions of dollars). For this purpose, laser companies were forced by further court cases to pay license fee.

The theoretical basis of the laser was the "old quantum mechanics" developed (1900–1917) by Planck, Bohr, and Einstein. The main results of the "old quantum mechanics" obtained a consequent founding by the quantum mechanics (developed 1925–1928). Why did it last about 40 years until maser and laser were operating? This question will be discussed in Sect. 9.13.

Nobel Prizes in the field of lasers

1964 Charles Townes, Nicolay Basov, Aleksandr Prokhorov: fundamental work in the field of quantum electronics, which has led to the construction of oscillators and amplifiers based on the maser-laser principle.
1966 Alfred Kastler: optical pumping.
1971 Dennis Gabor: holography.
1981 Nicolas Bloembergen, Arthur Schawlow: nonlinear optics and laser spectroscopy.
1989 Norman F. Ramsey: maser and atomic clocks.
1997 Steven Chu, Claude Cohen-Tannoudji, William D. Phillips: methods of cooling and trapping of atoms by using laser light.
1999 Ahmed Zewail (Chemistry): study of chemical reactions by using femtosecond laser pulses.

2000 Zhores Alferov, Herbert Kroemer: semiconductor heterostructures—the basis of lasers of high-speed opto-electronics.
2001 Wolfgang Ketterle, Eric Cornell, Carl Wieman: experimental realization of Bose Einstein condensation.
2002 John B. Fenn, Koichi Tanaka (chemistry); mass spectroscopic analysis of biomolecules by use of laser desorption.
2005 Roy Glauber: theory of coherence (the basis of modern quantum optics); Theodor Hänsch, John Hull: optical frequency analyzer.
2009 Charles K. Kao: ground braking achievements concerning the transmission of light in fibers of optical communications; Willard S. Boyle, George E. Smith: invention of an imaging semiconductor circuit—the CCD sensor.
2012 Serge Haroche, David Wineland: development of ground-breaking experimental methods that enable measurement and manipulation of individual quantum systems.
2014 Hiroshi Amano, Isamu Akasaki, Shuj Nakamura: invention of efficient blue light emitting diodes, which has enabled bright and energy-saving white light sources.
2014 Eric Belzig, Stefan Hell, William Moerner (Chemistry), development of super-resolved fluorescence microscopy.

References [1–25, 308, 309]

Problems

1.1 Physical constants. Remember numerical values of physical constants (in units of the international system, SI).

(a) c = speed of light.
(b) h = Planck's constant.
(c) $\hbar = h/(2\pi)$.
(d) e = elementary charge.
(e) m_0 = electron mass.
(f) μ_0 = magnetic field constant.
(g) ε_0 = electric field constant.
(h) k = Boltzmann's constant.
(i) N_A = Avogadro's number.
(j) R = gas constant.
(k) L_0 = Loschmidt's number.

[*Hint*: in examples in the text and in the Problems, an accuracy of several percent of a quantity is in most cases sufficient.]

1.2 Frequency, wavelength, wavenumber, and energy scale. It is helpful to characterize a radiation field on different scales: frequency ν; wavelength λ; wavenumber

$\tilde{\nu} = \nu/c = 1/\lambda$ (=number of wavelengths per unit of length = spatial frequency); and photon energy $h\nu$ in units of Joule or eV. Express each of the following quantities by the corresponding quantities on the three other scales.

(a) $\lambda = 1\,\mu\text{m}$.
(b) $\nu = 1\,\text{THz}$.
(c) $\lambda = 1\,\text{nm}$.
(d) $\tilde{\nu} = 1\,\text{m}^{-1}$.
(e) $h\nu = 1\,\text{eV}$.

1.3 Express the following values on different scales:

(a) kT for $T = 300\,\text{K}$ (T = temperature); (b) 1 meV; (c) $1\,\text{cm}^{-1}$; (d) $10\,\text{cm}^{-1}$.

1.4 Power of the sun light and of laser radiation. The intensity of the sun light on earth (or slightly outside of the atmosphere of the earth) is $1{,}366\,\text{W m}^{-2}$.

(a) Evaluate the power within an area of $1\,\text{cm}^2$.
(b) Estimate the power density (= intensity) if the radiation incident on a $1\,\text{cm}^2$ area is focused to an area of $100\,\mu\text{m}$ diameter; focusing to a smaller diameter is not possible because of the divergence (5 mrad) of the radiation from the sun.
(c) Determine the power density of the radiation of a helium-neon laser (power 1 mW, cross sectional area $1\,\text{cm}^2$) focused to an area that has a diameter of $1\,\mu\text{m}$.

1.5 Determine the time it takes light to propagate

(a) A distance that corresponds to the diameter of an atom.
(b) A distance of 1 cm.
(c) From a point of the surface of the earth to a point on the surface of the moon (at a distance of 174,000 km).

Chapter 2
Laser Principle

A laser (=laser oscillator) is a self-excited oscillator. A self-excited oscillator starts oscillation by itself and maintains an oscillation. Laser radiation is generated by stimulated transitions in an active medium. The active medium is a gain medium—propagation of radiation in the active medium results in an increase of the energy density of the radiation. The active medium in a laser experiences feedback from radiation stored in a laser resonator. A portion of radiation coupled out of the resonator represents the useful radiation.

In this chapter, we characterize an active medium by a population inversion in an ensemble of two-level atomic systems. We formulate the threshold condition of laser oscillation. We solve the resonator eigenvalue problem and find possible frequencies of laser oscillation. We also show that the buildup of steady state oscillation of a laser takes time—the oscillation onset time.

To describe a coherent electromagnetic wave, we make use of the model of a quasiplane wave—a parallel beam of coherent radiation. A quasiplane wave is characterized by a well-defined propagation direction, a finite spatial extension perpendicular to the propagation direction, and a constant field amplitude within the beam. The model is very useful for a basic description of the field in a laser oscillator. Later (in Chap. 11) we will introduce a modified description of a coherent electromagnetic wave.

In later chapters, we will specify the two-level atomic systems. A two-level atomic system can belong to various states: electronic states of an atom or an ion (in a gas, crystal, glass, or liquid); electronic, vibrational, or rotational states of a molecule; electronic states of electrons in a semiconductor or a semiconductor heterostructure.

We will introduce (Chap. 4) still another type of active medium—an active medium containing energy-ladder systems rather than two-level systems.

K.F. Renk, *Basics of Laser Physics*, Graduate Texts in Physics,
DOI 10.1007/978-3-319-50651-7_2

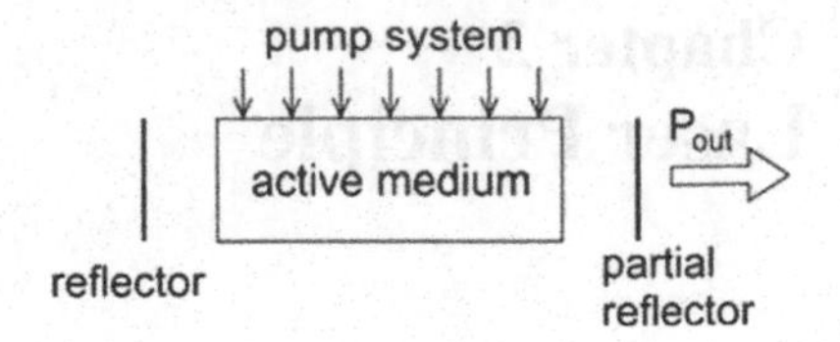

Fig. 2.1 A laser

2.1 A Laser

A laser (Fig. 2.1) emits coherent radiation of an output power P_{out}. A laser has the following parts.

- *Active medium* (=*gain medium* = *laser medium*). The active medium is able to amplify electromagnetic radiation. The active medium, located inside a resonator, fills out a resonator partly or completely.
- *Pump system*. It "pumps" the active medium. Methods of pumping are: optical pumping with another laser or a lamp; pumping with a gas discharge; pumping with a current through a semiconductor or a semiconductor heterostructure; chemical pumping.
- *Laser resonator*. The laser resonator has the task to store a coherent electromagnetic field and to enable the field to interact with the active medium—the active medium experiences feedback from the coherent field. We will describe resonators that consist of two mirrors—one is a reflector of a reflectivity R_1 near 1, and the other is a *partial reflector* serving as *output coupler*. The output coupling mirror has a reflectivity (R_2) that also can have a value near 1 but that can be much smaller; semiconductor lasers can have reflectors with $R_1 = R_2 \sim 0.3$. Each type of laser requires its own resonator design. There is a main criterion concerning reflectivities of resonators: a laser should be able to work at all. Depending on the task of a laser, other criteria can be chosen—for instance, that a laser should have optimum efficiency of conversion of pump power to laser output power.

2.2 Coherent Electromagnetic Wave

We describe a coherent electromagnetic wave generated by a continuous wave laser as a *quasiplane wave* (=parallel beam of coherent light),

$$E(z,t) = A\cos[\omega(t - t_0) - k(z - z_0)]. \quad (2.1)$$

E is the electric field at time t and location z, A is the amplitude, $\omega = 2\pi\nu$ the angular frequency, ν the frequency, and k the wave vector of the wave; t_0 defines a time coordinate and z_0 a spatial coordinate. The direction of E (and A) is perpendicular to the direction of the propagation direction (z direction). The dispersion relation,

$$\omega = ck, \tag{2.2}$$

relates the frequency and the wave vector. The quasiplane wave has a finite lateral extension. We suppose that the amplitude of the field does not vary, at a fixed z, over the cross section and that it is independent of z. The quasiplane wave is a section of a plane wave (which has infinite extensions in the plane perpendicular to the direction of propagation).

If we choose $t_0 = 0$ and $z_0 = 0$ to describe a quasiplane wave propagating in free space, we can write

$$E(z, t) = A\cos(\omega t - kz). \tag{2.3}$$

The instantaneous energy density, u_{inst}, in the electromagnetic field is

$$u_{\mathrm{inst}} = \varepsilon_0 A^2 \cos^2(\omega t - kz), \tag{2.4}$$

where ε_0 is the electric field constant. The *energy density* u of the electromagnetic field, that is, the instantaneous energy density averaged over a temporal period $T = 2\pi/\omega$, is equal to

$$u = \frac{1}{2}\varepsilon_0 A^2. \tag{2.5}$$

The quasiplane wave transports energy in z direction. The power P of the wave is

$$P = \frac{1}{2} c\varepsilon_0 A^2 a_1 a_2, \tag{2.6}$$

where $a_1 a_2$ is the cross-sectional area of a beam of rectangular shape. The *intensity* (= power per unit area = energy flux density) is

$$I = \frac{P}{a_1 a_2} = \frac{1}{2} c\varepsilon_0 A^2. \tag{2.7}$$

We interpret the transport of radiation energy as a flux of photons along the z direction and introduce the average number of photons per unit of volume, the *photon density* Z, by the relation

$$u = Zh\nu = Z\hbar\omega. \tag{2.8}$$

The energy density is equal to the photon density times the photon energy $h\nu$.

Simplifying further, we describe a light beam of laser light as a *parallel bundle of light rays* (Fig. 2.2). To characterize the propagation of light within a parallel light bundle, we introduce the *disk of light*. It is a section of a light bundle and has the length δz; we assume that δz is much larger than the wavelength of the radiation,

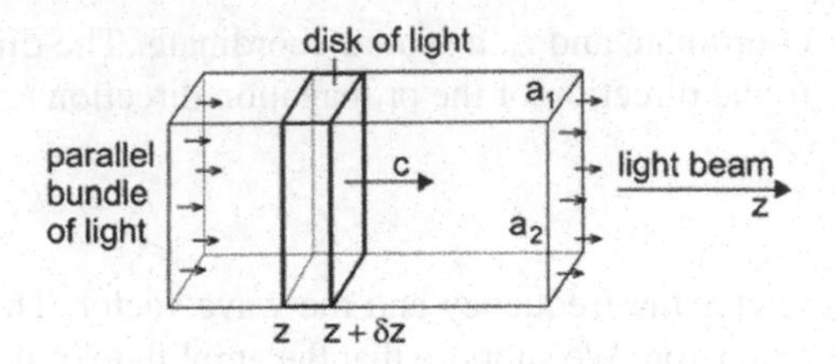

Fig. 2.2 Parallel light bundle and disk of light

$\delta z \gg \lambda$. The disk of light propagates along the z direction with the speed of light. The energy density in a disk of light is $u(\nu, z)$ and the photon density is equal to

$$Z(\nu, z) = \frac{u(\nu, z)}{h\nu}. \tag{2.9}$$

We will make use of the complex notation of the field. A complex field $\tilde{E}$ corresponds to a real field according to the relation

$$E = \mathrm{Re}[\tilde{E}] = \frac{1}{2}(\tilde{E} + \tilde{E}^*) = \frac{1}{2}\tilde{E} + c.c., \tag{2.10}$$

where $\mathrm{Re}[\tilde{E}]$ is the real part of $\tilde{E}$. The real field is equal to the sum of $\tilde{E}/2$ and its conjugate complex ($c.c.$) $\tilde{E}^*/2$.

The real part of a complex field, which is the product of a complex quantity $\tilde{A}$ and another complex quantity $\tilde{K}$, is

$$E = \mathrm{Re}[\tilde{E}] = \frac{1}{2}(\tilde{A}^*\tilde{K} + \tilde{A}\tilde{K}^*) = \frac{1}{2}\tilde{A}^*\tilde{K} + c.c. \tag{2.11}$$

The complex field $\tilde{E} = A\,\mathrm{e}^{\mathrm{i}(\omega t - kz)}$, with the real amplitude A, corresponds to the real field $E = A\cos(\omega t - kz)$. It follows that the energy density in an electromagnetic field is

$$u = \frac{\varepsilon_0}{2}\tilde{E}\tilde{E}^* = \frac{\varepsilon_0}{2}|\tilde{E}|^2 = \frac{\varepsilon_0}{2}A^2. \tag{2.12}$$

The photon density is given by

$$Z = \frac{\varepsilon_0}{2h\nu}\tilde{E}\tilde{E}^* = \frac{\varepsilon_0}{2h\nu}|\tilde{E}|^2 = \frac{\varepsilon_0}{2h\nu}A^2. \tag{2.13}$$

Accordingly, the amplitude of the field is

$$A = \sqrt{2h\nu Z/\varepsilon_0}. \tag{2.14}$$

More generally, we can characterize a quasiplane wave by

$$E = A\cos[\omega(t - t_0) - kz + \varphi_0], \tag{2.15}$$

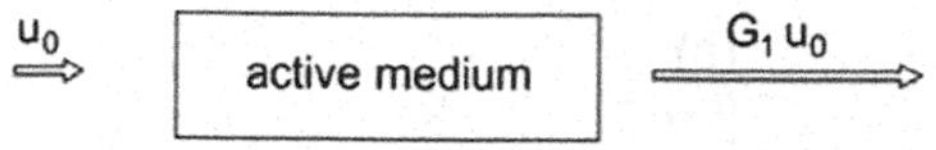

Fig. 2.3 Amplification of radiation in an active medium

where the time t_0 defines the time axis and φ_0 the z axis. The corresponding complex field is

$$\tilde{E} = A\, \mathrm{e}^{\mathrm{i}[\omega(t-t_0)-kz+\varphi_0]}. \tag{2.16}$$

If a wave propagates in a dielectric medium (dielectric constant ε), then ε_0 has to be replaced by $\varepsilon\varepsilon_0$ in the expressions concerning u, Z, A, and intensity.

The transit of coherent radiation through an active medium (Fig. 2.3) results in an increase of the photon density,

$$Z = G_1 Z_0, \tag{2.17}$$

and of the energy density,

$$u = G_1 u_0. \tag{2.18}$$

G_1 (>1) is the single-pass *gain factor*. Z_0 is the photon density and u_0 the energy density in the incident beam. Z is the photon density and u the energy density in the beam after passing through the active medium. We write

$$G_1 = \mathrm{e}^{\alpha L}, \tag{2.19}$$

where α is the *gain coefficient* of the active medium and L the length of the active medium. It follows that

$$\alpha = \frac{1}{L} \ln G_1 = \frac{1}{\ln 10} \frac{1}{L} \log G_1 = 0.43 \frac{1}{L} \log G_1. \tag{2.20}$$

The transit of radiation through an absorbing medium results in a decrease of energy density and of photon density,

$$u = \bar{G}_1 u_0, \tag{2.21}$$

$$Z = \bar{G}_1 Z_0, \tag{2.22}$$

where $\bar{G}_1 < 1$ is the absorption factor $(=Z/Z_0)$. We write

$$\bar{G}_1 = \mathrm{e}^{-\alpha_{\mathrm{abs}} L}. \tag{2.23}$$

α_{abs} is the *absorption coefficient* of a medium. It follows that

$$\alpha_{\mathrm{abs}} = \frac{1}{L} \ln \bar{G}_1 = 0.43 \frac{1}{L} \log \bar{G}_1. \tag{2.24}$$

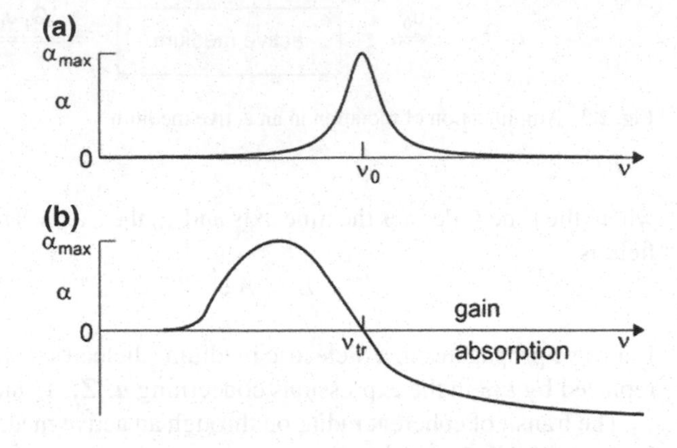

Fig. 2.4 Frequency dependence of the gain coefficient of an active medium. **a** Gain coefficient at frequencies around a resonance frequency. **b** Gain coefficient at frequencies around a transparency frequency

We can interpret the absorption coefficient as a negative gain coefficient; for units of gain, *see* Sect. 16.11.

An active medium can have a gain coefficient that has a maximum α_{max} at a resonance frequency ν_0 (Fig. 2.4a). The gain coefficient decreases toward smaller and larger frequencies and remains positive. However, an active medium can have a gain coefficient that changes sign (Fig. 2.4b). Such a medium has a transparency frequency ν_{tr}— the active medium is amplifying at frequencies $\nu < \nu_{tr}$ but absorbing at frequencies $\nu > \nu_{tr}$.

2.3 An Active Medium

An atom (or molecule) used in a laser has *two laser levels*, besides other energy levels. We ignore for the moment the other levels and describe an atom as a *two-level atomic system* (Fig. 2.5) and accordingly an ensemble of atoms as an *ensemble of two-level atomic systems*. We introduce the following notation.

- Level 2 (energy E_2) = upper laser level.
- Level 1 (energy E_1) = lower laser level.
- $E_{21} = E_2 - E_1$ = energy difference between the two laser levels = transition energy.
- N_2 = population of level 2 = density of excited two-level atomic systems = number density (number per unit volume) of excited two-level atomic systems.
- N_1 = population of level 1 = density of unexcited two-level atomic systems.
- $N_2 - N_1$ = *population difference*.
- $N_1 + N_2$ = density of two-level atomic systems.

Monochromatic electromagnetic radiation of frequency ν can interact with a two-level atomic system if Bohr's energy-frequency relation

$$h\nu = E_{21} = E_2 - E_1 \tag{2.25}$$

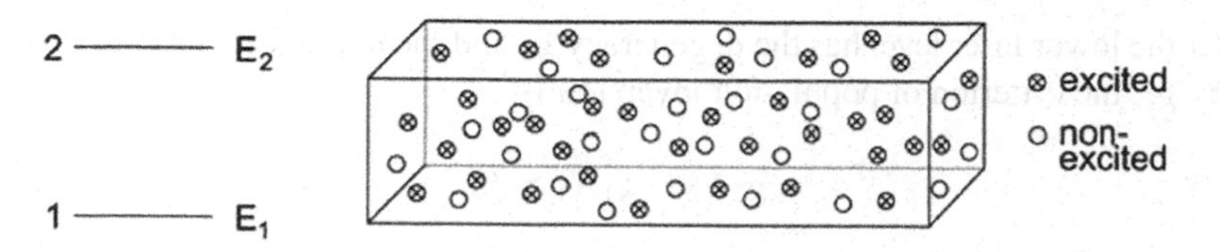

Fig. 2.5 Two-level atomic system and ensemble of two-level atomic systems

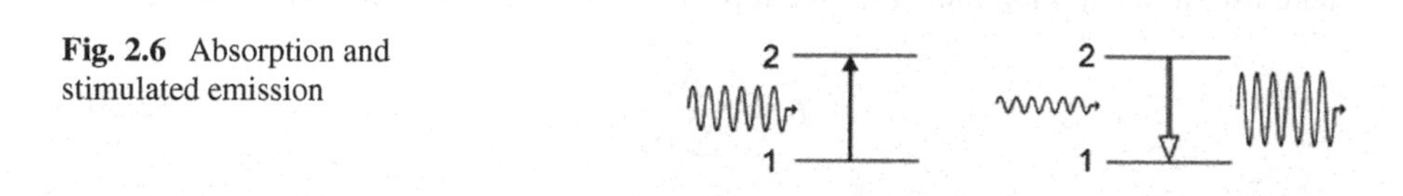

Fig. 2.6 Absorption and stimulated emission

holds, that is, if the photon energy $h\nu$ of the photons of the radiation field is equal to the energy difference E_{21}. But because of lifetime broadening of the upper level, a two-level atomic system can also interact if $h\nu$ is unequal to E_{21}.

Two processes are competing with each other in a laser, *absorption* and *stimulated emission* of radiation. In an absorption process (Fig. 2.6), a photon is converted to excitation energy of a two-level atomic system by a $1 \rightarrow 2$ transition. An excited two-level atomic system transfers by a stimulated emission process its excitation energy to the light field. Einstein showed: *Radiation created by stimulated emission has the same frequency, direction, polarization and phase as the stimulating radiation.*

If the active medium is an ensemble of two-level systems, the strength of stimulated emission is proportional to N_2, and the strength of absorption is proportional to N_1. We will later (Sect. 6.5) show, for an ensemble of identical two-level systems, that the factor of proportionality is the same for both processes. The net effect is proportional to the population difference $N_2 - N_1$. Stimulated emission prevails if $N_2 - N_1 > 0$ while absorption prevails if $N_2 - N_1 < 0$. In an active medium, the population difference $N_2 - N_1$ is larger than zero,

$$N_2 - N_1 > 0. \tag{2.26}$$

Alternatively, we can write:

$$N_2 > N_1; \tag{2.27}$$

in an active medium the population of the upper laser level is larger than the population of the lower laser level. Stimulated emission and absorption compensate each other if $N_2 = N_1$. In this case, the medium is transparent. We can formulate the *transparency condition*:

$$N_2 - N_1 = 0. \tag{2.28}$$

(If the lower laser level has the degeneracy g_1 and the upper laser level the degeneracy g_2, the criterion of population inversion is

$$g_2 N_2 - g_1 N_1 > 0. \tag{2.29}$$

We assume, for convenience, in the following that $g_1 = g_2$.)

It is useful, in particular with respect to the treatment of semiconductor lasers, to make use of *occupation numbers*. We introduce the *(relative) occupation number*

$$f_i = \frac{N_i}{\sum N_i}, \tag{2.30}$$

where N_i is the population of level i and $\sum N_i$ is the sum of the populations of all levels of an ensemble of atomic systems. The sum of the relative occupation numbers of an ensemble is unity, $\sum f_i = 1$. The relative occupation number f_i is equal to the probability that level i is occupied.

The relative occupation number of the upper laser level (Fig. 2.7) is equal to

$$f_2 = \frac{N_2}{N_1 + N_2}, \tag{2.31}$$

and the relative occupation number of the lower laser level is

$$f_1 = \frac{N_1}{N_1 + N_2}. \tag{2.32}$$

The sum of the relative occupation numbers is unity,

$$f_2 + f_1 = 1. \tag{2.33}$$

The *occupation number difference* (that is the difference between two probabilities) is

$$f_2 - f_1 = \frac{N_2 - N_1}{N_1 + N_2}. \tag{2.34}$$

The occupation number difference is the ratio of the population difference $N_2 - N_1$ and the density $N_1 + N_2$ of two-level atomic systems. Thus, the population difference is equal to the occupation number difference times the density of two-level atomic systems,

$$N_2 - N_1 = (f_2 - f_1)(N_1 + N_2). \tag{2.35}$$

Fig. 2.7 Relative occupation numbers of a two-level atomic system

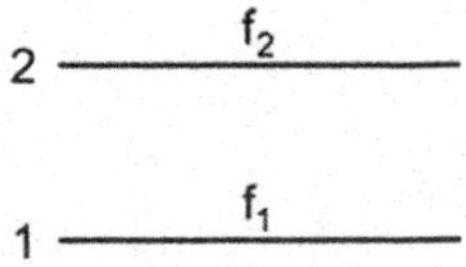

It follows that the strength of stimulated emission of radiation is proportional to f_2, and that the strength of absorption of radiation is proportional to f_1. The net effect—the difference between the strength of stimulated emission and absorption—is proportional to $f_2 - f_1$. Stimulated emission prevails if

$$f_2 - f_1 > 0, \tag{2.36}$$

while absorption prevails if $f_2 - f_1 < 0$. The condition $f_2 - f_1 > 0$ is again the condition of gain. We can write the transparency condition in the form:

$$f_2 - f_1 = 0; \tag{2.37}$$

a medium is transparent if the occupation number difference (more accurately: the relative occupation number difference) is zero. The corresponding density of two-level atomic systems in the upper laser level is the *transparency density* N_{tr} $(= N_{2,tr})$.

A population inversion corresponds to a nonequilibrium state of an ensemble of two-level atomic systems. At thermal equilibrium, the population N_2 of an ensemble of two-level atomic systems is always smaller than the population N_1.

Thermal equilibrium of many media containing an ensemble of atomic systems is governed by *Boltzmann's statistics*. If Boltzmann's statistics holds, the ratio of the population of the upper level and the population of the lower level is given by

$$\frac{N_2}{N_1} = e^{-(E_2-E_1)/kT}, \tag{2.38}$$

where k $(= 1.38 \times 10^{-23}\ \mathrm{J\,K^{-1}})$ is Boltzmann's constant and T the temperature of the ensemble. At thermal equilibrium, the population difference is always negative, $N_2 - N_1 < 0$; the net effect of stimulated emission and absorption of radiation results in damping of radiation at the frequency $\nu \sim (E_2 - E_1)/h$. For an ensemble of two-level atomic systems, which obeys Boltzmann's statistics, the occupation number of the upper level is equal to

$$f_2^{\mathrm{Boltz}} = \frac{N_2}{N_1 + N_2} = \frac{1}{\exp[(E_2 - E_1)/kT] + 1} \tag{2.39}$$

and the occupation number of the lower level is

$$f_1^{\mathrm{Boltz}} = \frac{N_1}{N_1 + N_2} = \frac{1}{\exp[-(E_2 - E_1)/kT] + 1}. \tag{2.40}$$

At thermal equilibrium, the relative occupation number of the lower level is always larger than the relative occupation number of the upper level, $f_1^{\mathrm{Boltz}} - f_2^{\mathrm{Boltz}} > 0$, that is, at thermal equilibrium, absorption always exceeds stimulated emission.

In the case that the energy levels of an ensemble governed by Boltzmann's statistics are degenerate, the relative occupation number of level i is given by

$$f_i^{\text{Boltz}} = \frac{g_i \exp[E_i/kT]}{\sum g_i \exp[E_i/kT]}. \tag{2.41}$$

In atomic physics and thermodynamics, the occupation number of an atomic level concerns the total number of atoms. To describe laser media, it is convenient to make use of number densities and of relative occupation numbers. In order to avoid a confusion, we mark total numbers by the suffix "tot." If an ensemble of two-level systems is distributed in the volume V, we are dealing with the following quantities:

- N_1 = density of atoms in level 1.
- N_2 = density of atoms in level 2.
- $N_{1,\text{tot}} = N_1 \times V$ = occupation number of level 1 = total number of two-level systems in level 1.
- $N_{2,\text{tot}} = N_2 \times V$ = occupation number of level 2 = total number of two-level systems in level 2.
- $N_{\text{tot}} = (N_1 + N_2) \times V$ = total number of two-level atomic systems.
- $f_2 = N_2/(N_2 + N_1) = N_{2,\text{tot}}/N_{\text{tot}} \times V$ = relative occupation number of level 2.
- $f_1 = N_1/(N_2 + N_1) = N_{1,\text{tot}}/N_{\text{tot}} \times V$ = relative occupation number of level 1.
- $f_2 - f_1 = (N_2 - N_1)/(N_2 + N_1) = (N_{2,\text{tot}} - N_{1,\text{tot}})/N_{\text{tot}}$ = *occupation number difference* (=difference of the relative occupation numbers).

We will use the notation "occupation number" instead of "relative occupation number."

2.4 Laser Resonator

The *Fabry–Perot resonator* (Fig. 2.8) consists of two plane mirrors arranged in parallel at a distance L; the Fabry–Perot resonator is an open resonator—it has no sidewalls. We consider a Fabry–Perot resonator with reflectors of rectangular shape. We choose cartesian coordinates with the z axis parallel to the resonator axis; laser

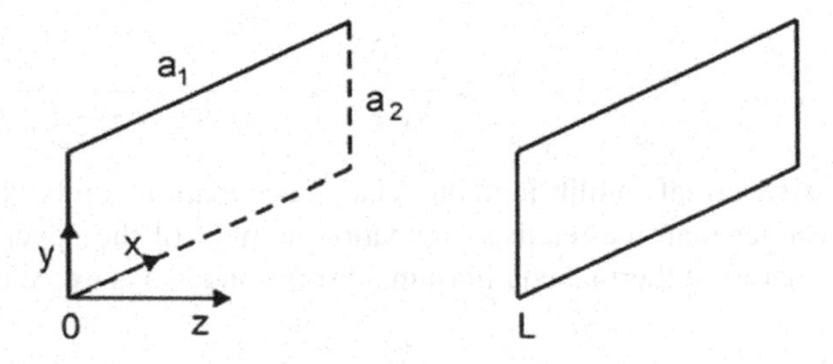

Fig. 2.8 Fabry–Perot resonator

radiation propagates along z. Characteristic quantities of a Fabry–Perot resonator are:

- a_1 = width of the resonator (along x).
- a_2 = height of the resonator (along y).
- L = length of the resonator (along z).
- $z = 0$ = location of mirror 1.
- $z = L$ = location of mirror 2.
- R_1 = reflectivity of mirror 1.
- R_2 = reflectivity of mirror 2.

We assume for the mirrors are perfectly reflecting ($R_1 = R_2 = 1$) and describe the laser field within a resonator as a *quasiplane standing wave* composed of two waves of equal amplitude and opposite propagation directions:

$$E = \frac{1}{2}A\cos[\omega t - (kz - \varphi_0)] + \frac{1}{2}A\cos[\omega t + (kz - \varphi_0)]. \qquad (2.42)$$

The field E and the amplitude A have an orientation along a direction (e.g., the x direction) perpendicular to the z direction. Using the relations $\cos(\alpha \pm \beta) = \cos\alpha\cos\beta \mp \sin\alpha\sin\beta$, we can write (2.42) in the form

$$E = A\cos(kz - \varphi_0)\cos\omega t, \qquad (2.43)$$

which describes a standing wave. To find k, ω, and φ_0, we make use of three conditions:

- The solution of the *resonator eigenvalue problem* provides discrete values of the wave vector.
- The dispersion relation for electromagnetic radiation then yields the resonance frequencies of a resonator.
- Two *boundary conditions* for electromagnetic fields provide the phase.

The *resonator eigenvalue problem reads*: after a round trip transit through the resonator, the field at a location z at time $t + T$ is the same as the field at time t,

$$E(z, t+T) = E(z, t). \qquad (2.44)$$

This leads to the condition

$$2kL = l \times 2\pi; \quad l = 1, 2, 3, \ldots. \qquad (2.45)$$

The integer l is the order of a resonance. The change of phase per round trip transit is $2kL = 2\pi$. Accordingly, the wave vector has discrete values,

$$k_l = l \times \frac{2\pi}{2L}. \qquad (2.46)$$

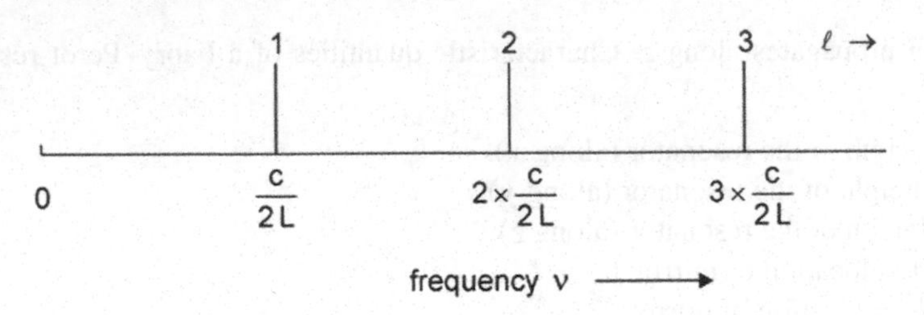

Fig. 2.9 Resonance frequencies of the Fabry–Perot resonator

We obtain, with $k = \omega/c$,

$$\omega_l = l \times \frac{2\pi c}{2L}, \tag{2.47}$$

or

$$\nu_l = l \times \frac{c}{2L}. \tag{2.48}$$

The resonance frequencies (= eigenfrequencies) ν_l of a Fabry–Perot resonator are multiples of $c/2L$. The resonance frequencies ν_l are equidistant. Next near resonance frequencies have the frequency distance (Fig. 2.9)

$$\nu_l - \nu_{l-1} = \frac{c}{2L}. \tag{2.49}$$

The round trip transit time—the time it takes the radiation to perform a round trip transit through the resonator—is

$$T = 1/\nu_1 = 2L/c. \tag{2.50}$$

The resonance wavelengths of radiation in a Fabry–Perot resonator are given by the relation

$$l \times \lambda_l/2 = L; \tag{2.51}$$

the length of the Fabry–Perot resonator is a multiple of $\lambda_l/2$. To determine the resonance wavelengths of a Fabry–Perot resonator containing a medium of refractive index n, we have to take into account that the speed of light in a medium is c/n and the wavelength is λ/n, where λ is the wavelength of light in vacuum.

Taking account of the boundary conditions and the dispersion relation, we obtain:

$$E = A\cos(k_l z - \varphi_0)\cos\omega_l t. \tag{2.52}$$

The *boundary conditions are*: $E = 0$ at $z = 0$ and $z = L$. We obtain the phase $\varphi_0 = \pi/2$. Thus, the standing wave has the form

$$E = A\sin k_l z\ \cos\omega_l t. \tag{2.53}$$

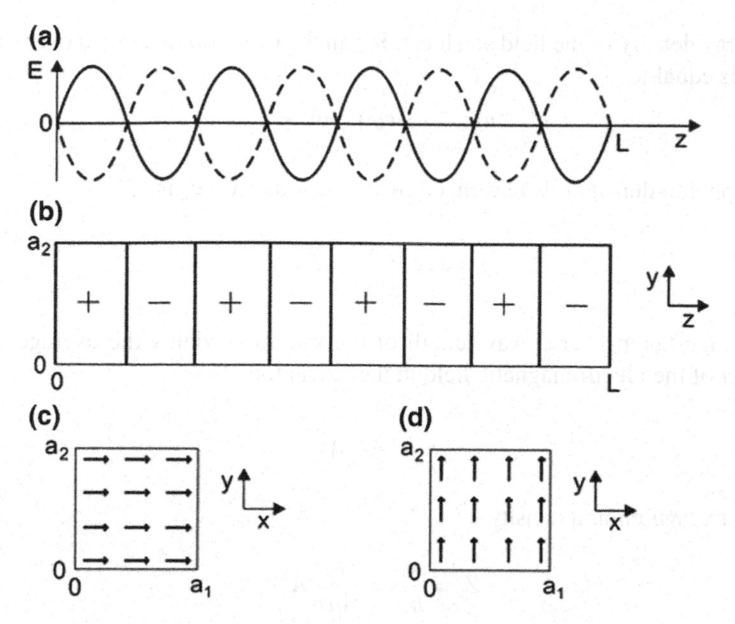

Fig. 2.10 Standing wave in a Fabry–Perot resonator. **a** Field $E(z)$. **b** Phase of the field in the zy plane. **c** Field lines in the xy plane for $E \parallel x$. **d** Field lines in the xy plane for $E \parallel y$

A resonance characterized by a frequency $\omega_l = 2\pi\nu_l$ and a wave vector k_l corresponds to a *resonator mode*, that is, to a particular pattern of the amplitude of the electromagnetic wave within the resonator. Figure 2.10a shows the electric field $E(z)$ for $t = 0$ and $T/2$. At a fixed time, the field varies sinusoidally along the resonator axis according to the variation of the phase kz. The sign of the field varies in z direction (Fig. 2.10b). The polarization of the electric field has a direction perpendicular to the z axis—along the x axis (Fig. 2.10c) or along the y axis (Fig. 2.10d). We now summarize the main properties of a quasiplane standing wave in a Fabry–Perot resonator.

- *Amplitude.* The amplitude A is a constant everywhere within the Fabry–Perot resonator.
- *Phase variation along the z axis.* The phase varies along the z axis.
- *Phase variation perpendicular to the z axis.* The phase does not vary in directions perpendicular to the z axis.
- *Polarization of the radiation.* The field is oriented perpendicular to z.

Two waves propagating in opposite directions add to the field, now written in complex form,

$$\tilde{E} = \frac{1}{2} A\, e^{i[\omega t - (kz - \varphi_0)]} + \frac{1}{2} A\, e^{i[\omega t + (kz - \varphi_0)]}. \tag{2.54}$$

The energy density of the field at a location z in the resonator averaged over a period of time is equal to

$$u(z) = \frac{1}{2}\varepsilon_0 A^2 \sin^2 k_l z \tag{2.55}$$

and the photon density (also averaged over a period of time) is

$$Z(z) = \frac{\varepsilon_0 A^2}{2h\nu} \sin^2 k_l z. \tag{2.56}$$

The average taken over a wavelength of the radiation yields the average energy density u of the electromagnetic field in the resonator

$$u = \frac{1}{4}\varepsilon_0 A^2 \tag{2.57}$$

and the average photon density

$$Z = \frac{u}{h\nu} = \frac{\varepsilon_0}{4h\nu} A^2. \tag{2.58}$$

If a resonator has two reflectors both with $R = 1$, light within the resonator travels without loss; it performs an infinite number of round trip transits. But the number of round trip transits is finite if one of the reflectors is a partial reflector acting as output coupling mirror. Then, a reflection at the output coupling mirror corresponds to a reduction of the energy density within the resonator. How long does a photon remain in a resonator? We consider the energy density u at a fixed location within the resonator (Fig. 2.11). After one round trip of the radiation, the energy density is Vu, where the V factor describes how much of the energy remained in the resonator after one round trip transit; accordingly, the photon density Z is reduced to VZ after one round trip of the radiation. The V factor is a measure of loss.

- *V factor* = fraction of radiation energy that remains in the resonator after a round trip transit = fraction of the number of photons remaining in the resonator after one round trip = survival probability of a photon after a round trip transit through the resonator.
- $V = 1$, there is no loss.
- $V < 1$, there is loss.

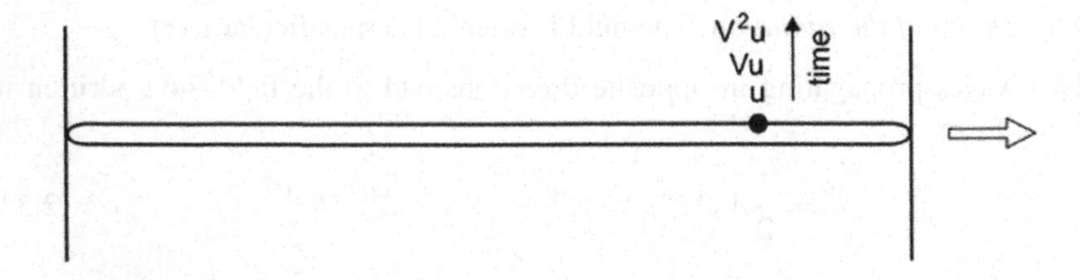

Fig. 2.11 Resonator with loss

The photon density develops as follows:
$t = 0$; Z_0.
$t = T$; one round trip transit, $Z = VZ_0$.
$t = sT$; s round trip transits,

$$Z(s) = V^s Z_0. \tag{2.59}$$

Replacing s by the continuous variable t/T, we write

$$Z(t) = Z_0 V^{t/T}. \tag{2.60}$$

Using the identity $a^x = \mathrm{e}^{x \ln a}$, we obtain

$$Z(t) = Z_0\, \mathrm{e}^{-\kappa t} = Z_0\, \mathrm{e}^{-t/\tau_p}, \tag{2.61}$$

where

$$\kappa = \frac{1}{\tau_p} = \frac{-\ln V}{T} \tag{2.62}$$

is the *loss coefficient* of the resonator and

$$\tau_\mathrm{p} = \frac{T}{-\ln V} \tag{2.63}$$

is the *photon lifetime* (= average lifetime of a photon in the resonator = decay time of the energy density of radiation in the resonator). We write the V factor as

$$V = \mathrm{e}^{-T/\tau_\mathrm{p}} = \mathrm{e}^{-\kappa T}. \tag{2.64}$$

The energy density decreases exponentially with the same decay constant as the photon density,

$$u = u_0\, \mathrm{e}^{-t/\tau_\mathrm{p}} = u_0\, \mathrm{e}^{-\kappa t}, \tag{2.65}$$

where u_0 is the initial energy density.

If the loss is due to both output coupling loss (described by V_out) and internal loss in the resonator (V_i), the total V factor is equal to

$$V = V_\mathrm{out} V_\mathrm{i}. \tag{2.66}$$

Then the loss coefficient of a resonator is

$$\kappa = \kappa_\mathrm{out} + \kappa_i, \tag{2.67}$$

where

- κ_{out} is the decay coefficient due to output coupling of radiation and
- κ_{i} is the loss coefficient due to internal loss.

Diffraction at the reflectors, for instance, causes internal loss.

The relative decrease of the energy density after one round trip transit is

$$(u - Vu)/u = 1 - V. \tag{2.68}$$

The quantity $1 - V$ is the *loss per round trip transit.*

Example A resonator has the V factor V =0.9. This means:

- There remain, after one round trip of the radiation, 90% of the photons in the resonator.
- The loss per round trip is 10%.
- The photon lifetime is $\tau_p = T/(-\ln V) = 9.5T$.
- The survival probability of a photon after a round trip through the resonator is 0.9.

2.5 Laser = Laser Oscillator

The laser (=laser oscillator) is a *self-excited oscillator* (=self-sustained oscillator). It is characteristic of a self-excited oscillator that it starts oscillation itself and maintains oscillation as long as pump energy is supplied by an external energy source.

We mention a classical self-excited oscillator. A string of a violin is excited to an oscillation during a continuous motion of the bow. The string together with the bow, which steadily delivers energy to the oscillation, is a self-excited oscillator. The length of the string determines the fundamental frequency.

We will now formulate the condition of laser oscillation and also show that the buildup of a steady state oscillation takes time.

2.6 Radiation Feedback and Threshold Condition

Radiation in a resonator containing an active medium is repeatedly propagating through the active medium. The active medium experiences feedback from the radiation that is stored in the resonator.

We now assume that population inversion and gain are suddenly turned on at time $t = 0$. At the start of a laser oscillation, the energy density of the radiation is u. The energy density is equal to VGu after one round trip (Fig. 2.12). G is the *gain*

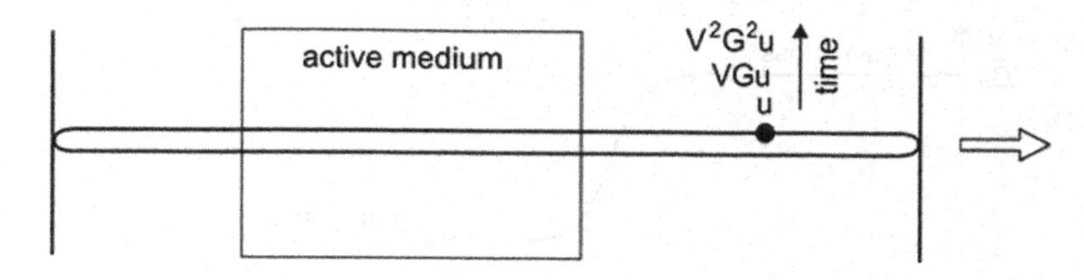

Fig. 2.12 Balance of energy in a laser

factor per round trip transit. It is equal to the product of the single-pass gain factors ($G = G_1^2$). The energy density increases if

$$VGu > u \tag{2.69}$$

or

$$GV > 1. \tag{2.70}$$

The energy in a resonator increases with time if the product of the gain factor and the V factor is larger than unity. Without loss ($V=1$), the energy density after one round trip is Gu. The relative increase of the energy density after one round trip is

$$\frac{Gu - u}{u} = G - 1. \tag{2.71}$$

The quantity $G - 1$ is the *gain per round trip*. If gain and loss are present, we can write the condition of net gain in the form

$$\frac{Gu - u}{u} > \frac{u - Vu}{u} \tag{2.72}$$

or

$$G - 1 > 1 - V, \tag{2.73}$$

gain per round trip > loss per round trip. We did not differ between radiation propagating in the resonator in clockwise or counterclockwise direction: we supposed that propagation in both directions leads to gain.

We assume that V does not change with time. Can we also assume that the gain factor G is independent of time? If G would be constant, the energy in the laser would permanently increase and reach an infinitely large value. But this is not the case if $V < 1$. Then the energy in a laser resonator becomes finite—G decreases during the buildup of the light field in the laser resonator. We have the condition that $VG_\infty u = u$ at steady state or

$$G_\infty V = 1; \tag{2.74}$$

at steady state oscillation, the product of the gain factor G_∞ and the V factor is 1.

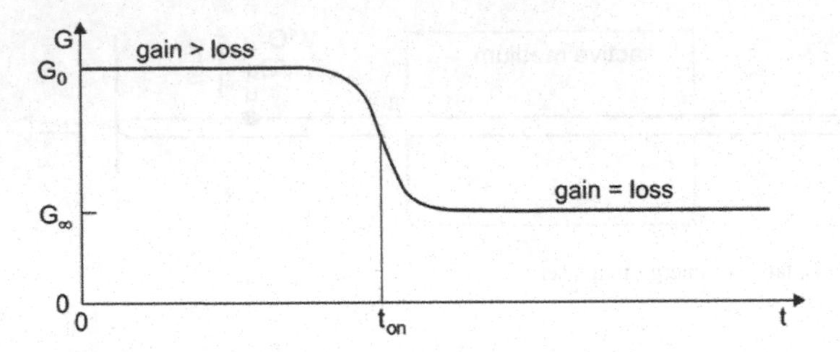

Fig. 2.13 Gain during onset of laser oscillation

We describe the following case: an active medium is suddenly turned on at time $t = 0$. The initial gain factor is G_0 (Fig. 2.13). The gain factor remains nearly constant and then decreases to the steady state value G_∞. The transition from G_0 to G_∞ occurs at the *onset time* t_{on}, which is a measure of the time it takes to build up a steady state oscillation. G_0 is the small-signal gain factor and G_∞ the large-signal gain factor. The two conditions lead to the *threshold condition of laser oscillation* (= laser condition):

$$GV \geq 1. \tag{2.75}$$

The condition implies that during the buildup of a laser field, G is larger than at steady state oscillation. We can also interpret the threshold condition as follows: an oscillation builds up if G_0 is only slightly larger than G_∞. In the extreme case that $G_0 \rightarrow G_\infty$, reaching a steady state takes infinitely long time ($t_{on} \rightarrow \infty$). In this sense, we introduce the threshold gain factor,

$$G_{th} = G_\infty. \tag{2.76}$$

The small-signal gain factor G_0 is always larger than the large-signal gain factor G_∞ ($= G_{th}$). At steady state oscillation, the gain factor is clamped at

$$G_\infty = V^{-1}. \tag{2.77}$$

We will treat onset of oscillation in more detail (Sects. 2.9, 8.4, and 9.7; Fig. 9.6). A laser is a regenerative amplifier: at steady state oscillation, radiation lost during a round trip transit through the resonator is regenerated after the round trip.

The energy density of radiation in a lossless resonator of a continuously pumped laser increases to infinitely large values—but optical damage limits the energy density (Sect. 16.10).

2.7 Frequency of Laser Oscillation

At steady state oscillation, the electric field at a fixed location in the laser resonator reproduces itself after each round trip transit through the resonator; this is the resonator eigenvalue problem in the case that a resonator contains an active medium. The electric field in a resonator (Fig. 2.14) has to obey the condition

$$\tilde{G}_{\mathrm{E}} \tilde{V}_{\mathrm{E}} \exp(-\mathrm{i}(2kL - \Delta\phi - \varphi_3 + \varphi_{\mathrm{R1}} + \varphi_{\mathrm{R2}}))\, \tilde{E} = \tilde{E}. \tag{2.78}$$

The quantities concern a round trip transit through the active medium.

- $\tilde{G}_{\mathrm{E}} = G_{\mathrm{E}}\mathrm{e}^{\mathrm{i}\varphi_1}$ =complex gain factor with respect to the field.
- $\tilde{V}_{\mathrm{E}} = V_{\mathrm{E}}\mathrm{e}^{\mathrm{i}\varphi_2}$ =complex loss factor with respect to the field.
- 2kL = geometric phase shift.
- φ_{R1} = phase shift due to reflection at one of the mirrors.
- φ_{R2} = phase shift due to reflection at the other mirror.
- φ_3 = phase shift due to dispersion in the resonator (Sect. 13.3).
- $\Delta\phi$ = *Gouy phase shift* = additional phase shift occuring for radiation propagating in a resonator with curved mirrors (Sect. 11.7); the Gouy phase shift of radiation in a Fabry–Perot resonator is zero.

We obtain the condition

$$G_{\mathrm{E}} V_{\mathrm{E}} \exp{-\mathrm{i}(2kL - \Delta\phi - \varphi_3 - \varphi_1 - \varphi_2 + \varphi_{\mathrm{R1}} + \varphi_{\mathrm{R2}})} = 1. \tag{2.79}$$

The factor to the exponential has to be equal to unity, and the sum of all phases has to be a multiple of 2π. It follows that $G_{\mathrm{E}} V_{\mathrm{E}} = 1$ and, with $G_{\mathrm{E}}^2 = G$ and $V_{\mathrm{E}}^2 = V$, that $GV = 1$ as already derived in the preceding section. The second condition is

$$2kL - \Delta\phi - \varphi_3 + \varphi_{\mathrm{R1}} + \varphi_{\mathrm{R2}} = l \times 2\pi, \tag{2.80}$$

where l is an integer. The sum of all changes of phase after a round trip transit has to be a multiple of 2π. Since $k = \omega/c$, the condition provides the eigenfrequencies $\omega_l = 2\pi\nu_l$. In the special case that all additional phases—but not the geometric phase shift kz—are zero, we obtain the eigenfrequencies $\nu_l = lc/(2L)$, with $l = 1, 2, \ldots$; otherwise the resonance frequencies are shifted.

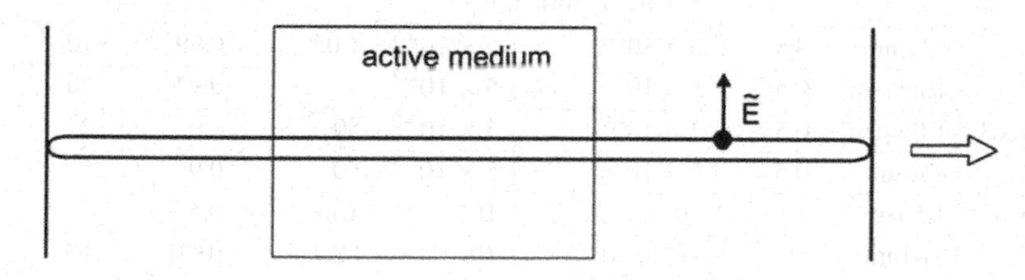

Fig. 2.14 Field in a laser

Electromagnetic radiation propagating in a medium of refractive index n has the wave vector $k = n\omega/c$. It depends on the properties of the resonator and of the active medium at which resonance frequency (or frequencies) a laser oscillates.

We will discuss the origin of phase shifts of electromagnetic wave propagating in an active medium in Chap. 9 and phase shifts of an electromagnetic wave in a laser resonator in Chap. 11 and Sect. 13.4.

2.8 Data of Lasers

Table 2.1 shows data of continuous wave lasers. The data concern the quantities (*see* also Fig. 2.15):

- L = resonator length.
- d = diameter in case of a circular resonator.
- a_1 = width and a_2 = height of a rectangular resonator.
- Resonator volume = $\pi(d/2)^2L$ of a circular-mirror resonator and a_1a_2L of a rectangular resonator.
- G = gain factor per round trip of the radiation; G_1, per single transit.
- V = V factor per round trip; it indicates the reduction of the photon density in the resonator per round trip transit, and V_1 loss per single transit.
- P_{out} = output power.

By modifying a laser (e.g., by choosing an other cross sectional areas), it can be possible to obtain a larger or a smaller output power. The lasers generate radiation at wavelengths listed in the table but also at other wavelengths. The length of the active medium of a gas laser is about equal to the length of the resonator. The length of the active medium of a solid state laser is smaller than the resonator length. The length of the active medium of a semiconductor laser is about equal to the resonator length. Semiconductor lasers have much smaller sizes than other lasers.

- *Helium–neon laser*. The gain is small. Therefore, the V factor has to be close to unity—the reflectivities of the resonator mirrors have to be near unity

Table 2.1 Data of lasers

Laser	λ	L(m)	Resonator d(m), or a_1 (m); a_2 (m)	Volume (m^3)	G [or G_1]	V [or V_1]	P_{out} (W)
HeNe	633 nm	0.5	2×10^{-3}	5×10^{-5}	1.02	0.99	10^{-2}
CO_2	10.6 μm	0.5	2×10^{-2}	5×10^{-3}	3	0.95	70
Nd:YAG	1.06 μm	0.5	2×10^{-2}	5×10^{-3}	50	0.9	2
TiS	830 nm	0.5	2×10^{-2}	5×10^{-3}	50	0.9	5
Fiber	1.5 μm	10	10^{-5}	10^{-9}	100	0.5	1
SC	810 nm	10^{-3}	10^{-6}; 10^{-4}	10^{-13}	[12]	[0.3]	10^{-1}
QCL	5 μm	10^{-3}	10^{-5}; 10^{-4}	10^{-12}	10	0.9	10^{-3}

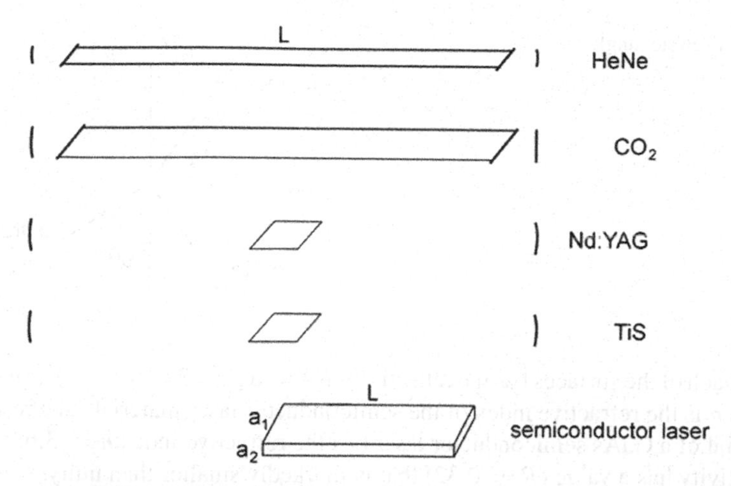

Fig. 2.15 Various lasers

(e.g., $R_1 = 0.998$ and $R_2 = 0.99$). A glass plate covered on the front surface with a highly-reflecting multilayer coating and on the back surface with anantireflecting dielectric coating is a high-reflectivity mirror (Sect. 25.7). The front surface with its coating acts as resonator mirror while the back surface is outside the resonator. The glass tube that contains the laser gas is closed by Brewster windows.

- *CO_2 laser.* The gain is large. One of the reflectors is a metal mirror; a metal mirror has a reflectivity near unity for radiation at wavelengths larger than about 5 μm. The output coupling mirror has a reflectivity that is noticeably smaller than unity (e.g., $R_1 = 1$; $R_2 = 0.95$). The output coupling mirror is a dielectric plate (e.g., a germanium plate) covered on the resonator side with a dielectric multilayer coating and on the other side with a dielectric antireflecting coating; a metal film is not suitable as a partial reflector because of a very high absorptivity for radiation passing through a metal film (Problem 25.18).
- *Neodymium YAG laser* (Nd:YAG laser). The gain is large. The length of the active medium is much smaller than the length of the resonator. Both mirrors (e.g., $R_1 \sim 1$ and $R_2 = 0.95$) consist of dielectric multilayers on glass plates. The Nd:YAG crystal surfaces are obliquely oriented relative to the beam axis so that the angle of incidence of the radiation is the Brewster angle and radiation traverses the crystal surfaces without loss.
- *Titanium–sapphire laser.* The active medium also fills a small portion of the resonator. The gain is large (mirror reflectivities are, e.g., $R_1 \sim 1$ and $R_2 \sim 0.95$). The titanium–sapphire crystal surfaces are obliquely oriented relative to the beam axis so that the angle of incidence of the radiation is the Brewster angle and radiation traverses the crystal surfaces without loss.
- *Fiber laser.* The active medium is a doped fiber of small diameter and large length. The output power can reach several hundred watt.
- *Bipolar semiconductor laser* (SC). The gain can be large already at a small length of an active medium. It is possible to use the semiconductor surfaces as reflectors.

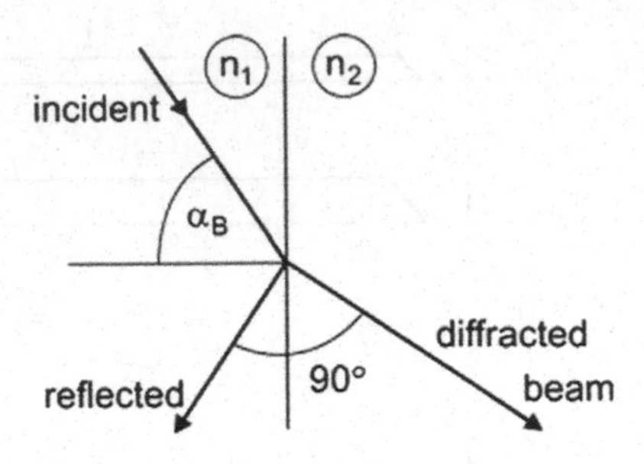

Fig. 2.16 Brewster angle

Then each of the surfaces has the reflectivity $R_1 = R_2 = R = (n-1)^2/(n+1)^2$, where n is the refractive index of the semiconductor laser material. The resonator material of a GaAs semiconductor laser has the refractive index $n = 3.6$, and the reflectivity has a value ($R = 0.32$) that is markedly smaller than unity.

- *Quantum cascade laser*. The gain is also large at small length of the gain medium.

Radiation passing at normal incidence through an interface between air and a medium experiences loss due to reflection, while radiation of the appropriate polarization direction passes the interface under the Brewster angle without reflection loss. The Brewster angle follows from *Snell's law*

$$\frac{\sin \alpha_2}{\sin \alpha_1} = \frac{n_1}{n_2}, \tag{2.81}$$

where α_1 is the angle of incidence, α_2 the angle of the transmitted beam, n_1 (~ 1) the refractive index of air, and n_2 the refractive index of the dielectric material. The reflectivity is zero if the electric field vector lies within the plane of incidence (p polarization) and if the angle of incidence is equal to the Brewster angle α_B (Fig. 2.16). The Brewster angle is determined by the relation

$$\tan \alpha_B = n_2/n_1. \tag{2.82}$$

The diffracted and the reflected beam (that has no power) are perpendicular to each other. Radiation of a polarization perpendicular to the plane of incidence (s polarization) is partly reflected and is therefore attenuated by a Brewster window. Accordingly, the radiation of a laser that contains Brewster windows is polarized.

2.9 Oscillation Onset Time

The threshold condition does not specify the value of the initial energy density of the electromagnetic field. However, there is a physical limit: if there is no electromagnetic energy in the resonator, nothing can be amplified. Already one photon in the resonator can initiate laser oscillation if the threshold condition is fulfilled. One photon in a resonator corresponds to a photon density of $Z_0 = (a_1 a_2 L)^{-1}$, where $a_1 a_2 L$ is the volume of the resonator. After the round trip transit time, the density of the photons is VG_0 and after s round trip transits, the density of the photons in the resonator is

$$Z(s) = Z_0 (VG_0)^s. \tag{2.83}$$

To estimate the onset time, we assume that the gain, described by the round trip gain factor G_0, is turned on at $t = 0$, then remains constant, and suddenly decreases at $t = t_{on}$ to G_∞ (Fig. 2.17, upper part). It follows that the photon density increases exponentially from the initial value Z_0 until it reaches at $t = t_{on}$ the steady state value Z_∞ (= density of photons in the resonator at steady state oscillation). We write

$$(VG_0)^{s_{on}} = Z_\infty / Z_0, \tag{2.84}$$

where s_{on} is the number of round trip transits necessary to reach the steady state. It follows that the oscillation onset time, $t_{on} = s_{on} T$, is given by

$$t_{on} = T \frac{\ln(Z_\infty / Z_0)}{\ln(VG_0)}. \tag{2.85}$$

The onset time is proportional to the round trip transit time and to the natural logarithm of Z_∞ / Z_0. And it is inversely proportional to the natural logarithm of the product VG_0.

We replace s by the continuous variable t/T and write

$$Z(t) = Z_0 \, (VG_0)^{t/T}. \tag{2.86}$$

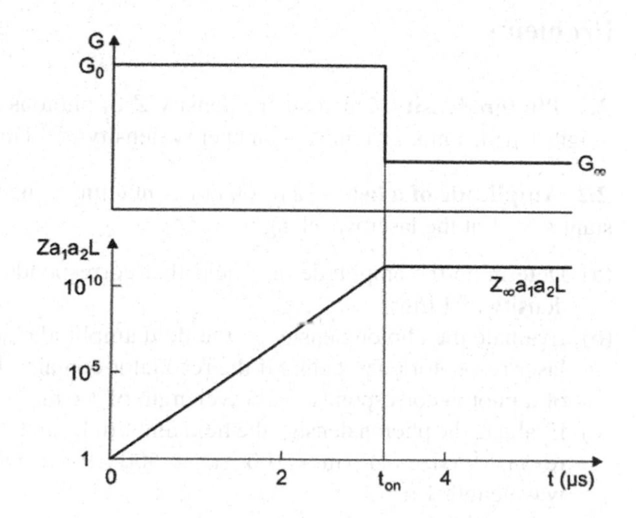

Fig. 2.17 Onset of laser oscillation: gain factor and photon number

With the identity $a^x = \mathrm{e}^{x \ln a}$, we obtain

$$Z = Z_0\, \mathrm{e}^{\ln(VG_0)t/T}. \tag{2.87}$$

The number of photons increases exponentially until, at $t = t_{\mathrm{on}}$, a steady state oscillation is established.

It follows, for $1 - V \ll 1$ and $G_0 - 1 \ll 1$, that $\ln(VG_0) = \ln V + \ln G_0 = (G_0 - 1) - (1 - V) =$ gain minus loss per round trip. Then we can write

$$t_{\mathrm{on}} = \frac{T}{(G_0 - 1) - (1 - V)} \ln(Z_\infty / Z_0). \tag{2.88}$$

If the gain is small compared to unity, $(G_0 - 1) \ll 1$, the oscillation onset time is large compared to the round trip time, $t_{\mathrm{on}} \gg T$.

Example A helium–neon laser (length 0.5 m; cross-sectional area 1 mm^2; output power 1 mW; gain $= G - 1 = 0.02$; loss $1 - V = 0.01$) starts with one photon in the laser mode ($Z_0 a_1 a_2 L = 1$) and contains, at steady state oscillation, $Z_\infty a_1 a_2 L = P_{\mathrm{out}} \tau_{\mathrm{p}} = 10^{10}$ photons, where $\tau_{\mathrm{p}} = T/(1 - V) = 3.3 \times 10^{-7}$ s is the photon lifetime. The density of photons in the resonator at $t = 0$ is $Z_0 = (a_1 a_2 L)^{-1} = 10^6\ \mathrm{m}^{-3}$. The density of photons at steady state oscillation is $Z_\infty \sim 10^{16}\ \mathrm{m}^{-3}$. The round trip transit time is $T = 3.3 \times 10^{-9}$ s. It follows that the onset time is $t_{\mathrm{on}} \sim 8\ \mu$s. The buildup of steady state oscillation of a helium–neon laser requires that the radiation performs about thousand round trip transits through the resonator. The photon density (Fig. 2.17, lower part) increases exponentially during the onset time ($t < t_{\mathrm{on}}$) and has a constant value for $t > t_{\mathrm{on}}$.

References [1–11, 26–28].

Problems

2.1 Photon density. Calculate the density Z of photons in a radiation field (wavelength 1 μm, 1 nm, or 1 mm) of an energy density of 1 J m^{-3}.

2.2 Amplitude of a field in a resonator containing a medium of the dielectric constant $\varepsilon = 1$ at the laser wavelength.

(a) Determine the amplitude of a field that corresponds to radiation of an energy density of 1 J/m^3.
(b) Evaluate the photon density Z, the field amplitude, and the energy density in a laser resonator (size 1 cm^3) if the resonator contains 1 photon and if the energy of a photon corresponds to a wavelength of 1 μm.
(c) Evaluate the photon density, the field amplitude, and the energy density in a laser resonator (size 0.4 μm × 100 μm × 500 μm) if it contains 1 photon (photon wavelength 1 μm).

2.3 Thermal occupation number of an atomic system governed by Boltzmann statistics.

(a) Show that $f_1^{\rm Boltz} - f_2^{\rm Boltz} > 0$ for an ensemble of two-level atomic systems in thermal equilibrium.
(b) Estimate the thermal occupation number difference $f_2^{\rm Boltz} - f_1^{\rm Boltz}$ for an ensemble of two-level atomic systems (at 300 K) at level separations that correspond to visible radiation (wavelength of 600 nm).
(c) Calculate the occupation number difference at level separations that correspond to far infrared radiation ($\lambda = 300\ \mu$m).

2.4 Threshold condition of laser oscillation.

(a) Show that the threshold condition (expressed for the power of radiation) is the same for light propagating in $-z$ direction as for light propagating in $+z$ direction.
(b) Show that the threshold condition is also the same for two electromagnetic fields propagating in $\pm$ z directions if $G_2 \neq G_1$, where the gain factors correspond to the single-pass gain factors for the two propagation directions.

2.5 Brewster angle. Determine the Brewster angles of materials used in lasers as windows or as active materials.

(a) Helium–neon laser (633 nm); quartz glass, $n = 1.4$.
(b) CO_2 laser; NaCl crystal, $n = 1.5$.
(c) Nd:YAG laser; YAG, $n = 1.82$.
(d) Titanium–sapphire laser; sapphire, $n = 1.76$.

2.6 Photon lifetime and oscillation onset time. Determine the photon lifetime and the oscillation onset time of lasers mentioned in Table 2.1.

2.7 Fresnel coefficients.
Derive the Brewster angle by use of the *Fresnel coefficients*:

$$r_\perp = \frac{E_\perp^{(r)}}{E_\perp^{(i)}} = \frac{n_1 \cos\theta_1 - n_2 \cos\theta_2}{n_1 \cos\theta_1 + n_2 \cos\theta_2}, \qquad (2.89)$$

$$t_\perp = \frac{E_\perp^{(t)}}{E_\perp^{(i)}} = \frac{2 n_1 \cos\theta_1}{n_1 \cos\theta_1 + n_2 \cos\theta_2}, \qquad (2.90)$$

$$r_\parallel = \frac{E_\parallel^{(r)}}{E_\parallel^{(i)}} = \frac{n_2 \cos\theta_1 - n_1 \cos\theta_2}{n_1 \cos\theta_1 + n_2 \cos\theta_2}, \qquad (2.91)$$

$$t_\parallel = \frac{E_\parallel^{(t)}}{E_\parallel^{(i)}} = \frac{2 n_1 \cos\theta_1}{n_1 \cos\theta_1 + n_2 \cos\theta_2}. \qquad (2.92)$$

- $r_\perp$ = Fresnel coefficient of reflection, with the electric field direction perpendicular to the plane of incidence.

- $r_{||}$ = Fresnel coefficient of reflection, with the electric field direction in the plane of incidence.
- θ_1 = angle of incidence.
- θ_2 = angle of the refracted beam.
- n_1 = refractive index of medium 1.
- n_2 = refractive index of medium 2.

The coefficients $r_\perp$ and $r_{||}$ are the corresponding Fresnel coefficients of transmission.

2.8 Fresnel coefficients of normal incidence.

(a) Show that $r_{||} = r_\perp = r = (n_1 - n_2)/(n_1 + n_2)$ and $t_{||} = t_\perp = t = 2n_1/(n_1 + n_2)$ and determine the reflectivity R and the transmissivity T.
(b) Show that $r_{21} = -r_{12}$ and that $t_{12}t_{21} - r_{12}r_{21} = 1$.

2.9 Relate the intensity of radiation to the photon density.

2.10 Photon flux. The photon flux is equal to the number of photons per second per unit area.

(a) Relate to other quantities that characterize a plane wave, namely photon density, energy density, intensity, and amplitude.
(b) Determine the photon flux for the radiation fields mentioned in Problem 2.1.
(c) Determine the photon flux for the output of the lasers mentioned in Table 2.1.

2.11 Determine the Brewster angle for laser materials mentioned in Table 6.1.

2.12 Radiation of a helium-neon laser (power 1 mW, wavelength 633 nm) is focused to an area of diameter 10 μm^2. Determine the intensity, the photon density, the energy density and the amplitude of the electric field in the focus.

2.13 Circularly polarized radiation.

(a) Characterize the field of circularly polarized radiation.
(b) Show that circularly polarized radiation can be obtained by sending a plane wave through a quarter-wave plate; a quarter-wave plate consists of an anisotropic crystal with different refractive indices for the ordinary and the extraordinary beam propagating in the same direction.

Chapter 3
Fabry–Perot Resonator

The main topics of this chapter concern the characterization of a resonator mirror, of the Fabry–Perot interferometer, and of the Fabry–Perot resonator.

In the 1890s, Charles Fabry and Alfred Perot (Marseille, France) introduced the Fabry–Perot *interferometer*. It consists, in principal, of two partial mirrors that have *infinitely* large lateral extensions. We will determine the transmissivity of a Fabry–Perot interferometer. The transmission curve (Airy curve) exhibits, at high reflectivities of the mirrors, narrow resonances with Lorentzian shape.

A Fabry–Perot *resonator* consists of two partial mirrors of *finite* lateral extensions. The spectral transmission curve of a Fabry–Perot resonator is—for quasiplane waves—the same as that of a Fabry–Perot interferometer. We will show that the transmission curve narrows when a gain medium is inserted into a Fabry–Perot resonator.

We mention different types of laser resonators and then discuss properties of a Fabry–Perot resonator. We introduce the ideal mirror and determine the transmission curves of the Fabry–Perot interferometer and of the Fabry–Perot resonator.

The Fabry–Perot resonator, together with the description of radiation as a quasiplane waves, represents a model resonator that is well suited to study basic properties of a resonator and of a laser. We will later (Sect. 11.6) show that the Fabry–Perot resonator differs from the Fabry–Perot interferometer, particularly for radiation that propagates at an angle to the resonator axis. There is another very important difference: radiation in a Fabry–Perot resonator experiences diffraction (Sect. 11.8) while diffraction plays no role in a Fabry–Perot interferometer.

3.1 Laser Resonators and Laser Mirrors

There are different types of resonators:

- *Resonator with curved mirrors as reflectors* (Chap. 11). The resonators are suitable for most of the gas and solid state lasers as well as for free-electron lasers.

K.F. Renk, *Basics of Laser Physics*, Graduate Texts in Physics,
DOI 10.1007/978-3-319-50651-7_3

- *Fabry–Perot resonator* (this chapter). It has two plane reflectors. This resonator represents an ideal model resonator suitable for the study of basic properties of a resonator and of a laser. This type of resonator is used in vertical-cavity surface-emitting lasers (Sect. 22.7).
- *Waveguide Fabry–Perot resonator.* It has two plane reflectors oriented parallel to each other. Within the Fabry–Perot resonator, the light is guided by an optical waveguide structure. This type of resonator is used in edge–emitting semiconductor lasers (Sect. 20.5) and fiber lasers (Sect. 15.7).
- *Cavity resonator* (Chap. 10). The cavity resonator has metallic walls. It is used as resonator of semiconductor lasers of the far infrared and of microwave lasers.
- *Photonic crystal resonator* (Chap. 25). A reflector of a photonic crystal resonator can consist of periodically arranged materials of different refractive indices. The photonic crystal resonator can be a Fabry–Perot-like resonator or a cavity-like resonator. Photonic crystal resonators are becoming more and more important of a variety of lasers, e.g., of microlasers and nanolasers.

The design of mirrors of a laser resonator depends on the availability of materials of mirrors. We mention a few types of mirrors:

- *Ideal mirror* (Sect. 3.4). The ideal mirror is lossless; its thickness is infinitely small; it is able to divide an incident wave in a reflected and a transmitted wave. We can choose the reflectivity of this model mirror. We will make use of the model mirror to describe main properties of a Fabry–Perot resonator (this chapter) and of resonators with curved mirrors (Chap. 11).
- *Crystal surface.* A crystal surface is able to divide, without loss, an optical wave in a reflected and a transmitted wave. However, the reflectivity at normal incidence is much smaller than unity. GaAs crystal surfaces are nevertheless suitable as reflectors of semiconductor lasers (Sect. 20.5).
- *Dielectric multilayer mirror* (Sect. 25.7). A stack of dielectric quarter-wavelength layers represents a dielectric multilayer mirror that can be designed as an almost lossless reflector with a reflectivity very near to unity (e.g., $R = 0.999$) for visible radiation or as a lossless partial reflector with a reflectivity that we can choose.
- *Metal mirror.* A metal mirror has a relatively low reflectivity and large absorptivity for visible radiation and is not suitable as mirror of a laser that generates visible radiation—a silver mirror shows a reflectivity of 0.95 and an absorptivity of 5% for red light. Metal reflectors are in use as reflectors of infrared radiation at wavelengths larger than several micrometers. Metal films are not suitable as partial reflectors because the absorptivity of a metal film is much larger than the transmissivity (*see* Problems 25.17 and 25.18).

In Chap. 25, we will discuss various mirrors and methods used for realization of feedback and we will describe a method which is suitable to calculate properties of multilayer reflectors. In this chapter we treat the Fabry–Perot resonator.

3.2 V Factor and Related Quantities

We first study an empty Fabry–Perot resonator that has a reflector (reflectivity = 1) and a partial reflector (reflectivity R) as described in Sect. 2.4. Radiation starting at time $t = 0$ and performing many round trip transits through a resonator propagates during the time t over a total distance $d = ct$. We can write

$$Z(d) = Z_0\, \mathrm{e}^{-d/l_\mathrm{p}}. \tag{3.1}$$

Z is density of photons after propagation over the distance d, Z_0 is the photon density at $d = 0$ and

$$l_\mathrm{p} = c\tau_\mathrm{p} = \frac{2L}{-\ln V} \tag{3.2}$$

is the average path length of a photon in the resonator. During its lifetime in the resonator, a photon propagates over the distance l_p. We introduce the effective number $s_\mathrm{eff} = l_\mathrm{p}/2L$ of round trip transits of a photon. After s_eff round trip transits, the photon density decreases to Z_0/e. We obtain the relations

$$s_\mathrm{eff} = \frac{l_\mathrm{p}}{2L} = \frac{1}{-\ln V} = \frac{\tau_p}{T}. \tag{3.3}$$

Besides Z and u, also the total number Za_1a_2L of photons in the resonator and the total energy $Za_1a_2Lh\nu$ in the resonator decrease exponentially with the photon lifetime τ_p.

The *quality factor* Q (=Q factor =Q value) of a resonator is equal to 2π times the ratio of the energy stored in the resonator and the energy loss per oscillation period. It follows that the Q factor of a Fabry–Perot resonator is given by

$$Q = 2\pi l \times \frac{u \times a_1a_2L}{u \times a_1a_2L\tau_\mathrm{p}^{-1} \times T} = \omega\tau_\mathrm{p} = \frac{\omega T}{-\ln V} = \frac{2\pi l}{-\ln V} = 2\pi l s_\mathrm{eff}. \tag{3.4}$$

The Q factor is equal to the product of the order l of the resonance and the effective number of round trip transits of a photon through the resonator, multiplied by 2π.

We summarize the relations of the V factor and other quantities:

- $\tau_\mathrm{p} = T/(-\ln V)$ = lifetime of a photon in the resonator.
- $l_\mathrm{p} = 2L/(-\ln V)$ = path length of a photon in the resonator
- $s_\mathrm{eff} = 1/(-\ln V)$ = number of round trip transits of a photon within the resonator.
- $Q = 2\pi l/(-\ln V)$ = Q factor.

T is the round trip transit time, L the length of the resonator, and l the order of resonance. If V has a value near unity ($1 - V \ll 1$), the expansion of $\ln V$ in a

Table 3.1 Lifetimes of photons in laser resonators

Laser	λ	L (m)	T (s)	V; [V_1]	τ_p (s)	$\frac{\tau_p}{T}\left[\frac{\tau_p}{T/2}\right]$
HeNe	632 nm	0.5	3.3×10^{-9}	0.99	3.3×10^{-7}	100
CO_2	10 μm	0.5	3.3×10^{-9}	0.9	3.3×10^{-8}	10
Nd:YAG	1.06 μm	0.5	3.3×10^{-9}	0.9	3.3×10^{-8}	10
TiS	830 nm	0.5	3.3×10^{-9}	0.9	3.3×10^{-8}	10
Fiber	1.5 μm	10	6.7×10^{-8}	0.5	10^{-7}	1.4
SC	840 nm	10^{-3}	2.4×10^{-11}	[0.33]	2.4×10^{-11}	[0.9]
QCL	5 μm	10^{-3}	2.4×10^{-11}	0.9	2.4×10^{-10}	10

Taylor series yields

$$-\ln V = 1 - V. \tag{3.5}$$

In the case that the output coupling loss is the main loss and that only a small portion of radiation is coupled out per round trip transit, $1 - R \ll 1$, we obtain the relations:

- $V = R$.
- $\tau_p = T/(-\ln R) = T/(1 - R)$ = lifetime of a photon in the resonator.
- $l_p = 2L/(-\ln R) = 2L/(1 - R)$ = path length of a photon in a resonator.
- $s_{\text{eff}} = 1/(-\ln R) = 1/(1 - R)$ = number of round trip transits.
- $Q = 2\pi l/(-\ln R) = 2\pi l/(1 - R)$ = Q factor.

Table 3.1 shows data of lifetimes of photons in different laser resonators. The lifetimes differ by several orders of magnitude and correspond to ten to hundred round trip transits through a resonators. There is an exception: a semiconductor laser is, in principle, able to operate if the surfaces of a semiconductor act as laser mirrors. A reflectivity of $R = 0.33$ of each of the surfaces (of a GaAs semiconductor crystal) corresponds to a photon lifetime $\tau_p = 0.9T/2$. The V factor at a single transit is $V_1 = 0.33$ and the laser threshold condition $V_1 G_1 = 1$ requires a threshold gain factor $G_1 = 3$ at a single transit of the photons through the resonator.

3.3 Number of Photons in a Resonator Mode

The energy of radiation in a mode of a resonator is quantized (Fig. 3.1) and assumes the values

$$E_n = n \times h\nu, \tag{3.6}$$

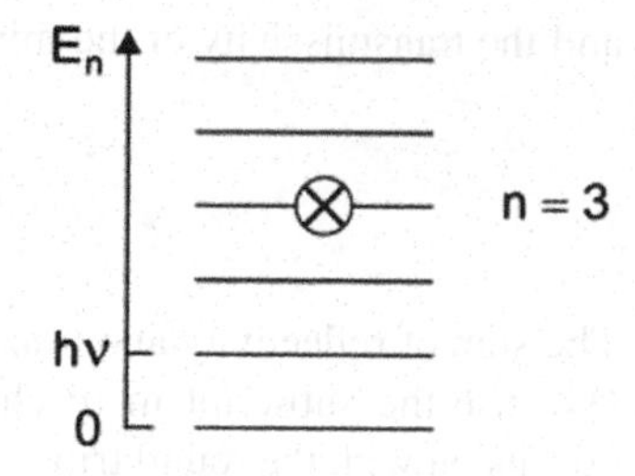

Fig. 3.1 Number of photons in a mode

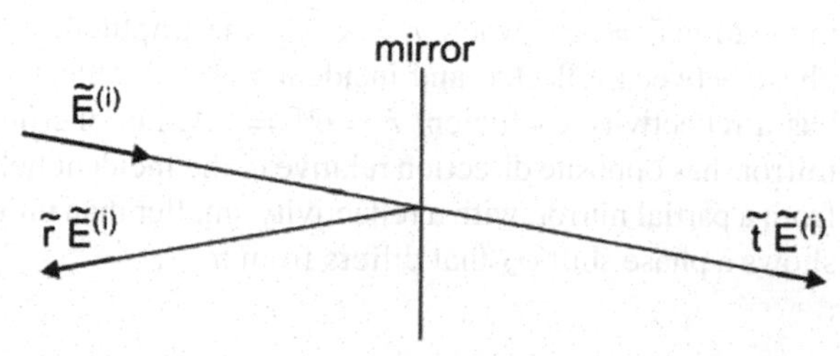

Fig. 3.2 Laser mirror

with $n = 0, 1, 2, \ldots$; we neglect the zero point energy. The occupation number n is equal to the total number of photons in a single-mode resonator. The photon occupation number of a mode of a rectangular Fabry–Perot resonator is Za_1a_2L, where Z is the average photon density.

3.4 Ideal Mirror

We characterize a mirror by the complex reflection coefficient $\tilde{r}$ and the complex transmission coefficient $\tilde{t}$ (Fig. 3.2). A plane wave

$$\tilde{E}^{(i)} = A\, e^{i(\omega t - kz)} \tag{3.7}$$

incident on a mirror is partly reflected and partly transmitted. The reflected field is

$$\tilde{E}^{(r)} = \tilde{r} A\, e^{i(\omega t + kz)} \tag{3.8}$$

and the transmitted field is

$$\tilde{E}^{(t)} = \tilde{t} A\, e^{i(\omega t - kz)}. \tag{3.9}$$

We introduce an *ideal mirror* as a mirror of negligible thickness. The reflectivity of the mirror is

$$R = \frac{|\tilde{E}^{(r)}|^2}{|\tilde{E}^{(i)}|^2} = |\tilde{r}|^2 \tag{3.10}$$

and the transmissivity of the mirror is

$$T_{\mathrm{m}} = \frac{|\tilde{E}^{(\mathrm{t})}|^2}{|\tilde{E}^{(\mathrm{i})}|^2} = |\tilde{t}|^2. \tag{3.11}$$

The sum of reflectivity and transmissivity of a lossless mirror is unity, $R + T_{\mathrm{m}} = 1$. (We use the subscript m to characterize the mirror transmissivity T_{m} to avoid a confusion with the round trip transit time T.) We can write the reflection coefficient in the form $\tilde{r} = r\mathrm{e}^{\mathrm{i}\varphi}$, where $r = \sqrt{R}$ is the amplitude reflection coefficient and φ the phase between reflected and incident wave. A reflector with the reflectivity $R = 1$ has a reflectivity coefficient $\tilde{r} = \mathrm{e}^{\mathrm{i}\pi} = -1$, i.e., the field at $z = 0$ (location of the mirror) has opposite direction relative to the incident field. Radiation that is reflected from a partial mirror, with a reflectivity smaller than unity and a finite transmissivity, shows a phase shift φ_{R} that differs from π.

3.5 Fabry–Perot Interferometer

A Fabry–Perot interferometer (Fig. 3.3) consists of two plane parallel partial reflectors—of infinite lateral extensions—at a distance L. We consider an interferometer with two equal mirrors at $z = 0$ and $z = L$. Due to multiple reflections within the interferometer, a plane wave incident on the interferometer is split into an infinite number of plane waves. We treat the case that the propagation direction of the radiation is parallel to the interferometer axis. Then, the partial waves transmitted by the interferometer add to the transmitted field

$$\tilde{E}^{(\mathrm{t})} = \tilde{E}_1 + \tilde{E}_2 + \tilde{E}_3 + \ldots = \mathrm{e}^{\mathrm{i}kL}\tilde{t}\tilde{t}\left(1 + \tilde{r}^2\mathrm{e}^{\mathrm{i}\delta} + \tilde{r}^4\mathrm{e}^{2\mathrm{i}\delta} + \ldots\right) A\mathrm{e}^{\mathrm{i}(\omega t - kz)}, \tag{3.12}$$

where

$$\delta = k \times 2L + 2\varphi = 2\omega L/c + 2\varphi \tag{3.13}$$

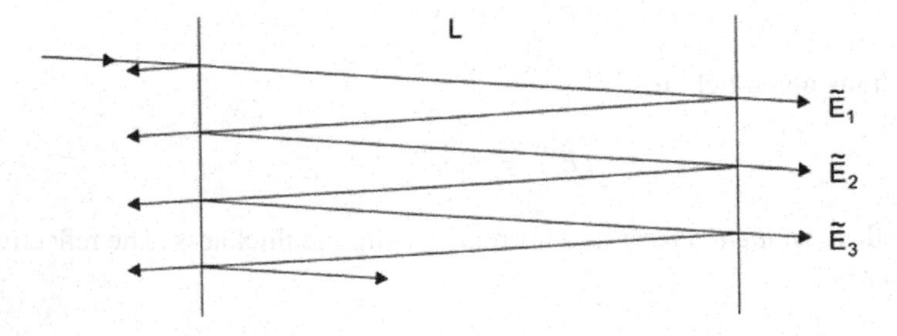

Fig. 3.3 Fabry–Perot interferometer

is the phase difference between two successive partial waves and φ is the phase change due to reflection at a mirror. The partial waves (described by a geometric series) add to the plane wave

$$\tilde{E}^{(t)} = \frac{\tilde{t}\tilde{t}}{1-\tilde{r}^2 e^{i\delta}} A\, e^{i[\omega t - k(z-L)]}. \tag{3.14}$$

The transmissivity of the Fabry–Perot interferometer is given by

$$T_{FP} = \frac{|\tilde{E}^{(t)}|^2}{|\tilde{E}^{(i)}|^2} = \frac{T_m^2}{1+R^2-2R\cos\delta}, \tag{3.15}$$

where $E^{(i)} = A\, e^{i(\omega t - kz)}$ is the incident plane wave, T_m is the transmissivity, and R the reflectivity of a mirror. With $\cos\delta = 1 - 2\sin^2\delta/2$, we obtain the *Airy formula*

$$T_{FP} = \frac{T_m^2}{(1-R)^2}\,\frac{1}{1+4R(1-R)^{-2}\sin^2(\delta/2)}. \tag{3.16}$$

The maximum transmissivity is equal to

$$T_{FP,max} = \frac{T_m^2}{(1-R)^2}, \tag{3.17}$$

obtained for

$$\delta_l = l \times 2\pi;\quad l = 1,\ 2,\ \ldots. \tag{3.18}$$

The maxima appear at the frequencies

$$\nu_l = l \times \frac{c}{2L} + \frac{\varphi}{2\pi}\frac{c}{2L};\quad l = 1,\ 2,\ \ldots. \tag{3.19}$$

The frequency distance between next–near maxima is the free spectral range

$$\Delta\nu_1 = \frac{c}{2L}. \tag{3.20}$$

The halfwidth (full width at half maximum; FWHM) of the Airy curve is equal to

$$\Delta\nu_{res} = \frac{\nu_l}{lF} = \frac{1-R}{\pi\sqrt{R}} \times \frac{c}{2L}. \tag{3.21}$$

The quantity

$$F = \frac{\pi\sqrt{R}}{1-R} \tag{3.22}$$

is the *finesse* of the Fabry–Perot interferometer. The halfwidth is independent of the order of resonance.

In the case that the Fabry–Perot mirrors are lossless, $T_m = 1 - R$, the maximum transmissivity is unity and the transmissivity of a Fabry–Perot interferometer is given by

$$T_{FP} = \frac{1}{1 + 4R(1-R)^{-2}\sin^2\delta/2}. \tag{3.23}$$

Figure 3.4 shows the transmissivity of the Fabry–Perot interferometer for different values of the reflectivity R.

- $R = 0.1$. This case corresponds to a simple plane parallel plate, $R = (n-1)^2 (n+1)^{-2} = 0.1$ for $n = 1.9$; application: low-Q resonator, suitable for wavelength selection of a single-mode laser (Sect. 12.3).
- $R = 0.3$. This case corresponds to a GaAs plate for radiation at a wavelength of 800 nm ($n = 3.6$; $R = 0.33$); application: low-Q resonator of semiconductor lasers.
- $R = 0.8$. Such a Fabry–Perot interferometer shows already Lorentzian resonance curves (*see* next section).

A Fabry–Perot interferometer can be used as an optical frequency analyzer. The Fabry–Perot interferometer mainly transmits radiation of frequencies around ν_l, ν_{l+1}, ... and mainly reflects radiation at frequencies in the ranges between the resonances. At resonance, the transmissivity is unity. The resolving power,

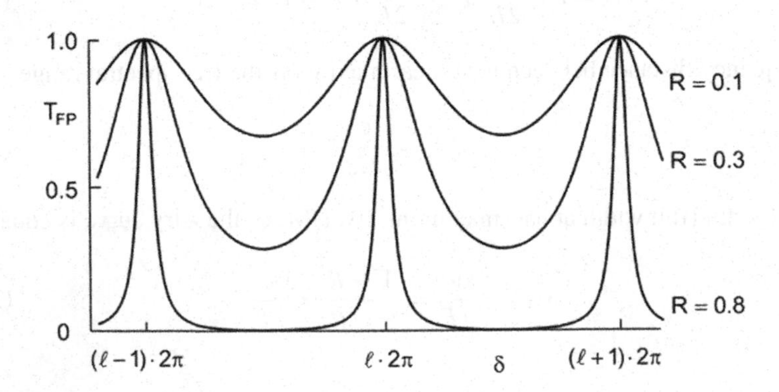

Fig. 3.4 Transmissivity of a Fabry–Perot interferometer

$$\frac{\nu_{\mathrm{res}}}{\Delta\nu_{\mathrm{res}}} = \frac{\lambda_{\mathrm{res}}}{\Delta\lambda_{\mathrm{res}}} = lF, \tag{3.24}$$

is equal to the product of the order of resonance and the finesse; ν_{res} is a resonance frequency, $\Delta\nu_{\mathrm{res}}$ the halfwidth of the transmission curve on the frequency scale, $\lambda_{\mathrm{res}} = \nu_{\mathrm{res}}/c$ the resonance wavelength, and $\Delta\lambda_{\mathrm{res}}$ the halfwidth on the wavelength scale. The free spectral range (on the frequency scale) is equal to $c/2L$.

3.6 Resonance Curve of a Fabry–Perot Resonator

We can consider a Fabry-Perot interferometer as a Fabry-Perot resonator with two equal mirrors. The resonance frequencies are

$$\nu_l = l \times \frac{c}{2L}; \quad l = 1, 2, \ldots. \tag{3.25}$$

The halfwidth

$$\Delta\nu_{\mathrm{res}} = \frac{\nu_l}{lF} = \frac{1-R}{\pi\sqrt{R}} \times \frac{c}{2L} \tag{3.26}$$

of the resonance curve is, as already mentioned, independent of the order of resonance. If the reflectivity is near unity, $\sqrt{R} \approx 1$, the halfwidth is given by

$$\Delta\nu_{\mathrm{res}} = \frac{1-R}{\pi T} = \frac{1}{2\pi\tau_{\mathrm{p}}}. \tag{3.27}$$

$T = 2L/c$ is the round trip transit time of the radiation and

$$\tau_{\mathrm{p}} = \frac{T}{2(1-R)} \tag{3.28}$$

is the photon lifetime. If $R \approx 1$, the Q factor is equal to

$$Q_{\mathrm{res}} = \frac{\nu_l}{\Delta\nu_{\mathrm{res}}} = \frac{\pi l}{1-R} = \omega_l\tau_{\mathrm{p}}. \tag{3.29}$$

The values of τ_{p} and of Q_{res} are half the corresponding values of a Fabry–Perot resonator that consists of a reflector with a reflectivity of unity and another reflector with a reflectivity R.

We consider the transmissivity of a Fabry–Perot resonator at frequencies in the vicinity of a resonance. We expand the sine function in the denominator of (3.23), $\sin(\delta/2) = l\pi(\nu - \nu_l)/\nu_l$. We find

$$T_{\text{FP}} = \frac{\Delta\nu_{\text{res}}^2}{4} \frac{1}{(\nu - \nu_l)^2 + (\Delta\nu_{\text{res}}^2/4)}, \tag{3.30}$$

which we write as

$$T_{\text{FP}} = \frac{\pi\,\Delta\nu_{\text{res}}}{2}\, g_L(\nu_l), \tag{3.31}$$

where

$$g_L(\nu) = \frac{\Delta\nu_{\text{res}}}{2\pi} \frac{1}{(\nu - \nu_{\text{res}})^2 + (\Delta\nu_{\text{res}}/2)^2} \tag{3.32}$$

is the Lorentzian function (=Lorentz resonance function). It has the properties:

- ν_{res} = resonance frequency.
- $\Delta\nu_{\text{res}}$ = halfwidth of the resonance curve.
- $g_{\text{L}}(\nu_0) = 2(\pi\,\Delta\nu_{\text{res}})^{-1} \approx 0.64/\Delta\nu_{\text{res}}$.
- $\int_0^\infty g_{\text{L}}(\nu)\text{d}\nu = 1$.

Our treatment shows that the resonance curve of a Fabry–Perot resonator (for R near unity) is a Lorentzian, and that the halfwidth of the resonance curve is $\Delta\nu_{\text{res}} = 1/2\pi\,\tau_{\text{p}}$. The width of the resonance curve is determined by the lifetime of a photon in the resonator.

There is an essential difference between a Fabry–Perot interferometer and a Fabry–Perot resonator:

- A Fabry–Perot interferometer has an infinite lateral extension.
- A Fabry–Perot resonator has finite lateral extensions, with the consequence that specific mode patterns occur (Sect. 11.6) and that diffraction plays an important role (Sect. 11.8).

3.7 Fabry–Perot Resonator Containing a Gain Medium

The field transmitted by a Fabry–Perot resonator containing a gain medium is given by

$$\tilde{E}^{(\text{t})} = \text{e}^{ikL}\tilde{t}\tilde{t}G_{1,\text{E}}\left(1 + G_{1,\text{E}}^2 r^2\text{e}^{\text{i}\delta} + G_{1,\text{E}}^4 r^4\text{e}^{2\text{i}\delta} + \cdots\right) A\text{e}^{\text{i}(\omega t - kz)}. \tag{3.33}$$

$G_{1,\text{E}}$ is the single-pass gain factor of the field. We assume that $G_{1,\text{E}}$ is real and that $G_1 R = G_{1,\text{E}}^2 rr^* = GR$ is smaller than 1. We obtain the transmissivity of a Fabry–Perot resonator containing an active medium:

$$T_{\text{FP}}^* = \frac{G(1-R)^2}{(1-GR)^2} \times \frac{1}{1 + 4GR\left[(1-GR)^{-2}\,\sin^2(\delta/2)\right]}. \tag{3.34}$$

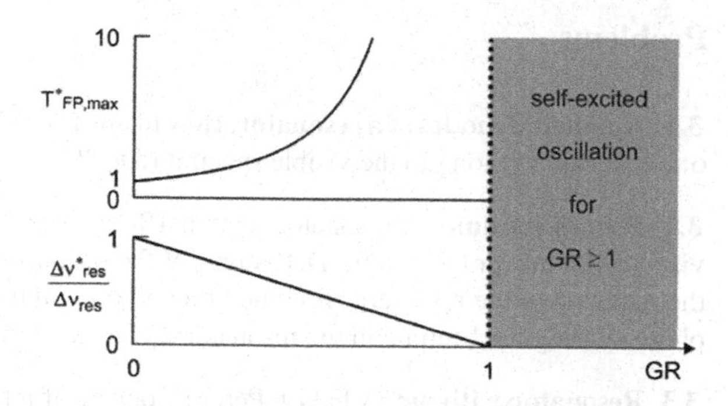

Fig. 3.5 Fabry–Perot resonator containing an active medium

The transmissivity (for $1 - R \ll 1$) is given by

$$T^*_{\mathrm{FP}} = T^*_{\mathrm{FP,max}} \frac{(\Delta \nu^*_{\mathrm{res}}/2)^2}{(\nu - \nu_l)^2 + (\Delta \nu^*_{\mathrm{res}}/2)^2}. \tag{3.35}$$

The transmission curve has Lorentzian shape. The halfwidth is equal to

$$\Delta \nu^*_{\mathrm{res}} = \frac{1 - GR}{\pi} \frac{c}{2L} = \frac{1 - GR}{\pi T}. \tag{3.36}$$

The maximum transmissivity (for $\delta = l \times 2\pi$) is given by

$$T^*_{\mathrm{FP,max}} = \frac{G(1 - R)^2}{(1 - GR)^2}. \tag{3.37}$$

The maximum transmissivity of a Fabry–Perot resonator containing an active medium (Fig. 3.5) is larger than unity. Incident radiation is amplified. The transmissivity becomes infinitely large for $GR \to 1$. The halfwidth of the resonance curve decreases with increasing GR and approaches zero for $GR \to 1$. If $GR > 1$, a Fabry–Perot resonator with an active medium is able to perform a self-excited oscillation—an oscillation starts from noise and is self-sustained. We will show later (Sect. 8.9) that the halfwidth of the spectral distribution of radiation generated in a resonator containing an active medium can be very small but is always nonzero.

We mention that a quasiclassical oscillator can also generate radiation with Lorentzian lineshape (Chap. 31, Fig. 31.3b).

Example Fabry–Perot resonator ($R_1 = R_2 = R = 0.9$) containing a gain medium ($G = 1.1$). The peak transmissivity is $T^*_{\mathrm{FP,max}} = 120$ and the width $\Delta \nu^*_{\mathrm{res}} = 0.01/\pi T$ in comparison to $\Delta \nu_{\mathrm{res}} = 0.1/\pi T$.

References [1, 8, 9, 26, 28–30, 310].

Problems

3.1 Number of modes of a resonator. How many modes of a Fabry–Perot resonator of 1 cm length belong to the visible spectral range?

3.2 Photon lifetime. A resonator (length 10 cm) has loss due to output coupling via a partial mirror ($R = 0.9$). Determine: V factor; number of round trip transits of the radiation in the resonator; lifetime of a photon in the resonator; path length of a photon during the lifetime in the resonator; Q factor of the resonator.

3.3 Resonator with air. A Fabry–Perot resonator of a length $L = 0.5$ m operates, for visible radiation near 600 nm, in vacuum or in air (refractive index $n = 1.00027$).

(a) Evaluate the frequency difference between next–near resonances.
(b) Determine the change of the mode separation if the resonator, originally in vacuum, is flooded with air.

3.4 Energy density of radiation in a Fabry–Perot resonator.

(a) Evaluate the average energy density of a quasiplane standing wave in a Fabry–Perot resonator.
(b) Show that the average energy density is twice the energy density (averaged over time and space) of two beams of light propagating in opposite directions.

3.5 Laser with two output coupling mirrors. A resonator with two output coupling mirrors (reflectivities $R_1 = R_2 = R$, with $R - 1 \ll 1$) emits radiation in two opposite directions. Determine the V factor and the photon lifetime.

3.6 Photon density. Relate the average density of photons in a Fabry–Perot resonator and the average density of photons in the light beam outside the resonator for the following two cases.

(a) $R_1 = 1$ and $R_2 = R$.
(b) $R_1 = R_2 = R$.

3.7 Evaluate the reflectivity of a symmetric lossless Fabry–Perot interferometer ($R_1 = R_2 = R$).

3.8 Determine the transmissivity (Airy formula) of an asymmetric Fabry–Perot interferometer.

3.9 Fabry–Perot interferometer with absorbing mirrors.

(a) Determine the transmissivity (Airy formula) and the maximum transmissivity of a symmetric Fabry–Perot interferometer with two absorbing mirrors.
(b) What is the condition, with respect to the absorptivity of the mirrors, that the maximum transmissivity of the Fabry–Perot interferometer is larger than 0.98? [*Hint*: the sum of the transmissivity T_m, the reflectivity R, and the absorptivity A_m of a mirror is unity.]

3.10 Fabry–Perot interferometer at obliquely incident radiation.

(a) Determine the resonance condition of a symmetric Fabry–Perot interferometer if the direction of incident radiation has an angle θ relative to the interferometer axis.
(b) Determine the resonance condition assuming that the interferometer is a plane parallel plate.

3.11 Determine the reflectivity of a plane surface of a dielectric medium from Fresnel's formulas of normal incidence (*see* Problems to Chap. 2).

3.12 Determine the mode spacing for different Fabry–Perot resonators: $L = 1\,\mathrm{m}$ $(n = 1)$; $L = 10\,\mathrm{cm}$ $(n = 1)$; $L = 10\,\mu\mathrm{m}$ $(n = 3.6)$.

3.13 A Fabry–Perot interferometer with nonabsorbing mirrors has, for radiation at a resonance frequency, a transmissivity of 1 although the entrance mirror can have, for itself, a reflectivity of, for instance, 99%. Why do the boundary conditions of the incident wave not require that a portion of radiation is reflected at the entrance mirror? [*Hint*: sketch the incident field and the field that propagates within the interferometer for a moment of strong field strength at the position of the entrance mirror and compare the situation with the case that radiation is reflected at a single mirror rather than at a resonator.]

3.14 Resonance frequencies of a Fabry–Perot interferometer. Determine the resonance frequencies of a Fabry–Perot interferometer, taking account of phase changes at the mirrors; suppose that radiation is propagating along the axis of the interferometer.

(a) Show that the resonance condition leads to the frequencies $\nu_l' = \nu_l(1 + \varphi_\mathrm{R}/2\pi)$, where $\varphi_\mathrm{R} = 2\pi - (\varphi_\mathrm{R1} + \varphi_\mathrm{R2})$ and φ_R1 and φ_R2 are the phase changes due to reflection at the two mirrors (R1 and R2).
(b) Show that the frequency distance between next-nearest modes is the same for $\varphi_\mathrm{R} \neq 0$ as for $\varphi_\mathrm{R} = 0$.
(c) Show that the resonance wavelengths are equal to multiples of λ.

Chapter 4
The Active Medium: Energy Levels and Lineshape Functions

We present a characterization of active media with respect to energy levels and line broadening. We make a distinction between two-level based lasers and energy-ladder based lasers.

In a two-level based laser, stimulated transitions occur between two levels of an atomic system. A two-level system of a particular atom or molecule is a subsystem of the energy levels of the atom or the molecule. We characterize the two-level based lasers as: four-level lasers; three-level lasers; two-level lasers; two-band lasers; and quasiband lasers. All presently operating lasers except free-electron lasers can, in principle, be described as two-level based lasers.

In an energy-ladder based laser, stimulated transitions occur between levels of energy-ladder systems. The (yet hypothetical) Bloch laser (Chap. 32) belongs to this type. We will make use (Chap. 19) of the concept of an energy-ladder based laser to illustrate properties of free-electron lasers.

Line broadening can be due to homogeneous or inhomogeneous broadening. We discuss the Lorentzian and Gaussian lineshape functions. We describe the classical oscillator model of an atom and the natural line broadening.

Finally, we introduce low-dimensional active media. In a low-dimensional medium, the free motion of electrons is spatially restricted. There are two-dimensional media, with electrons moving along a plane, or one-dimensional media, with electrons moving along a line. The strongest restriction occurs in a zero-dimensional medium—all three dimensions are restricted (like in an atom). The classification as three-, two-, one-, and zero-dimensional media concerns semiconductor lasers. The most important semiconductor lasers—quantum well lasers—operate with two-dimensional active media.

K.F. Renk, *Basics of Laser Physics*, Graduate Texts in Physics,
DOI 10.1007/978-3-319-50651-7_4

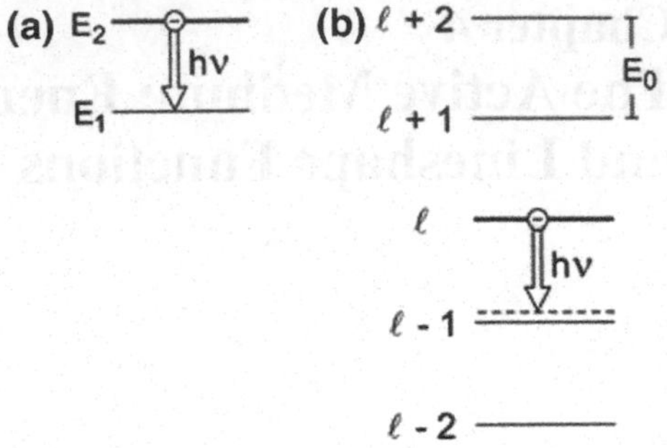

Fig. 4.1 Types of laser. **a** Two-level based laser. **b** Energy-ladder based laser

4.1 Two-Level Based and Energy-Ladder Based Lasers

A *two-level based laser* contains two-level atomic systems. Coherent radiation is generated by stimulated transitions between the upper and the lower laser levels of two-level systems (Fig. 4.1a). Quantities characterizing a two-level system are:

- E_2 = energy of the upper laser level
- E_1 = energy of the lower laser level
- $E_2 - E_1$ = transition energy
- $\nu_0 = (E_2 - E_1)/h$ = transition frequency = atomic resonance frequency

The laser frequency has a value at or near the transition frequency,

$$\nu_L \sim \nu_0. \tag{4.1}$$

We will characterize the two-level based lasers according to the number of different levels involved in transitions in a laser medium (Sects. 4.2 and 4.3). A two-level system is a subsystem of the energy level system of an atom or a molecule. All presently operating lasers—except free-electron lasers—are two-level based lasers.

An *energy-ladder based laser* contains energy-ladder systems. A free-electron in a spatially periodic field executes oscillations (free-electron oscillations). According to an energy level description, an oscillating free-electron forms an energy-ladder system and occupies one of the levels of the energy-ladder system. Stimulated emission of a photon by the oscillating electron corresponds to a stimulated transition between next-near energy levels of the energy-ladder system (Fig. 4.1b). The energy levels are equidistant,

$$E_l = l\, E_0, \tag{4.2}$$

where l is an integer, Quantities characterizing an energy-ladder system are:

- E_0 = energy distance between next-near levels = transition energy.
- $\nu_0 = E_0/h$ = transition frequency = resonance frequency of the electron oscillation.

The laser frequency ν_L of an energy-ladder based laser is slightly smaller than the resonance frequency,

$$\nu_L < \nu_0. \tag{4.3}$$

We will describe the (yet hypothetical) superlattice Bloch laser (Sect. 32.8) as an energy-ladder based laser. We will furthermore show that the free-electron laser can, in principle, be interpreted as an energy-ladder based laser (Sect. 19.17).

4.2 Four-Level, Three-Level, and Two-Level Lasers

In a four-level laser (Fig. 4.2), a pump excites atoms, molecules, or other atomic systems from the ground state level (level 0) to an excited state level (level 3 = pump level). Relaxation leads to population of the upper laser level (level 2). Stimulated emission by 2 → 1 transitions results in a population of the lower laser level (level 1). Depopulation of the lower level occurs by relaxation to the ground state. We have three relaxation processes, namely 3 → 2, then 2 → 1, and 1 → 0. We assume that the relaxation 3 → 2 is very fast. We ignore other relaxation processes (e.g., 2 → 0). Continuous pumping maintains a *permanent population inversion* ($N_2 > N_1$) if the *relaxation time* τ_{rel}^* of the upper laser level is larger than the relaxation time τ_{rel} of the lower laser level,

$$\tau_{\mathrm{rel}}^* > \tau_{\mathrm{rel}}. \tag{4.4}$$

Without stimulated emission, the population of the upper level is equal to the product of the *pump rate* r and the relaxation time τ_{rel}^*,

$$N_2 = r\tau_{\mathrm{rel}}^*. \tag{4.5}$$

Pumping of a four-level laser medium creates two-level atomic systems; a two-level atomic system is either in its ground state (level 1) or in its excited state (level 2).

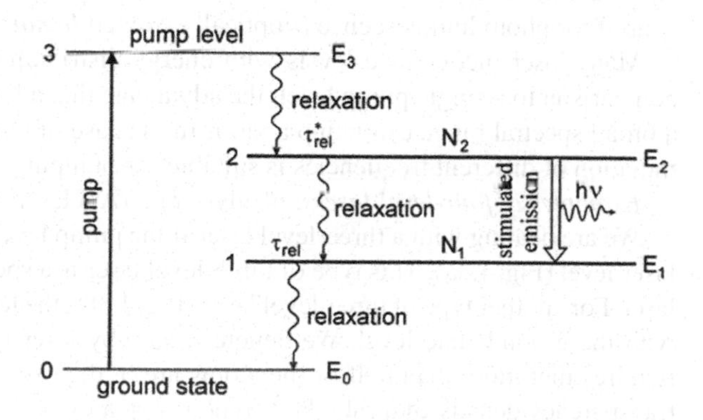

Fig. 4.2 Four-level laser (*principle*)

Table 4.1 Relaxation times of laser levels

Laser	λ	τ_{rel}^{*}	τ_{rel}
HeNe	633 nm	100 ns	10 ns
CO_2	10.6 μm	5 s	< 5 s
Nd:YAG	1.06 μm	230 μs	≪ 230 μs
TiS	790 nm	3.8 μs	10^{-12} s
QCL	5 μm	10^{-11} s	10^{-12} s

The density of two-level atomic systems, $N_2 + N_1$, increases with increasing pump strength. Population inversion in the active medium of a four-level laser medium occurs already at the smallest pump rate.

The maximum efficiency of conversion of a pump quantum (energy $E_3 - E_0$) to a laser light quantum (energy $h\nu$) is the *quantum efficiency*

$$\eta_{\mathrm{q}} = \frac{h\nu}{E_3 - E_0} = \frac{E_2 - E_1}{E_3 - E_0}. \tag{4.6}$$

The quantum efficiency is very small ($\eta_{\mathrm{q}} \ll 1$) if $E_2 - E_1 \ll E_3 - E_0$. Then, a large portion of the pump energy is converted to relaxation energy and therefore to heat (or to radiation produced by spontaneous emission). The quantum efficiency is near unity if the pump level lies only slightly above the upper laser level and if, at the same time, the lower laser level lies only slightly above the ground state level.

Table 4.1 shows values of relaxation times of upper and lower laser levels of a few laser materials. The relaxation times differ by many orders of magnitude. The relaxation $2 \rightarrow 1$ can be due to spontaneous emission of radiation or due to nonradiative relaxation. Relaxation of the lower level can also be due to spontaneous emission of radiation or due to nonradiative relaxation, that is, by a radiationless transition. We will specify the relaxation processes later in connection with the discussion of specific lasers. The lasers mentioned in the table can operate as continuous wave lasers ($\tau_{\mathrm{rel}}^{*} > \tau_{\mathrm{rel}}$).

Optical transitions between two discrete energy levels lead to fluorescence lines and absorption lines that have finite linewidths (Sect. 4.4); the notation *fluorescence* is used for photo luminescence (=optically excited *luminescence*).

Many laser media have levels with energy distributions. A pump band has, in comparison to a single pump level, the advantage that a lamp that emits radiation in a broad spectral range can pump a laser. In the case of pumping with another laser, radiation of different frequencies is suitable for pumping.

Examples of four-level lasers: neodymium YAG laser; titanium–sapphire laser.

We are dealing with a three-level laser if the pump level coincides with the upper laser level (Fig. 4.3a). This type of three-level laser is a special case of the four-level laser. For another type of three-level laser (Fig. 4.3b), the lower laser level is identical with the ground state level. We denote it as ruby laser type. Population inversion requires that more than half of the atoms are in the excited state. Accordingly, the transparency density of a ruby laser type is given by

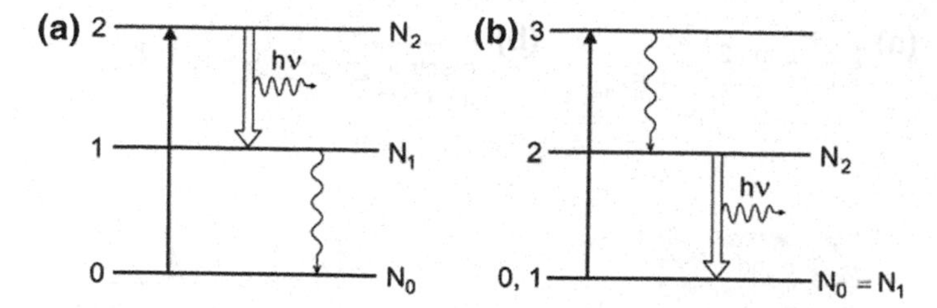

Fig. 4.3 Three-level lasers. **a** Three-level laser with coinciding pump and upper laser level. **b** Ruby laser type

$$N_{\text{tr}} = N_0/2. \tag{4.7}$$

N_0 is the density of impurity ions in a crystal.

In a *two-level laser* only two atomic energy levels play a role.

Example of a two-level laser: the ammonia (NH_3) maser [32]. To obtain population inversion, long-lived excited molecules are permanently injected into a resonator where stimulated emission processes occur. The molecules leave the resonator spatially. The upper laser level is 35 μ eV above the ground state level. Ammonia gas in a box contains NH_3 molecules in the ground state and in the excited state. A hole in a box with NH_3 gas at room temperature is the source of a molecular beam consisting of excited and nonexcited NH_3 molecules. The molecular beam traverses an atomic filter that separates the excited molecules and the nonexcited molecules. The atomic filter consists of an inhomogeneous field (an electric quadrupole field) that exerts forces on the molecules due to their electric dipole moments. The magnitude of the dipole moment of a molecule in the excited state differs from that in the ground state. Therefore, the forces lead to a spatial separation of the molecules. The excited molecules pass the resonator and deliver, via stimulated emission, the excitation energy to the laser field in the resonator. The ammonia laser was the first microwave maser (frequency near 24 GHz).

4.3 Two-Band Laser and Quasiband Laser

A *two-band laser* medium has (besides other energy levels or energy bands) a lower energy band (band 1) and an upper energy band (band 2), separated by an energy gap (Fig. 4.4a). The gap energy is E_g. Without pumping, almost all energy levels belonging to the lower band are full and all energy levels of the upper band are empty. We suppose that E_g is much larger than kT so that thermal excitation from the lower to the upper band can be ignored.

Pumping—injection of electrons into the upper band and extraction of electrons from the lower band (Fig. 4.4b)—results in a quasithermal population of energy levels in the upper band and in empty levels of the lower band. The electrons in the

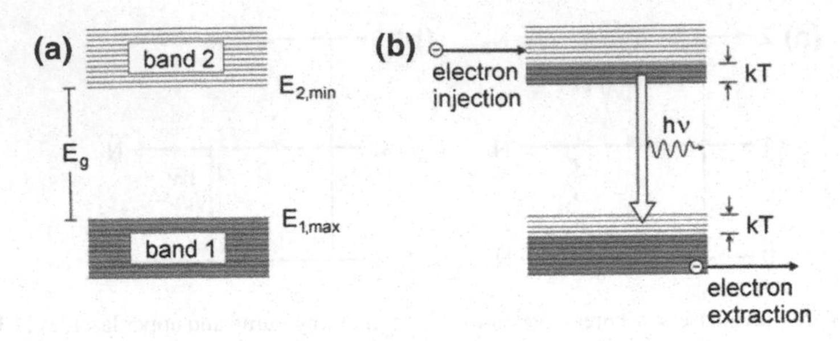

Fig. 4.4 Two-band laser. **a** Energy bands. **b** Laser principle

Table 4.2 Relaxation times of laser levels of two-band laser media and quasiband laser media

Laser	λ (μm)	τ^*_{rel}	τ_{intra} (s)
Semiconductor	0.4–2	1–5 ns	10^{-13}
Fiber	1–2	10^{-2} s	10^{-13}

upper band undergo fast intraband relaxation. The population in the upper band is in a quasithermal equilibrium, which is determined by the lattice temperature of the active medium. The populated levels have energies near the minimum of the upper band. The width of the energy distribution of populated levels is $\sim kT$ (or larger at strong pumping). Energy levels near the energy minimum of the upper band have the largest population. The energy distribution of populated levels in the upper band is governed by Fermi's statistics.

The population in the lower band is also in a quasithermal equilibrium, corresponding to the lattice temperature of the active medium. The empty levels have energies near the maximum of the lower band. The width of the energy distribution of empty levels is $\sim kT$ (or larger at strong pumping). Energy levels near the maximum of the lower band have the lowest population. Fermi's statistics determines the energy distribution of populated levels in the lower band too.

Quasithermal means that the population within an energy band has a thermal distribution according to the lattice temperature of the active medium—but that the population of the upper energy band is, relative to the population in the lower energy band, far out of equilibrium. Stimulated transitions from occupied levels in the upper band to empty levels in the lower band are the source of laser radiation. Establishment of a quasiequilibrium in the upper band and establishment of a quasiequilibrium in the lower band are due to the interaction of electrons with phonons, that is, due to electron–phonon scattering.

Table 4.2 shows relaxation times: the intraband relaxation time τ_{intra} is much smaller than the interband relaxation time τ^*_{rel}. Interband relaxation in an active medium is mainly due to spontaneous emission of radiation.

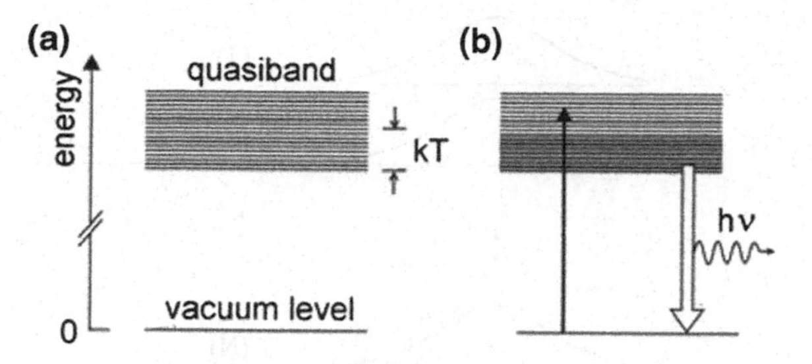

Fig. 4.5 Quasiband laser. **a** Quasiband. **b** Laser principle

The *quasiband laser* represents a model of a glass fiber laser. The active medium (Fig. 4.5a) contains excited-impurity quasiparticles in a *quasiband*. The quasiband lies ~ 1 eV above a vacuum level. The width of a quasiband of an impurity-doped glass can have a value of 10–100 meV. Optical pumping via transitions from the vacuum level to the quasiband creates quasiparticles (Fig. 4.5b). Annihilation of quasiparticles via stimulated transitions from the quasiband to the vacuum level is the origin of laser radiation. The quasiparticles in the quasiband have a quasithermal distribution determined by Fermi's statistics; for relaxation times, see Table 4.2. The quasiband model will be described in Chap. 18.

Examples Two-band lasers: all bipolar semiconductor lasers.
Quasiband lasers: erbium-doped fiber laser; other fiber lasers and fiber amplifiers (Sect. 15.7 and Chap. 18).
Two-quasiband lasers (with the active medium having a lower and an upper quasiband): organic and polymer lasers (Sect. 34.4).

4.4 Lineshape: Homogeneous and Inhomogeneous Line Broadening

We use the notation "lineshape" in different ways:

- Lineshape of a luminescence line.
- Lineshape of an absorption line = absorption profile = shape of an absorption coefficient = slope of $\alpha_{abs}(\nu)$.
- Lineshape of a gain curve = gain profile = shape of a gain coefficient = shape of $\alpha(\nu)$.

We characterize the lineshape of a line that is due to transitions between two levels of an atomic system by:

- ν_0 = center frequency of a line.
- $\Delta\nu_0$ = linewidth = halfwidth = full width at half maximum (FWHM).
- The lineshape function $g(\nu)$ with the normalization $\int_0^\infty g(\nu)d\nu = 1$ or, alternatively, $\bar{g}(\nu)$ with the normalization $\bar{g}(\nu_0) = 1$.

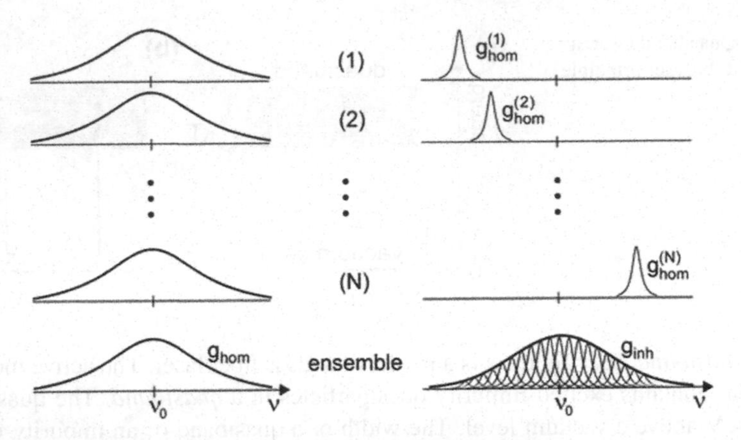

Fig. 4.6 Homogeneous and inhomogeneous line broadening

There are many different mechanisms responsible for lineshapes. Accordingly, there is a large number of different lineshapes. We divide the lineshapes as lineshapes due to homogeneous broadening and lineshapes due to inhomogeneous broadening.

A homogeneous line broadening occurs if all two-level atomic systems have the same lineshape function $g_{\rm hom}(\nu) = g^{(1)} = g^{(2)} = \ldots = g^{(N)}$, where N is the number of atomic systems (Fig. 4.6, left). The atomic resonance frequency ν_0 is the same for all two-level atomic systems.

In the case that a line is inhomogeneously broadened (Fig. 4.6, right), each two-level atomic system of an ensemble has its own resonance frequency. The linewidth of the ensemble is larger than the transition linewidth of a single two-level atomic system. We can regard an inhomogeneously broadened line (with the lineshape function $g_{\rm inh}$ and the center frequency ν_0) as composed of homogeneously broadened lines with the frequencies $\nu_{0,1}, \nu_{0,2}, \ldots, \nu_{0,N}$ of different two-level systems.

Examples of homogeneous line broadening: collision broadening in gases (Sect. 14.2) and vibronic line broadening of the transition in Ti^{3+}:Al_2O_3 used for operation of the titanium–sapphire laser (Chap. 5; Sects. 7.6 and 15.2; Chap. 17).

Example of inhomogeneous broadening: Doppler broadening of transition lines in gases (Sect. 14.1).

4.5 Lorentz Functions

An important lineshape function is the *Lorentz resonance function* (Fig. 4.7),

$$g_{\rm L}(\nu) = g_{\rm L,res}(\nu) = \frac{\Delta\nu_0}{2\pi} \frac{1}{(\nu_0 - \nu)^2 + \Delta\nu_0^2/4}, \tag{4.8}$$

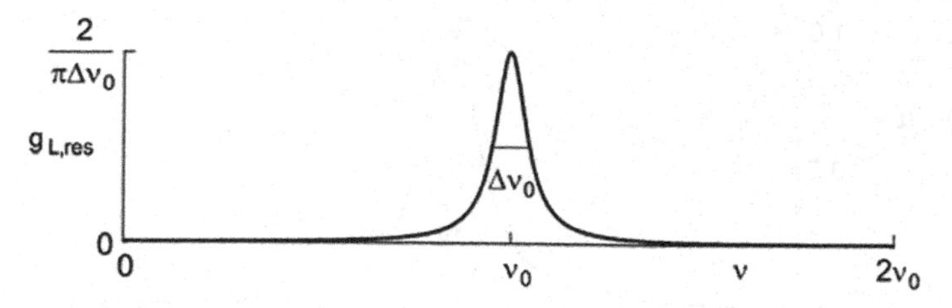

Fig. 4.7 Lorentz resonance function (Lorentzian lineshape function)

where ν_0 is the resonance frequency and $\Delta\nu_0$ the halfwidth (full width at half maximum, FWHM). We suppose that we are dealing with a narrow line. Then, the integral over the Lorentz resonance function, from zero to infinite, is approximately equal to unity,

$$\int_0^\infty g_{\mathrm{L,res}}(\nu)\mathrm{d}\nu = 1. \tag{4.9}$$

The maximum value is equal to

$$g_{\mathrm{L,res}}(\nu_0) = \frac{2}{\pi \Delta\nu_0} \approx \frac{0.64}{\Delta\nu_0}. \tag{4.10}$$

The peak value of the Lorentz resonance curve is equal to the inverse of the halfwidth of the curve (times $2/\pi$); with decreasing halfwidth, the Lorentz resonance curve narrows.

$$g_{\mathrm{L,disp}}(\nu) = \frac{1}{\pi}\frac{\nu_0 - \nu}{(\nu_0^2 - \nu^2)^2 + \Delta\nu_0^2/4} = \frac{\nu_0 - \nu}{\Delta\nu_0/2}\, g_{\mathrm{L,res}}(\nu). \tag{4.11}$$

We write the Lorentz resonance function, normalized to unity at the line center, in dimensionless units:

$$\bar{g}_{\mathrm{L,res}}(\nu/\nu_0) = \frac{(\Delta\nu_0/\nu_0)^2/4}{(1 - \nu/\nu_0)^2 + (\Delta\nu_0/\nu_0)^2/4}. \tag{4.12}$$

The corresponding Lorentz dispersion function is equal to

$$\bar{g}_{\mathrm{L,disp}}(\nu/\nu_0) = \frac{(\nu_0 - \nu)(\Delta\nu_0/2)}{(\nu_0 - \nu)^2 + (\Delta\nu_0)^2/4} = \frac{1 - \nu/\nu_0}{\Delta\nu_0/2\nu_0}\, \bar{g}_{\mathrm{L,res}}(\nu/\nu_0). \tag{4.13}$$

The Lorentz resonance curve (Fig. 4.8, upper part) is symmetric with respect to the resonance frequency ν_0, while the Lorentz dispersion curve (Fig. 4.8, lower part) is antisymmetric. The Lorentz dispersion curve is zero at $\nu = \nu_0$ and has extrema at the frequencies $\nu_0 \pm \Delta\nu_0/2$. The extrema of $\bar{g}_{\mathrm{L,disp}}$ are equal to $\pm$ 0.5.

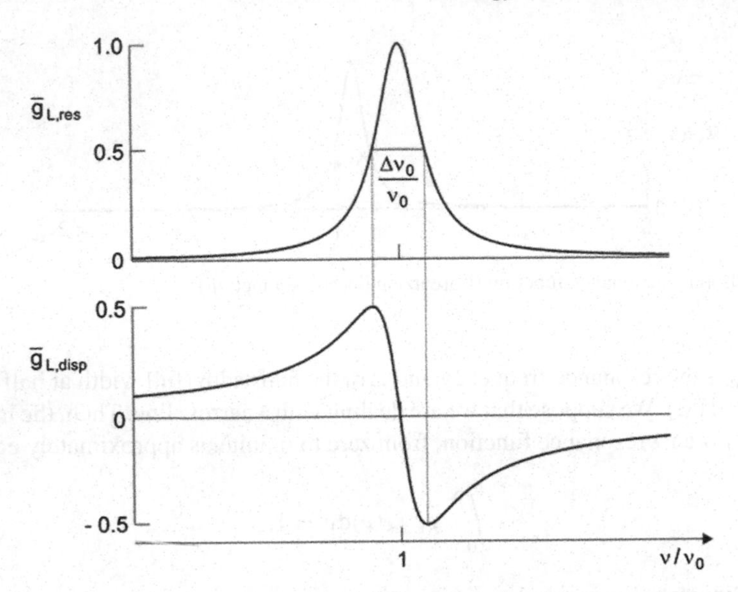

Fig. 4.8 Lorentz functions. **a** Lorentz resonance function. **b** Lorentz dispersion function

The Lorentz resonance function on the ω scale is given by

$$g_{\mathrm{L,res}}(\omega) = \frac{\Delta\omega_0}{2\pi} \frac{1}{(\omega_0 - \omega)^2 + \Delta\omega_0^2/4}. \tag{4.14}$$

We have the relation, because of $g(\omega)\mathrm{d}\omega = g(\nu)\mathrm{d}\nu$, $\omega = 2\pi\nu$ and $\int g(\omega)\mathrm{d}\omega = 1$,

$$g_{\mathrm{L,res}}(\nu) = 2\pi\, g_{\mathrm{L,res}}(\omega). \tag{4.15}$$

On the ω scale, the Lorentz resonance function, normalized to unity at the line center, has the form

$$\bar{g}_{\mathrm{L,res}}(\omega) = \frac{\pi\,\Delta\omega_0}{2}\, g_{\mathrm{L,res}}(\omega) = \frac{\Delta\omega_0^2/4}{(\omega_0 - \omega)^2 + \Delta\omega_0^2/4}. \tag{4.16}$$

The corresponding Lorentz dispersion function is

$$\bar{g}_{\mathrm{L,disp}}(\omega) = \frac{\pi\,\Delta\omega_0}{2}\, g_{\mathrm{L,disp}}(\omega) = \frac{(\omega_0 - \omega)\Delta\omega_0/2}{(\omega_0 - \omega)^2 + \Delta\omega_0^2/4}. \tag{4.17}$$

The Lorentz resonance function on the energy scale is given by

$$g_{\mathrm{L,res}}(h\nu) = \frac{\Delta E_0}{2\pi} \frac{1}{(E_{21} - h\nu)^2 + \Delta E_0^2/4}, \tag{4.18}$$

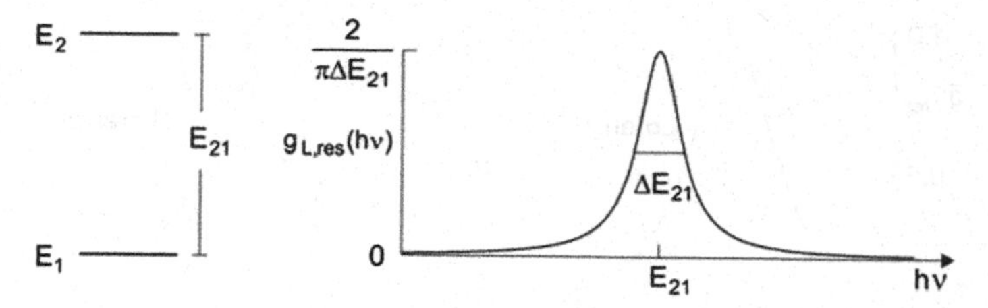

Fig. 4.9 Lorentz resonance function on the energy scale

where we have the quantities:

- $\Delta E_0 = \Delta E_{21} = h\Delta\nu_0 =$ halfwidth of the line on the energy scale.
- $E_{21} = E_2 - E_1 =$ transition energy.
- $\nu_0 = E_{21}/h = (E_2 - E_1)/h =$ transition frequency.
- $h\nu =$ quantum energy of the photons in a radiation field of frequency ν.

The relation $g_{\mathrm{L,res}}(\nu)\mathrm{d}\nu = g_{\mathrm{L,res}}(h\nu)\mathrm{d}(h\nu)$ leads to

$$g_{\mathrm{L,res}}(h\nu) = \frac{1}{h} g_{\mathrm{L,res}}(\nu). \tag{4.19}$$

Because of line broadening, the photon energy $h\nu$ (Fig. 4.9) does not need to coincide with the transition energy E_{21}.

The Lorentz functions we presented describe narrow resonance lines, $\Delta\omega_0 \ll \omega_0$. Otherwise we have functions that we call *general Lorentz functions*. The general Lorentz resonance function is given by

$$G_{\mathrm{L,res}}(\omega) = \frac{\omega\Delta\omega_0}{(\omega_0^2 - \omega^2)^2 + (\omega\Delta\omega_0)^2}. \tag{4.20}$$

The general Lorentz resonance function normalized to 1 at the line center is equal to

$$\bar{G}_{\mathrm{L,res}}(x) = \frac{a^2 x}{(1 - x^2)^2 + a^2 x^2}, \tag{4.21}$$

where $x = \omega/\omega_0 = \nu/\nu_0$ and $a = \Delta\omega_0/\omega_0 = \Delta\nu_0/\nu_0 = \Delta E_0/E_0$. The Lorentz resonance function increases proportionally to frequency at small frequencies, $\omega \ll \omega_0$, and decreases inversely proportional to the third power of the frequency at large frequencies, $\omega \gg \omega_0$ (Fig. 4.10, upper part); the relative halfwidth of the curves in the figure is $a = 0.03$.

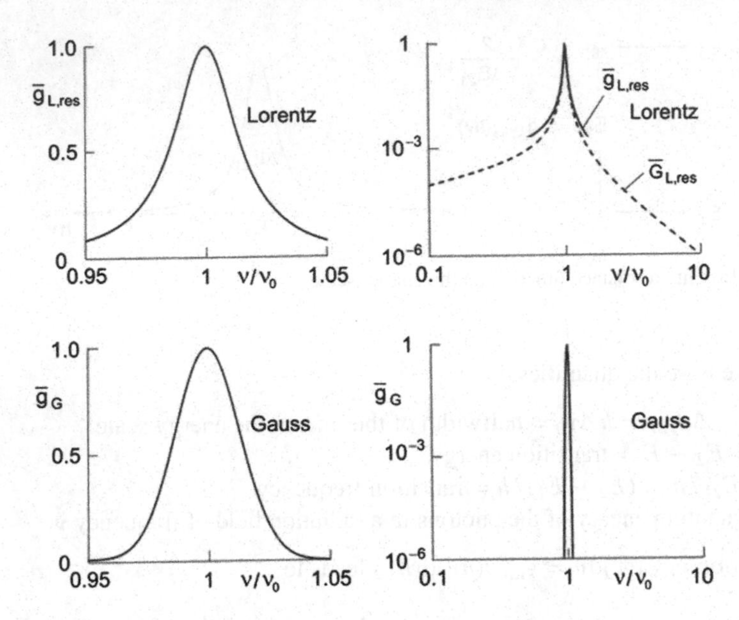

Fig. 4.10 Lorentz resonance function and Gaussian distribution function

4.6 Gaussian Lineshape Function

Line broadening can lead to a *Gaussian lineshape* described by

$$g_G(\nu) = \frac{2}{\Delta\nu_0}\left(\frac{\ln 2}{\pi}\right)^{1/2} \exp\left[-\frac{\ln 2\,(\nu-\nu_0)^2}{\Delta\nu_0^2/4}\right], \tag{4.22}$$

where ν_0 is the center frequency and $\Delta\nu_0$ the half width. The maximum value is

$$g_G(\nu_0) = \frac{2}{\Delta\nu_0}\sqrt{\frac{\ln 2}{\pi}} \approx \frac{0.94}{\Delta\nu_0}. \tag{4.23}$$

The lineshape function is normalized, $\int_0^\infty g_G(\nu)\mathrm{d}\nu = 1$. The Gaussian lineshape function normalized to unity at the line center is

$$\bar{g}_G(\nu) = \exp\left[-\frac{\ln 2\,(\nu-\nu_0)^2}{\Delta\nu_0^2/4}\right]. \tag{4.24}$$

A Gaussian line (Fig. 4.10, lower part) and a Lorentzian line of the same relative halfwidth shows only small differences to at frequencies around the line center. But there are essential differences in the wings. The Gaussian line decreases exponentially and has negligibly small values at frequencies a few halfwidths away from the center frequency; see the double logarithmic plots (Fig. 4.10, right). The Gaussian line has

Table 4.3 Linewidths

Laser	λ	ν (THz)	$\Delta\nu_0$ (GHz)	$\Delta\nu_{nat}$	τ_{sp}
HeNe	633 nm	474	1.6	1.2 MHz	100 ns
CO_2	10.6 μm	28	(0.07–500)	0.03 Hz	5 s
Nd:YAG	1.06 μm	280	140	1 kHz	230 μs

finite values around the center frequency, while the Lorentzian line (with the same linewidth) extends far into the wings.

Examples of Gaussian lines: Doppler broadened lines (Sect. 14.1) are inhomogeneously broadened; the line of Ti^{3+}:Al_2O_3 used for operation of the titanium–sapphire laser is homogeneously broadened (Sect. 17.4).

4.7 Experimental Linewidths

Table 4.3 shows values of the linewidth $\Delta\nu_0$ of $2 \rightarrow 1$ transition lines together with values of the natural linewidth $\Delta\nu_{nat}$. The halfwidth $\Delta\nu_{nat}$ of the upper laser level follows from the relation $\Delta\nu_{nat} = (2\pi\tau_{sp})^{-1}$, where τ_{sp} is the lifetime with respect to $2 \rightarrow 1$ spontaneous transitions (Sect. 4.9). Various methods of determination of linewidths are available. We mention a few methods.

- *Helium–neon laser.* The fluorescence line is inhomogeneously broadened (due to Doppler broadening). The linewidth can be calculated by use of the expression of Doppler broadening (Sect. 14.1).
- *CO_2 laser.* The $2 \rightarrow 1$ fluorescence line is Doppler broadened at low gas pressure and collision broadened at high pressure.
- *Nd:YAG laser.* A fluorescence experiment provides the linewidth.

A lower limit of the linewidth of a transition is the natural linewidth. Active media of lasers operated at room temperature show linewidths of the atomic transitions that are always larger than the natural linewidth as a study of the specific lasers shows (see the chapters beginning with Chap. 14).

4.8 Classical Oscillator Model of an Atom

An atom consists of a nucleus and an electron cloud (Fig. 4.11a). In the *classical oscillator model of an atom*, the electron cloud is replaced by an electron located at the center of the electron cloud, that is, at the position of the nucleus. In equilibrium, the electron does not move. When it is brought out of its equilibrium position (Fig. 4.11b), it performs an oscillation with a displacement $x(t)$ and an amplitude x_0. The oscillation of the electron (charge $q = -e$) corresponds to an oscillation of

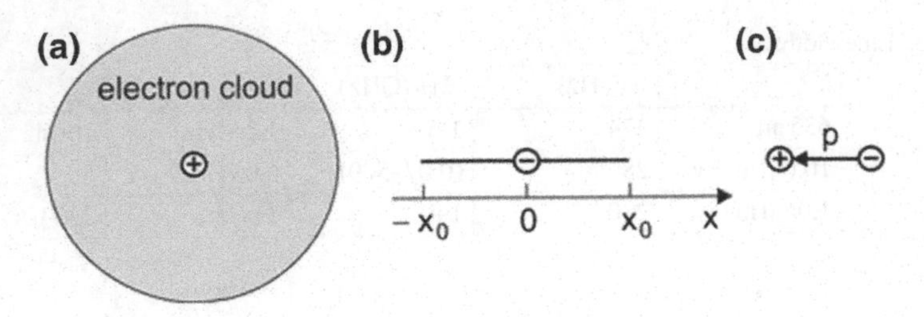

Fig. 4.11 Classical oscillator model of an atom. **a** Electron cloud and nucleus of an atom. **b** Oscillation of an electron around the equilibrium position $x = 0$. **c** Excited atom as an oscillating dipole

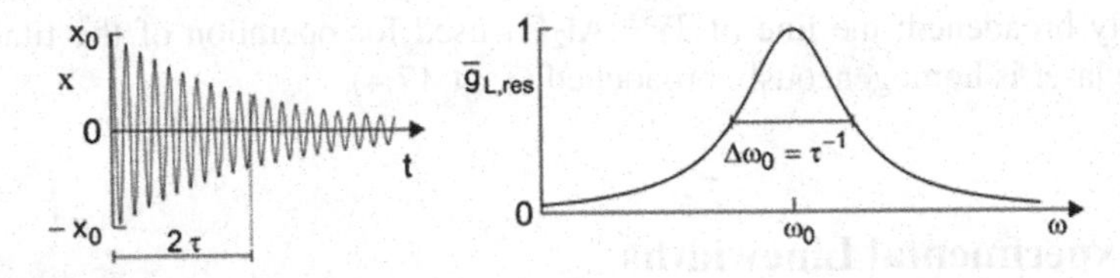

Fig. 4.12 Damped oscillator (*left*) and shape of the fluorescence line (*right*)

a dipole with the dipole moment $p = qx = -ex$ (Fig. 4.11c). An oscillation of a classical electric dipole (= *Hertzian dipole*) gives rise to emission of radiation. Here, we describe the model of the atomic oscillator (= dipole oscillator model = classical oscillator model of an atom = Lorentz model of an atom); later (in Chap. 9), we will make use of the model to derive the gain coefficient of radiation propagating in an active medium.

We consider the atomic oscillation as a damped oscillation. The equation of motion is given by

$$\ddot{x} + \beta\dot{x} + \omega_0^2 x = 0, \tag{4.25}$$

where x is the displacement from the equilibrium position ($x = 0$), ω_0 is the resonance frequency, and β the damping constant. We chose the resonance frequency so that $\hbar\omega_0 = E_2 - E_1$, where E_1 is the energy of the ground state, E_2 the energy of the excited state of the two-level atom, and $E_{21} = E_2 - E_1$ is the transition energy.

The solution to the equation of motion has, for $\beta \ll \omega_0$, the form

$$\tilde{x} = x_0 \mathrm{e}^{-\frac{1}{2}\beta t} \mathrm{e}^{-\mathrm{i}\omega_0 t}, \tag{4.26}$$

where x_0 is a displacement at $t = 0$; the displacement is 0 for $t < 0$. The time of decay of the amplitude of the oscillation is 2τ (Fig. 4.12, left), and the time of decay of the energy of the atomic oscillation is the lifetime $\tau = 1/\beta$.

Connected with an oscillation is an electric field

$$\tilde{E} = A\,\mathrm{e}^{-\frac{1}{2}\beta t}\mathrm{e}^{-\mathrm{i}\omega_0 t}. \tag{4.27}$$

The initial value A of the amplitude of the field corresponds to the initial value of the displacement. The field is not monochromatic but has a frequency distribution (Fig. 4.13, right) that follows by Fourier transformation,

$$\tilde{E}(\omega) = \frac{A}{\sqrt{2\pi}}\int_{-\infty}^{\infty} \tilde{E}(t)\mathrm{e}^{\mathrm{i}\omega t}\mathrm{d}t, \tag{4.28}$$

which leads, with $\beta = \Delta\omega_0$, to

$$\begin{aligned}\tilde{E}(\omega) &= \frac{A}{\sqrt{2\pi}}\int_{-\infty}^{\infty} \exp\mathrm{i}[(\omega-\omega_0)t + i(\beta/2)t]\mathrm{d}t \\ &= \frac{-A}{\sqrt{2\pi}}\frac{1}{\mathrm{i}(\omega-\omega_0)+\Delta\omega_0/2}.\end{aligned} \tag{4.29}$$

Accordingly, the oscillating electron emits an electromagnetic wave with an intensity distribution that corresponds to a Lorentzian line (Fig. 4.12, right),

$$I(\omega)/I_0 = \bar{g}_{\mathrm{L,res}}(\omega). \tag{4.30}$$

I_0 is the intensity at the line center. The linewidth $\Delta\omega_0$ is equal to τ^{-1}.

In the classical oscillator model description of an atom, the nonoscillating state corresponds to the atomic ground state, and an oscillating state corresponds (independent of the value of the amplitude of the oscillation) to the excited state of the atom.

4.9 Natural Line Broadening

The finite lifetime of the upper level of a two-level atomic system with respect to spontaneous emission of radiation by $2 \to 1$ transitions leads to a line broadening described by the natural lineshape function

$$g_{\mathrm{nat}}(\nu) = \frac{\Delta\nu_{\mathrm{nat}}}{2\pi}\frac{1}{(\nu_0-\nu)^2 + \Delta\nu_{\mathrm{nat}}^2/4}, \tag{4.31}$$

where ν_0 is the resonance frequency,

$$\Delta\nu_{\mathrm{nat}} = \frac{1}{2\pi\tau_{\mathrm{sp}}} \tag{4.32}$$

is the *natural linewidth*, and τ_{sp} the lifetime of the upper level with respect to spontaneous emission of radiation by $2 \rightarrow 1$ transitions. We designate τ_{sp} as the *spontaneous lifetime* of level 2. The maximum value of the lineshape function,

$$g_{nat}(\nu_0) = 4\tau_{sp} = \frac{2}{\pi \Delta\nu_{nat}}, \tag{4.33}$$

is proportional to the spontaneous lifetime, that is, inversely proportional to the natural linewidth.

4.10 Energy Relaxation

We describe energy relaxation in the classical model of an atom. An atomic dipole loses energy due to damping. The amplitude of the oscillation decreases exponentially (Fig. 4.13a). The decay time of the amplitude is $2T_1$ and the decay time of the energy content in the oscillator is equal to the *energy relaxation time*. The linewidth of the frequency distribution of radiation emitted by a dipole oscillator is equal to

$$\Delta\omega_0 = \frac{1}{T_1}. \tag{4.34}$$

The halfwidth of the frequency distribution of radiation emitted by a dipole is equal to the reciprocal of the energy relaxation time.

We will later study an ensemble of dipole oscillators. The decay of the energy contained in an ensemble of dipole oscillators occurs also with the energy relaxation time T_1.

[In an ensemble of electrons, the decay of the energy content is joined with the decay of the *polarization* with the time T_1, which is then called *longitudinal relaxation time* of the polarization; see next section and Sect. 9.9].

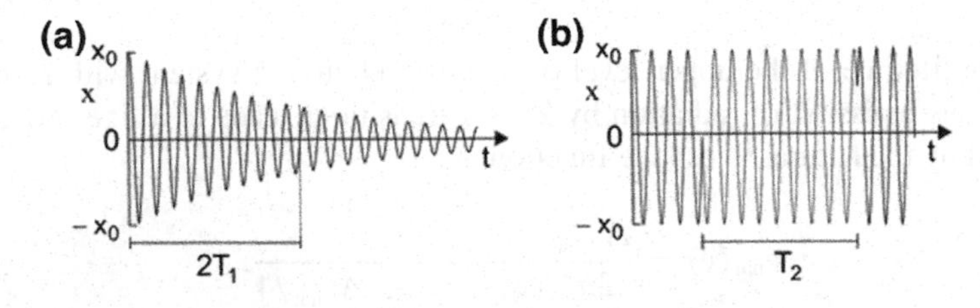

Fig. 4.13 Change of an oscillation. **a** Energy relaxation. **b** Dephasing

4.11 Dephasing

External forces can have the effect that a dipole oscillator randomly changes the phase while the amplitude x_0 and the resonance frequency ω_0 remain unchanged. The average time between two successive changes of the phase is the *dephasing time* T_2 (Fig. 4.13b). Radiation emitted by an oscillator consists of a series of subsequent wave trains $E(t)$ of constant amplitude A. The average duration of a wave train is equal to T_2. A Fourier analysis shows that the power spectrum has a Lorentzian lineshape. The center frequency is an average frequency, the lineshape function is a Lorentz resonance function and the linewidth (halfwidth) is given by

$$\Delta\omega_0 = \frac{1}{T_2/2} = \frac{2}{T_2}. \tag{4.35}$$

Examples Line broadening due to collisions of atoms and molecules in gases (Sect. 14.2); line broadening due to an elastic collision of an ion in a crystal with a phonon (Sect. 15.9).

We will later study an ensemble of dipole oscillators that are prepared in such a way that all elementary atomic oscillators are oscillating with the same phase. The sum of all atomic dipole moments per unit volume is the *dielectric polarization*. Due to randomly occurring dephasing processes of the atomic dipole oscillators, the dielectric polarization decays also with the dephasing time T_2. In connection with the decay of the polarization, T_2 is called *phase relaxation time* or *transverse relaxation time* (Sect. 9.8).

We will (in Chaps. 19 and 32) find dephasing processes characteristic of *monopole oscillators* (next section).

4.12 Dipole Oscillator and Monopole Oscillator

A dipole oscillator (Sect. 4.8) consists of a negative charge and a positive charge oscillating against each other. Interaction of a dipole oscillator with the surrounding can lead to loss of energy and to change of the phase of the oscillation. We are using the dipole oscillator as a classical model of a two-level atomic system (Sect. 4.8).

A *monopole oscillator* consists of an oscillating single charge. Interaction of the charge with the surroundings can change the phase of the oscillation but not the amplitude. An electron traversing a spatially periodic magnetic field represents an example of a monopole oscillator. An electron crossing a spatially periodic electric field is another example.

The monopole oscillation of an electron propagating at a relativistic velocity through a transverse periodic magnetic field is the elementary excitation occurring in a free-electron laser (Chap. 19); interaction of a high frequency electric field with an ensemble of the electron-monopole oscillators leads to gain for the high frequency field in the free-electron laser.

Monopole oscillation of an electron occurs also in the (hypothetical) Bloch laser (Chap. 32). An electron in a periodic potential formed by a semiconductor superlattice executes, under the action of a static electric field, an oscillation (Bloch oscillation); a superlattice consists, in the simplest case, of two different semiconductor layers in turn. The Bloch oscillation is the elementary excitation of a Bloch laser.

A laser based on monopole oscillations as the elementary excitations in an active medium can be regarded as an *inversionless laser*.

An electron that performs a Bloch oscillation can, alternatively, be described as an electron that occupies an energy level of an *energy-ladder system*. The energy ladder consists of equidistant energy levels. In the picture of the energy-ladder system, the Bloch laser is a laser with population inversion in the active medium (Sect. 32.7). Radiation is generated by stimulated transitions between energy levels of the energy-ladder systems.

It is a question whether an electron propagating through a transverse periodic magnetic field can also be characterized by an energy-ladder system. If such a description would be possible, a free-electron laser would be describable as a laser with a population inversion, alternatively to the description as an inversionless laser (Chap. 19).

Interaction of radiation with an active medium based on dipole oscillators leads to a gain coefficient that has the shape of a Lorentz resonance curve (see Chap. 9). Interaction of radiation with a monopole oscillator results in a gain coefficient that has the shape of a Lorentz dispersion curve (Chaps. 19 and 32).

4.13 Three-Dimensional and Low-Dimensional Active Media

We classify media used in lasers as three-dimensional (3D) and low-dimensional media. A low-dimensional active medium is either two-dimensional (2D), one-dimensional (1D), or zero-dimensional (0D). This classification is useful with respect to semiconductor lasers. Semiconductor lasers make use of low-dimensional media. Low-dimensional media are realized by means of semiconductor heterostructures.

The dimensionality concerns solely the question whether electrons *move freely* (between two collisions) in three dimensions, in two dimensions, in one dimension, or cannot move freely at all. In this sense, an atom is a 0D system.

Figure 4.14 gives a survey of active media of different dimensionality:

- *3D active medium.* The electrons move freely in three dimensions. The unit of the density N of electrons is m^{-3}.
  Example electrons in a 3D bipolar semiconductor.
- *2D active medium.* The electron motion is bound to a plane. The unit of the two-dimensional density (area density = sheet density) N^{2D} of electrons is m^{-2}.
  Example electrons in a quantum film (in a quantum well laser).

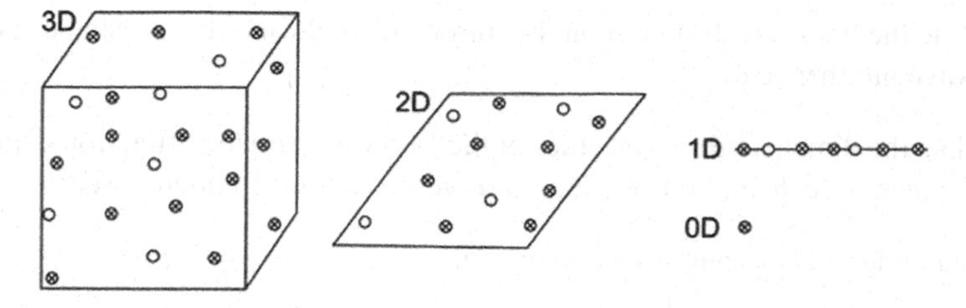

Fig. 4.14 Three-dimensional and low-dimensional active media

- *1D active medium.* The electron motion is bound to a line. The unit of the one-dimensional density (=line density) N^{1D} of electrons is m^{-1}.
  Example Electrons in a quantum wire (in a quantum wire laser).
- *0D active medium.* The electrons are imprisoned. An ensemble of 0D active media forms a 3D active medium.
  Example electrons in a quantum dot (in a quantum dot laser); we can regard a quantum dot as an artificial atom.

Ensembles of atoms, molecules, or ions are three-dimensional media. Each atom (or molecule or ion) is for itself a quantum system. Boltzmann's statistics governs the populations of the energy levels in an ensemble of atoms because the interaction (for instance, by collisions in gases) between the atoms is weak. An electron gas (=ensemble of the electrons) in the upper band in a bipolar semiconductor laser forms a quantum system. The electron gas (=ensemble of the electrons) in the lower band forms another quantum system. The two coexisting electron gases obey, each for itself, Fermi's statistics and have different Fermi energies (called *quasi-Fermi energies*).

References [1–4, 31, 32].

Problems

4.1 Lineshape functions. At which frequency distance from the central line ($\nu_0 = 4 \times 10^{14}$ Hz; $\Delta\nu_0 = 1$ GHz) does the lineshape function decrease by a factor of 100 (a) if the line has Lorentzian shape and (b) if the line has Gaussian shape?

4.2 Absolute number of two-level atomic systems. Determine the absolute number N_{tot} of two-level atomic systems for systems of different dimensionality.

(a) Three-dimensional medium with a density $N = 10^{24}\,m^{-3}$ of two-level systems and a volume of $1\,mm \times 1\,mm \times 1\,mm$.
(b) Two-dimensional medium with an area density $N^{2D} = 10^{16}\,m^{-2}$ and an area of $1\,mm \times 1\,mm$.
(c) One-dimensional medium with a line density $N^{1D} = 10^{7}\,m^{-1}$ and a length of 1 mm.

4.3 Relate the lineshape function on the frequency scale and the lineshape function on the wave number scale.

4.4 Relate the dimensionless variables of the Lorentz resonance function expressed on the frequency scale and those expressed on the angular frequency scale.

4.5 Area under a Gaussian or Lorentzian curve.

(a) Show that the width of a rectangular curve, which has the same height as a Gaussian curve and encloses the same area, is equal to

$$\Delta\nu_0/(2\sqrt{\ln 2}) \approx 1.06 \times \Delta\nu_0 \approx \Delta\nu_0.$$

(b) Show that the width of a rectangular curve, which has the same height as a Lorentzian curve and encloses the same area, is approximately equal to

$$(\pi/2)\Delta\nu_0 \approx 1.57 \times \Delta\nu_0.$$

4.6 Show that the integral over a narrow Lorentzian curve is approximately unity.

4.7 Derive the maximum value, Eq. (4.10), for the Lorentz resonance function.

4.8 Show that the width of a rectangular shape that has the same height and the same area as a Gaussian $\Delta\nu_0/(2\sqrt{\ln 2}) \approx 1.06 \times \Delta\nu_0$, and, correspondingly, for a Lorentzian shape by $(\pi/2)\Delta\nu_0 \approx 1.57 \times \Delta\nu_0$.

Chapter 5
Titanium–Sapphire Laser

As an example of a laser, we describe the titanium–sapphire laser.

Titanium–sapphire (=titanium-doped sapphire) has a broad pump band, a long-lived upper laser level, and a distribution of short-lived lower laser levels. The broad pump band allows for pumping with a lamp or a laser. The broad distribution of lower laser levels makes it possible to operate the laser as a tunable continuous wave laser or as a femtosecond laser.

A continuous wave titanium–sapphire laser is tunable over a large frequency range—extending from the red to the infrared spectral region. A titanium–sapphire femtosecond laser generates ultrashort light pulses with a duration between about 5 and 100 fs (depending on the special arrangement).

In a simplified description, we characterize the titanium–sapphire laser as a four-level laser with a broad distribution of pump levels, a sharp upper laser level, a broad distribution of lower laser levels, and a sharp ground state level.

In this chapter, we discuss the principle of the titanium–sapphire laser, and we will give a short description of the design. Additionally, we will present absorption and fluorescence spectra of titanium–sapphire. We will obtain more information about the titanium–sapphire laser in later sections and chapters (particularly in Sect. 7.6, Chap. 13, Sect. 15.2, and Chap. 17).

5.1 Principle of the Titanium–Sapphire Laser

A titanium–sapphire (Ti^{3+}:Al_2O_3) crystal contains Ti^{3+} ions replacing Al^{3+} ions in a sapphire (Al_2O_3) crystal. The density (number density) of Ti^{3+} ions in Al_2O_3 is typically 1×10^{25} m^{-3}, corresponding to a doping concentration of 0.03% by weight Ti_2O_3 in Al_2O_3.

Figure 5.1a shows the energy level diagram of a Ti^{3+} ion in Al_2O_3. Above the ground state level (energy $E = 0$), there is a continuum of energy levels (the vibronic levels of the ground state). Above the lowest excited state level (at $E \sim 2.0$ eV), there is another continuum of energy levels (the vibronic levels of the excited state). The

K.F. Renk, *Basics of Laser Physics*, Graduate Texts in Physics,
DOI 10.1007/978-3-319-50651-7_5

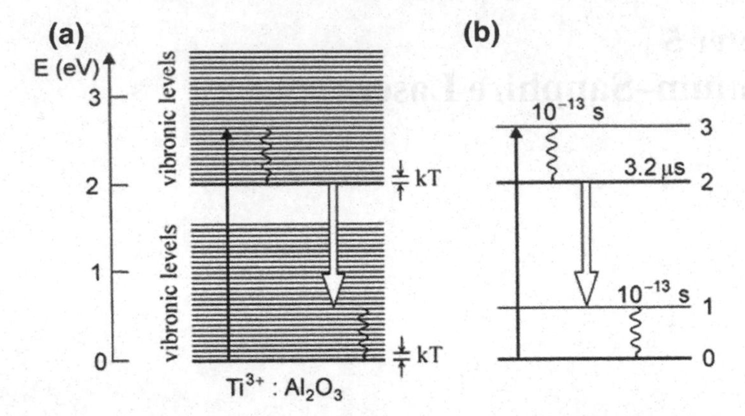

Fig. 5.1 Titanium–sapphire laser. **a** Energy level diagram and transitions. **b** Simplified energy level diagram

lowest excited state level and the vibronic energy levels of the excited state have a long lifetime (~3.2 μs) with respect to relaxation to the ground state level or to vibronic levels of the ground state; the lifetime is mainly due to spontaneous emission of radiation (spontaneous lifetime ~3.8 μs). Optical pumping and fast nonradiative relaxation lead to population of the lowest excited state level. Laser radiation is generated by stimulated transitions to vibronic levels of the ground state—to levels well above the ground state level. The vibronic levels of the ground state then relax by nonradiative relaxation processes. Relaxation within the excited state levels and within the ground state levels occurs in relaxation times of the order of 10^{-13} s. Absorption processes are mainly due to transitions from the ground state to levels that lie, in comparison to the value of kT at room temperature, far above the lowest excited state level.

We describe the titanium–sapphire laser, which is a vibronic laser, for simplicity, as a four-level laser (Fig. 5.1b)—with a pump level (that has a broad energy distribution), a sharp upper laser level, a lower laser level (that has a broad energy distribution too), and a sharp ground state level.

We will discuss vibronic systems in more detail in Sect. 15.2 and Chap. 17. All Ti^{3+} ions in Al_2O_3 have the same energy level distribution: optically pumped titanium–sapphire is an active medium with a homogeneously broadened $2 \rightarrow 1$ fluorescence line (Sect. 17.4).

The titanium–sapphire laser operates as a cw laser or as femtosecond laser. The laser frequency of a cw laser—and thus the energy of the lower laser level—is mainly determined by the resonance frequency of the laser resonator, which itself is adjustable by the use of appropriate frequency selective elements within the resonator. In a femtosecond laser, with a broadband resonator, transitions occur at the same time into a large number of lower laser levels; in this case, the resonator contains elements controlling the phases of the electromagnetic waves of different frequencies that are present in the resonator at the same time.

5.2 Design of a Titanium–Sapphire Laser

Pumping of a titanium–sapphire laser is possible by use of a discharge lamp or of another laser. If a laser is pumped by use of a lamp, light of the lamp is focused on the crystal leading to an almost homogeneous excitation of the crystal. Pumping with a laser is possible by transverse or longitudinal pumping. In the case of transverse pumping, the pump radiation irradiates the titanium–sapphire crystal from the side.

In the case of longitudinal pumping (Fig. 5.2), the pump radiation passes a dichroitic mirror and is then absorbed in the titanium–sapphire crystal. The dichroitic mirror is transparent for the pump radiation but is a reflector for the laser radiation.

By inserting into the resonator an element indicated as "black box" in the figure, the laser operates in different ways.

If the black box is a dispersive element (e.g., a prism or a diffraction grating), the laser generates

- cw radiation; tuning range from about 650–1200 nm on the wavelength scale (250–460 THz on the frequency scale).

If the black box is a mode coupler, the laser generates

- ultrashort pulses; pulse duration 5–100 fs.

Heat produced by nonradiative relaxation processes leaves the titanium–sapphire crystal by heat transfer mainly via the mechanical support of the crystal. Heating effects can strongly be reduced if only a small portion of a Ti^{3+}:Al_2O_3 crystal rod is optically pumped. The pump volume is, for example, a cylindrical volume of a diameter of 0.5 mm in a Ti^{3+}:Al_2O_3 rod of 1 cm diameter. The heat produced in the optically pumped region is distributed over the whole crystal. This results in a much smaller temperature enhancement than in the case that the crystal is homogeneously pumped. Thus, much larger populations of the upper laser level or a much larger output power can be obtained. Al_2O_3 has a large heat conductivity. This favors a fast heat escape from the active volume.

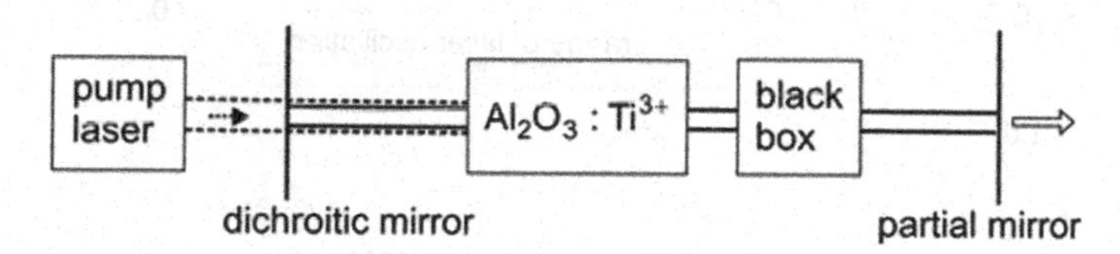

Fig. 5.2 Titanium–sapphire laser

5.3 Absorption and Fluorescence Spectra of Titanium–Sapphire

Spectra of the absorption cross section and fluorescence spectra of $Ti^{3+}:Al_2O_3$ are known from experimental studies [33]. Figure 5.3 (upper part, left) shows the absorption cross section $\sigma_{abs}(\lambda)$ of a Ti^{3+} ion in sapphire. The absorption cross section follows from the relation

$$\alpha_{abs}(\lambda) = N_0\,\sigma_{abs}(\lambda), \tag{5.1}$$

where α_{abs} is the experimental absorption coefficient and N_0 the density of Ti^{3+} ions. The absorption band extends from the blue to the green spectral region. Sapphire is an anisotropic crystal. Therefore, the optical properties depend on the orientation of the direction of the electric field E of the electromagnetic wave relative to the direction of the optic axis (c axis) of the crystal. The absorption lines for $E \parallel c$ (π polarization) and $E \perp c$ (σ polarization) have different strengths but the same shape.

Figure 5.3 (lower part) shows the fluorescence spectrum. $S_\lambda(\lambda)$ is the spectral distribution of the fluorescence radiation on the wavelength scale and $S_{\lambda,max}$ is the maximum of the spectral distribution. The fluorescence spectrum extends from the red to the near infrared. The fluorescence lines for $E \parallel$ c and $E \perp$ c have different strengths but the same shape. The large linewidth and especially the long infrared tail of the fluorescence band are the basis of the broadband tunability of a titanium–sapphire laser and of the operation as femtosecond laser. The range of laser oscillation (dashed in Fig. 5.3) extends far into the range of the infrared tail of the fluorescence curve.

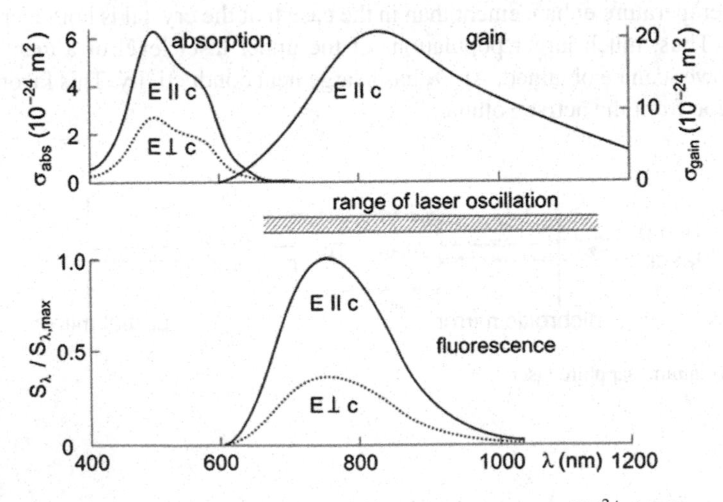

Fig. 5.3 Absorption and gain cross sections and fluorescence spectra of $Ti^{3+}:Al_2O_3$

The occurrence of a broad range of laser oscillation is a consequence of the special properties of Ti^{3+} in Al_2O_3. We will discuss these properties later (in Sect. 15.2). We will also show (in Sect. 17.4) that almost all excited Ti^{3+} ions contribute to gain, and that each of the excited Ti^{3+} ions contributes equally, i.e., all excited Ti^{3+} ions contribute equally to generation of radiation in a continuous-wave laser or to generation of femtosecond pulses in a femtosecond titanium–sapphire laser.

There remains the question: how can we determine, from the fluorescence spectrum, the gain cross section and especially the frequency dependence of the gain cross section? This will be discussed in Sect. 7.6. We anticipate the result. Figure 5.3 (upper part, right) shows the gain cross section σ_{gain} of an excited Ti^3+ ion as derived from the fluorescence spectrum and by taking into account the spontaneous lifetime of an excited Ti^3+ ion. The gain cross section has a maximum near 830 nm. The maximum value of the gain cross section of excited Ti^3+ is about 4 times larger than the absorption cross section of Ti^3+; a possible reason for the difference will be discussed at the end of Sect. 17.2. In comparison to the gain profile, the fluorescence curve decreases strongly with wavelength because of the wavelength dependence of spontaneous emission.

5.4 Population of the Upper Laser Level

Without laser oscillation, the population of the upper laser level (Fig. 5.4) increases with increasing pump power P linearly with P at small P (dashed line) and less than linearly at large P. The population saturates at very large P, where the N_2 population approaches the density N_0 of Ti^{3+} ions. The population difference varies according to $N_2 - N_1 = K(P/P_{sat})(1 + P/P_{sat})^{-1}$, where K is a constant and P_{sat} the saturation-pump power. At the saturation pump power the absorption coefficient $\alpha_{abs}(\nu)$ is half of its value at weak pumping. A population inversion ($N_2 > N_1$) occurs already at the smallest pump power.

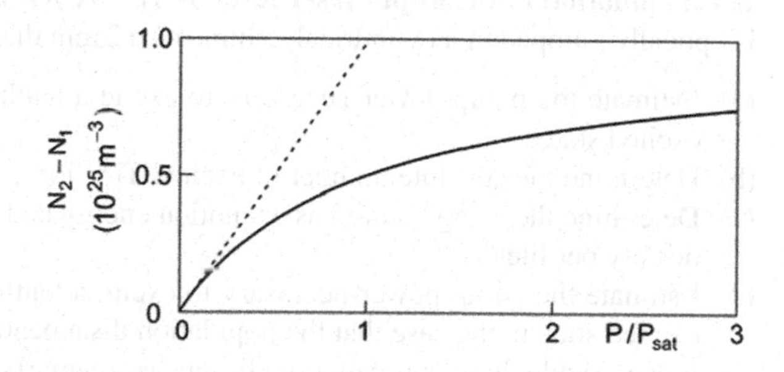

Fig. 5.4 Population of the upper laser level of $Ti^{3+}:Al_2O_3$

5.5 Heat and Phonons

Nonradiative relaxation processes in active media produce heat. Therefore, cooling of an active medium is necessary. Cooling occurs via heat conduction and heat transfer. In solids, phonons play an important role in relaxation processes; phonons are quanta of lattice vibrations, i.e., of vibrations of the atoms of a solid. After a relaxation process, a phonon decays into heat. Phonons are treated in textbooks on solid state physics. As a standard book of heat conduction and heat transfer, *see* [34].

References [1–4, 7, 33, 34]

Problems

5.1 Geometrical length of the resonator. Determine the optical length of a resonator (distance between reflector and an output coupling mirror 50 cm) that contains a titanium–sapphire crystal (length 1 cm; refractive index $n = 1.76$).

5.2 Photon density. The diameter of a laser beam at the output coupling mirror is 10 cm. The laser generates visible radiation of a power of 1 W and fluorescence radiation of a power of 1 W too.

(a) Estimate the power of laser radiation from the active medium of the laser passing through an area of 1 cm diameter in a distance 10 m away from a laser that emits radiation into a cone with a cone angle of 0.1 mrad.
(b) Estimate the power of fluorescence radiation from the active medium of the laser passing through the same area.

5.3 What is the prescription of conversion of the shape of a narrow fluorescence spectrum on the wavelength scale into the shape of the spectrum on the frequency scale? [*Hint*: for the answer, *see* Sect. 7.6.]

5.4 Population of the upper laser level. A $Ti^{3+}:Al_2O_3$ crystal of a length of 1 cm is optically pumped in a cylindrical volume of 0.2 mm diameter.

(a) Estimate the pump power necessary to excite a tenth of the Ti^{3+} ions into the excited state.
(b) Determine the absolute number of excited Ti^{3+} ions.
(c) Determine the energy stored as excitation energy and the corresponding energy density per liter.
(d) Estimate the pump power necessary to excite a tenth of the Ti^{3+} ions into the excited state in the case that the population disappears every 300 ns. [Why may it be possible that the population disappears regularly?]

Part II
Theoretical Basis of the Laser

Chapter 6
Basis of the Theory of the Laser: The Einstein Coefficients

According to Bohr's atomic model (1911), which is based on spectroscopic investigations, transitions between discrete energyd levels of an atom can lead to emission or absorption of radiation of a frequency that fulfills Bohr's energy-frequency relation. In an absorption process, a photon is absorbed. In an emission process, a photon is emitted. Einstein found that the emission of a photon is possible by two different processes, spontaneous and stimulated emission, and that the coefficients describing the three processes—absorption, stimulated and spontaneous emission—are related to each other (Einstein relations).

Making use of Planck's radiation law, we derive the Einstein relations. We also show that stimulated emission of radiation is a process that occurs permanently around us. There remains the question: what is, in addition to the stimulated emission, a specific property of a laser?

Einstein coefficients can be extracted from results of experimental studies of optical properties of matter at thermal equilibrium. In this chapter, we consider an ensemble of two-level systems in thermal equilibrium determined by Boltzmann's statistics. Later (in Chap. 21), we will treat ensembles that obey Fermi's statistics.

6.1 Light and Atoms in a Cavity

How does light interact with a two-level atomic system? We will study this question in three steps:

- We describe the thermal equilibrium between the radiation in a cavity and the walls of the cavity.
- We describe the thermal equilibrium between an ensemble of two-level atomic systems in a cavity and the walls of the cavity.
- We consider a cavity that contains an ensemble of two-level atomic systems *and* radiation.

K.F. Renk, *Basics of Laser Physics*, Graduate Texts in Physics,
DOI 10.1007/978-3-319-50651-7_6

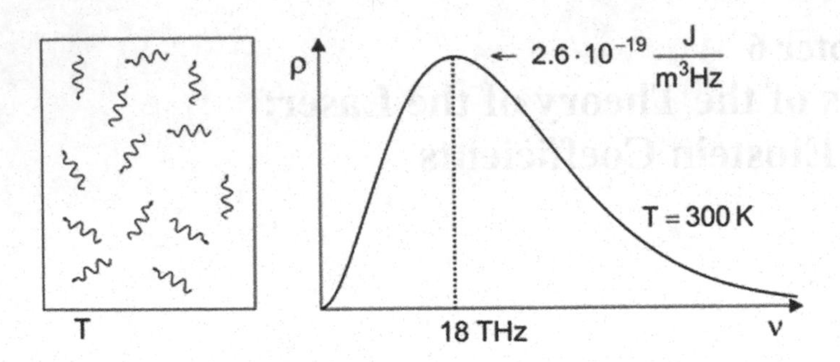

Fig. 6.1 A cavity and Planckian distribution of radiation

A cavity (Fig. 6.1, left) contains blackbody radiation. The spectral distribution of the energy density $\rho(\nu)$ of the radiation depends on the temperature T of the walls of the cavity. The spectral energy density is determined by *Planck's radiation law*

$$\rho(\nu) = \frac{8\pi\nu^2}{c^3}\frac{h\nu}{e^{h\nu/kT}-1}, \tag{6.1}$$

where k is Boltzmann's constant. The frequency distribution is shown in Fig. 6.1 (right). The frequency ν_{max} of the maximum of the distribution is directly proportional to the temperature according to the relation

$$h\nu_{max} \sim 2.8\,kT. \tag{6.2}$$

If the walls are at room temperature ($T = 300\,\mathrm{K}$), the maximum of the distribution lies in the infrared ($\nu_{max} = 1.8 \times 10^{13}$ Hz). The spectral density increases as ν^2 at small frequency ($\nu \ll \nu_{max}$) and decreases as $\nu^3 e^{-h\nu/kT}$ at large frequency. Thermal equilibrium is established by absorption of radiation by the walls of the cavity and by emission of radiation from the walls into the cavity.

The energy density of radiation in the frequency interval ν, $\nu + d\nu$ is

$$u(\nu) = \rho(\nu)\,d\nu. \tag{6.3}$$

We now treat a cavity containing an ensemble of two-level atomic systems in thermal equilibrium, which is determined by Boltzmann's statistics,

$$N_2/N_1 = e^{-(E_2-E_1)/kT}. \tag{6.4}$$

The population ratio is near unity if $E_2 - E_1 \ll kT$. It decreases exponentially with the energy difference $E_2 - E_1$. Thermal equilibrium is established by collisions of the two-level atomic systems with each other and with the walls of the cavity.

Impurity ions in a solid have fixed locations. The populations of the energy levels of different ions are in thermal equilibrium with the solid due to absorption and emission of phonons. The populations are governed by Boltzmann's statistics.

Einstein showed [38] that the thermal equilibrium in a gas of atoms can also be established by the direct interaction of the radiation with the atoms and that three processes of interaction between radiation and atoms must occur: absorption, spontaneous emission and stimulated emission.

Using Bohr's energy-frequency relation,

$$h\nu_0 = E_2 - E_1, \tag{6.5}$$

where ν_0 is the transition frequency, we can write

$$N_2/N_1 = \mathrm{e}^{-h\nu_0/kT}. \tag{6.6}$$

We will now characterize the three processes by the three Einstein coefficients.

6.2 Spontaneous Emission

Excited atoms (Fig. 6.2) can emit photons spontaneously, i.e., without external cause. The radiation emitted spontaneously is incoherent and the emission occurs into all spatial directions. The change $\mathrm{d}N_2$ of the population N_2 of the upper level, within a time interval $\mathrm{d}t$, is proportional to N_2 and to $\mathrm{d}t$,

$$\mathrm{d}N_2 = -A_{21} N_2 \mathrm{d}t. \tag{6.7}$$

A_{21} is the *Einstein coefficient of spontaneous emission.* The population of the upper level decays exponentially,

$$N_2(t) = N_2(0)\,\mathrm{e}^{-A_{21}t} = N_2(0)\,\mathrm{e}^{-t/\tau_{\mathrm{sp}}}. \tag{6.8}$$

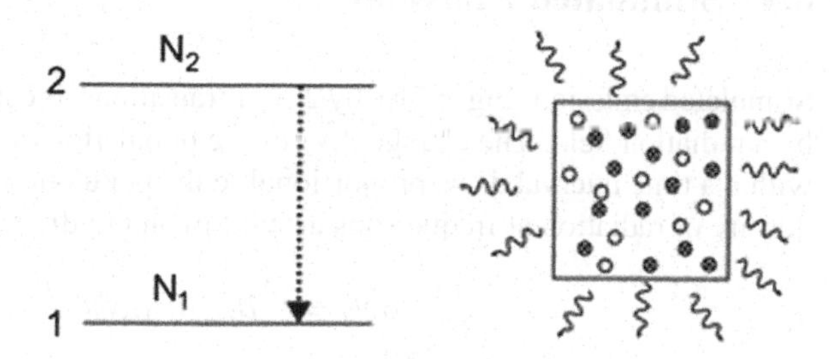

Fig. 6.2 Spontaneous emission

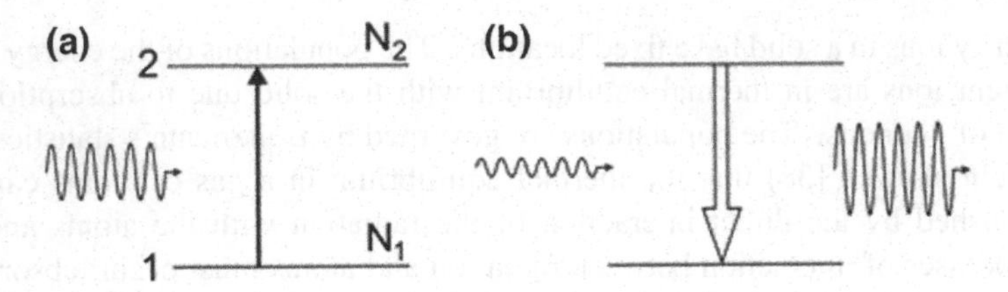

Fig. 6.3 Stimulated transitions. **a** Absorption and **b** stimulated emission

$N_2(0)$ is the density of excited two-level atomic systems at $t = 0$ and τ_{sp} is the average *lifetime of an excited two-level atomic system with respect to spontaneous emission* (=*spontaneous lifetime*). We have the simple relation

$$A_{21} = 1/\tau_{sp}. \tag{6.9}$$

The Einstein coefficient A_{21} is equal to the reciprocal of the spontaneous lifetime.

6.3 Absorption

Photons of a light field can be absorbed (Fig. 6.3a) by $1 \rightarrow 2$ transitions. The change $\mathrm{d}N_1$ of the population N_1 of the ground state, within a time interval $\mathrm{d}t$, is proportional to the population of the ground state itself, to the spectral energy density ρ of the radiation field and to $\mathrm{d}t$,

$$\mathrm{d}N_1 = -B_{12}\rho(\nu_0)N_1\mathrm{d}t. \tag{6.10}$$

B_{12} is the *Einstein coefficient of absorption* and $\rho(\nu_0)$ is the spectral energy density of radiation at frequencies around ν_0. Absorption is only possible in the presence of a field—the absorption is a stimulated process.

6.4 Stimulated Emission

Stimulated emission (Fig. 6.3b), by $2 \rightarrow 1$ transitions, is caused (stimulated, induced) by a radiation field. The change $\mathrm{d}N_2$ of the population of atoms in the excited state, within a time interval $\mathrm{d}t$, is proportional to the population N_2, to the spectral energy density of radiation at frequencies around ν_0 and to $\mathrm{d}t$,

$$\mathrm{d}N_2 = -B_{21}\rho(\nu_0)N_2\mathrm{d}t. \tag{6.11}$$

B_{21} is the *Einstein coefficient of stimulated emission.* The radiation created by stimulated emission has the same frequency, direction, polarization and phase as the stimulating radiation.

6.5 The Einstein Relations

We are looking for relations between the Einstein coefficients. As discussed, the interaction of a two-level atomic system with radiation occurs (Fig. 6.4) via absorption, stimulated and spontaneous emission. We describe the three processes by rate equations that correspond to differential equations of first order:

- The rate of change of the population N_1 due to absorption is given by

$$(\mathrm{d}N_1/\mathrm{d}t)_{\mathrm{abs}} = -B_{12}\,\rho(\nu_0)\,N_1; \tag{6.12}$$

  the temporal change of the population N_1 due to absorption is proportional to $\rho(\nu_0)$ and to N_1.
- The rate of change of the population N_2 due to stimulated emission is equal to

$$(\mathrm{d}N_2/\mathrm{d}t)_{\mathrm{stim}} = -B_{21}\,\rho(\nu_0)\,N_2; \tag{6.13}$$

  the temporal change of the population N_2 due to stimulated emission is proportional to $\rho(\nu_0)$ and to N_2.
- The rate of change of the population N_2 due to spontaneous emission is

$$(\mathrm{d}N_2/\mathrm{d}t)_{\mathrm{sp}} = -A_{21}N_2; \tag{6.14}$$

  the temporal change of the population N_2 due to spontaneous emission of radiation is proportional to N_2.

We consider a cavity with an ensemble of two-level atomic systems and radiation in thermal equilibrium. In the time average, the ratio N_2/N_1 is a constant. Therefore, the absorption rate has to be equal to the emission rate,

$$(\mathrm{d}N_1/\mathrm{d}t)_{\mathrm{abs}} = (\mathrm{d}N_2/\mathrm{d}t)_{\mathrm{sp}} + (\mathrm{d}N_2/\mathrm{d}t)_{\mathrm{stim}}. \tag{6.15}$$

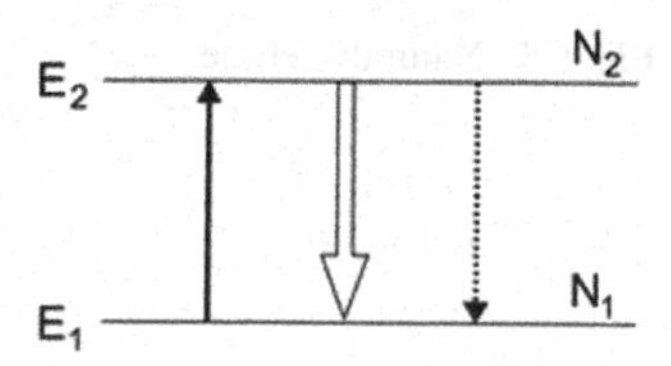

Fig. 6.4 Absorption, stimulated emission, and spontaneous emission

This leads to the relation

$$B_{12}\rho(\nu_0)N_1 = A_{21}N_2 + B_{21}\rho(\nu_0)N_2. \tag{6.16}$$

It follows that

$$\rho(\nu_0) = \frac{A_{12}/B_{21}}{(B_{12}/B_{21})N_1/N_2 - 1}. \tag{6.17}$$

The Boltzmann factor determines the ratio N_1/N_2. The comparison with Planck's radiation law provides the *Einstein relations*

$$B_{21} = B_{12}, \tag{6.18}$$

$$A_{21} = \frac{8\pi\nu^2}{c^3} h\nu B_{21}. \tag{6.19}$$

The frequency ν (replacing ν_0) follows from Bohr's relation $h\nu = E_2 - E_1$. We have the result:

- *The same Einstein coefficient governs both stimulated emission and absorption.*
- *There is a connection between the coefficients of spontaneous and stimulated emission.*
- *The Einstein coefficient A_{21} increases strongly with frequency.*

Figure 6.5 shows the spontaneous lifetime for different transition frequencies $\nu = (E_2 - E_1)/h$ at a fixed value of B_{21} ($= 10^{18}\,\mathrm{m^3\,J^{-1}\,s^{-2}}$); the spontaneous lifetime is of the order of 10^{-6} s at a transition frequency (5×10^{14} Hz) in the visible, 100 s at a transition frequency (10^{12} Hz) in the far infrared, and 10^{-15} s at a transition frequency (10^{17} Hz) in the X-ray range. Spontaneous lifetimes at X-ray transition frequencies are very short. Therefore, operation of an X-ray laser is difficult (Sect. 16.4).

If energy levels are degenerate, Boltzmann's statistics yields

$$\frac{N_2}{N_1} = \frac{g_2}{g_1} e^{-(E_2-E_1)/kT}, \tag{6.20}$$

where g_1 is the degree of degeneracy of level 1 and g_2 the degree of degeneracy of level 2. The treatment of the equilibrium between the atomic populations and the radiation in a cavity leads to the relations

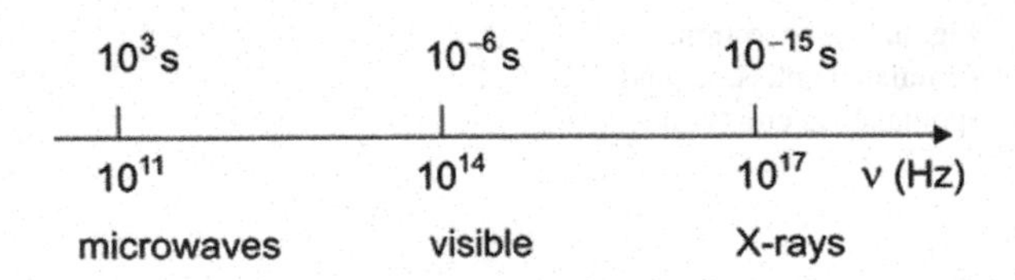

Fig. 6.5 Natural lifetime

$$B_{12} = \frac{g_2}{g_1} B_{21} \tag{6.21}$$

and (as in the nondegenerate case)

$$A_{21} = \frac{8\pi \nu^2}{c^3} h\nu B_{21}. \tag{6.22}$$

In the following, we will treat the case of nondegenerate energy levels ($g_1 = g_2 = 1$).

If two-level atomic systems are embedded in a medium of refractive index n, the speed of light in vacuum has to be replaced by the speed of light in the medium. The Einstein relations then are

$$B_{21} = B_{12}, \tag{6.23}$$

$$A_{21} = \frac{8\pi \nu^2}{(c/n)^3} h\nu B_{21}, \tag{6.24}$$

and

$$B_{21} = \frac{(c/n)^3}{8\pi h\nu^3} A_{21}. \tag{6.25}$$

In this form, the Einstein relations are valid if a medium is optically isotropic. If a medium is optically anisotropic, the relation between A_{21} and B_{21} has to be modified.

Table 6.1 shows values of Einstein coefficients determined by the use of experimental or theoretical methods. A few methods are mentioned in the following:

- Measurement of τ_{sp} (by a luminescence experiment) provides A_{21} and (via the Einstein relations) B_{21} too. *Example* Nd:YAG.
- Measurement of the absorption coefficient provides (Chap. 7) B_{21} and (via the Einstein relations) A_{21}.
- An analysis of the luminescence spectrum yields A_{21}; *Example* bipolar semiconductor lasers (*see* chapters on semiconductor lasers).
- Theoretical studies of the transition rates provide B_{21}; *Example* QCL.

Table 6.1 Einstein coefficients

Laser	λ	n	τ_{sp}	A_{21} (s^{-1})	B_{21} ($m^3\ J^{-1}\ s^{-2}$)
HeNe	633 nm	1	100 ns	10^7	1.5×10^{20}
CO_2	10.6 μm	1	5 s	0.2	1.4×10^{16}
Nd:YAG	1.06 μm	1.82	230 μs	4.3×10^3	5.1×10^{16}
TiS (E ∥ c)	830 nm	1.74	3.8 μs	2.6×10^5	1.7×10^{18}
Fiber	1.5 μm	1.5	10 ms	10^2	6.6×10^{15}
Semiconductor	810 nm	3.6		3×10^9	3.7×10^{21}
QCL	5 μm	3.6			4×10^{21}

The Einstein coefficients of different systems differ by many orders of magnitude.

If the spectral energy density is given on the angular frequency scale, $\rho = \rho(\omega)$, the Einstein coefficient B_{21}^{ω} is smaller by the factor 2π, $B_{21}^{\omega} = B_{21}/2\pi$. The Einstein relations then are $B_{12}^{\omega} = B_{21}^{\omega}$ and $A_{21} = \left(\hbar\omega^3/\pi^2 c^3\right) B_{21}^{\omega}$.

6.6 Einstein Coefficients on the Energy Scale

If the spectral energy density is given on the energy scale,

$$\rho(h\nu) = \rho(\nu)/h, \tag{6.26}$$

the Einstein coefficients of stimulated and spontaneous emission are

$$\bar{B}_{21} = B_{21}^{h\nu} = hB_{21}, \tag{6.27}$$

$$\bar{A}_{21} = A_{21} = \frac{8\pi(h\nu)^3}{h^3(c/n)^3}\bar{B}_{21} = \frac{8\pi\nu^3}{(c/n)^3}\bar{B}_{21}. \tag{6.28}$$

$\bar{B}_{21}$ is given in units of $m^3\ s^{-1}$.

6.7 Stimulated Versus Spontaneous Emission

Stimulated emission is a general phenomenon that does not only occur in lasers. Stimulated emission is a permanent process, for instance, in the lecture hall.

The thermal occupation number of a mode of a cavity is given by the Bose-Einstein factor (Fig. 6.6)

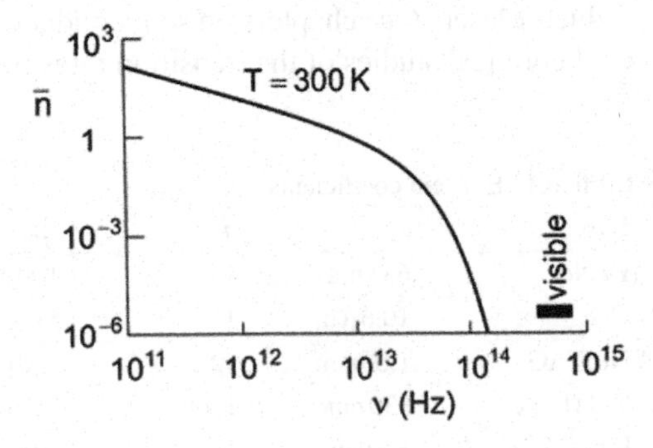

Fig. 6.6 Thermal occupation number of a photon mode at frequency ν

$$\bar{n} = \frac{1}{e^{h\nu/kT} - 1}. \tag{6.29}$$

At room temperature, $T = 300\,\mathrm{K}$, the thermal occupation number has very different values in different spectral regions:

- $\bar{n} \ll 1$ for $h\nu \gg kT$ (visible).
- $\bar{n} \sim 1$ for $h\nu \approx kT$ (infrared).
- $\bar{n} \gg 1$ for $h\nu \ll kT$ (far infrared and microwaves).

The approximation for small frequencies, $\bar{n} = kT/h\nu$, shows that $\bar{n}$ increases to infinitely large values ($\bar{n} \to \infty$) for $\nu \to 0$). At a frequency $\nu = 10^{11}$ Hz, the thermal occupation number is large, $\bar{n} \approx 100$.

We return to Planck's radiation law and write it in the form

$$\rho(\nu)\mathrm{d}\nu = D(\nu)\mathrm{d}\nu \times \bar{n}(\nu) \times h\nu, \tag{6.30}$$

where we have the quantities:

- $\rho(\nu)$ = spectral energy density of radiation at the frequency ν.
- $\rho(\nu)\mathrm{d}\nu$ = energy density in the frequency interval ν, $\nu + \mathrm{d}\nu$.
- $D(\nu) = 8\pi\nu^2/c^3$ =mode density =density of states of photons =density of modes per unit of volume and unit of frequency (Sect. 10.4).
- $D(\nu)\mathrm{d}\nu$ = number of modes per unit of volume in the frequency interval ν, $\nu + \mathrm{d}\nu$.
- $h\nu$ = quantum energy of a photon (=photon energy).

We can formulate Planck's radiation law as follows: *the energy density of blackbody radiation is equal to the product of mode density, thermal occupation number, and energy of a photon.*

A two-level atomic system (Fig. 6.7a) in equilibrium with thermal radiation can emit radiation either by spontaneous emission at an emission rate per excited atomic system of $\tau_{\mathrm{sp}}^{-1} = A_{21}$ or by stimulated emission at an emission rate per atomic system of $\tau_{\mathrm{stim}}^{-1} = \rho(\nu)B_{21}$. The stimulated emission dominates

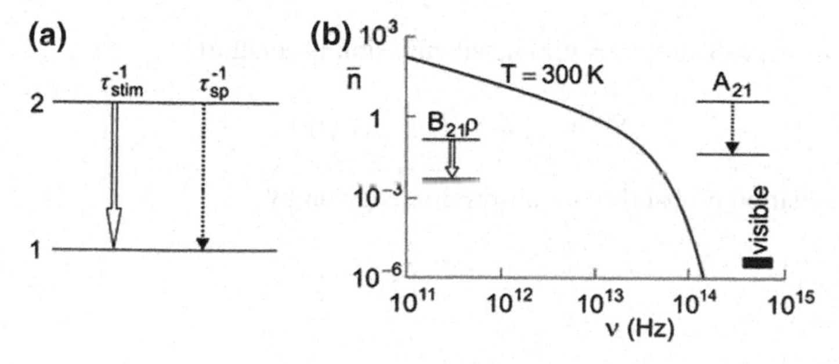

Fig. 6.7 Stimulated and spontaneous emission of radiation. **a** The two processes and **b** regions of their dominance

$$\frac{\tau_{\mathrm{stim}}^{-1}}{\tau_{\mathrm{sp}}^{-1}} = \frac{\rho(\nu)B_{21}}{A_{21}} = \frac{\rho(\nu)}{D(\nu)h\nu} = \bar{n}(\nu) > 1, \tag{6.31}$$

if the occupation number of the modes at the frequency ν is larger than unity. In thermal equilibrium of an ensemble of two-level atomic systems with radiation (Fig. 6.7b), the transitions $2 \to 1$ are:

- Mainly due to stimulated emission at small frequencies ($h\nu \ll kT$).
- Mainly due to spontaneous emission at large frequencies ($h\nu \gg kT$).

Stimulated transitions between energy levels at a transition frequency of 10^{11} Hz occur, at room temperature, almost 100 times faster than spontaneous transitions! Not only the walls of a lecture hall but also the persons in the hall permanently emit 10^{11}-Hz radiation mainly by stimulated emission.

What is specific about a laser? A thermal system contains radiation with portions in all spatial directions. Stimulated emission of radiation propagating in a direction compensates absorption of radiation propagating in exactly the opposite direction. In a lecture hall, we have permanently stimulated emission due to the interaction of the thermal radiation with the persons in the hall and with the walls. The situation is completely different in a laser. Pumping of produces an active medium, which is in a nonequilibrium state. The active medium experiences feedback from radiation stored in the laser resonator and therefore emits, by stimulated emission, radiation only in the direction of the stimulating radiation—i.e., the active medium is able to emit radiation in a single mode only.

6.8 Transition Probabilities

We express the transition rates by transition probabilities. The transition probability for spontaneous emission, i.e., the transition rate per two-level system per second, is given by

$$w_{21,\mathrm{sp}} = A_{21}. \tag{6.32}$$

The transition probability for stimulated emission is equal to

$$w_{21,\mathrm{stim}} = B_{21,\mathrm{stim}}\,\rho(\nu_0) \tag{6.33}$$

and the transition probability for absorption is given by

$$w_{21,\mathrm{abs}} = B_{12,\mathrm{abs}}\,\rho(\nu_0). \tag{6.34}$$

The transition probability per unit time for a stimulated emission process in an atomic two-level system is given by *Fermi's golden rule*

$$w_{21,\mathrm{stim}} = \frac{2\pi}{\hbar}|\mu_{21}|^2, \tag{6.35}$$

where μ_{21} is the matrix element for the stimulated transition $2 \rightarrow 1$ (Sect. 6.9).

6.9 Determination of Einstein Coefficients from Wave Functions

The stimulated emission of radiation by a two-level atomic system, characterized by the wave function ψ_1 of the lower level and ψ_2 of the upper level, is determined (for an electric dipole transition) by the dipole matrix element

$$\boldsymbol{\mu}_{21} = -\int \psi_2^* e\boldsymbol{r}\psi_1 \mathrm{d}V, \tag{6.36}$$

where $\mathrm{d}V$ denotes a volume element. The Einstein coefficient of stimulated emission is equal to

$$B_{21} = \frac{2\pi^2|\boldsymbol{\mu}_{21}|^2}{3\varepsilon_0 h^2}, \tag{6.37}$$

where the spectral energy density of the field that stimulates the transition is given on the frequency scale, $\rho = \rho(\nu)$. The Einstein coefficient of absorption is

$$B_{12} = B_{21}. \tag{6.38}$$

Quantum mechanics taking account of the quantization of the electromagnetic field shows that spontaneous emission is caused by vacuum fluctuations of the electromagnetic field. Theory yields the Einstein coefficient of spontaneous emission

$$A_{21} = \frac{16\pi^3\nu^3|\boldsymbol{\mu}_{21}|^2}{3\varepsilon_0 h c^3}. \tag{6.39}$$

The Einstein relations are satisfied. Quantum theory thus provides a foundation of the old quantum mechanics.

References [1–4, 6, 31, 35–38].

Problems

6.1 Photon density. Estimate the density of photons, present in a lecture room, in a frequency interval of 1 MHz

(a) At a microwave frequency of 1 GHz.
(b) At a terahertz frequency of 1 THz.
(c) At a frequency (500 THz) in the visible.

6.2 Number of thermal photons in a mode of a laser resonator. Calculate the average number of thermal photons in a mode of a laser resonator at room temperature, for different lasers.

(a) Titanium–sapphire laser (frequency 400 THz).
(b) CO_2 laser (30 THz).
(c) Far infrared laser (1 THz).

6.3 Einstein coefficients. Determine the Einstein coefficients from spontaneous lifetimes of laser media mentioned in Table 6.1:

(a) Helium–neon laser; $\tau_{sp} = 100\,\text{ns}$.
(b) CO_2 laser; $\tau_{sp} = 5\,\text{s}$.
(c) Nd:YAG laser; $\tau_{sp} = 230\,\mu\text{s}$.

6.4 Einstein coefficient. Relate the Einstein coefficients B_{21}^{ω} (for ρ on the ω scale) and B_{21}^{ν} (for ρ on the ν scale).

6.5 Write Planck's radiation law on the wavelength scale.

6.6 Radiation laws. Derive from Planck's radiation law other laws:

(a) Wien's displacement law on the frequency scale; $\nu_{max}(T)$
(b) Wien's displacement law on the wavelength scale. $\lambda_{max}(T)$
(c) Rayleigh–Jeans law ($h\nu \ll kT$).
(d) Wien's law ($h\nu \gg kT$).
(d) Stefan–Boltzmann law. [*Hint*: $\int x^3(\mathrm{e}^x - 1)^{-1}\mathrm{d}x = \pi^4/15$.]

6.7 Maximum of the Planckian distribution.

(a) The spectrum of the cosmic background radiation has a Planck distribution corresponding to a temperature of 2.7 K. Determine the frequency ν_{max} of the maximum of the distribution on the frequency scale and the wavelength λ_{max} of the distribution on the wavelength scale. [*Hint*: $\lambda_{max} \neq \nu_{max}/c$.]
(b) Determine ν_{max} and λ_{max} for blackbody radiation emitted by a blackbody at a temperature of 300 K.

6.8 Determine the number of photons contained in a cavity with walls at temperature T. [*Hint*: $\int_0^\infty x^2(\mathrm{e}^x - 1)^{-1}\mathrm{d}x = 2.40$.]

6.9 Derive (6.2) from (6.1).

6.10 Determine the matrix elements for optical transitions of media characterized in Table 6.1.

Chapter 7
Amplification of Coherent Radiation

In the preceding chapter, we discussed the interaction of *broadband* radiation with an ensemble of two-level atomic systems. Here, we treat the interaction of *monochromatic* radiation with an ensemble of two-level atomic systems. We will show that the photon density in a disk of light traveling in an active medium increases exponentially with the traveling path length. The gain coefficient of an active medium is proportional to the Einstein coefficient of stimulated emission and to the population difference. We express the gain coefficient as the product of the gain cross section of a two-level system and the population difference.

The largest gain cross section is obtainable for an active medium with a naturally broadened $2 \rightarrow 1$ fluorescence line. Then the gain cross section at the line center is equal to the square of the wavelength of the radiation divided by 2π; we assume that the medium is optically isotropic. The broadening of a transition line of an active medium operated at room temperature is always due to another mechanism (and not by natural broadening). Therefore, the gain cross section of atoms in an active medium at room temperature is smaller than the square of the wavelength of the radiation divided by 2π.

In the case that an active medium is a two-band medium, it is convenient to introduce an effective gain coefficient. It is related to the difference of the density of electrons in the upper band and the transparency density.

We compare gain coefficients and gain cross sections of different active media and discuss, in particular, the gain coefficient of titanium–sapphire.

A two-dimensional active medium can interact with a light beam, which is three dimensional, in two ways: it can propagate along the two-dimensional medium or it can cross the two-dimensional medium. In the case that radiation *propagates along* an active medium, it is useful to introduce a modal gain coefficient, which is related to the average density of atomic two-level systems within a photon mode—the populations of atomic two-level systems still underly the laws governing the two-dimensional medium. In the case that radiation *crosses* an active medium, a description by use of the gain factor of radiation (rather than a gain coefficient of the active medium) is adequate.

K.F. Renk, *Basics of Laser Physics*, Graduate Texts in Physics,
DOI 10.1007/978-3-319-50651-7_7

7.1 Interaction of Monochromatic Radiation with an Ensemble of Two-Level Systems

In a laser, *monochromatic radiation* acts on an ensemble of two-level atomic systems. Which are, in this case, the rate equations?

We characterize monochromatic radiation (Fig. 7.1) by a spectral energy density $\rho(\nu)$ that has a constant value within a frequency interval $\nu, \nu + \mathrm{d}\nu$ and is zero outside this interval. The energy density $u(\nu)$ of the monochromatic radiation is

$$u(\nu) = \rho(\nu)\mathrm{d}\nu. \tag{7.1}$$

We assume that the spectral width of the monochromatic radiation is small compared to the linewidth $\Delta\nu_0$ of the atomic transition,

$$\mathrm{d}\nu \ll \Delta\nu_0. \tag{7.2}$$

If only natural line broadening is present, then $\mathrm{d}\nu \ll \Delta\nu_{\mathrm{nat}}$.

(We treat $\mathrm{d}\nu$ as a small but finite physical quantity; $\mathrm{d}\nu$ appears also as a differential in differential equations or integrals. The two aspects—to consider $\mathrm{d}\nu$ as a finite quantity or as a differential—are compatible with each other, *see* [20].)

We ignore, for the moment, spontaneous emission. Stimulated emission processes depopulate the upper level and absorption processes populate it. The temporal change of the population of the upper level is

$$\mathrm{d}N_2/\mathrm{d}t = -B_{21}\rho(\nu)g(\nu)\mathrm{d}\nu N_2 + B_{12}\rho(\nu)g(\nu)\mathrm{d}\nu N_1, \tag{7.3}$$

where $g(\nu)\mathrm{d}\nu$ is the portion of the transition probability in the interval $\nu, \nu + \mathrm{d}\nu$ and $g(\nu)$ is the lineshape function.

Justification: Broadband radiation with a constant spectral density in the frequency region of the spectral line leads to the temporal change of population of the upper level.

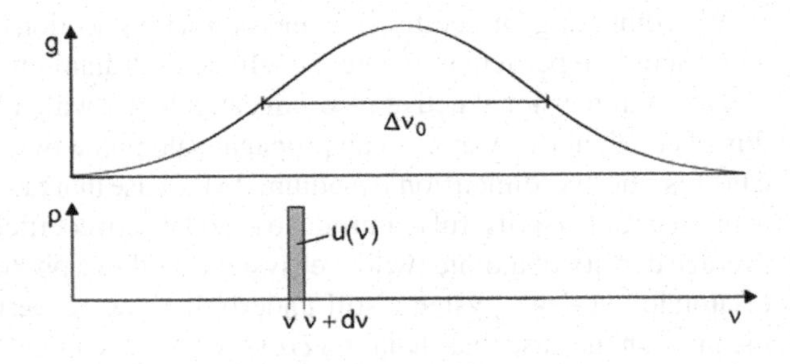

Fig. 7.1 Monochromatic radiation

$$\frac{dN_2}{dt} = -B_{21}\rho(\nu)N_2 \int_0^\infty g(\nu')d\nu' + B_{12}\rho(\nu)N_1 \int_0^\infty g(\nu')d\nu'. \quad (7.4)$$

This is, because of $\int g d\nu = 1$, equal to the result of the preceding chapter. (We assumed that B_{21} is independent of ν.)

We continue the discussion of the interaction of monochromatic radiation with an ensemble of two-level atomic systems and write, with $B_{12} = B_{21}$, the decay rate in the form

$$dN_2/dt = -B_{21}\rho(\nu)g(\nu)d\nu(N_2 - N_1). \quad (7.5)$$

It follows, with $u = u(\nu) = \rho(\nu)d\nu$, that the temporal change of the population of the upper level is given by

$$dN_2/dt = -B_{21}u\, g(\nu)(N_2 - N_1). \quad (7.6)$$

The transitions $2 \to 1$ dominate ($dN_2/dt < 0$) if $N_2 - N_1 > 0$. The change dN_2 of the population N_2 is connected with a change du of the energy density of the radiation,

$$du = -\, dN_2 \times h\nu. \quad (7.7)$$

Thus, we obtain

$$du/dt = h\nu g(\nu)B_{21}(N_2 - N_1)\, u. \quad (7.8)$$

We replace the energy density u by the photon density, $Z = u/h\nu$, and obtain the temporal change of the population of the upper level

$$dN_2/dt = -\, h\nu\, g(\nu)B_{21}(N_2 - N_1)\, Z, \quad (7.9)$$

and the change of the photon density

$$\frac{dZ}{dt} = -\frac{d(N_2 - N_1)}{dt}. \quad (7.10)$$

It follows that

$$dZ/dt = b_{21}(N_2 - N_1)Z, \quad (7.11)$$

where

$$b_{21} = h\nu B_{21}g(\nu) \quad (7.12)$$

is the *growth rate constant*, which is a measure of the strength of stimulated emission. *The growth rate constant $b_{21}(\nu)$ is equal to the product of the photon energy $h\nu$, the Einstein coefficient of stimulated emission and the value of the lineshape function at frequency ν.* We have the result: the growth rate (dZ/dt) of the photon density is proportional to the population difference and the photon density.

According to the equation

$$dN_2/dt = -b_{21} Z (N_2 - N_1), \tag{7.13}$$

we also can interpret b_{21} as the decay rate for stimulated decay of the population N_2, per unit of the photon density and per unit of population difference.

We replace the population difference by the occupation number difference, $N_2 - N_1 = (N_2 + N_1)(f_2 - f_1)$, and write

$$dN_2/dt = -b_{21}(N_1 + N_2)(f_2 - f_1) Z. \tag{7.14}$$

The decay rate of the population of the upper laser level is proportional to the density $N_1 + N_2$ of two-level atomic systems and to the occupation number difference $f_2 - f_1$. The net decay rate of the decay of a single two-level atomic system in an ensemble of two-level atomic systems is equal to

$$r_{21}(\nu) = -h\nu B_{21} g(\nu)(f_2 - f_1) Z. \tag{7.15}$$

Alternatively, we can write

$$r_{21}(h\nu) = r_{21}(\nu) = -h\nu \bar{B}_{21} g(h\nu)(f_2 - f_1) Z, \tag{7.16}$$

where the lineshape function g is now expressed on the energy scale and where $\bar{B}_{21} = h B_{21}$.

7.2 Growth and Gain Coefficient

The temporal change of the photon density is equal to

$$dZ/dt = \gamma Z, \tag{7.17}$$

where

$$\gamma = h\nu B_{21} g(\nu)(N_2 - N_1) = b_{21}(N_2 - N_1) \tag{7.18}$$

is the *growth coefficient* of an active medium.

Fig. 7.2 Monochromatic radiation in an active medium

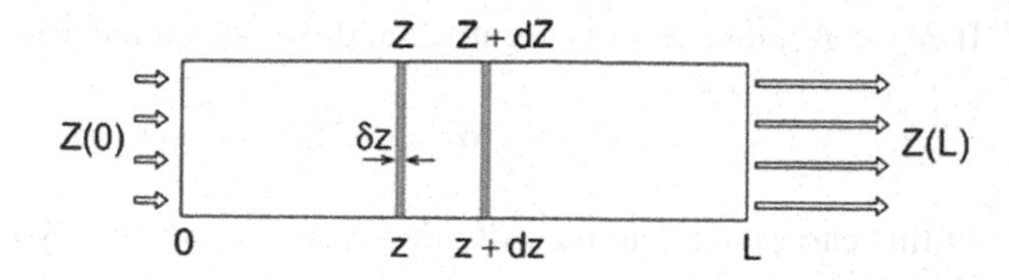

If we suddenly turn on, at time $t = 0$, the population inversion, then the density of photons increases exponentially,

$$Z = Z_0 e^{\gamma t}. \tag{7.19}$$

Z_0 is the photon density at $t = 0$.

The radiation in a disk of light (thickness δz) propagating in an active medium (Fig. 7.2) is amplified. On the path from z to $z + dz$, the change of the photon density within the disk of thickness $\delta z \ll dz$ is

$$dZ = b_{21}(N_2 - N_1)Z dt, \tag{7.20}$$

where

$$dt = dz/(c/n) \tag{7.21}$$

is the time the disk of light takes to travel the distance dz. We can write

$$dZ = \alpha(\nu)\, Z\, dz, \tag{7.22}$$

where

$$\alpha(\nu) = \frac{\gamma(\nu)}{c/n} = \frac{b_{21}(\nu)}{c/n}(N_2 - N_1) = \frac{h\nu}{c/n} B_{21} g(\nu)(N_2 - N_1) \tag{7.23}$$

is the *gain coefficient* (=small-signal gain coefficient) of an active medium. The gain coefficient is proportional to b_{21} and to the population difference.

(If the energy levels are degenerate, the gain coefficient is given by

$$\alpha(\nu) = \frac{\gamma(\nu)}{c/n} = \frac{b_{21}(\nu)}{c/n}\left(N_2 - N_1\frac{g_2}{g_1}\right) = \frac{h\nu}{c/n} B_{21} g(\nu)\left(N_2 - N_1\frac{g_2}{g_1}\right). \tag{7.24}$$

In the following, we consider nondegenerate two-level systems.)

The photon density increases exponentially with the traveling path length,

$$Z(z) = Z(z_0)\, e^{\alpha(\nu)(z - z_0)}. \tag{7.25}$$

If $N_2 < N_1$, then $\alpha(\nu)$ is negative, and we obtain the absorption coefficient

$$\alpha_{\mathrm{abs}} = -\alpha = (n/c)h\nu B_{21}g(\nu)(N_1 - N_2). \tag{7.26}$$

In this case, the photon density decreases exponentially according to the Lambert–Beer law

$$Z(z) = Z(z_0)\mathrm{e}^{-\alpha_{\mathrm{abs}}(z-z_0)}. \tag{7.27}$$

We continue the discussion of gain. We can replace the population difference by the product of the occupation number difference and the density of two-level atomic systems, $N_2 - N_1 = (f_2 - f_1)(N_1 + N_2)$, and obtain

$$\alpha(\nu) = (n/c)h\nu B_{21}g(\nu)\,(N_1 + N_2)\,(f_2 - f_1). \tag{7.28}$$

The gain coefficient is proportional to the density of two-level atomic systems and to the occupation number difference $f_2 - f_1$. We introduce the *gain bandwidth* $\Delta\nu_{\mathrm{g}}$ as the halfwidth of the gain curve $\alpha(\nu)$. If B_{21} is independent of frequency, $\Delta\nu_{\mathrm{g}}$ is determined by the halfwidth of the lineshape function $g(\nu)$.

If the lineshape function is given on the energy scale, then

$$\alpha(\nu) = \alpha(h\nu) = (n/c)h\nu \bar{B}_{21}g(h\nu)\,(N_1 + N_2)\,(f_2 - f_1). \tag{7.29}$$

If a line has Lorentzian shape, we can write the gain coefficient in the form

$$\alpha(\nu) = \frac{h\nu}{c/n}B_{21}\,\frac{2}{\pi\,\Delta\nu_0}\,\bar{g}_{\mathrm{L,res}}(\nu)\,(N_2 - N_1), \tag{7.30}$$

or

$$\alpha(\nu) = (n/c)h\nu B_{21}\frac{2}{\pi\,\Delta\nu_0}\,\bar{g}_{\mathrm{L,res}}(N_1 + N_2)\,(f_2 - f_1), \tag{7.31}$$

where $\bar{g}_{\mathrm{L,res}}$ is the lineshape function normalized to unity at the line center.

In a light beam propagating through an active medium of length L, the photon density increases from the value Z_0 to the value

$$Z = Z_0\mathrm{e}^{\alpha(\nu)L}. \tag{7.32}$$

The *single-path gain factor* is equal to

$$G_1(\nu) = \mathrm{e}^{\alpha(\nu)L}. \tag{7.33}$$

The single-path gain is

$$\frac{Z - Z_0}{Z_0} = \frac{u - u_0}{u_0} = G_1(\nu) - 1 = e^{\alpha(\nu)L} - 1. \tag{7.34}$$

If $\alpha(\nu)L \ll 1$, then

$$G_1(\nu) - 1 = \alpha(\nu)L, \tag{7.35}$$

i.e., the single-path gain $G_1 - 1$ is equal to the product of the gain coefficient and the length of the active medium.

7.3 Gain Cross Section

We write the gain coefficient of an active medium (containing two-level atomic systems) in the form

$$\alpha(\nu) = N_2\sigma_{21} - N_1\sigma_{12}, \tag{7.36}$$

where σ_{21} is the *gain cross section* of a two-level atomic system and σ_{12} the *absorption cross section*. The cross sections are equal, $\sigma_{12} = \sigma_{21}$. It follows that the gain coefficient of an active medium containing an ensemble of two-level atomic systems is given by

$$\alpha = (N_2 - N_1)\,\sigma_{21}, \tag{7.37}$$

where

$$\sigma_{21}(\nu) = \frac{b_{21}}{c/n} = \frac{h\nu}{c/n} B_{21} g(\nu). \tag{7.38}$$

The gain cross section at the frequency ν is proportional to ν, to the Einstein coefficient of stimulated emission, and to the value of the lineshape function at the frequency ν.

In the case that a line is naturally broadened, we can use the Einstein relations and the relation $A_{21} = \tau_{sp}^{-1}$. We then find

$$\sigma_{21}(\nu) = \frac{c^2 A_{21}}{8\pi n^2 \nu^2} g_{nat}(\nu) = \frac{c^2}{8\pi n^2 \nu^2} \frac{1}{\tau_{sp}} g_{nat}(\nu). \tag{7.39}$$

The largest gain cross section of an isotropic two-level atomic system in an active medium is obtainable if the $2 \rightarrow 1$ line is naturally broadened. Then $g(\nu_0) = 4\tau_{sp}$ and

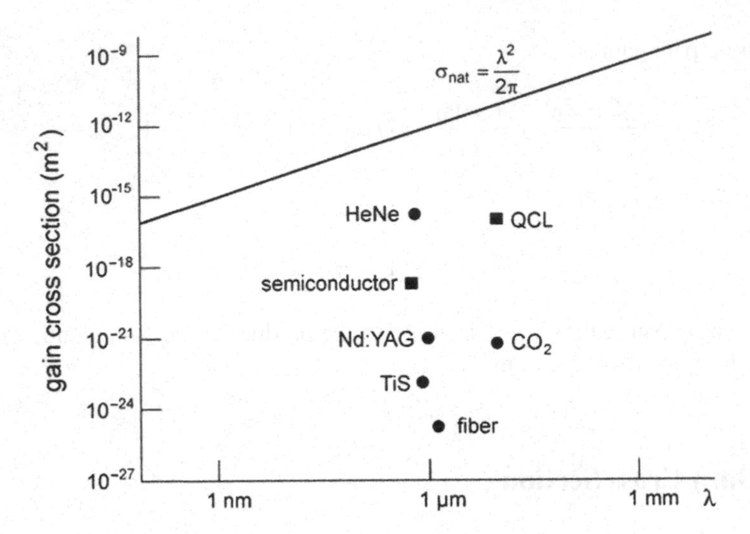

Fig. 7.3 Gain cross sections

$$\sigma_{\rm nat} = \sigma_{21,\rm nat}(\nu_0) = \frac{c^2}{2\pi n^2 \nu_0^2} = \frac{(\lambda/n)^2}{2\pi}, \tag{7.40}$$

where $\lambda = c/\nu_0$ is the wavelength of the radiation in vacuum, n is the refractive index of the active medium at the frequency ν_0, and $\nu_0 = (E_2 - E_1)/h$. The gain cross section of a two-level system with a naturally broadened line increases with the square of the wavelength (Fig. 7.3, solid line).

We will see later, when we will discuss specific lasers (Chaps. 14–16 and chapters on semiconductor lasers), that the active media of all lasers operated at room temperature show $1 \rightarrow 2$ absorption lines and the $2 \rightarrow 1$ fluorescence lines that are not broadened by natural broadening, but that other mechanisms dominate the line broadening. Therefore, the gain cross section of radiation propagating in an active (isotropic) medium at room temperature is smaller than $(\lambda/n)^2/2\pi$.

If the gain coefficient curve is a Lorentz resonance curve, we can write

$$\sigma_{21}(\nu) = \frac{h\nu}{c/n} B_{21}\, g_{\rm L,res}(\nu) = \frac{2}{\pi\Delta\nu_0}\frac{h\nu}{c/n} B_{21}\, \bar{g}_{\rm L,res}(\nu). \tag{7.41}$$

The gain cross section at the center of a Lorentzian gain curve (with the halfwidth $\Delta\nu_0$) is

$$\sigma_{21,\rm L}(\nu_0) = \frac{\Delta\nu_{\rm nat}}{\Delta\nu_0}\sigma_{\rm nat}(\nu_0) = \frac{\Delta\nu_{\rm nat}}{\Delta\nu_0}\frac{(\lambda/n)^2}{2\pi}. \tag{7.42}$$

The gain cross section of radiation at the line center of a Lorentzian line is by the factor $\Delta\nu_{\rm nat}/\Delta\nu_0$ smaller than in case of a naturally broadened line.

Table 7.1 Gain bandwidths and gain cross sections

Laser	λ	n	$\Delta\nu_g$	$\Delta\nu_g/\nu_0$	σ_{21} (m^2) [σ_{eff} (m^2)]
HeNe	633 nm	1	1.5 GHz	3×10^{-6}	1.4×10^{-16}
CO_2	10.6 μm	1	69 MHz–500 GHz	2.5×10^{-6}–1.7×10^{-2}	1.2×10^{-20}
Nd:YAG	1.06 μm	1.82	140 GHz	1.4×10^{-4}	8.1×10^{-22}
TiS ($E \parallel$ c)	830 nm	1.74	110 THz	0.3	2.3×10^{-23}
TiS ($E \perp$ c)					8×10^{-24}
Fiber	1.5 μm	1.5	5 THz–12 THz	2.5–6 $\times 10^{-2}$	2×10^{-25}–[6×10^{-25}]
Semiconductor	840 nm	3.6	10 GHz–1 THz	3×10^{-5}–3×10^{-3}	[3×10^{-19}]
QCL	5 μm	3.4	10 GHz–1 THz	2×10^{-4}–1.6×10^{-2}	10^{-16}

The gain cross section of radiation at the center of a Gaussian line is

$$\sigma_{21,G}(\nu_0) = \frac{\Delta\nu_{nat}}{\Delta\nu_0}\sqrt{\pi \ln 2}\,\sigma_{nat}(\nu_0) \sim 1.48\,\frac{\Delta\nu_{nat}}{\Delta\nu_0}\frac{(\lambda/n)^2}{2\pi}. \quad (7.43)$$

Table 7.1 shows values of gain bandwidths and gain cross sections (and of effective gain cross sections, *see* next section and Sect. 18.7). Different halfwidths, mentioned in the following, are: $\Delta\nu_g$ = gain bandwidth = halfwidth of the gain curve; $\Delta\nu_0$ = halfwidth of an absorption line; $\Delta\nu_{fluor}$ = halfwidth of the fluorescence line, measured on the frequency scale.

- *Helium–neon laser*. The gain bandwidth is equal to the halfwidth of the 2 → 1 fluorescence line, $\Delta\nu_g = \Delta\nu_{fluor} = \Delta\nu_0$.
- *CO_2 laser*. $\Delta\nu_g = \Delta\nu_{fluor} = \Delta\nu_0$.
- *Nd:YAG laser*. $\Delta\nu_g = \Delta\nu_{fluor} = \Delta\nu_0$.
- *Titanium–sapphire laser*. The gain bandwidth is very large (Sect. 7.6). In the table, the crystal anisotropy of Ti^{3+}:Al_2O_3 is taken into account. An average gain cross section is $\sigma_{21} = \frac{1}{3}(\sigma_1 + 2\sigma_2)$, where σ_1 is the gain cross section for $E \parallel$ c and σ_2 is the gain cross section for $E \perp$ c. The experimental fluorescence curves indicate that $\sigma_2 \sim 3\sigma_1$ and that therefore $\sigma_1 \sim 1.8\,\sigma_{21}$ and $\sigma_2 \sim 0.6\,\sigma_{21}$.
- *Fiber laser*. The gain bandwidth depends on the pump strength (Chap. 18).
- *Bipolar semiconductor laser*. The gain bandwidth changes if the strength of the pumping (i.e., the current flowing through the active semiconductor medium) changes (Chaps. 21 and 22).
- *Quantum cascade laser*. The gain bandwidth varies if the strength of the pumping changes. The gain bandwidth of an active medium of a specific quantum cascade laser can be obtained by a detailed analysis of the properties of the active medium (Chap. 29).

The gain bandwidths of the different active media differ by five orders of magnitude. The titanium–sapphire laser has by far the largest gain bandwidth, corresponding to 30% of the center frequency. The gain cross section (*see* also Fig. 7.3) differs by nine orders of magnitude. The helium–neon laser and the quantum cascade laser have large values of the gain cross section. The bipolar semiconductor lasers have smaller values. CO_2 lasers and solid state lasers have still smaller values.

7.4 An Effective Gain Cross Section

We consider a two-band laser with an active medium containing N_0 two-level atomic systems per unit volume. Without pumping, all levels in the lower band are occupied and all levels of the upper band are empty (Fig. 7.4a); we assume that $E_{2,\min} - E_{1,\max} \gg kT$. Pumping leads to a population in the upper band and to empty levels in the lower band. The population in the upper band and the population in the lower band are in a nonequilibrium relative to each other. At weak pumping, the population N_2 (=N) in the upper band is small and the density N_1 (=N) of empty levels in the lower band is also small. Accordingly, the relative occupation number f_2 (at energies near the minimum of the upper band) is small ($f_2 \ll 1$), the relative occupation number f_1 (at energies near the maximum of the lower band) is only slightly smaller than unity and absorption of radiation prevails. The width of the energy distributions of each of the populations is of the order of kT. With increasing N, f_2 increases and f_1 decreases until N reaches the transparency density N_{tr}, where $f_2 - f_1 = 0$ (Fig. 7.4b). The width of the energy distributions of each of the populations is still of the order of kT. The largest population in the upper band occurs at energies near the band minimum (energy $E_{2,\min}$) and the smallest population in the lower band at energies near the band maximum (energy $E_{1,\max}$).

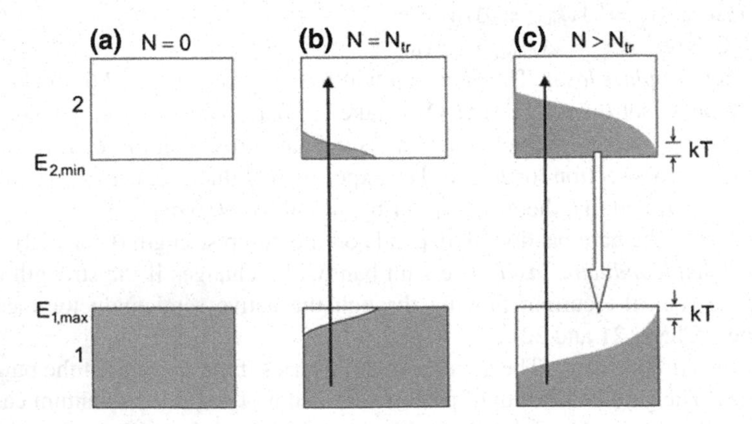

Fig. 7.4 Two-band laser. **a** Population without pumping. **b** Quasi-thermal distributions of the populations for $N = N_{tr}$. **c** Quasi-thermal distributions of the populations for $N > N_{tr}$

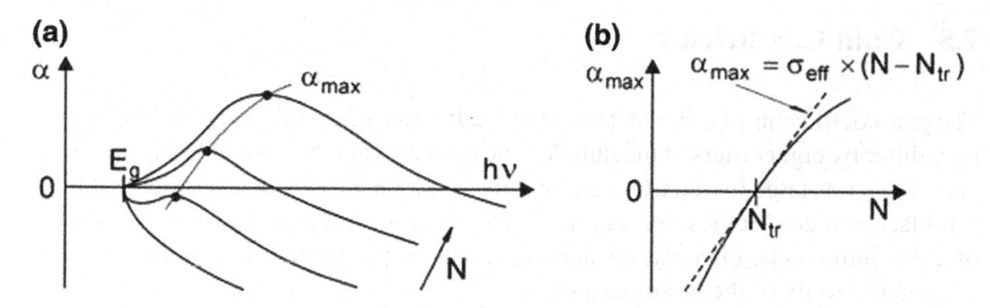

Fig. 7.5 Gain coefficient of a two-band laser medium. **a** Frequency dependence of the gain coefficient. **b** Dependence of the maximum gain coefficient on the population N in the upper band

If $N > N_{tr}$ (i.e., $f_2 > f_1$), the medium is a gain medium and the gain increases with increasing N. The gain bandwidth increases with increasing band filling and becomes larger than kT at large filling (Fig. 7.4c); then the widths of the distributions are larger than $k\,T$.

The gain coefficient α depends on different parameters: temperature of the active medium; Einstein coefficient B_{21}; electron density N; energy distributions in the levels of the lower and the levels in the upper band. The $\alpha(\nu)$ curve can have a complicated shape (Fig. 7.5a). Gain occurs if the maximum α_{max} of the $\alpha(\nu)$ curve becomes (for $N = N_{tr}$) positive. With increasing N, α_{max} increases and the gain bandwidth increases too. Figure 7.5b (solid line) shows α_{max} versus N for values of N around the transparency density. The expansion of α_{max} leads to

$$\alpha_{max} = \sigma_{eff} \times (N - N_{tr}), \tag{7.44}$$

where the differential cross section

$$\sigma_{eff} = (\partial\alpha_{max}/\partial N)_{N=N_{tr}} \tag{7.45}$$

is an *effective gain cross section*. The effective gain cross section (Fig. 7.5b, dashed) corresponds to the slope of $\alpha_{max}(N)$ near N_{tr}. The effective gain cross section is the gain cross section related to the density of two-level systems excited in addition to the two-level systems that are, at the transparency density, already in the excited state. We will discuss later (in Chaps. 21 and 22) how we can determine the effective gain cross sections and α_{max} as well as gain bandwidths of bipolar semiconductor lasers; a value of an effective gain cross section is given in Table 7.1.

We will introduce two other effective gain cross sections in connection with the discussion of gain coefficient of a doped fiber (Sect. 18.7).

7.5 Gain Coefficients

The gain coefficients of different laser media differ markedly (Table 7.2 and Fig. 7.6—they differ by eight orders of magnitude. The gain coefficient is small for the helium–neon laser, it is large for the CO_2 laser and for solid-state lasers. It is very large for the (bipolar) semiconductor laser and the quantum cascade laser. The gain coefficient of a medium can be obtained by an analysis of the properties of an active medium or from the study of the laser thershold. The length L' of an active medium is about 1 mm or smaller for a semiconductor laser and lies between several centimeters and about 1 m for the other lasers. We can ask whether it is possible to increase $N_2 - N_1$ in order to obtain larger gain coefficients. The answer is different for the different lasers.

Table 7.2 Gain coefficients

Laser	λ	α (m^{-1})	L' (m)	G (G_1)	σ_{21} [σ_{eff}] (m^2)	$N_2 - N_1$ [$N - N_{tr}$] (m^{-3})
HeNe	633 nm	0.014	0.5	1.014	1.4×10^{-16}	10^{14}
CO_2	10.6 μm	5	0.5	3	1.2×10^{-20}	1.5×10^{18}
Nd:YAG	1.06 μm	20	0.1	50	8.1×10^{-22}	1×10^{23}
TiS $E\|\|c$	830 nm	20	0.1	50	2.3×10^{-23}	8×10^{23}
Fiber	1.5 μm	0.7	10	(10^3)	[6×10^{-25}]	1.2×10^{24}
Semiconductor	840 nm	1,500	10^{-3}	(4.5)	[3×10^{-19}]	[6×10^{21}]
QCL	5 μm	1,000	10^{-3}	2.7	10^{-16}	10^{19}

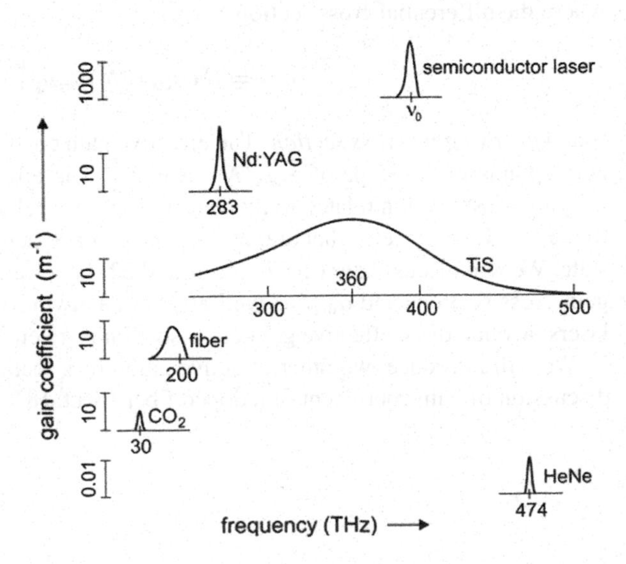

Fig. 7.6 Gain coefficient of different laser media

- *Helium–neon laser.* The excited neon atoms have a large gain cross section. However, the population difference $N_2 - N_1$ that can be reached is very small.
- *CO_2 laser.* The value of the population difference given in the table is obtained at a discharge in a gas of 5 mbar pressure. Higher population differences are obtainable at higher gas pressures (e.g., at a gas pressure of 1 bar in pulsed lasers), which emit pulses of high power. The gain coefficient α at the line center does not change with $N_2 - N_1$ because of collision broadening (Sects. 14.2 and 14.8).
- *Nd:YAG laser.* The population difference $N_2 - N_1$ increases in case of stronger pumping. However, it is preferable to make use of stronger pumping to increase the laser output power rather than to enhance the population difference (Chap. 8).
- *Titanium–sapphire laser.* The population difference cannot increase much further, it has already a value near 10% of the density of Ti^{3+} ions (at a doping level of 10^{25} m^{-3}). A further increase of $N_2 - N_1$ leads to saturation of the pump rate (Sect. 5.4).
- *Fiber laser.* An increase of the population difference (at stronger pumping) by a factor of 10 is possible; the impurity concentration, $N_0 = 7 \times 10^{25}\,\mathrm{m}^{-3}$, is by about an order of magnitude larger than for crystals. The transparency density is $N_{tr} \sim N_0/2$.
- *Bipolar semiconductor laser.* An increase of the population difference $N - N_{tr}$ and of α by less than an order of magnitude is possible.
- *Quantum cascade laser.* Whether the population difference can be increased depends on the specific design.

While the gain coefficient curves of gas lasers follow directly from atomic properties of gases and of solid-state laser media from atomic properties of impurity ions in solids, the situation is completely different for semiconductor and quantum cascade lasers: it is possible to choose the center frequency ν_0 of a gain coefficient curve through the choice of an appropriate semiconductor material and an appropriate heterostructure. Designing bipolar semiconductor lasers is possible for almost all frequencies of radiation in the near UV, the visible and the near infrared (150–800 THz). Designing quantum cascade lasers is possible for all frequencies in the range 11–150 THz or, as cooled quantum cascade lasers operating at a temperature of 80 K, in the range 1–5 THz.

7.6 Gain Coefficient of Titanium–Sapphire

In the preceding section, we presented gain data of titanium–sapphire. Here, we show how we can obtain the data. The fluorescence band extends over a large wavelength range. Therefore, we have to take into account that the Einstein coefficient of spontaneous emission varies strongly with frequency. We now determine the spectral distribution $S_\nu(\nu)$ on the frequency scale from the spectral distribution $S_\lambda(\lambda)$ on the wavelength scale, represented in Fig. 5.3 (Sect. 5.3). We use the relation

$$S_\nu(\nu)|d\nu| = S_\lambda(\lambda)|d\lambda|, \tag{7.46}$$

where $d\nu$ is a frequency range near the frequency ν and $d\lambda$ the corresponding wavelength range near the wavelength $\lambda = c/\nu$. With $\nu = c/\lambda$ and $|d\nu| = |d\lambda|/\lambda^2$, we obtain

$$S_\nu = \frac{1}{|d\nu/d\lambda|}\, S_\lambda = \frac{\lambda^2}{c}\, S_\lambda. \tag{7.47}$$

S_ν is proportional to $g(\nu)$ and to $A_{21}(\nu)$, i.e.,

$$g(\nu) = K_1 \frac{S_\nu(\nu)}{A_{21}(\nu)} = K_2\, \lambda^5\, S_\lambda. \tag{7.48}$$

K_1 and K_2 are constants. Multiplying S_λ (Fig. 5.3) by λ^5 and normalizing the maximum of the lineshape function to 1, we obtain the gain profile $\bar{g}(\nu)$ shown in Fig. 7.7 (solid line). The curve yields the center frequency ν_0 (~ 360 THz) and the gain bandwidth $\Delta\nu_g$ ($\sim$110 THz). The ratio of the gain bandwidth and the center frequency is about 0.3. The gain coefficient $\alpha(\nu)$ has a Gaussian-like profile. At small frequencies ($\nu < \nu_0$), the decrease of the $\bar{g}(\nu)$ curve is less steep than for a Gaussian lineshape (dotted). A Gaussian-like gain profile, with a deviation from a Gaussian profile at small frequencies, is consistent with the vibronic character of the energy levels as we will discuss later (in Chap. 17)—the deviation is a consequence of the anharmonicity of the lattice vibrations of sapphire. We attribute the line broadening to homogeneous broadening (Sect. 17.4).

Figure 7.8 (upper part) shows the absorption coefficient of Ti^{3+}:Al_2O_3 and, furthermore, the gain coefficient of excited Ti^{3+}:Al_2O_3 in the case that 8% of the titanium ions (in a crystal containing 10^{25} Ti^{3+} ions per m^3) are in the excited state. The maximum gain coefficient follows from the relation $\alpha_{max} = N\sigma_{max}$, where N_2 is the density of excited Ti^{3+} ions. We find that the maximum cross section of stimulated emission is equal to

$$\sigma_{max} = a\sqrt{\pi \ln 2}\, \frac{\Delta\nu_{nat}}{\Delta\nu_0}\, \sigma_{nat}(\nu_0) \sim 1.48a\, \frac{\Delta\nu_{nat}}{\Delta\nu_0}\, \frac{(\lambda/n)^2}{2\pi}, \tag{7.49}$$

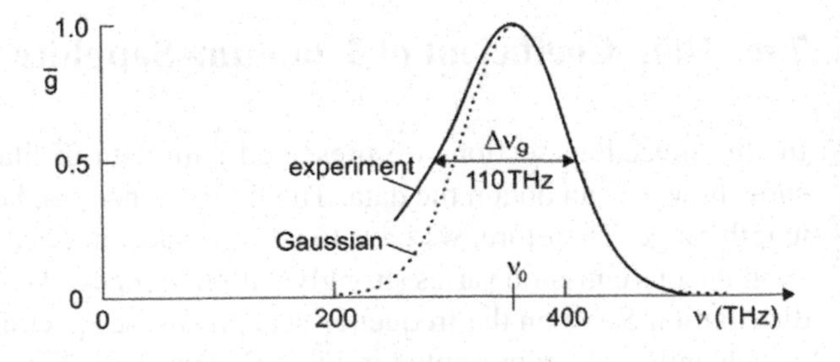

Fig. 7.7 Gain profile of Ti^{3+}:Al_2O_3

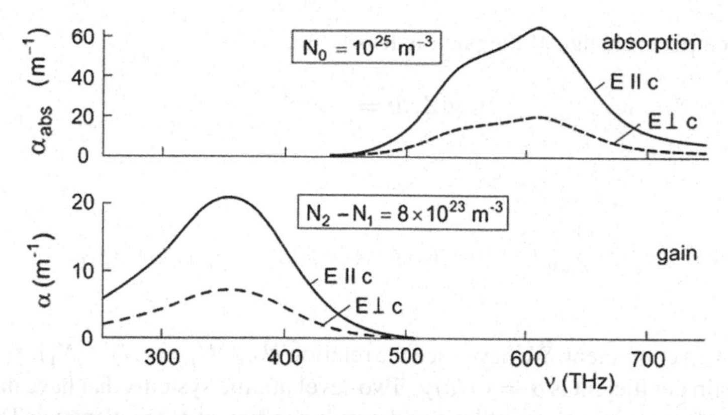

Fig. 7.8 Absorption and gain coefficients of Ti^{3+}:Al_2O_3

where $\Delta\nu_{nat}$ is the natural linewidth, $\Delta\nu_0$ the gain bandwidth, $\sigma_{nat}(\nu_0)$ the cross section corresponding to a naturally broadened line and n (=1.74) the refractive index of sapphire; the factor $\sqrt{\pi \ln 2}$ takes account of the difference of a Gaussian and a Lorentzian profile and the factor a ($\sim$2) of crystal anisotropy.

7.7 Gain Coefficient of a Medium with an Inhomogeneously Broadened Line

We can decompose an inhomogeneously broadened line, e.g., a Gaussian line, into homogeneously broadened lines. We introduce:

- ν_c = center frequency of an inhomogeneous broadened line.
- $g_{inh}(\nu, \nu_c)$ lineshape function describing the inhomogeneous broadening.
- ν_0 = resonance frequency of a specific two-level atomic system; each two-level atomic system has its own resonance frequency.
- $g_{hom}(\nu, \nu_0)$ = lineshape function describing the homogeneous broadening of a two-level atomic system that has the resonance frequency ν_0.
- $dN_2^{d\nu_0} = N_2 g_{inh}(\nu_0, \nu_c)d\nu_0$ = density of two-level atomic systems in the upper laser level that have the resonance frequency in the frequency interval ν_0, $\nu_0 + d\nu_0$.
- $dN_1^{d\nu_0} = N_1 g_{inh}(\nu_0, \nu_c)d\nu_0$ = density of two-level atomic systems in the lower laser level that have the resonance frequency in the frequency interval ν_0, $\nu_0 + d\nu_0$.

The temporal change of the population difference is given by

$$d(N_2 - N_1)/dt = -\gamma\, Z, \tag{7.50}$$

and the temporal change of the photon density is

$$\mathrm{d}Z/\mathrm{d}t = \gamma\, Z, \tag{7.51}$$

where

$$\gamma(\nu) = \int_0^\infty h\nu g_{\mathrm{hom}}(\nu - \nu_0) B_{21}(N_2 + N_1)(f_2 - f_1) g_{\mathrm{inh}}(\nu_0)\mathrm{d}\nu_0 \tag{7.52}$$

is the growth coefficient. We have used the relation $N_2 - N_1 = (N_2 + N_1)(f_2 - f_1)$.

The gain coefficient is $\alpha = (n/c)\gamma$. Two-level atomic systems that have different resonance frequencies ν_0 contribute to the gain coefficient at frequency ν. The Einstein coefficient B_{21} can depend on frequency. A special case of (7.52) is the Voigt profile (Problem 14.2b).

7.8 Gain Characteristic of a Two-Dimensional Medium

There are two possibilities to arrange a two-dimensional active medium in a light beam. The propagation direction of the light can be parallel to the plane of the medium or perpendicular. We treat here the first case and the second case in the next section.

We consider the propagation of a parallel light beam that contains a two-dimensional active medium (Fig. 7.9). The propagation direction of the light is parallel to the plane of the medium. We introduce the *average density of two-level atomic systems in the light beam*,

$$N_{\mathrm{av}} = \frac{N^{2\mathrm{D}}}{a_2}, \tag{7.53}$$

where a_2 is the extension of the beam perpendicular to the film plane—the height of the photon mode—and $N^{2\mathrm{D}}$ is the two-dimensional density of two-level systems in the two-dimensional medium. If $N_2^{2\mathrm{D}} - N_1^{2\mathrm{D}}$ is the population difference, the average population difference in the light beam is

$$(N_2 - N_1)_{\mathrm{av}} = \frac{N_2^{2\mathrm{D}} - N_1^{2\mathrm{D}}}{a_2}. \tag{7.54}$$

The average density is *independent* of the thickness of the two-dimensional active medium. It follows that the temporal change of the photon density in a disk of light is equal to

$$\frac{\mathrm{d}Z}{\mathrm{d}t} = b_{21}\frac{N_2^{2\mathrm{D}} - N_1^{2\mathrm{D}}}{a_1}\, Z. \tag{7.55}$$

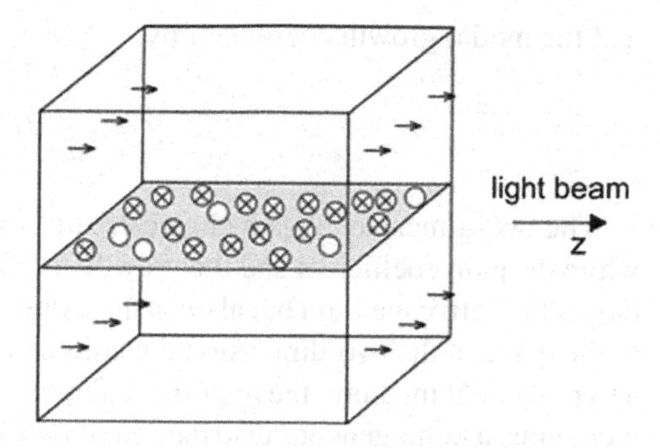

Fig. 7.9 Two-dimensional medium in a light beam

Our procedure is justified because it does not matter at which position within the photon mode the two-level atomic systems are located. As an essential condition, we assumed that the photons in the light beam belong to a single mode.

The growth coefficient is equal to

$$\gamma = b_{21} \frac{N_2^{2D} - N_1^{2D}}{a_2}. \tag{7.56}$$

Taking into account that $b_{21} = (c/n)\sigma_{21}$, we find the gain coefficient

$$\alpha = \sigma_{21} \frac{N_2^{2D} - N_1^{2D}}{a_2}. \tag{7.57}$$

The gain coefficient α is inversely proportional to the extension of the photon mode perpendicular to the plane of the two-dimensional medium and is called *modal gain coefficient*. The single-path gain factor of radiation transversing a medium of length L is

$$G_1 = e^{\alpha L}. \tag{7.58}$$

We introduce the two-dimensional gain characteristic

$$H^{2D}(\nu) = \sigma_{21}(\nu)(N_2^{2D} - N_1^{2D}), \tag{7.59}$$

as the product of the gain cross section and the difference of the two-dimensional populations. The modal gain coefficient is given by

$$\alpha = \frac{1}{a_2} H^{2D} \tag{7.60}$$

and the modal growth coefficient by

$$\gamma = \frac{c}{na_2} H^{2D}. \quad (7.61)$$

The two-dimensional gain characteristic completely describes the active medium while the gain coefficient and the growth coefficient depends not only on the properties of the active medium but also on the extension of the photon mode perpendicular to the plane of the two-dimensional medium. Thus, a modal gain coefficient refers to a hypothetical medium: the hypothetical medium has the height of the photon mode; it contains a homogeneous distribution of two-level systems; the density of two-level systems in the hypothetical medium is equal to the density of two-level systems in the two-dimensional medium divided by the height of the photon mode.

Example GaAs quantum well in a light beam in a quantum well laser (Chaps. 21 and 22).

7.9 Gain of Light Crossing a Two-Dimensional Medium

In the case that a light beam is crossing a two-dimensional active medium (Fig. 7.10), it is convenient to make use of the gain factor rather than the gain coefficient. The average population difference in a disk of light is

$$(N_2 - N_1)_{av} = \frac{N_2^{2D} - N_1^{2D}}{\delta z}, \quad (7.62)$$

where δz is the length of the disk. The change of the photon density within the interaction time $\delta t = n\delta z/c$ is

$$\delta Z = \frac{N_2^{2D} - N_1^{2D}}{\delta z} b_{21}\, Z\, \delta t. \quad (7.63)$$

It follows that

$$\frac{\delta Z}{Z} = \sigma_{21}(N_2^{2D} - N_1^{2D}) \quad (7.64)$$

and, with $G_1 - 1 = \delta Z/Z$, that

$$G_1 - 1 = \sigma_{21}(N_2^{2D} - N_1^{2D}) = H^{2D}(\nu). \quad (7.65)$$

The single-path gain $G_1 - 1$ of radiation crossing a two-dimensional medium is equal to the two-dimensional gain characteristic.

Example a light beam crossing a GaAs quantum well in a quantum well laser (Chaps. 21 and 22).

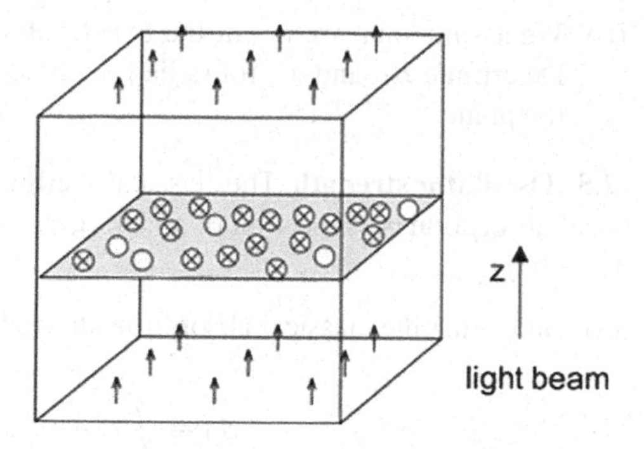

Fig. 7.10 Light beam crossing a two-dimensional active medium

References [1–4, 6, 31, 35–37].

Problems

7.1 Amplification of radiation in titanium–sapphire. Given is an active titanium–sapphire medium with a population difference $N_2 - N_1 = 10^{24}\ \mathrm{m}^{-3}$.

(a) Determine the gain coefficient at the frequency of maximum gain.
(b) Determine the single-path gain factor at the frequency of maximum gain when the crystal has a length of 10 cm.
(c) Determine the gain coefficient and the single path gain factor of radiation at a wavelength in vacuum of 1 μm.

7.2 Gain cross section of Ti^{3+} in titanium–sapphire. Compare the gain cross section of an excited Ti^{3+} ion with the gain cross section of a two-level system that has a naturally broadened line at the frequency of maximum gain coefficient of titanium–sapphire.

7.3 Two-dimensional gain medium. The two-level systems of a two-dimensional gain medium have a gain cross section $\sigma_{21} = 1.5 \times 10^{-19}\,\mathrm{m}^2$. The population difference is equal to $N_2^{2D} - N_1^{2D} = 10^{16}\,\mathrm{m}^{-2}$.

(a) Estimate the modal gain coefficient in the case that radiation propagates along the active medium and that the mode has a height of 800 nm.
(b) Estimate the gain for radiation traversing the medium.

7.4 Anisotropic media.

(a) We manipulate the two-level atomic systems of an active medium (for example, by applying a magnetic field, so that the atomic dipoles have an orientation mainly in one direction instead of a random orientation); we assume that A_{21} does not change. Determine B_{21} and σ_{21} for radiation of different orientations of the electric field vector of electromagnetic radiation.

(b) We assume that we orient the two-level systems with their dipoles in a plane. Determine B_{21} and σ_{21} for radiation polarized either parallel or perpendicular to the plane.

7.5 Oscillator strength. The classical oscillator model of an atom provides the classical absorption cross section of an atom, $\sigma_{\rm cl}(\nu) = e^2/(4\varepsilon_0 m_0 c) g_{\rm L,res}(\nu)$, according to (9.67)

(a) Show that the classical absorption strength is

$$S_{\rm cl} \equiv \int \sigma_{\rm cl}(\nu)\mathrm{d}\nu = \frac{e^2}{4\,\epsilon_0\, m_0 c}. \tag{7.66}$$

(b) Show that the quantum mechanical absorption strength is equal to

$$S \equiv \int \sigma_{21}(\nu)\mathrm{d}\nu = \frac{n}{c} h\nu B_{21} = \frac{c^2 A_{21}}{8\pi n^2 \nu^2} = \frac{c}{8\pi n^2 \nu^2 \tau_{\rm sp}}. \tag{7.67}$$

(c) We introduce the oscillator strength f via the relation

$$S = S_{\rm cl} \times f. \tag{7.68}$$

Estimate the oscillator strengths, which correspond to the absorption and to the gain cross sections of titanium–sapphire.

(d) Show that in case of a narrow line

$$S = \frac{\lambda_0^2}{8\pi n^3 \tau_{\rm sp}}, \tag{7.69}$$

where $\lambda_0 = c/\nu_0$.

7.6 Fluorescence line and absorption cross section.

(a) Show that we can write, in case of a narrow line caused by transitions in an ensemble of two-level atomic systems,

$$\frac{1}{\tau_{\rm sp}} \approx \frac{8\pi c n^2}{\lambda_0^4} \int \sigma_{12}(\lambda)\mathrm{d}\lambda, \tag{7.70}$$

where λ_0 is the center wavelength, n the refractive index and σ_{12} the absorption cross section.

(b) Show that this leads to the relation

$$\sigma_{12}(\lambda) = \frac{\lambda_0^4}{8\pi n^2 \tau_{\rm sp}} \frac{S(\lambda)\mathrm{d}\lambda}{\int S(\lambda)\mathrm{d}\lambda}, \tag{7.71}$$

where $S(\lambda)\mathrm{d}\lambda$ is the fluorescence intensity in the wavelength interval $\mathrm{d}\lambda$ at the wavelength λ and $\int S(\lambda)\mathrm{d}\lambda$ is the total fluorescence intensity. This relation is sometimes called Füchtbauer-Ladenburg relation; in the 1920s, Füchtbauer studied absorption lines [48] and Ladenburg (*see* Sect. 9.10) fluorescence lines of atomic gases.

7.7 Gain saturation. We consider a four-level laser medium and take into account both pumping and relaxation. Instead of (7.13), we write

$$\mathrm{d}N_2/\mathrm{d}t = r - b_{21}Z(N_2 - N_1) - N_2/\tau^{\star}_{\mathrm{rel}}, \tag{7.72}$$

where r is the pump rate (per unit of volume). We assume that $\tau_{\mathrm{rel}} \ll \tau^{\star}_{\mathrm{rel}}$ and therefore $N_2 \ll N_2$, and find

$$N_2 = N_{2,0}(1 + b_{21}\tau^{\star}_{\mathrm{rel}}Z). \tag{7.73}$$

We introduce the intensity $I = cZh\nu$. It follows that the large-signal gain coefficient is

$$\alpha_{\mathrm{I}} = \alpha/(1 + I/I_s), \tag{7.74}$$

where

$$I_{\mathrm{s}} = c/\big(B_{21}g(\nu)\tau^{\star}_{\mathrm{rel}}\big) \tag{7.75}$$

is the saturation intensity.

(a) Sketch gain curves for $I/I_{\mathrm{s}} = 0; 1; 10$. [*Hint*: in the case of homogeneous broadening, the whole line saturates.]
(b) Determine the saturation intensity For Nd:YAG.
(c) Determine the saturation intensity For titanium–sapphire.

7.8 Saturation of absorption.

(a) Consider an ensemble of two-level atomic systems and show that the large-signal absorption coefficient is

$$\alpha_{\mathrm{abs,I}} = \alpha_{\mathrm{abs}}/(1 + I/I_{\mathrm{s}}), \tag{7.76}$$

where $\alpha_{\mathrm{abs}} = -(n/c)h\nu B_{12}g(\nu)(N_2 - N_1)$ is the small-signal absorption coefficient, $I = cZh\nu$ the intensity of radiation, and

$$I_{\mathrm{s}} = c/\left(2B_{12}g(\nu)\tau^{\star}_{\mathrm{rel}}\right) \tag{7.77}$$

the saturation intensity. [*Hint*: begin with (7.26); take into account that the total population density $N_{\mathrm{tot}} = N_2 + N_1$ is constant; introduce the population difference ΔN, with $\Delta N = N_2 - N_1$; then derive the differential equation for

$d(\Delta N)/dt$ and determine the steady state solution; because the lower level remains populated, the saturation intensity is smaller (by a factor two) than in case of a four-level system with a short lifetime of the lower laser level (Problem 7.7).]

(b) Determine ΔN, N_2, and N_1 for $I = I_s$.
(c) Sketch absorption curves of a transition for $I/I_s = 0; 1; 10$.
(d) Why is the saturation intensity in case of saturation of absorption and in case of gain saturation independent of the populations of the two-level systems?

7.9 Show that the transition probability for stimulated emission induced by monochromatic radiation in a frequency band $d\nu$ is given by
$w_{21\text{stim}} = B_{21}\rho(\nu)g(\nu)d\nu$.

7.10 The photon flux in a beam of monochromatic radiation is $\Phi = cZ$, where Z is the photon density.

(a) Show that the transition probability for stimulated emission for an atom in the beam is equal to $w_{21}(\nu) = \sigma(\nu)\Phi$.
(b) Determine the photon flux that is necessary to reach $w_{21} = 10^{-9}$ s for the laser materials mentioned in Table 7.1.

7.11 Relate the transition probability for stimulated emission to the growth coefficient and to the gain coefficient.

7.12 Test of equations. Compare the dimensions of left and right side of the following equations:

(7.18), (7.23), (7.38), and (7.52).
[*Hint*: for the dimension of B_{21}, *see* (6.11) or Table 6.1.]

7.13 Determine B_{12} for the transition that is responsible for the absorption band of titanium–sapphire for E ‖ c (Fig. 7.7).

7.14 Optical thickness and self-absorption. The optical thickness of a material is defined as the product αL, where α is the absorption coefficient and L the length of the material; a material is optically thick if $\alpha L \gg 1$ and optically thin if $\alpha L \ll 1$.

(a) Determine the length of a titanium-sapphire crystal for which $\alpha L = 1$ in the center of the pump band of a TiS laser. [*Hint*: make use of Fig. 7.8]
(b) Determine the thickness of a TiS crystal at which self-absorption of fluorescence radiation strongly influences the fluorescence spectrum.

Chapter 8
A Laser Theory

In this chapter, we present simple laser equations describing the dynamics of laser oscillation. The equations are coupled rate equations relating the populations of the laser levels and the photon density.

The laser equations provide the threshold condition, the pump threshold, and the threshold population difference. The solutions to the equations indicate that, at steady state oscillation, clamping of the population difference occurs. Pumping with a pump power exceeding the pump threshold results in generation of laser radiation. The analysis of the laser equations allows us, furthermore, to determine the oscillation onset time and to calculate the optimum output coupling efficiency of a laser.

During the onset of laser oscillation, the interplay of the active medium with the field in a laser resonator can lead to oscillations (relaxation oscillations) of both the density of photon in the resonator and the population difference. We derive a criterion of the occurrence of relaxation oscillations. The relaxation oscillations have frequencies in the GHz range.

We perform an estimate of the laser linewidth. It is finite because of the influence of noise on laser oscillation. Amplification of radiation, which is either due to spontaneous emission by the active medium or due to thermal radiation in the laser resonator is the origin of the finite linewidth of laser radiation.

In the next chapter, we will extend the theory taking into account that laser oscillation is joined with a high frequency polarization of the active medium.

8.1 Rate Equations

To describe dynamical processes occurring in a laser, we make use of a rate equation theory; the rate equations correspond to differential equations of first order. We treat the four-level laser. The theory applies, without modification, also to the three-level laser (with the pump level coinciding with the upper laser level).

In the center of the four-level laser (Fig. 8.1) is a two-level atomic system with the upper laser level 2 and the lower laser level 1. An ensemble of two-level atomic

K.F. Renk, *Basics of Laser Physics*, Graduate Texts in Physics,
DOI 10.1007/978-3-319-50651-7_8

Fig. 8.1 Four-level laser

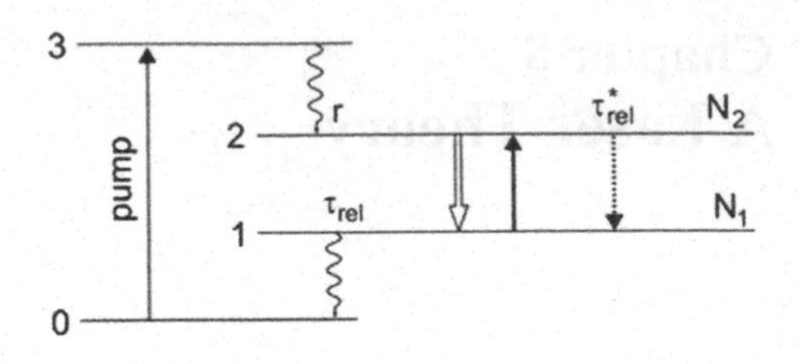

systems interacts with the laser radiation by stimulated emission and absorption of radiation. The upper laser level can relax (relaxation time τ_{rel}^*) by transitions to the lower laser level. Pumping into the pump level (level 3) and fast relaxation leads to a population of the upper laser level. The population rate r is a measure of the pump strength. The lower laser level is depopulated by relaxation (relaxation time τ_{rel}). We assume that further processes, like the relaxation $3 \rightarrow 1$ or $2 \rightarrow 0$, are negligibly week.

We describe the dynamics of the four-level laser by the *laser rate equations:*

$$\frac{dN_2}{dt} = r - \frac{N_2}{\tau_{rel}^*} - b_{21}Z(N_2 - N_1), \tag{8.1}$$

$$\frac{dN_1}{dt} = -\frac{N_1}{\tau_{rel}} + \frac{N_2}{\tau_{rel}^*} + b_{21}Z(N_2 - N_1), \tag{8.2}$$

$$\frac{dZ}{dt} = b_{21}Z(N_2 - N_1) - \frac{Z}{\tau_p}. \tag{8.3}$$

These laser equations take into account the following processes:

- The upper laser level is populated by pumping with the pump rate r. It is depopulated by relaxation with the relaxation rate N_2/τ_{rel}^* and by the net effect of stimulated emission and absorption with the rate $b_{21}Z(N_2 - N_1)$.
- The lower laser level is depopulated by relaxation to the ground state with the rate N_1/τ_{rel}. It is populated by the relaxation of the upper laser level and the net effect of stimulated emission and absorption.
- The photon density increases according to the net effect of stimulated $2 \rightarrow 1$ transitions and absorption processes and decreases due to loss of photons in the resonator.

The three equations are nonlinear differential equations relating N_1, N_2, and Z. We list the quantities used for description of the four-level laser:

- N_2 = population of the upper laser level = density of two-level atomic systems in the upper laser level = number density (=number per m^3) of two-level atomic systems in the upper laser level.
- N_1 = population of the lower laser level = density of two-level atomic systems in the lower laser level.
- $N_2 - N_1$ = population difference.

- $N_1 + N_2$ = density of two-level atomic systems.
- $\tau^*_{\rm rel}$ = lifetime of the upper laser level with respect to $2 \rightarrow 1$ relaxation.
- $\tau_{\rm rel}$ = lifetime of the lower laser level with respect to $1 \rightarrow 0$ relaxation.
- r = pump rate (per unit volume) = number of two-level atomic systems in the upper laser level that are excited per m^3 and s.
- $E_{21} = E_2 - E_1$ = energy difference of the laser levels = transition energy.
- ν = frequency of the laser radiation.

Because of line broadening effects, the quantum energy $h\nu$ of a laser photon is not necessarily equal to the transition energy E_{21}. We are describing a laser that oscillates on *one* mode. We characterize the light in the laser resonator by the quantities:

- Z = photon density (=number of photons per m^3).
- $\tau_{\rm p}$ = photon lifetime = average lifetime of a photon in the resonator.
- $\kappa = \kappa_{\rm i} + \kappa_{\rm out}$ $(=1/\tau_{\rm p})$ = photon loss coefficient of the resonator.
- $\kappa_{\rm i}$ = internal loss coefficient describing loss of photons within the resonator.
- $\kappa_{\rm out}$ = loss coefficient describing loss of photons by output coupling of radiation.
- $b_{21}(\nu) = h\nu B_{21} g(\nu)$ = growth rate constant.
- $\sigma_{21} = n b_{21}/c$ = gain cross section.
- n = refractive index of the active medium at the laser frequency.
- c = speed of light in vacuum.

8.2 Steady State Oscillation of a Laser

At steady state oscillation, the populations and the photon density are independent of time,

$$dN_2/dt = 0; \quad dN_1/dt = 0; \quad dZ/dt = 0. \tag{8.4}$$

We obtain the three laser equations

$$r - N_2/\tau^*_{\rm rel} - b_{21} Z (N_2 - N_1) = 0, \tag{8.5}$$

$$-N_1/\tau_{\rm rel} + N_2/\tau^*_{\rm rel} + b_{21} Z (N_2 - N_1) = 0, \tag{8.6}$$

$$b_{21} Z (N_2 - N_1) - Z/\tau_{\rm p} = 0. \tag{8.7}$$

(In the case that the laser levels are degenerate, we have to replace $N_2 - N_1$ by $N_2 - N_1 g_2/g_1$. Population inversion then corresponds to the condition $N_2 > N_1 g_2/g_1$; g_1 = degree of degeneracy of level 1 and g_2 = degree of degeneracy of level 2. In the following, we treat an ensemble of two-level atomic systems, $g_1 = g_2 = 1$.)

At steady state oscillation, the photon density is unequal to zero ($Z \neq 0$) and we can eliminate Z from the first two equations and find

$$N_{1,\infty}/\tau_{\rm rel} = r. \tag{8.8}$$

The relaxation rate of the lower laser level is equal to the pump rate. This result is obvious (*see* Fig. 8.1): at steady state, the pumping compensates the loss of two-level atomic systems.

Equation (8.7) yields the threshold condition:

$$(N_2 - N_1)_{\mathrm{th}} = (N_2 - N_1)_\infty = \frac{1}{b_{21}\tau_{\mathrm{p}}} = \frac{\kappa_{\mathrm{i}} + \kappa_{\mathrm{out}}}{h\nu B_{21} g(\nu)}. \tag{8.9}$$

$(N_2 - N_1)_{\mathrm{th}}$ is the *threshold population difference*. The population difference at steady state oscillation is equal to the threshold population difference,

$$(N_2 - N_1)_\infty = (N_2 - N_1)_{\mathrm{th}}. \tag{8.10}$$

The population difference is independent of the pump rate and is "clamped" to the threshold population difference $(N_2 - N_1)_{\mathrm{th}}$. The population difference is equal to the reciprocal of the product of the growth rate constant and the lifetime of a photon in the resonator. The threshold decreases with increasing growth rate constant and with increasing photon lifetime. The threshold condition is also discussed in the next section.

We find, with

$$(N_2 - N_1)_\infty = (N_{2,\infty} - N_{1,\infty}), \tag{8.11}$$

that

$$N_{2,\infty} = N_{1,\infty} + \frac{1}{b_{21}\tau_{\mathrm{p}}}. \tag{8.12}$$

Both $N_{2,\infty}$ and $N_{1,\infty}$ increase linearly with the pump rate while the difference experiences clamping.

It follows from (8.5) and (8.6) that the photon density is given by

$$Z_\infty = r\left(1 - \frac{\tau_{\mathrm{rel}}}{\tau_{\mathrm{rel}}^*}\right)\tau_{\mathrm{p}} - \frac{(N_2 - N_1)_\infty \tau_{\mathrm{p}}}{\tau_{\mathrm{rel}}^*}. \tag{8.13}$$

The photon density at steady state oscillation increases linearly with the pump rate (Fig. 8.2). The *threshold pump rate* r_{th} follows from the last equation, for $Z_\infty = 0$,

$$r_{\mathrm{th}} = \frac{(N_2 - N_1)_\infty}{\tau_{\mathrm{rel}}^*(1 - \tau_{\mathrm{rel}}/\tau_{\mathrm{rel}}^*)}. \tag{8.14}$$

The threshold pump rate (=pump rate at laser threshold) compensates the loss of N_2 population that is due to $2 \rightarrow 1$ relaxation processes.

Equation (8.13) shows that the photon density Z_∞ becomes infinitely large if $\tau_{\mathrm{p}} = \infty$, i.e., if the lifetime of the photons is not limited by loss of photons by escape from the resonator or by loss within the resonator. A laser without any loss, $\kappa_{\mathrm{i}} = \kappa_{\mathrm{out}} = 0$, would contain an infinitely large number of photons.

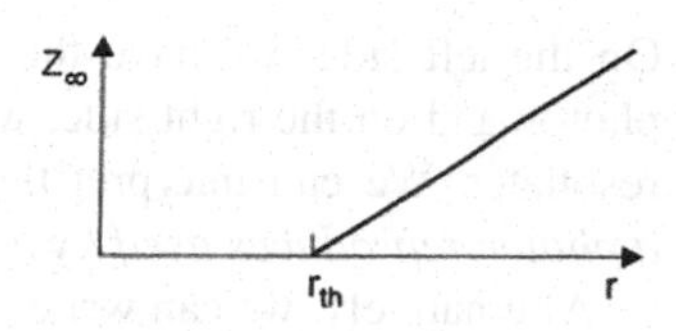

Fig. 8.2 Photon density at steady state oscillation

We now consider the case that the relaxation time of the lower level is small compared with the relaxation time of the upper laser level, $\tau_{\rm rel} \ll \tau^*_{\rm rel}$. Then the threshold pump rate is equal to

$$r_{\rm th} = \frac{(N_2 - N_1)_\infty}{\tau^*_{\rm rel}} = \frac{1}{b_{21}\tau^*_{\rm rel}\tau_{\rm p}} = \frac{\kappa_{\rm i} + \kappa_{\rm out}}{b_{21}\tau^*_{\rm rel}} \tag{8.15}$$

and the photon density at steady state oscillation is

$$Z_\infty = (r - r_{\rm th})\tau_{\rm p} = \left(\frac{r}{r_{\rm th}} - 1\right)\frac{1}{b_{21}\tau^*_{\rm rel}}. \tag{8.16}$$

Without output coupling loss ($\kappa_{\rm out} = 0$) but with internal loss, the threshold pump rate is equal to

$$r_{\rm th,i} = \frac{\kappa_{\rm i}}{b_{21}\tau^*_{\rm rel}} \tag{8.17}$$

and the photon density is

$$Z_{\infty,\rm i} = \frac{r}{\kappa_{\rm i}} - \frac{1}{b_{21}\tau^*_{\rm rel}}. \tag{8.18}$$

8.3 Balance Between Production and Loss of Photons

We will express the threshold condition in different ways. All formulations are equivalent.

The condition of steady state oscillation is the following: *the rate of photon production is equal to the rate of photon loss*:

$$b_{21}(N_2 - N_1)_\infty Z_\infty = \frac{Z_\infty}{\tau_{\rm p}}. \tag{8.19}$$

Dividing by Z_∞, we obtain

$$\frac{1}{b_{21}(N_2 - N_1)_\infty} = \tau_{\rm p}. \tag{8.20}$$

On the left side, we have the time it takes, in the time average, to produce one photon and on the right side, we have the average lifetime of a photon in the laser resonator. We can interpret the steady state: *during its lifetime in the resonator, a photon reproduces itself by a stimulated emission process exactly once.*

Alternatively, we can write

$$(N_2 - N_1)_\infty = \frac{1}{c\tau_p\sigma_{21}} = \frac{1}{l_p\sigma_{21}}. \tag{8.21}$$

The threshold population difference is inversely proportional to the product of the path length l_p of a photon in the resonator and of the gain cross section σ_{21}. We can also write

$$(N_2 - N_1)_\infty \sigma_{21} l_p = 1. \tag{8.22}$$

This means: On its multiple path through the resonator, a photon induces exactly one photon by a stimulated emission process. Or, on its multiple path through the resonator, a photon reproduces itself before it leaves the resonator. Finally, we write

$$(N_2 - N_1)_\infty = \frac{1}{\sigma_{21} l_p}. \tag{8.23}$$

By replacing the photon path length $l_p = 2nL/(-\ln V)$, we obtain the threshold population difference

$$(N_2 - N_1)_\infty = \frac{-\ln V}{2nL\sigma_{21}}. \tag{8.24}$$

The threshold population difference tends to zero if the V factor approaches unity.

8.4 Onset of Laser Oscillation

We assume that we suddenly, at time $t = 0$, turn on a population difference $(N_2 - N_1)_0$. The temporal change of the photon density Z in the laser resonator is, for small Z, given by the equation

$$\frac{1}{Z}\frac{dZ}{dt} = b_{21}(N_2 - N_1)_0 - \frac{1}{\tau_p}. \tag{8.25}$$

The solution is

$$Z(t) = Z_0\, e^{(\gamma_0 - \kappa)t}. \tag{8.26}$$

Z_0 is the photon density at $t = 0$,

$$\gamma_0 = b_{21}(N_2 - N_1)_0 \tag{8.27}$$

is the small-signal growth coefficient, and $\kappa = 1/\tau_p$ is the decay coefficient of the resonator with respect to the decay of a photon.

To estimate the onset time t_{on}, we now assume that the population difference remains constant during the buildup of laser oscillation and changes suddenly to the steady state value $(N_2 - N_1)_\infty$. Under this assumption, the density Z of photons in the resonator increases exponentially until it reaches the steady value Z_∞. Accordingly, we find

$$Z_\infty = Z_0\, e^{(\gamma_0 - \kappa)t_{on}}. \tag{8.28}$$

It follows that the oscillation onset time is equal to

$$t_{on} = \frac{\ln(Z_\infty/Z_0)}{\gamma_0 - \kappa}. \tag{8.29}$$

This is the same result as derived earlier (in Sect. 2.9) since the gain factor is $G_0 = e^{\gamma_0 T}$ and the V factor is $V = e^{-\kappa T}$, where T is the round trip transit time; *see* (2.85).

According to our description of the buildup of laser oscillation, the photon density increases exponentially (Fig. 8.3) until it reaches, at the onset time t_{on}, the steady state value Z_∞. The population density decreases at the onset time t_{on} from $(N_2 - N_1)_0$ to

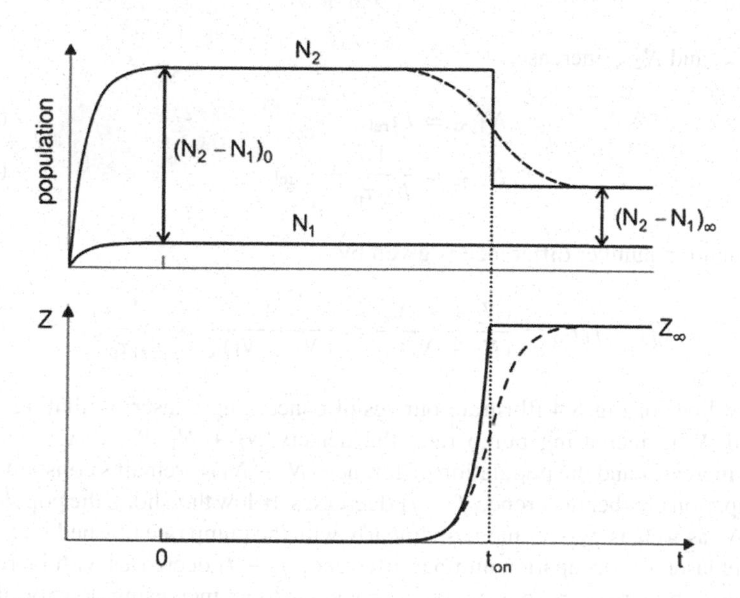

Fig. 8.3 Onset of laser oscillation

$(N_2 - N_1)_\infty$. We will later (in Sect. 9.7) show that the population difference $N_2 - N_1$ and the photon density Z are smoothly going over into their steady state values (dashed curves in Fig. 8.3).

In our discussion of the onset of laser oscillation, we assume that the relaxation time of the upper laser level is much smaller than the onset time, $\tau_{\mathrm{rel}}^* \ll t_{\mathrm{on}}$. In this case, the population N_2 reaches a constant value at $t = 0$, immediately after the start of the pumping, as indicated in Fig. 8.3. Together with the population N_2, the population N_1 reaches a constant value immediately after the start of pumping too.

The helium–neon laser belongs to the lasers that fulfill the condition of a fast relaxation in comparison with the oscillation onset time. The relaxation time τ_{rel}^* of many other laser media (for instance, of titanium–sapphire) is much larger than the oscillation onset time. Then the population and the photon density show dynamic effects, which we will discuss later (Sects. 8.8 and 9.8).

8.5 Clamping of Population Difference

The population difference at steady state oscillation is clamped to the threshold population difference. What does this mean with respect to the occupation number difference $f_2 - f_1$? It follows from the laser equations of the steady state that the density of two-level atomic systems increases with increasing pump rate according to

$$(N_1 + N_2)_\infty = \frac{1}{b_{21}\tau_{\mathrm{p}}} + 2r\tau_{\mathrm{rel}}. \tag{8.30}$$

Both $N_{1,\infty}$ and $N_{2,\infty}$ increase,

$$N_{1,\infty} = r\tau_{\mathrm{rel}}, \tag{8.31}$$

$$N_{2,\infty} = \frac{1}{b_{21}\tau_{\mathrm{p}}} + r\tau_{\mathrm{rel}}. \tag{8.32}$$

The occupation number difference is given by

$$(f_2 - f_1)_\infty = \frac{(N_2 - N_1)_\infty}{(N_2 + N_1)_\infty} = \frac{1}{(N_2 + N_1)_\infty} \times \frac{1}{b_{21}\tau_{\mathrm{p}}}. \tag{8.33}$$

The solid lines of Fig. 8.4 illustrate our result concerning a laser oscillating above threshold. With increasing pump rate, the density $N_2 + N_1$ of two-level atomic systems increases and the population difference $(N_2 - N_1)_\infty$ remains constant while the occupation number difference $f_2 - f_1$ decreases; below threshold, the populations N_2 and N_1 as well as $f_2 - f_1$ increase linearly with the pump rate (dashed lines). In a four-level laser, the occupation number difference $f_2 - f_1$ decreases with increasing pump rate. An increaasing pump rate corresponds to an increasing density of two-level systems in the active medium.

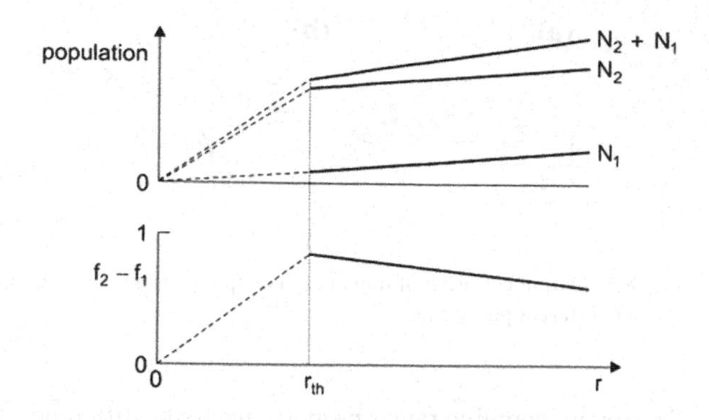

Fig. 8.4 Populations and occupation number difference of a four-level laser

If the lifetime of the lower laser level is very small, $\tau_{\mathrm{rel}} \ll \tau_{\mathrm{rel}}^*$, the two-level atomic systems are mainly in their excited states because the population of level 1 is small compared to the population of level 2 ($N_1 \ll N_2$). Then the population difference is nearly equal to the density of the two-level atomic systems, $N_2 - N_1 \approx N_2 + N_1$. Accordingly, the occupation number difference is near unity, $(f_2 - f_1)_\infty \sim 1$.

We will later find (Chaps. 21 and 22) that for a two-band laser, clamping occurs for the occupation number difference $f_2 - f_1$. The reason is that the density of twolevel systems in a two-band medium is constant and does not depend on the pump strength.

8.6 Optimum Output Coupling

How can we obtain optimum laser output? We have two limiting cases:

- If we choose an output coupling mirror of reflectivity $R = 1$, a strong laser field builds up. The laser, however, does not emit radiation.
- If we choose an output coupling mirror of a reflectivity allowing laser oscillation to occur just at threshold, then the laser field in the resonator is extremely weak. The output power of the laser is negligibly small too.

Optimum output corresponds to an intermediate case. We can choose the reflectivity of the output coupling mirror of a laser and thus the output coupling coefficient κ_{out} (Fig. 8.5a). We are now looking for the value of κ_{out} that leads to optimum output; we assume that $\tau_{\mathrm{rel}} \ll \tau_{\mathrm{rel}}^*$. We introduce the photon output coupling rate r_{out}. At steady state, the output coupling rate (=number of photons coupled out from the resonator per m^3 and per s) is

$$r_{\mathrm{out}} = \kappa_{\mathrm{out}} Z_\infty = \frac{\kappa_{\mathrm{out}}}{\kappa_{\mathrm{i}} + \kappa_{\mathrm{out}}}(r - r_{\mathrm{th}}). \tag{8.34}$$

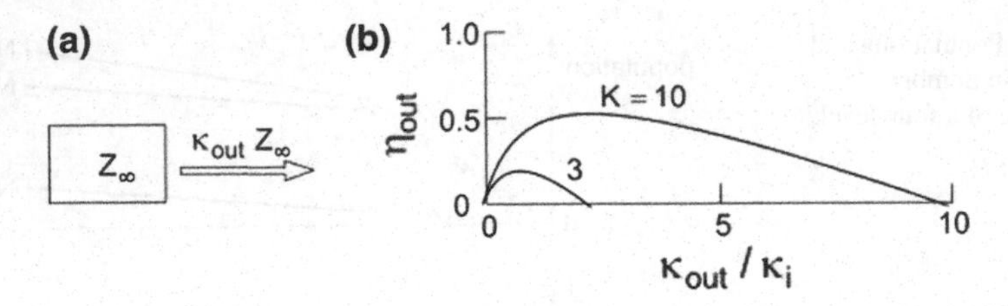

Fig. 8.5 Output coupling of radiation. **a** Output coupling coefficient. **b** Output coupling efficiency for two different pump rates

The output coupling rate is proportional to the difference of pump rate and threshold pump rate. We can write

$$r_{out} = Z_\infty \kappa_{out} = \frac{r\kappa_{out}}{\kappa_{out} + \kappa_i} - \frac{\kappa_{out}}{b_{21}\tau_{rel}^*}. \tag{8.35}$$

We define the output coupling efficiency by

$$\eta_{out} = r_{out}/r, \tag{8.36}$$

where r is the pump rate. A straightforward calculation yields

$$\eta_{out} = \frac{1}{K} \frac{(K-1)\kappa_{out}/\kappa_i - \kappa_{out}^2/\kappa_i^2}{1 + \kappa_{out}/\kappa_i}, \tag{8.37}$$

where the parameter

$$K = r/r_{th,i} \tag{8.38}$$

is a measure of the pump rate. By differentiating η_{out} with respect to κ_{out} and equating to zero, we find that *optimum output coupling* occurs if

$$(\kappa_{out}/\kappa_i)_{opt} = \sqrt{K} - 1. \tag{8.39}$$

The output coupling efficiency depends on the pump rate parameter K (Fig. 8.5b). At a fixed pump rate (e.g., corresponding to $K = 3$), the efficiency increases at weak output coupling ($\kappa_{out} < \kappa_i$) linearly with κ_{out}, reaches a maximum and decreases to zero at the threshold value of κ_{out}.

The maximum output coupling efficiency (Fig. 8.6) increases, for $K > 1$, with K according to

$$\eta_{out,max} = (\sqrt{K} - 1)^2/K. \tag{8.40}$$

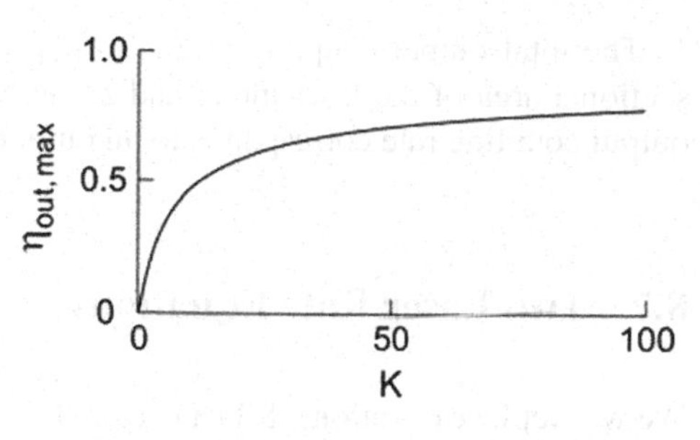

Fig. 8.6 Dependence of the maximum output coupling efficiency on the pump rate parameter K

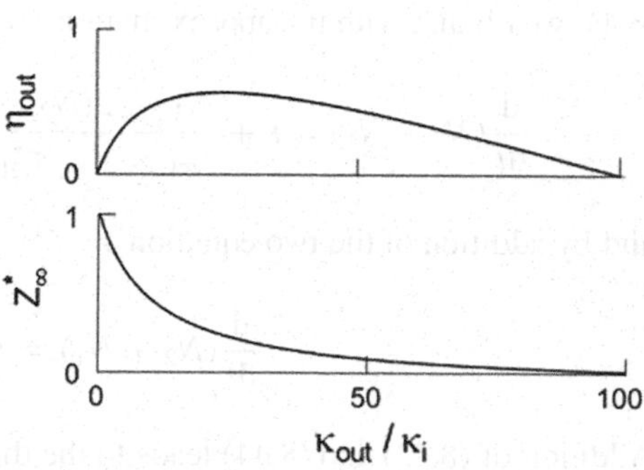

Fig. 8.7 Output coupling efficiency and density of photons in a laser resonator (for $K = 10$)

When a laser is pumped far beyond threshold, $\sqrt{K} \gg 1$, the optimum efficiency η_{out} approaches unity. In this case, the pump power is converted into energy of relaxation and energy of photons in the laser mode; almost all photons are coupled out. The intrinsic loss of photons (e.g., due to diffraction or due to absorption of radiation within the laser resonator) becomes negligibly small.

Which is the density Z_∞ of photons in the laser resonator? Using the relation

$$\kappa_{out} Z_\infty = r_{out} = \eta_{out} r, \tag{8.41}$$

we obtain, after a simple calculation,

$$Z^*_\infty = \frac{Z_\infty}{r_{th,i}/\kappa_i} = \frac{(K-1)}{1 + \kappa_{out}/\kappa_i}, \tag{8.42}$$

where the ratio $r_{th,i}/\kappa_i = (b_{21}\tau^*_{rel})^{-1}$ is a quantity that characterizes a two-level atomic system and where Z^*_∞ is the photon density in units of this quantity.

Figure 8.7 shows, for $K = 10$, the output coupling efficiency and the density of photons in the laser resonator. At optimum output coupling, the number of photons in the resonator is by far smaller than at weak output coupling. The analysis shows that optimum output coupling corresponds to a compromise between a high density of photons in the resonator and a large output coupling efficiency.

The total output coupling rate is $r_{out,tot} = r_{out}a_1a_2L$, where a_1a_2 is the cross-sectional area of the laser mode and L the length of the active medium. The total output coupling rate corresponds to an output power $P_{out} = r_{out,tot}h\nu$.

8.7 Two Laser Rate Equations

We will replace equations (8.1) and (8.2) by one equation. By subtracting (8.2) from (8.1), we obtain, with the approximation $N_1 \ll N_2$,

$$\frac{d}{dt}(N_2 - N_1) = r + \frac{N_1}{\tau_{rel}} - \frac{2(N_2 - N_1)}{\tau^*_{rel}} - 2b_{21}Z(N_2 - N_1) \tag{8.43}$$

and by addition of the two equations,

$$\frac{d}{dt}(N_2 + N_1) = r - \frac{N_1}{\tau_{rel}} = 0. \tag{8.44}$$

Addition of (8.43) and (8.44) leads to the differential equation

$$\frac{d}{dt}(N_2 - N_1) = 2r - \frac{2(N_2 - N_1)}{\tau^*_{rel}} - 2b_{21}Z(N_2 - N_1). \tag{8.45}$$

We investigate the case that the population difference is suddenly turned on. At $t = 0$, immediately after the production of the population inversion, the photon density Z is negligibly small. It follows that

$$(N_2 - N_1)_0 = r\tau^*_{rel}. \tag{8.46}$$

The population difference $(N_2 - N_1)_0$ at time $t = 0$ is equal to the pump rate multiplied by the lifetime of the upper laser level. By replacing r, we obtain (instead of originally three equations) *two laser rate equations*:

$$\frac{d}{dt}(N_2 - N_1) = \frac{2(N_2 - N_1)_0}{\tau^*_{rel}} - \frac{2(N_2 - N_1)}{\tau^*_{rel}} - 2b_{21}Z(N_2 - N_1), \tag{8.47}$$

$$\frac{dZ}{dt} = b_{21}(N_2 - N_1)Z - \frac{Z}{\tau_p}. \tag{8.48}$$

At steady state, $d(N_2 - N_1)/dt = 0$, the population difference is given by

$$(N_2 - N_1)_\infty = \frac{(N_2 - N_1)_0}{1 + b_{21}\tau^*_{rel}}Z_\infty = \frac{(N_2 - N_1)_0}{1 + Z_\infty/Z_s}, \tag{8.49}$$

where

$$Z_s = \frac{1}{b_{21}\tau^*_{\rm rel}} \tag{8.50}$$

and where Z_s is the saturation density. The decrease of the population difference during onset of laser oscillation corresponds to a decrease of gain (*gain saturation*). At the steady state, the large-signal gain coefficient is equal to

$$\alpha_\infty = \frac{b_{21}}{c/n}\frac{(N_2 - N_1)_0}{1 + Z_\infty/Z_s} = \frac{\alpha_0}{1 + Z_\infty/Z_s}, \tag{8.51}$$

where α_0 is the small-signal gain coefficient, i.e., the gain coefficient in at $Z \ll Z_s$.

We relate the initial population, the population at steady state and the photon density at steady state. We assume again that $\tau_{\rm rel} \ll \tau^*_{\rm rel}$. We find, from (8.9), (8.15) and (8.46) and with $c\sigma_{21} = b^0_{21}$, the relations

$$\frac{(N_2 - N_1)_0}{(N_2 - N_1)_\infty} = \frac{r}{r_{\rm th}} = 1 + c\tau^*_{\rm rel}\sigma_{21} Z_\infty = 1 + \tau^*_{\rm rel} b^0_{21} Z_\infty. \tag{8.52}$$

Now, the questions remain how $N_2 - N_1$ develops in the time region $t \approx t_{\rm on}$ from the initial value $(N_2 - N_1)_0$ to $(N_2 - N_1)_\infty$ and how the photon density changes from the exponential increase at $t \ll t_{\rm on}$ to the constant value Z_∞. To study the transition from the initial to the steady state, it is necessary to know more about the role of the active medium. This question is a topic of the next chapter.

We obtain a connection to the next chapter by considering the energy content of a laser. A laser contains three forms of energy: energy of excitation of two-level atomic systems $u_{\rm ex} = (N_2 - N_1)_\infty(E_2 - E_1) = h\nu/c\sigma_{21}\tau_{\rm p}$; electromagnetic field energy $u = \varepsilon_0 A^2_\infty/4$; and polarization energy of density $u_{\rm pol}$. A goal of the discussion presented in Chap. 9 will be to find out the relations between the three forms of energy during the buildup of laser oscillation as well as at steady state—we will find that, at steady state, the polarization energy density is equal to the field energy density.

8.8 Relaxation Oscillation

A *relaxation oscillation* can occur during the buildup of a laser oscillation. We search for an oscillation of the population difference and of the photon density at time $t = t_{\rm on}$. We use the ansatz

$$(N_2 - N_1) = (N_2 - N_1)_\infty + N_{\rm osc}, \tag{8.53}$$

$$Z = Z_\infty + Z_{\rm osc}. \tag{8.54}$$

$N_{osc} \equiv (N_2 - N_1)_{osc}$ is the oscillating portion of the population difference and Z_{osc} the oscillating portion of the photon density. We assume, for simplicity, that $N_{osc} \ll (N_2 - N_1)_\infty$ and $Z_{osc} \ll Z_\infty$. It follows from the laser equations that

$$\frac{dN_{osc}}{dt} = -\frac{2}{\tau_p} Z_{osc} - 2rb_{21}\tau_p N_{osc} \tag{8.55}$$

and

$$\frac{dZ_{osc}}{dt} = \left(\frac{r}{r_{th}} - 1\right)\frac{N_{osc}}{\tau^*_{rel}}. \tag{8.56}$$

We neglected the terms with the product $N_{osc}Z_{osc}$ and made use of the relations $(N_2 - N_1)_\infty = 1/(b_{21}\tau_p)$ and $Z_\infty = (r/r_{th} - 1)/(b_{21}\tau^*_{rel})$, and supposed that $\tau^*_{rel} \ll \tau_p$. By differentiating the first of the two equations and using the second equation, we find

$$\frac{d^2N_{osc}}{dt} + \frac{2}{\tau^*_{rel}}\frac{r}{r_{th}}\frac{dN_{osc}}{dt} + \frac{2}{\tau_p\tau^*_{rel}}\left(\frac{r}{r_{th}} - 1\right)N_{osc} = 0. \tag{8.57}$$

This is the equation of a damped harmonic oscillation with the solution

$$N_{osc} = N_{osc}(0)\, e^{(t-t_{on})/\tau_{damp}} \cos\left(\omega_{osc}(t - t_{on})\right). \tag{8.58}$$

$N_{osc}(0)$ is an initial value of the oscillating portion of the population difference at time $t = t_{on}$. The ansatz yields the frequency of the relaxation oscillation

$$\omega_{osc} = \frac{1}{\tau^*_{rel}}\sqrt{\frac{2\tau^*_{rel}}{\tau_p}\left(\frac{r}{r_{th}} - 1\right) - \left(\frac{r}{r_{th}}\right)^2} \tag{8.59}$$

and the damping (relaxation) time

$$\tau_{damp} = \frac{r_{th}}{r}\tau^*_{rel}. \tag{8.60}$$

A relaxation oscillation occurs if

$$\left(\frac{r}{r_{th}}\right)^2 < \left(\frac{r}{r_{th}} - 1\right)\frac{2\tau^*_{rel}}{\tau_p}. \tag{8.61}$$

Otherwise, the relaxation oscillation is overdamped, i.e., there is no relaxation oscillation.

At a pump rate $r = 2 \times r_{th}$, a relaxation oscillation is expected if $\tau_p \leq \tau^*_{rel}$. This is plausible. An instantaneously large population difference leads to a large

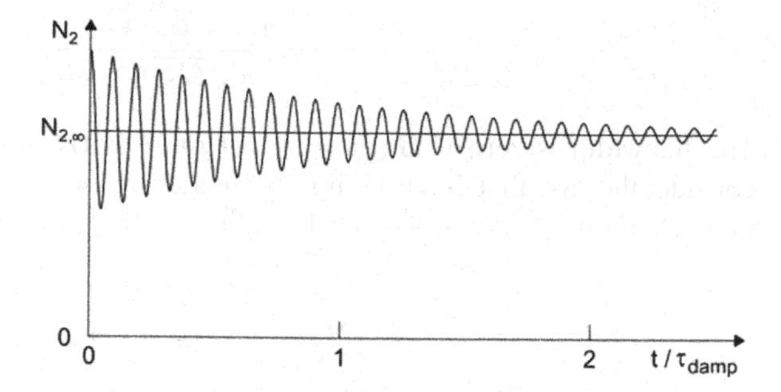

Fig. 8.8 Relaxation oscillation

photon density. This causes a strong decrease of the population difference. Since the photons leave the resonator quickly, the population can build up again and the process of decrease and enhancement of population is repeats until the damping suppresses relaxation oscillation. If the photons have a long lifetime ($\tau_p \geq \tau_{rel}^*$), an instantaneous accumulation of a population difference is not possible and there is no relaxation oscillation.

A helium–neon laser fulfills the condition $\tau_p \geq \tau_{rel}^*$; it shows no relaxation oscillation. The photon lifetime of solid-state lasers and semiconductor lasers is shorter than the relaxation time of the upper laser level, which is a condition of occurrence of relaxation oscillations.

Example Relaxation oscillation in a bipolar semiconductor laser (Fig. 8.8).

- $r/r_{th} = 2$; $\tau_{rel}^* = 4\,\text{ns}$; $\tau_p = 10^{-11}\,\text{s}$.
- $\omega_{osc} = 7 \times 10^9\,\text{Hz}$; $\nu_{osc} = 1.1\,\text{GHz}$; $\nu_{osc}^{-1} = 0.9\,\text{ns}$.
- $\tau_{damp} = 8\,\text{ns}$.

A bipolar semiconductor laser driven by a modulated current emits radiation pulses. The modulation frequency has to be smaller than the oscillation frequency. Otherwise, instabilities can occur. We will discuss in Sect. 9.9 how we can calculate the dynamics of such instabilities.

8.9 Laser Linewidth

Due to spontaneous emission, a laser line has a finite spectral width. To estimate the laser linewidth, we make use of results with respect of a Fabry–Perot resonator containing an active medium (Sect. 3.7). The power P of radiation emitted by a Fabry–Perot resonator (containing an active medium) at a frequency ν in the vicinity of a resonance frequency ν_l is given by

$$P(\nu) = \frac{K}{(1 - \sqrt{G_\infty V})^2 + 4\sqrt{G_\infty V}\sin^2[2\pi(\nu - \nu_l)L/c]}. \tag{8.62}$$

K is a measure of the maximum power of a particular laser. It follows that the linewidth (=*laser linewidth* [FWHM] is given by) is

$$\Delta\nu_{\mathrm{L}} = \frac{1-\sqrt{G_\infty V}}{\pi\sqrt{G_\infty V}}\,\frac{c}{2L}. \tag{8.63}$$

The linewidth is zero if $G_\infty V = 1$ but finite if $G_\infty V$ is smaller than unity. We consider the case that $G_\infty V$ is slightly smaller than unity. We can approximate, with $(1-\sqrt{a})(1+\sqrt{a}) = 1-a$ and $1+\sqrt{a} \approx 2$, the laser linewidth by

$$\Delta\nu_L = \frac{1-G_\infty V}{2\pi}\,\frac{c}{2L}. \tag{8.64}$$

We will now show that $G_\infty V$ is slightly smaller than unity because of spontaneous emission. We modify the rate equation of the photon density. The change of the density of photons during a round trip transit is the sum of the change due to stimulated and spontaneous emission,

$$\frac{\mathrm{d}Z}{\mathrm{d}t} = \frac{GV-1}{T}Z + A_{21}g(\nu)\Delta\nu\frac{1}{D(\nu)\,\Delta\nu_1 a_1 a_2 L}N_2. \tag{8.65}$$

The frequency range of spontaneous emission, covered by a mode, is $\Delta\nu_1 = c/2L = 1/T$ and the probability of spontaneous emission of radiation into one mode is the inverse of the density of states of photon modes in the resonator volume a_1a_2L within the frequency interval $\Delta\nu_1$. The condition of steady state oscillation, $\mathrm{d}Z/\mathrm{d}t = 0$, leads to the relation

$$(1-G_\infty V)\,Z_\infty = Tb_{21}N_{2,\mathrm{th}}\,\frac{1}{a_1a_2L} \tag{8.66}$$

or

$$\Delta\nu_L = \frac{b_{21}N_{2,\mathrm{th}}}{2\pi a_1a_2LZ_\infty}. \tag{8.67}$$

We relate the photon density and the power of the laser radiation,

$$\frac{Z_\infty a_1a_2Lh\nu}{\tau_\mathrm{p}} = P_\mathrm{out}. \tag{8.68}$$

It follows, with the resonator linewidth $\Delta\nu_\mathrm{res} = (2\pi\tau_\mathrm{p})^{-1}$, that

$$\Delta\nu_L = \frac{h\nu}{P_\mathrm{out}}\,\Delta\nu_\mathrm{res}b_{21}N_{2,\mathrm{th}}. \tag{8.69}$$

Using the threshold condition that is still approximately valid,

$$(N_{2,\mathrm{th}}-N_{1,\mathrm{th}})\,b_{21} = N_{2,\mathrm{th}}\frac{N_{2,\mathrm{th}}-N_{1,\mathrm{th}}}{N_{2,\mathrm{th}}}\,b_{21} = \frac{1}{\tau_\mathrm{p}}, \tag{8.70}$$

we find the *Schawlow-Townes formula* [39]

$$\Delta\nu_{\mathrm{L}} = \frac{2\pi h\nu(\Delta\nu_{\mathrm{res}})^2}{P_{\mathrm{out}}} \frac{N_{2,\mathrm{th}} - N_{1,\mathrm{th}}}{N_{2,\mathrm{th}}}. \tag{8.71}$$

We define the quality factor Q_L of the laser radiation as the ratio of the laser frequency and the halfwidth of the laser line. It follows, with $P_{\mathrm{out}} = Z_{\mathrm{tot}} \times h\nu/\tau_{\mathrm{p}}$, where Z_{tot} is the total photon number in the resonator, that

$$Q_{\mathrm{L}} = \frac{\nu}{\Delta\nu_{\mathrm{L}}} = Z_{\mathrm{tot}}\, Q_{\mathrm{res}}\, \frac{N_{2,\mathrm{th}} - N_{1,\mathrm{th}}}{N_{2,\mathrm{th}}} = Z_{\mathrm{tot}}\, Q_{\mathrm{res}}\, \frac{f_2 - f_1}{f_2}. \tag{8.72}$$

$Q_{\mathrm{res}} = \nu/\Delta\nu_{\mathrm{res}} = 2\pi\nu\tau_{\mathrm{p}}$ is the Q factor of the resonator and τ_{p} the lifetime of a photon in the resonator. The quality factor of the laser radiation is proportional to: the number of photons in the resonator, the quality factor of the resonator, and the occupation number difference divided by the relative occupation number of the upper laser level. If $f_1 \ll f_2$, we have $(f_2 - f_1)/f_1 = 1$. Then the quality factor of the laser radiation is equal to the product of the number of photons in the resonator and the quality factor of the resonator,

$$Q_{\mathrm{L}} = Z_{\mathrm{tot}}\, Q_{\mathrm{res}}. \tag{8.73}$$

Since we are considering a single mode laser, Z_{∞} is equal to the occupation number n of the photons in the laser resonator mode.We obtain the simple relationship:

$$\Delta\nu_{\mathrm{L}} = \frac{\Delta\nu_{\mathrm{res}}}{n} \tag{8.74}$$

the halfwidth of the laser line is equal to the halfwidth of the resonance curve of the laser resonator divided by the occupation number of the mode that is excited in the laser resonator!

In the case that a laser is started by thermal radiation, i.e., if $h\nu \ll kT$, the quality factor of the laser radiation is smaller by the factor $h\nu/kT$,

$$Q_{\mathrm{L}} = Z_{\mathrm{tot}}\, Q_{\mathrm{res}}\, \frac{h\nu}{kT}. \tag{8.75}$$

Example helium–neon laser; $\nu = 5 \times 10^{14}$ Hz; $\Delta\nu_{\mathrm{res}} = 1$ MHz; $P_{\mathrm{out}} = 1$ mW; theoretical laser linewidth $\Delta\nu_L = 10^{-3}$ Hz. A laser linewidth of 0.1 Hz has been realized by thermal and mechanical stabilization. The experimental laser linewidth corresponded to a relative frequency width of the laser radiation of $\Delta\nu_L/\nu \sim 10^{-14}$.

The frequency width of laser radiation can be very narrow as a consequence of the feedback an active medium experiences from the radiation in the laser resonator. Because of thermal and mechanical fluctuations, stabilization of a continuous-wave laser, in order to use it as a frequency standard, is extremely difficult. Femtosecond lasers are more suited to develop a frequency standard (Sect. 13.7).

References [1–4, 6, 8, 31, 35–37, 39].

Problems

8.1 Threshold condition. Evaluate the threshold condition of a titanium–sapphire laser operated as cw laser. The data of the laser: Fabry–Perot resonator $L = 10\,\text{cm}$, filled with the active titanium-sapphire crystal (gain cross section $\sigma_{21} = 3 \times 10^{-23}\,\text{m}^2$; frequency $\nu = 360\,\text{THz}$); reflectivity of the output coupling mirror $R = 0.98$; cross-sectional area of the laser beam $a_1 a_2 = 0.5\,\text{mm}^2$.

8.2 Photon density, output power and efficiency. Determine the density of photons in the laser resonator and the laser output power of the laser described in Problem 8.1, for a pump power that is 10 times larger than the threshold pump power. Evaluate the efficiency of conversion of a pump photon into a laser photon.

8.3 Oscillation onset time.

(a) Show that the oscillation onset time is always large compared to the period $2\pi/\omega$ of the laser field. [*Hint*: make use of the data of Table 7.1.]
(b) Estimate the oscillation onset time of the titanium-sapphire laser (described in Problem 8.1).

8.4 Formulate the threshold condition in the case that the length L' of the active medium is smaller than the length of the laser resonator. Is the condition $GV = 1$ still valid?

8.5 Estimate the laser linewidth of a semiconductor laser of a wavelength of $0.8\,\mu\text{m}$ and an output power of $1\,\text{mW}$; loss factor $V_1 = 0.3$ and volume of the active medium $= 10^{-13}\,\text{m}^3$.

8.6 Coherence length. Monochromatic laser radiation consists of radiation of a line (halfwidth $\Delta\lambda$) at the laser wavelength λ.

(a) Determine the coherence length l_{coh}. [*Hint*: use as criterion that the number of wavelengths of radiation at $\lambda - \Delta\lambda/2$ and $\lambda + \Delta\lambda/2$ differs by 1.]
(b) Determine the coherence length of radiation generated by a semiconductor laser.
(c) Determine l_{coh} of radiation generated by a highly stabilized helium–neon laser.
(d) Determine the coherence length of the radiation of a hypothetical continuous wave laser at a frequency of 4×10^{14} Hz that is stabilized with a relative accuracy of 10^{-16}.

Chapter 9
Driving a Laser Oscillation

We investigate the role of electric polarization of a laser medium in order to obtain further insight into dynamical processes occurring in a laser. The reader, who does not wish to interrupt the description of a laser and its operation, may skip over this chapter.

We study interaction of a medium with a high frequency field by use of Maxwell's equations. We derive five coupled differential equations of second order. Applying the slowly varying amplitude approximation, we can reduce the equations to five nonlinear differential equations of first order. The equations relate: population difference; amplitude of the field; phase of the field; amplitude of the polarization; and phase of the polarization.

We can reduce the five differential equations to three in the case that transverse relaxation of the polarization is absent and that the three relevant frequencies—laser frequency, atomic transition frequency, and resonance frequency of the laser resonator—coincide with each other. The three equations relate population difference, amplitude of the field, and amplitude of the polarization. The solutions yield the temporal development of population difference, amplitude of the field, and amplitude of the polarization during onset of laser oscillation.

We finally derive, in the slowly varying amplitude approximation, the laser equations in the case that transverse relaxation is present and that the laser frequency is equal to the resonance frequency of the resonator but differs from the atomic transition frequency. We obtain five nonlinear coupled differential equations of first order (Lorenz–Haken equations).

This chapter begins with an introduction of the electric polarization of a medium. We make use of the classical oscillator model of an atom. We derive a classical expression of the dielectric susceptibility, which relates the field in a medium and the polarization of the medium. We determine the classical absorption coefficient of a medium. Comparing the classical absorption coefficient with the absorption coefficient derived earlier by quantum mechanical arguments, we obtain a quantum mechanical expression of the dielectric susceptibility. The susceptibility of an active medium depends

K.F. Renk, *Basics of Laser Physics*, Graduate Texts in Physics,
DOI 10.1007/978-3-319-50651-7_9

linearly on the population difference. We mention the Kramers–Kronig relations, which relate real and imaginary part of a physical response function—the polarization of a medium is the response to an external field and the dielectric susceptibility is the corresponding response function.

According to Maxwell's equations, a laser oscillation can be driven either by a high frequency electric polarization or by a high frequency current.

9.1 Maxwell's Equations

To describe the response of a medium to an electromagnetic field, we make use of Maxwell's equations,

$$\nabla \times \boldsymbol{H} = \boldsymbol{j} + \frac{\partial \boldsymbol{D}}{\partial t}, \tag{9.1}$$

$$\nabla \times \boldsymbol{E} = -\frac{\partial \boldsymbol{B}}{\partial t}, \tag{9.2}$$

$$\nabla \cdot \boldsymbol{E} = 0, \tag{9.3}$$

$$\nabla \cdot \boldsymbol{B} = 0. \tag{9.4}$$

$\nabla = (\partial/\partial x, \partial/\partial y, \partial/\partial z)$ is the *del operator* (=*Nabla operator*), $\boldsymbol{E}$ the electric field, $\boldsymbol{D}$ the displacement field, $\boldsymbol{j}$ the electric current density of a high frequency electric current carried by free-electrons, $\boldsymbol{H}$ the magnetic field and $\boldsymbol{B}$ the magnetic induction. We exclude, with $\nabla \cdot \boldsymbol{E} = 0$, local charge accumulations. Material equations provide further relations:

$$\boldsymbol{D}(\boldsymbol{E}) = \varepsilon_0 \boldsymbol{E} + \boldsymbol{P}(\boldsymbol{E}), \tag{9.5}$$

$$\boldsymbol{j} = \boldsymbol{j}(\boldsymbol{E}), \tag{9.6}$$

$$\boldsymbol{B} = \mu_0 \boldsymbol{H}, \tag{9.7}$$

where $\varepsilon_0 = 8.86 \times 10^{-12}\,\mathrm{A\,s\,V^{-1}\,m^{-1}}$ is the electric field constant, $\mu_0 = 4\pi \times 10^{-7}\,\mathrm{V\,s\,A^{-1}\,m^{-1}}$ the magnetic field constant and $\varepsilon_0\mu_0 = c^{-2}$. We ignore, using the relation $\boldsymbol{B} = \mu_0 \boldsymbol{H}$, magnetic effects. The displacement field is the sum of the field (times ε_0) and the high frequency polarization $\boldsymbol{P}$, which itself depends on the field. The current density depends on the field. The quantities $\boldsymbol{E}$, $\boldsymbol{P}$ and $\boldsymbol{j}$ can depend on time and location. We can write the first Maxwell equation in the form

$$\nabla \times \boldsymbol{H} = \boldsymbol{j} + \varepsilon_0 \frac{\partial \boldsymbol{E}}{\partial t} + \frac{\partial \boldsymbol{P}}{\partial t}. \tag{9.8}$$

We now assume that the response is linear and introduce the *dielectric susceptibility* $\tilde{\chi}$, the *dielectric constant* $\tilde{\varepsilon}$ and the *conductivity* $\tilde{\sigma}$ by the relations:

$$P = \varepsilon_0 \, \tilde{\chi} \, E, \tag{9.9}$$
$$D = \varepsilon_0 \, \tilde{\varepsilon} \, E, \tag{9.10}$$
$$\tilde{\varepsilon} = 1 + \tilde{\chi}, \tag{9.11}$$
$$j = \tilde{\sigma} \, E. \tag{9.12}$$

The function $\tilde{\chi}$ is a *linear response function.* It characterizes the linear response of the polarization of a dielectric medium to a high frequency electric field. Correspondingly, $\tilde{\varepsilon}$ is the linear response function for the dielectric displacement while $\tilde{\sigma}$ is the linear response function for the current density.

The first Maxwell equation describing linear response is given by

$$\nabla \times H = j + \varepsilon_0 \frac{\partial(\tilde{\varepsilon}\, E)}{\partial t} = \tilde{\sigma}\, E + \varepsilon_0 \frac{\partial \tilde{\varepsilon}}{\partial t} E + \varepsilon_0 \, \tilde{\varepsilon} \frac{\partial E}{\partial t}. \tag{9.13}$$

When a medium is in thermal equilibrium, then, $\partial\tilde{\varepsilon}/\partial t = 0$. But, the dielectric constant can be time-dependent, $\partial\tilde{\varepsilon}/\partial t \neq 0$, for an active medium—which is always in a nonequilibrium state.

The first Maxwell equation contains the displacement current density

$$j_{\mathrm{d}} = \frac{\partial D}{\partial t} = \varepsilon_0 \frac{\partial E}{\partial t} + \frac{\partial P}{\partial t}, \tag{9.14}$$

which is the sum of the displacement current density $\varepsilon_0 \partial E/\partial t$ and the polarization current density $\partial P/\partial t$. The polarization current density corresponds to the portion of the displacement current density that is due to polarization of a medium.

We assume that high frequency field, polarization and current are oriented along x. A field

$$\tilde{E} = A\, \mathrm{e}^{\mathrm{i}\omega t}, \tag{9.15}$$

gives rise to a polarization

$$\tilde{P} = \varepsilon_0 \tilde{\chi}\, \tilde{E} = P_1 - \mathrm{i}\, P_2 = \varepsilon_0 \chi_1 A \cos\omega t - \mathrm{i}\, \varepsilon_0 \chi_2 A \sin\omega t, \tag{9.16}$$

with

$$\tilde{\chi} = \chi_1 - \mathrm{i}\, \chi_2. \tag{9.17}$$

The real part of the polarization,

$$P_1 = \varepsilon_0 \chi_1 \, A \, \cos\omega t, \tag{9.18}$$

has the same phase as the field. The imaginary part $-\mathrm{i}P_2$, where

$$P_2 = \varepsilon_0 \chi_2 \, A \, \sin \omega t \tag{9.19}$$

has a phase of 90° relative to the field. We will see that gain occurs if $\chi_2(\omega)$ is negative. *Note*: throughout the book, we discuss, for convenience, the negative imaginary part (e.g., χ_2) of a complex quantity, which characterizes a material property, rather than the imaginary part ($-\chi_2$) itself.

The polarization is phase shifted relative to the field,

$$P = \varepsilon_0 \chi \, A \, \cos\,[\omega t + \varphi(\omega)], \tag{9.20}$$

where

$$\chi = \sqrt{\chi_1^2 + \chi_2^2} \tag{9.21}$$

is the absolute value of the susceptibility and where φ is the phase between polarization and field. The phase is given by the relation

$$\tan \varphi = \chi_2 / \chi_1. \tag{9.22}$$

We now describe the linear response of a conductive gain medium. A high frequency field

$$\tilde{E}(\omega) = A \, \mathrm{e}^{\mathrm{i}\omega t} \tag{9.23}$$

gives rise to a high frequency current of current density

$$\tilde{j} = \tilde{\sigma} \, \tilde{E}, \tag{9.24}$$

where $\tilde{\sigma}$ is the complex high frequency conductivity,

$$\tilde{\sigma} = \sigma_1 - \mathrm{i}\,\sigma_2. \tag{9.25}$$

The current density,

$$\tilde{j} = j_1 - \mathrm{i}\, j_2 = \sigma_1 A \cos \omega t - \mathrm{i}\sigma_2 A \sin \omega t, \tag{9.26}$$

has a real part

$$j_1 = \sigma_1 \, A \, \cos \omega t \tag{9.27}$$

that has the same phase as the field. We will see that gain occurs if $\sigma_1(\omega)$ is negative. The (negative) imaginary part

$$j_2 = \sigma_2 \, A \, \sin \omega t \tag{9.28}$$

has a phase of 90° relative to the field and corresponds to a lossless current. The current is phase shifted relative to the field,

$$j = \sigma\ A\ \cos\left[\omega t + \varphi(\omega)\right], \tag{9.29}$$

where

$$\sigma = \sqrt{\sigma_1^2 + \sigma_2^2} \tag{9.30}$$

is the absolute value of the conductivity and where φ is the phase between current density and field. The phase is given by the relation

$$\tan\varphi = \sigma_2/\sigma_1. \tag{9.31}$$

We can introduce a *generalized dielectric constant* $\tilde{\varepsilon}_{\text{gen}}$ and a *generalized conductivity* $\tilde{\sigma}_{\text{gen}}$. We write the first Maxwell equation in different ways:

$$\nabla \times \boldsymbol{H} = \boldsymbol{j} + \frac{\partial \boldsymbol{D}}{\partial t} = \tilde{\sigma}\tilde{\boldsymbol{E}} + \varepsilon_0\tilde{\varepsilon}\frac{\partial \boldsymbol{E}}{\partial t} = (\tilde{\sigma} + \mathrm{i}\omega\varepsilon_0\tilde{\varepsilon})\ \boldsymbol{E} = \tilde{\sigma}_{\text{gen}}\ \boldsymbol{E} = \mathrm{i}\omega\varepsilon_0\tilde{\varepsilon}_{\text{gen}}\ \boldsymbol{E}. \tag{9.32}$$

We obtain the relation

$$\tilde{\sigma}_{\text{gen}} = \sigma_{\text{gen},1} - \mathrm{i}\sigma_{\text{gen},2} = \mathrm{i}\ \omega\varepsilon_0\tilde{\varepsilon}_{\text{gen}} \tag{9.33}$$

or, alternatively,

$$\tilde{\varepsilon}_{\text{gen}} = \varepsilon_{\text{gen},1} - \mathrm{i}\varepsilon_{\text{gen},2} = \frac{\tilde{\sigma}_{\text{gen}}}{\mathrm{i}\omega\varepsilon_0}. \tag{9.34}$$

It follows that

$$\varepsilon_{\text{gen},1} = \varepsilon_1 - \frac{\sigma_2}{\omega\varepsilon_0}, \tag{9.35}$$

$$\varepsilon_{\text{gen},2} = \varepsilon_2 + \frac{\sigma_1}{\omega\varepsilon_0}. \tag{9.36}$$

This formulation is useful for determination of optical constants and other optical properties of a medium in thermal equilibrium or of an active medium too. The complex refractive index $n_1 - in_2$ follows from the relation

$$(n_1 - \mathrm{i}n_2)^2 = \varepsilon_{\text{gen},1} - \mathrm{i}\varepsilon_{\text{gen},2}. \tag{9.37}$$

We obtain

$$n_{1,2} = \sqrt{\frac{1}{2}\left(\varepsilon_{\mathrm{gen},1} \pm \sqrt{\varepsilon_{\mathrm{gen},1}^2 + \varepsilon_{\mathrm{gen},2}^2}\right)}. \tag{9.38}$$

9.2 Possibilities of Driving a Laser Oscillation

It follows from Maxwell's equations, with

$$\nabla \times (\nabla \times \boldsymbol{E}) = \nabla \cdot \nabla \cdot \boldsymbol{E} - \nabla^2 \boldsymbol{E}, \tag{9.39}$$

that

$$\frac{\partial^2 \boldsymbol{E}}{\partial t^2} - \frac{1}{\mu_0 \varepsilon_0} \nabla^2 \boldsymbol{E} = -\frac{1}{\varepsilon_0}\frac{\partial^2 \boldsymbol{P}}{\partial t^2} - \frac{1}{\varepsilon_0}\frac{\partial \boldsymbol{j}}{\partial t}. \tag{9.40}$$

On the right side, we have two terms, the second derivative of the polarization and the derivative of the electric current density with respect to time. There are two possibilities to obtain gain:

- A high frequency polarization can be the origin of gain or
- A high frequency electric current can be the origin of gain.

In the following sections (Sects. 9.3–9.6), we will present a model of a dielectric medium. The model is suited to study basic properties of lasers (Sects. 9.7–9.10). The model describes a dielectric medium that shows a homogeneously broadened narrow line.

We will begin with a derivation of the classical susceptibility and the classical absorption coefficient α_{cl} of an ensemble of classical oscillators. A comparison of the classical absorption coefficient with the quantum mechanical expression of the absorption coefficient α_{abs} (Sect. 7.2) will lead to a procedure that allows us to change from classical expressions of the susceptibility to quantum mechanical expressions. The model yields the complex susceptibility of a dielectric medium consisting of an ensemble of two-level atomic systems.

We will show later (in Chaps. 19 and 32) how a high frequency electric current carried by free-electrons can give rise to gain.

9.3 Polarization of an Atomic Medium

We make use of the classical oscillator model to describe the interaction of an atom with an electromagnetic field. An electric field

$$E = \text{Re}\,[\tilde{E}] = \frac{1}{2}\,(A\text{e}^{\text{i}\omega t} + c.c.) = A\cos\omega t, \tag{9.41}$$

excites an oscillator to a forced oscillation described by the equation of motion

$$\frac{\text{d}^2x}{\text{d}t^2} + \beta\frac{\text{d}x}{\text{d}t} + \omega_0^2 x = \frac{q}{m_0}\,E, \tag{9.42}$$

where $q = -e$ is the electron charge (e = elementary charge), m_0 the electron mass, and β the damping constant with respect to the energy; the decay constant with respect to the amplitude is $\beta/2$. We write the displacement as a complex quantity

$$x = \text{Re}[\tilde{x}] = \frac{1}{2}(\tilde{x}\text{e}^{\text{i}\omega t} + c.c.), \tag{9.43}$$

where $\tilde{x}(\omega)$ is a frequency-dependent complex amplitude of the oscillation. We assume that the amplitude (envelope) is slowly varying, $|\text{d}\tilde{x}/\text{d}t| \ll \omega|\tilde{x}|$ (slowly varying envelope approximation, SVEA) and find, with $\beta = \Delta\omega_0$, the solution

$$\tilde{x} = -\frac{e}{m_0}\,\frac{1}{\omega_0^2 - \omega^2 + \text{i}\omega\Delta\omega_0}\,A. \tag{9.44}$$

(We can write

$$\tilde{x} = -\frac{e}{m_0}\,\tilde{G}_\text{L}(\omega)\,A, \tag{9.45}$$

where

$$\tilde{G}_\text{L}(\omega) = \frac{1}{\omega_0^2 - \omega^2 + \text{i}\omega\Delta\omega_0} \tag{9.46}$$

is the complex Lorentz response function in general form; *see* Sect. 9.11.

In the following, we assume that $\beta \ll \omega_0$, i.e., that $\Delta\omega_0 \ll \omega_0$ and we restrict the frequency ω to a range around ω_0 so that $|\omega - \omega_0| \ll \omega_0$. We obtain from (9.44), with $\omega_0^2 - \omega^2 = 2\omega_0(\omega_0 - \omega)$, the solution

$$\tilde{x} = -\frac{e}{2m_0\omega_0}\,\frac{1}{\omega_0 - \omega + \text{i}\Delta\omega_0/2}\,A. \tag{9.47}$$

An oscillating electron is connected with an oscillating electric dipole moment

$$\tilde{p} = -e\tilde{x}. \tag{9.48}$$

It follows that

$$\frac{d^2p}{dt^2} + \beta\frac{dp}{dt} + \omega_0^2 p = \frac{e^2}{m_0}E. \tag{9.49}$$

The ansatz

$$p = \frac{1}{2}\,(\tilde{p}\,e^{i\omega t} + c.c.), \tag{9.50}$$

with the complex amplitude $\tilde{p}$ of the dipole moment, leads to

$$\tilde{p} = \frac{e^2}{2m_0\omega_0}\,\frac{1}{\omega_0 - \omega + i\Delta\omega_0/2}\,A. \tag{9.51}$$

The dipole moment shows, as the displacement, a resonance at the frequency ω_0.

A medium consisting of an ensemble of two-level atomic systems of density N experiences, under the action of an electric field, the electric polarization

$$\boldsymbol{P} = \sum_{i=1}^{N} \boldsymbol{p}_i, \tag{9.52}$$

where $\boldsymbol{p}_i$ is the electric dipole moment of the ith two-level atomic system and N the number of two-level systems per unit volume. Without a high frequency electric field, the dipole moments are zero, $\boldsymbol{p}_i = 0$, and there is no polarization. Under the action of a high frequency field, atomic dipole moments and polarization can oscillate synchronously to the field.

The differential equation

$$\frac{d^2P}{dt^2} + \beta\frac{dP}{dt} + \omega_0^2 P = \frac{Ne^2}{m_0}E, \tag{9.53}$$

that follows from the equation of motion of a single dipole relates a high frequency polarization P and a high frequency electric field. The ansatz

$$P = \text{Re}\,[\tilde{P}] = \frac{1}{2}(\tilde{P}\,e^{i\omega t} + c.c.), \tag{9.54}$$

where $\tilde{P}$ is the complex amplitude of the polarization, yields

$$\tilde{P} = \frac{Ne^2}{2m_0\omega_0}\,\frac{1}{\omega_0 - \omega + i\Delta\omega_0/2}\,A. \tag{9.55}$$

The polarization has the same frequency as the electric field. The amplitude is proportional to the amplitude of the field that produces the polarization. The polarization has a resonance at ω_0. We find the electric susceptibility

$$\tilde{\chi} = \frac{Ne^2}{2\varepsilon_0 m_0 \omega_0} \frac{1}{\omega_0 - \omega + \mathrm{i}\Delta\omega_0/2}. \tag{9.56}$$

The real part and the (negative) imaginary part of the susceptibility are

$$\chi_1(\omega) = \frac{Ne^2}{2\varepsilon_0 m_0 \omega_0} \frac{\omega_0 - \omega}{(\omega - \omega_0)^2 + (\Delta\omega_0)^2/4} = \frac{\omega_0 - \omega}{\Delta\omega_0/2} \chi_2(\omega), \tag{9.57}$$

$$\chi_2(\omega) = \frac{Ne^2}{2\varepsilon_0 m_0 \omega_0} \frac{\Delta\omega_0/2}{(\omega - \omega_0)^2 + (\Delta\omega_0)^2/4} = \frac{N\pi e^2}{2\varepsilon_0 m_0 \omega_0} g_{\mathrm{L,res}}(\omega), \tag{9.58}$$

where $g_{\mathrm{L,res}}(\omega)$ is the Lorentz resonance function. To obtain the susceptibility on the frequency scale, we replace $g_{\mathrm{L,res}}(\omega)$ by $(1/2\pi)g_{\mathrm{L,res}}(\nu)$ and ω_0 by $2\pi\nu_0$. We find

$$\chi_1(\nu) = \frac{\nu_0 - \nu}{\Delta\nu_0/2} \chi_2(\nu), \tag{9.59}$$

$$\chi_2(\nu) = N\frac{e^2}{8\pi\varepsilon_0 m_0 \nu_0} g_{\mathrm{L,res}}(\nu). \tag{9.60}$$

We assumed in our derivation of the susceptibility that all atomic dipoles have the same resonance frequency and the same damping constant. This corresponds to homogeneous line broadening. The damping can be due to emission of radiation (resulting in natural line broadening) or due to other energy relaxation processes that lead to a Lorentzian line.

9.4 Quantum Mechanical Expression of the Susceptibility of an Atomic Medium

We characterize a dielectric medium, which we assume to be optically isotropic, by the complex displacement field

$$\tilde{D} = \varepsilon_0 \tilde{E} + \tilde{P} = \varepsilon_0 \tilde{E} + \varepsilon_0 \tilde{\chi} \tilde{E} = \tilde{\varepsilon} \tilde{E}, \tag{9.61}$$

where $\varepsilon = \varepsilon_1 - \mathrm{i}\varepsilon_2$ is the complex dielectric constant. The real part is $\varepsilon_1 = 1 + \chi_1$ and the imaginary part $\varepsilon_2 = \chi_2$. We obtain the complex refractive index $n_1 - \mathrm{i}n_2$ from the relation

$$(n_1 - \mathrm{i}n_2)^2 = \varepsilon_1 - \mathrm{i}\varepsilon_2. \tag{9.62}$$

The wave vector of a plane wave travelling in z direction is

$$k = (n_1 - in_2)\,\frac{\omega}{c}, \tag{9.63}$$

where $n_1 \equiv n$ is the *refractive index*. The complex field of the plane wave is

$$\tilde{E} = A\,\mathrm{e}^{\mathrm{i}(\omega t - kz)} = A\,\mathrm{e}^{-(n_2\omega/c)z}\,\mathrm{e}^{\mathrm{i}[\omega t - (n_1\omega/c)z]}. \tag{9.64}$$

In the vicinity of the center of a Lorentzian line, the imaginary part of the refractive index is $n_2 \approx \frac{1}{2}\chi_2$. At frequencies around the resonance frequency ω_0, the energy density in the wave is

$$u = u_0\mathrm{e}^{-\alpha_{\mathrm{cl}}z}, \tag{9.65}$$

where $u_0 = \frac{1}{2}\varepsilon_0 A^2$ is the energy density at $z = 0$ and

$$\alpha_{\mathrm{cl}}(\omega) = \frac{\pi e^2 N}{2\varepsilon_0 m_0 c}\,g_{\mathrm{L,res}}(\omega) \tag{9.66}$$

is the classical absorption coefficient. The absorption coefficient on the frequency scale is given by

$$\alpha_{\mathrm{cl}}(\nu) = N\frac{e^2}{4\varepsilon_0 m_0 c}\,g_{\mathrm{L,res}}(\nu). \tag{9.67}$$

We now compare this formula with the quantum mechanical expression for the absorption coefficient, which follows from (7.23),

$$\alpha_{\mathrm{abs}}(\nu) = \frac{1}{c}h\nu B_{12}g_{\mathrm{L,res}}(\nu)(N_1 - N_2). \tag{9.68}$$

The two expressions of $\alpha_{\mathrm{cl}}(\nu)$ and $\alpha_{\mathrm{abs}}(\nu)$ are in accord with each other if we replace

$$\frac{Ne^2}{4\varepsilon_0 m_0} \rightarrow (N_1 - N_2)h\nu B_{12},\ \text{or, respectively,}\ (N_1 - N_2)h\omega B_{12}^{\omega} \tag{9.69}$$

on the ω scale. The replacement in the expressions of the susceptibility leads to

$$\chi_1(\omega) = \frac{\omega_0 - \omega}{\Delta\omega_0/2}\,\chi_2(\omega), \tag{9.70}$$

$$\chi_2(\omega) = (N_1 - N_2)\frac{2\pi}{\omega_0}\,b_{21}(\omega), \tag{9.71}$$

where $\Delta\omega_0$ is the halfwidth of the atomic transition and where

$$b_{21}(\omega) = \hbar\omega_0 B_{21}^{\omega} g_{\mathrm{L,res}}(\omega). \tag{9.72}$$

The complex susceptibility $\tilde{\chi} = \tilde{\chi}_1 - \mathrm{i}\tilde{\chi}_2$ is equal to

$$\tilde{\chi}(\omega) = 2\hbar B_{21}^{\omega}(N_1 - N_2)\,\frac{1}{\omega_0 - \omega + \mathrm{i}\Delta\omega_0/2}. \tag{9.73}$$

The susceptibilities on the frequency scale are given by

$$\chi_1(\nu) = \frac{\nu_0 - \nu}{\Delta\nu_0/2}\,\chi_2(\nu) = hB_{21}(N_1 - N_2)\,g_{\mathrm{L,disp}}(\nu), \tag{9.74}$$

$$\chi_2(\nu) = (N_1 - N_2)\frac{1}{\nu_0}\,b_{21}(\nu) = hB_{21}(N_1 - N_2)\,g_{\mathrm{L,res}}(\nu), \tag{9.75}$$

where $b_{21}(\nu) = h\nu B_{21} g_{\mathrm{L,res}}(\nu)$.

The susceptibilities of an inactive medium (Fig. 9.1a) indicate that the interaction of an electric field with a medium is strongest at resonance, where χ_2 has a maximum and χ_1 is zero. Absorption of radiation is strongest at the transition frequency ω_0. In comparison with a nonactive medium, the susceptibilities of an active medium (Fig. 9.1b) have opposite signs. The gain coefficient of an active medium will be largest if the frequency of the radiation is equal to the transition frequency.

We have the important result: the imaginary part of the susceptibility of a medium that shows a homogeneously broadened transition caused by energy relaxation has the shape of a Lorentz resonance function while the real part has the shape of a Lorentz dispersion function.

The replacement of the classical expressions of the susceptibilities by the quantum mechanical expressions in (9.53) leads to the differential equation

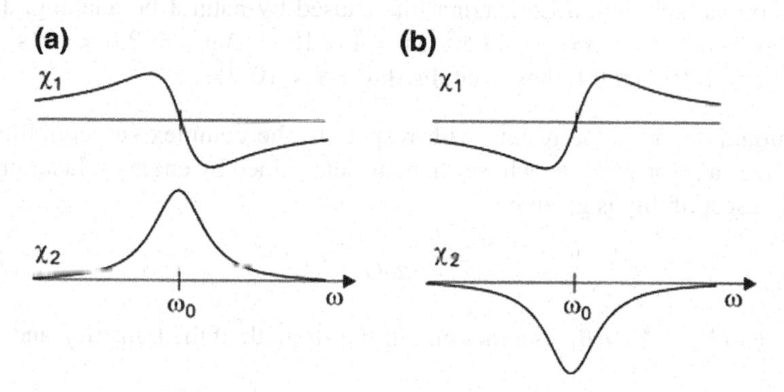

Fig. 9.1 Dielectric susceptibilities **a** of an inactive medium and **b** of an active medium

$$\frac{d^2P}{dt^2} + \Delta\omega_0 \frac{dP}{dt} + \omega_0^2 P = \frac{2}{\pi}\varepsilon_0 \hbar\omega B_{21}^{\omega}(N_1 - N_2)E. \quad (9.76)$$

P and E are real quantities.

At the center of a resonance line, at the frequency ω_0, the real part of the susceptibility is zero. In the vicinity of ω_0, the real part of the susceptibility is proportional to the frequency difference,

$$\chi_1 = \frac{8}{\Delta\omega_0}\hbar B_{21}^{\omega}(N_1 - N_2)(\omega_0 - \omega). \quad (9.77)$$

It follows that the refractive index varies with frequency—i.e., the medium shows *dispersion*—and that the refractive index is approximately given by

$$n(\omega) = n(\omega_0) + (dn/d\omega)_{\omega_0}(\omega_0 - \omega). \quad (9.78)$$

Differentiation of $n_1^2 - n_2^2 = 1 + \chi_1$ leads to $2n_1 dn_1/d\omega = \varepsilon_0 d\chi_1/d\omega$ and where we wrote n instead of n_1. Differentiation of $n^2 = 1 + \chi_1$ leads to $2n\, dn/d\omega = d\chi_1/d\omega$ and to

$$\left(\frac{dn}{d\omega}\right)_0 = -\frac{4}{n\Delta\omega_0^2}\hbar B_{21}^{\omega}(N_1 - N_2), \quad (9.79)$$

where $\Delta\omega_0$ is the atomic linewidth.

On the frequency scale, the change of the refractive index is

$$\left(\frac{dn}{d\nu}\right)_0 = \frac{-2}{n\pi\Delta\nu_0^2} hB_{21}(N_1 - N_2). \quad (9.80)$$

Example Dispersion of an active medium. We estimate the dispersion of optically pumped titanium–sapphire ($N_2 - N_1 = 10^{24}$ m^3; $B_{21}^{\omega} = 2\pi \times 1.7 \times 10^{18}$ m^3 J^{-1} s^{-2}). In comparison with a Lorentzian line caused by natural broadening, $dn/d\omega$ is reduced by a factor $\Delta\omega_0/(1.48\Delta\omega_{nat}) = 4 \times 10^{10}$; $\Delta\omega_{nat} = 2.6 \times 10^5\,s^{-1}$ and $\Delta\omega_0 = 2\pi \times 100$ THz. It follows that $dn/d\omega = 4 \times 10^{-11}$ s.

We summarize here the results with respect to the complex susceptibility that characterizes a resonance line whose shape is determined by energy relaxation. The complex susceptibility is given by

$$\tilde{\chi} = a\omega\tilde{G}_L, \quad (9.81)$$

where $a = 4(N_1 - N_2)\hbar B_{12}$ is a measure of the strength of the transition and where

$$\tilde{G}_L = G_{L,disp} - iG_{L,res} = \frac{1}{\omega_0^2 - \omega^2 + i\omega\Delta\omega_0} \quad (9.82)$$

is the (general) complex Lorentz function. At frequencies around a narrow Lorentzian resonance, the susceptibility is $\tilde{\chi} = a'\tilde{g}_L$, where $a' = a/2$ and

$$\tilde{g}_{\rm L} = g_{\rm L,disp} - {\rm i}g_{\rm L,res} = \frac{1}{\omega_0 - \omega + {\rm i}\Delta\omega_0/2}. \quad (9.83)$$

Many textbooks treat the theory of classical dispersion (*see* preceding section) or present quantum mechanical derivations of the susceptibility; *see*, for instance, [5, 6].

9.5 Polarization of an Active Medium

A field $E = A\cos\omega t$ in a medium produces a polarization. In the special case that the frequency of the field is equal to the resonance frequency of the atomic transition ($\omega = \omega_0$), we obtain

$$P = \varepsilon_0\chi_2 \sin(\omega_0 t)A = \frac{2\pi\,\varepsilon_0}{\omega_0}(N_1 - N_2)\,b_{21}(\omega_0)\sin(\omega_0 t)A. \quad (9.84)$$

The polarization of an inactive medium (Fig. 9.2a) is delayed (by $\pi/2$ for $\omega = \omega_0$) with respect to the field that creates the polarization while the polarization of an active medium (Fig. 9.2b) is advanced (by $\pi/2$) with respect to the field.

An external electric field in a nonactive medium delivers energy to an ensemble of two-level atomic systems. The power transferred to a single two-level system is equal to

$$\frac{\text{force times path}}{\text{time}} = qE\dot{x} = E\dot{p}. \quad (9.85)$$

The power, averaged over time, a field (frequency $\omega = \omega_0$) delivers to a medium of polarization $P = (N_1 - N_2)p$ is equal to

$$W = \left< E\frac{{\rm d}P}{{\rm d}t} \right>_t = \frac{1}{2}\varepsilon_0\chi_2(\omega_0)A^2 = (N_1 - N_2)\frac{\pi\,\varepsilon_0}{\omega_0}b_{21}(\omega_0)A^2. \quad (9.86)$$

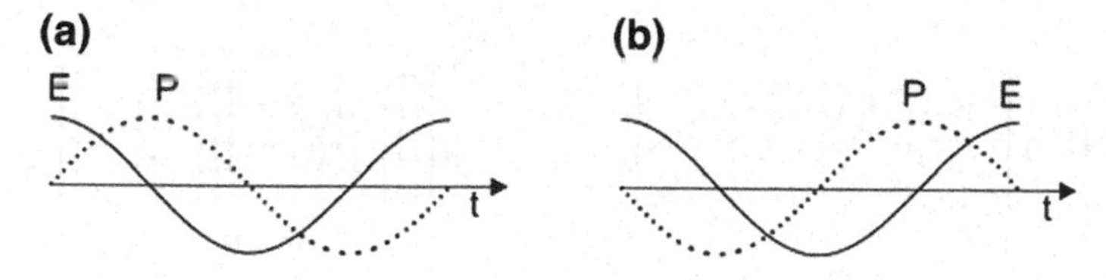

Fig. 9.2 Polarization and electric field at resonance **a** of an inactive medium and **b** of an active medium

The power is negative, $W < 0$, if a medium is active: then the polarization delivers energy to the field. The polarization is maintained by pumping of the active medium. Thus, the polarization mediates gain of the field. Pump power is converted—via the polarization—to power of the high frequency field.

If an active medium interacts with radiation at a frequency that is not the resonance frequency ($\omega \neq \omega_0$), the phase shift between field and polarization is

$$\tan \varphi(\omega) = \chi_2(\omega)/\chi_1(\omega) \tag{9.87}$$

and the power is

$$W(\omega) = W_0 \sin \varphi, \tag{9.88}$$

where

$$W_0 = (N_1 - N_2) \frac{\pi \varepsilon_0}{\omega_0} b_{21}(\omega_0) A^2 \tag{9.89}$$

is the power delivered by the field for $\omega = \omega_0$. Accordingly, the power transfer is smaller at frequencies outside the resonance frequency, that is, the gain in a laser medium is largest for radiation at the resonance frequency.

In an active medium of a laser, after a sudden turning on of the population difference (at $t = 0$), the atomic dipole moments oscillate with arbitrary phases relative to each other and therefore the polarization is zero (Fig. 9.3, left). An electric field of small amplitude $A(t = 0)$ produces a weak polarization and this enhances the field. The amplitude of the field and the amplitude of the polarization grow together by the mutual interaction of field and polarization until a steady state oscillation is established (right).

The interplay of the radiation and the atomic dipoles results in the growth of both the field and the polarization. With increasing field, the atomic dipole oscillations become more and more synchronized to the field. Accordingly, the polarization becomes more and more able to deliver energy to the field. The energy necessary for the buildup of the field originates from the excitation energy of the ensemble of two-level atomic systems. The initial field that starts oscillation can be due to spontaneous emission of radiation by the ensemble of two-level atomic systems (in the visible, UV, and X-ray range) or due to thermal radiation (in the far infrared spectral region).

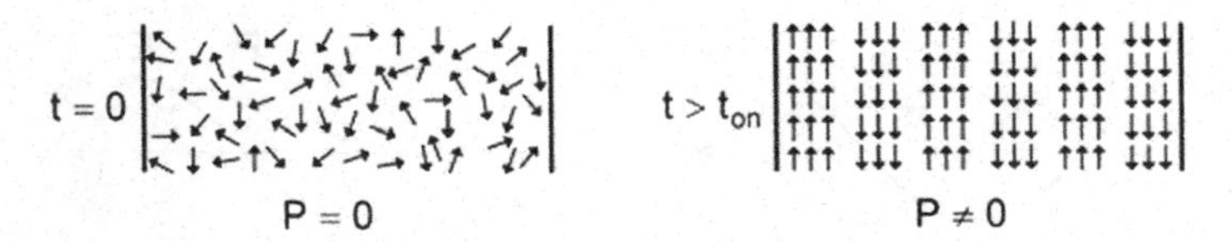

Fig. 9.3 Polarization of an active medium in a laser immediately after generation of population inversion and at steady state oscillation of the laser

At steady state oscillation, synchronization of the field and the polarization is maintained. Loss of polarization is compensated by pumping. The field synchronizes the atomic dipole oscillations that are produced by the pumping.

A high frequency polarization is characteristic of a large variety of excitations used in lasers:

- Electronic excitations of atoms, molecules, or ions.
- Electronic excitations by interband transitions in bipolar semiconductor lasers.
- Electronic excitations by intersubband transitions in quantum cascade lasers.
- Vibrational excitations of molecules.
- Rotational excitations of molecules.

The origin of electric dipole moments are oscillating charges in electronic and vibrational excitations and rotating charges in rotational excitations.

Before we treat the question how the amplitudes of the field and of the polarization build up during the onset of laser oscillation, we introduce the polarization current.

9.6 Polarization Current

If transverse relaxation is absent, we can introduce the *polarization current* characterized by the polarization current density

$$j_{\mathrm{pol}} = Nq\dot{x} = \mathrm{d}P/\mathrm{d}t. \tag{9.90}$$

The polarization current density is equal to the rate of change of the polarization. By differentiation of (9.53) with respect to time and multiplication of the equation by Nq, we obtain (with $q = -e$)

$$\frac{\mathrm{d}^2 j_{\mathrm{pol}}}{\mathrm{d}t^2} + \beta \frac{\mathrm{d} j_{\mathrm{pol}}}{\mathrm{d}t} + \omega_0^2 j_{\mathrm{pol}} = \mathrm{Re}\left[\mathrm{i}\omega \frac{Ne^2}{m_0} A e^{\mathrm{i}\omega t}\right]. \tag{9.91}$$

With $j_{\mathrm{pol}} = \mathrm{Re}[\tilde{j}_{\mathrm{pol}}]$ and the ansatz

$$\tilde{j}_{\mathrm{pol}} = \tilde{\sigma}^{\mathrm{pol}} \tilde{E}, \tag{9.92}$$

we find the complex *polarization conductivity*

$$\tilde{\sigma}^{\mathrm{pol}} = \frac{Ne^2}{2m_0} \frac{1}{\omega_0 - \omega + \mathrm{i}\Delta\omega_0/2}. \tag{9.93}$$

We write

$$\tilde{\sigma}^{\mathrm{pol}} = \sigma_1^{\mathrm{pol}} - \mathrm{i}\sigma_2^{\mathrm{pol}}. \tag{9.94}$$

Making use of (9.69), we obtain quantum mechanical expressions of the real and imaginary parts of the polarization conductivity,

$$\sigma_1^{\rm pol}(\omega) = \varepsilon_0 \omega \chi_2(\omega) = (N_1 - N_2) \times 2\pi \varepsilon_0 b_{21}(\omega), \tag{9.95}$$

$$\sigma_2^{\rm pol}(\omega) = \varepsilon_0 \omega \chi_1(\omega) = \frac{\omega_0 - \omega}{\Delta\omega_0/2}\, \sigma_1(\omega). \tag{9.96}$$

The polarization conductivities on the frequency scale are

$$\sigma_1^{\rm pol}(\nu) = (N_1 - N_2) \times \varepsilon_0 b_{21}(\nu), \tag{9.97}$$

$$\sigma_2^{\rm pol}(\nu) = \frac{\nu_0 - \nu}{\Delta\nu_0/2} \sigma_1(\nu). \tag{9.98}$$

We can also write

$$\sigma_1^{\rm pol}(\omega) = 2\pi \varepsilon_0 \omega_0 B_{21}^{\omega} (N_1 - N_2)\, g_{\rm L,res}(\omega), \tag{9.99}$$

$$\sigma_2^{\rm pol}(\omega) = \frac{\omega_0 - \omega}{\Delta\omega_0/2}\, \sigma_1(\omega) = 2\pi \varepsilon_0 \omega B_{21}^{\omega} (N_1 - N_2)\, g_{\rm L,disp}(\omega), \tag{9.100}$$

or

$$\sigma_1^{\rm pol}(\nu) = 2\pi \varepsilon_0 \nu_0 B_{21} (N_1 - N_2)\, g_{\rm L,res}(\nu), \tag{9.101}$$

$$\sigma_2^{\rm pol}(\nu) = \frac{\nu_0 - \nu}{\Delta\nu_0/2}\, \sigma_1(\nu) = 2\pi \varepsilon_0 \nu_0 B_{21} (N_1 - N_2)\, g_{\rm L,disp}(\nu). \tag{9.102}$$

The real part of the polarization conductivity of a nonactive medium (Fig. 9.4a) is positive and has a Lorentzian shape. The (negative) imaginary part is positive for $\omega < \omega_0$ and negative for $\omega > \omega_0$. The polarization conductivities are dynamical conductivities (=high frequency conductivities). The real part of the polarization conductivity shows a resonance curve with the maximum at ω_0. The signs of the polarization conductivities are reversed in case of an active medium (Fig. 9.4b).

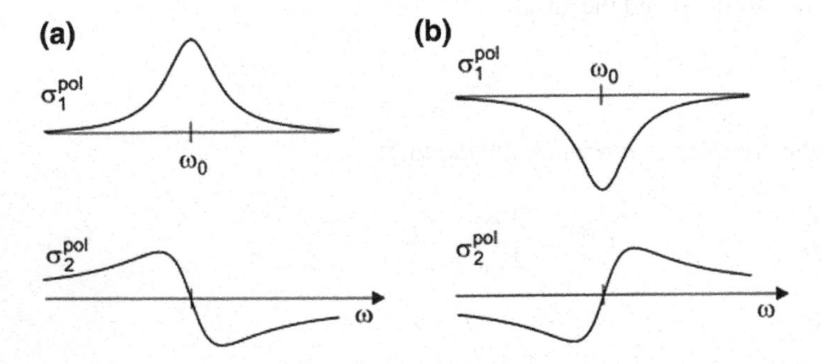

Fig. 9.4 Polarization conductivities **a** of an inactive medium and **b** of an active medium

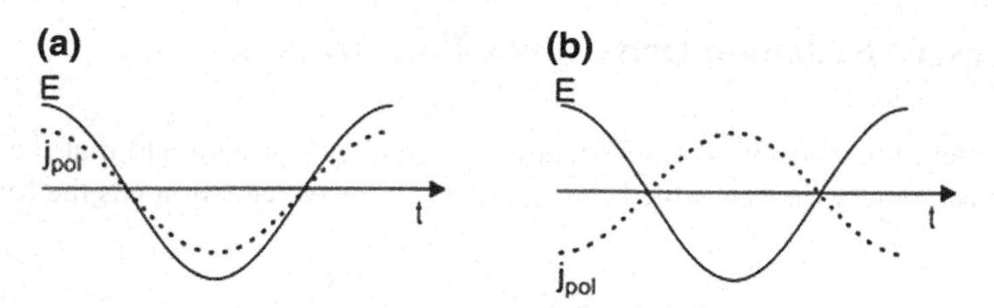

Fig. 9.5 Polarization current and field at resonance **a** of an inactive medium and **b** of an active medium

Thus, we found that the real part of the polarization conductivity of a medium with a homogeneously damped resonance transition has the shape of a Lorentz resonance function and the imaginary part has the shape of a Lorentz dispersion function.

We consider the special case that $\omega = \omega_0$ and that therefore $\sigma_2 = 0$. With $E = A\cos\omega_0 t$, we obtain

$$j_{\mathrm{pol}} = \sigma_1^{\mathrm{pol}}(\omega_0)E = (N_1 - N_2)\times 2\pi\varepsilon_0 b_{21}(\omega_0)\cos(\omega_0 t)\,A. \tag{9.103}$$

The polarization current of a nonactive medium has the same phase as the field (Fig. 9.5a) and $j_{\mathrm{pol}}E$ is positive at any moment. The field experiences damping. Power of the field is converted to heat due to relaxation of the polarization. The polarization current of an active medium has a phase of π relative to the field (Fig. 9.5b). The product $j_{\mathrm{pol}}E$ is negative at any moment. The field is amplified. Pump power is converted to power of the field. Amplification is mediated by the polarization current.

An electric field delivers to an oscillating charge q the

$$\text{power} = \frac{\text{work}}{\text{time}} = qE\dot{x} \tag{9.104}$$

and to an ensemble of electrons the power $j_{\mathrm{pol}}E$. The time average of the power transferred from the field to the polarization current is given by

$$(N_1 - N_2)\sigma_1(\omega_0)A < \cos^2\omega_0 t >_t = (N_1 - N_2)\times \pi\varepsilon_0 b_{21}(\omega_0)A^2. \tag{9.105}$$

To describe the resonance interaction of a medium with radiation, we will make use of both the polarization and the polarization current. The polarization is the fundamental quantity. The use of the polarization current density—in the case that transverse relaxation can be neglected—has the advantage that we can write some expressions in a simpler form, particularly in the special case that $\omega = \omega_0$; then the phase between current and field is either zero or π.

9.7 Laser Oscillation Driven by a Polarization

We now treat the case that the polarization is the origin of gain and that the active medium is a purely dielectric medium ($j = 0$). The wave equation has the form

$$\frac{1}{c^2}\frac{\partial^2 \boldsymbol{E}}{\partial t^2} - \nabla^2 \boldsymbol{E} = -\frac{1}{\varepsilon_0}\frac{\partial^2 \boldsymbol{P}}{\partial t^2}. \tag{9.106}$$

We assume, for simplicity, that the active medium is optically isotropic and fills a Fabry–Perot resonator completely. We also assume that the field does not vary over the cross-sectional area of the resonator. We can write

$$\frac{\partial^2 E(z,t)}{\partial t^2} - c^2\frac{\partial^2 E(z,t)}{\partial z^2} = -\frac{1}{\varepsilon_0}\frac{\partial^2 P(z,t)}{\partial t^2}. \tag{9.107}$$

The field in the resonator represents a standing wave

$$E(z,t) = E(t)\sin kz. \tag{9.108}$$

This leads, with $k^2 = \omega_{\mathrm{res}}^2/c^2$, where ω_{res} is the resonance frequency of the resonator, to the differential equation

$$\left(\frac{\mathrm{d}^2 E}{\mathrm{d}t^2} - \omega_{\mathrm{res}}^2 E\right)\sin kz = -\frac{1}{\varepsilon_0}\frac{\mathrm{d}^2 P(z,t)}{\mathrm{d}t^2}. \tag{9.109}$$

The polarization has the same z dependence as the field. Therefore, we can divide by $\sin kz$, except at the positions where $E(z) = 0$. Thus, we obtain, with $P(z,t) = P(t)\sin kz$ and $P = P(t)$, the differential equation:

$$\frac{\mathrm{d}^2 E}{\mathrm{d}t^2} - \omega_{\mathrm{res}}^2 E = -\frac{1}{\varepsilon_0}\frac{\mathrm{d}^2 P}{\mathrm{d}t^2}. \tag{9.110}$$

In our derivation of the wave equation, we did not include damping of the field that is, for instance, due to output coupling of radiation. We now introduce damping by use of the differential equation of the empty resonator:

$$\frac{\mathrm{d}^2 E}{\mathrm{d}t^2} + \kappa\frac{\mathrm{d}E}{\mathrm{d}t} + \omega_{\mathrm{res}}^2 E = 0, \tag{9.111}$$

where κ is the damping coefficient of the resonator. We assume that $\kappa \ll \omega_{\mathrm{res}}$, i.e., that the field in the empty resonator is given by

$$E = A_0 \mathrm{e}^{-\frac{1}{2}\kappa t}\cos \omega_{\mathrm{res}} t. \tag{9.112}$$

A_0 is the amplitude at $t = 0$. Now, the differential equation describing a field in a Fabry–Perot resonator containing an active medium is given by

$$\frac{d^2E}{dt^2} + \kappa\frac{dE}{dt} + \omega_{res}^2 E = -\frac{1}{\varepsilon_0}\frac{d^2P}{dt^2}. \tag{9.113}$$

We now assume that transverse relaxation of the polarization is absent and treat the particular case that all three frequencies—laser frequency, transition frequency and resonance frequency of the resonator—coincide with each other,

$$\omega = \omega_{res} = \omega_0. \tag{9.114}$$

We therefore can replace the second derivative of the polarization by the first derivative of the polarization current,

$$\frac{d^2P}{dt^2} = \frac{dj_{pol}}{dt}, \tag{9.115}$$

and obtain the differential equation

$$\frac{d^2E}{dt^2} + \kappa\frac{dE}{dt} + \omega^2 E = -\frac{1}{\varepsilon_0}\frac{dj_{pol}}{dt}. \tag{9.116}$$

At a sudden turning on of an initial population difference, the amplitude of the field is time dependent,

$$E(t) = A(t)\cos\omega t. \tag{9.117}$$

Immediately after starting the pumping, the polarization current density is

$$j_{pol} = \sigma_{1,0}^{pol} E, \tag{9.118}$$

where

$$\sigma_{1,0}^{pol} = -\varepsilon_0 b_{21}^0 (N_2 - N_1)_0 \tag{9.119}$$

is the small-signal polarization conductivity, $(N_2 - N_1)_0$ is the initial population difference, and

$$b_{21}^0 = b_{21}(\omega_0) = \hbar\omega_0 B_{21}^{\omega} g_{L,res}(\omega_0) \tag{9.120}$$

is the growth rate constant at the transition frequency ω_0. It follows that

$$\frac{d^2E}{dt^2} + (-\gamma_0 + \kappa)\frac{dE}{dt} + \omega_0^2 E = 0, \tag{9.121}$$

where

$$\gamma_0 = \gamma(\omega_0) = -\frac{1}{\varepsilon_0}\sigma_{1,0}^{\text{pol}} = b_{21}^0 (N_2 - N_1)_0 \tag{9.122}$$

is the growth coefficient of the active medium at ω_0 and

$$\alpha_0 = \alpha(\omega_0) = \frac{\gamma_0}{c/n} = -\frac{n}{c\varepsilon\varepsilon_0}\sigma_{1,0}^{\text{pol}} = \frac{n}{c}\, b_{21}^0\, (N_2 - N_1)_0 \tag{9.123}$$

is the gain coefficient at ω_0. If $\gamma_0 \ll \omega$, the solution of (9.121) is given by

$$E = A_0\, \mathrm{e}^{\frac{1}{2}(\gamma_0 - \kappa)t} \cos \omega t. \tag{9.124}$$

A_0 is the amplitude of the starting field. A very small field e.g., a field corresponding to one photon in the resonator mode, can initiate the oscillation. The amplitude of the field increases exponentially.

We now investigate the large-signal behavior. We assume that the amplitude $A(t)$ is a slowly varying function, i.e., that the envelope of the function $E(t)$ varies slowly (*slowly varying envelope approximation*). This means: the temporal change of the amplitude during one period of the oscillation period is negligibly small,

$$|\mathrm{d}A/\mathrm{d}t| \ll \omega |A(t)|, \tag{9.125}$$

$$|\mathrm{d}^2 A/\mathrm{d}t^2| \ll \omega |\mathrm{d}A/\mathrm{d}t| \ll \omega^2 |A(t)|. \tag{9.126}$$

Thus, we obtain

$$\frac{\mathrm{d}E}{\mathrm{d}t} = \frac{\mathrm{d}A}{\mathrm{d}t}\cos \omega t - \omega A \sin \omega t, \tag{9.127}$$

$$\frac{\mathrm{d}^2 E}{\mathrm{d}t^2} = -2\omega \frac{\mathrm{d}A}{\mathrm{d}t}\sin \omega t - \omega^2 A \cos \omega t. \tag{9.128}$$

In the last equation, we omitted the term $(\mathrm{d}^2 A/\mathrm{d}t^2)\cos \omega t$. Making use of the last two equations, we find from (9.116) the differential equation

$$-2\omega \frac{\mathrm{d}A}{\mathrm{d}t}\sin \omega t - \kappa \omega A \sin \omega t = -\frac{1}{\varepsilon_0}\frac{\mathrm{d}j_{\text{pol}}}{\mathrm{d}t}. \tag{9.129}$$

We neglected the term $\kappa(\mathrm{d}A/\mathrm{d}t)\cos \omega t$ in accordance with the condition $|\mathrm{d}A/\mathrm{d}t| \ll \omega |A(t)|$.

We write the polarization current density in the form

$$j_{\text{pol}} = -J(t)\cos \omega t. \tag{9.130}$$

$J(t)$ is the time-dependent amplitude of the polarization current density. This leads to the differential equation

$$\frac{\mathrm{d}A}{\mathrm{d}t} + \frac{\kappa}{2}A = \frac{1}{2\varepsilon_0}J. \tag{9.131}$$

This differential equation relates the amplitude of the field and the amplitude of the polarization current density. From $j_{\mathrm{pol}} = \sigma_1^{\mathrm{pol}}E$, we find immediately a relation between the amplitude of the current density and the population difference,

$$J = \varepsilon_0 b_{21}^0 (N_2 - N_1)A. \tag{9.132}$$

Another relation follows from the energy conservation law: the change of the population difference by stimulated emission and by relaxation from level 2 to level 1 compensates the power, which the polarization current transfers to the field. We assume that the population of the lower level is negligibly small compared with the population of the upper level. Then we can write

$$\left[\frac{\mathrm{d}}{\mathrm{d}t}(N_2 - N_1) + \frac{(N_2 - N_1)}{\tau_{\mathrm{rel}}^*} - \frac{(N_2 - N_1)_0}{\tau_{\mathrm{rel}}^*}\right] h\nu = j_{\mathrm{pol}}E. \tag{9.133}$$

Neglecting the rapidly varying term in $j_{\mathrm{pol}}E$ and averaging over the temporal and spatial variation,

$$< j_{\mathrm{pol}}E >_{t,z} = JA < \cos^2 \omega t >_t < \sin^2 kz >_z = \frac{1}{4}JA, \tag{9.134}$$

we obtain *three laser equations*

$$\frac{\mathrm{d}A}{\mathrm{d}t} + \frac{\kappa}{2}A = \frac{1}{2\varepsilon_0}J, \tag{9.135}$$

$$J = \varepsilon_0 b_{21}^0 (N_2 - N_1)A, \tag{9.136}$$

$$\frac{\mathrm{d}}{\mathrm{d}t}(N_2 - N_1) + \frac{(N_2 - N_1)}{\tau_{\mathrm{rel}}^*} - \frac{(N_2 - N_1)_0}{\tau_{\mathrm{rel}}^*} = -\frac{1}{4h\nu}JA. \tag{9.137}$$

These relate the amplitude of the field, the amplitude of the current density and the population difference.

By eliminating J and $N_2 - N_1$ from the two first equations and from the first and the third equation, respectively, we obtain

$$\frac{\mathrm{d}A}{\mathrm{d}t} + \frac{\kappa}{2}A = \frac{b_{21}^0}{2}(N_2 - N_1)A, \tag{9.138}$$

$$\frac{\mathrm{d}}{\mathrm{d}t}(N_2 - N_1) + \frac{N_2 - N_1}{\tau_{\mathrm{rel}}^*} - \frac{(N_2 - N_1)_0}{\tau_{\mathrm{rel}}^*} = -\frac{\varepsilon_0}{2h\nu}A\frac{\mathrm{d}A}{\mathrm{d}t} - \frac{\varepsilon_0\kappa}{4h\nu}A^2. \tag{9.139}$$

From the first of these equations, we obtain

$$N_2 - N_1 = \frac{2}{b_{21}^0}\frac{1}{A}\frac{\mathrm{d}A}{\mathrm{d}t} + \frac{\kappa}{b_{21}^0} \tag{9.140}$$

and, by differentiation,

$$\frac{\mathrm{d}}{\mathrm{d}t}(N_2 - N_1) = \frac{2}{b_{21}^0}\frac{\mathrm{d}}{\mathrm{d}t}\left(\frac{1}{A}\frac{\mathrm{d}A}{\mathrm{d}t}\right). \tag{9.141}$$

By elimination of $\mathrm{d}(N_2 - N_1)/\mathrm{d}t$ and of $N_2 - N_1$ from (9.138), (9.139), and (9.141), we find

$$a\,\frac{\mathrm{d}A}{\mathrm{d}t} + \left[\frac{\kappa}{2} - \frac{b_{21}^0(N_2 - N_1)_0}{2}\right]A + \frac{\varepsilon_0 b_{21}^0 \tau_{\mathrm{rel}}^* \kappa}{8h\nu}A^3 = 0. \tag{9.142}$$

The abbreviation

$$a = 1 + \frac{\varepsilon_0 b_{21}^0 \tau_{\mathrm{rel}}^*}{4h\nu}A^2 + \tau_{\mathrm{rel}}^*\frac{1}{A}\frac{\mathrm{d}A}{\mathrm{d}t}, \tag{9.143}$$

contains two terms that are small compared to 1, so that we obtain

$$\frac{\mathrm{d}A}{\mathrm{d}t} + \frac{1}{2}(-\gamma_0 + \kappa)A + \frac{\varepsilon_0 b_{21}^0 \tau_{\mathrm{rel}}^* \kappa}{8h\nu}A^3 = 0. \tag{9.144}$$

The amplitude A increases at small times exponentially and approaches at large times the steady state value

$$A_\infty = 2\sqrt{\frac{(\gamma_0 - \kappa)h\nu}{\varepsilon_0 b_{21}^0 \tau_{\mathrm{rel}}^* \kappa}}. \tag{9.145}$$

The differential (9.144) has the solution

$$A(t) = \frac{A_\infty}{\sqrt{1 + (A_\infty/A_0)^2 \mathrm{e}^{-(\gamma_0 - \kappa)t}}}. \tag{9.146}$$

$A_0 = A(t = 0)$ is the initial amplitude of the field. According to (9.131), the amplitude of the polarization current density is

$$J = \varepsilon_0 \kappa A + 2\varepsilon_0 \frac{\mathrm{d}A}{\mathrm{d}t}. \tag{9.147}$$

It follows from (9.138) that the population difference is equal to

$$N_2 - N_1 = \frac{1}{\varepsilon_0 b_{21}^0} \frac{J}{A}. \tag{9.148}$$

Multiplication of the differential (9.144) by $2A$ leads to

$$\frac{\mathrm{d}}{\mathrm{d}t}(A^2) + (-\gamma_0 + \kappa)A^2 + \frac{\varepsilon_0 b_{21}^0 \tau_{\mathrm{rel}}^* \kappa}{4h\nu} A^4 = 0. \tag{9.149}$$

Taking into account that the energy density of the field is $Zh\nu = \frac{1}{4}\varepsilon_0 A^2$, we obtain

$$\frac{\mathrm{d}Z}{\mathrm{d}t} = (\gamma_0 - \kappa)Z - b_{21}^0 \tau_{\mathrm{rel}}^* \kappa Z^2. \tag{9.150}$$

The equation describes the initial exponential increase of Z as well as transition to the steady state. The photon density at the steady state is

$$Z_\infty = \frac{\gamma_0 - \kappa}{b_{21}^0 \tau_{\mathrm{rel}}^* \kappa} = (r - r_{\mathrm{th}})\tau_p, \tag{9.151}$$

where $\tau_p = \kappa^{-1}$ is the lifetime of a photon in the resonator, $r_{\mathrm{th}} = (N_2 - N_1)_{\mathrm{th}}/\tau_{\mathrm{sp}}$ is the threshold pump rate and $r = (N_2 - N_1)_0/\tau_{\mathrm{sp}}$ is the pump rate. The expression of (9.151) is the same as (8.16), derived earlier.

The differential equation (9.150) has, with $Z(t = 0) = Z_0 \ll Z_\infty$, the solution

$$Z(t) = \frac{Z_\infty}{1 + Z_\infty/Z_0\, \mathrm{e}^{-(\gamma_0 - \kappa)t}}. \tag{9.152}$$

We define the oscillation onset time as the time where

$$Z(t_{\mathrm{on}}) = Z_\infty/2 \tag{9.153}$$

and find a value,

$$t_{\mathrm{on}} = \frac{\ln(Z_\infty/Z_0)}{\gamma_0 - \kappa}, \tag{9.154}$$

which we derived in Sects. 2.9 and 8.4 by simple arguments; *see* (2.85) and (8.29).

Figure 9.6 shows the buildup of laser oscillation at a sudden turning on of the population difference; the numbers concern a helium–neon laser (*see* next example). The curves of the figure indicate the following:

- The initial population difference (produced at $t = 0$) remains almost constant and decreases near t_{on} smoothly to the steady state value.
- The amplitude of the polarization current increases exponentially, shows a maximum at the time $t = t_{\mathrm{on}}$ and then decreases to the steady state value.

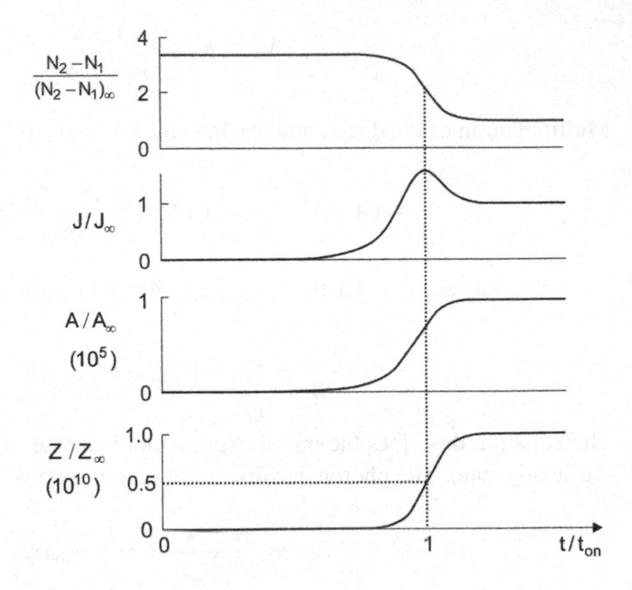

Fig. 9.6 Onset of laser oscillation: population difference; amplitude of the polarization current density; amplitude of the electric field; and photon density

- The amplitude of the field increases exponentially at $t < t_{\mathrm{on}}$ and reaches, at $t > t_{\mathrm{on}}$, the steady state value.
- The photon density reaches half the steady state value at $t = t_{\mathrm{on}}$.

It follows from the preceding equations that the steady state amplitude of the polarization current density is equal to

$$J_\infty = \varepsilon_0 A_\infty / \tau_p \tag{9.155}$$

and that the energy density of the polarization is

$$u_{\mathrm{pol}} = \frac{1}{4} J_\infty A_\infty \tau_p = \frac{\varepsilon_0}{4} A_\infty^2 = u. \tag{9.156}$$

At steady state oscillation, the polarization energy density is equal to the energy density of the electric field. During the buildup of laser oscillation, the polarization energy exceeds the electric field energy; the polarization energy is largest at the onset time t_{on}.

The ratio of the initial population difference and the steady state population difference is given by

$$\frac{(N_2 - N_1)_0}{(N_2 - N_1)_\infty} = 1 + b_{21}^0 \tau_{\mathrm{rel}}^* Z_\infty. \tag{9.157}$$

During the buildup of laser oscillation, the polarization conductivity changes from the small-signal value

$$\sigma_{1,0}^{\text{pol}} = -(N_2 - N_1)_0 \varepsilon_0 b_{21}^0 \tag{9.158}$$

to the large-signal value

$$\sigma_{1,\infty}^{\text{pol}} = -(N_2 - N_1)_\infty \varepsilon_0 b_{21}^0. \tag{9.159}$$

The ratio of the small-signal and the large-signal polarization conductivities is equal to the corresponding ratios of the susceptibilities and gain coefficients,

$$\frac{\sigma_{1,0}^{\text{pol}}}{\sigma_{1,\infty}^{\text{pol}}} = \frac{\chi_{1,0}}{\chi_{1,\infty}} = \frac{\alpha_0}{\alpha_\infty} = \frac{(N_2 - N_1)_0}{(N_2 - N_1)_\infty} = 1 + b_{21}^0 \tau_{\text{rel}}^* Z_\infty. \tag{9.160}$$

Example Helium–neon laser: output power 3 mW; $\pi(d/2)^2 L = 10^{-6}\,\text{m}^3$; L = 0.5 m; gain cross section $\sigma_{21} = 1.4 \times 10^{-16}\,\text{m}^2$; $\tau_{\text{rel}}^* = 100\,\text{ns}$; $\tau_p = 1.8 \times 10^{-7}\,\text{s}$. We find the following values:

- $P_{\text{out}} = Z_\infty \pi (d/2)^2 L h\nu/\tau_p$; $Z_\infty = 1.6 \times 10^{15}\,\text{m}^{-3}$.
- $Z_\infty = \varepsilon_0 A_\infty^2/4h\nu$; $A_\infty = 1.5 \times 10^4\,\text{V}\,\text{m}^{-1}$.
- $(N_2 - N_1)_0/(N_2 - N_1)_\infty = 1 + c\tau_{\text{rel}}^* \sigma_{21} Z_\infty = 4.1$.
- $(N_2 - N_1)_\infty = (\tau_p b_{21}^0)^{-1} = 1.2 \times 10^{14}\,\text{m}^{-3}$; $b_{21}^0 = c\sigma_{21}$.
- $\sigma_{1,0}^{\text{pol}}/\sigma_{1,\infty}^{\text{pol}} = \chi_{1,0}/\chi_{1,\infty} = \alpha_0/\alpha_\infty = (G_0 - 1)/(G_\infty - 1) = 4.1$.
- $G_\infty = 1.02$; $G_0 = 1.08$.
- $J_\infty = \varepsilon_0 A_\infty/\tau_p = 0.68\,\text{A}\,\text{m}^{-2}$.
- $Z_0 = 10^6\,\text{m}^{-3}$; $Z_\infty/Z_0 = 1.6 \times 10^9$.
- $t_{\text{on}} = T \ln(Z_\infty/Z_0)/\ln(G_0 V) = 22\,T = 720\,\text{ns}$.

A current density-field curve (Fig. 9.7a) is a straight line with a negative slope described by the relation

$$j_{\text{pol}}(t) = \sigma_1^{\text{pol}} E(t). \tag{9.161}$$

It follows from the negative slope that $j_{\text{pol}}(t)$ and $E(t)$ have opposite phases. During the onset of laser oscillation, the negative polarization conductivity varies with time (Fig. 9.7b). The variation is very slow, i.e., σ_1^{pol} is nearly constant during a cycle of the field. The absolute value of the polarization conductivity is large at $t = 0$ and decreases with increasing amplitude of the field until it reaches the steady state value $|\sigma_{1,\infty}^{\text{pol}}| = \varepsilon_0 \kappa$.

That the magnitude of the polarization conductivity σ_1^{pol} changes with time is the consequence of the *quantum mechanical origin of gain*: during the buildup of the laser field, the population difference decreases and therefore the magnitude of σ_1^{pol} decreases.

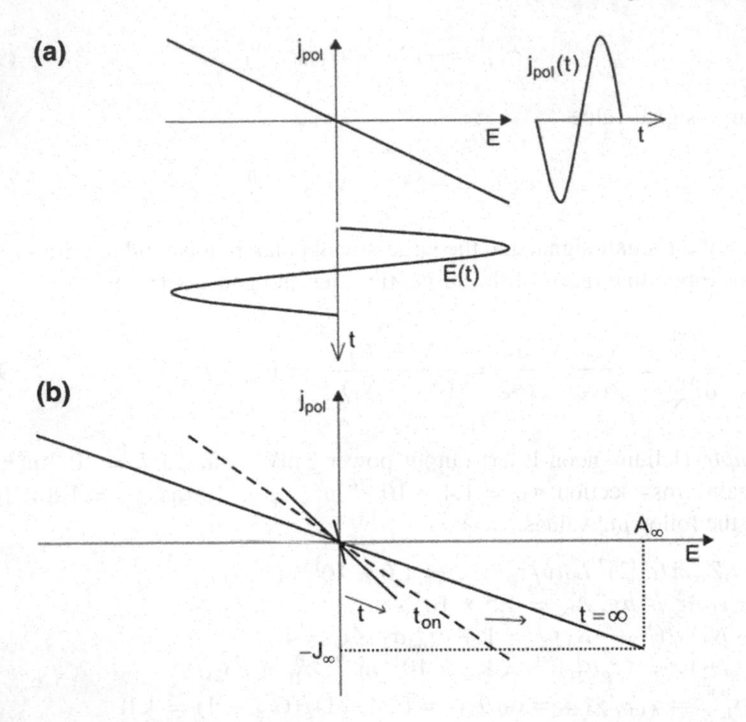

Fig. 9.7 Interplay of polarization current and field. **a** A current density-field curve and time-dependent current and field. **b** Current density-field curves during buildup of laser oscillation

9.8 Relaxation of the Polarization

Relaxation of the polarization occurs in ensembles of excited two-level systems. We consider the case that all atomic oscillators in a medium oscillate, under the action of a laser field, at the same frequency and with the same phase. We suppose that energy relaxation is absent. Then, the dipole moment of each of the oscillators is

$$p_i = p_0 \cos(\omega t + \varphi_0), \tag{9.162}$$

where

$$p_0 = q\, x_0 \tag{9.163}$$

is the amplitude of the dipole moment, $+q$ and $-q$ the oscillating charges, and x_0 the amplitude of the oscillation. The polarization is defined as the dipole moment density,

$$P = \sum\nolimits_{i=1}^{N_0} p_i = P_0 \cos(\omega t + \varphi_0). \tag{9.164}$$

N_0 is the number of atomic oscillators per unit volume and

$$P_0 = N_0\, p_0 = N_0\, q_0 x_0 \tag{9.165}$$

is the amplitude of the polarization.

We now assume that we suddenly turn off the laser field. Then, dephasing processes can destroy the polarization. We assume that dephasing processes occur randomly in time. The number of oscillators that undergo dephasing in a small time interval dt is

$$\mathrm{d}N = \frac{N}{T_2}\,\mathrm{d}t. \tag{9.166}$$

N is the number of coherently oscillating dipoles per unit volume and T_2 is the dephasing time. It follows that

$$N(t) = N_0\,\mathrm{e}^{-t/T_2}. \tag{9.167}$$

Accordingly, the polarization decreases exponentially,

$$P(t) = P_0\,\mathrm{e}^{-t/T_2}. \tag{9.168}$$

The polarization decays with the dephasing time T_2, which is called, in connection with the decay of the polarization, *transverse relaxation time* or *phase relaxation time*.

We assumed that the population of excited two-level systems remained constant, i.e., that T_2 was much smaller than the energy relaxation time T_1 of the upper energy level, $T_2 \ll T_1$. In the case that $T_2 \gg T_1$, the change of coherently oscillating dipoles is

$$\mathrm{d}N = \frac{N}{2T_1}\,\mathrm{d}t. \tag{9.169}$$

Now, the polarization is related to the amplitude of the oscillation that decays with the time constant $2T_1$ (while the energy content in the oscillator decays with T_1).

In the intermediate case, $T_2 \approx T_1$, the polarization decays according to

$$P(t) = P_0 \exp\left[\left(-\frac{1}{2T_1} + \frac{1}{T_2}\right) t\right]. \tag{9.170}$$

In connection with the polarization, the time T_1 is called *longitudinal relaxation time* (of the polarization).

9.9 Laser Equations

We study the laser oscillation of a more general case (assuming again that the population difference is suddenly turned on):

- $\omega \neq \omega_{\text{res}} \neq \omega_0$; i.e., laser frequency, resonance frequency of the resonator and transition frequency have different values.
- The polarization undergoes both longitudinal and transverse relaxation.

Our goal is to find differential equations that relate the field, the polarization and the population difference. We make use of the quantities:

- ω = laser frequency.
- ω_{res} = resonance frequency of the laser resonator.
- ω_0 = transition frequency; $\omega_0 = (E_2 - E_1)/\hbar$.
- $\tilde{A}$ = amplitude of the field.
- $\tilde{B}$ = amplitude of the polarization.
- $\Delta N = N_2 - N_1$ = population difference.
- $\Delta N_0 = (N_2 - N_1)_0$ = initial population difference.
- $\kappa/2 = 1/(2\tau_p)$; $2/\kappa$ = lifetime of the field in the laser resonator.
- T_1 = spontaneous lifetime of the population difference = longitudinal relaxation time.
- T_2 = transverse relaxation time.
- $\Delta\omega_0/2 = 1/(2T_1) + 1/T_2$; $2/\Delta\omega_0$ = relaxation time of the polarization.
- $b_{21}^0 = \hbar\omega_0 B_{21}^{\omega} g(\omega_0)$ = growth rate constant at ω_0.

The energy conservation law requires: the change of the population difference times the quantum energy of a photon is equal to the change of the average density of the energy contained in the field and the polarization,

$$\left(\frac{\mathrm{d}\Delta N}{\mathrm{d}t} + \frac{\Delta N - \Delta N_0}{T_1}\right)\hbar\omega = - < E\frac{\mathrm{d}P}{\mathrm{d}t} >_{\mathrm{t,z}}; \tag{9.171}$$

the average is taken over a temporal and a spatial period of the field. Atomic excitation energy is converted to field and polarization energy.

We have seen (in Sect. 9.7) that a polarization can drive a laser oscillation. We now assume that we inject into the laser resonator of an oscillating laser an external high frequency electric field E_1. The external field acts as an additional source term $\kappa_1 \mathrm{d}E_1/\mathrm{d}t$, where κ_1 is a coupling constant.

We make use of laser equations of second order, namely of (9.113) with the additional source term and of (9.76). We then have, together with (9.171), the laser equations:

$$\frac{d^2E}{dt^2}+\kappa\frac{dE}{dt}+\omega_{res}^2 E=-\frac{1}{\varepsilon_0}\frac{d^2P}{dt^2}+\kappa_1\frac{dE_1}{dt}, \tag{9.172}$$

$$\frac{d^2P}{dt^2}+\Delta\omega_0\frac{dP}{dt}+\omega_0^2 P=-\frac{2}{\pi}\varepsilon_0\hbar\omega B_{21}^{\omega}(N_2-N_1)E, \tag{9.173}$$

$$\left(\frac{d\Delta N}{dt}+\frac{\Delta N-\Delta N_0}{T_1}\right)=-\frac{1}{\hbar\omega}< E\frac{dP}{dt} >_{t,z}. \tag{9.174}$$

The equations are the *semiclassical laser equations* (also called neoclassical laser equations): the atomic states are quantized while the field is treated classically. The equations are suited to describe the dynamics of a laser oscillator.

We make use of the ansatz:

$$E=\frac{1}{2}[\tilde{A}\,e^{i\omega t}+c.c.],\quad E_1=\frac{1}{2}[\tilde{F}\,e^{i\omega_1 t}+c.c.],\quad P=\frac{1}{2}[\tilde{B}\,e^{i\omega t}+c.c.]. \tag{9.175}$$

The slowly varying envelope approximation

$$\left|\frac{d\tilde{A}}{dt}\right|\ll\omega\left|\tilde{A}\right|\quad\text{and}\quad\left|\frac{d^2\tilde{A}}{dt^2}\right|\ll\omega\left|\frac{d\tilde{A}}{dt}\right|, \tag{9.176}$$

$$\frac{d\tilde{B}}{dt}\ll\omega\left|\tilde{B}\right|\quad\text{and}\quad\left|\frac{d^2\tilde{B}}{dt^2}\right|\ll\omega\left|\frac{d\tilde{B}}{dt}\right|, \tag{9.177}$$

and the restriction to frequencies around ω_0 so that

$$\omega^2-\omega_0^2=(\omega+\omega_0)(\omega-\omega_0)\approx 2\omega(\omega-\omega_0), \tag{9.178}$$

leads to the laser equations:

$$\frac{d\tilde{A}}{dt}+\left[\frac{\kappa}{2}+i(\omega-\omega_{res})\right]\tilde{A}=-\frac{i\omega}{2\varepsilon_0}\tilde{B}+\frac{\kappa_1}{2}\tilde{F}, \tag{9.179}$$

$$\frac{d\tilde{B}}{dt}+\left[\frac{\Delta\omega_0}{2}+i(\omega-\omega_0)\right]\tilde{B}=\varepsilon_0 b_{21}^0\Delta N\tilde{A}, \tag{9.180}$$

$$\frac{d\Delta N}{dt}+\frac{\Delta N-\Delta N_0}{T_1}=-\frac{1}{4\hbar}(\tilde{A}\tilde{B}^*-\tilde{A}^*\tilde{B}). \tag{9.181}$$

The laser equations are coupled differential equations relating: amplitude of the field, phase of the field; amplitude of the polarization, phase of the polarization; population difference. The equations take into account the dephasing of the polarization by transverse relaxation.

We introduce amplitudes and phases,

$$\tilde{A}(t) = A(t)\mathrm{e}^{\mathrm{i}\varphi(t)}, \quad \tilde{F}(t) = F(t)\mathrm{e}^{\mathrm{i}\varphi_1(t)}, \quad \tilde{P}(t) = [C(t) - \mathrm{i}S(t)]\mathrm{e}^{\mathrm{i}\varphi(t)}. \tag{9.182}$$

and obtain five laser equations (in slowly varying envelope approximation):

$$\frac{\mathrm{d}A}{\mathrm{d}t} + \frac{\kappa}{2}A = \frac{\omega}{2\varepsilon_0}S + \frac{\kappa_1}{2}F\cos(\varphi - \varphi_1), \tag{9.183}$$

$$\frac{\mathrm{d}\varphi}{\mathrm{d}t} + \omega - \omega_0 = -\frac{\omega}{2\varepsilon_0}\frac{C}{A} - \frac{\kappa_1}{2}\frac{F}{A}\sin(\varphi - \varphi_1), \tag{9.184}$$

$$\left(\frac{\mathrm{d}}{\mathrm{d}t} + \frac{\Delta\omega_0}{2}\right)C + \left(\frac{\mathrm{d}\varphi}{\mathrm{d}t} + \omega - \omega_0\right)S = 0, \tag{9.185}$$

$$\left(\frac{\mathrm{d}}{\mathrm{d}t} + \frac{\Delta\omega_0}{2}\right)S - \left(\frac{\mathrm{d}\varphi}{\mathrm{d}t} + \omega - \omega_0\right)C = -\,\varepsilon_0 b_{21}^0 \Delta N A, \tag{9.186}$$

$$\frac{\mathrm{d}\Delta N}{\mathrm{d}t} + \frac{\Delta N - \Delta N_0}{T_1} = -\frac{1}{4\hbar}AS. \tag{9.187}$$

These five equations describe dynamical processes in laser oscillators.

Example Injection locking of a laser. Injection locking (=frequency locking = phase locking) of a self-excited oscillator means that an external high frequency field, injected into the resonator of the oscillator, forces the oscillator to assume the frequency of the external field rather than to execute an oscillation at its "natural" frequency of the free running laser. At the same time, the field in the resonator has a fixed phase relative to the external field. Injection of an external field (frequency ω_L) of a small amplitude can force a laser (natural frequency ω_L) to oscillate at ω_1. The natural frequency is not necessarily the frequency of maximum gain but is the frequency that is determined by the corresponding condition (2.80). The small-power laser acts as "seed" laser (=master oscillator) of the large-power laser (=slave laser) as shown in Fig. 9.8a. The frequency of the seed laser can be chosen in a range around the frequency of maximum gain (Fig. 9.8b). The power of the frequency-locked laser is the same as that of the free running laser. Injection of a monochromatic field results in a large occupation number of photons of frequency ω_1. Thus, the laser starts oscillation at ω_1 rather than at ω_0. If a laser is already oscillating at ω_L and the seed laser starts, the laser frequency "jumps" to ω_1. Injection locking by use of a frequency stabilized small-power laser can result in a stabilization of the large-power laser with respect to frequency and output power. The minimum power necessary for phase locking and the locking range can be derived by use of the five laser equations; a detailed treatment of injection locking can be found in [1].

Frequency locking of an oscillator is a general phenomenon. Two mechanical oscillators that are weakly coupled can force themselves to oscillate at a fixed phase. In 1865, Huygens observed phase locking of two pendulum clocks fixed at a wall and mechanically coupled via the wall. Injection locking of an electrical classical self-excited oscillator (Sect. 31.8) is well-known: a high frequency current flowing through the active element of a self-excited classical oscillator is superimposed with

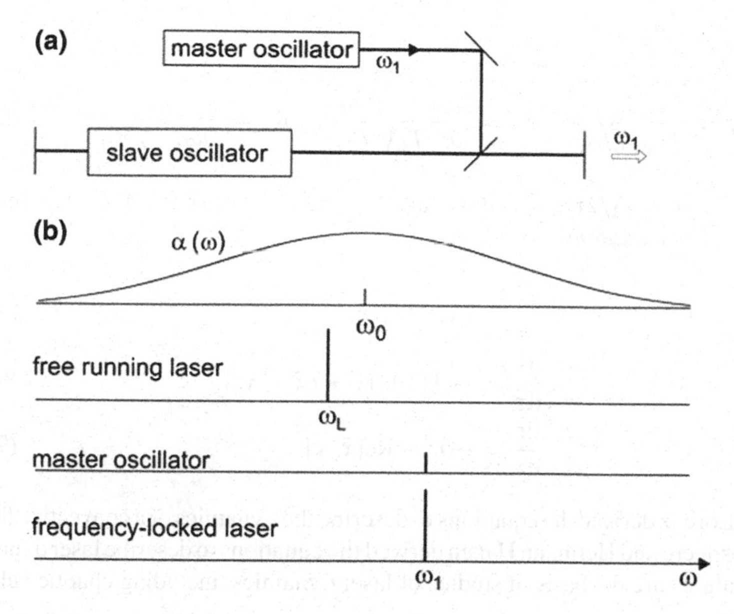

Fig. 9.8 Injection locking. **a** Seed laser (*master oscillator*) and laser. **b** Gain curve, free-running laser, external source, and frequency-locked laser

an external high frequency current. Frequency locking of a laser oscillator occurs via the influence of the electric field on the polarization of the active medium according to (9.172).

In the following, we consider the case that an external field is absent and that the laser frequency is equal to the resonance frequency of the resonator ($\omega = \omega_{\mathrm{res}}$) and, furthermore, that the phase relaxation time is much smaller than the energy relaxation time. We obtain the equations

$$\frac{\mathrm{d}\tilde{A}}{\mathrm{d}t} = -\frac{\tilde{A}}{2\tau_p} - \frac{\mathrm{i}\omega}{2\varepsilon_0}\tilde{B}, \tag{9.188}$$

$$\frac{\mathrm{d}\tilde{B}}{\mathrm{d}t} = -\frac{\tilde{B}}{T_2} + \mathrm{i}(\omega - \omega_0)\tilde{B} + \varepsilon_0 b_{21}^0 \Delta N \tilde{A}, \tag{9.189}$$

$$\frac{\mathrm{d}}{\mathrm{d}t}\Delta N = -\frac{1}{T_1}(\Delta N - \Delta N_0) - \frac{\mathrm{i}}{4\hbar}(\tilde{A}\tilde{B}^* - \tilde{A}^*\tilde{B}). \tag{9.190}$$

We introduce dimensionless variables.

- $\tau = t/T_2$ = dimensionless time.
- $\tilde{x} = \tilde{A}/K_{\mathrm{E}}$ = dimensionless amplitude of the field.
- $\tilde{y} = \tilde{B}/K_{\mathrm{P}}$ = dimensionless amplitude of the polarization.
- $z = (\Delta N_0 - \Delta N)/K_{\mathrm{N}}$; $\Delta N_0/K_{\mathrm{N}}$ = dimensionless initial population inversion; $\Delta N/K_3$ = dimensionless population inversion.

With

$$K_{\mathrm{E}} = \frac{1}{T_2\sqrt{\varepsilon_0 B_{21}}}, \quad K_{\mathrm{P}} = \frac{1}{\omega\tau_p T_2}\sqrt{\frac{\varepsilon_0}{B_{21}}}, \quad K_{\mathrm{N}} = \frac{1}{\hbar\omega T_2\tau_p B_{21}}, \tag{9.191}$$

$b = T_2/T_1$, $\sigma = T_2/2\tau_p$, $\delta = \mathrm{i}(\omega - \omega_0)T_2$, and $r = \omega T_2\tau_p b_{21}^0 \Delta N_0$, we find the *Lorenz-Haken equations*

$$\frac{\mathrm{d}\tilde{x}}{\mathrm{d}\tau} = -\sigma\tilde{x} + \sigma\tilde{y}, \tag{9.192}$$

$$\frac{\mathrm{d}\tilde{y}}{\mathrm{d}\tau} = -(1 - \mathrm{i}\delta)\tilde{y} + r\tilde{x} - \tilde{x}z, \tag{9.193}$$

$$\frac{\mathrm{d}z}{\mathrm{d}\tau} = -bz + \mathrm{Re}[\tilde{x}^*\tilde{y}]. \tag{9.194}$$

Edward Lorenz derived the equations to describe the dynamics of a convective fluid of the atmosphere and Hermann Haken derived the equations to describe laser dynamics. The equations are the basis of studies of laser dynamics, including chaotic behavior [41–45].

9.10 Laser-van der Pol Equation

According to the last section, we can characterize laser oscillation (if transverse relaxation is absent) in slowly varying envelope approximation by the equations:

$$E(t) = A(t)\cos\omega t, \tag{9.195}$$

$$\frac{\mathrm{d}A}{\mathrm{d}t} - \frac{1}{2}(\gamma_0 - \kappa)A + \frac{\varepsilon_0 b_{21}^0 \tau_{\mathrm{rel}}^* \kappa}{8h\nu}A^3 = 0. \tag{9.196}$$

This is the *laser-van der Pol equation*. It is a nonlinear differential equation of first order for the amplitude A.

We have found in Sect. 9.7 that the laser equations lead, in slowly varying envelope approximation, to analytical expressions for the time dependences of the amplitude of field, (9.161) and, furthermore of the amplitude of the polarization and of the polarization-current density

$$j_{\mathrm{pol}} = -\varepsilon_0\kappa A(t)\cos\omega t \tag{9.197}$$

and of the population difference and the population difference

$$N_2 - N_1 = \frac{\kappa}{b_{21}^0} \frac{1}{A(t)}. \tag{9.198}$$

Thus, the analytical solution of (9.169) for $A(t)$, equation (9.146), provides also analytical expressions for the polarization current (and therefore of the polarization) and of the population difference.

The laser-van der Pol equation is always applicable for description of a laser oscillation (if transverse relaxation is absent). The reason is the following: for all lasers, the gain per period T of the laser field is small compared with unity, $(\gamma_0 - \kappa)T \ll 1$. This is the condition for the applicability of the slowly varying envelope approximation.

We have derived the laser-van der Pol equation by using the condition that laser frequency, atomic transition frequency, and resonance frequency of the resonator coincide with each other. The equation is, however, also applicable for a high frequency field composed of fields of different frequencies if the frequencies have values near the frequency of maximum gain and near the eigenfrequency of the laser resonator, i.e., if the gain curve is sufficiently broad and the Q value of the laser resonator is sufficiently small. The van der Pol equation of the laser is suitable to describe, for instance, mode beating in a laser that oscillates on two modes at the same time; *see*, for instance, [5, 40–42]. It can also be applied to treat frequency locking.

The van der Pol equation in slowly varying amplitude approximation,

$$\frac{dA}{dt} - \frac{1}{2}(\gamma_0 - \kappa)A + \beta A^3 = 0, \tag{9.199}$$

is a differential equation of first order. It describes an oscillation of the quantity $y = A(t)\cos\omega t$, where $A(t)$ is a slowly varying amplitude, γ_0 is a growth coefficient, κ is a damping term, and β is a term that follows from the mechanism responsible for oscillation. The equation has two solutions. One of the solutions is $A = 0$. It follows that the occurrence of oscillation supposes an initial amplitude that is nonzero. A small initial amplitude causes exponential growth of the amplitude. The A^3 term is responsible for limitation of the amplitude. In a laser, the initial amplitude of the laser field stems from spontaneous emission of radiation by the active medium. The equation holds only if the net gain per oscillation period is small compared with 1, that is, if $(\gamma_0 - \kappa)T \ll 1$, as already mentioned.

The three equations (9.196) through (9.198) are the most simple laser equations but represent a complete set of laser equations.

The differential equation (9.196) is also suitable, to analyze oscillation of a classical model oscillator if $(\gamma_0 - \kappa)T \ll 1$. In a classical oscillator model, the parameters γ_0, κ, and β represent quantities, which characterize mechanisms that differ completely from the mechanisms in a laser (*see* Chap. 31). The equation (9.196) is the slowly varying envelope approximation of the (general) van der Pol equation that is a nonlinear differential equation of second order (*see* Sect. 31.8 and Problems to this chapter). In contrast to laser oscillators, classical oscillators work in most cases with a much larger gain per period. Thus, a laser oscillator is *the* application of the classical van der Pol equation in slowly varying envelope approximation. (The van der Pol oscillator has been reported by *van der Pol* [243] long before the invention of the laser.)

9.11 Kramers–Kronig Relations

The Kramers–Kronig relations relate the real part of a linear response function and the imaginary part. If the real part of a linear response function is known for all frequencies, then the imaginary part can be calculated for all frequencies. And if the imaginary part of a quantity is known for all frequencies, the real part can be calculated for all frequencies. The Kramers–Kronig relations for the complex susceptibility are

$$\chi_2(\omega) = -\frac{2\omega}{\pi}\int_0^\infty \frac{\chi_1(\omega')}{\omega'^2-\omega^2}\,\mathrm{d}\omega', \tag{9.200}$$

$$\chi_1(\omega) = -\frac{2}{\pi}\int_0^\infty \frac{\omega'\chi_2(\omega')}{\omega'^2-\omega^2}\,\mathrm{d}\omega'. \tag{9.201}$$

Example If a susceptibility has the form of a complex (general) Lorentz function

$$\tilde{\chi}(\omega) = \chi_1 - \mathrm{i}\chi_2 = a\omega\tilde{G}_\mathrm{L}(\omega), \tag{9.202}$$

where a is a measure of the strength of the corresponding transition and

$$\begin{aligned}\tilde{G}_\mathrm{L}(\omega) &= \frac{1}{(\omega_0^2-\omega^2)+\mathrm{i}\omega\Delta\omega_0}\\ &= \frac{\omega_0^2-\omega^2}{(\omega_0^2-\omega^2)^2+(\omega\Delta\omega_0)^2}+\mathrm{i}\frac{\omega\Delta\omega_0}{(\omega_0^2-\omega^2)^2+(\omega\Delta\omega_0)^2},\end{aligned} \tag{9.203}$$

is the complex general Lorentz function. An electric field $\tilde{E} = A\mathrm{e}^{\mathrm{i}\omega t}$ causes a polarization $\tilde{P} = \varepsilon_0(\chi_1 - \mathrm{i}\chi_2)\tilde{E}$. The susceptibilities χ_1 and χ_2 are related according to (9.200) and (9.201).

The Kramers–Kronig relations are a consequence of causality. The Dutch physicists Kramers [46] and Kronig [47] derived the relations independently from each other. The relations are treated in many textbooks on Solid State Physics and Optics; *see*, for instance, [59, 177, 179, 180, 184, 297, 302].

9.12 Lorentz Functions: A Survey

As already mentioned, in mathematics, the Lorentzian function is defined as

$$f_{\mathrm{L}}(x) = \frac{1}{\pi} \frac{\Delta x_0/2}{(x_0 - x)^2 + \Delta x_0^2/4}, \tag{9.204}$$

$$\int_{-\infty}^{\infty} f_{\mathrm{L}}(x) \mathrm{d}x = 1. \tag{9.205}$$

Besides the normalization condition, the values of x_0 and Δx_0 are not restricted.

For a narrow resonance function, with $\Delta x_0 \ll x_0$, the integral from zero to infinity is approximately unity. We use this approximation of the Lorentzian function to describe narrow resonances in physical systems and designate the function as *Lorentz resonance function*, $g_{\mathrm{L,res}}(x)$.

In the case that a resonance is **not** narrow, we use the function that we call *general Lorentz resonance function*. The Kramers–Kronig relations connect the general Lorentz resonance function with the *general Lorentz dispersion function*.

In the book, we use the following functions.

- $\tilde{G}_{\mathrm{L}} = G_{\mathrm{L,res}} - \mathrm{i}\, G_{\mathrm{L,disp}}$, general complex Lorentz function.
- $G_{\mathrm{L,res}}$, general Lorentz resonance function (=real part of the general Lorentz function).
- $G_{\mathrm{L,disp}}$, general Lorentz dispersion function (=imaginary part of the general Lorentz function).
- $\bar{G}_{\mathrm{L,res}}$, normalized general Lorentz resonance function.
- $\bar{G}_{\mathrm{L,disp}}$, "normalized" general Lorentz dispersion function; normalized is the corresponding real part.
- $\tilde{g}_{\mathrm{L}} = g_{\mathrm{L,res}} - \mathrm{i}\, g_{\mathrm{L,disp}}$, complex Lorentz function.
- $g_{\mathrm{L,res}}$, Lorentz resonance function.
- $g_{\mathrm{L,disp}}$, Lorentz dispersion function.
- $\bar{g}_{\mathrm{L,res}}$, normalized Lorentz resonance function.
- $\bar{g}_{\mathrm{L,disp}}$, "normalized" Lorentz dispersion function; normalized is the corresponding Lorentz resonance function.

Table 9.1 gives the functions on the ω scale.

The Lorentz resonance function describes the frequency dependence of the gain coefficient of active media based on dipole oscillators (Table 9.2), i.e., of conventional lasers containing two-level atomic systems as the elementary systems. The Lorentz dispersion function describes the gain coefficient of an active medium in a free-electron laser containing monopole oscillators, or in a quantum mechanical description, containing energy-ladder systems. The monopole oscillations in a free-electron laser (Chap. 19) are narrow-band oscillations.

The elementary systems in a Bloch laser (Chap. 32) are energy-ladder systems, or, in a classical description, monopole oscillations of electrons. The gain function

Table 9.1 Lorentz functions

	Resonance function	Dispersion function
G_L	$\frac{\omega\,\Delta\omega_0}{(\omega_0^2-\omega^2)^2+(\omega\,\Delta\omega_0)^2}$	$\frac{\omega_0^2-\omega^2}{(\omega_0^2-\omega^2)^2+(\omega\,\Delta\omega_0)^2}$
$\bar{G}_L$	$\frac{\omega^2\Delta\omega_0^2}{(\omega_0^2-\omega^2)^2+(\omega\,\Delta\omega_0)^2}$	$\frac{(\omega_0^2-\omega^2)\,\omega\,\Delta\omega_0}{(\omega_0^2-\omega^2)^2+(\omega\,\Delta\omega_0)^2}$
g_L	$\frac{1}{\pi}\frac{\Delta\omega_0/2}{(\omega_0-\omega)^2+\Delta\omega_0^2/4}$	$\frac{1}{\pi}\frac{(\omega_0-\omega)}{(\omega_0-\omega)^2+\Delta\omega_0^2/4}$
$\bar{g}_L$	$\frac{\Delta\omega_0^2/4}{(\omega_0-\omega)^2+\Delta\omega_0^2/4}$	$\frac{(\omega_0-\omega)\,\Delta\omega_0/2}{(\omega_0-\omega)^2+\Delta\omega_0^2/4}$

Table 9.2 Shape of gain curves of different types of lasers

	Type of laser	Elementary system	Classical model
$g_{L,res}$	Conventional lasers	Two-level systems	Dipole oscillation
$\bar{g}_{L,disp}$	free-electron laser	Energy-ladder system	Monopole oscillations of electrons
$\bar{g}_{L,disp} - K(\omega)$	Bloch laser	Energy-ladder system	Monopole oscillations of electrons

is a modified Lorentz dispersion function, $\bar{g}_{L,\,disp} - K(\omega)$. This function takes into account that the response of an active medium of a Bloch laser is composed of a dynamic part, described by $\bar{g}_{L,\,disp}$, and a term, $K(\omega)$, which takes into account that a direct current is always present in a Bloch laser.

9.13 A Third Remark About the History of the Laser

We come back to the question: why did it take—after the discovery of stimulated emission by Einstein—40 years until the maser and the laser were invented? The laser is an apparatus developed by experimentalists. We consider the development of spectroscopy after stimulated emission became known. In the time from 1900 to 1930, Berlin was a center of spectroscopy, with two outstanding spectroscopists: Heinrich Rubens (1865–1922) and Rudolf Ladenburg (1882–1952). Rubens developed methods suited to study the far infrared spectral range. His result (1900) with respect to the spectral distribution of radiation emitted by a thermal radiation source was a basis of the derivation of Planck's radiation law. One of Rubens' Ph.D students, Marianus Czerny (1896–1985), studied the (far infrared) rotational spectrum of HCl [49]; he found that the positions of the absorption lines did not agree with predictions of classical physics. Richard Tolman (Caltech, Pasadena) analyzed in 1924 [50] Czerny's data (that he knew before publication by Czerny) with respect to the strength of absorption, taking into account thermal populations of energy levels, which had transition energies of the order of kT. Tolman included in his analysis stimulated emission (then called "negative absorption"). Tolman discussed in his lectures about quantum mechanics the three processes described by Einstein. Thus, stimulated emission was known in the physics community (*see* [22]).

Based on the quantum theory of Einstein and making use of the correspondence principle, Hendrik A. Kramers [51] developed a theory of the refractive index of gases taking account of stimulated emission. He showed that the refractive index in the vicinity of a resonance line of an atomic gas is expected to decrease if a portion of atoms is in the excited state, i.e., if the population difference is reduced. The effect that is due to stimulated emission was called "negative dispersion". In 1928, Rudolf Ladenburg and coworkers [52–56] found experimentally that stimulated emission resulted in a reduction of the refractive index of excited neon. Making use of an interferometric technique combined with a spectral analysis, Ladenburg and coworkers measured the change of refractive index of a gas (neon), contained in a long tube, that was excited by a gas discharge. The refractive index in the neighborhood of lines decreased, at strong current, with increasing discharge current, i.e., with increasing excitation. Thus, Ladenburg and coworkers performed experimental studies of energy levels of gases in nonequilibrium states. Ladenburg studied (beginning in 1908) atomic gases by analyzing emission spectra excited by gas discharges. He emigrated in 1928 to the USA (becoming professor at Princeton University).

In Germany, the activities in the field of experimental spectroscopy were strongly reduced after the stock market crash in 1929 and when Hitler came to power. In the U.S.A. and other countries, the Great Depression resulted in a reduced investment in physics. It seems that atomic physics was in principal understood at the end of the 1920s. During the 1930s and especially after the discovery of nuclear fission, the field of nuclear physics became most attractive for physicists. Great interest in spectroscopy (including microwave spectroscopy) began with the discovery of masers and lasers.

Townes writes in his memoirs [22]: "By the 1950s, then, the idea of getting amplification by stimulated emission of radiation was already recognized here and there, but for one reason or another, nobody really saw the idea's potency or published it, except for me and the Russians [Basov and Prokhorov], whose work was then unknown for me." The essential new idea, besides the idea to make use of atomic transitions in a system with population inversion, was the idea—introduced by Townes—to use a resonator in order to realize a self-excited oscillator. The next step, toward the optical maser (laser), was the idea to use an optical resonator, i.e., a resonator without sidewalls.

Was there a chance to invent the laser already in the late 1920s? In his first paper mentioning negative dispersion [52], Ladenburg reported the formula of the refractive index of a gas near a resonance line but outside the range of absorption,

$$n_1 - 1 = \frac{\lambda_0^5}{\lambda - \lambda_0} \frac{1}{16\pi c} A_{21} \left(\frac{g_2}{g_1} N_1 - N_2 \right), \tag{9.206}$$

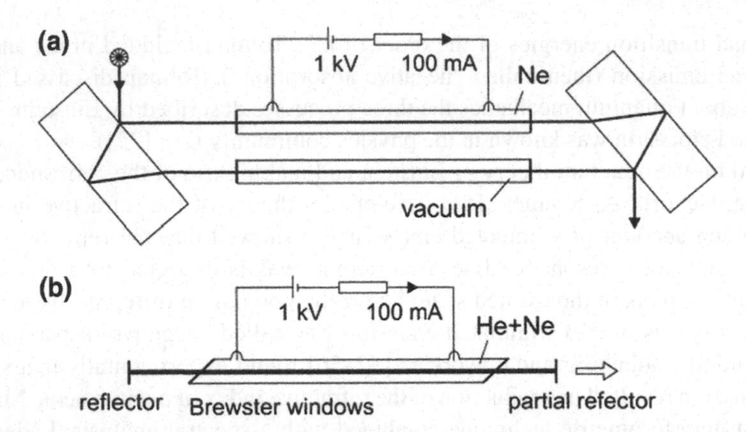

Fig. 9.9 A comparison. **a** Ladenburg's arrangement used for measurement of negative dispersion of a neon gas. **b** Helium–neon laser

where g_1 is the degeneracy of the lower level, g_2 the degeneracy of the upper level, λ the wavelength of the radiation, and $\lambda_0 = c/\nu_0$. Equation (9.206) follows from (9.77) if degeneracy of energy levels is taken into account. Ladenburg wrote (translated from German): "It is one of the most important tasks to detect experimentally Kramers' negative dispersion, whose theoretical importance is unquestionable."

Ladenburg and coworkers used an arrangement [53] shown in Fig. 9.9a. The optical arrangement consisted of a *Jamin interferometer*. One arm of the interferometer contained a tube filled with neon gas (at low pressure) and the other arm contained an evacuated reference tube. A beam of white light from an arc discharge lamp was divided into two beams (one traversing the tube with the gas and the other traversing the empty tube). Then the two beams were superimposed and passed a diffraction grating. The interference pattern on a photo plate contained information on the frequency dependence of the refractive index. At large current, both the lower and the upper level of a resonance transition were occupied. The results [54–56] indicated that the refractive index decreased as predicted by theory. The relative population difference $(N_1 - N_2 g_1/g_2)/N_1$ was most likely between ten and forty percent. The arrangement had similarities to that of a helium–neon laser (Fig. 9.9b). The dimension of the helium tube, the gas pressure, and the strength of current were similar. There are two important differences: the helium–neon laser contains, in addition to neon, also helium and the gas tube is enclosed between highly reflecting mirrors; additionally, the gas tube of a laser is closed by Brewster windows. The helium is essential to obtain a population inversion and the high-reflectivity mirrors are necessary to reach laser threshold. The effect of energy transfer from helium to neon atoms was not known at Ladenburg's time and highly reflecting dielectric mirrors were not available. A helium–neon laser generating visible light was realized in 1962. If Ladenburg or somebody else would have had the idea of a neon laser, a more elaborated investigation of gas discharges and of laser mirrors would have been necessary.

In the 1920s and the beginning 1930s, there was not yet much knowledge about optical properties of dielectric crystals, of doped dielectric crystals and of semiconductors. Therefore, a solid state laser would not have been a reachable goal at the time. Research in the field of solid state physics grew strongly between 1920 and 1960 [57].

References [1, 4–6, 12, 13, 22, 35, 36, 40–58, 177–181, 184, 243, 297, 302].

Problems

9.1 Susceptibilities and polarization conductivities. Instead of our ansatz of an electromagnetic wave, $\tilde{E} = A\mathrm{e}^{\mathrm{i}(\omega t - kz)}$, we could use the ansatz $\tilde{E} = A\mathrm{e}^{\mathrm{i}(kz - \omega t)}$. Show that the imaginary part of the susceptibility changes sign but that the real field and the real polarization are the same in both cases. Discuss the corresponding polarization conductivity.

9.2 Linear dispersion.

(a) Determine the linear dispersion $\mathrm{d}n/\mathrm{d}\omega$ of titanium–sapphire at a population difference $N_2 - N_1 = 10^{24}\,\mathrm{m}^{-3}$.
(b) Determine the shift of the resonance frequencies of a Fabry–Perot resonator (length 0.5 m) due to optical pumping of a crystal of 1 cm length, i.e., at a change of the population difference $N_2 - N_1 = 0$ to $N_2 - N_1 = 10^{24}\,\mathrm{m}^{-3}$.

9.3 Nonlinear dispersion of optically pumped titanium–sapphire.

(a) Determine the nonlinear dispersion $\mathrm{d}^2n/\mathrm{d}\nu^2$ around the center frequency ω_0.
(b) Determine the nonlinear dispersion in the case that $N_2 - N_1 = 10^{24}\,\mathrm{m}^{-3}$.
(c) How large is the change of the refractive index in the frequency range $\nu_0 - \Delta\nu_0/2$, ν_0 and in the range ν_0, $\nu_0 + \Delta\nu_0/2$?
(d) Determine the shift of the resonance frequencies (due to nonlinear dispersion of a crystal of 1 cm length of a Fabry-Perot resonator (length 0.5 m) due to optical pumping, i.e., at a change of the population difference $N_2 - N_1 = 0$ to $N_2 - N_1 = 10^{24}\,\mathrm{m}^{-3}$.

9.4 Drude theory. We obtain the Drude theory of the electric transport if we treat the electrons in a solid as free-electrons, i.e., if we set $\omega_0 = 0$ in (9.42) and introduce the electron velocity $\mathrm{v} = \mathrm{d}x/\mathrm{d}t$. Then $\beta^{-1} = \tau$ is the relaxation time of an electron: an electron (accelerated at time $t = 0$) by an electric field loses its energy after the time τ.

(a) Derive the high frequency conductivities $\sigma_1(\omega)$ and $\sigma_2(\omega)$.
(b) Determine the real part and imaginary of the high frequency mobility $\tilde{\mu}(\omega)$; $\tilde{\mathrm{v}}(\omega) = \tilde{\mu}(\omega)\tilde{E}(\omega)$.

(c) Determine the corresponding frequency-dependent susceptibilities and dielectric constants (=dielectric functions).

9.5 Perfect conductor of high frequency currents. We define a perfect conductor of high frequency currents as a conductor with free-electrons that have an infinitely long relaxation time. [A superconductor at temperatures that are small compared to its superconducting transition temperature T_c can be a perfect conductor of high frequency currents at frequencies where $h\nu < 2\Delta$; 2Δ is the superconducting energy gap; $T_c = 7\,\mathrm{K}$ for lead and 90 K for the high temperature superconductor $YBa_2Cu_3O_7$.]

(a) Derive the high frequency conductivity of a perfect conductor.
(b) Determine the dielectric function.
(c) Calculate the values of σ_2 of an ideal conductor that contains free-electrons of a density $N = 10^{28}\,\mathrm{m}^3$; $N = 10^{25}\,\mathrm{m}^3$; $N = 10^{22}\,\mathrm{m}^3$.

9.6 Show that the slowly varying amplitude approximation is valid if the change of the amplitude within a quarter of the period of a high frequency field is small compared to the amplitude of a high frequency field.

9.7 Rabi oscillation. An ensemble of two-level atomic systems that interact with a strong electric field can show an oscillation of the population inversion and, synchronously, an oscillation of the polarization. We assume that the frequency of the field is equal to the atomic resonance frequency and that transverse relaxation is absent. We furthermore assume that the only relaxation process is spontaneous emission of radiation but that the spontaneous lifetime T_1 is much larger than the period of the field. We describe the dynamics of the polarization by,

$$\frac{d^2P}{dt^2} + \Delta\omega_0 \frac{dP}{dt} + \omega_0^2 P = \frac{2}{\pi} \varepsilon_0 \hbar\omega_0 B_{21}^{\omega} \Delta N E. \tag{9.207}$$

An electric field $E = A\cos\omega_0 t$ causes a polarization $P = B\sin\omega_0 t$ that is 90° phase shifted relative to the field. We find, in slowly varying amplitude approximation, the equation

$$\frac{dB}{dt} + \frac{\Delta\omega_0}{2} B(t) = \frac{1}{\pi} \varepsilon_0 \hbar\omega_0 B_{21}^{\omega} A \Delta N(t). \tag{9.208}$$

The time dependence of the population difference is determined by the differential equation:

$$\frac{d\Delta N(t)}{dt} + \frac{\Delta N(t) - \Delta N_0}{T_1} = \frac{1}{2\hbar\omega_0} A B(t); \tag{9.209}$$

the change of the population difference averaged over a period of the field, multiplied by the energy of a photon, is equal to $AB/2$.

(a) Show that, under certain conditions, these two equations are equivalent to two second-order differential equations,

$$\mathrm{d}^2 B/\mathrm{d}t^2 + \omega_\mathrm{R}^2 B = 0, \tag{9.210}$$

$$\mathrm{d}^2 \Delta N/\mathrm{d}t^2 + \omega_\mathrm{R}^2 \Delta N = 0, \tag{9.211}$$

where

$$\omega_\mathrm{R}^2 = \varepsilon_0 b_{21}^0/(4\hbar\omega_0)A^2 \tag{9.212}$$

and ω_R is the Rabi frequency and, furtheremore, $\omega_\mathrm{R} \ll \omega_0$; this condition allows for application of the slowly varying envelope approximation. The differential equations are approximately valid if $\omega_\mathrm{R} \gg \omega_0$. The solutions are

$$B = B_0 \sin \omega_\mathrm{R} t, \tag{9.213}$$

$$\Delta N = \Delta N_0 \cos \omega_\mathrm{R} t. \tag{9.214}$$

The amplitude of the polarization and the population difference oscillate with the Rabi frequency. The Rabi frequency is proportional to the amplitude of the electric field.

(b) Make a draft of the time dependences of the population difference ΔN and the amplitude of the polarization.

(c) Calculate the Rabi frequencies for a medium with a naturally broadened line, with $T_1 = 10^{-2}$ s. What is the minimum field amplitude necessary for the occurrence of a Rabi oscillation? (For more information about Rabi oscillations, *see*, for instance, [1, 5, 40]).

9.8 Start of laser oscillation.
Show, by use of the van der Pol (vdP) equation of a laser, that laser oscillation cannot start without an initial field [*Hint*: the vdP equation has two different solutions, depending on the initial conditions].

9.9 Write the van der Pol equation of a laser in dimensionless units, as well as the solution for the electric field.

9.10 Derive the van der Pol equation from the Lorenz-Haken equations.

9.11 The van der Pol equation.
Derive the van der Pol equation of a laser from the (general) van der Pol equation

$$\frac{\mathrm{d}^2 y}{\mathrm{d}\tau^2} + \varepsilon(-1 + y^2)\frac{\mathrm{d}y}{\mathrm{d}\tau} + y = 0,$$

where y is the dimensionless field, ε (>0) is a parameter, and τ the dimensionless time [*Hint*: Make use of SVEA; *see* Problem 31.4].

9.12 Determine the amplitude of the polarization of a helium–neon laser medium that carries a polarization-current of 0.68 A m^{-2} (*see Example* to Fig. 9.6).

9.13 Phase portrait.

(a) Characterize onset of laser oscillation by a phase portrait. *Hint*: Make use of the solution $A(t)$ of the van der Pol equation of a laser. The phase portrait is obtained for a plot of $\dot{A}$ (on the y axis) versus A (on the x axis), with the time t as a parameter that varies from $t = 0$ to $t \rightarrow \infty$.

(b) Draw the phase portrait of a laser oscillation for the case that the gain is suddenly turned off.

Part III
Operation of a Laser

Chapter 10
Cavity Resonator

After a basic description of a laser in the first parts of the book, we now are dealing with the question how we can operate a laser. For this purpose, we will first discuss laser resonators. In this chapter, we treat the cavity resonator, which is a closed resonator. In the next chapter, we will study the open resonator.

We solve the wave equation for electromagnetic radiation in a metallic rectangular cavity and determine the eigenfrequencies and the field distributions of modes of a cavity resonator. A cavity resonator has a low frequency cutoff. The cutoff frequency, determined by the geometry of the resonator, corresponds to a resonance of lowest order. The field of a mode is a standing wave.

Standing waves composed of two waves that propagate in opposite directions along one of the three axes of a rectangular resonator are forbidden modes. A long resonator has modes that are composed of waves that propagate nearly parallel to the long axis. We express the frequency separation of these modes in a simple way by the use of the Fresnel number; we will later see that Fresnel numbers are important parameters of the theory of diffraction.

We finally calculate the mode density that corresponds to frequencies, which are large compared to the cutoff frequency. This leads to the expression of the mode density we used in connection with the discussion of Planck's radiation law and the Einstein coefficients (Sect. 6.7).

10.1 Cavity Resonators in Various Areas

Max Planck used the model of a cavity resonator (="hohlraum" resonator) to derive the radiation law.

Microwave oscillators (Chap. 31) and far infrared semiconductor lasers (Chap. 29) make use of cavity resonators.

K.F. Renk, *Basics of Laser Physics*, Graduate Texts in Physics,
DOI 10.1007/978-3-319-50651-7_10

High-Q microwave cavities that are able to store electromagnetic fields of large amplitude are suited to accelerate particles in accelerators.

10.2 Modes of a Cavity Resonator

We discuss properties of a rectangular metallic cavity resonator (Fig. 10.1). All walls are metallic. We assume that the walls are ideal conductors, i.e., that reflection of radiation at the walls occurs without absorption loss. The extensions of the cavity resonator are: a_1 = width (along x axis); a_2 = height (along y); L = length (along z). Coupling of radiation into a resonator is possible, for instance, by means of a hole in one of the walls. We treat the interior of the resonator as a vacuum space, ignoring the effect of air (or of another medium).

To describe the electromagnetic field in the resonator, we make use of Maxwell's equations

$$\nabla \times \boldsymbol{H} = \partial \boldsymbol{D}/\partial t, \tag{10.1}$$

$$\nabla \times \boldsymbol{E} = -\partial \boldsymbol{B}/\partial t, \tag{10.2}$$

$$\nabla \cdot \boldsymbol{E} = 0, \tag{10.3}$$

$$\nabla \cdot \boldsymbol{B} = 0. \tag{10.4}$$

$\boldsymbol{E}$ is the electric field, $\boldsymbol{B}$ the magnetic induction, $\boldsymbol{H}$ the magnetic field, $\boldsymbol{D}$ the dielectric displacement, ε_0 the electric field constant, and μ_0 the magnetic field constant. A field in vacuum is characterized by

$$\mu_0 \boldsymbol{H} = \boldsymbol{B}, \tag{10.5}$$

$$\boldsymbol{D} = \varepsilon_0 \boldsymbol{E}. \tag{10.6}$$

The boundary conditions for electromagnetic fields require continuity of the tangential component E_t of the electric field at a boundary and continuity of the normal component H_n of the H field. The tangential component of the electric field is zero

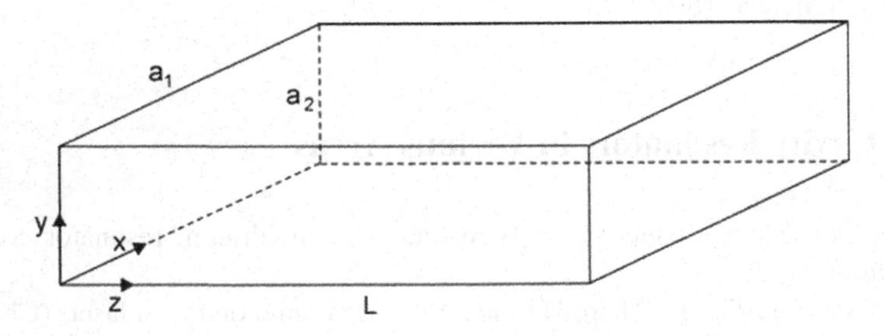

Fig. 10.1 Rectangular cavity resonator

everywhere on the walls of an ideal conductor,

$$E_t(\text{wall}) = 0. \tag{10.7}$$

Forming $\nabla \times (\nabla \times \boldsymbol{E} + \partial \boldsymbol{B}/\partial t) = 0$, and with $\nabla \times (\nabla \times \boldsymbol{E}) = \nabla \times \nabla \times \boldsymbol{E} - \nabla^2 \boldsymbol{E}$, we obtain the wave equation

$$\nabla^2 \boldsymbol{E} - \frac{1}{\mu_0 \varepsilon_0} \frac{\partial^2 \boldsymbol{E}}{\partial t^2} = 0. \tag{10.8}$$

This corresponds to three equations, concerning the field component E_x, E_y and E_z:

$$\left(\frac{\partial^2}{\partial x^2} + \frac{\partial^2}{\partial y^2} + \frac{\partial^2}{\partial z^2}\right) E_x - \frac{1}{c^2} \frac{\partial^2 E_x}{\partial t^2} = 0, \tag{10.9}$$

$$\left(\frac{\partial^2}{\partial x^2} + \frac{\partial^2}{\partial y^2} + \frac{\partial^2}{\partial z^2}\right) E_y - \frac{1}{c^2} \frac{\partial^2 E_y}{\partial t^2} = 0, \tag{10.10}$$

$$\left(\frac{\partial^2}{\partial x^2} + \frac{\partial^2}{\partial y^2} + \frac{\partial^2}{\partial z^2}\right) E_z - \frac{1}{c^2} \frac{\partial^2 E_z}{\partial t^2} = 0, \tag{10.11}$$

where $c = 1/\sqrt{\varepsilon_0 \mu_0}$ is the speed of light in vacuum. The ansatz

$$E_x = f(x)\, g(y)\, h(z) \cos \omega t \tag{10.12}$$

leads to the equation

$$\frac{1}{f} \frac{\partial^2 f}{\partial x^2} + \frac{1}{g} \frac{\partial^2 g}{\partial y^2} + \frac{1}{h} \frac{\partial^2 h}{\partial z^2} + \frac{\omega^2}{c^2} = 0. \tag{10.13}$$

By separation of the variables,

$$f^{-1} \partial^2 f / \partial x^2 = k_x^2, \tag{10.14}$$

$$g^{-1} \partial^2 g / \partial y^2 = k_y^2, \tag{10.15}$$

$$h^{-1} \partial^2 h / \partial z^2 = k_z^2, \tag{10.16}$$

we obtain

$$\omega = c \sqrt{k_x^2 + k_y^2 + k_z^2}. \tag{10.17}$$

The solutions concerning the field components, the wave vector $\mathbf{k}$, and the eigenfrequency ω are given by the following equations:

$$E_x(\boldsymbol{r},t) = A_x \cos k_x x \sin k_y y \sin k_z z \cos \omega t, \tag{10.18}$$

$$E_y(\boldsymbol{r},t) = A_y \sin k_x x \cos k_y y \sin k_z z \cos \omega t, \tag{10.19}$$

$$E_z(\boldsymbol{r},t) = A_z \sin k_x x \sin k_y y \cos k_z z \cos \omega t, \tag{10.20}$$

$$\boldsymbol{k} = \boldsymbol{k}_{\rm mnl} = \left(m\frac{\pi}{a_1},\ n\frac{\pi}{a_2},\ l\frac{\pi}{L}\right), \tag{10.21}$$

$$\omega = \omega_{\rm mnl} = c\sqrt{\left(m\frac{\pi}{a_1}\right)^2 + \left(n\frac{\pi}{a_2}\right)^2 + \left(l\frac{\pi}{L}\right)^2}. \tag{10.22}$$

A_x, A_y, and A_z are the amplitudes of the three field components, $\boldsymbol{r} = (x, y, z)$ is a location, m, n, and l are integers. With the exception that at least two of the three numbers are nonzero, these can have the values

$$m = 0, 1, 2...;\quad n = 0, 1, 2...;\quad l = 0, 1, 2, \ldots .$$

We thus obtained the modes of a resonator. A mode is characterized by the number triple mnl and a discrete eigenfrequency $\omega_{\rm mnl}$. The electric field fulfills the condition of transversality, $\nabla \times \boldsymbol{E} = 0$, or

$$\boldsymbol{k} \times \boldsymbol{E} = 0 \tag{10.23}$$

corresponding to

$$k_x A_x + k_y A_y + k_z A_z = 0. \tag{10.24}$$

The 101 mode has the frequency

$$\nu_{101} = \frac{c}{2}\sqrt{\frac{1}{a_1^2} + \frac{1}{L^2}}. \tag{10.25}$$

The electric field of the 101 mode is

$$E_y = A_y \sin\frac{\pi x}{a_1} \sin\frac{\pi z}{L} \cos \omega t. \tag{10.26}$$

The field (Fig. 10.2a) is oriented along the y axis. The field strength has the largest value in the center of the cavity.

The 011 mode has the frequency

$$\nu_{011} = \frac{c}{2}\sqrt{\frac{1}{a_2^2} + \frac{1}{L^2}}. \tag{10.27}$$

The field of the 011 mode is oriented along x. In the case that $a_1 = a_2$, the 011 mode is degenerate with the 101 mode, i.e., the modes have the same frequency but different field patterns. Microwave cavities often have a side ratio close to $a_2/a_1 = 1/2$.

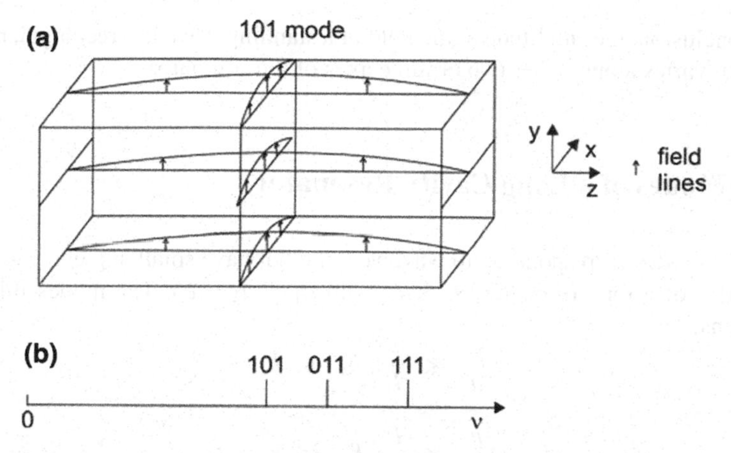

Fig. 10.2 Modes of a cavity resonator. **a** Field lines of the 101 mode at a fixed time. **b** Frequency distribution

The frequency distribution of the modes (Fig. 10.2b) shows that there is a forbidden frequency range between $\nu = 0$ and ν_{101} (for $a_2 < a_1$). The lowest frequency is the *cutoff frequency* of the resonator. The corresponding cutoff wavelength (of free-space radiation) is $\lambda = c/\nu_{101}$. The cutoff wavelength of a cubic cavity resonator is $\lambda = a\sqrt{2}$.

Example $a_1 = 1\,\text{cm}$, $a_2 = 0.5\,\text{cm}$, $L = 1\,\text{cm}$; $\nu_{101} = \omega_{101}/2\pi = (c/2)\sqrt{a_1^{-2} + L^{-2}}$ $\sim 21.2\,\text{GHz}$. The corresponding free-space wavelength of the radiation is about 1.42 cm.

The solutions (10.18) through (10.20) describe standing waves. As an example, we consider a n01 mode

$$E_y = A_y \sin k_x x \sin k_z z \cos \omega t. \tag{10.28}$$

We can consider the field as composed of two waves propagating in $+z$ and $-z$ direction,

$$E_y = \frac{1}{2} A_y \sin(k_x x) \left(\cos\left(\omega t - k_z z - \frac{\pi}{2}\right) + \cos\left(\omega t + k_z z + \frac{\pi}{2}\right)\right). \tag{10.29}$$

The two waves have the same amplitude, but the amplitude varies along the x direction. Alternatively, we can describe the n01 wave as composed of two waves propagating in x and $-x$ direction,

$$E_y = \frac{1}{2} A_y \sin(k_z z) \left(\cos\left(\omega t - k_x x - \frac{\pi}{2}\right) + \cos\left(\omega t + k_x x + \frac{\pi}{2}\right)\right). \tag{10.30}$$

In this description, the amplitude varies along the z direction.

In conclusion, the amplitude of the field of a standing wave in a rectangular cavity resonator varies along either two or three axes of the resonator.

10.3 Modes of a Long Cavity Resonator

We study modes corresponding to wave vectors that have small angles with respect to the axis of a long resonator ($L \gg a$, with $a_1 = a_2 = a$). The modes fulfill the conditions:

$$\frac{m}{a} \ll \frac{l}{L}; \; k_x \ll k_z, \tag{10.31}$$

$$\frac{n}{a} \ll \frac{l}{L}; \; k_y \ll k_z.$$

The frequency of a mode is

$$\nu_{\mathrm{mnl}} = \frac{c}{2}\frac{l}{L}\left(1 + \frac{m^2 + n^2}{l^2}\frac{L^2}{a^2}\right)^{1/2}. \tag{10.32}$$

A Taylor expansion yields

$$\nu_{\mathrm{mnl}} \approx \frac{c}{2L}\left(l + m^2\frac{L^2}{2la^2} + n^2\frac{L^2}{2la^2}\right). \tag{10.33}$$

The frequency distance of the modes with high order in l and low order in m and n (Fig. 10.3) is equal to

$$\Delta\nu_m = \nu_{\mathrm{l,m+1,n}} - \nu_{\mathrm{l,m,n}} = \frac{c}{2L}\left(m + \frac{1}{2}\right)\frac{L^2}{4la^2}. \tag{10.34}$$

A simple calculation yields

$$\Delta\nu_m = \Delta\nu_l\,\frac{m + \frac{1}{2}}{F}, \tag{10.35}$$

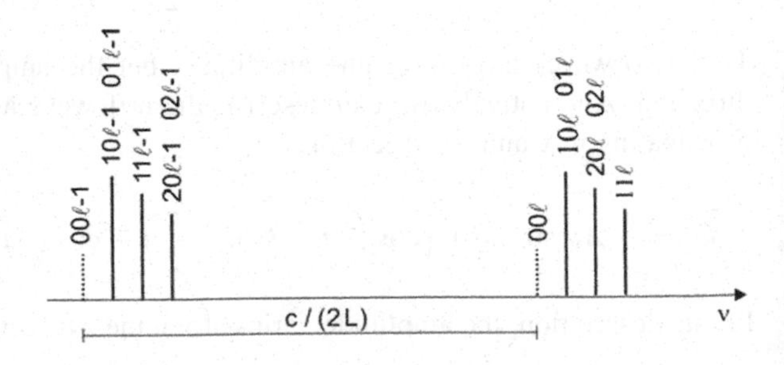

Fig. 10.3 Frequency spectrum of modes of a long resonator

where $\Delta\nu_l = c/2L$ is the frequency distance between neighboring 00l modes (that are *forbidden modes*) at the frequencies $\nu_l = l \times c/(2L)$ and where

$$F = \frac{(a/2)^2}{L\lambda} \tag{10.36}$$

is the *Fresnel number*. If $F = 1$, the frequency separation between the 0nl mode and the 1nl mode is $\Delta\nu_{\mathrm{m}=0} = \Delta\nu_l/2$. If $F \gg 1$, the frequency distance, $\Delta\nu_{\mathrm{m}=0} = \Delta\nu_l/(2F)$, is small compared to $\Delta\nu_l$. The Fresnel number combines geometric quantities (the area of the reflectors and the distance L between the reflectors) and the wavelength of the radiation; *see* also Sect. 10.5.

10.4 Density of Modes of a Cavity Resonator

The k vectors of the modes of a cavity resonator have discrete values. In k space (Fig. 10.4), the k values are

$$\boldsymbol{k} = \left(m\frac{\pi}{a_1}, n\frac{\pi}{a_2}, l\frac{\pi}{L}\right), \tag{10.37}$$

where m, n, and l are integers. The numbers are positive, the k vectors lie in one quadrant of the k space. There is one allowed k point in the k-space volume

$$V_1 = \frac{\pi}{a_1} \times \frac{\pi}{a_2} \times \frac{\pi}{L}. \tag{10.38}$$

A spherical shell of radius k and thickness dk has the k-space volume

$$V = \frac{1}{8} \times 4\pi k^2 \mathrm{d}k. \tag{10.39}$$

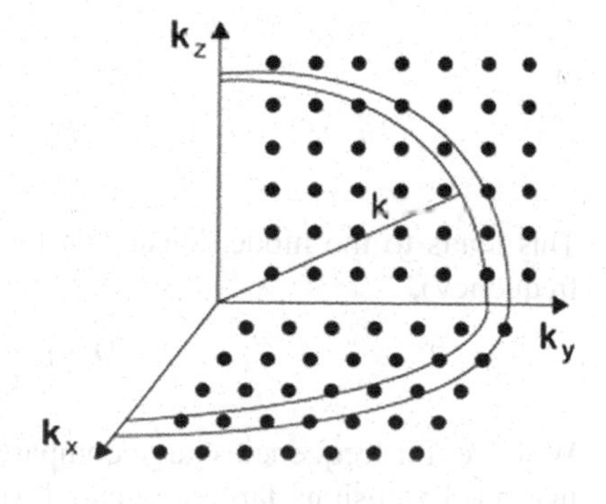

Fig. 10.4 Modes in k space

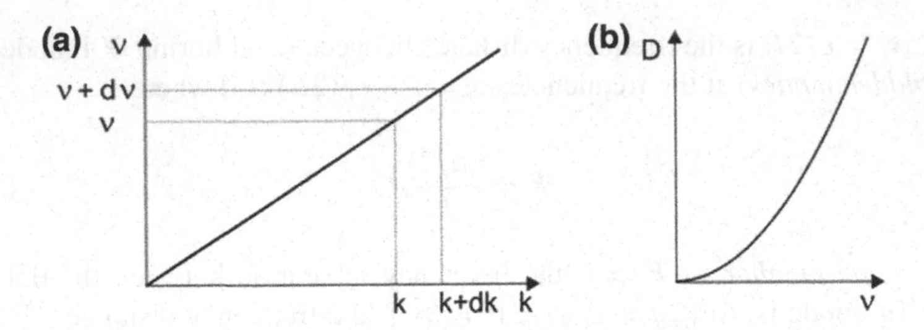

Fig. 10.5 Electromagnetic waves in a cavity of large volume. **a** Dispersion relation. **b** Mode density

At large k, the number of k values in the interval $k, k + \mathrm{d}k$ is equal to

$$\bar{D}(k)\mathrm{d}k = \frac{V}{V_1} = \frac{a_1 a_2 L}{2\pi^2} k^2 \mathrm{d}k. \tag{10.40}$$

The density of modes in k space, i.e., the density of allowed k values, is equal to

$$\bar{D}(k) = \frac{a_1 a_2 L}{2\pi^2} k^2. \tag{10.41}$$

It follows from the dispersion relation for light (Fig. 10.5a),

$$\nu = \frac{c}{2\pi} k, \tag{10.42}$$

that the frequency interval

$$\mathrm{d}\nu = \frac{c}{2\pi} \mathrm{d}k \tag{10.43}$$

contains as many modes as the corresponding interval $\mathrm{d}k$. Taking into account that there are two waves of different polarization for each k vector, we obtain

$$\bar{D}(\nu)\mathrm{d}\nu = 2\bar{D}(k)\mathrm{d}k \tag{10.44}$$

or

$$\bar{D}(\nu) = \frac{2\bar{D}(k)}{\mathrm{d}\nu/\mathrm{d}k}. \tag{10.45}$$

This leads to the mode density on the frequency scale (=number modes per unit frequency),

$$\bar{D}(\nu) = a_1 a_2 L \, \frac{8\pi \nu^2}{c^3}. \tag{10.46}$$

We have, for frequencies large compared to the cutoff frequency, i.e., for a resonator that has extensions large compared to the wavelength of radiation, the result: *the density of the modes of a cavity resonator increases proportionally to the square of the frequency* (Fig. 10.5b). It also increases proportional to the volume of the cavity. The mode density per unit volume is given by

$$D(\nu) = \frac{1}{a_1 a_2 L}\bar{D}(\nu) = \frac{8\pi\nu^2}{c^3}. \tag{10.47}$$

Accordingly, $D(\nu)\mathrm{d}\nu$ is the number of modes (per unit volume) in the frequency interval $\nu, \nu + \mathrm{d}\nu$.

The density of modes in a cavity resonator containing an optically isotropic medium (refractive index n) is equal to

$$D(\nu) = \frac{8\pi\nu^2}{(c/n)^3}. \tag{10.48}$$

Due to the smaller wavelength, i.e., the larger wave vector of radiation of frequency ν, the density of modes of a cavity containing a dielectric medium with the refractive index n is by the factor n^3 larger than the density of modes of the cavity without a medium.

The density of modes (per unit volume) in free space is the same as the density of modes (per unit volume) in a cavity resonator (Problem 10.7).

10.5 Fresnel Number

The Fresnel number (Sect. 10.3)

$$F = \frac{a^2}{4\lambda L} = \frac{a/2}{\lambda} \times \frac{a/2}{L} \tag{10.49}$$

is a combination of resonator extensions (width a, height a, length L) and wavelength λ. The Fresnel number is dimensionless. Many properties of optic apparatus of different size depend solely on F; the Fresnel number plays an important role in the characterization of diffraction occurring in laser resonators (Sect. 11.8).

We introduce the Fresnel number in a different way. We consider (Fig. 10.6) the widening of a light beam by diffraction at an iris diaphragm (diameter a). The angle of diffraction is approximately given by

$$\theta \approx \frac{\lambda}{a}. \tag{10.50}$$

A mirror (also of diameter a) at distance L from the iris reflects a portion of the radiation. The Fresnel number is

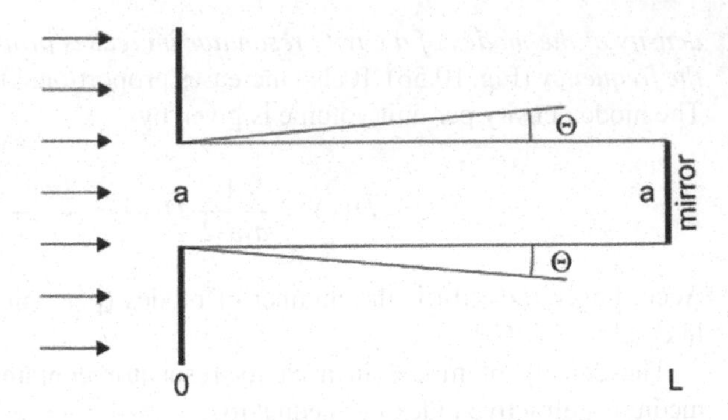

Fig. 10.6 Fresnel number

$$F = \frac{\text{intensity of the reflected light}}{\text{intensity of the unreflected light}}$$
$$= \frac{\pi a^2/4}{\pi(a/2 + L\theta)^2 - \pi a^2/4} = \frac{a^2/4}{a^2/4 + aL\theta - a^2/4} = \frac{a^2}{4\lambda L}. \tag{10.51}$$

The mirror reflects half of the light if $F = 1$ and most of the light if $F \gg 1$. The intensity of the reflected beam is

$$I_r = I_0 \frac{F}{1 + F}, \tag{10.52}$$

where I_0 is the intensity of the incident beam.

10.6 TE Waves and TM Waves

We choose the z axis of a rectangular cavity resonator as a preferred axis. Then we can divide the waves, with respect to the z axis, in TE and TM waves:

- TE wave (=transverse electric wave). The electric field is transverse to z. The magnetic field has a z component as well as an x or a y component. Or it has z, x, and y components. (A TE wave is also called H wave or magnetic wave.)
- TM wave (=transverse magnetic wave). The magnetic field is transverse to z. The electric field has a z component and an x or a y component. Or it has z, x, and y components. (A TM wave is also called E wave or electric wave.)

To calculate the fields of different modes, we have to take into account that the boundary conditions for electromagnetic fields at the boundary of a perfect conductor are $E_t(\text{wall}) = 0$ and $H_n(\text{wall}) = 0$. It turns out that a mode of a rectangular cavity resonator is either a TE mode or a TM mode; i.e., if a mode is excited, the corresponding wave is a TE wave or a TM wave. (For calculations of amplitudes of electric and magnetic fields, *see* Problems).

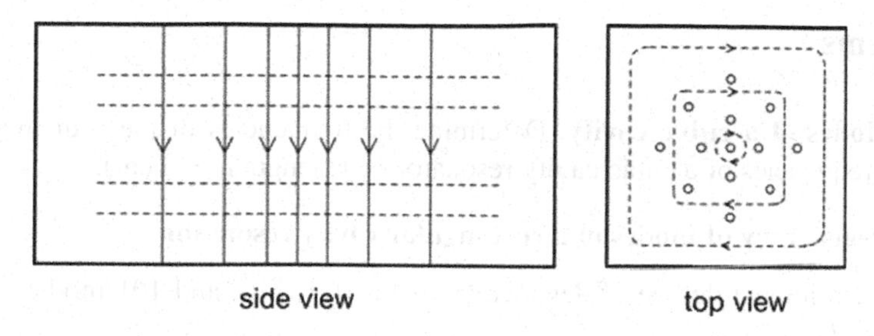

Fig. 10.7 Rectangular cavity resonator: electric field lines (*solid*) and magnetic field lines (*dashed*)

Example (Fig. 10.7). The magnetic field of the 101 mode can be calculated by use of (10.2) and (10.5). We obtain (with $A_y = E_0$):

$$E_y = E_0 \sin\frac{\pi x}{a_2}\sin\frac{\pi z}{L}\cos\omega t, \tag{10.53}$$

$$H_x = \sqrt{\frac{\varepsilon_0}{\mu_0}}\frac{\lambda}{2L}E_0\sin\frac{\pi x}{a_2}\cos\frac{\pi z}{L}\sin\omega t, \tag{10.54}$$

$$H_z = \sqrt{\frac{\varepsilon_0}{\mu_0}}\frac{\lambda}{a_2}E_0\cos\frac{\pi x}{a_2}\sin\frac{\pi z}{L}\sin\omega t, \tag{10.55}$$

where $\lambda = c/\nu = 2\pi c/\omega$ is the free-space wavelength; the parameter $\sqrt{\mu_0/\varepsilon_0} = 377\,\Omega$ is the impedance of free space. The 101 mode is a TE mode, which we can design as TE_{101} mode.

10.7 Quasioptical Arrangement

Microwave radiation generated by a microwave oscillator (Chap. 31) can be guided by means of waveguides. Emission of radiation into free space is possible by the use of an antenna. An antenna mediates the excitation of a wave in free space. A free-space wave excited by an antenna consists of a mixture of radiation belonging to different modes of free space. The strongest portion of radiation can belong to the fundamental (Gaussian) mode (next chapter). However, radiation belonging to other modes (corresponding to a power of radiation of the order of 1%) cannot be avoided. Vice versa, it is not possible to completely convert radiation belonging to a single mode of the free space to radiation of a single mode of a waveguide. The combination of microwave and optical techniques leads to quasioptical arrangements [63].

References [59–62].

Problems

10.1 Modes of a cubic cavity. Determine the frequencies of the four modes of lowest frequencies of a cubic cavity resonator (side length $a = 1\,\mathrm{cm}$).

10.2 Degeneracy of modes of a rectangular cavity resonator.

(a) Determine the degree of degeneracy of the 011, 110, and 101 modes if $a_1 = a_2 = L$.
(b) Determine the degree of degeneracy of the 011, 110, and 101 modes if $a_1 = a_2 \neq L$.
(c) Determine the degree of degeneracy of the 011, 110, and 101 modes if $a_1 \neq a_2 \neq L$.
(d) Determine the degree of degeneracy of the 111 mode.

10.3 Density of modes of a cavity resonator. Determine the density of modes of a cubic cavity resonator of 1 cm side length at a frequency corresponding to a vacuum wavelength $\lambda = 700\,\mathrm{nm}$ for the following cases.

(a) If the cube (with metallic walls) is empty,
(b) If the cube contains an Al_2O_3 crystal (refractive index n = 1.8) and fills the cavity completely,
(c) If the cube contains a GaAs crystal (n = 3.65) that completely fills out the cube.

10.4 Number of modes. Determine the number of modes of a cubic cavity (side length 1 cm) in the frequency interval $1 \times 10^{14}\,\mathrm{Hz}$, $1.1 \times 10^{14}\,\mathrm{Hz}$.

10.5 Mode density on different scales. Determine the relations between the mode density on the frequency scale and on different other scales:

(a) scale of photon energy $h\nu$; (b) ω scale; (c) scale of vacuum wavelength λ.

10.6 Variation of the resonance frequency of a mode. By changing the length L of a resonator, the resonance frequencies change. Determine the dependence of the frequency ν of the 101 mode on the change δL of the length L of a long resonator ($L \gg a_1$).

10.7 Density of modes in free space. Determine the density of modes of electromagnetic waves in free space. [*Hint*: make use of periodic boundary conditions.]

10.8 Energy of a field in a cavity resonator.

(a) Determine the energy of a field in the 101 mode of a rectangular cavity resonator (width a_1, height a_2, length L; field of amplitude A).
(b) Determine the energy content in the case that $a_1 = 1\,\mathrm{cm}, a_2 = 0.5\,\mathrm{cm}, L = 2\,\mathrm{cm}$, $A = 1\,\mathrm{V\,cm^{-1}}$.

10.9 Magnetic field in a rectangular cavity resonator.

(a) Derive the wave equations describing the H field.
(b) Solve the wave equations. [*Hint*: the normal component H_{n} of the magnetic field is zero everywhere on the walls, H_{n} (wall) $= 0$.]

10.10 TE$_{\mathrm{mnl}}$ modes of a rectangular cavity resonator.

(a) Determine the fields of a TE$_{\mathrm{mnl}}$ mode.
(b) Express the amplitudes of the field components by the amplitude of the z component of the H field. [*Hint*: take into account that $\boldsymbol{k} \times \boldsymbol{E} = 0$ and $\boldsymbol{k} \times \boldsymbol{H} = 0$.]

10.11 TM mode of a rectangular cavity resonator.

(a) Determine the fields of a TM$_{\mathrm{mnl}}$ mode.
(b) Express the amplitudes of the field components by the amplitude of the z component of the E field.

10.12 Field components of different modes of a rectangular cavity resonator.

(a) Determine the magnetic field components of the 101 mode. [Solutions are given in (10.54) and (10.55).]
(b) Determine the electric and magnetic field components of the 011 mode.
(c) Determine the electric and magnetic field components of the TE$_{111}$ mode and the TM$_{111}$ mode.
(d) Show that the 101, 011 and 110 modes exist only as TE modes.
(e) Show that the TE$_{\mathrm{mnl}}$ and TM$_{\mathrm{mnl}}$ modes are degenerate if none of the three numbers is zero.

10.13 Rectangular waveguide. If we omit in a rectangular resonator the two walls perpendicular to the z axis, we obtain a rectangular waveguide.

(a) Characterize the TE mode of lowest order.
(b) Characterize the TM mode of lowest order.

10.14 Show that the number of (short-wavelength) cavity modes in a frequency interval $\mathrm{d}\nu$ for a rectangular cavity is given by $(8\pi/\lambda^3)\ V_c\ \mathrm{d}\nu/\nu$, where V_c is the cavity volume and λ the free-space wavelength. Show that the number of modes in a spherical cavity is given by the same expression.

Chapter 11
Gaussian Waves and Open Resonators

A large number of gas and solid state lasers as well as free-electron lasers make use of an open resonator.

Before discussing open resonators, we introduce the Gaussian wave (=Gaussian beam). It is a kind of a natural mode of electromagnetic radiation in free space. A Gaussian wave is a paraxial wave, that is a wave with a well-defined propagation direction along the beam axis (z axis) and a small divergence. The amplitude of the field perpendicular to the beam axis has a Gaussian distribution. A Gaussian beam traveling from $z = -\infty$ to $z = \infty$ has a beam waist. Accordingly, the diameter of the beam shows a minimum at the beam waist.

A Gaussian wave is a solution of the wave equation—which we use in the form of the Helmholtz equation—and an appropriate boundary condition: the energy transported by the wave through a plane perpendicular to the propagation direction is finite. Besides the Gaussian mode (=fundamental Gaussian mode), the wave equation provides higher order Gaussian modes.

A Gaussian wave fits to a resonator with spherical mirrors—a longitudinal mode of an open resonator is a standing wave composed of two Gaussian waves propagating in opposite directions. Higher order Gaussian modes lead to transverse modes of a resonator. A laser with a spherical-mirror resonator is able to generate a Gaussian wave.

The analysis of resonators having mirrors of various curvature shows that there are stable and unstable resonators. The confocal, the concentric, and the plane parallel resonator are three special types of resonators.

We describe the effect of diffraction that can be used to suppress laser oscillation on transverse modes and to operate a laser on longitudinal modes only.

We introduce the ray matrix (ABCD matrix) to describe the propagation of paraxial optical rays in free space and in optical systems. We show that a Gaussian beam can be focused by a lens to an area of a diameter that is equal to about a wavelength of the radiation.

K.F. Renk, *Basics of Laser Physics*, Graduate Texts in Physics,
DOI 10.1007/978-3-319-50651-7_11

The wavelength of a monochromatic Gaussian wave is a constant far outside the beam waist but shows a (small) variation in the range of the waist. As a consequence, the resonance frequencies of a resonator with spherical mirrors are not multiples of a minimum frequency but are shifted toward higher frequencies. The change of wavelength in a beam waist corresponds to a change of phase that has been predicted and experimentally demonstrated by L. G. Gouy in 1891 and experimentally demonstrated also recently by the use of femtosecond pulses. The Gouy phase shift influences the frequency spectrum of optical frequency combs (Sect. 13.4).

We begin this chapter with a characterization of laser radiation generated by the use of a resonator with spherical mirrors.

11.1 Open Resonator

Figure 11.1a shows a design of a laser (e.g., of a titanium-sapphire laser). The laser resonator consists of spherical mirrors (diameter 1 cm) at a distance of 1 m. The active medium has a diameter of 1 cm. The diameter of the laser wave is about 1 mm. The spherical mirrors have the extraordinary property to concentrate the radiation within the resonator at the resonator axis. The radiation circulates within the resonator. A portion of radiation, coupled out via the partial reflector, has a small beam divergence

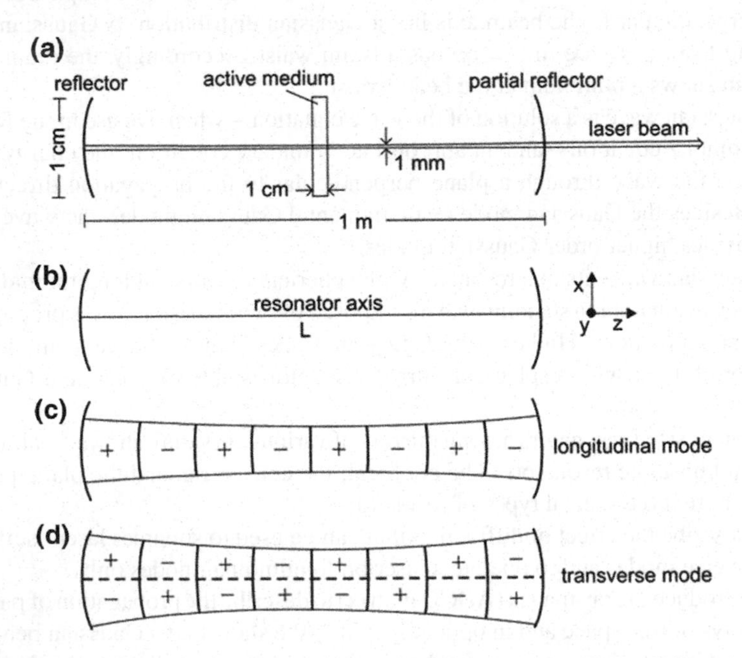

Fig. 11.1 Laser with a spherical-mirror and modes of the resonator. **a** Laser. **b** Open resonator. **c** Longitudinal mode. **d** Transverse mode

(e.g., 1 mrad). Diffraction of the wave at the reflector and the partial mirror has (in the laser design shown in the figure) almost no effect on the wave—except that diffraction plays an important role with respect to elimination of radiation belonging to unwanted modes (Sect. 12.2).

We give here a short characterization of a spherical-mirror resonator: it is an open resonator—it has no sidewalls (Fig. 11.1b). The length L of the resonator is much larger than the wavelength of the radiation. We characterize the modes of the resonator by the use of a cartesian coordinate system; we choose the direction of the resonator axis as z axis. We now blow up the lateral extension of the wave in order to visualize different modes. The simplest type is a *longitudinal mode* (Fig. 11.1c). The phase of the field varies along the resonator axis. The field amplitude has the largest value on the resonator axis and decreases in directions perpendicular to the resonator axis. Figure 11.1d shows a *transverse mode*: the phase of the field varies in axial direction (as for a longitudinal mode); however, the amplitude of the field is zero at the resonator axis and changes the sign in x direction. There are many other types of transverse modes as we will see.

We will show that longitudinal and transverse modes of a resonator with spherical mirrors correspond to standing waves in accord with the wave equation and with appropriate boundary conditions. A standing wave in a spherical-mirror resonator consists of two Gaussian waves propagating in $+z$ and $-z$ direction. The waves are Gaussian waves (=Gaussian beams). Before treating resonators, we will introduce Gaussian waves as solutions of the wave equation describing electromagnetic waves in free space (*see* the next two sections). A Gaussian wave is a paraxial wave: it has a well-defined propagation direction and a small beam divergence. We will see that Gaussian waves in free space can also be divided into longitudinal and transverse modes. A Gaussian (or higher-order Gaussian) mode of the free space is characterized by the propagation direction (z direction) and a number pair mn, where m is the number of changes of the sign of the amplitude in x direction and n is the number of changes in y direction.

The electric field of a Gaussian wave in free space is transverse or nearly transverse to the z direction. The direction of the magnetic field (that has always a direction perpendicular to the electric field) is also transverse or nearly transverse to the z direction. A Gaussian wave characterized as

$$\mathrm{TEM}_{\mathrm{mn}} \text{ wave}$$

means that the electric and magnetic fields of the wave are transverse or nearly transverse to the z direction (Problem 11.6). A $\mathrm{TEM}_{\mathrm{mn}}$ wave can be a longitudinal or a transverse mode.

- *Longitudinal mode* = 00 mode = axial mode = Gaussian mode = fundamental Gaussian mode = lowest-order Gaussian mode—the longitudinal mode appears under different names which will become clear during this chapter. The phase of the field in a longitudinal mode varies in the axial (=longitudinal) direction. The sign of the amplitude does not change in the directions perpendicular to the resonator axis.

- *Transverse mode* (=higher-order Gaussian mode). The phase of a transverse mode varies in the axial direction and the sign of the amplitude varies in one or two directions perpendicular to the resonator axis. We will introduce the Hermite–Gaussian modes.

Paraxial electromagnetic waves are transverse electromagnetic (TEM) waves (=transversely polarized electromagnetic waves), whether they belong to longitudinal or transverse modes. The active medium of a laser resonator is able to excite a standing Gaussian mode in the resonator and—if one of the spherical mirrors is a partial mirror—also a Gaussian wave propagating in free space.

A Gaussian mode or a higher-order Gaussian mode within a resonator is a

$$\mathrm{TEM}_{mnl} \text{ mode.}$$

The index l indicates the number of half wavelengths of the field in a resonator. The electric and magnetic fields of a Gaussian wave in a resonator are transverse or nearly transverse to the z direction. Each number triple corresponds to a mode of the electromagnetic field, i.e., to a particular pattern of the field in a resonator.

A polarized electromagnetic wave in a mode mn of free space or on a mode mnl of a resonator can be polarized in one of two directions perpendicular to each other (and perpendicular to the propagation direction). The direction of the polarization of laser radiation can be chosen by inserting a polarizer or other elements (for instance, a Brewster window) into the laser resonator.

The characterization of modes as longitudinal or transverse modes is of practical interest: most lasers generate radiation belonging mainly to longitudinal modes.

11.2 Helmholtz Equation

We make use of a simple wave optics, first described by Helmholtz. We start with the equation

$$\nabla^2 \boldsymbol{E} - \frac{1}{c^2}\frac{\partial^2 \boldsymbol{E}}{\partial t^2} = 0, \tag{11.1}$$

which represents three wave equations, one for each of the three components of the field vector $\boldsymbol{E}$. Ignoring the polarization of the electric field, we can reduce the three wave equations to one equation,

$$\nabla^2 E - \frac{1}{c^2}\frac{\partial^2 E}{\partial t^2} = 0. \tag{11.2}$$

E is the field treated as a scalar quantity. We consider a monochromatic wave

$$E(x, y, z, t) = \psi(x, y, z)\, \mathrm{e}^{i\omega t}. \tag{11.3}$$

The time independent part of the field, $\psi(x, y, z)$, obeys the *Helmholtz equation*

$$\nabla^2\psi + k^2\psi = 0, \quad (11.4)$$

where $k = \omega/c$. The energy density of the field is $u(x, y, z) = \frac{1}{2}\varepsilon_0|\psi(x, y, z)|^2$. Among the many solutions of the Helmholtz equation are two simple cases.

Plane wave. The solution is

$$E = A\,\mathrm{e}^{\mathrm{i}(\omega t - kz)}, \quad (11.5)$$

where

$$k = \omega/c \quad (11.6)$$

is the wave vector and c the speed of light. The wave vector is independent of x, y, and z. The phase of the wave assumes constant values,

$$\varphi(t, z) = \omega t - kz = \text{const.} \quad (11.7)$$

The condition $\partial\varphi/\partial t = 0$ yields the phase velocity $\mathrm{v_{ph}} = \mathrm{d}z/\mathrm{d}t = \omega/k = c$. The condition $\partial\varphi/\partial z = 0$ yields the group velocity $\mathrm{v_g} = \mathrm{d}z/\mathrm{d}t = c$. Group and phase velocities are equal to the speed of light. The wavelength $\lambda = 2\pi/k$, i.e., the spatial period, is independent of x, y, and z. The amplitude A of a plane wave is the same everywhere in space. The phase kz is a constant in planes of fixed z.

The wave has no angular spread. We can decompose the phase,

$$\varphi(t, z) = \varphi_t(t) - \varphi_z(z), \quad (11.8)$$

where $\varphi_t(t)$ is the time-dependent portion of the phase and $\varphi_z(z)$ is the position-dependent portion. The temporal change of $\varphi_t(t)$ is the angular frequency,

$$\mathrm{d}\varphi_t/\mathrm{d}t = \omega, \quad (11.9)$$

and spatial change of $\varphi_z(z)$ is the wave vector ($= 2\pi \times$ saptial frequency),

$$\mathrm{d}\varphi_z/\mathrm{d}z = k. \quad (11.10)$$

Spherical wave. A spherical wave has the form

$$E = \frac{K}{s}\,\mathrm{e}^{\mathrm{i}(\omega t - ks)}. \quad (11.11)$$

K is a measure of the strength of a wave. Inserting (11.11) in (11.4) yields $k = \omega/c$. The amplitude decreases inversely proportional to the distance s from a point source. The phase ks is a constant on spheres around the origin s=0. The phase and group velocities are equal to the speed of light. The direction of the phase and of the group velocity is radial away from the source point $s=0$. The wavelength $\lambda=2\pi/k$ is independent of x, y, and z.

The plane wave and the spherical wave cannot be realized experimentally. We will now look for *paraxial waves*. These have a well-defined propagation direction (along z) and a small angular spread. We describe the waves by the ansatz

$$\psi = f(x, y, z)\, e^{-ikz}. \tag{11.12}$$

We suppose that f changes only weakly with z. We can therefore neglect the second derivative of f with respect to z and obtain the Helmholtz equation of paraxial waves,

$$\frac{\partial^2 f}{\partial x^2} + \frac{\partial^2 f}{\partial y^2} - 2ik\frac{\partial f}{\partial z} = 0. \tag{11.13}$$

Gaussian waves and waves in optical resonators are described in many textbooks; *see* REFERENCES at the end of the chapter. Studies of mode patterns began in 1961 [67–69]. We will study various aspects of Gaussian waves. We will begin with the discussion of a solution of the Helmholtz equation, following [40].

11.3 Gaussian Wave

A Gaussian wave (=Gaussian beam) is a paraxial wave. We solve the Helmholtz equation of paraxial waves by use of the ansatz

$$f(x, y, z) = G(z)\, e^{-(x^2+y^2)/F(z)}. \tag{11.14}$$

G and F are complex functions that change only weakly with z.

Differentiation yields

$$\frac{d^2 f}{dx^2} = \left(-\frac{2G}{F} + 4x^2\frac{G}{F^2}\right) e^{-(x^2+y^2)/F}, \tag{11.15}$$

$$\frac{df}{dz} = \left(\frac{dG}{dz} + \frac{x^2+y^2}{F^2}G\frac{dF}{dz}\right) e^{-(x^2+y^2)/F}. \tag{11.16}$$

The Helmholtz equation leads to an equation

$$-\frac{2}{F(z)} - ik\frac{1}{G(z)}\frac{dG(z)}{dz} + \frac{x^2+y^2}{F^2(z)}\left(2 - ik\frac{dF}{dz}\right) = 0, \tag{11.17}$$

which includes two conditions,

$$2 - \mathrm{i}k\frac{\mathrm{d}F}{\mathrm{d}z} = 0, \tag{11.18}$$

$$-\frac{2}{\mathrm{i}kF} - \frac{1}{G}\frac{\mathrm{d}G}{\mathrm{d}z} = 0. \tag{11.19}$$

Integrating (11.18) yields

$$F(z) = \frac{2}{\mathrm{i}k}(z + C_1). \tag{11.20}$$

The integration constant C_1 is a complex quantity. We suppose that the wave front at $z = z_0$ is a plane, i.e., that the phase of $f(x, y, z)$ is independent of x and y. Then $F(z_0)$ is real. By writing $F(z_0)=w_0^2$, we find

$$C_1 = \frac{\mathrm{i}k}{2}w_0^2 - z_0 \tag{11.21}$$

and

$$F = w_0^2 + \frac{2}{\mathrm{i}k}(z - z_0). \tag{11.22}$$

We separate $1/F$ in real and imaginary part,

$$\frac{1}{F} = \frac{k^2w_0^2 + 2\mathrm{i}k(z - z_0)}{k^2w_0^4 + 4(z - z_0)^2} = \frac{1}{w^2} + \frac{\mathrm{i}k}{2R}, \tag{11.23}$$

where

$$w = w_0\sqrt{1 + \frac{4(z - z_0)^2}{k^2w_0^4}}, \tag{11.24}$$

$$R = z - z_0 + \frac{k^2w_0^4}{4(z - z_0)}. \tag{11.25}$$

We obtain, with $r^2 = x^2 + y^2$, the solution

$$f(z, r) = G(z)\,\mathrm{e}^{-r^2/w^2(z)}\,\mathrm{e}^{-\mathrm{i}kr^2/2R(z)}. \tag{11.26}$$

We will see that $w = w(z)$ is the beam radius and $R = R(z)$ is the radius of curvature of the beam at the location z. The beam radius has the smallest value for $z = z_0$, i.e., the beam has a waist at $z = z_0$, where the radius of the beam is equal to w_0. That R is the radius of curvature follows from the relation (Fig. 11.2):

$$\frac{2(z - z_0)r^2}{k^2w_0^4 + 4(z - z_0)^2} = \frac{r^2}{2R}. \tag{11.27}$$

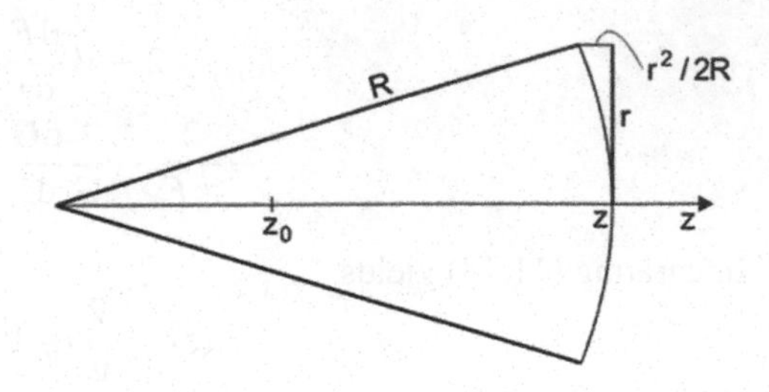

Fig. 11.2 Curvature of the wave front of a Gaussian wave

Making use of (11.22) and (11.19) we obtain the differential equation

$$\frac{1}{G}\frac{\mathrm{d}G}{\mathrm{d}z} = -\frac{1}{z - z_0 + \mathrm{i}kw_0^2/2}. \tag{11.28}$$

We write

$$G(z) = K(z)\,\mathrm{e}^{\mathrm{i}\phi(z)}. \tag{11.29}$$

$K{=}|G|$ is the absolute value of G and ϕ is a phase, the *Gouy phase*. We obtain the differential equation

$$\frac{1}{G}\frac{\mathrm{d}G}{\mathrm{d}z} = \frac{1}{K}\frac{\mathrm{d}K}{\mathrm{d}z} + \mathrm{i}\frac{\mathrm{d}\phi}{\mathrm{d}z} = -\frac{z - z_0}{(z - z_0)^2 + k^2w_0^4/4} + \frac{\mathrm{i}kw_0^2/2}{(z - z_0)^2 + k^2w_0^4/4}. \tag{11.30}$$

Separation of real and imaginary part provides two differential equations,

$$\frac{1}{K}\frac{\mathrm{d}K}{\mathrm{d}z} = \frac{z - z_0}{(z - z_0)^2 + k^2w_0^4/4}, \tag{11.31}$$

$$\frac{\mathrm{d}\phi}{\mathrm{d}z} = \frac{kw_0^2/2}{(z - z_0)^2 + k^2w_0^4/4}. \tag{11.32}$$

The solutions are

$$K = \frac{2C_2}{kw_0^2}, \tag{11.33}$$

$$\phi(z) = \tan^{-1}\frac{2(z - z_0)}{kw_0^2} \quad \text{or} \quad \tan\phi(z) = \frac{2(z - z_0)}{kw_0^2}. \tag{11.34}$$

C_2 is an integration constant, which is real. (Instead of the notation tan^{-1}, the notation $arctg$ can be used). It follows that the field is

$$\psi(z, r) = \frac{C_3}{w(z)}\,\mathrm{e}^{-r^2/w^2(z)}\,\mathrm{e}^{-\mathrm{i}[kz - \phi(z) + kr^2/2R(z)]}. \tag{11.35}$$

$C_3 = 2C_2/kw_0$ is a constant. The phase shows a change according to propagation and, additionally, due to the Gouy phase shift $\phi(z)$. The amplitude of the field decreases in propagation direction inversely proportional to the beam radius. The field amplitude at a fixed z decreases from its value on the axis ($r = 0$) to 1/e at the beam radius $w(z)$. The solution contains two integration constants, w_0 (contained in the expression of w) and C_3. The values of w_0 and C_3 of a particular Gaussian wave can be determined experimentally—for instance by determination of the beam diameter of the intensity distribution at a fixed location z (e.g., at z_0) and determination of the power of the wave. In the *beam waist*, i.e., at the location of minimum beam diameter, the field distribution is equal to

$$\psi(z_0, r) = \frac{C_3}{w_0}\, \mathrm{e}^{-r^2/w_0^2}\, \mathrm{e}^{-\mathrm{i}[kz_0 - \phi(z_0)]}. \tag{11.36}$$

We can write

$$G(z) = \frac{C_2}{z + C_1} = \frac{C_2}{\mathrm{i}kF(z)/2} \tag{11.37}$$

and therefore

$$\psi(z, r) = \frac{2C_2}{\mathrm{i}kF(z)}\, \mathrm{e}^{-r^2/w^2(z)}\, \mathrm{e}^{-\mathrm{i}[kz - \phi(z) + kr^2/2R(z)]}, \tag{11.38}$$

where

$$\frac{1}{F(z)} = \frac{1}{w^2(z)} + \frac{ik}{2R(z)}. \tag{11.39}$$

The propagation of a Gaussian wave is completely described by the beam radius $w(z)$ and the radius of curvature $R(z)$ or, alternatively, by the complex beam parameter $F(z)$. It is convenient to introduce another complex beam parameter

$$\tilde{q}(z) = \frac{ik}{2} F(z). \tag{11.40}$$

(We omit in this section the tilde sign of complex quantities, except of the beam parameter $\tilde{q}$). It follows that

$$\frac{1}{\tilde{q}(z)} = \frac{1}{R(z)} - \frac{2i}{kw^2(z)} = \frac{1}{R(z)} - \frac{i\lambda}{\pi w^2(z)}. \tag{11.41}$$

We can write the field in the form

$$\psi(z, r) = \frac{2C_2}{\mathrm{i}kF} \exp\left(-ikz - \frac{r^2}{F}\right) = \frac{C_2}{\tilde{q}} \exp\left(-ikz - \frac{r^2}{2\tilde{q}}\right). \tag{11.42}$$

We now discuss the solution in more detail. The field is equal to

$$E(z, r) = \frac{C_3}{w(z)} \mathrm{e}^{-r^2/w^2(z)} \mathrm{e}^{\mathrm{i}(\omega t - [kz - \phi(z) + kr^2/2R])}. \tag{11.43}$$

The spatially dependent part is

$$\psi(z, r) = \frac{C_3}{w(z)} \mathrm{e}^{-r^2/w^2(z)} \mathrm{e}^{-\mathrm{i}[kz - \phi(z) + kr^2/2R]}. \tag{11.44}$$

We write

$$\psi(z, r) = A(z, r) \, \mathrm{e}^{-i\varphi(z,r)}. \tag{11.45}$$

The amplitude of the Gaussian wave is

$$A(z, r) = \frac{C_3}{w(z)} \mathrm{e}^{-r^2/w^2(z)} \tag{11.46}$$

and the phase is

$$\varphi(z, r) = kz - \phi(z) + \frac{r}{2R(z)} kr. \tag{11.47}$$

The expression contains two terms that depend on z only and another term that depends additionally on r; this term vanishes on the beam axis. The field distribution has the following properties (Fig. 11.3 and Table 11.1):

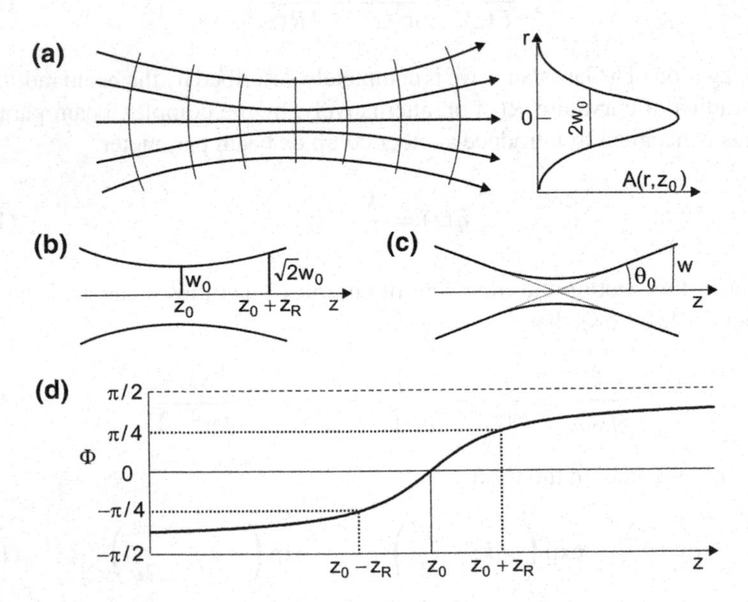

Fig. 11.3 Gaussian wave. **a** Rays and lateral distribution of the amplitude. **b** Rayleigh range. **c** Divergence. **d** Gouy phase

Table 11.1 Properties of field and energy density of a Gaussian beam propagating from $z=-\infty$ to $z=+\infty$

	$z=-\infty$	$z=z_0$	$z=(z_0+z_R)$	$z\to\infty$
Radius (field)		w_0	$\sqrt{2}\,w_0$	$w_0(z-z_0)/z_R$
$A(r=0)$		A_0	$A_0/\sqrt{2}$	$A_0 z_R/(z-z_0)$
ϕ	$-\pi/2$	0	$\pi/4$	$\pi/2$
Wave front	Spherical	Plane	Curved	Spherical
Radius (energy)		$r_{u,0}=w_0/\sqrt{2}$	$\sqrt{2}r_{u,0}$	$r_{u,0}(z-z_0)/z_R$
$u(r=0)$	$u_0 z_0^2/(z-z_0)^2$	u_0	$u_0/2$	$u_0 z_0^2/(z-z_0)^2$

- $\psi(r,z)$ is circularly symmetric around the beam axis.
- The amplitude of the wave decreases laterally according to the Gaussian function and is, on the beam axis, inversely proportional to the beam radius $w(z)$.
- The wave has a waist. The beam radius of the waist is w_0. In the waist, the wave front is a plane and the field amplitude distribution is given by

$$A(z_0,r)=\frac{C_3}{w_0}\,e^{-r^2/w_0^2}=A_0\,e^{-r^2/w_0^2}. \tag{11.48}$$

  A_0 is the amplitude of the wave on the beam axis (r=0) at z=z_0.
- If $z\neq z_0$, the wave front is curved and the beam radius increases with increasing $|z-z_0|$.
- For large $|z-z_0|$, namely for $|z-z_0|\gg kw_0^2/2$, the curvature is $R(z)$=$z-z_0$ and the beam radius is $w(z)$=$2kw_0(z-z_0)$. Both the curvature and the beam radius increase linearly with $|z-z_0|$.
- The *Rayleigh range* is equal to

$$z_R=kw_0^2/2=\pi w_0^2/\lambda. \tag{11.49}$$

The beam diameter increases in the range z_0, z_0+z_R by the factor $\sqrt{2}$. In the range z_0-z_R, z_0+z_R the Gaussian wave remains almost parallel. This range is the near-field (or Fresnel) range. The range $|z-z_0|>z_R$ is the far-field (Fraunhofer) range. We can express the three parameters w, R, and ϕ of a Gaussian wave by the beam waist w_0 and the Rayleigh range z_R,

$$w(z)=w_0\sqrt{1+\frac{(z-z_0)^2}{z_R^2}}, \tag{11.50}$$

$$R(z)=z-z_0+\frac{z_R^2}{z-z_0}, \tag{11.51}$$

$$\phi(z)=\tan^{-1}\frac{z-z_0}{z_R}. \tag{11.52}$$

The distance $2z_R$ between the points $z_0 - z_R$ and $z_0 + z_R$ is the *confocal parameter* or *depth of focus*.

- *Gouy phase.* The Gouy phase—inherent to a Gaussian wave—describes a phase that is associated with the spatial and the temporal change of the curvature of the wave front (Sect. 11.7).
- *Change of phase.* When a wave front with the field distribution $\psi(z_1, r)$ propagates from z_1 to z_2, the phase φ changes according to

$$\varphi(z_2) - \varphi(z_1) = [kz_2 - \phi(z_2)] - [kz_1 - \phi(z_1)]. \tag{11.53}$$

$\phi(z_2) - \phi(z_1)$ is the Gouy phase shift. When the wave front propagates through the beam waist from a far-field location $z_1 \ll z_0$ to a far-field location $z_2 \gg z_0$, the phase φ changes by

$$\varphi(z_2) - \varphi(z_1) = kz_2 - kz_1 - \pi. \tag{11.54}$$

The propagation through the beam waist changes the phase of the wave by $-\pi$, in addition to the geometrical phase change $kz_2 - kz_1$. When the wave front travels from z_1=$-z_R$ to z_2=z_R, the change of phase is (for $z_0 = 0$) equal to

$$\varphi(z_R) - \varphi(-z_R) = kz_R - \pi/2. \tag{11.55}$$

- The field of a Gaussian wave is given by

$$E(z, r, t) = A_0 \frac{w_0}{w} \, e^{-r^2/w^2(z)} \cos\left[\omega t - (kz - \phi + kr^2/2R)\right]. \tag{11.56}$$

A_0=C_3/w_0 is the amplitude in the center of the beam waist (at $z = 0$ and $r = 0$).

- The energy density, averaged over a temporal period, is

$$u(z, r) = \frac{1}{2}\varepsilon_0 A_0^2 \, \frac{r_{u,0}^2}{r_u^2} \, e^{-r^2/r_u^2}, \tag{11.57}$$

where r_u=$w/\sqrt{2}$ is the beam radius with respect to the energy density distribution. The radius of the energy distribution at the beam waist is $r_{u,0}$=$w_0/\sqrt{2}$. At the beam waist, the energy density decreases within the radius $r_{u,0}$ to 1/e relative to the energy density on the beam axis.

- *Divergence.* At large $|z - z_0|$, the angle of divergence of the field is given by

$$\theta_0 = \frac{w_0}{z_R} = \frac{\lambda}{\pi w_0}. \tag{11.58}$$

The product of the far-field aperture angle θ_0 and the diameter at the beam waist is a constant,

$$2w_0 \times \theta_0 = 2\lambda/\pi. \tag{11.59}$$

The angle of divergence, with respect to the energy density, is $\theta_{u,0} = \theta_0/\sqrt{2}$. The product

$$2r_{u,0}\theta_{u,0} = \lambda/\pi \tag{11.60}$$

is by a factor of two smaller than the product with respect to the field amplitude.

- *Radiance of a Gaussian wave.* The radiance of a paraxial beam is defined as the power of radiation passing through an area (oriented perpendicular to the beam direction) divided by the area and the solid angle of the beam. To estimate the radiance of a Gaussian beam, we approximate $\exp(-r^2/r_u^2)$ by a rectangular radial distribution of diameter $2r_u$ and find that the power of radiation passing through the beam waist is approximately given by

$$P = L_u \times \text{area} \times \Omega, \tag{11.61}$$

which is the product of the area of the beam in the beam waist, the solid angle Ω of the beam, and the radiance (=brightness) L_u. We can write:

$$L_u = \frac{P}{\text{area} \times \Omega}. \tag{11.62}$$

For small values of $\theta_{u,0}$, the solid angle of the beam is $\Omega = \pi\theta_{u,0}^2$. This leads to

$$L_u = \frac{P}{\pi r_{u,0}^2 \times \pi\theta_{u,0}^2}. \tag{11.63}$$

Taking into account the relation (11.60), we find that the radiance of a Gaussian beam is equal to the power divided by $(\lambda/2)^2$.
Example P=1 W; λ=0.5 μm; L_u=1.6 × 10^{13} W m^{-2} sr^{-1}.

Is it possible to realize a Gaussian wave? The answer is: a laser with appropriately arranged spherical mirrors as resonator mirrors is able to produce a Gaussian wave. We will begin the discussion of spherical-mirror resonators by treating a particular spherical-mirror resonator, the symmetric confocal resonator.

11.4 Confocal Resonator

A confocal resonator consists of two spherical mirrors, which have the same focus. We discuss the symmetric confocal resonator. It has two equal mirrors.

Two spherical mirrors arranged at a distance L=R form a (symmetric) *confocal resonator* (Fig. 11.4, left). We choose z=0 as the location of the center of one of the mirrors. We will show that a Gaussian wave can fit to a confocal resonator. The symmetry of the arrangement requires that the beam waist lies at $z_0 = L/2$. We choose $z_R = L/2$ and obtain, by using (11.49) and (11.50), the radius of curvature

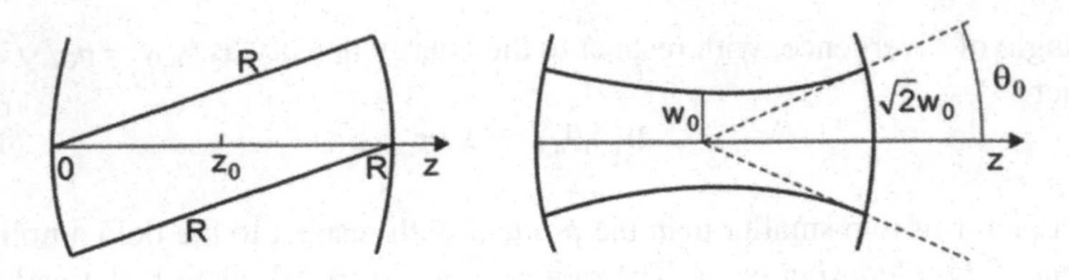

Fig. 11.4 Confocal resonator

$$R = \frac{L}{2}\left(1+\frac{k^2 w_0^4}{L^2}\right) \tag{11.64}$$

and the beam radius (=mode radius) with respect to the field amplitude in the waist

$$w_0 = \sqrt{\frac{L\lambda}{2\pi}}. \tag{11.65}$$

With respect to the energy density, the mode radius in the beam waist is $r_{\mathrm{u},0} = \sqrt{L\lambda/4\pi}$.

The Gaussian mode within the resonator (Fig. 11.4, right) has the beam waist w_0. The beam radius on each of the mirrors is $w_0\sqrt{2}$. The distance between the center of the resonator and a mirror is equal to the Rayleigh range. In the far-field range, outside the resonator—with one of the spherical mirrors being a partial reflector—the field has the divergence angle θ_0=$2\lambda/(\pi w_0)$.

The field in the resonator corresponds to a Gaussian standing wave, i.e., to two Gaussian waves propagating in opposite directions,

$$E(z,r) = \frac{1}{2}A(z,r)\,\mathrm{e}^{\mathrm{i}[\omega t-\varphi(z,r)+\varphi_0]} + \frac{1}{2}A(z,r)\,\mathrm{e}^{\mathrm{i}[\omega t+\varphi(z,r)-\varphi_0]}. \tag{11.66}$$

The field at the axis is equal to

$$E = A_0\,\frac{w_0}{w}\,\mathrm{e}^{-r^2/w^2}\,\cos[kz-\phi(z)-\varphi_0]\cos\omega t. \tag{11.67}$$

We obtain the resonance frequencies by the use of the resonance condition (2.80), namely that the change of the phase of a field propagating in the resonator is, at a round trip transit, a multiple of 2π (that is the resonator eigenvalue problem). The Gouy phase shift per round trip transit is

$$\Delta\phi = \phi(z_2)-\phi(z_1)+\phi(z_2)-\phi(z_1) = 2\,(\phi(z_2)-\phi(z_1)) = \pi \tag{11.68}$$

This leads to the condition

$$2kL - \Delta\phi = 2kL - \pi = l \times 2\pi \tag{11.69}$$

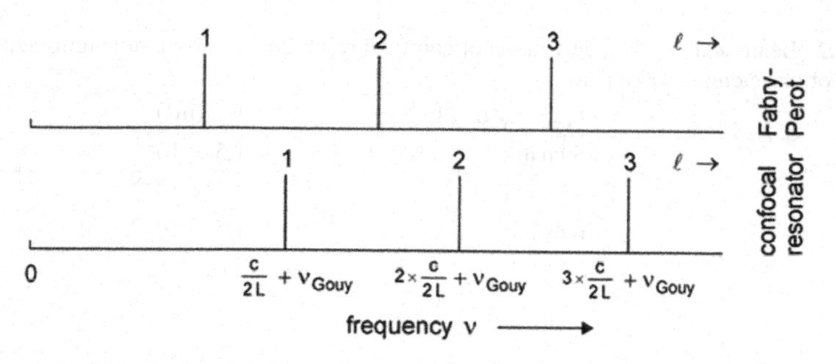

Fig. 11.5 Resonance frequencies

or

$$k_l L = \left(l + \frac{1}{2}\right)\pi; \ l = 1, 2, ... \tag{11.70}$$

The resonance frequencies are (Fig. 11.5)

$$\nu_l = \frac{c}{2L}\left(l + \frac{1}{2}\right) = l \times \frac{c}{2L} + \nu_{\mathrm{Gouy}}, \tag{11.71}$$

where $\nu_{\mathrm{Gouy}} = c/4L$ is the Gouy frequency shift for the symmetric confocal resonator. In comparison with a Fabry–Perot interferometer of the same length, the resonance frequencies are shifted towards higher frequencies. However, the frequency separation between adjacent modes is the same,

$$\nu_{l+1} - \nu_l = \frac{c}{2L}. \tag{11.72}$$

The phase φ_0 is determined by the choice of the origin (z=0) of the z axis. We have chosen the position of one of the reflectors as z = 0. The boundary conditions, namely that the field on the mirror (assumed to have a reflectivity near 1), has to be zero, requires that

$$\varphi_0 + \phi(0) = \varphi_0 - \pi/4 = 3\pi/4 \tag{11.73}$$

and therefore $\varphi_0 = \pi/2$. Thus, we obtain the field at the axis:

$$\psi = A_0 \frac{w_0}{w} \mathrm{e}^{-r^2/w^2} \sin[kz - \phi(z)] \cos \omega t. \tag{11.74}$$

The energy density, averaged over both a temporal period and a spatial period, is

$$u = \frac{\varepsilon_0}{4} A_0^2 \frac{r_{\mathrm{u},0}^2}{r_{\mathrm{u}}^2} \mathrm{e}^{-r^2/r_{\mathrm{u}}^2} = u_0 \frac{r_{\mathrm{u},0}^2}{r_{\mathrm{u}}^2} \mathrm{e}^{-r^2/r_{\mathrm{u}}^2}, \tag{11.75}$$

Table 11.2 Beam waist and mode volume of confocal resonators of different lengths suitable for radiation of wavelength λ=0.6 μm

L(m)	$r_{\mathrm{u},0} = \sqrt{L\lambda/4\pi}$	V_{00} (m^3)
0.1	69 μm	1.5×10^{-8}
0.5	155 μm	3.5×10^{-8}
10	0.69 mm	1.5×10^{-5}

where $u_0 = (\varepsilon_0/4)A_0^2$ is the energy density at r=0 at the beam waist (z=z_0).

The energy contained in a mode is given by

$$\int_0^L \mathrm{d}z \int_0^\infty u(r) \times 2\pi r \mathrm{d}r = u_0 \int_0^L \mathrm{d}z \frac{r_{\mathrm{u},0}^2}{r_{\mathrm{u}}^2} \int_0^\infty 2\pi r \mathrm{d}r e^{-r^2/r_{\mathrm{u}}^2} = u_0 \pi r_{\mathrm{u},0}^2 L. \tag{11.76}$$

What is the volume V_0 of a (hypothetical) mode, which contains the same radiation energy as the 00 mode, but with a constant energy density? We can write $u_0 V_{00}$=$u_0 \pi r_{\mathrm{u},0}^2 L$ and interpret V_{00} as the *mode volume*. Thus, the mode volume of the 00 mode of a confocal resonator is equal to

$$V_{00} = \pi r_{\mathrm{u},0}^2 L. \tag{11.77}$$

The mode volume of the 00 mode of a confocal resonator is equal to the product of the cross sectional area of the beam waist (with respect to the energy distribution) and the length of the resonator. Table 11.2 shows values of beam waists and mode volumes for radiation of a fixed wavelength.

The confocal resonator is suitable as resonator of, for instance, a helium–neon laser or a free-electron laser.

- *Helium–neon laser* (λ=633 nm). The small gain of the active medium requires a large length (typically 0.5 m) of the resonator. The mechanism of the relaxation of the excited neon atoms makes it necessary to use a tube with a small diameter; the neon atoms relax by collisions with the walls (Sect. 14.3).
- *Free-electron laser*. A free-electron laser requires a resonator of a length of typically 10 m. The Gaussian wave in a confocal resonator has a large overlap with an electron wave that propagates along the axis of the resonator (Sect. 19.1).

Other resonator configurations will be discussed in the next section.

11.5 Stability of a Field in a Resonator

There are a large number of different resonators. However, not all have stable modes. We treat a resonator with two spherical mirrors (Fig. 11.6) that have different radii (R_1 and R_2) of curvature. A Gaussian mode, fitting to the resonator, can be determined

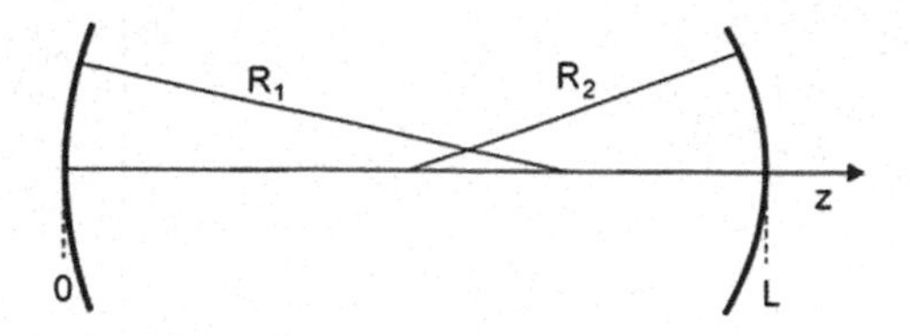

Fig. 11.6 Resonators with spherical mirrors

by means of the conditions for the radii of curvature:

$$R_1 = z_0\left(1 + \frac{z_R^2}{z_0^2}\right) = -R_1(0), \tag{11.78}$$

$$R_2 = (L - z_0)\left(1 + \frac{z_R^2}{(L - z_0)^2}\right), \tag{11.79}$$

where z_0 is the location of the beam waist; z=0 and z=L are the positions of the two mirrors. From the two relations, we find

$$z_0 = \frac{(1 - g_1)g_2 L}{g_1 + g_2 - 2g_1g_2}, \tag{11.80}$$

$$z_R^2 = \frac{(1 - g_1g_2)g_1g_2L^2}{(g_1 + g_2 - 2g_1g_2)^2}, \tag{11.81}$$

$$w_0^2 = \frac{L\lambda}{\pi}\frac{\sqrt{(1 - g_1g_2)g_1g_2}}{g_1 + g_2 - 2g_1g_2}, \tag{11.82}$$

$$w_i^2 = \frac{L\lambda}{\pi g_i}\sqrt{\frac{g_1g_2}{1 - g_1g_2}}, \tag{11.83}$$

where w_i (i=1, 2) is the beam radius at the mirror 1 or 2, respectively, and where

$$g_1 = 1 - L/R_1, \tag{11.84}$$

$$g_2 = 1 - L/R_2 \tag{11.85}$$

are the *mirror parameters*. The mode diameters w_i are infinitely large if the product $g_1g_2 = 1$. There is no real solution if $g_1g_2(1 - g_1g_2)^{-1}$ is negative. This leads to the *stability criterion*: a stable mode can be realized if

$$0 \leq g_1g_2 \leq 1. \tag{11.86}$$

Fig. 11.7 shows the resonator stability diagram. Stable resonators have mirror parameters in the shadowed regions.

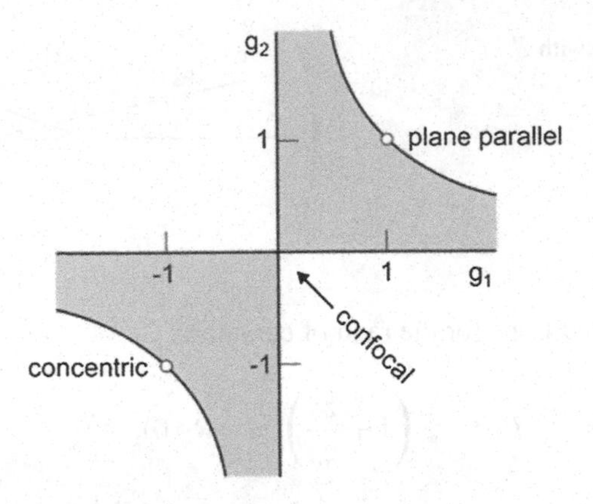

Fig. 11.7 Resonator stability diagram

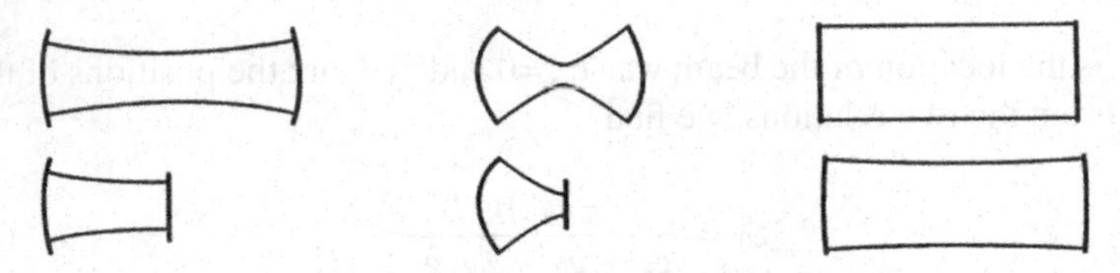

Fig. 11.8 Types of resonators; confocal and semiconfocal (*left*); near-concentric and semiconcentric (*center*); planar and near-planar (*right*)

There are limiting cases.

- $R_1 = R_2 = \infty$; $g_1 = g_2 = 1$; plane parallel (=Fabry-Perot) resonator.
- $R_1 + R_2 = 2L$; $g_1 + g_2 - g_1 g_2 = 0$; confocal resonator (general case).
- $R_1 = R_2 = L$; $g_1 = g_2 = 0$; symmetric confocal resonator.
- $R_1 + R_2 = L$; $g_1 g_2 = 1$; concentric resonator (general case).
- $R_1 + R_2 = L$; $g_1 = g_2 = -1$; symmetric concentric resonator.

The different types of resonators (Fig. 11.8) have advantages and disadvantages.

- The *confocal resonator*. It has the lowest diffraction loss (Sect. 11.8). Corresponding to the stability criterion, the confocal resonator is at the limit of stability. To reach stability, the distance between the mirrors should be slightly smaller than the radius of curvature of the mirrors. In comparison with other resonators, this resonator can easily be adjusted.
- A *semiconfocal resonator* consists of a spherical mirror and a plane mirror at the distance $R/2$, where R is the radius of curvature of the spherical mirror—the plane mirror is located at the position of the beam waist of a corresponding confocal resonator.
- The *concentric resonator* (=spherical resonator) has a beam waist of $w_0 = 0$ and an infinitely large divergence. It is therefore not realizable.
- The *near-concentric resonator* has the smallest mode volume.

- A *semiconcentric resonator* (=hemispheric resonator) consists of a spherical mirror and a plane mirror at the distance $R/4$—the plane mirror is located at the position of the beam waist of a corresponding confocal resonator.
- The *plane parallel (=Fabry-Perot) resonator* has, in comparison with all other resonators, the largest mode volume. It is difficult to adjust.
- The *near-planar resonator* (=superconfocal resonator) has one or two mirrors with a radius of curvature that is much larger than the length of the resonator. This resonator has the advantage, in comparison with the plane parallel resonator, that it is easier to adjust. In comparison with the confocal resonator, the near-planar resonator has a larger mode volume at the same resonator length. In special cases, larger laser output power is obtainable. The beam radius of a near-planar resonator, with R_1=R_2=$R \geq L$, is almost constant along the resonator axis,

$$w_0^2 = w_1^2 = w_2^2 = \frac{\lambda L}{\pi}\sqrt{\frac{R}{2L}}. \tag{11.87}$$

The change of the phase of a Gaussian wave propagating in a near-planar resonator during a single transit through a resonator is small, since $\phi(z{=}0) \approx \phi(z_0) \approx \phi(z{=}L) \sim 0$. This follows from the expression $\phi = \tan^{-1}(z - z_0)/z_{\mathrm{R}}$.

In the general case of a stable resonator, the change of the Gouy phase shift per round trip transit is given by

$$\Delta\phi = 2[\phi(L) - \phi(0)] = 2\left(\tan^{-1}\frac{L - z_0}{z_{\mathrm{R}}^2} - \tan^{-1}\frac{-z_0}{z_{\mathrm{R}}^2}\right) = 2\cos^{-1}(\pm\sqrt{g_1 g_2}). \tag{11.88}$$

We made use of the relations $\tan^{-1} x + \tan^{-1} y = \tan^{-1}\left([x + y][1 - xy]^{-1}\right)$ and $\tan^{-1} x{=}\cos^{-1}(1/\sqrt{1 + x^2})$. (Note that the inverse trigonometric function $\cos^{-1} x \equiv \arccos x$.) The condition

$$2kL - \Delta\phi = l \times 2\pi, \quad l = 1, 2, ... \tag{11.89}$$

leads to the resonance frequencies

$$\nu_l = \frac{c}{2L}\left[l + \frac{1}{\pi}\cos^{-1}(\pm\sqrt{g_1 g_2})\right], \tag{11.90}$$

where the plus sign has to be chosen if g_1 and g_2 are positive while the minus sign has to to be chosen if g_1 and g_2 are negative. Limiting cases are as follows:

- Fabry–Perot resonator. $g_1, g_2 \to 1$; $\cos^{-1}\sqrt{g_1 g_2} \to 0$.
- Confocal resonator; $g_1, g_2 \to 0$; $\cos^{-1}\sqrt{g_1 g_2} \to \pi/2$.
- Concentric resonator; $g_1, g_2 \to -1$; $\cos^{-1}\sqrt{\pm g_1 g_2} \to \pi$.

The frequency difference between adjacent modes is

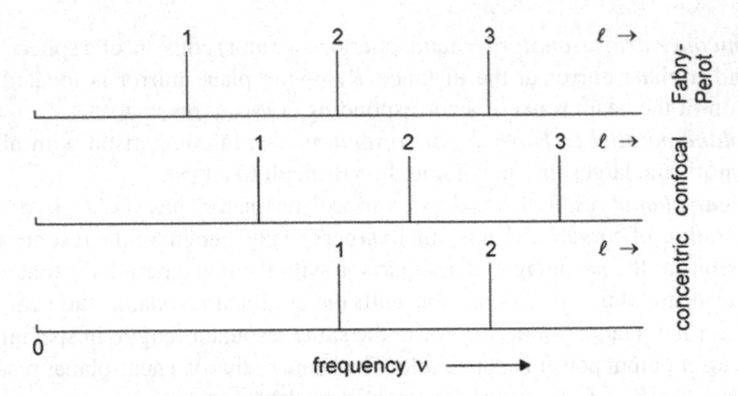

Fig. 11.9 Resonance frequencies of different resonators

$$\nu_{l+1} - \nu_l = c/(2L). \tag{11.91}$$

This is an important result: the Gouy phase shift of radiation propagating within a resonator causes a shift of the resonance frequencies of the resonator toward higher frequencies. The shift is the same for all resonance frequencies. But the frequency distance between neighboring resonances remains uninfluenced by the Gouy phase shift.

Figure 11.9 shows the frequencies of the modes of different resonators. The Gouy frequency is zero for a Fabry-Perot resonator, $c/4L$ for a symmetric confocal resonator, and $c/2L$ for a symmetric concentric resonator.

11.6 Transverse Modes

We use the ansatz

$$f(x, y, z) = X(x)\, Y(y)\, G(z)\, \mathrm{e}^{-(x^2+y^2)/F(z)}, \tag{11.92}$$

where X depends on x only, Y on y only, F and G on z only. Differentiation yields

$$\frac{\partial^2 f}{\partial x^2} = \left(X'' - \frac{4x}{F}X' - \frac{2}{F}X + \frac{4x^2}{F^2}\right) \mathrm{XYGe}^{-r^2/F}, \tag{11.93}$$

$$\frac{\partial f}{\partial z} = \left[G' + \frac{\mathrm{GF}'}{F^2}(x^2 + y^2)\right] \mathrm{XYe}^{-r^2/F}, \tag{11.94}$$

with $X' = \mathrm{d}X/\mathrm{d}x$, $Y' = \mathrm{d}Y/\mathrm{d}y$ and $G' = \mathrm{d}G/\mathrm{d}z$. The Helmholtz differential equation leads to

$$\frac{X''}{X} - \frac{4x}{F}\frac{X'}{X} + \frac{Y''}{Y} - \frac{4y}{F}\frac{Y'}{Y} - \frac{4}{F} - 2\mathrm{i}k\frac{G'}{G} + 2\frac{x^2 + y^2}{F^2}(2 - \mathrm{i}kF') = 0. \tag{11.95}$$

We obtain again the condition that $2 - ikF'$=0. With the same arguments used earlier, we find again that $F(z)$=$w_0^2 + 2/(ik)(z - z_0)$. Furthermore, we obtain differential equations for X and Y,

$$\frac{X''}{X} - \frac{4x}{F}\frac{X'}{X} + \frac{4m}{F} = 0, \tag{11.96}$$

$$\frac{Y''}{Y} - \frac{4y}{F}\frac{X'}{X} + \frac{4n}{F} = 0, \tag{11.97}$$

where m and n are dimensionless numbers. We introduce the dimensionless variable $\zeta = x\sqrt{2/F}$. Then the differential equation for X becomes

$$\frac{d^2X}{d\zeta^2} - 2\zeta\frac{dX}{d\zeta} + 2mX = 0. \tag{11.98}$$

This is Hermite's differential equation. The solutions are the Hermite polynomials $H_m(\zeta)$; H_m is the Hermite polynomial of mth order. We list a few Hermite polynomials.

- $m = 0$; $H_0(\zeta) = 1$.
- $m = 1$; $H_1(\zeta) = 2\zeta$.
- $m = 2$; $H_2(\zeta) = 4\zeta^2 - 2$.
- $m = 3$; $H_3(\zeta) = 8\zeta^3 - 12\zeta$.

The Hermite function H_0 is an even function with respect ζ, H_1 is an odd function, H_2 an even function and so on. The Hermite polynomials obey the recursion formula

$$H_{m+1}(\zeta) = 2\zeta H_m(\zeta) - 2mH_{m-1}(\zeta). \tag{11.99}$$

The solutions of (11.96) and (11.97) are

$$X(x) = H_m\left(\sqrt{\frac{2}{F}}\,x\right) \quad \text{and} \quad Y(y) = H_n\left(\sqrt{\frac{2}{F}}\,y\right). \tag{11.100}$$

We obtain from (11.95) by separation of the variables the differential equation

$$\frac{G'}{G} = \frac{2i}{k}\frac{1+m+n}{F} = -\frac{1+m+n}{z+C_1}. \tag{11.101}$$

We write

$$G_{\mathrm{mn}}(z) = |G_{\mathrm{mn}}(z)|\, e^{i\phi_{\mathrm{mn}}(z)}. \tag{11.102}$$

Separation in real and imaginary part leads to two differential equations for $|G|$ and ϕ. The solution for $|G|$ is

$$|G(z)| = |G_{\mathrm{mn}}(z)| = \frac{C_{3,\mathrm{mn}}}{w}, \tag{11.103}$$

where

$$C_{3,\mathrm{mn}} = \frac{C_2}{(kw_0/2)^{1+\mathrm{m}+\mathrm{n}}} \tag{11.104}$$

and where C_2 is an integration constant, which is real. The solution for ϕ is

$$\phi_{\mathrm{mn}}(z) = (1 + m + n)\,\phi(z). \tag{11.105}$$

$\phi(z)$ is the same as in the case $m = n = 0$. Thus, we have the solution

$$\psi_{\mathrm{mn}}(x, y, z) = H_m\left(\sqrt{\frac{2}{F}}\,x\right) H_n\left(\sqrt{\frac{2}{F}}\,y\right) \frac{C_{3,\mathrm{mn}}}{w(z)}\,\mathrm{e}^{-\mathrm{i}(kz-\phi_{\mathrm{mn}}+kr^2/2R)}. \tag{11.106}$$

The Gouy phase increases with increasing m and increasing n. Each number pair mn corresponds to a mode of radiation in free space. We design the paraxial modes with mn = 00 (=fundamental Gaussian modes = Gaussian modes) as *longitudinal modes* and the paraxial modes with mn ≠ 00 as *transverse modes* (=Hermite-Gaussian modes = higher-order Gaussian modes).

Figure 11.10a shows the amplitudes of the fields of a few modes together with field lines. The amplitudes of longitudinal and transverse modes have different spatial distributions.

- 00 mode. The field amplitude ψ has the largest value at the beam axis.
- 10 mode. In x direction, the amplitude changes once the sign and has two extrema—according to the Hermite polynomial $H_1(x)$. The field amplitude is zero at the beam axis.
- 20 mode. In x direction, the amplitude changes twice the sign and has three extrema.

The transverse modes mnl have the same beam diameter and the same radius of curvature as the longitudinal mode 001. A transverse mnl mode has, along the z axis, the same number of field maxima as the longitudinal 00l mode (compare with Fig. 11.1c and d). Figure 11.10b shows different mode patterns as they can be observed for the intensity distribution of laser radiation outside a laser resonator; special filters placed in a laser resonator can select a particular mode at which a laser oscillates.

The phase shift per round trip transit of radiation in a mode mnl has to obey the resonator eigenvalue condition

$$2kL - (1 + m + n)\Delta\phi = l \times 2\pi, \tag{11.107}$$

where

$$\Delta\phi = \phi(z_2) - \phi(z_1) + \phi(z_2) - \phi(z_1) = 2\,(\phi(z_2) - \phi(z_1)), \tag{11.108}$$

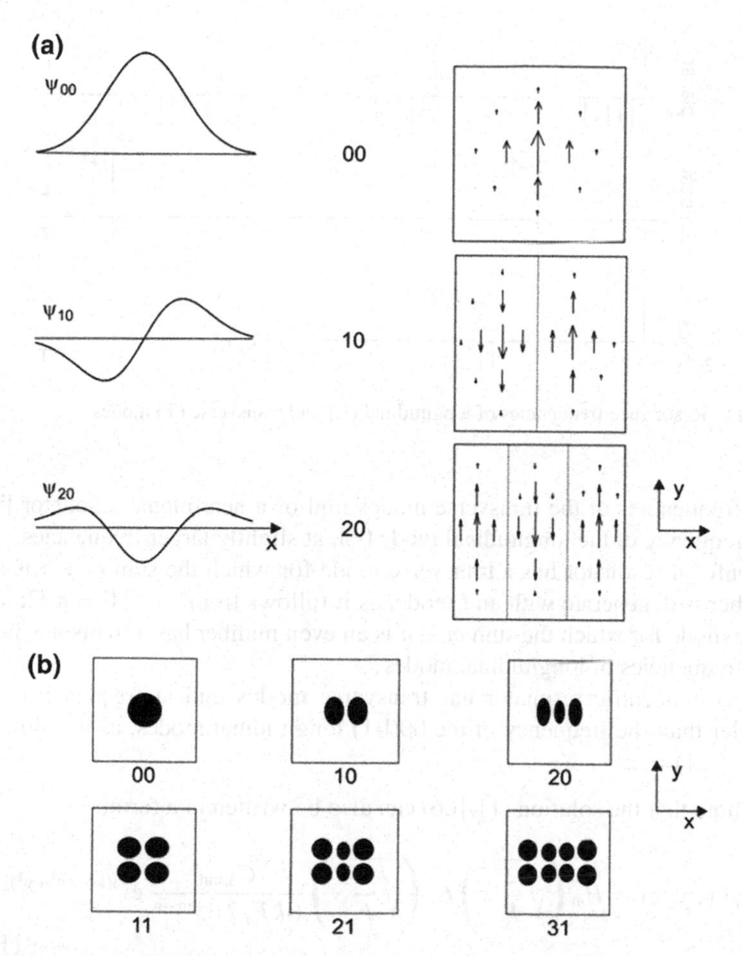

Fig. 11.10 Longitudinal and transverse modes. **a** Amplitude distributions and field lines. **b** Intensity distributions

$$\phi(z_i) = \tan^{-1}\frac{z_i - z_0}{z_R}, \tag{11.109}$$

and where z_1=0 and z_2=L. We obtain the resonance frequencies

$$\nu_{lmn} = \frac{c}{2L}\left[l + \frac{1+m+n}{\pi}\cos^{-1}(\pm\sqrt{g_1 g_2})\right]. \tag{11.110}$$

The frequency separations between longitudinal and transverse modes depend on the values of g_1 and g_2. The frequency separation between two neighboring modes mn(l + 1) and mnl is always $c/(2L)$.

Figure 11.11 shows special cases that follow from (11.110).

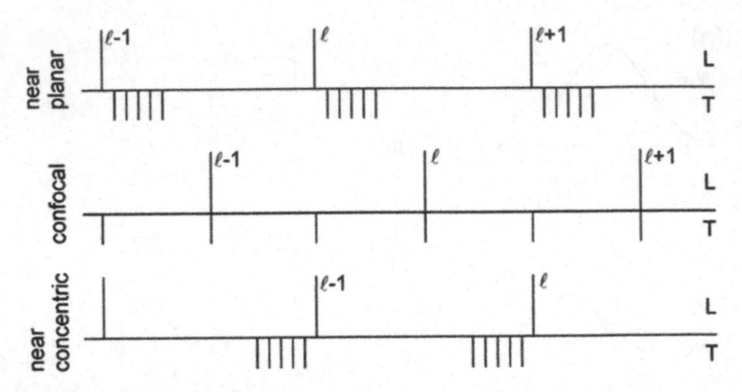

Fig. 11.11 Resonance frequencies of longitudinal (L) and transverse (T) modes

- The frequencies of the transverse modes mnl of a near-planar resonator lie near the frequency of the longitudinal mode 00l, at slightly larger frequencies.
- A confocal resonator has a transverse mode for which the sum $m + n$ is an odd number is degenerate with an l mode, as it follows from $\cos^{-1} 0 = \pi/2$; a transverse mode for which the sum $m + n$ is an even number has a frequency between two frequencies of longitudinal modes.
- A near-concentric resonator has transverse modes mnl at frequencies slightly smaller than the frequency of the 00(l-1) longitudinal modes, as it follows from $\cos^{-1}(-1) = \pi$.

We note that the solution (11.106) can also be written in a form,

$$\psi_{\mathrm{mn}}(x, y, z) = H_m\left(\sqrt{\frac{2}{F}}\, x\right) H_n\left(\sqrt{\frac{2}{F}}\, y\right) \frac{C_{3,\mathrm{mn}}}{(\mathrm{i}kF/2)^{1+\mathrm{m}+\mathrm{n}}}\, \mathrm{e}^{-\mathrm{i}[kz-(x^2+y^2)/F]}, \tag{11.111}$$

that contains the complex beam parameter $F(z)$, which is the same as for a fundamental Gaussian wave.

In our study of Gaussian waves, we have made use of Cartesian coordinates. The solution to the Helmholtz equation of paraxial waves provides the fundamental Gaussian waves and the Hermite-Gaussian waves. The number pair mn describes the variation of the sign of the amplitude along the x and the y axis. Solutions to the Helmholtz equation are Laguerre-Gaussian modes too. These are also characterized by a number pair mn, however m now describes the variation of the sign of the amplitude in radial direction and n the variation in azimuthal direction. The waves are also TEM waves. The Laguerre-Gaussian modes are obtained by solving the Helmholtz equation written in cylinder coordinates. Depending on the experimental arrangement of a laser, either type of higher-order Gaussian mode can be observed.

11.7 The Gouy Phase

The field of a Gaussian wave is given by

$$E(z, r, t) = A_0 \frac{w_0}{w} \mathrm{e}^{-r^2/w^2} \cos\left(\omega t - [k(z - z_0) - \phi + kr^2/2R]\right). \qquad (11.112)$$

A_0 is the amplitude, w_0 the beam radius in the beam waist, z_0 the position of the beam waist, $w(z)$ the beam radius at the position z, k the wave vector of the radiation, ϕ the Gouy phase and $kr^2/2R(z)$ a phase in lateral direction that is zero on the beam axis; we have chosen the phase φ_0 so that the beam waist lies at z_0. The beam radius is

$$w(z) = w_0 \sqrt{1 + \frac{(z - z_0)^2}{z_\mathrm{R}^2}}, \qquad (11.113)$$

where

$$z_\mathrm{R} = kw_0^2/2 = \pi w_0^2/\lambda \qquad (11.114)$$

is the Rayleigh range. The Gouy phase (Fig. 11.12, upper part) is given by

$$\phi(z) = \tan^{-1} \frac{z - z_0}{z_\mathrm{R}}. \qquad (11.115)$$

The curvature of the wave front is

$$R(z) = z - z_0 + \frac{z_\mathrm{R}^2}{z - z_0}. \qquad (11.116)$$

The derivative of the time-dependent portion φ_t of the phase yields the frequency,

$$\mathrm{d}\varphi_t/\mathrm{d}t = \omega. \qquad (11.117)$$

From the position-dependent portion of the phase,

$$\varphi_z = k(z - z_0) - \phi, \qquad (11.118)$$

we obtain, by differentiation, the effective wave vector

$$k_\mathrm{eff} = k \quad \mathrm{d}\phi/\mathrm{d}z. \qquad (11.119)$$

The effective wavelength is

$$\lambda_\mathrm{eff} = \frac{2\pi}{k_\mathrm{eff}} = \frac{\lambda}{1 - k^{-1}\mathrm{d}\phi/\mathrm{d}z} = \frac{\lambda}{1 - \lambda/(2\pi)\mathrm{d}\phi/\mathrm{d}z}, \qquad (11.120)$$

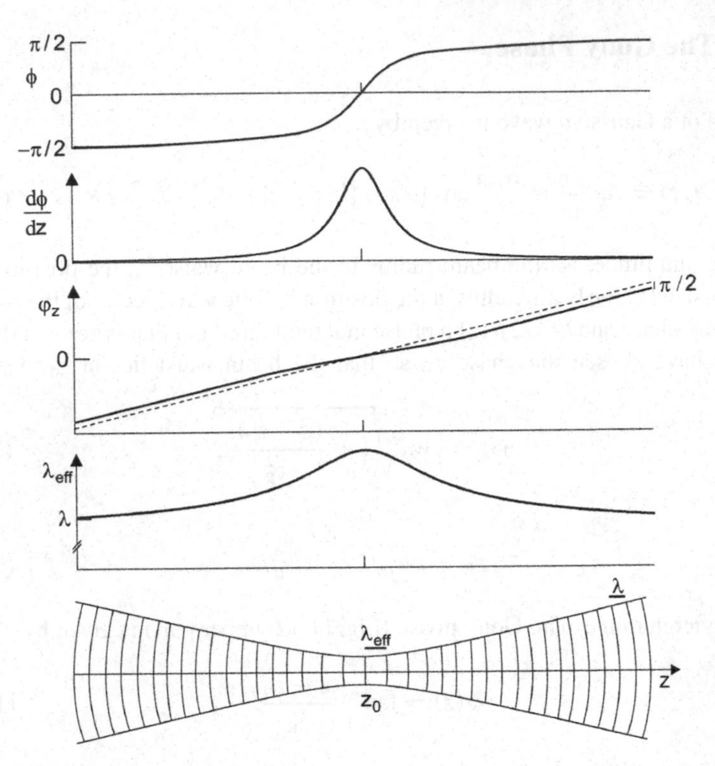

Fig. 11.12 Gaussian beam: Gouy phase; variation of the Gouy phase; spatial part (*dashed*) of the phase and total phase (*solid*); effective wavelength; wavefronts

where $\lambda = 2\pi/k$ is the wavelength of the radiation far outside the beam waist. The effective wave vector and the effective wavelength depend on the position z. The Gouy phase shows the strongest change in the Rayleigh range and the derivative

$$\frac{\mathrm{d}\phi}{\mathrm{d}z} = \frac{1/z_R}{1 + (z - z_0)^2/z_R^2} \tag{11.121}$$

has a maximum at the center of the beam waist. We consider a wave front, which propagates through the beam waist from a far-field location z_1 to a far-field location z_2 (Fig. 11.12, third panel). The phase far outside the beam waist, at $z_1 - z_0 < 0$, is given by

$$\varphi_{(z_1)} = k \times (z_1 - z_0) + \pi/2, \tag{11.122}$$

and for the range far outside the beam waist at $z_2 - z_0 > 0$ by

$$\varphi_{(z_2)} = k \times (z_2 - z_0) - \pi/2. \tag{11.123}$$

The Gouy phase shift at a transit of radiation through the beam waist, from a far-field location to a location in the other far-field is equal to $-\pi$. In the range of the beam waist, the effective wavelength (Fig. 11.12, lower panels) is larger than λ. In the center of the beam waist, the effective wavelength is equal to $\lambda_{\text{eff}} = \lambda + \lambda/(4\pi)$. The wave fronts have the largest distance in the center of the beam waist. The sum of all differences $\lambda_{\text{eff}} - \lambda$ at a transit of radiation through the beam waist is equal to $\lambda/2$.

We now discuss the Gouy phase shift of radiation in a resonator. The field of a standing wave of an open resonator is

$$E(z, r, t) = A_0 \frac{w_0}{w} \mathrm{e}^{-r^2/w^2} \cos[k(z - z_0) - \phi + kr^2/2R] \cos \omega t. \tag{11.124}$$

The resonance condition requires that

$$2kL - \Delta\phi = l \times 2\pi; \quad l = 1, 2, ..., \tag{11.125}$$

where l is the order of resonance and $\Delta\phi$ the Gouy phase shift per round trip. At resonance, the phase change $2kL$ per round trip transit is larger than 2π because of the Gouy phase shift,

$$2kL = l \times 2\pi + \Delta\phi. \tag{11.126}$$

It follows that the resonance frequencies are

$$\nu_l = l \times \frac{c}{2L} + \nu_{\text{Gouy}}, \tag{11.127}$$

where ν_{Gouy} is the *Gouy frequency*,

$$\nu_{\text{Gouy}} = \frac{\Delta\phi}{2\pi} \times \frac{c}{2L}. \tag{11.128}$$

The resonance frequencies of an open resonator (Fig. 11.13) are multiples of $c/2L$ but shifted toward higher frequencies by the Gouy frequency.

Figure 11.14 and Table 11.3 show values of Gouy frequencies of stable resonators. The Gouy frequency is zero for a Fabry–Perot resonator and has the largest value

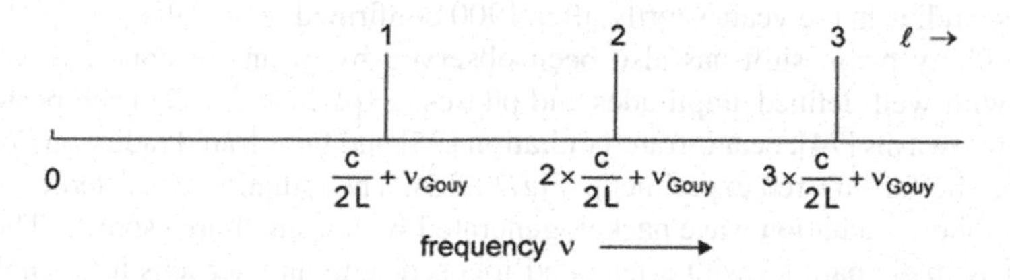

Fig. 11.13 Low-order resonance frequencies of an open resonator

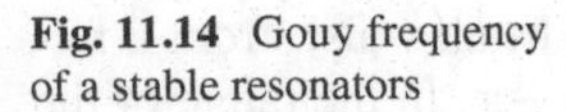
Fig. 11.14 Gouy frequency of a stable resonators

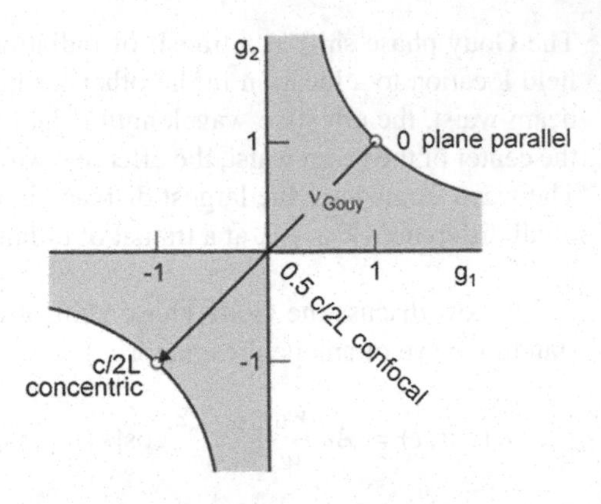

Table 11.3 Gouy phase shift and Gouy frequency of resonators

Resonator	$\Delta\phi$	ν_{Gouy}
Fabry-Perot	0	0
Symmetric confocal	π	$0.5\ c/(2L)$
Semiconfocal	$\pi/2$	$0.25\ c/(2L)$
Symmetric concentric	2π	$c/(2L)$

for a concentric resonator. The Gouy frequency of a symmetric confocal resonator is $\nu_{\text{Gouy}} = (1/2)c/2L$.

Example The Gouy phase shift of a symmetric confocal resonator of a length of 0.5 m is $\Delta\phi = \pi$ and the Gouy frequency is $0.5c/(2L) = 150$ MHz; for a semiconfocal resonator of the same length, the Gouy phase shift is $\pi/2$ and the Gouy frequency is 75 MHz.

In 1891, Louis Gouy (Lyon, France) found that an electromagnetic wave changes the phase by π if it propagates through a focus point—besides the phase change due to spatial propagation [70–72]. Gouy studied an interference pattern of two beams (arising from the same white light source), which were reflected from two plane mirrors, and observed an additional phase shift when he replaced one of the mirrors by a spherical mirror that produced a focus point in one of the beams; Gouy derived the phase shift from an analysis of the focusing process by use of Huygens' principle. Various studies in the years shortly after 1900 confirmed the results (*see* [73]).

The Gouy phase shift has also been observed by means of coherent waves—waves with well-defined amplitudes and phases. Experiments have been performed with microwaves [74], near infrared radiation [75] and far infrared radiation [76]. We describe the far infrared experiment (Fig. 11.15). The radiation consisted of single-cycle terahertz radiation wave packet, generated with a small-area source. The radiation was made parallel with a lens and focused with another lens to a small-area detector. The detector monitored the time dependent amplitude and phase of the radiation at the position of the detector. Alternatively, two additional lenses produced

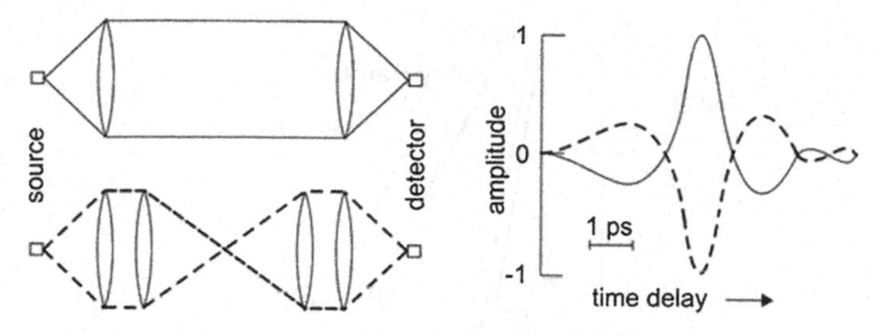

Fig. 11.15 Measurement of the Gouy phase shift

a focus point between source and detector. Signals were measured at different time delays relative to the starting time of a pulse. The experiment showed that the phase of a wave packet changed by π when the wave propagated through the focus; we will describe the method of measuring phases and amplitudes of electromagnetic fields in Sect. 13.5.

We can ask: is the group velocity of radiation traversing a beam waist smaller than the speed of light? The answer is no: the change of phase by π causes a reversal of the direction of the electric field of the wave. The group velocity remains therefore unchanged.

11.8 Diffraction Loss

Up to now, we have neglected loss by diffraction at the resonator mirrors. The diffraction loss depends on the resonator type. Figure 11.16 shows examples of the diffraction loss δ (=loss per round trip) for resonators of different Fresnel numbers (F).

- The diffraction loss of radiation in a confocal resonator is (for $F \geq 0.5$) much smaller than the diffraction loss of radiation in a planar resonator.
- Longitudinal modes have a smaller diffraction loss than transverse modes.
- The diffraction loss decreases strongly with increasing Fresnel number.

The theory of Kirchhoff (1882) allows for determination of diffraction loss. We give here a short sketch of the theory. We are looking for a solution of the wave equation in the form of the Helmholtz equation,

$$\nabla^2 \psi + k^2 \psi = 0. \quad (11.129)$$

Originally, Kirchhoff formulated the theory assuming that a parallel light wave is incident on an iris diaphragm (Fig. 11.17a). The boundary condition is the following: in the open part of the iris, the field has the same value ψ as without iris. According to Huygens' principle, spherical waves are leaving from each point in the open part of

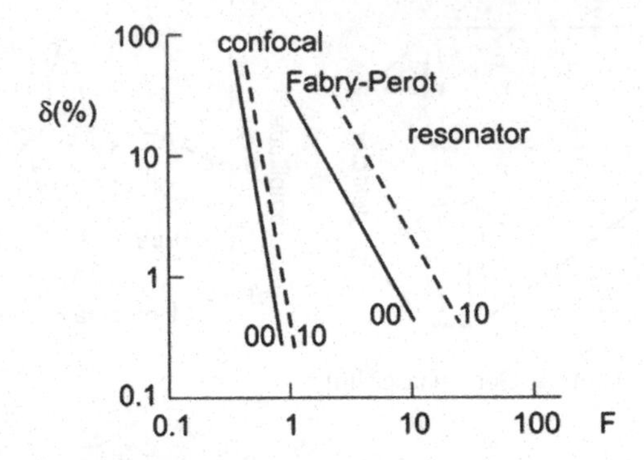

Fig. 11.16 Diffraction loss per round trip

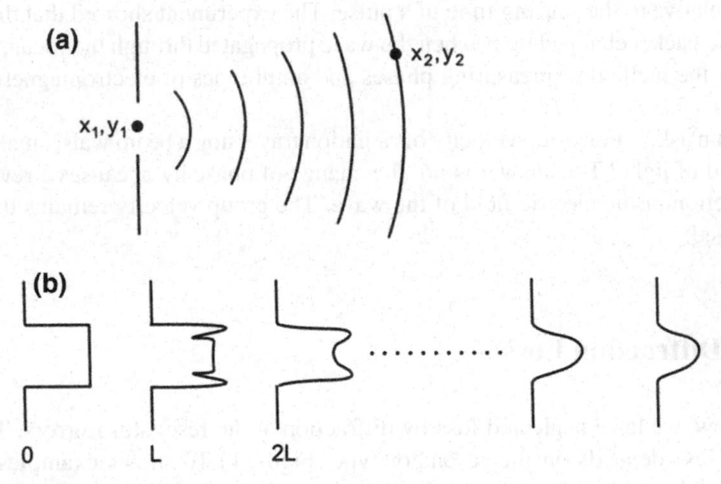

Fig. 11.17 Diffraction. a Diffraction at an iris diaphragm. b Multiple diffraction at iris diaphragms in series

the iris. The field amplitude at a point (x_2, y_2) is the sum of all partial waves arriving from all points x_1, y_1 in the open part of the iris. The summation yields

$$\psi(x_2, y_2) = \frac{ik}{4\pi} \int (1 + \cos\theta) \frac{e^{iks}}{s} \psi(x_1, y_1) \, dx_1 dy_1. \tag{11.130}$$

The amplitude depends on the distance s between the iris and the point (x_2, y_2) and on the angle θ between the central axis and the direction between the iris and the point; s is large compared to the diameter of the open part of the iris. The solution obeys the Helmholtz equation. The factor i to the integral (11.130) implies the occurrence of the Gouy phase shift.

Radiation in a resonator undergoes *multiple reflection with diffraction*, illustrated in Fig. 11.17b for iris diaphragms in series. The calculation starts with an arbitrarily assumed field distribution $\psi_1(x_1, x_2)$, for instance, a constant distribution over one of the mirrors. A first integration provides the distribution at the second mirror—at a single transit through the resonator. The numerical calculation of ψ_{n+1} from ψ_n, where n is the number of passes through the resonator, leads to the following results.

- *Stable resonator.* After a field has performed a certain number of transits through the resonator, the field obeys the relation

$$\psi_{n+1}(x, y) = \eta\, \psi_n(x, y), \tag{11.131}$$

  where η (< 1) is a number. The shape of the field distribution is reproduced, but there is a loss at each reflection.
- *Instable resonator*. The distribution $\psi(x, y)$ does not stabilize.

We have seen that in the far field of a fundamental Gaussian wave the product of the beam diameter and the angle of aperture is a constant,

$$D_0 \times \theta_0 = \frac{4}{\pi}\lambda. \tag{11.132}$$

D_0=$2w_0$ is the diameter of the field distribution at the beam waist. If diffraction at the output coupling mirror or at another optical element in a laser resonator enhances the angle of aperture, the beam diameter D can be larger than D_0 and the angle θ can be larger than θ_0. The product can be written as

$$D \times \theta = M^2 D_0 \theta_0. \tag{11.133}$$

The M factor is a measure of the quality of a beam. $M = 1$ corresponds to a Gaussian beam.

11.9 Ray Optics

We characterize an optical ray (Fig. 11.18) at a the point z, r by the vector

$$\boldsymbol{r} = \begin{pmatrix} r \\ r' \end{pmatrix}, \tag{11.134}$$

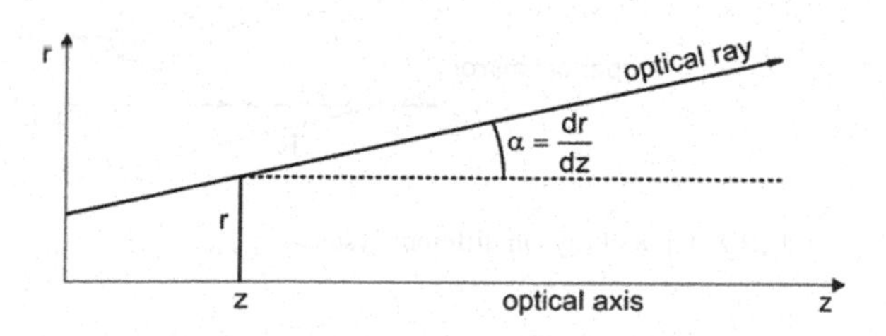

Fig. 11.18 Paraxial optical beam

where r is the distance of the ray from the beam axis and $r' = \mathrm{d}r/\mathrm{d}z$ is the slope of the ray. The slope of a paraxial ray is approximately equal to the angle between the ray and the optical axis. Therefore, we can make use of the approximation $\mathrm{d}r/\mathrm{d}z = \sin\alpha \approx \alpha$.

We describe the trajectory of an optical ray propagating from a location $\boldsymbol{r}_1$ to a location $\boldsymbol{r}_2$ by

$$\boldsymbol{r}_2 = \begin{pmatrix} A & B \\ C & D \end{pmatrix} \boldsymbol{r}_1, \tag{11.135}$$

where $\begin{pmatrix} A & B \\ C & D \end{pmatrix}$ is the *ray matrix* (=ABCD matrix).

The propagation of an optical ray in an optical system with s optical elements in series is described by the matrix product

$$\begin{pmatrix} A & B \\ C & D \end{pmatrix} = \begin{pmatrix} A_s & B_s \\ C_s & D_s \end{pmatrix} \cdots \begin{pmatrix} A_2 & B_2 \\ C_2 & D_2 \end{pmatrix} \begin{pmatrix} A_1 & B_1 \\ C_1 & D_1 \end{pmatrix}. \tag{11.136}$$

We illustrate the method by various examples (Fig. 11.19).

- *Propagation in free space*; $\begin{pmatrix} A & B \\ C & D \end{pmatrix} = \begin{pmatrix} 1 & L \\ 0 & 1 \end{pmatrix}$.

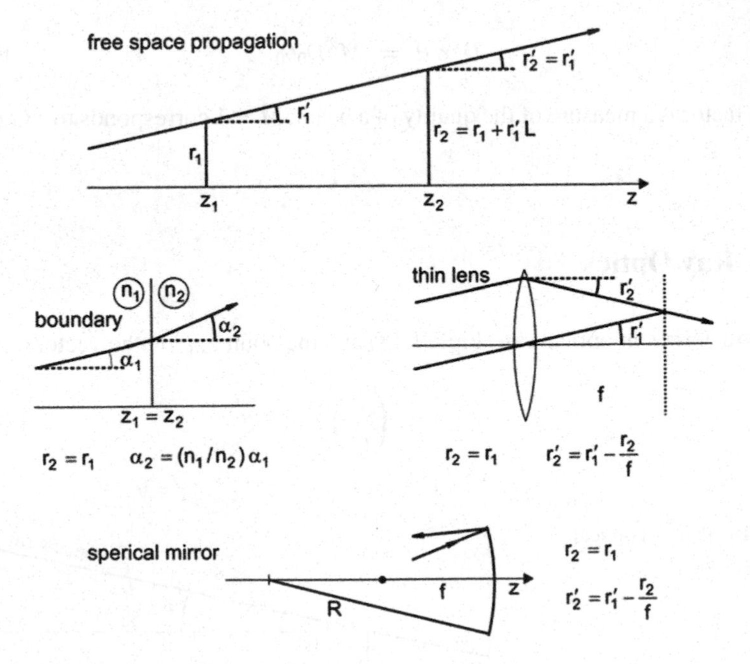

Fig. 11.19 Optical rays in different systems

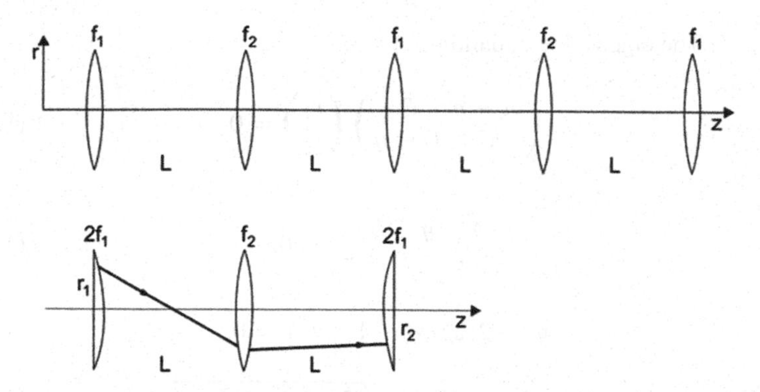

Fig. 11.20 Optical beam passing lenses in series (*upper part*) and periodicity interval (*lower part*)

- *Snell's law*; $\begin{pmatrix} A & B \\ C & D \end{pmatrix} = \begin{pmatrix} 1 & 0 \\ 0 & n_1/n_2 \end{pmatrix}$.
- *Thin lens* with a focus length f; $\begin{pmatrix} A & B \\ C & D \end{pmatrix} = \begin{pmatrix} 1 & 0 \\ -f^{-1} & 1 \end{pmatrix}$.
- *Spherical mirror*; $\begin{pmatrix} A & B \\ C & D \end{pmatrix} = \begin{pmatrix} 1 & 0 \\ -2R^{-1} & 1 \end{pmatrix} = \begin{pmatrix} 1 & 0 \\ f^{-1} & 1 \end{pmatrix}$. The radius has a positive sign ($R > 0$) for a concave mirror and a negative sign ($R < 0$) for a convex mirror.

We derive, by the use of ray optics, the stability criterion for resonators. A spherical mirror and a thin lens are equivalent optical elements. Accordingly, we can replace a two-mirror resonator by a series of lenses with the focus lengths $f_1=R_1/2$ and $f_2=R_2/2$ (Fig. 11.20). The periodicity interval of the series of lenses includes a half-lens with the focus length $2f_1=R_1$, a lens with the focus length $f_2 = R_2/2$ and another half lens with the focus length $2f_1 = R_1$. A round trip through the resonator corresponds to the path through the periodicity interval in the lens system from $\boldsymbol{r}_1$ to $\boldsymbol{r}_2$, where

$$\boldsymbol{r}_2 = \begin{pmatrix} 1 & 0 \\ -(2f_1)^{-1} & 1 \end{pmatrix} \begin{pmatrix} 1 & L \\ 0 & 1 \end{pmatrix} \begin{pmatrix} 1 & 0 \\ -f_2^{-2} & 1 \end{pmatrix} \begin{pmatrix} 1 & L \\ 0 & 1 \end{pmatrix} \begin{pmatrix} 1 & 0 \\ -(2f_1)^{-1} & 1 \end{pmatrix} \boldsymbol{r}_1. \quad (11.137)$$

We obtain, with the mirror parameters $g_1 = 1 - L/R_1$ and $g_2 = 1 - L/R_2$, the ABCD matrix

$$\boldsymbol{r}_2 = \begin{pmatrix} 2g_1g_2 - 1 & 2g_2L \\ -2g_1(g_1g_2 - 1)L^{-1} & 2g_1g_2 - 1 \end{pmatrix} \boldsymbol{r}_1. \quad (11.138)$$

We are looking for rays that remain unchanged after the propagation through a periodicity interval. A stable trajectory requires that

$$\begin{pmatrix} A & B \\ C & D \end{pmatrix} \boldsymbol{r}_1 = \eta\, \boldsymbol{r}_1 \quad (11.139)$$

and $|\eta| = 1$. The eigenvalue equation

$$\begin{pmatrix} A-\eta & B \\ C & D-\eta \end{pmatrix} \begin{pmatrix} r_1 \\ r_1' \end{pmatrix} = 0 \tag{11.140}$$

leads to

$$\begin{vmatrix} A-\eta & B \\ C & D-\eta \end{vmatrix} = 0, \tag{11.141}$$

$$\eta^2 - 2(2g_1g_2 - 1)\eta + 1 = 0, \tag{11.142}$$

$$\eta_{\mathrm{a,b}} = 2g_1g_2 - 1 \pm \sqrt{(2g_1g_2 - 1)^2 - 1}. \tag{11.143}$$

There are two possibilities.

- η_a and η_b are real if $g_1g_2 \geq 1$. This corresponds to instable resonators because, after N round trip transits through the resonator and $N \to \infty$, the vector

$$\begin{pmatrix} r_N \\ r_N' \end{pmatrix} = \eta^N \begin{pmatrix} r_1 \\ r_1' \end{pmatrix} \tag{11.144}$$

  diverges.
- η_a and η_b are imaginary if $g_1g_2 < 1$. This is the stability criterion. We obtain $\eta_{\mathrm{a,b}} = \exp(\pm\varphi)$, where $\cos\varphi = 2g_1g_2 - 1$. After N round trip transits, the vectors

$$\begin{pmatrix} r_N \\ r_N' \end{pmatrix} < |\eta_{\mathrm{a,b}}| \mathrm{e}^{\pm i\varphi} \begin{pmatrix} r_1 \\ r_1' \end{pmatrix} \tag{11.145}$$

  remain stable because $|\eta_a| = |\eta_b| < 1$. In our derivation of the stability criterion, we did not specify the values of r_1 and r_1'. Thus, the result is valid for all paraxial rays.

It is possible to describe the propagation of a Gaussian beam through an optical system by the use of the ABCD matrix of the optical system. We have found, *see* (11.41) and (11.42), that a Gaussian beam can be characterized by the complex beam parameter $\tilde{q}(z)$. We now make use of this complex beam parameter: if the complex beam parameter $\tilde{q}_1(z_1)$ is known, then $\tilde{q}_2(z_2)$ follows from the relation

$$\tilde{q}_2 = \frac{A\tilde{q}_1 + B}{C\tilde{q}_1 + D} \tag{11.146}$$

or

$$\frac{1}{\tilde{q}_2} = \frac{C + D/\tilde{q}_1}{A + B/\tilde{q}_1} = \frac{1}{R(z_2)} - \frac{\mathrm{i}\lambda}{\pi w^2(z)}. \tag{11.147}$$

We mention two examples.

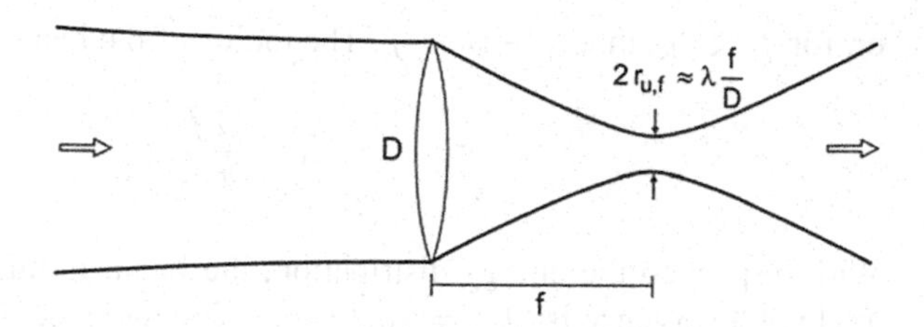

Fig. 11.21 Focusing of a Gaussian beam by a lens

Example propagation of a Gaussian beam from the location z_0 (beam waist) to a location z. The elements of the ABCD matrix for propagation in free space are $A = 1$, $B = z - z_0$, $C = 0$, and $D = 1$. At the beam waist, we have $R(z_0) = \infty$, $w(z_0) = w_0$, and $1/\tilde{q}_1 = -\mathrm{i}\lambda/\pi w_0^2$. It follows, for $z = z_2$, that

$$\frac{-\mathrm{i}\lambda(\pi w_0^2)^{-1}}{1 - \mathrm{i}\lambda(\pi w_0^2)^{-1}(z - z_0)} = \frac{1}{R(z)} - \frac{\mathrm{i}\lambda}{\pi w^2(z)}. \tag{11.148}$$

Equating real and imaginary parts leads to expressions for $w(z)$ and $R(z)$ that we derived (in Sect. 11.3) by the use of the Helmholtz equations; *see* (11.24) and (11.25). The agreement may be seen as a justification of the relation (11.146).

Example focusing a Gaussian beam by a thin lens (Fig. 11.21). A thin lens is located in the beam waist of a Gaussian beam. The ABCD matrix describing propagation through a thin lens at z_0 and then over a distance $z - z_0$ is

$$\begin{pmatrix} A & B \\ C & D \end{pmatrix} = \begin{pmatrix} 1 & z - z_0 \\ 0 & 1 \end{pmatrix} \begin{pmatrix} 1 & 0 \\ -f^{-1} & 1 \end{pmatrix} = \begin{pmatrix} 1 - (z - z_0)f^{-1} & z \\ -f^{-1} & 1 \end{pmatrix}. \tag{11.149}$$

We find, with $1/\tilde{q}_1 = -\mathrm{i}/z_R = -\mathrm{i}\lambda/(\pi w_0^2)$ and

$$\frac{-f^{-1} - \mathrm{i}z_R^{-1}}{1 - (z - z_0)f^{-1} - \mathrm{i}(z - z_0)z_R^{-1}} = \frac{1}{R(z)} - \mathrm{i}\frac{\lambda}{\pi w^2(z)}, \tag{11.150}$$

the values

$$\frac{1}{R(z)} = \frac{-f^{-1} + (z - z_0)(f^{-2} + z_R^{-2})}{1 - (z - z_0)^2 f^{-2} + (z - z_0)^2 z_R^{-2}}, \tag{11.151}$$

$$\frac{\lambda}{\pi w^2(z)} = \frac{z_R^{-1}}{(1 - (z - z_0)^2 f^{-2} + (z - z_0)^2 z_R^{-2}}. \tag{11.152}$$

At the focus point of the lens, $z = z_f$, the curvature $R(z_f)$ is infinitely large. It follows that

$$z_f - z_0 = \frac{f}{1 + f^2 z_R^{-2}} \tag{11.153}$$

or, for $f \ll z_R$, that $z_f - z_0 \approx f$. The radius of the beam in the focus of the lens is

$$w_f = \frac{\lambda f}{\pi w_0}. \tag{11.154}$$

With respect to the energy distribution, the beam radius is $r_{u,f} = w_f/\sqrt{2}$ and the angle of divergence is $\Theta_{u,f} = w_0/(f\sqrt{2})$. The radiance in the focus of the lens,

$$L_u = \frac{P}{\pi^2 r_{u,f}^2 \theta_{u,f}^2}, \tag{11.155}$$

is the same as in the incident Gaussian beam, namely $P/(\lambda/2)^2$ in units of W m^{-2} sr^{-1}. When a Gaussian beam traverses more than one optical element and all optical elements in the beam produce ideal images (without optical aberration and without diffraction), the radiance is the same at any location along the beam.

The diameter of the wave (with respect to the energy density) is

$$2r_{u,f} = \frac{\lambda}{\pi}\frac{f}{w_0}. \tag{11.156}$$

A lens of focal length $f = \pi w_0$ focuses the radiation of a Gaussian beam to an area with a diameter that is about equal to the wavelength of the radiation. If we choose a lens of diameter $D = 2w_0$, the diameter of the focused beam is equal to

$$2r_{u,f} = \frac{2\lambda}{\pi}\frac{f}{D} \sim \lambda\,\frac{f}{D}. \tag{11.157}$$

The beam diameter is λ, i.e., $2r_{u,f} = \lambda$, if the *f-number* of the focusing lens is $f/D = \pi/2 \sim 1.6$. It follows that a lens with an f-number of 1.6 can focus a Gaussian beam to an area $\pi(\lambda/2)^2 \approx \lambda^2$. The light intensity in the focus is

$$I_f = \frac{P}{\lambda^2}, \tag{11.158}$$

where P is the power of the radiation.

We mention another radiometric quantity, the *brilliance* of a beam:

$$B = \frac{r_p}{\Omega\,\Delta\nu}. \tag{11.159}$$

The brilliance of an optical beam is equal to the photon flux r_p (number of photons per second and m^2) divided by the solid angle of the beam and by the bandwidth of the radiation. For a detailed discussion of radiometric (physical) quantities and photometric quantities (how the human eye records radiation), *see* [29].

References [1–4, 6–11, 26, 29, 40, 64–76].

Problems

11.1 Gaussian wave.

(a) Determine the energy that is contained in a sheet (perpendicular to the beam axis) of thickness δz at the position z.
(b) Calculate the portion of power of radiation passing an area that has the beam radius $r_{u,0} = r_0$.
(c) Evaluate the radius r_p of the area passed by radiation of a portion p of the total power of the Gaussian wave.
(d) Evaluate r_p if $p = 95\%$.
(e) Evaluate r_p if $p = 99\%$.
(f) Determine the power of the radiation that passes an area of radius $r_p \ll r_0$.

11.2 Determine the minimum diameter of the tube of a helium–neon laser ($\lambda_L = 633$ nm) that is necessary to keep, per round trip, 99% of the radiation within a confocal resonator ($L = 0.5$ m).

11.3 Angle of divergence. Determine the angle of divergence of a Gaussian beam generated by a helium–neon laser (resonator length 0.5 m; radius of the energy density distribution at the beam waist $r_{u,0}$=0.16 mm; wavelength 633 nm).

11.4 Photon density in a Gaussian wave. An argon ion laser (length 1 m; radius of the beam waist 1 cm; wavelength 480 nm; power 1 Watt) emits a Gaussian wave. By the use of a telescope, the angle of aperture diminishes by a factor of 10. Estimate the number of photons arriving each second at a detector of 2 cm diameter at different distances between laser and detector.

(a) 100 km.
(b) 374,000 km (distance earth-moon).

11.5 ABCD matrix. Determine the effective focal length of an arrangement of two thin lenses (focal lengths f_1 and f_2) in contact.

11.6 Transversality of the radiation of a Gaussian wave. If a polarizer is located in a parallel beam of polarized radiation, the amplitude of the field transmitted by the polarizer is $A = A_0 \cos\theta$, where θ is the angle between the direction of polarization of the incident wave and the direction of the radiation for which the polarizer is transparent. (We assume that the transmissivity of the polarizer is 1 for $\theta = 0$.) Determine the loss of power of a Gaussian wave passing a polarizer (that is assumed to be thin compared to the Rayleigh range z_0 if the polarizer is located at different positions.

(a) In the beam waist at z_0.
(b) At $z = z_0/2$.
(c) At $z = z_0$.
(d) At $z \gg z_0$.

(e) Estimate the contribution of the polarizer to the V factor of a confocal laser resonator of 1 m length if the polarizer has a thickness of 1 cm and is located in the center of the resonator.

11.7 Hermite-Gaussian wave. Given is a 101 Hermite–Gaussian wave.

(a) Determine the radius of the wave at the beam waist and the angles of divergence in the far-field.
(b) Compare the results with corresponding values of a 001 Gaussian wave.

11.8 Calculate the Gouy phase of a Gaussian wave (λ=0.6 μm; $w_0 = 1$ mm) for propagation from the center of the beam waist over a distance of one wavelength; 1 mm; 1 cm; and 1 m.

11.9 Calculate the Gouy phase per round trip transit through a resonator of a Gaussian wave (λ=0.6 μm).

(a) If the resonator is a near-planar resonator with two mirrors (radius of curvature $R_1 = R_2 = 7$ m; resonator length = 1 m).
(b) If the resonator is a near-confocal resonator (radius of curvature $R_1 = R_2 =$ 1.10 m; resonator length = 1 m).

11.10 Show that a concentric resonator is not realizable. [*Hint*: consider the beam waist and the angle of divergence.]

11.11 Show that (11.44) is a solution of the Helmholtz equation.

11.12 Derive ray matrices for different optical arrangements.

(a) Reflection of radiation at a plane surface of a dielectric medium.
(b) Propagation of radiation through a thin lens.
(c) Focusing of radiation by a spherical mirror. [*Hint*: for solutions, *see* Sect. 11.9.]

11.13 Show that the intensity of radiation in a Gaussian beam averaged over an optical period is $I = c\,\epsilon_0\, A^2 \pi w_0^2/2$ and that

$$I(z,r) = \frac{2P}{\pi w^2(z)}\, \mathrm{e}^{-2r^2/w^2(z)}, \tag{11.160}$$

where $P = 2\pi \int I(z,r) r\mathrm{d}r$ is the power of the radiation.

11.14 Estimate radiance and brilliance of radiation of a helium-neon laser (power 10 mW; angle of divergence 1 mrad; beam waist in the laser 0.5 mmm; bandwidth 1 kHz) and compare the values with those of a light bulb (electric power 10 W).

11.15 Heisenberg uncertainty principle
Show that a photon in a Gaussian beam obeys the Heisenberg uncertainty relation $\Delta y \Delta p_y \geq \hbar$, where Δy is the uncertainty of the position y and Δp_y the uncertainty of the momentum p_y of the photon. [*Hint*: Show that the full width at half maximum

of the lateral energy density in the waist is equal to $D = \sqrt{2 \ln 2}\, w_0$ and that the full angle of divergence, related to the full width at half maximum of the lateral energy density in the far-field, is equal to $\vartheta = (\sqrt{2 \ln 2}/\pi)\lambda/w_0$. It follows that the product is $D\vartheta = (2 \ln 2/\pi)\lambda$. Now, determine the wave vector spread, according to $\vartheta = \Delta k_y/k_x$. It follows, with $\Delta y = D$ and $\Delta p_y = \hbar \Delta k_y$ that $\Delta y \Delta p_y = (4 \ln 2)\, \hbar$.

11.16 Gaussian beam in a medium.
Show that a Gaussian beam in a medium with the refractive index n (> 1) has a smaller divergence than in free space if the beam radius in the waist is the same.

11.17 Confined Gaussian beam.
A medium with a radial dependence of the refractive index of the form $n(r) = n_0 - a\, r^2$, with $a > 0$, is able to guide a wave without divergence. Show that the Helmholtz equation has the solution $\psi(z, r) = \psi_0 \exp\, (-r^2/w_1^2 + \mathrm{i}\, \lambda\, z/w_1^2)$, where the beam radius w_1 is given by $w_1^2 = \lambda/(\pi\sqrt{2a})$ and where λ is the vacuum wavelength. [*Hint*: make use of (11.4) and (11.13), with the relation $k = n\, \omega/c$.]

Chapter 12
Different Ways of Operating a Laser

In this chapter, we describe techniques used to operate lasers as continuous wave lasers or as pulsed lasers—in the next chapter we will treat femtosecond lasers.

We discuss single mode lasers. We mention spectral hole burning, occurring in lasers that operate with inhomogeneously broadened transitions. We give a short introduction to various methods of Q-switching of lasers used to generate laser pulses.

Furthermore, we describe two applications of continuous wave lasers—optical tweezers and gravitational wave detector.

12.1 Possibilities of Operating a Laser

Lasers can operate as continuous wave lasers, as pulsed lasers, or as femtosecond lasers. Lasers operated in different ways at different wavelengths have various applications in physics, chemistry [77, 78], biology, and medicine [79–82, 127–129, 332]. An early application was the holography [83, 84].

There are continuous wave lasers and different types of pulsed lasers:

- The *cw (continuous wave) laser*. Continuous pumping maintains the laser oscillation.
- *Pulsed laser*. A pump pulse generates a population inversion. Or more general: each laser that delivers pulses is a pulsed laser.
- *Q-switched laser*. In the Q-switched laser, the quality factor Q of the laser resonator varies with time. The Q factor is small for most of the time and large for a short time. During the time of small Q, population in the upper laser level is collected. During the time of large Q, laser oscillation occurs and the population of the upper laser level is strongly reduced.

K.F. Renk, *Basics of Laser Physics*, Graduate Texts in Physics,
DOI 10.1007/978-3-319-50651-7_12

- *Giant pulse laser*. This is a Q-switched laser with an upper laser level that has a very long lifetime (for instance 1ms); pumping leads to a large concentration of atoms in the upper laser level.
- *Femtosecond laser* (Chap. 13). The pumping is continuous. For most of the time, the Q factor of the resonator is small but it is large during short time intervals that follow each other periodically. The laser emits a coherent pulse train.

12.2 Operation of a Laser on Longitudinal Modes

A *mode diaphragm* eliminates transverse modes. The transverse modes suffer stronger diffraction than longitudinal modes and cannot reach the threshold condition. This mode selection makes it possible to operate a laser on longitudinal modes, 00l, 00(l + 1), ...

There are different possibilities of the operation of a laser on longitudinal modes:

- *Single line laser*, operated on a few neighboring longitudinal modes at frequencies in a narrow frequency range.
- *Single mode laser*, operated on a single longitudinal mode.
- *Mode-locked laser*, operated on a large number of longitudinal modes with frequencies in a large frequency range—the phases of the electromagnetic fields of different longitudinal modes are coupled (locked) to each other (Chap. 13).

12.3 Single Mode Laser

Many lasers (e.g., the helium–neon laser) have narrow gain profiles, but oscillate on a few modes (Fig. 12.1). By inserting an *etalon* (a plane parallel plate) into the resonator, selection of a single mode is possible. An etalon represents a low-Q resonator of the resonance wavelength

$$\lambda_s = \frac{2nd}{s} \cos \theta, \tag{12.1}$$

where n is the refractive index, d the thickness of the etalon, and θ the angle between the laser beam within the etalon and the normal to the etalon; s (an integer) is the order of resonance. Rotation of the etalon changes the angle θ and the resonance wavelength λ_s.

Example Without an etalon in the resonator, a helium–neon laser oscillates on about three modes at once (Fig. 12.1, lower part) but with an etalon on a single mode.

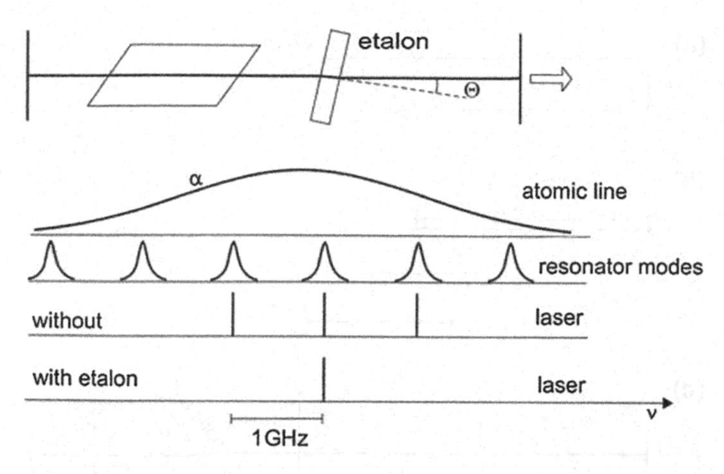

Fig. 12.1 Mode selection with an etalon; arrangement and result for a helium–neon laser ($\lambda =$ 633 nm, $\nu = 474$ THz)

12.4 Tunable Laser

A laser with a broadband gain profile oscillates on a single line if the laser resonator contains an appropriate wavelength selective element. We mention three wavelength selective elements suited to force a laser with a broad gain profile to emit a single line.

- *Prism* (Fig. 12.2a). A prism in the resonator selects the laser wavelength. The laser is tunable; rotation of the prism changes the wavelength. With an etalon additionally inserted into the resonator, a laser is able to oscillate on a single mode.
- *Echelette grating* (Fig. 12.2b). An echelette grating acts as one of the two reflectors of a laser resonator. In the *Littrow arrangement*, radiation that is incident on the echelette grating is diffracted in first order. The diffracted beam has the reverse direction relative to the incident beam. A rotation of the grating changes the wavelength of the backward diffracted radiation and thus the laser wavelength. A mirror telescope extends the diameter of the beam in order to obtain a higher resolving power and, furthermore, to reduce the field strength at the surface of the echelette grating thus avoiding damage of the grating. With an etalon that is additionally inserted into the resonator, a laser is able to oscillate on a single mode.
- *Birefringent filter* (Fig. 12.2c) The frequency selective element is a birefringent plate (e.g., a crystal of KDP = potassium dihydrogen phosphate) located between two polarizers in the laser resonator. The optic axis of the birefringent plate is oriented along the surface of the plate. The birefringent plate splits a beam of polarized radiation, incident under the Brewster angle, into an ordinary and an extraordinary beam. Behind the plate, the radiation has elliptical polarization and

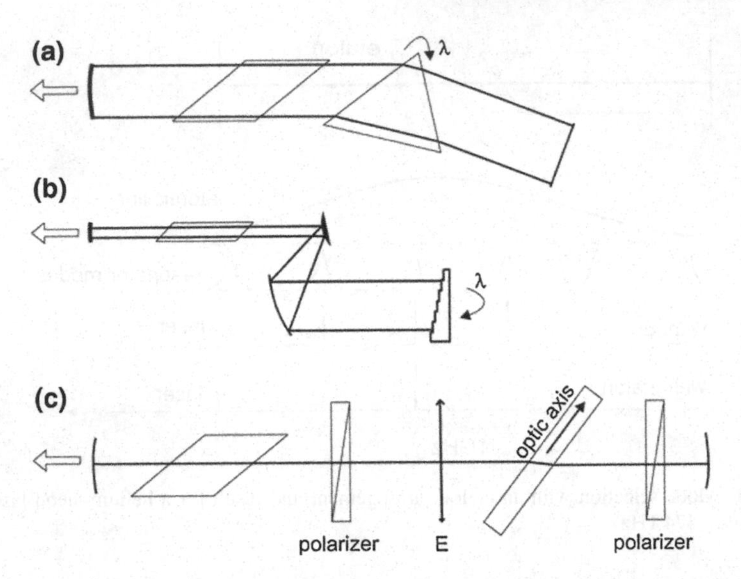

Fig. 12.2 Tunable laser. **a** Line selection with a prism. **b** Line selection with a grating. **c** Line selection with a birefringent filter

the second polarizer causes loss. There is no loss if the change of the phase between the ordinary and extraordinary beam is a multiple of π,

$$\frac{2\pi}{\lambda}(n_e - n_o)l_p = s \times \pi, \tag{12.2}$$

where n_e is the refractive index of the extraordinary beam, n_o the refractive index of the ordinary beam, l_p the length of the plate along the beam direction, and s is an integer. By rotating the plate while keeping the angle of incidence at the Brewster angle, the direction of the optic axis relative to the direction of the electric field vector (E) changes, leading to changes of n_e and of λ.

12.5 Spectral Hole Burning in Lasers Using Inhomogeneously Broadened Transitions

The oscillation behavior of a continuous wave laser depends on the type of line broadening.

A cw laser based on a homogeneously broadened line oscillates at the frequency of maximum gain (Fig. 12.3). When the population inversion begins, laser oscillation at the line center—where the gain coefficient α has its maximum—builds up. The onset of laser oscillation leads to a reduction of the population difference, from $(N_2 - N_1)_0$ to $(N_2 - N_1)_{th}$ for frequencies at the line center. Accordingly, the gain coefficient

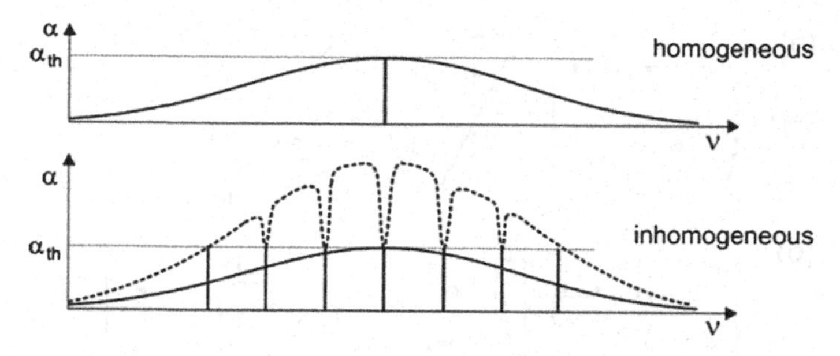

Fig. 12.3 Continuous wave laser based on a homogeneously broadened transition (*upper part*) or an inhomogeneously broadened transition (*lower part*)

changes from the small-signal gain coefficient to the threshold gain coefficient α_{th} for frequencies at the line center. Then the population difference and the gain coefficient are not sufficient for laser oscillation at frequencies in the wings of the line.
Examples of continuous wave lasers based on transitions with homogeneous line broadening: Nd:YAG laser; titanium–sapphire laser.

If laser oscillation is based on an inhomogeneously broadened line, a cw laser can oscillate on all modes that reach the threshold gain. Laser oscillation on one mode does not directly influence the population of two-level atomic systems that contribute to oscillation on other modes.
Examples of lasers based on transitions with inhomogeneous broadening: helium–neon laser; cw CO_2 laser.

The gain curve $\alpha(\nu)$ of a laser operated with an inhomogeneous broadened line shows 'holes" (*see* Fig. 12.3)—the effect is a manifestation of *spectral hole burning*. Irradiation of a medium with laser radiation can lead to a hole in the absorption spectrum of a medium. (Generation of a spectral hole in a medium by using a pulsed laser and the probing of the spectrum with cw radiation or with probe pulses allows for the measurement of the lifetime of a spectral hole; different methods of spectral hole burning are widely used in physics and chemistry for studying spectral properties of various media.)

12.6 Q-Switched Lasers

We discuss a few methods of Q-switching.

- *Mechanical Q-switching* (Fig. 12.4a). The reflector of the laser resonator rotates (for example with an angular frequency of 100 turns per second). The resonator has a high Q factor only during a short time in which the rotating mirror is oriented parallel to the output coupling mirror. During the time of a low Q factor, the upper laser level is populated and during the time of large Q, it is depopulated.

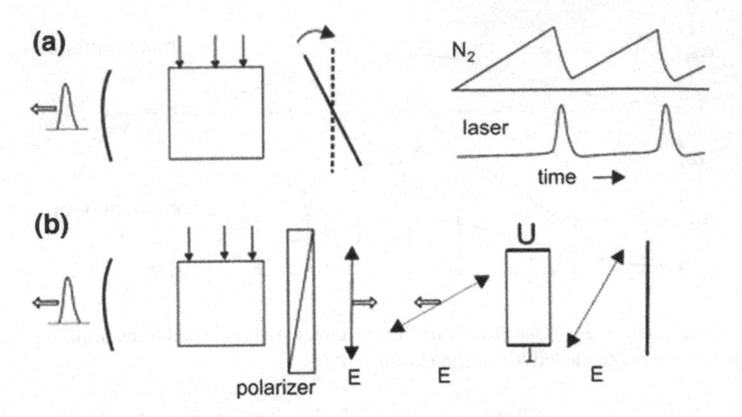

Fig. 12.4 Q-switching. **a** Mechanical and **b** electro-optic Q-switching

Example Generation of pulses (duration 100 ns) with a Q-switched CO_2 laser.

- *Electro-optic Q-switch making use of the Pockels effect* (Fig. 12.4b). A Pockels cell switches the Q factor of the laser resonator from a low to a high value. An optically isotropic crystal becomes birefringent when a static voltage (U) across the crystal is applied and produces a static field E_s in the crystal. Then the refractive indices for a light field E are different for the polarization directions parallel and perpendicular to the static field. The difference of the refractive indices is given by

$$n(\boldsymbol{E}||\boldsymbol{E}_s) - n(\boldsymbol{E}\perp\boldsymbol{E}_s) = aU, \tag{12.3}$$

where a is a material constant. A Pockels cell with applied voltage rotates the polarization direction of the light after two transits through the cell by $\pi/2$. A polarizer blocks the radiation. When the voltage is quickly turned off, the crystal is no longer blocking the radiation—a laser pulse builds up in the resonator. The Pockels effect is large for KDP (for a specific crystal orientation). A voltage of about 25kV is necessary for Q-switching with a crystal of 5mm height and 5 cm length.

- *Electro-optic Q-switch making use of the Kerr effect.* In a Kerr cell, an isotropic medium becomes birefringent under the action of a static field. The difference between the refractive indices of the ordinary and the extraordinary beam varies quadratically with the voltage,

$$n(\boldsymbol{E}||\boldsymbol{E}_s) - n(\boldsymbol{E}\perp\boldsymbol{E}_s) = b\,U^2. \tag{12.4}$$

A static field orients the molecules in a Kerr cell giving rise to birefringence. The effect is especially large for liquid nitrobenzene ($C_6H_5NO_2$). A voltage of about 10 kV is necessary at a cell size of 1 cm height and 1 cm length.

- *Q-switching with a saturable absorber* (Fig. 12.5a). As an example of Q-switching with a saturable absorber, we discuss Q-switching with a dye solved in a liquid. The

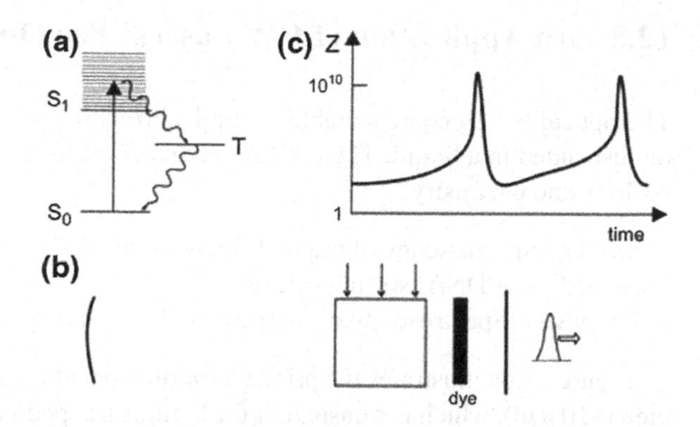

Fig. 12.5 Q-switching with a saturable absorber. **a** Saturation process for dye molecules. **b** Arrangement. **c** Number of photons in the laser resonator

ground state of a dye molecule is a singlet state (S_0). The two lowest excited states are a singlet state (S_1) and a triplet state (T), at a smaller energy. By transitions $S_0 \rightarrow S_1$, laser radiation is absorbed and by nonradiative transitions $S_1 \rightarrow$ T, molecules are transferred into the triplet state. A triplet state has a long lifetime (0.1–1 μ s); the decay of a triplet state occurs mainly via nonradiative transitions. A Q-switched laser based on a saturable absorber (Fig. 12.5b) is continuously pumped. At the start of the pumping, a laser field begins to build up. The buildup of a laser field occurs slowly because of absorption of radiation by the dye molecules. The absorption saturates the dye molecules, then almost all dye molecules are in the triplet state and the dye cell becomes transparent giving rise to generation of a strong laser pulse (Fig. 12.5c). During the buildup of a pulse, the population of the upper laser level becomes reduced to a low value. Due to relaxation of the dye molecules to their ground state, absorption sets in and the Q value becomes small. The buildup of a laser field begins again. Pumping is possible with radiation of another laser.

We will later (in Sect. 13.2) discuss other methods of Q-switching. Dye molecules are discussed in more detail in Sect. 16.1.

12.7 Longitudinal and Transverse Pumping

An active medium can be obtained by longitudinal or transverse pumping.

- *Examples of longitudinal pumping*: pumping with a gas discharge, with the electric field being parallel to the laser beam (Chap. 14); optical pumping with laser radiation whose propagation direction is along the laser beam (Sect. 5.2).
- *Examples of transverse pumping*: pumping with a gas discharge, with the electric field, which drives the discharge, oriented perpendicular to the laser beam (Sect. 14.8); optical pumping with a gas discharge lamp.

12.8 An Application of CW Lasers: The Optical Tweezers

The optical tweezers are suitable for optical trapping of single biomolecules solved or suspended in a liquid. The optical tweezers have many applications especially in biology and chemistry.

- *Biology*. Spectroscopy of trapped single molecules (e.g., red blood cells); study of properties of DNA; sorting of cells.
- *Chemistry*. Spectroscopy of trapped single organic macromolecules.

Figure 12.6a illustrates the principle of the optical tweezers. A glass pearl (diameter 1–10 μm), which is transparent for light, is trapped in the focus of a laser beam. The beam (power $\sim$ 1 mW) is strongly focused. When the glass pearl has a position below the focus, the pearl acts as a lens and the light leaves the pearl at an angle of aperture that is smaller than the angle under which it enters the pearl. Accordingly, the light gains momentum in the direction of the light beam, namely downward, and therefore the pearl gains momentum toward the focus. When the pearl has a position above the focus, it acts as a diverging lens and the force on the pearl is downward. The light leaves the pearl under the same angular distribution as it enters the pearl and there is no momentum transfer. When the glass pearl is located at the height of the focus but shifted to the left, the light beam is deflected to the left and the pearl moves toward the focus. Finally, when the pearl is on the right side of the focus, there is a force toward the left. Thus, the stable location of the pearl is the focus of the lens.

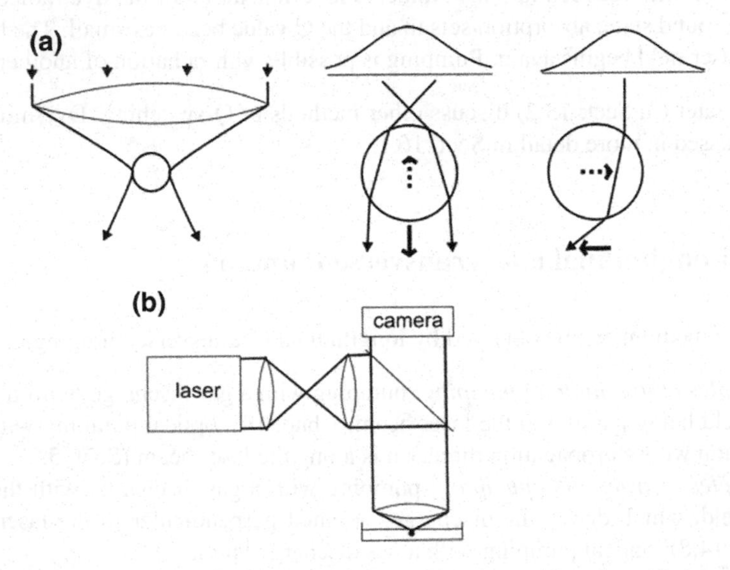

Fig. 12.6 Optical tweezers. **a** Trapping of a glass pearl in a focused laser beam. **b** Arrangement

An optical tweezers system can be realized as a modified microscope (Fig. 12.6b). The object lens produces a strongly focused beam of laser radiation. A camera monitors reflected radiation. The optical tweezers are able to trap a macromolecule in a solvent. Investigation of a trapped molecule is possible by using spectroscopic techniques, e.g., by studying fluorescence radiation. Applications in biology are described in [79].

12.9 Another Application: Gravitational Wave Detector

An extraordinary ambitious project concerns the goal to detect gravitational waves (*see*, for instance, [85]). The collapse of a big star generates (according to the general theory of gravitation) a gravitational wave, propagating with the speed of light. The wavelength of a gravitational wave is of the order of the extension of the collapsing star. A gravitational wave is expected to compress the space in a direction perpendicular to the propagation direction and to dilate the space in the other direction perpendicular to the propagation direction. At present, gravitational wave detectors based on the Michelson interferometer are built and tested at many places.

The center of the gravitational wave detector (Fig. 12.7) is a Michelson interferometer with a laser light source. The Michelson interferometer has two arms arranged perpendicular to each other. A beam splitter divides the laser beam into two beams. A gravitational wave pulse traversing the Michelson interferometer is expected to shorten one arm and to lengthen the other arm. Estimates of the effect suggest that the path length difference may only be of the order of 10^{-22} m for $L_1 = L_2 = 1$ km. The experiment requires an extremely high stability of the laser and of the arrangement and, furthermore, an extremely high sensitivity of detection. To have a chance to observe a signal, the arms have to be very long (of the order of 1 km or much longer). Making use of satellites, very large arms are realizable. The arms have slightly different lengths ($L_1 - L_2 = \lambda/4$, where λ is the wavelength of the laser radiation) in order to have a high sensitivity of detection.

In 2016, observation of a gravitational wave has been reported [311].

References [1–4, 6, 77–85, 127–129, 310, 311].

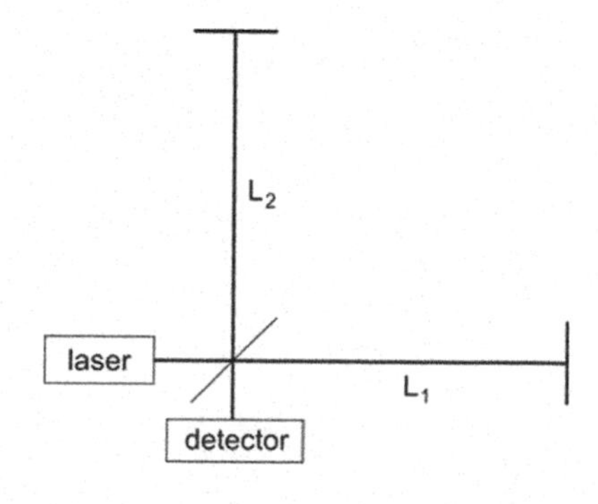

Fig. 12.7 Gravitational wave detector

Problems

12.1 Resonance condition for a planparallel plate. Derive the resonance condition (12.1) for a planparallel plate. [*Hint*: Determine the difference of the optical path of a beam directly reflected at the surface of the plate and a beam reflected at the backside of the plate.]

12.2 Michelson interferometer. A Michelson interferometer operates with a parallel laser beam (wavelength $\lambda = 580$ nm; P_0 = power of the laser radiation).

(a) Calculate the intensity $I(x)$ at the detector for a length difference x when one arm has the length L and the other arm the length $L + x$.
(b) Determine the path difference δx for values of x in the interval $x_0 \leq x \leq x_0 + \lambda$ for $x_0 = 1$ km that leads to the largest signal-to-noise ratio for the signal.
(c) Estimate the change of the signal if one of the arms changes its length by $\delta L/L = 10^{-15}$ and the other arm by $\delta L/L = -10^{-15}$. [*Hint*: The beam splitter in the Michelson interferometer splits an incident electromagnetic field into two fields.]

12.3 It is possible to reduce suddenly the reflectivity of the output mirror of a laser. Show, qualitatively, that this "cavity dumping" results in a much stronger laser pulse than without cavity dumping.

Chapter 13
Femtosecond Laser

Mode locking allows a laser with a broad gain bandwidth to generate femtosecond pulses. A mode-locked laser oscillates at the same time on a large number of modes. The fields of all modes are phase-locked to each other.

We describe the principle of mode locking, techniques of mode locking, and a method of determination of the duration of femtosecond pulses. We explain the pump-probe method, which is suited to take ultrashort snapshots during dynamical processes in an atomic system. Additionally, we study the onset of oscillation of a femtosecond laser.

The femtosecond laser is the basis of a great variety of new areas of research and applications. We discuss: femto-chemistry; optical frequency analyzer; terahertz time domain spectrometer; and attosecond pulses—that is an area of nonlinear optics with very strong optical fields. Other applications concern surgery and material processing.

An optical frequency analyzer makes use of an optical frequency comb. A titanium–sapphire laser generates a frequency comb with a frequency distribution that extends over about an octave. We show in this chapter that the exact position of the frequencies of a frequency comb generated with a femtosecond laser are determined by: the optical length of the resonator; the Gouy phase shift; dispersion of the active medium; and dispersion of the optical elements. Making use of methods of nonlinear optics, the distribution can be broadened—a frequency comb can consist of fields at equally spaced frequencies corresponding to radiation from the near infrared to the near ultraviolet. The position of the frequencies generated by a particular laser can be determined with a very high accuracy (Sect. 35.7).

We will introduce (Sect. 13.1) the mode locking without taking account of Gouy phase shift and dispersion. In Sect. 13.4, we will discuss the role of the Gouy phase shift and of dispersion.

K.F. Renk, *Basics of Laser Physics*, Graduate Texts in Physics,
DOI 10.1007/978-3-319-50651-7_13

13.1 Mode Locking

The secret of the femtosecond laser is the mode locking: the laser oscillates at the same time on a large number of longitudinal modes—with equal frequency separation between next-near modes—and all oscillations have fixed phases relative to each other.

We choose an active medium with a broad gain coefficient profile (Fig. 13.1a) that has a Gaussian shape. The spectral profile $F^2(\omega)$ for a particular femtosecond laser depends on parameters of the laser. For a first treatment of a femtosecond laser, we assume that the profile has a rectangular shape and that the width is equal to the gain bandwidth (Fig. 13.1b). Accordingly, the amplitude A of the field components is constant within the frequency range $\omega_0 - \Delta\omega_g/2$, $\omega_0 + \Delta\omega_g/2$ and zero outside this range. The frequency distribution represents an *optical frequency comb*: the frequency distribution consists of equally spaced peaks. The frequency separation between next-near peaks is equal to

$$\Omega = 2\pi/T = 2\pi c/2L = \pi c/L, \tag{13.1}$$

where $T = 2L/c$ is the round trip transit time. The number of modes with frequencies in the gain bandwidth $\Delta\omega_g$ is

$$N = \Delta\omega_g/\Omega. \tag{13.2}$$

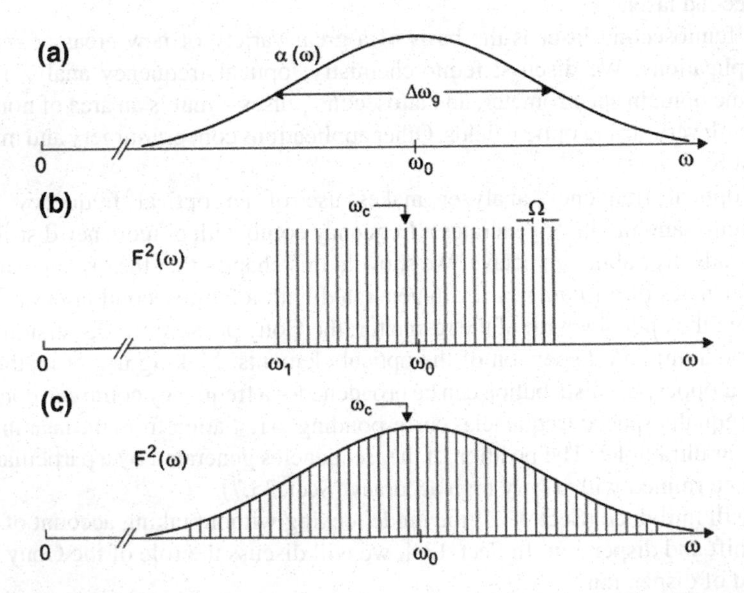

Fig. 13.1 Mode locking. **a** Gain coefficient profile. **b** Spectral intensity profile with a rectangular shape. **c** Spectral intensity profile with a Gaussian shape

The instantaneous electric field at a fixed location z in the laser resonator is given by

$$\tilde{E}(t) = A \sum_{s=0}^{N-1} e^{i[(\omega_1 + s\Omega)t + \varphi_s]}, \tag{13.3}$$

where φ_s is the phase of the mode s and $\omega_1 = s_1\omega$ is the lowest frequency of the oscillating modes; s_1 is an integer. Without mode coupling, the fields of the different modes have different phases (which fluctuate with time). Therefore, the field fluctuates very strongly. The laser emits laser radiation in a broad frequency band ($\Delta\omega_g$); the radiation propagates along the resonator axis. The average intensity of the laser radiation is

$$I_{\text{incoh}} = \frac{1}{2} c\varepsilon_0 \text{NA}^2; \tag{13.4}$$

A is the amplitude of the field components.

Mode locking forces the fields to oscillate in phase,

$$\varphi_s(t) = \varphi_s = \varphi. \tag{13.5}$$

We choose the timescale so that

$$\varphi_s(0) = \varphi_s = 0. \tag{13.6}$$

A round trip transit changes the phase of the fields at fixed z by 2π (Fig. 13.2). The field at a fixed location z in the resonator is given by

$$\tilde{E} = A \sum_{s=0}^{N-1} e^{i(\omega_1 + s\Omega)t} = A \sum_{s=0}^{N-1} e^{is\Omega t} e^{i\omega_1 t}. \tag{13.7}$$

This is a geometric series. With $r = e^{i\Omega t}$, we have

$$\tilde{E} = A(1 + r + r^2 + ... + r^{N-1})e^{i\omega_1 t} = A\,\frac{1 - r^N}{1 - r} e^{i\omega_1 t} = A\,\frac{1 - e^{iN\Omega t}}{1 - e^{i\Omega t}}\, e^{is_1\Omega t}. \tag{13.8}$$

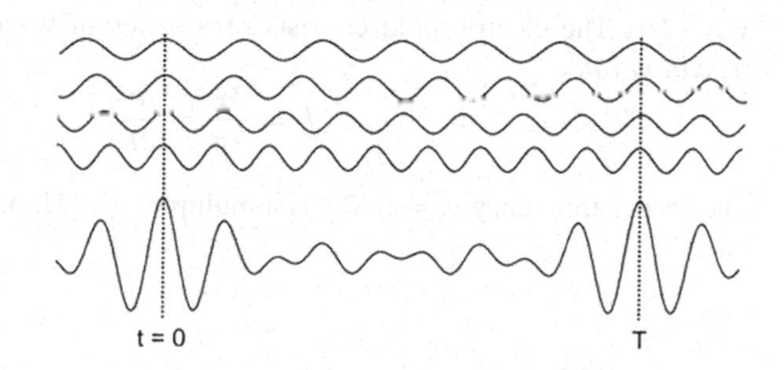

Fig. 13.2 Field of different modes and total field at a fixed position z in a laser

We introduce the carrier frequency ω_c. The carrier frequency is a multiple of the round trip transit frequency (*see* Fig. 13.1c) and lies in the vicinity of the center frequency ω_0. We can write

$$\tilde{E} = A\,\frac{e^{-i\frac{1}{2}N\Omega t} - e^{i\frac{1}{2}N\Omega t}}{e^{-i\frac{1}{2}\Omega t} - e^{i\frac{1}{2}\Omega t}}\,e^{i[\omega_1 t + (N/2-1/2)\Omega t]} = A\,\frac{\sin\left(\frac{1}{2}N\Omega t\right)}{\sin\left(\frac{1}{2}\Omega t\right)}\,e^{i\omega_c t}. \tag{13.9}$$

The carrier frequency is given by

$$\omega_c = \omega_1 + (N/2 - 1/2)\,\Omega \tag{13.10}$$

if N is an odd number and by

$$\omega_c = \omega_1 + (N/2)\,\Omega \tag{13.11}$$

if N is an even number. The carrier frequency is a multiple of the frequency separation between next-near peaks,

$$\omega_c = l_c \times \Omega, \tag{13.12}$$

where l_c is an integer. The femtosecond pulse train does not change if we add another integer to l_c (or subtract another integer from l_c) as long as this number is small compared to l_c.

The real part of the field is

$$E = \mathrm{Re}[\tilde{E}] = A\,\frac{\sin\left(\frac{1}{2}N\Omega t\right)}{\sin\left(\frac{1}{2}\Omega t\right)}\,\cos\omega_c t. \tag{13.13}$$

We write

$$E = A(t)\cos\omega_c t, \tag{13.14}$$

where

$$A(t) = A\,\frac{\sin\left(\frac{1}{2}N\Omega t\right)}{\sin\left(\frac{1}{2}\Omega t\right)} \tag{13.15}$$

is a time-dependent amplitude. The amplitude has the form $\sin(Nx)/\sin x$, where $x = \frac{1}{2}\Omega t$. The electric field consists of a series of wave packets with the *pulse repetition rate*

$$f_r = \frac{\Omega}{2\pi} = \frac{c}{2L}. \tag{13.16}$$

The carrier frequency $\nu_c = \omega_c/2\pi$ is a multiple of $c/2L$, i.e., of the pulse repetition rate.

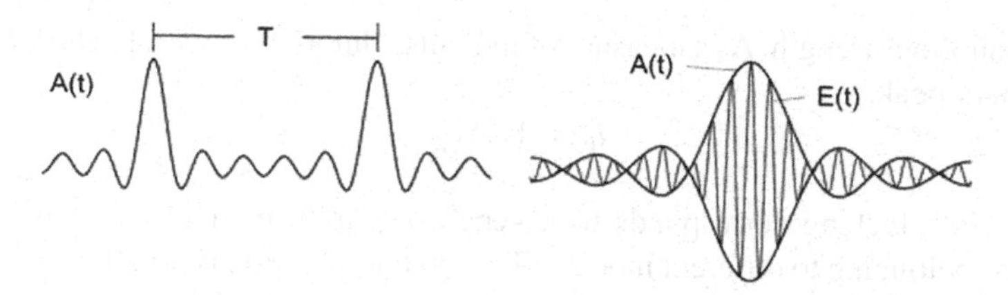

Fig. 13.3 Amplitude of the wave train of femtosecond pulses and field of a pulse

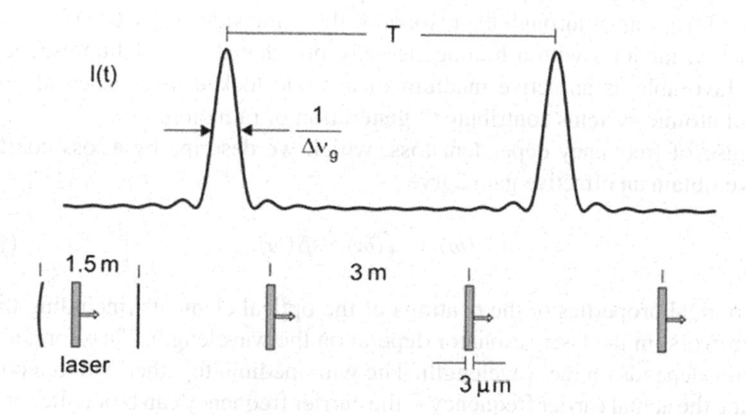

Fig. 13.4 Femtosecond pulse train on the timescale (*at a fixed location*) and in space (*at a fixed time*)

The amplitude of the field (Fig. 13.3) shows main maxima and side maxima.

- *Main maxima* occur for $\Omega t/2 = s \times \pi$. Main maxima appear at the times

$$t = sT \text{ with } s = 0, 1, 2, \ldots. \tag{13.17}$$

The temporal distance between next-near pulses is equal to the round trip transit time of radiation in the laser.

- The first point, t_1, of zero amplitude follows from the relation $N\Omega t_1/2 = \pi$, leading to $t_1 = 2\pi/(N\Omega) = T/N$.

A femtosecond pulse train is coherent. It consists of periodically repeated wave packets. The amplitude $A(t)$ is the envelope of the electric field curve $E(t)$.

The intensity $I(t)$ of a femtosecond pulse train (Fig. 13.4, upper part) has main maxima and side maxima. The peak intensity,

$$I_{\text{peak}} = \frac{1}{2} c \varepsilon_0 A^2 N^2, \tag{13.18}$$

is proportional to the square of the number of oscillating modes; in case of a measurement (outside the laser resonator), the peak intensity is smaller according to the output coupling strength. As a measure of the pulse duration t_p, we take the halfwidth of the main peak,

$$t_p = 1/\Delta\nu_g. \tag{13.19}$$

The mode locking corresponds to a synchronization of fields of different frequencies (belonging to different modes). The synchronization is possible because all frequencies are multiples of the same fundamental frequency $\Omega = 2\pi/T$ according to the resonance condition $\omega_l = l \times \Omega$; the phase of each field component has, after each round trip transit through the resonator, the same value ($\varphi_l = \varphi = 0$).

An active medium with a homogeneously broadened $2 \rightarrow 1$ fluorescence line is most favorable as an active medium of a mode locked laser. Then all excited two-level atomic systems contribute to generation of radiation.

Because of frequency dependent loss, which we describe by a loss coefficient $\beta(\omega)$, we obtain an effective gain curve

$$\alpha_{\text{eff}}(\omega) = \alpha(\omega) - \beta(\omega). \tag{13.20}$$

The optical properties of the coatings of the optical elements (including the resonator mirrors) in the laser resonator depend on the wavelength. Therefore, also the loss factor depends on the wavelength. The gain medium together with the coatings determine the actual carrier frequency—the carrier frequency can be smaller or larger than ω_0. Thus, the carrier frequency can be chosen by making use of appropriate optical elements.

Mode locking is possible by active mode locking or passive mode locking. We will describe techniques of mode locking in the next section.

Before, we should mention that the field $E(t)$ is the *Fourier transform* of the frequency spectrum $F(\omega)$ and vice versa:

$$E(t) = \frac{1}{2\pi}\int_{-\infty}^{+\infty} F(\omega)\mathrm{e}^{\mathrm{i}\omega t}\,\mathrm{d}\omega \tag{13.21}$$

and

$$F(\omega) = \int_{-\infty}^{+\infty} E(t)\mathrm{e}^{-\mathrm{i}\omega t}\,\mathrm{d}t. \tag{13.22}$$

Figure 13.5 (upper part) shows a rectangular spectral intensity profile $F^2(\nu)$. The lower part shows $E(t)$, obtained by a Fourier transformation of $F(t)$, together with the envelope function $A(t)$. The product of the pulse duration and the gain bandwidth is equal to unity, $t_p\Delta\nu_g = 1$.

Example titanium–sapphire laser (*see* Fig. 13.4, lower part). Width of the gain profile $\Delta\nu_g = 1.1 \times 10^{14}$ Hz (Sect. 7.6); length of the resonator 1.5 m; pulse duration $t_p = 9$ fs; spatial length of a single pulse $ct_p = 3\,\mu$m; distance between subsequent pulses ~ 3 m. The number of phase-locked modes is $N \sim 1 \times 10^{14}$ Hz/10^8 Hz $\sim 10^6$. The round trip transit time of a light pulse is $T = 10$ ns and the pulse repetition rate $f_r = 100$ MHz.

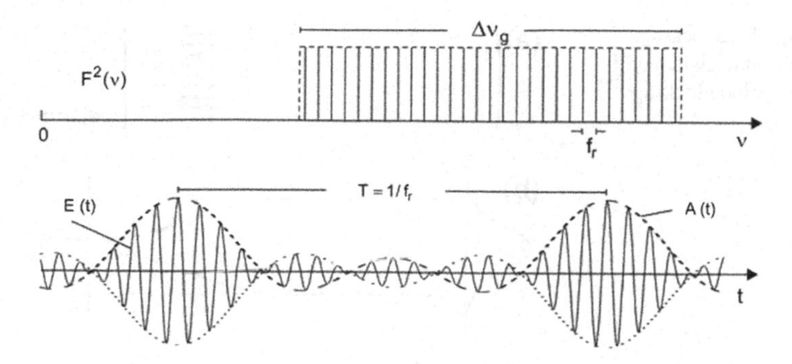

Fig. 13.5 A frequency comb: the spectral intensity profile (of rectangular shape) and the field

We mention here also *Gaussian pulses* [86]. A Gaussian pulse is characterized by the following quantities.

- Spectral intensity profile $F^2(\omega) = \exp[-4 \ln 2(\omega - \omega_0)^2/\Delta\omega_g^2]$; $\Delta\omega_g =$ gain bandwidth (FWHM).
- Fourier transformation of $E(\omega)$ yields the time-dependent amplitude $A(t)$ and thus the temporal intensity profile $I(t)/I_p = A^2(t)/A^2$, where I_p is the peak intensity.
- Temporal intensity profile $I(t)/I_p = \exp[-4 \ln 2t^2/t_p^2]$; $t_p = 4 \ln 2/\Delta\omega_g =$ pulse duration (FWHM)

For a Gaussian profile of the spectral-intensity envelope, the pulse duration bandwidth product is, with $\Delta\nu_g = \Delta\omega_g/2\pi$ equal to

$$t_p \Delta\nu_g = \frac{2 \ln 2}{\pi} = 0.441. \tag{13.23}$$

According to the gain bandwidth of titanium–sapphire, pulses with a pulse duration as short as ~4 fs should be attainable; pulses of a duration of ~5 fs have indeed been observed [87]. (Note: in comparison with a rectangular shape of the intensity profile, a Gaussian shape with the same gain bandwidth has a broader spectral distribution of the amplitudes and leads therefore to shorter pulses.)

13.2 Active and Passive Mode Locking

An acousto-optic modulator (Fig 13.6a) is suitable for active mode locking. The switch consists of a periodically in time-varying diffraction grating. An ultrasonic wave in a crystal modulates spatially the mass density of the crystal and therefore the refractive index. Every half period of the ultrasonic field, the modulation disappears during a short moment. Therefore, there is no diffraction pattern for a short moment and a light pulse passes the modulator without diffraction loss. Laser pulses pass the modulator without diffraction at twice the ultrasonic frequency f_s. The frequency of

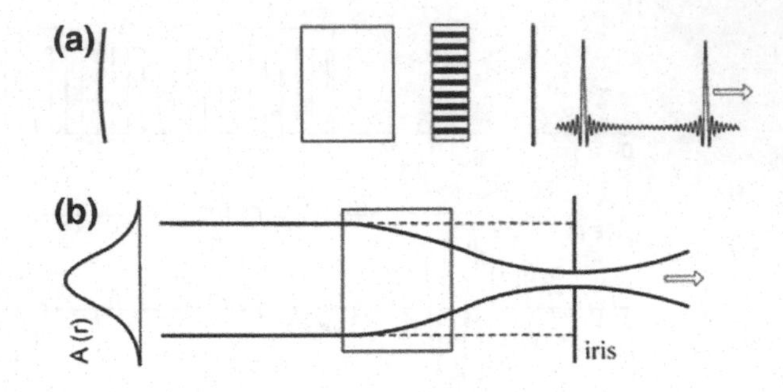

Fig. 13.6 Mode locking. **a** Acousto-optic switch and **b** Kerr lens mode locking

the ultrasonic wave is $2f_s = 1/T$. A laser of length 1.5 m, $1/T = c/2L = 10^8$ Hz (100 MHz), requires an ultrasonic wave of a frequency of 50 MHz.

Kerr lens mode locking is a method of passive mode locking. The refractive index of a material depends on the radiation intensity I,

$$n(x, y, t) = n_0 + n_2 I(x, y, t), \tag{13.24}$$

where n_0 is the refractive index and n_2 the Kerr coefficient.

Figure 13.6b shows the principle of the Kerr lens mode locking: a strong laser pulse produces a Kerr lens in the Kerr medium due to self-focusing of the radiation (Sect. 35.6). The Kerr lens is a transient lens; it exists only during the passage of the laser pulse through the Kerr medium. Focusing does not occur for radiation belonging to the wings of the temporal distribution of the intensity. Therefore, the Kerr lens cuts radiation in the wings of the temporal distribution of the intensity. At steady state oscillation of a femtosecond laser, a pulse lengthening during a round trip transit through the laser resonator is compensated by the action of the Kerr lens.

The active medium itself, a Ti^{3+}:Al_2O_3 crystal, is suitable as a Kerr lens in a titanium–sapphire laser. The Kerr coefficient of sapphire has the value $n_2 = 3 \times 10^{-20}$ m^2 W^{-1}; for an estimate of n_2, *see* Sect. 35.6 and Problem 35.3. To reach a change of the refractive index ($n = 1.74$) by an appreciable amount (for example, by 0.3), the power density of the radiation has to be very large ($\sim 10^{19}$ W m^{-2}).

The titanium–sapphire femtosecond laser (Fig. 13.7a) contains a chirped mirror that compensates different optical path lengths of radiation of different wavelengths; a chirped mirror consists of an antireflecting surface layer and of multilayers composed of layers of different thicknesses. The orientation of the crystal surfaces of the titanium–sapphire crystal correspond to the Brewster angle. A pump laser produces population inversion. The spectral distribution of the radiation emitted by a femtosecond titanium–sapphire laser (Fig. 13.7b) corresponds (for pulses with a duration of about 5 fs) to a frequency width (110 THz), which is about a third of the carrier frequency (360 THz). Femtosecond pulse operation is possible at different carrier frequencies; however, a modification that corresponds to an effective spectral gain coefficient leads to a narrowing of the spectral gain coefficient profile and therefore to a lengthening of the femtosecond pulses.

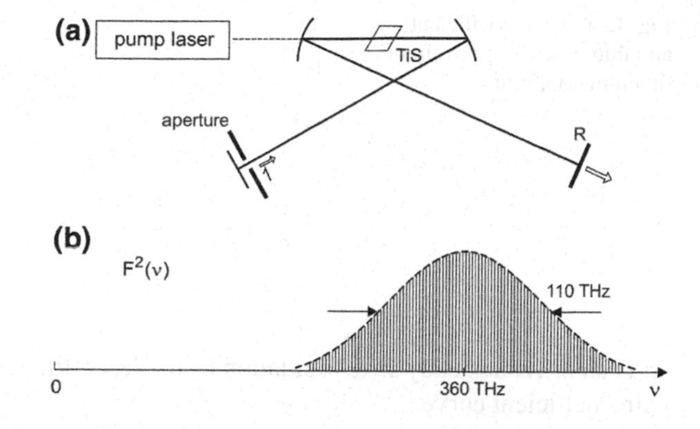

Fig. 13.7 Titanium–sapphire laser. **a** Arrangement; R, chirped mirror. **b** Spectral intensity distribution of laser radiation of a mode locked titanium–sapphire laser

Another technique of mode locking makes use of a saturable absorber. A laser pulse saturates the absorption. After each transit of a pulse through the saturated absorber, the pulse is amplified in the gain medium and the population of the upper laser level is strongly reduced. After each transit of a pulse, the population builds up again. Dye molecules solved in a solvent are suitable as saturable absorbers of visible radiation (Sect. 12.6).

Later, we will discuss a further technique of passive mode locking, namely mode locking via an intensity-dependent reflectivity of a mirror (Sect. 15.6).

13.3 Onset of Oscillation of a Mode-Locked Titanium–Sapphire Laser

We now discuss onset of oscillation of a mode-locked titanium–sapphire femtosecond laser. After turning on the optical pumping of the titanium–sapphire crystal, spontaneous emission of radiation (fluorescence radiation) initiates oscillation of the laser.

The gain coefficient of titanium–sapphire has its maximum at a frequency $\nu_0 \approx$ 350 THz (Fig. 13.8), while the maximum of the spectrum of spontaneous emission lies at a higher frequency; gain curve and fluorescence curve of titanium–sapphire are studied in Sects. 5.3 and 7.6, and in Chap. 17. Spontaneously emitted radiation emitted in the low frequency wing of the fluorescence spectrum initiates oscillation. The shape of the fluorescence spectrum deviates strongly from a Gaussian: emission of radiation is still very strong in the low frequency wing of the fluorescence line. Therefore, initial radiation is available in the whole range in which the gain coefficient is sufficiently large for building up laser oscillation. Finally, a radiation field is build up that corresponds to the shape of the gain coefficient (supposed that the elements in the laser resonator work appropriately in a very broad frequency band). At the onset of oscillation of the laser, spontaneous emission produces an initial photon distribution centered at $\nu_0 + \Delta\nu_0/2$. Amplification shifts the spectrum toward smaller frequencies

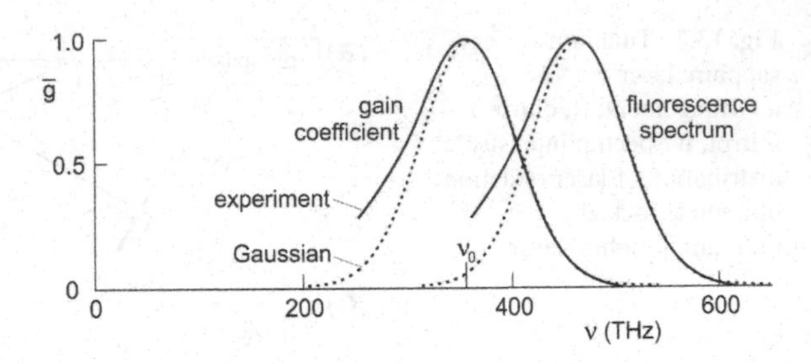

Fig. 13.8 Gain coefficient and fluorescence spectrum of titanium–sapphire

and can have, at steady state oscillation of the laser, the same Gaussian shape as the gain coefficient curve.

A femtosecond laser, which operates with an active medium whose gain coefficient has a Gaussian profile, can have a spectral intensity profile that can also have a Gaussian shape (*see* Fig. 13.1). How is this possible? In order to answer this question, we study a titanium–sapphire femtosecond laser containing a Kerr lens.

We discuss the corresponding requirement with respect to the Q-switch. In the steady state of an oscillating Q-switched laser, a field component E_s (oscillating on mode s) reproduces itself after a round trip transit of the radiation through the resonator. This leads to the condition

$$V_{\mathrm{out}}\ V_{\mathrm{s}}(\nu)\ G(\nu) = 1,$$

where $G(\nu)$ is the gain factor for the field at frequency ν, V_{out} is the V factor describing output coupling loss, and $V_{\mathrm{s}}(\nu)$ is the V factor describing intrinsic loss (Sect. 2.4). $V_{\mathrm{s}}(\nu)$ should have a value near unity at frequencies of large gain. (While the gain factor is large during the time a pulse passes through the active medium and small in the time between two subsequent pulses, the loss factors do not depend on time.) If the gain factor is only slightly larger than unity, $2\alpha(\nu)L_{\mathrm{ac}} \ll 1$, then we can write $G(\nu) = 1 - 2\alpha(\nu)L_{\mathrm{ac}}$ and find

$$V_{\mathrm{out}}\ V_{\mathrm{i}}(\nu) = 1 - 2\alpha(\nu)L_{\mathrm{ac}}.$$

L_{ac} is the length of the active medium. Thus, in the special case described by the last equation, the gain curve has the same shape as the gain coefficient curve.

13.4 Optical Frequency Comb

In the preceding section, we assumed that the mode-locked fields have frequencies that are multiples of the repetition rate. We neglected three effects—Gouy phase shift; phase shift due to dispersion of the active medium; and phase shift due to dispersion of the optical elements.

We first discuss the influence of the Gouy phase shift on the wave packets generated by a femtosecond laser. Instead of (13.7), the field is equal to

$$\tilde{E} = A\sum_{s=0}^{N-1} e^{i[(\omega_1+s\Omega)t+\omega_{Gouy}t]} = A\sum_{s=0}^{N-1} e^{is\Omega t} e^{i(\omega_1+\omega_{Gouy})t}, \tag{13.25}$$

where $\omega_{Gouy} = 2\pi\nu_{Gouy}$ is the Gouy angular frequency. It follows, from a calculation according to (13.10)–(13.15), that the field is given by

$$E = A(t)\cos(\omega_c + \omega_{Gouy})t, \tag{13.26}$$

$$A(t) = A\,\frac{\sin(\frac{1}{2}N\Omega t)}{\sin(\frac{1}{2}\Omega t)}. \tag{13.27}$$

While ω_c is still a multiple of the repetition rate Ω, the frequency $\omega_c + \omega_{Gouy}$ is not a multiple of Ω (except for the Gouy phase zero). Taking into account that $\omega_{Gouy} \ll \omega_c$, we can write

$$E = A(t)\cos[\omega_c t + \varphi_{ce}(t)], \tag{13.28}$$

wherc

$$\varphi_{ce}(t) = 2\pi\nu_{Gouy}t = \Delta\phi\, t/T \tag{13.29}$$

is a time-dependent phase, the *carrier envelope phase*, and $\Delta\phi$ the Gouy phase shift per round trip transit. The carrier envelope phase varies slowly in comparison with the phase $\omega_c t$. A variation by 2π occurs in a time distance that corresponds to many periods of the carrier frequency.

A frequency comb (Fig. 13.9a) is characterized by:

- $\nu_c = \omega_c/2\pi$ = carrier frequency (near the frequency ν_0 of maximum gain);
- $f_r = c/2L = 1/T = \Omega/2\pi$ = pulse repetition rate (=pulse repetition frequency);
- f_o = offset frequency;
- $\nu_l = l \times f_r + f_o$ = frequencies of the frequency comb;
- $\nu_{l+1} - \nu_l = \Omega/2\pi$ = c/2L = frequency distance between next-near peaks.

The halfwidth of a peak is determined by the pulse duration. The field (Fig. 13.9b) shows that there is a jitter between the $A(t)$ curve and the $\cos(\omega_c t + \varphi_{ce})$ curve. $A(t)$ is periodic with the period $1/T$. The $\cos(\omega_c t + 2\pi\nu_{Gouy}t)$ term changes continuously its phase from $2\pi\nu_{Gouy}t = 0$ to 2π but remains, in the time average, synchronous to $A(t)$; the fields $E_1(t)$, $E_2(t)$, $E_3(t)$and $E_4(t)$ (Fig. 13.9b) are the fields at the times $t, t + f_o^{-1}/4, t + f_o^{-1}/2$ and $t + 3f_o^{-1}/4$, respectively.

Example of an offset frequency due to the Gouy phase shift. A femtosecond titanium–sapphire laser (confocal resonator of length $L = 0.5$ m; $f_r = 300$ MHz; Gouy phase shift $\phi_{Gouy} = \pi$) shows an offset frequency $f_o = \nu_{Gouy} = 150$ MHz.

We now discuss the influence of dispersion of an active medium. We characterize the active medium by the gain coefficient $\alpha(\omega)$ (Fig. 13.10a). We assume that

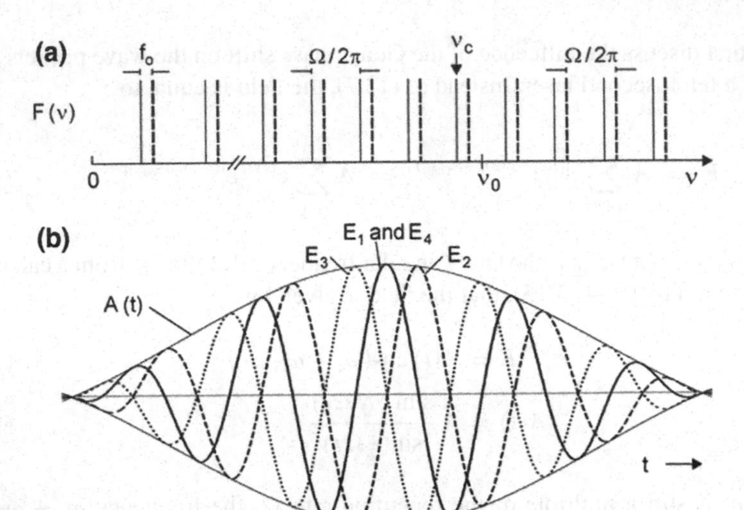

Fig. 13.9 A frequency comb influenced by the Gouy phase shift. **a** Frequency offset. **b** Amplitude of the femtosecond pulses and field curves at different times

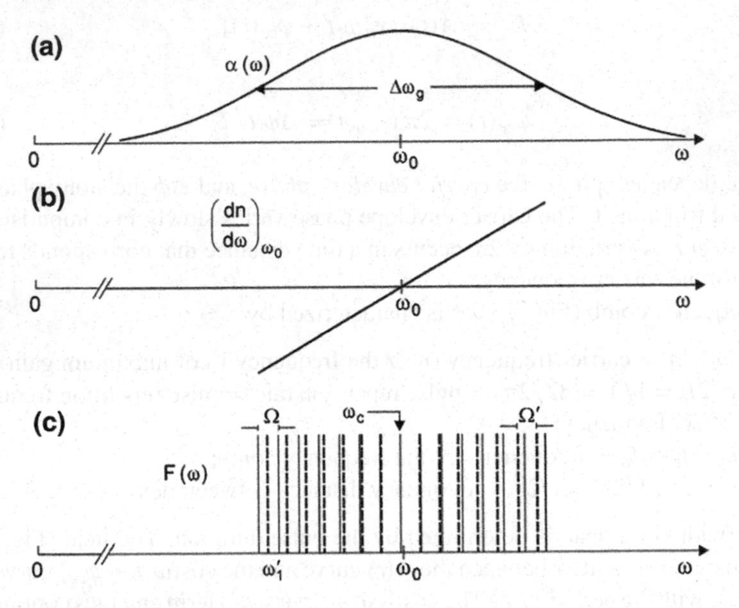

Fig. 13.10 Influence of dispersion in the active medium on a frequency comb. **a** Gain coefficient. **b** Change of refractive index. **c** Frequency comb (*dashed*) with a frequency shift due to dispersion of the active medium

the refractive index increases linearly with frequency (Fig. 13.10b), as expected for frequencies around the frequency ω_0 of maximum gain (Sect. 9.4, *Example*). Above the center frequency ω_0, the change of the refractive index is positive, causing an increase of the resonance frequencies of the laser resonator. Below ω_0, the change of the refractive index is negative, causing a decrease of the resonance frequencies of the resonator. The frequency separation between next-near modes has a constant value (Fig. 13.10c) because the frequency shift is proportional to $\omega - \omega_0$. Without dispersion, the separation between next-near modes is

$$\Omega = \frac{2\pi c}{2L}. \tag{13.30}$$

With dispersion of the active medium (of length L'), the separation between next-near modes follows from

$$\Omega' = \frac{2\pi c}{2L + 2L'\mathrm{d}n/\mathrm{d}\omega\Omega} = \frac{2\pi c}{2L}\left(1 - \frac{L'}{L}\frac{\mathrm{d}n}{\mathrm{d}\omega}\Omega\right) = \Omega\left(1 - \frac{L'}{L}\frac{\mathrm{d}n}{\mathrm{d}\omega}\Omega\right). \tag{13.31}$$

The frequency difference is

$$\Omega - \Omega' = \Omega\,\frac{L'}{L}\frac{\mathrm{d}n}{\mathrm{d}\omega}\,\Omega. \tag{13.32}$$

Dispersion of the active medium reduces the frequency distance between next-near resonances. The field at a fixed position in the laser resonator is given by

$$\tilde{E} = A\sum_{s=0}^{N-1} \mathrm{e}^{\mathrm{i}(\omega_1 + s\Omega')t} = A\sum_{s=0}^{N-1} \mathrm{e}^{\mathrm{i}s\Omega' t}\mathrm{e}^{\mathrm{i}\omega_1 t}. \tag{13.33}$$

In comparison with (13.7), Ω' replaces Ω. The summation leads to

$$E = A(t)\cos\omega_{\mathrm{c}} t, \tag{13.34}$$

where

$$A(t) = A\,\frac{\sin(\frac{1}{2}N\Omega' t)}{\sin(\frac{1}{2}\Omega' t)} \tag{13.35}$$

is the time-dependent amplitude. The repetition rate of the pulses is

$$f_{\mathrm{r}} = 2\pi/\Omega'. \tag{13.36}$$

The carrier frequency ω_{c} is not a multiple of the repetition rate but of $\Omega = 2\pi c/2L$,

$$\omega_{\mathrm{c}} = l_{\mathrm{c}}\Omega, \tag{13.37}$$

where l_c is an integer. We write

$$\omega_c = \omega_c' + \omega_o, \tag{13.38}$$

where

$$\omega_c' = l_c'\Omega' \tag{13.39}$$

now is a carrier frequency that is a multiple of the repetition rate Ω', l_c' is an integer and where

$$\omega_o = l_c\Omega - l_c'\Omega' \tag{13.40}$$

is a carrier offset angular frequency. It follows that the field is

$$E = A\,\frac{\sin(\frac{1}{2}N\Omega' t)}{\sin(\frac{1}{2}\Omega' t)}\,\cos\left[\omega_c' t + \varphi_{ce}(t)\right], \tag{13.41}$$

where

$$\varphi_{ce}(t) = \omega_o t \tag{13.42}$$

is the carrier envelope phase. We find the carrier offset angular frequency

$$\omega_o' = l_c\Omega - l_c'\Omega' = l_c'\Omega\frac{L'}{L}\frac{dn}{d\omega}\Omega - (l_c' - l_c)\Omega. \tag{13.43}$$

We obtain, with $l_c'\Omega \sim \omega_0$, an estimate of the offset angular frequency,

$$\omega_o' = \Omega\frac{L'}{L}\frac{dn}{d\omega}\omega_0 - (l_c' - l_c)\Omega, \tag{13.44}$$

and thus of the offset frequency,

$$f_o = \left(\frac{L'}{L}\frac{dn}{d\omega}\omega_0 - (l_c' - l_c)\right)\frac{c}{2L}. \tag{13.45}$$

We choose l_c' so that f_o is positive but not larger than $c/2L$,

$$f_o \le \frac{c}{2L}. \tag{13.46}$$

Example of an offset frequency due to dispersion of an active medium. A femtosecond titanium–sapphire laser (active medium: population difference $N_2 - N_1 = 10^{22}\,\mathrm{m}^{-3}$, $dn/d\omega = 1 \times 10^{-13}\,\mathrm{s}^{-1}$, crystal length $L' = 1.5\,\mathrm{cm}$; resonator length $L = 0.5\,\mathrm{m}$; pulse repetition rate $f_r = 100\,\mathrm{MHz}$) shows an offset frequency due to dispersion of $f_o = 400\,\mathrm{MHz}$; the offset frequency due to dispersion of the active medium depends on the density of excited two-level states and depends therefore on the pump strength.

A third effect contributes to the carrier envelope offset phase: dispersion of the optical elements in the resonator of a femtosecond laser. Depending on the sign of $dn/d\omega$, of an optical element (for instance, of a reflector), the corresponding change of phase can lead to an increase or a decrease of the frequency separation between next-near peaks of the frequency comb.

Thus, the offset frequency of a femtosecond laser is the sum of the offset frequencies that are caused by Gouy phase shift, dispersion of the active medium, and dispersion of the optical elements.

The field of a frequency comb has peaks the frequencies

$$\nu_l = l f_r + f_o. \tag{13.47}$$

To find exact values of the frequencies of a frequency comb, we have to determine three parameters: order l; pulse repetition rate f_r; and frequency offset f_o. For a particular laser, all three quantities can be determined experimentally with a high accuracy (Sect. 35.7) by the use of techniques based on nonlinear optics.

Nonlinear dispersion, occurring in addition to linear dispersion, disturbs each pulse during a round trip transit through the laser resonator. This disturbance would continuously deform the shapes of the pulses. However, it is strongly suppressed from pulse to pulse by the Kerr lens. Due to the pulse shaping by the Kerr lens, the round trip transit time is strictly periodic: the pulses propagate within the resonator with the group velocity described by the envelope function $A(t+T) = A(t)$. The carrier wave (at the frequency ω_c), on the other hand, propagates with the phase velocity.

It is possible to broaden a frequency comb in the frequency space by the use of techniques of nonlinear Optics (Sect. 35.7). A broad frequency comb, extending over the entire visible spectral range, represents white light. Focusing ultrashort light pulses onto a transparent material (a solid, a liquid or a gas) can lead to generation of a white light continuum (with a spectral super broadening of the femtosecond pulses—the origin is the interplay of short pulses with a dispersive medium; for more information, *see* for instance [86]).

13.5 Optical Correlator

How can we determine the duration of femtosecond pulses? In an autocorrelator (Fig. 13.11a), a beam splitter divides the laser beam into two beams. The paths of the two beams are different but join each other after passing another beam splitter. A lens focuses the beams to a frequency doubler, which produces second harmonic radiation. A filter behind the frequency doubler blocks the radiation of the fundamental frequency. Another lens focuses the second harmonic radiation on a detector. The detector signal is a measure of the strength of the second harmonic radiation. The

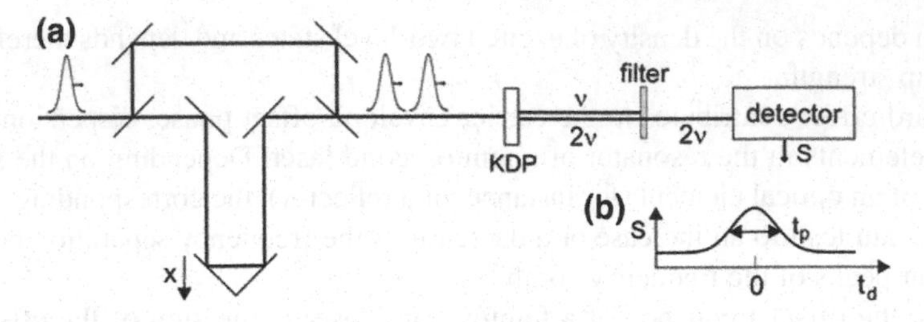

Fig. 13.11 Optical autocorrelator. **a** Arrangement. **b** Signal

two second-harmonic pulses arriving at the frequency doubler have equal strength. The detector has a large response time and monitors the average power of the second harmonic radiation. It is not necessary that the detector is able to resolves the single pulses temporally. A KDP crystal, which has a high nonlinearity of second harmonic generation, is suitable as a frequency doubler. The delay time between the two pulses is

$$t_d = 2x/c, \tag{13.48}$$

where x is the shift of the movable mirror and where $x = 0$ corresponds to equal path lengths of the two pulses. The detector signal (Fig. 13.11b) has a maximum if the delay time is zero. The pulse duration follows from the shape of the signal curve. The intensity of the second harmonic radiation increases quadratically with the intensity,

$$I(2\omega, t, t_d) = K\,|I_1(t) + I_2(t + t_d)|^2 . \tag{13.49}$$

K is a constant and I_1 and I_2 are the intensities of the two beams. The integration with respect to time yields the signal as a correlation

$$S(t_d) = \frac{K}{\tau_{det}} \int_0^{\tau_{det}} I(2\omega, t_d)\mathrm{d}t, \tag{13.50}$$

where τ_{det} is the integration time of the detector (or of a following electronic monitoring device). If the intensities are equal, we expect the signal

$$S(t_d) = \frac{K}{\tau_{det}} \left[2 < I^2(t) > +4 < I(t)I(t + t_d) >\right]. \tag{13.51}$$

The signal caused by a beam with a Gaussian shape of the temporal distribution is, for $t_d = 0$, about three times the signal caused by the corresponding two pulses arriving at large delay ($t_d \gg t_p$).

We will treat the mechanism of frequency doubling later (Sect. 35.3). Besides the measurement of the intensity autocorrelation, which we described here, there are various other techniques of autocorrelation measurements (e.g., measurement of the field autocorrelation; *see* books on femtosecond lasers).

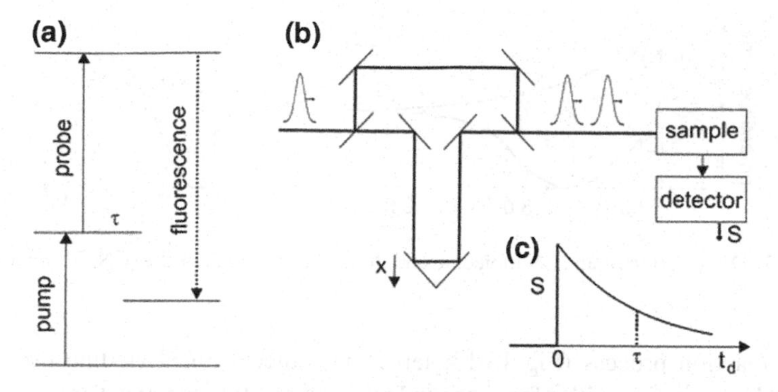

Fig. 13.12 Pump-probe experiment. **a** Principle. **b** Arrangement. **c** Signal

13.6 Pump-Probe Method

The femtosecond pulses are suited to investigate the dynamics of fast processes. Examples are the studies of short-lived excited states of atoms, molecules or solids. Applications lie in fields of physics, chemistry, biology and medicine.

In a pump-probe experiment (Fig. 13.12a), a pump pulse (of large pulse energy) and a probe pulse (of small pulse energy) are passing a sample containing, for example, molecules. The pump pulse excites molecules into an excited state. The probe pulse excites molecules further to an energetically higher lying state, which decays by emission of fluorescence radiation. The second excitation is only possible during the lifetime (τ) of the excited state. A pump-probe arrangement (Fig. 13.12b) consists of a femtosecond laser, a beam splitter and a delay section. To measure the fluorescence radiation, a detector with a large response time (large compared to the temporal separation of two subsequent pulses) is suitable. The detector signal $S(t_d)$, determined for different time delays t_d, yields the lifetime τ of the first excited state (Fig. 13.12c); the delay time is $t_d = 2x/c$, where $x = 0$ corresponds to the situation that probe and pump beam passed the same path length when they reach the sample.

The *pump-probe method* provides ultrashort snapshots.

13.7 Femtosecond Pulses in Chemistry

In 1999, the Egyptian scientist Ahmed H. Zewail at the California Institute of Technology in Pasadena (USA) received the Nobel Prize in chemistry for his "Outstanding research on the transition states of chemical reactions with the femtosecond spectroscopy." Zewail and coworkers and other research groups developed methods (*femtochemistry*) allowing for an investigation of the dynamics of chemical reactions. Here, we discuss an experiment.

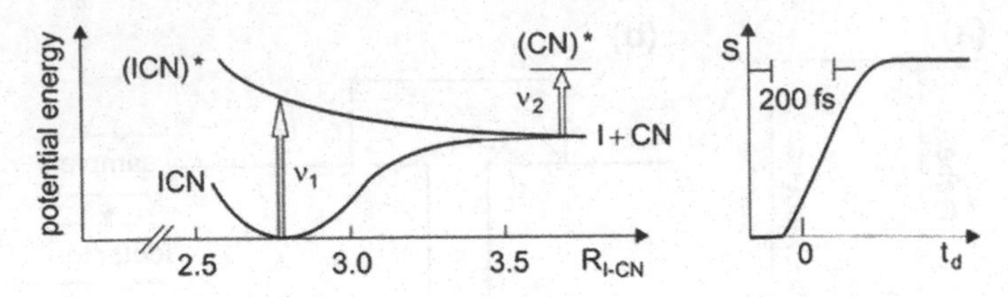

Fig. 13.13 Dissociation of an ICN molecule and fluorescence signal of the (CN)* fragment

In a reaction process (Fig. 13.13, left), the molecule ICN (iodine cyanide) is brought, by excitation with a femtosecond pump pulse (frequency ν_1), into an excited state (ICN)*. This state is antibonding and decays into I (iodine) atom and a CN radical. How long does it take until the CN radical forms by the dissociation of ICN in I and CN? To study this question, a probe pulse (frequency ν_2) following the first pulse excites CN radicals. Fluorescence radiation (Fig. 13.13, right) from excited CN radicals (CN)* indicates that it takes about 200 fs until CN forms.

13.8 Optical Frequency Analyzer

In 2005, the Nobel Prize in physics was donated to Roy Glauber, John Hall, and Theodor Hänsch for pioneering work in quantum optics and laser spectroscopy. Glauber performed theoretical investigations in the field of quantum optics. Hall and Hänsch received the Nobel Prize "for their contribution to the development of laser-based precision spectroscopy, including the optical frequency comb". The frequency comb is the basis of an *optical frequency analyzer* and, in future, most likely of a new frequency standard.

The main part of an optical frequency analyzer (Fig. 13.14a) is a frequency comb. It consists of radiation at discrete frequencies (ν_l, ν_{l+1}, ν_{l+2}, ...). The frequency distribution extends from the near infrared to the near ultraviolet. The optical frequency analyzer serves for determination of the frequency ν of a monochromatic radiation source (for instance of a highly stabilized continuous wave laser). The measurement of the beat frequency

$$f_{\mathrm{beat}} = \nu_l - \nu, \tag{13.52}$$

yields the value of ν. The beat frequency is measured by frequency mixing of radiation at the frequencies ν_l and ν in a photodiode (Fig. 13.14b). The beat frequency can be in the range of 1–10 GHz.

The frequencies ν_l are equally spaced. The exact position of a frequency ν_l is influenced by the Gouy phase and by dispersion effects (Sect. 13.4). In order to determine ν_l of a particular laser and to reach a high accuracy (1:10^{-16}) of the frequency measurement, the operation of an optical frequency analyzer makes use of nonlinear optical effects (Sect. 35.7).

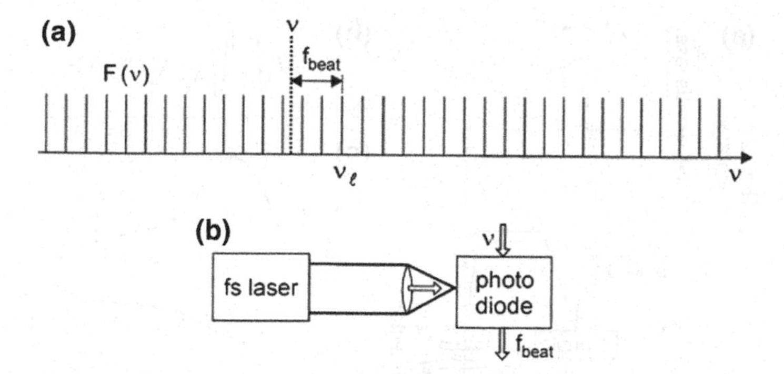

Fig. 13.14 Optical frequency analyzer. **a** Frequencies involved in the frequency analyzer. **b** Arrangement

13.9 Terahertz Time Domain Spectroscopy

Thetime domain spectroscopy [92–96] is a new spectroscopic method—it allows for the simultaneous measurement of the time dependences of both the amplitude and phase of an electromagnetic wave. The method is particularly suited to perform spectroscopic investigations with coherent terahertz and sub-terahertz waves, in the frequency range from about 0.1 to 100 THz (wavelength range 3 μm–0.3 mm). Applications lie in fields of physics, chemistry and biophysics. Optical properties of materials such as solids, liquids, chemicals and biomaterials can be determined.

A THz time domain spectrometer (Fig. 13.15a) consists of a THz field generator and a THz field detector both operated by the use of the same femtosecond laser (pulse duration 10 fs; repetition rate 50 MHz). The beam of the femtosecond laser is split into a main beam (used for generation of the THz field) and a reference beam (for detection). The main beam passes a generator crystal and the reference beam passes, after a time delay, a detector crystal. The generator crystal emits coherent THz radiation pulses; their duration is much larger than the duration of the femtosecond pulses. THz radiation reflected from a sample is focused to the detector crystal. The reference beam serves for the measurement of the instantaneous strength of the THz field at the location of the detector crystal. Under the action of the THz field, the detector crystal becomes birefringent and rotates the polarization direction of the optical radiation. Therefore, the radiation is able to pass a polarizer and to give rise to a signal by a photodetector; without THz field, the polarizer blocks the optical radiation.

The signal $S(x)$ of the photodetector corresponds, with $t_d = 2x/c$, to the signal $S(t_d)$ for different time delays t_d (Fig. 13.15b). $S(t_d)$ is a measure of the time dependence of the THz field. A Fourier analysis of the $S(t_d)$ curve yields (Fig. 13.15c) both the spectrum $F(\nu)$ of the amplitude and the spectrum $\varphi(\nu)$ of the phase of the THz field. From these informations, the complex reflectivity coefficient and thus real and imaginary parts of the dielectric response function of the sample can be extracted.

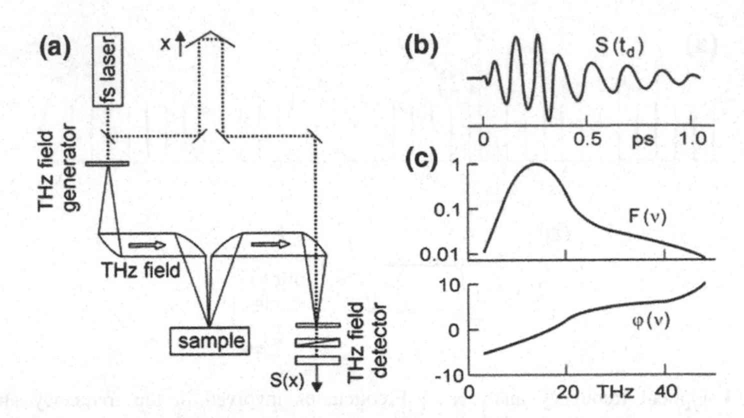

Fig. 13.15 Time domain THz spectroscopy. (**a**) Time domain THz spectrometer. (**b**) Detector signal. (**c**) Amplitude and phase spectra

The Fourier coefficients provide the connection between $F(\nu)$ and $\phi(\nu)$, on one side, and the real part $\chi_1(\nu)$ and the imaginary part $\chi_2(\nu)$ of the susceptibility, on the other side.

The origin of generation of a THz field is the difference frequency generation: nonlinear frequency mixing of the field components contained in a femtosecond pulse results in generation of a THz field. Femtosecond pulses with a spectral width of 100 THz lead to difference frequencies of all different field components from zero frequency to 100 THz. The difference frequency generation makes use of the nonlinear polarization (Sect. 35.4). GaSe has a large nonlinear coefficient for difference frequency generation.

Electrooptic crystals with large coefficients of THz field induced birefringence (used for detection) are GaSe and ZnTe.

The method makes it possible, as mentioned, to determine amplitudes and phases of THz fields. Thus, the real and imaginary parts of the susceptibility of materials can be determined. Almost all solid or liquid materials have excitations in the frequency range 1–100 THz. In this range, electrons, phonons and magnetic excitations determine optical properties of solids and biomolecules.

Time-domain spectroscopy began with a fast switch, the Auston switch [97–100] (Fig. 13.16a). Irradiation of semi-insulating GaAs with a 100-fs pulse results in generation of charge carriers. A static field produced with a static voltage (for instance 80 V across a GaAs crystal of 50 μm thickness) accelerates the electrons giving rise to generation of radiation. The spectrum of the radiation is determined by the temporal change of the current, dI/dt. The device acts as a Hertzian dipole. The spectrum of the radiation extends from ~100 GHz to several THz with a maximum at a wavelength around 1 mm (i.e., the radiation covers a range of sub-THz and THz frequencies). Another switch can be used as an antenna for measuring the instantaneous strength of a THz field (Fig. 13.16b). The THz field accelerates Free-electrons that are created by means of a femtosecond pulse. Variation of the delay between the femtosecond pulse and the THz pulse at the detector makes it possible to determine

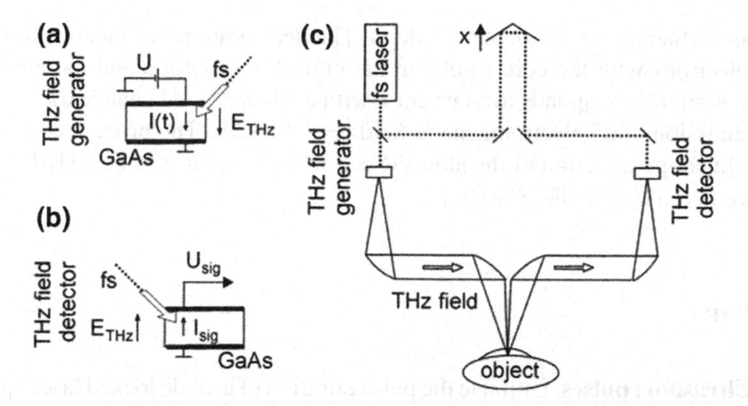

Fig. 13.16 Time domain sub-THz/THz spectrometer. **a** THz field generator. **b** THz field detector. **c** Arrangement

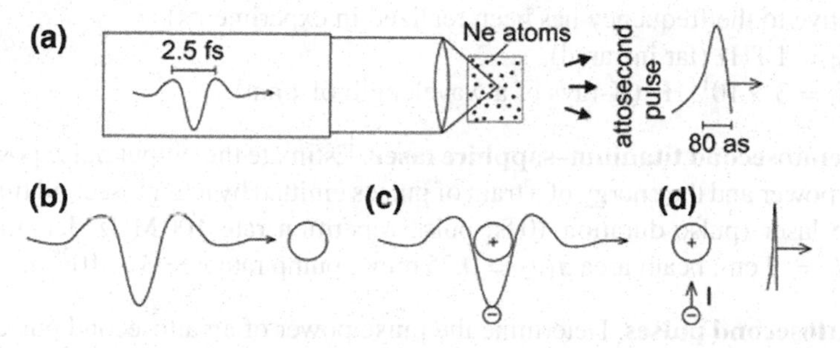

Fig. 13.17 Attosecond pulses. **a** Arrangement for generation of attosecond pulses. **b** Femtosecond pulse and a neon atom. **c** A neon atom and electrons in the femtosecond field. **d** Accelerated electrons corresponding to an instantaneous current I and attosecond pulse

amplitude and phase of the THz field. THz radiation reflected by an object can be detected (Fig. 13.17c). The signal obtained from the detector contains information on the surface region of an object.

13.10 Attosecond Pulses

Figure 13.17a illustrates a method of generation of attosecond pulses. An intense femtosecond pulse (duration 2.5 fs) of visible radiation focused on a box containing noble gas atoms (for instance neon) generates an attosecond pulse. The attosecond pulse (duration 80 as) represents an X-ray flash. The spectral distribution of the radiation lies mainly in the 10–20 nm range. Figure 13.17b shows an optical field pulse and a neon atom. The field excites electrons so strongly that they separate from the positive core (Fig. 13.17c). The field of a femtosecond pulse accelerates the electrons further and then decelerates them. The decelaration and the recombination of the electrons with the core results in the emission of attosecond radiation. The electron motion corresponds to a current I with a fast temporal change $\mathrm{d}I/\mathrm{d}t$ giving rise to emission of an electromagnetic field (Fig. 13.17d). The process corresponds to a nonlinear polarization of the atoms (Sect. 35.2); *see*, for instance, [101–103].

References [1–4, 7, 10, 86–105]

Problems

13.1 Ultrashort pulses. Estimate the pulse duration of a mode locked laser operated in a spectral range from ν_0 to 1.1 ν_0 for lasers in different frequency ranges.

(a) If $\nu_0 = 30\,\mathrm{THz}$ (range of the CO_2 laser; only a frequency region of 5% width relative to the frequency has been realized in experiments).
(b) If $\nu_0 = 1\,\mathrm{THz}$ (far infrared).
(c) If $\nu_0 = 3\times10^{17}\,\mathrm{Hz}$ (X-rays of a wavelength of 1 nm).

13.2 Femtosecond titanium–sapphire laser. Estimate the output pulse power, the average power and the energy of a train of pulses emitted by a femtosecond titanium–sapphire laser (pulse duration 10 fs; pulse repetition rate 100 MHz; length of the crystal $L' = 1\,\mathrm{cm}$; beam area $a_1a_2 = 0.25\,\mathrm{mm}^2$; pump rate $r = 3\times10^{28}\,\mathrm{m}^{-3}\,\mathrm{s}^{-1}$).

13.3 Attosecond pulses. Determine the pulse power of an attosecond pulse (duration 100 as) consisting of 10^8 photons of radiation at an average wavelength of 10 nm.

13.4 Unstabilized femtosecond laser. A femtosecond laser that is highly stabilized generates a train of pulses of duration of 10 fs. In the case that the laser is not sufficiently stabilized, the temporal separation of subsequent pulses varies due to fluctuations of the length of the laser resonator. The pulses can be described as pulses with an average amplitude $A(t)$ that has a Gaussian shape on the timescale.

(a) Give an expression of frequency spectrum.
(b) Determine the frequency spectrum if the pulse duration is equal to 100 fs.

13.5 Stabilization of a femtosecond laser. Determine the requirement of length stabilization of a femtosecond laser that produces pulses of a duration of 5 fs (repetition rate 100 MHz).

13.6 Acousto-optic switch.

(a) Relate the frequency of the ultrasonic wave and the length of the optical resonator.
(b) What is the condition that determines the length of the quartz plate?

13.7 Heisenberg's uncertainty principle.

(a) Show that a photon in a femtosecond pulse that has a Gaussian temporal profile obeys Heisenberg's uncertainty relation $\Delta x \Delta p_{\mathrm{x}} \geq \hbar$, where Δx is the uncertainty of the position x and Δp_x is the uncertainty of the momentum p_x. The pulse propagates along the x direction. [*Hint*: make use of (13.23); the result is $\Delta x \Delta p_x = (4 \ln 2)\, \hbar$.]

(b) Compare the result with the result of an analysis of a Gaussian beam of monochromatic radiation (Problem 11.15).

Part IV
Types of Lasers (Except Semiconductor Lasers)

Chapter 14
Gas Lasers

A gas laser contains atoms or molecules. Stimulated transitions occur in atoms between electronic states and in molecules between rotational, vibrational or electronic states. We describe various gas discharge lasers: helium–neon laser; metal vapor laser; argon ion laser; excimer laser; nitrogen laser; CO_2 laser; and optically pumped gas lasers.

The excimer laser and the CO_2 laser are two important industrial lasers. The excimer laser generates intense UV radiation pulses. The CO_2 laser is a source of infrared radiation. It has a high efficiency of conversion of electric power to power of laser radiation. The CO_2 laser is very versatile—it operates as continuous wave laser or as pulsed laser. Optically pumped gas lasers (pumped with CO_2 laser radiation) are suitable for generation of far infrared radiation.

We first treat two line broadening mechanisms that play a role in gas lasers: the Doppler and the collision broadening. Then we discuss different gas lasers.

14.1 Doppler Broadening of Spectral Lines

Doppler broadening is a main broadening mechanism of spectral lines for gases at low pressure. The frequency of the radiation that is due to transitions between two discrete energy levels of an atom (or a molecule) is

$$\nu = \nu_0 + (\mathrm{v}_z/c)\,\nu_0, \qquad (14.1)$$

where ν_0 is the frequency of the radiation emitted by the atom at rest and v_z is the velocity component in z direction. The atoms in a gas have a Maxwellian velocity distribution

$$f(\mathrm{v}_x, \mathrm{v}_y, \mathrm{v}_z) = \left(\frac{m}{2\pi kT}\right)^{3/2} \exp\left(-\frac{m}{2kT}(\mathrm{v}_x^2 + \mathrm{v}_y^2 + \mathrm{v}_z^2)\right). \qquad (14.2)$$

K.F. Renk, *Basics of Laser Physics*, Graduate Texts in Physics,
DOI 10.1007/978-3-319-50651-7_14

T is the temperature of the gas, m the mass of an atom (or molecule), and $f(\mathrm{v}_x, \mathrm{v}_y, \mathrm{v}_z)\mathrm{dv}_x\mathrm{dv}_y\mathrm{dv}_z$ is the probability to find an atom with a velocity $\mathrm{v}_x, \mathrm{v}_y, \mathrm{v}_z$ in the velocity element $\mathrm{dv}_x\mathrm{dv}_y\mathrm{dv}_z$. The integral over the distribution is equal to unity,

$$\int_{-\infty}^{\infty}\int_{-\infty}^{\infty}\int_{-\infty}^{\infty} f\,\mathrm{dv}_x\mathrm{dv}_y\mathrm{dv}_z = 1. \tag{14.3}$$

How large is the probability to find an atom in the velocity interval v_z, $\mathrm{v}_z + \mathrm{dv}_z$ (Fig. 14.1a)? It is

$$f(\mathrm{v}_z)\mathrm{dv}_z = \left(\frac{a}{\pi}\right)^{3/2} \mathrm{e}^{-a\mathrm{v}_z^2}\mathrm{dv}_z \int_{-\infty}^{\infty} \mathrm{e}^{-a\mathrm{v}_x^2}\mathrm{dv}_x \int_{-\infty}^{\infty} \mathrm{e}^{-a\mathrm{v}_y^2}\mathrm{dv}_y, \tag{14.4}$$

where $a = m/2kT$ is an abbreviation. It follows, with

$$\int_{-\infty}^{\infty} \mathrm{e}^{-a\mathrm{v}_x^2}\mathrm{dv}_x = (\pi a)^{-1/2}, \tag{14.5}$$

that

$$f(\mathrm{v}_z) = \sqrt{\frac{m}{2\pi kT}}\,\exp\left(-\frac{m}{2kT}\mathrm{v}_z^2\right). \tag{14.6}$$

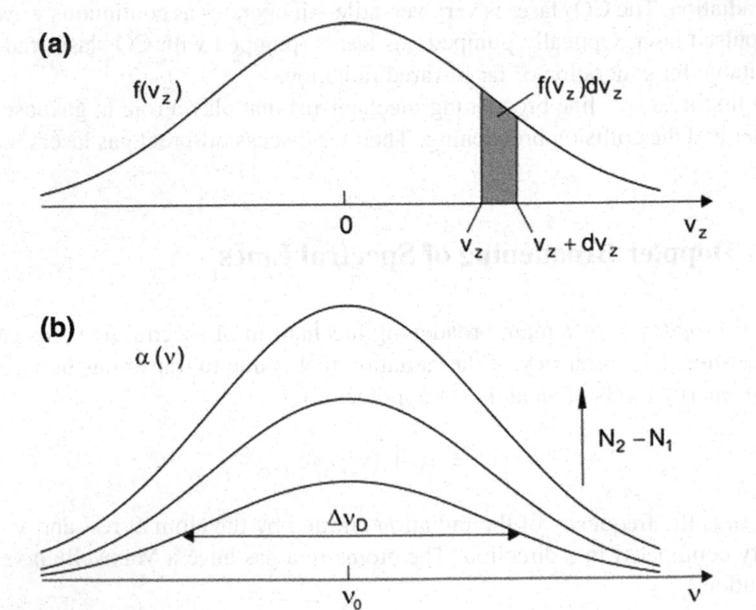

Fig. 14.1 Doppler broadening. **a** Maxwellian velocity distribution. **b** Gain coefficient of a medium with a Doppler broadened line

How large is the probability $g(\nu)\mathrm{d}\nu$ of a transition in the frequency interval ν, $\nu + \mathrm{d}\nu$? The answer is

$$g(\nu)\mathrm{d}\nu = f(\mathrm{v}_z)\mathrm{d}\mathrm{v}_z. \tag{14.7}$$

With

$$\mathrm{d}\mathrm{v}_z = \frac{c}{\nu_0}\,\mathrm{d}\nu, \tag{14.8}$$

we obtain

$$g(\nu) = \frac{2}{\Delta\nu_\mathrm{D}}\left(\frac{\ln 2}{\pi}\right)^{1/2}\exp\left[-\ln 2\,\frac{(\nu-\nu_0)^2}{(\Delta\nu_\mathrm{D}/2)^2}\right], \tag{14.9}$$

where

$$\Delta\nu_\mathrm{D} = 2\nu_0\sqrt{\frac{2kT\ln 2}{mc^2}} \tag{14.10}$$

is the *Doppler linewidth*. It depends on the temperature and the atomic mass of the atoms and is independent of the gas pressure. The Doppler broadening leads to a Gaussian line. The Doppler broadening is an inhomogeneous broadening mechanism because atoms of different velocities have emission lines (and absorption lines) at different frequencies.

The gain coefficient $\alpha(\nu)$ of an active medium with a Doppler broadened transition is proportional to the population difference $N_2 - N_1$ (Fig. 14.1b). The halfwidth of the gain curve is independent of $N_2 - N_1$.

Example Helium–neon laser; $\lambda = 633$ nm; $m_\mathrm{Ne} = 20\ m_\mathrm{p}$; m_p = proton mass; $k = 1.38 \times 10^{-23}\,\mathrm{J\,K^{-1}}$; $\Delta\nu_\mathrm{D} = 1.5 \times 10^9$ Hz.

14.2 Collision Broadening

According to a classical description of *collision broadening* (= pressure broadening) in gases, a collision of an excited atom with another (nonexcited) atom changes the phase of the sinusoidal oscillation of the excited atom. Therefore, collisions change the phase of radiation emitted by the atom (Fig. 14.2a); *see* also Sect. 4.11. The time τ_c between two collisions is a dephasing time. Between two collisions, an electron of the excited atom performs, in the picture of the classical oscillator model, (Sect. 4.8), an oscillation with the transition frequency.

A Fourier analysis of the electric field emitted by the atom leads to a Lorentzian line

$$g_{\mathrm{L,res}}(\nu) = \frac{\Delta\nu_c}{2\pi}\,\frac{1}{(\nu_0-\nu)^2 + \Delta\nu_c^2/4}. \tag{14.11}$$

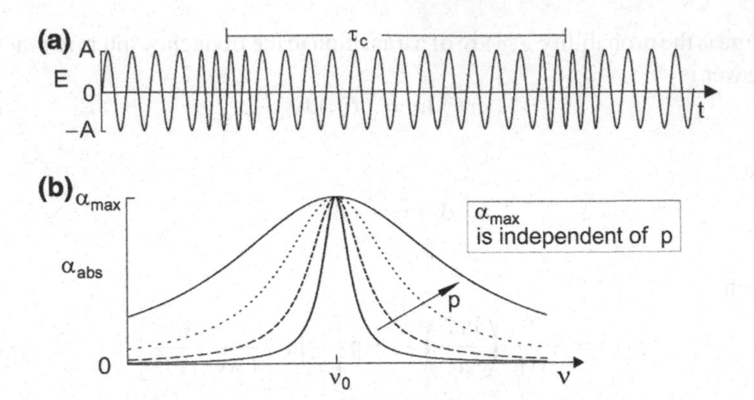

Fig. 14.2 Collision broadening. **a** Collision time. **b** Absorption coefficient

The linewidth is equal to

$$\Delta\nu_c = \frac{1}{\pi\tau_c}. \tag{14.12}$$

The theory of collision broadening provides, in accordance with experimental results, a relation between linewidth and collision time,

$$\Delta\nu_c = \sqrt{\frac{8}{\pi}}\,\frac{\sigma_c^2}{\sqrt{mkT}} \times p = K \times p. \tag{14.13}$$

The linewidth is proportional to the pressure. K is a characteristic constant of a gas, σ_c is the cross section of collisions, m the mass of the gas molecules (or atoms) and p the gas pressure. At room temperature, $K \sim 1\,\text{GHz}/p$, where p is measured in units of bar; K has values between 0.3 and $2.5\,\text{GHz}/p$, depending on the atoms or molecules.

The collision broadening corresponds to a homogeneous broadening mechanism because all atoms are submitted to collisions. The absorption coefficient of radiation interacting with a pressure broadened transition is equal to

$$\alpha_{\text{abs}}(\nu) = (h\nu/c)B_{12}g_L(\nu)N_1. \tag{14.14}$$

N_1 is the (number) density of molecules (atoms). The density is proportional to pressure. The maximum of the lineshape function is inversely proportional to pressure. Therefore, the absorption coefficient at line center is independent of pressure while the linewidth increases linearly with pressure (Fig. 14.2b).

The gain coefficient of an active medium consisting of molecules with a collision broadened line is equal to

$$\alpha(\nu) = (h\nu/c)B_{12}g_L(\nu) \times (N_2 - N_1). \tag{14.15}$$

Example CO_2 laser operated at large gas pressure (Sect. 14.8).

14.3 Helium–Neon Laser

The helium–neon laser belongs, besides the ruby laser, to the two first lasers and is still in use. In the helium–neon laser, Ne atoms are excited into s states (Fig. 14.3a). Laser transitions are s → p transitions. The helium–neon laser is a three-level laser type. Accidentally, the second lowest excited state of a helium atom (2^1S state) has almost the same energy as the 5s state of Ne. This coincidence allows for a selective excitation of the 5s state of Ne:

- In a gas discharge, electrons excite helium atoms; the excited helium atoms have very long lifetimes.
- Atomic collisions between excited He and Ne atoms lead to a transfer of excitation energy from He to Ne atoms.
- Stimulated 5s → 3p transitions result in generation of laser radiation of a wavelength of 633 nm.
- The 3p levels are depopulated by spontaneous emission of radiation (wavelength near 450 nm) by 3p → 3s transitions. The 3s state has a very long lifetime. Relaxation is possible via collisions of the neon atoms in the 3s state with the wall of the tube that contains the gas; it is a process of nonradiative relaxation. To obtain a sufficiently fast relaxation, a narrow gas tube is favorable.
- The lifetime of the 5s state is about 100 ns and the lifetime of the 3p state about 10 ns.
- Stimulated 5s → 4p transitions lead to generation of laser radiation of a wavelength of 3.4 μm.

The lowest excited state level of He (2^3S state) almost coincides with the 4s level of neon; the energy difference (~40 meV) is equal to $\sim 2kT$. Helium atoms, excited by electron collisions to their lowest excited state, transfer the excitation energy to neon atoms resulting in a population of the 4s level of neon. Stimulated 4s → 3p transitions lead to generation of laser radiation at a wavelength of 1.15 μm.

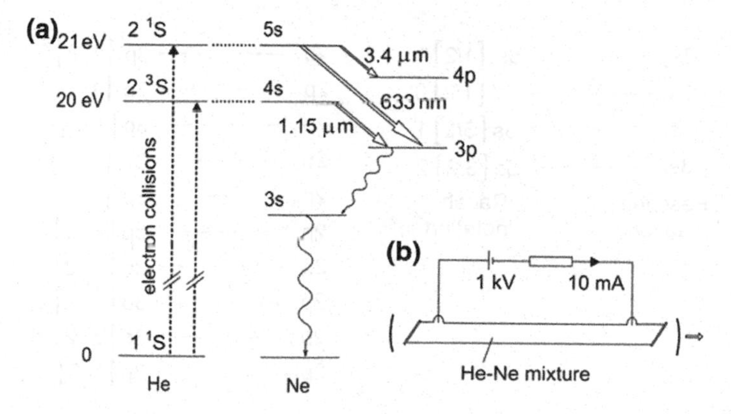

Fig. 14.3 Helium–neon laser. **a** Principle. **b** Arrangement

The helium–neon laser (Fig. 14.3b) contains a gas mixture of helium and neon (ratio 5:1; pressure ~5 mbar) in a glass tube (typical length 0.5 m; diameter 1–2 mm). Brewster windows close the tube. Radiation of the appropriate polarization passes the windows without reflection loss (*see* Fig. 2.16). A gas discharge (voltage ~2 kV; current ~10 mA) leads to a laser output power (~1 mW at 633 nm), which corresponds to an efficiency of the order of 0.01%. There are different reasons that the efficiency is small: the quantum efficiency is small and the pump process is not very efficient in the helium–neon gas. The laser resonator (especially the coating on the dielectric reflectors) determines the wavelength of a helium–neon laser.

Table 14.1 shows data of different helium–neon lasers; $\Delta\nu_g$ is the gain bandwidth.

The electronic configuration of Ne is $1s^2 2s^2 2p^6$. Excited states have the configurations $1s^2 2s^2 2p^5$—3s, 3p, 4s etc. The 3s, 3p, ... levels are split because of the interaction of an excited electron with the hole in the 2p shell (spin-orbit interaction). The s levels are split into 4 sublevels and the p levels into 10 sublevels. Due to the level splitting, a large number of transitions are available as laser transitions—about a hundred laser lines (many of them in the infrared and far infrared) are known. The first helium–neon laser operated in the infrared (wavelength 1.15 μm).

Figure 14.4 indicates a possible labeling of the energy levels of Ne. The $3s_2$ sublevel is the highest 3s level; the sublevels have the numbers 2 ... 5. The highest 2p

Table 14.1 Helium–Neon lasers

λ	Transition	$\Delta\nu_g$ (GHz)	α (m^{-1})	Power (mW)
543 nm	$3s_2 \rightarrow 2p_{10}$	1.75	0.005	1
594 nm	$3s_2 \rightarrow 2p_8$	1.60	0.005	1
612 nm	$3s_2 \rightarrow 2p_6$	1.55	0.017	1
633 nm	$3s_2 \rightarrow 2p_4$	1.50	0.1	1–10
1.15 μm	$2s_2 \rightarrow 2p_4$	0.83		1
1.52 μm	$2s_2 \rightarrow 2p_1$	0.63		1
3.39 μm	$3s_2 \rightarrow 3p_4$	0.28	100	10

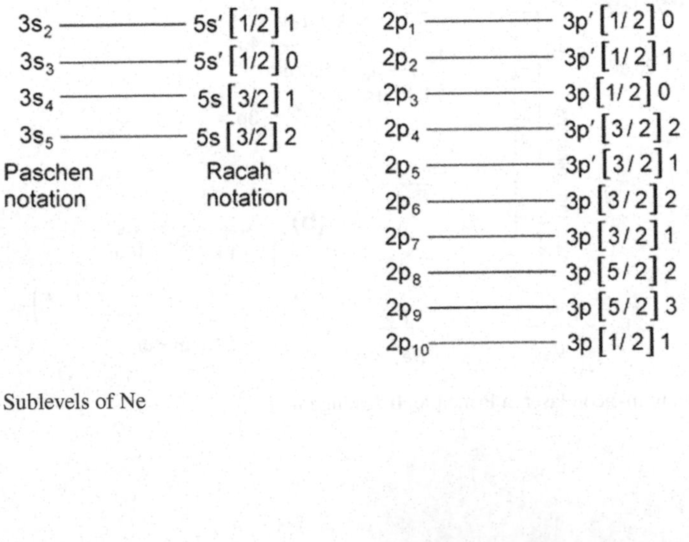

$3s_2$ — 5s′ [1/2] 1
$3s_3$ — 5s′ [1/2] 0
$3s_4$ — 5s [3/2] 1
$3s_5$ — 5s [3/2] 2
Paschen notation — Racah notation

$2p_1$ — 3p′ [1/2] 0
$2p_2$ — 3p′ [1/2] 1
$2p_3$ — 3p [1/2] 0
$2p_4$ — 3p′ [3/2] 2
$2p_5$ — 3p′ [3/2] 1
$2p_6$ — 3p [3/2] 2
$2p_7$ — 3p [3/2] 1
$2p_8$ — 3p [5/2] 2
$2p_9$ — 3p [5/2] 3
$2p_{10}$ — 3p [1/2] 1

Fig. 14.4 Sublevels of Ne

sublevel is $2p_1$ and the lowest 2p level is $2p_{10}$. In this notation (*Paschen* notation), the Ne^+ core is considered as an effective potential and the states of the additional electron are 1s, 2s, 2p, etc. An alternative, more detailed analysis uses the *Racah* notation: an excited neon atom has the configuration $1s^2 2s^2 2p^5$ plus an additional state with one electron (the outer electron). In the Racah notation, an energy level (for instance a 5s sublevel) is characterized by 5s[K]J or 5s′[K]J, where the symbols indicate the following:

- 5s or 5s′; configuration of the outer electron.
- K; quantum number of the sum of the total angular momentum J_c (quantum number j) of the core electrons and the orbital momentum L (quantum number l) of the outer electron.
- $J = K \pm \frac{1}{2}$, where $\frac{1}{2}$ is the quantum number of the spin of the outer electron.

The coupling leads to 4 sublevels of s states (Fig. 14.4); s is attributed to a state with K = 3/2 and s′ to a state with K = 1/2. A p state has 10 sublevels; 3p[K] configurations (j = 3/2) are possible with K = 1/2, 3/2 and 5/2 while 3p′[K] configurations are possible with K = 1/2 and 3/2. The coupling corresponds to intermediate coupling ($j - l$ coupling). The energy levels (energy values, lifetimes, and assignment to appropriate quantum states) have been studied long before the arrival of the laser; for discussions of Ne levels used in lasers, *see* [116–118].

Applications. The helium–neon laser generates monochromatic radiation with a small beam divergence. The laser serves for various applications (e.g., holography), which need a high coherence and low beam divergence.

14.4 Metal Vapor Laser

A metal vapor laser operates with copper, gold, lead, or cadmium vapor. In the copper vapor laser (Fig. 14.5), Cu atoms are excited by electron collisions from the ground state $3d^{10}4s$ to the $3d^{10}4p$ state, giving rise to stimulated transitions to $3d^9 4s^2$ states.

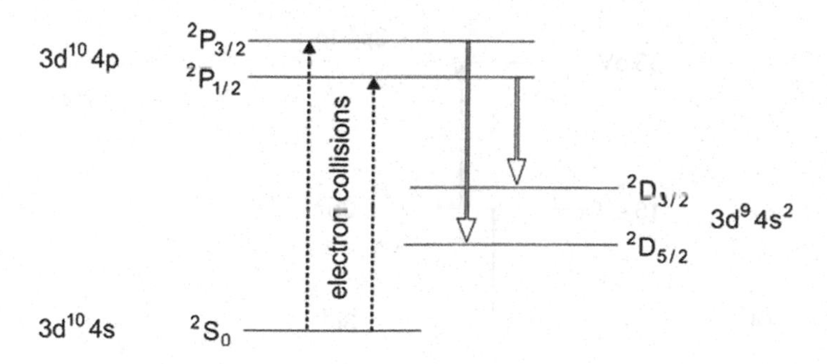

Fig. 14.5 Copper vapor laser

The level splitting is due to spin–orbit interaction. The levels are labeled according to the LS coupling (Russel-Saunders coupling); a level $^{2S+1}L_J$ corresponds to a state with the quantum number L of the orbital momentum, the quantum number S of the spin, and the quantum number J = L + S of the total momentum. 2S + 1 is the spin multiplicity and the S, P, D states correspond to states with L = 0, 1, 2.

A copper vapor laser consists of a ceramic tube (with Brewster windows) in the laser resonator. The tube contains a little piece of metallic copper. The laser oscillation depends very sensitively on the gas pressure and therefore on the temperature. There is only a narrow temperature window (1,500 °C ± 20 °C) in which the laser operates. Population inversion is produced by electric pulses (duration 20 ns; pulse energy 10 mJ; repetition rate 3 kHz). An electric pulse causes a pulsed discharge and excitation of copper atoms via electron collisions. The lifetime of the upper laser level is smaller than the lifetime of the lower laser level. Therefore, continuous oscillation is not possible; the laser is a *self-terminating laser*.

The copper vapor laser has a large gain coefficient (7 m^{-1}), and it has an excellent beam quality because of a large diameter of the active medium and of the resonator. The efficiency of conversion of electric pump energy to energy of laser radiation is about 1%.

Copper vapor lasers generate radiation at the wavelengths 510 and 578 nm and gold vapor lasers at 628 and 312 nm.

Applications lie in medicine, particularly in the detection and destruction of tumors by the photodynamic therapy. Today, metal vapor lasers are competing with semiconductor lasers.

14.5 Argon Ion Laser

In the argon ion laser (Fig. 14.6), subsequent electron collisions lead to ionization of argon atoms and to excitation of argon ions. The electron configurations are the following:

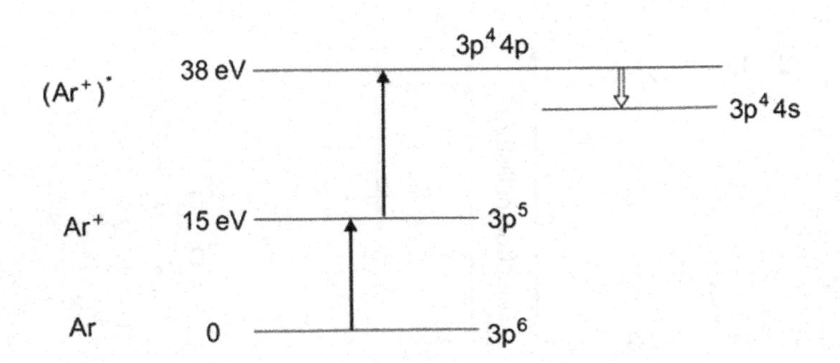

Fig. 14.6 Argon ion laser

- Ar $1s^2\,2p^6\,3s^2\,3p^6$; argon.
- Ar^+ $1s^2\,2p^6\,3s^2\,3p^5$; argon ion.
- $(Ar^+)^*$ $1s^2\,2p^6\,3s^2\,3p^4\,4p$; excited argon ion.

Different 4p → 4s transitions between the $3p^4 4p$ and $3p^4 4s$ levels (split due to spin–orbit interaction) give rise to cw laser emission in the blue and green, with strong emission lines at 488 and 514.5 nm.

A gas discharge in a ceramic tube (diameter 1–2 mm; length 1 m; cooled with water) containing the argon gas (pressure 0.1 mbar) pumps the argon ion laser. Because of the twofold excitation, the efficiency of the argon ion laser is proportional to the square of the current density in the gas discharge. At a high electric power (current 10 A; voltage 5 kV), the output power is large (20 W). The efficiency of the laser is small ($\leq 0.1\%$).

The krypton ion laser operates in the same way as the argon ion laser; it emits radiation at other wavelengths (between 406 and 676 nm). An important application of the argon and the krypton ion lasers is the optical pumping of other lasers, especially of the titanium–sapphire laser (and before this laser existed, the optical pumping of dye lasers). Today, semiconductor lasers serve as pump lasers.

14.6 Excimer Laser

We now treat an important industrial laser. The *excimer laser* makes use of the KrF excimer or of other excimers. The following processes occur in a KrF excimer laser (Fig. 14.7).

- A gas discharge in a mixture of krypton and fluorine gas produces $(KrF)^*$ molecules, i.e., KrF molecules in an excited electronic state. The lifetime of the excited state is of the order of 10^{-9} s.
- Stimulated transitions take place to nonbonding KrF states. After a transition, the Kr atom and the F atom repel each other and separate spatially. Therefore, the lower laser level has a shorter lifetime than the upper laser level. During an optical transition in a KrF excimer, the nuclear distance R_{Kr-F} between the nucleus of Kr and the nucleus of F does not change (Franck–Condon principle)—the transition corresponds to a vertical line in the energy-nuclear distance diagram.

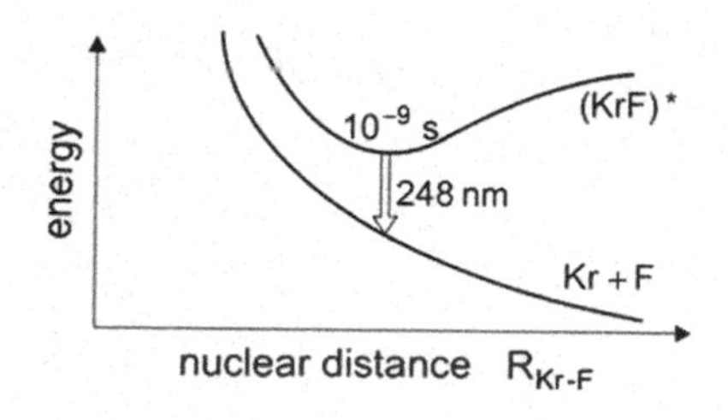

Fig. 14.7 KrF excimer laser

- The excitation occurs by electron collisions with Kr and by a chemical reaction, respectively, $Kr + e^- \rightarrow Kr^* + e^-$ and $Kr^* + F_2 \rightarrow (KrF)^* + F$:

An excimer (excited dimer) is a molecule with two equal atoms, which undergo chemical bonding in the excited state but not in the ground state.
Examples of excimers: Ar_2^* (emission at 126 nm); Kr_2^* (146 nm); Xe_2^* (172 nm). An exciplex (excited state complex) is denoted as excimer too.
Examples of exciplexes (excimers) and laser lines: ArF (193 nm); KrF (248 nm); XeCl (308 nm); XeF (351 nm); KrBr (206 nm); ArBr (161 nm); NeF (108 nm).

The excimer laser is a TEA laser (transversely excited atmospheric laser). We will describe a TEA laser arrangement in connection with the CO_2 laser (Sect. 14.8). The laser gas of a krypton fluoride excimer laser has the composition: He (= buffer gas, pressure ~1 bar); Kr (10%); and F_2 (0.1%). At a large pump power density (200 MW per liter gas volume), the gain is about 10% per cm (gain coefficient $\alpha = 10\,\mathrm{m}^{-1}$).

Data of an excimer laser: pumping by electric discharge pulses (voltage ~1 MV, current 10 kA, pulse duration 30 ns, electric energy per pulse 100 J); laser pulse energy 1 J; efficiency 1%; repetition rate 1–50 Hz.

Applications of the excimer laser are: labelling (of semiconductor chips, glasses, polymers, etc.) during mass production; structuring of materials by means of UV lithography—in 2011, semiconductor structures of lateral size of 45 nm are prepared by the use of the ArF laser (wavelength 193 nm).

14.7 Nitrogen Laser

The nitrogen laser is a prototype of a vibronic laser (Fig. 14.8). The electronic energy depends on the distance R_{N-N} between the nitrogen nuclei. A vibronic energy level of N_2 has electronic and vibrational energy,

$$E_{\mathrm{n,v}} = E_{\mathrm{n}} + \left(\mathrm{v} + \frac{1}{2}\right) h\nu_{\mathrm{vib}}. \tag{14.16}$$

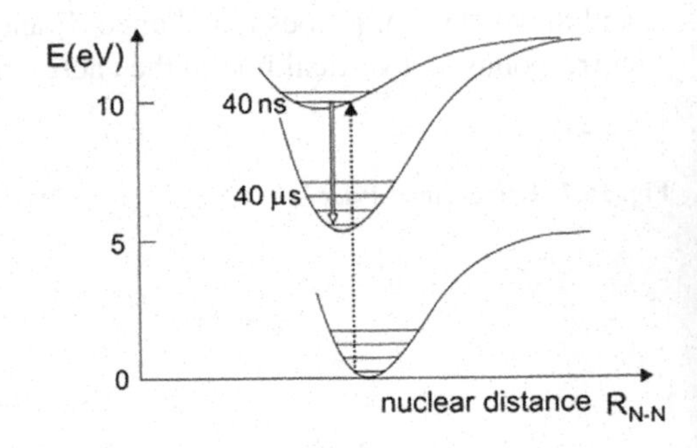

Fig. 14.8 Nitrogen laser

E_n is the electronic energy in the nth state; n = 1, ground state; n = 2, 3, ..., excited states; $(v + 1/2)h\nu_{vib}$ is the vibrational energy; v = 0, 1, 2, ... are the vibrational quantum numbers; and ν_{vib} (= 70.8 THz) is the vibrational frequency. Electron collisions in a gas discharge excite N_2 molecules to vibronic states belonging to the n = 3 electronic state. Stimulated transitions to vibronic levels of the n = 2 electronic level (energies $E_{2,v}$) produce laser radiation in the near UV (near 337 nm). The lifetime (40 ns) of the upper laser level is shorter than the lifetime of the lower laser level. Therefore, continuous operation is not possible; the laser is a self-terminating laser. Suitable for pumping are very short gas discharge pulses (duration 1 ns). Optical transitions obey the Franck–Condon principle.

14.8 CO_2 Laser

The CO_2 laser is of great importance:

- It has a high efficiency (10–50%) for conversion of electrical power to power of laser radiation.
- Different ways of operation are possible; in particular, cw operation, pulsed operation, and TEA laser operation.
- The cw CO_2 laser generates cw radiation of a large power (100 W at a length of the active medium of about 1 m, and up to 1,000 W or even more at very large length of the active medium).
- The TEA (transversely excited atmospheric) CO_2 laser produces pulses (duration ~100 ns) of high peak power (100 kW).

Applications of CO_2 lasers concern material processing (cutting, welding, hardening of metal surfaces, shock hardening at power densities of 10^9 W/cm^2) and medicine.

The CO_2 laser (Fig. 14.9a) makes use of vibrational-rotational levels,

$$E = E_{vib}(v_1, v_2, v_3) + E_{rot}(J). \tag{14.17}$$

E_{vib} is the vibrational energy and E_{rot} the rotational energy (J = quantum number of the rotation). The vibrational energy is

$$E_{vib} = \left(v_1 + \frac{1}{2}\right) h\nu_1 + \left(v_2 + \frac{1}{2}\right) h\nu_2 + \left(v_3 + \frac{1}{2}\right) h\nu_3, \tag{14.18}$$

where the oscillation frequency ν_1 (= 41.6 THz) corresponds to the *symmetric valence vibration*, ν_2 (=20.0 THz) to the *bending vibration*, and ν_3 (=70.5 THz) to the *antisymmetric valence vibration* (Fig. 14.9b); v_1, v_2, and v_3 are the vibrational quantum numbers; v_1 = 0, 1, 2, ...; v_2 = 0, 1, 2, ...; v_3 = 0, 1, 2, ... We denote a state with the quantum numbers v_1, v_2, and v_3 as $v_1v_2v_3$ state.

Electron collisions in a gas discharge can excite CO_2 molecules. More efficient is the indirect excitation. Electron collisions produce excited N_2 molecules

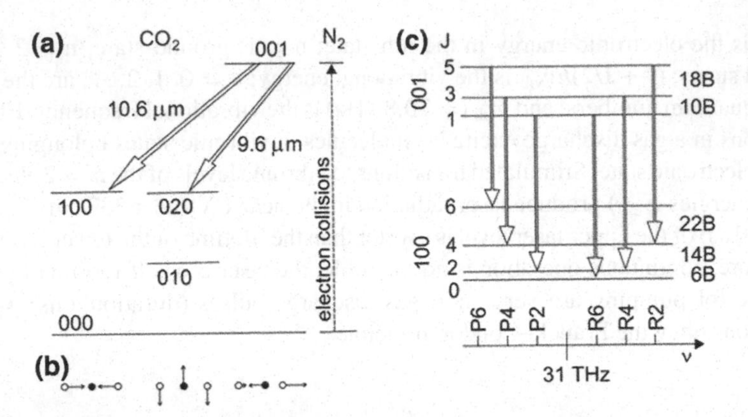

Fig. 14.9 CO_2 laser. **a** Vibrational levels of CO_2 and N_2. **b** Vibrations of the CO_2 molecule. **c** Vibrational-rotational transitions in CO_2

(in the lowest vibrational state); an excited N_2 molecule in the lowest vibrational state has a very large lifetime. Energy transfer processes by collisions between excited N_2 molecules and nonexcited CO_2 molecules lead to population of the 001 state of CO_2 molecules. This state has a very long lifetime (~4 s) with respect to spontaneous emission of radiation that is due to 001 → 100 and 001 → 100 transitions. There are two groups of laser transitions corresponding to two wavelength regions:

- 10.6 μm; transitions 001 → 100; frequencies near 28 THz.
- 9.6 μm; transitions 001 → 020; frequencies near 31 THz.

The transitions between different types of vibrations are allowed due to the anharmonicity of the vibrations. The depopulation of the lower states occurs by collisions of the molecules with walls (nonradiative relaxation). A vibrational transition in a CO_2 molecule is associated with a change of the rotational energy (Fig. 14.9c), where one of the selection rules

$$\Delta J = \pm 1 \tag{14.19}$$

must be fulfilled. The selection rule $\Delta J = +1$ corresponds to laser lines in the P branch and the selection rule $\Delta J = -1$ in laser lines in the R branch. The rotational energy is (approximately)

$$E_{\text{rot}} = BJ(J+1); \quad B = \frac{\hbar^2}{2\Theta}. \tag{14.20}$$

B (~15 GHz times h) is the rotational constant that is a measure of the rotational energy and Θ is the moment of inertia of a CO_2 molecule; each J state is $2J+1$ fold degenerate.

Not all rotational quantum numbers lead to allowed states. The CO_2 molecule is a Boson (more exactly, the $^{12}C^{16}O_2$ molecule). Interchange of the two O atoms must leave the total wave function of the molecule unchanged—the wave function must

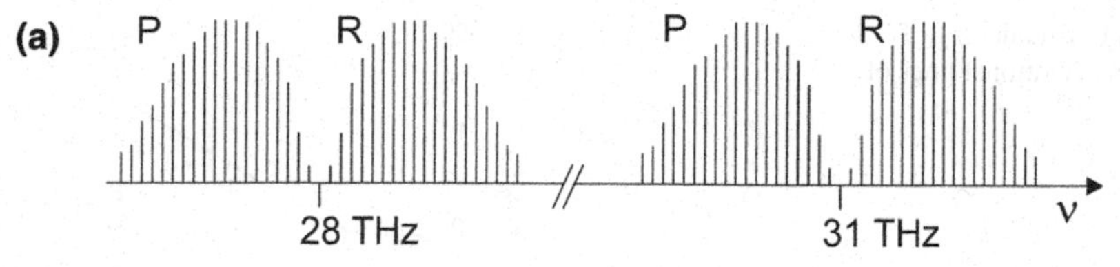

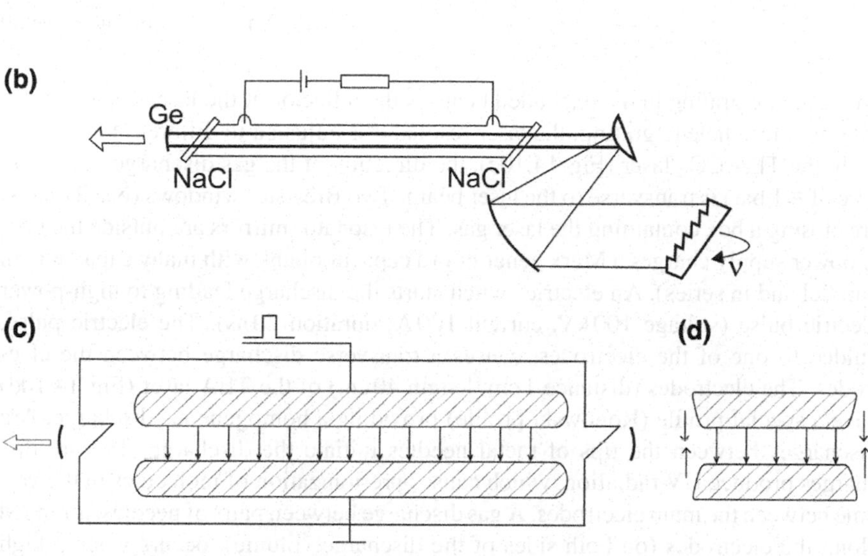

Fig. 14.10 CO_2 laser. **a** Laser lines. **b** Continuous wave CO_2 laser. **c** TEA (*transversely excited atmospheric*) CO_2 laser. **d** Profile of the electrodes of a TEA laser, together with discharge needles causing UV pre-ionization at the arrival of an electric pulse

be an even function. The electronic wave function of the electronic ground state of the molecule is even as well as the wave function of the nuclei (the nuclear spins of ^{12}C and of ^{16}O are zero). It follows: J is odd for an antisymmetric vibration, J is even for a symmetric vibration.

The frequency distance between two neighboring lines is $2 \times 2B/h = 4B/h$. Because of centrifugal distortion, the distance between two neighboring lines is not exactly $4B/h$ but depends on the vibrational quantum number and on the rotational quantum number. About twenty discrete laser lines belong to each of the four branches (Fig. 14.10a). The distance between next—near lines is ~60 GHz, or less because of the centrifugal distortion.

A gas discharge pumps the cw CO_2 laser (Fig. 14.10b). The gas, a mixture of CO_2, N_2, and He (at a ratio of about 1:1:8), can have a pressure of about 1 mbar. The glass tube (diameter 1 cm) that contains the laser gas is closed by Brewster windows (NaCl crystal plates). The spherical output coupling mirror consists of crystalline germanium. The outer side of the germanium mirror is covered with an antireflecting dielectric multilayer coating. Thus, standing waves in the output coupling mirror are avoided. The other surface, covered with another dielectric multilayer coating, has a reflectivity (~95%) that is appropriate to reach optimum output coupling.

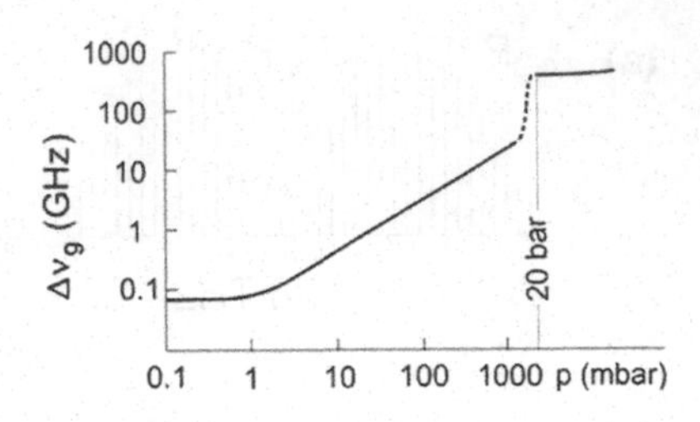

Fig. 14.11 Broadening of vibrational-rotational lines of CO_2

An echelette grating in the expanded beam is the reflector of the laser resonator. By rotating the echelette grating, the laser resonator is adjusted to different lines.

In the TEA CO_2 laser (Fig. 14.10c), the direction of the gas discharge (at a pressure of $\approx$1 bar) is transverse to the laser beam. Two Brewster windows (NaCl plates) are closing a box containing the laser gas. The resonator mirrors are outside the box. A power supply charges a Marx generator (a capacitor bank with many capacitors in parallel and in series). An electric switch starts the discharge leading to high-power electric pulse (voltage 100 kV; current 100 A; duration 20 ns). The electric pulse, guided to one of the electrodes, causes a transverse discharge between the electrodes. The electrodes (distance 1 cm, length 40 cm) of the TEA laser (Fig. 14.10d) have a special profile (Rogowski profile) providing a homogeneous discharge. Arc discharges between the tips of metal needles initiate the discharge. The arc discharges produce UV radiation, which causes pre-ionization of molecules in the volume between the main electrodes. A gas discharge between pairs of needles, arranged along the electrodes (on both sides of the discharge volume), occurs when a high voltage pulse arrives at the electrodes. The TEA laser is a multi-mode laser; a single pulse consists of radiation at several modes (longitudinal and transverse modes).

At small gas pressure, Doppler broadening of the vibrational-rotational lines of CO_2 determines the width of the lines (Fig. 14.11). Collision broadening dominates at pressures between 5 mbar and about 1,000 mbar; in this pressure range, the gain bandwidth increases proportionally to pressure. At still higher pressure, the vibrational-rotational lines overlap partly and above a pressure of 20 bar the single vibrational-rotational lines overlap completely. Then the gain profile of each of the four branches is continuous and has a width of about 500 GHz. A mode locked high-pressure CO_2 laser operating on one of the four branches produces picosecond pulses (duration $\sim$1 ps) consisting of radiation around a frequency of 30 THz.

14.9 Other Gas Discharge Lasers and Optically Pumped Far Infrared Lasers

Beside CO_2 lasers, there are other infrared and far infrared gas discharge lasers (Fig. 14.12). Laser oscillation is due to stimulated emission of radiation by transitions between vibrational-rotational levels (CO laser) or between rotational levels (D_2O and HCN lasers) in the vibrational ground state or an excited vibrational state.

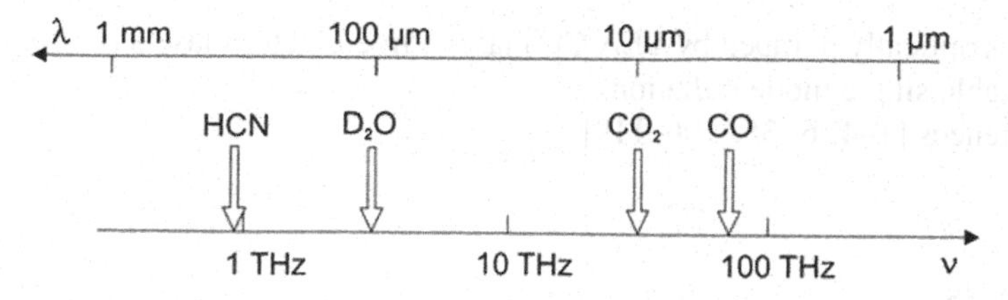

Fig. 14.12 Gas discharge lasers in the 1–100 THz range (*far infrared range*)

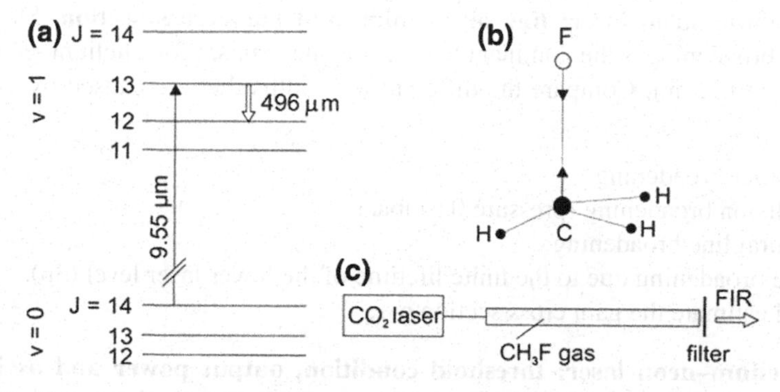

Fig. 14.13 Optically pumped CH_3F laser. **a** Principle. **b** CH_3F vibration. **c** Arrangement

The CO_2 laser is suitable for optical pumping of other gas lasers. Lasers operated with gases of CH_3F, D_2O, and alcohol molecules (and many other organic molecules) emit far infrared radiation at a large number of wavelengths.

Figure 14.13a shows an example of an optically pumped gas laser. Radiation of a CO_2 laser excites CH_3F from the vibrational ground state to an excited vibrational state. The vibration of the CH_3F molecule corresponds to a vibration of CH_3 against F (Fig. 14.13b). Stimulated rotational transitions ($J = 13 \rightarrow J = 12$) generate far infrared laser radiation (wavelength 496 μm, frequency near 605 GHz). A lens focuses the radiation of a CO_2 laser into a glass tube that contains the gas (Fig. 14.13c). A filter absorbs CO_2 laser radiation passing the tube.

Optical pumping is also possible if the CO_2 laser line and the absorption line of CH_3F do not completely coincide. Then stimulated Raman scattering (Sect. 35.8) results in generation of far infrared radiation. A variation of the CO_2 laser frequency leads to a variation of the frequency of the far infrared laser. The tuning range, however, is small (about 0.1% relative to a far infrared laser line).

The optically pumped gas lasers emit, depending on the gas and the wavelength of the CO_2 pump laser, radiation at a very large number of frequencies (about ten thousand laser lines have been reported). A gas laser pumped by a TEA laser generates intense far infrared radiation pulses (pulse power about 1 kW [119]).

In comparison with optically pumped cw far infrared gas lasers, quantum cascade lasers (Chap. 29) are becoming important alternatives. In comparison with far infrared gas lasers optically pumped by TEA CO_2 lasers, free-electron lasers (Chap. 19) produce tunable single mode radiation.

References [1–4, 6, 35, 106–119].

Problems

14.1 Helium–neon laser: line broadening and gain cross section. Show that Doppler broadening is the dominant broadening mechanism for a helium–neon laser operated at 633 nm. Compare the different linewidths that are caused by different effects.

(a) Doppler broadening.
(b) Collision broadening (pressure 0.5 mbar).
(c) Natural line broadening.
(d) Line broadening due to the finite lifetime of the lower laser level (3p).
(e) And estimate the gain cross section σ_{21}.

14.2 Helium–neon laser: threshold condition, output power and oscillation onset time. A helium–neon laser is characterized by: length of the active medium L = 0.5 m; cross section $a_1 a_2 = 4\,\text{mm}^2$; reflectivity of the output coupling mirror $R = 0.98$; reflectivity of the reflector $R = 0.998$. Determine the following quantities:

(a) Threshold population difference per m^3.
(b) Absolute value of the threshold population difference.
(c) Output power at a pump rate that is 10 times stronger than at threshold.
(d) Oscillation onset time.

14.3 Doppler effect in the helium-neon laser and Lamb dip.

(a) Calculate the frequency difference of the emission line at 633 nm for a neon atom that moves with a velocity of 500 m/s toward an observer and of an atom that moves with the same velocity away from the observer.
(b) In which velocity range do the emission lines overlap?
(c) Discuss the consequence for the gain in a helium–neon laser: the gain shows a minimum at the line center of the gain curve (= Lamb dip, according to W. Lamb).

14.4 CO_2 laser (length $L = 1$ m; cross-sectional area $a_1 a_2 = 1\,\text{cm}^2$; reflectivity of the output coupling mirror $R = 0.7$; lifetime of the upper laser level with respect to spontaneous emission of radiation by $2 \rightarrow 1$ transitions, $\tau_{\text{rel}}^* = 4\text{s}$; gas pressure 10 mbar).

(a) Calculate: Doppler linewidth; gain cross section; threshold condition; pump rate (relative to the threshold pump rate) that is necessary to obtain an output power $P_{\text{out}} = 60\,\text{W}$.

(b) Discuss the onset of laser oscillation taking into account that the upper laser level has a long lifetime with respect to spontaneous emission and that there are many rotational levels belonging to the excited state.
(c) Estimate the maximum gain coefficient of an excited CO_2 gas and the corresponding small-signal gain factor of radiation in a cw CO_2 laser. [*Hint*: the maximum gain coefficient is determined by the density of CO_2 molecules that are available in a gas at low pressure.]
(d) Show that the gain coefficient of an excited CO_2 gas in a TEA laser or in a high pressure CO_2 laser (pressure 20 bar) is about the same as in a cw laser at a gas pressure of 10 mbar. Why is the pulse power of a TEA laser or of a high pressure laser much larger than the power of the cw laser? Estimate the radiation energy of a pulse within a TEA laser.
(e) Estimate the oscillation onset time of a TEA laser.

14.5 Optical radar.
Determine the frequency difference between the frequency of radiation emitted by a helium–neon laser and the frequency of radiation reflected by a car traveling at a velocity of 60 km per hour.

14.6 CO molecule.

(a) Estimate the isotope shift of the vibrational frequency of CO (frequency $\tilde{\nu} = 2{,}170\,\mathrm{cm}^{-1}$) if ^{16}O is replaced by ^{18}O.
(b) Next-near lines that are due to transitions between vibrational-rotational levels have a frequency separation of $3.86\,\mathrm{cm}^{-1}$. Determine the rotational constant $\tilde{B} = B/(hc)$. Which of the rotational levels has the highest occupancy at room temperature?

14.7 Rotational levels at thermal equilibrium.

(a) Which of the J levels of a CO_2 molecule in the vibrational ground state has the largest occupancy in a gas at room temperature?
(b) Determine the excitation energy and the occupancy of the $\mathrm{v} = 0$, $J = 1$ state of a nitrogen molecule (N–N distance = 0.1 nm) in a gas at room temperature.

14.8 Estimate the density of neon atoms, the density of excited neon atoms, and the corresponding absolute numbers of nonexcited and excited neon atoms in a helium–neon laser.

14.9 Explain the nomenclature used to characterize: (a) the two lowest excited states of He; (b) the ground state and the four lowest excited states of Cu; (c) the states of Ar, Ar^+ and $(Ar^+)^\star$.

14.10 Spatial hole burning and diffusion of excited molecules in a CO_2 laser.
On the one hand, the excitation of CO_2 molecules in a gas discharge occurs homogeneously in the gas discharge tube. On the other hand, the amplitude of the standing wave field in the laser resonator shows a sin$z$ dependence along the resonator

axis. The stimulated emission is therefore spatially inhomogeneous. Show that—nevertheless—all excited CO_2 molecules can contribute to stimulated emission. Study the problem in case of a cw CO_2 laser (length 0.7 m; gas pressure 5 mbar; ratio He:Ne:CO_2 = 6:1:1; spontaneous lifetime of an excited CO_2 molecule $\tau_{sp} \sim$ 5 s; reflectivity of the output mirror $R = 0.95$; efficiency of conversion of pump power to power of laser radiation $\sim 20\%$).

(a) Determine the density of CO_2 molecules.
(b) Determine the density of excited CO_2 molecules.
(c) Estimate the diffusion constant $D = \bar{v}\lambda_m/3$, where $\bar{v}$ is an average velocity and λ_m ($\sim 100\,\mu$m) the mean free path of a CO_2 molecule with respect to a collision with another atom or molecule in the gas mixture of a CO_2 laser.
(d) Estimate the time τ_{esc} it takes an excited CO_2 molecules to escape from a region of weak field strength to a region of large field strength. [*Hint*: replace the cosine squared field distribution by a rectangular distribution and apply a one-dimensional diffusion equation to describe the dynamics of the local density N_{loc} of excited CO_2 molecules, $\mathrm{d}N_{loc}/\mathrm{d}t = D\mathrm{d}^2N_{loc}/\mathrm{d}x^2$.]

14.11 Collision cross sections of molecules in a CO_2 laser. Estimate the cross sections of collisions of CO_2 molecules with other CO_2 molecules, with N_2 molecules and with helium atoms. [*Hint*: Use the hard-sphere approximation of the cross section, $\sigma_c = \pi/4(d_1 + d_2)^2$, where d_1 and d_2 are the diameters of the two colliding molecules; treat the CO_2 molecule as a sphere (diameter 0.4 nm) as well as the N_2 molecule (diameter 0.2 nm).]

14.12 Voigt profile. A Voigt profile is observed when collision and Doppler broadening influence the spectral broadening of an optical transition. Atoms of velocity v have a transition frequency $\nu = \nu_0 + \nu_0 \mathrm{v}/c$. The lineshape function describing optical transitions in these atoms with the transition frequency ω_0' is

$$g(\omega_0', \omega) = \frac{\Delta\omega_0}{2\pi} \frac{1}{(\omega_0' - \omega)^2 + \Delta\omega_0^2/4}. \tag{14.21}$$

where $\Delta\omega_0$ is the halfwidth of a transition. The probability of a transition in the frequency interval $\omega_0', \omega_0' + \mathrm{d}\omega_0'$ is equal to

$$P(\omega_0')\mathrm{d}\omega_0' = \frac{2\sqrt{\ln 2}}{\sqrt{\pi}\,\Delta\omega_c} \exp\left(-\frac{\ln 2(\omega_0' - \omega_0)^2}{\Delta\omega_c^2/4}\right)\mathrm{d}\omega_0'. \tag{14.22}$$

where $\Delta\omega_c$ is the halfwidth and ω_0 the center frequency of the Gaussian profile. We obtain the spectral profile of a line by averaging,

$$S(\omega) = \int_0^\infty g(\omega_0', \omega) P(\omega_0')\mathrm{d}\omega_0'. \tag{14.23}$$

It follows that

$$S(\omega) = \frac{\Delta\omega_0}{\pi^{3/2}\Delta\omega_c}\int_0^{\infty}\frac{1}{(\omega_0' - \omega_0)^2 + \Delta\omega_0^2/4}\exp(-\frac{\text{In}2(\omega_0' - \omega_0)^2}{\Delta\omega_c^2/4})d\omega_0'. \quad (14.24)$$

The equation has to be solved numerically.

(a) Show that the limits of the Voigt profile are the Lorentzian or the Gaussian profile, depending on the ratio of the two halfwidth $\Delta\omega_0$ and $\Delta\omega_c$.
(b) Show that (14.24) is consistent with (7.52).

14.13 Characterize a carbon dioxide laser that operates with $^{12}C^{17}O_2$ or with $^{13}C^{16}O_2$.

14.14 A cw CO_2 laser beam (power 100 W, diameter 10 mm) hits a stone. How long does it take until the stone is glowing? Assume that the hot range has an extension of 1 mm.

14.15 Diffusion.
Describe diffusion of particles in an infinitely long rectangular slab. At $t = 0$, the particles are homogeneously distributed over the cross section at $x = 0$, with the two-dimensional particle density N_0. [*Hint*: Apply the one-dimensional diffusion equation $\partial\rho/\partial t = D\partial^2\rho/\partial x^2$ where ρ is the three-dimensional particle density.]

(a) Show that it has the solution $N_0/(2\sqrt{2Dt})\exp(-x^2/4Dt)$.
(b) Determine the variance and the halfwidth (FWHM) of the distribution.
(c) Choose as an example $N_0 = 10^{22}\,\text{m}^{-2}$ and $D = 10^{-2}\,\text{m}^2\,\text{s}^{-1}$. Determine the halfwidth of the distribution for the time at which the two-dimensional particle density at $x = 0$ decreased to one half of its original value.

Chapter 15
Solid State Lasers

We discuss solid state lasers that make use of electronic states of impurity ions in a dielectric crystals or in glasses—other types of solid state lasers, namely semiconductor lasers that are based on electrons in energy bands of semiconductors, will be treated in later chapters.

We describe the principle of the ruby laser. We treat the titanium–sapphire laser in more detail than in an earlier chapter. We mention other broadband solid state lasers. Then we present a description of the neodymium-doped YAG laser, of other neodymium lasers, and of other YAG lasers. We describe disk lasers and fiber lasers. We give a short survey of solid state lasers with respect to host materials and impurities. Finally, we describe line broadening processes occurring in solid state laser media.

The active medium of a disk laser has the form of a disk rather than the form of a rod. A disk laser pumped with a semiconductor laser has a high beam quality.

Glass lasers are used for generation of near infrared radiation of different wavelengths. The neodymium-doped glass laser can produce intense radiation pulses at a wavelength at 1.05 μm. Doped glass fiber lasers generate radiation in the wavelength range 0.7–3 μm. Fiber lasers are robust and flexible. They are suitable for applications in many areas (material processing, biophysics, medicine); fiber lasers are able to generate continuous wave radiation or picosecond pulses.

15.1 Ruby Laser

A ruby laser (Fig. 15.1) uses Cr^{3+} ions in an Al_2O_3 (sapphire) crystal with a doping concentration of typically 0.05% by weight Cr_2O_3; the density of Cr^{3+} ions is $N_0 = 1.6 \times 10^{25}$ m^{-3}. An excited Cr^{3+} ion in Al_2O_3 has two long-lived energy levels with a small energy separation. The lifetime of the levels with respect to spontaneous emission of radiation is about 3 ms. Two broad energy bands are suited as pump bands. The optical transitions between the long-lived levels and the ground state level occur at two slightly different wavelengths (R_1 fluorescence line at 694.3 nm

K.F. Renk, *Basics of Laser Physics*, Graduate Texts in Physics,
DOI 10.1007/978-3-319-50651-7_15

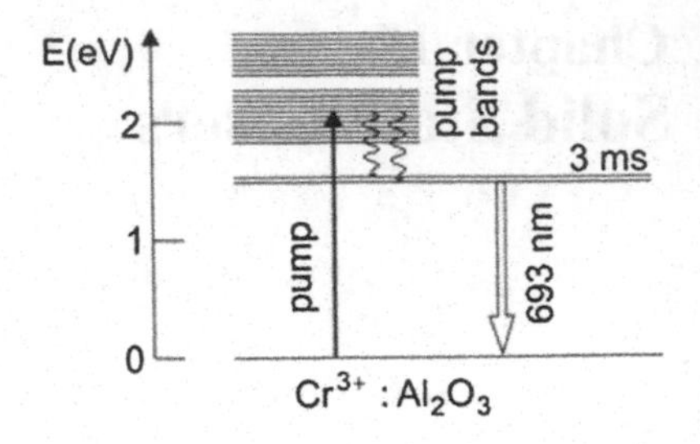

Fig. 15.1 Ruby laser (*principle*)

and R_2 line at 692.8 nm). The level splitting is due to a week trigonal crystalline field that is present, in addition to a cubic crystalline field, at the sites of the Cr^{3+} impurity ions in sapphire). The ground state level is identical with the lower laser level. By optical pumping into a pump band and fast nonradiative relaxation, the upper laser levels are populated. At sufficiently strong pumping, the populations of the upper laser levels are larger than the population of the ground state level. The gain cross section (at the center frequencies of the two lines) has the value $\sigma_{21} = 2.5 \times 10^{-24}\ \mathrm{m}^2$.

The further development of the ruby laser, after its first operation (in 1960 [120]), stimulated the development of special high-power discharge lamps (continuously working lamps and pulsed flash lamps too). Today, pumping of a ruby laser is possible with radiation of another laser. The long lifetime of the upper laser level makes it possible to excite almost all Cr^{3+} ions in a ruby crystal and to produce, by Q-switching, pulses of very large pulse energy.

15.2 More About the Titanium–Sapphire Laser

In an earlier chapter we have already introduced the titanium-sapphire laser ($Ti{:}Al_2O_3$ laser). Here, we discuss the laser in more detail.

We can describe the energy level diagram of Ti^{3+} in Al_2O_3 (Fig. 15.2, left) in a formal way. We introduce the *configuration coordinate* Q. It describes an average distance between a Ti^{3+} ion and neighboring ions. The energy of a level depends on Q. The energy curve $E(Q)$ indicates that the ground state is accompanied by vibronic levels. The energy of a vibronic level is composed of electronic and vibrational energy. Q_0 is the configuration coordinate at which the energy minimum of the electronic ground state occurs. Correspondingly, the $E^*(Q)$ curve indicates that the excited state of Ti^{3+} is accompanied by vibronic energy levels too. The configuration coordinate Q_0^* at which the energy minimum of excited Ti^{3+} occurs is larger than Q_0.

Figure 15.2 (right) illustrates the four-level description of the titanium-sapphire laser:

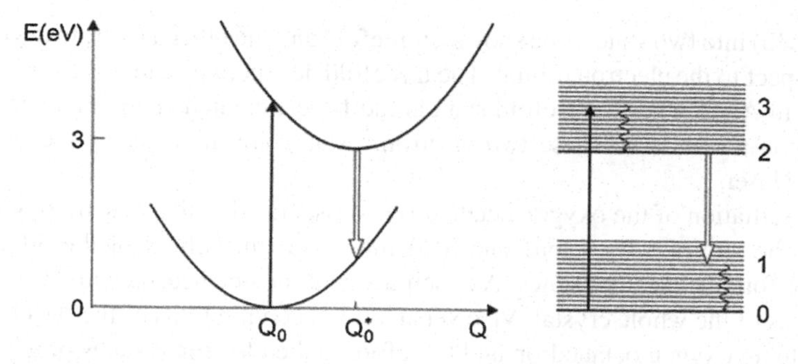

Fig. 15.2 Titanium–sapphire laser (*principle*)

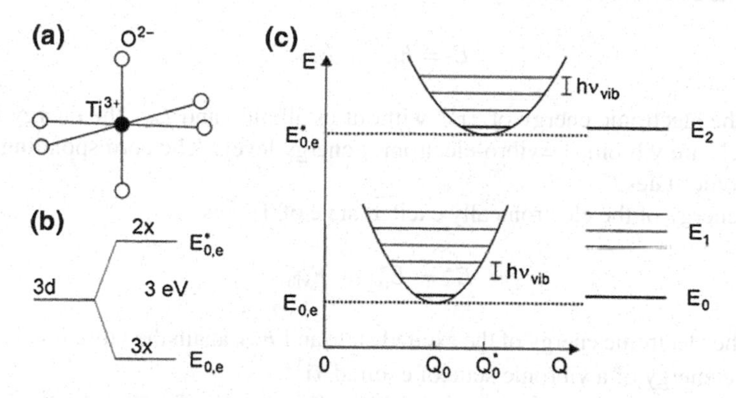

Fig. 15.3 Ti^{3+} in Al_2O_3. **a** Surrounding of a Ti^{3+} ion. **b** Crystal field splitting of the 3d state of Ti^{3+}. **c** Vibronic energy levels due to coupling to a phonon

- In an optical absorption process, a Ti^{3+} ion is excited from the ground-state level (level 0) to a vibronic level (level 3) of the electronically excited state.
- Fast nonradiative relaxation (relaxation time $\sim 10^{-13}$ s) leads to population of the lowest excited-state level (level 2) of Ti^{3+}.
- An optical transition occurs to a vibronic level of the electronic ground state.
- After fast relaxation (relaxation time $\sim 10^{-13}$ s), the Ti^{3+} ion is in its ground state.

Optical transitions are governed by the Franck–Condon principle: optical transitions occur without a change of the atomic distances.

We now discuss the origin of the vibronic energy levels of Ti^{3+} in Al_2O_3. The Ti^{3+} ion has the electron configuration $1s^2 2s^2 2p^6 3s^2 3p^6 3d$. It has filled shells (like an argon atom) and an external electron in the 3d shell. The 3d state of the free Ti^{3+} ion is fivefold degenerate according to the quantum number ($l = 2$) of the orbital momentum.

A Ti^{3+} ion in an Al_2O_3 crystal (Fig. 15.3a) is surrounded by an octahedron of oxygen ions (O^{2-} ions). In the field of the ions (crystal field), the 3d state splits

(Fig. 15.3b) into two states, one shows a threefold and the other a twofold degeneracy with respect to the electron orbital. The threefold degenerate state is the ground state of Ti^{3+} in Al_2O_3 and the twofold state is the lowest excited state; the energy level splitting is about 3 eV. These two electronic states are the basis of the titanium–sapphire laser.

An oscillation of the oxygen octahedron is associated with an oscillating electric field at the site of a Ti^{3+} ion. The field influences the orbital of the 3d electron and therefore the electron states. An oscillation of the octahedron couples to lattice vibrations of the whole crystal. Vice versa, all lattice vibrations of the Al_2O_3 crystal couple to an oxygen octahedron and therefore to the electronic states of a Ti^{3+} ion. The coupling gives rise to a distribution of electronic ground state levels as well as of excited state levels. The energy of the electronic ground state of Ti^{3+} is (Fig. 15.3c)

$$E = E_{0,\mathrm{e}} + E_{\mathrm{vib}}. \tag{15.1}$$

$E_{0,\mathrm{e}}$ is the electronic energy of Ti^{3+} without oscillation and E_{vib} the energy levels. The levels are vibronic (=vibro-electronic) energy levels. The corresponding states are vibronic states.

The energy of the electronically excited state of Ti^{3+} is

$$E^* = E^*_{0,\mathrm{e}} + E_{\mathrm{vib}}. \tag{15.2}$$

$E^*_{0,\mathrm{e}}$ is the electronic energy of the excited state and E_{vib} again the vibrational energy. E^* is the energy of a vibronic state of excited Ti^{3+}.

A single vibration of an Al_2O_3 crystal has the vibrational energy

$$E_{\mathrm{vib}} = \left(\mathrm{v} + \frac{1}{2}\right) h\nu_{\mathrm{vib}}, \tag{15.3}$$

where ν_{vib} is a vibrational frequency, v the vibrational quantum number of this vibration, and $\frac{1}{2}h\nu_{\mathrm{vib}}$ the zero point energy of the vibration.

An Al_2O_3 crystal has a large number of vibrational frequencies; the number of different lattice vibrations is of the order of 10^{22} for a crystal volume of 1 cm^3. Therefore, the vibronic levels have a continuous energy distribution. The different energy levels of Ti^{3+} in Al_2O_3 are:

- $E_0 = E_{0,\mathrm{e}}$+ zero point energy of all vibrations = energy of the ground state level.
- $E_2 = E^*_{0,\mathrm{e}}$+ zero point energy of all vibrations = lowest energy of the excited state.
- $E_1 = E_0 + E_{\mathrm{vib}}$ = lower laser levels, having a broad energy distribution.

The spontaneous lifetime of a vibronic level of an excited Ti^{3+} ion is ~3.8 μs.

Our discussion shows that the occurrence of a broad distribution of pump levels and of a broad distribution of lower laser levels in titanium–sapphire is a consequence of the vibronic character of the energy levels of Ti^{3+} in Al_2O_3. We will derive the gain profile of $Ti^{3+}:Al_2O_3$ in Sect. 17.2. Vibronic energy levels are the basis of many other lasers.

15.3 Other Broadband Solid State Lasers

We compare the titanium–sapphire laser with other broadband tunable solid state lasers (Table 15.1): alexandrite laser; Cr:LiSAF laser (chromium-doped lithium strontium aluminum fluoride laser); and Cr:LiCaF laser (chromium-doped lithium calcium fluoride laser). The wavelength λ given in the table is the wavelength of maximum gain coefficient. Titanium–sapphire has the largest gain bandwidth $\Delta\nu_g$.

The alexandrite laser was the first solid state laser that was tunable over a wide wavelength range (700–820 nm). In alexandrite ($BeAl_2O_4$ crystal doped with Cr^{3+}), the Cr^{3+} ions (concentration 3×10^{25} m^{-3}) replace about 0.1% of the Al^{3+} ions. The energy levels of Cr^{3+} in alexandrite, used in the laser, are (Fig. 15.4) the following:

- 0; ground state. Q_0 is the configuration coordinate of the energy minimum of the vibronic ground state levels.
- 1A; vibronic band of the ground state.
- 2A; vibronic band of excited Cr^{3+}; spontaneous lifetime 1.5 ms.
- 2B; another vibronic band of excited Cr^{3+}, 70 meV above the 2 A band; spontaneous lifetime of 1.5 μs. $Q_0^\star$ is the configuration coordinate of the energy minimum of this vibronic band.

Table 15.1 Tunable lasers

Lasers	λ (nm)	τ_{sp} (μs)	σ_{21} (m^2)	$\Delta\nu_g$ (THz)	Tuning range (nm)
TiS	790	3.8	3×10^{-23}	110	660–1,180
alexandrite	760	260	10^{-24}	50	700–820
Cr:LiSAF	850	70	5×10^{-24}	80	780–1,010
Cr:LiCaF	780	170	13×10^{-24}	60	720–840

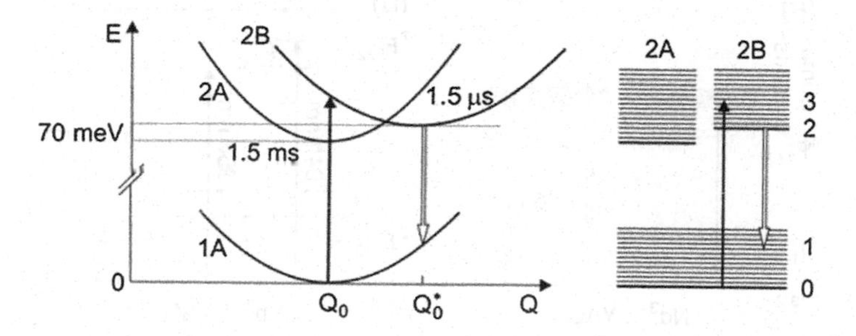

Fig. 15.4 Alexandrite laser

- 2; upper laser level (belonging to 2B); spontaneous lifetime 1.5 μs.
- 1; lower laser level (vibronic level belonging to 1A).
- 3; pump levels (belonging to 2B), pumped with radiation around 680 nm.

Optical pumping and fast relaxation result in populations of the 2A and the 2B vibronic bands. The population of each of the vibronic bands is in thermal equilibrium and the populations of the two bands are in thermal equilibrium with each other. The equilibrium is determined by the crystal temperature. To obtain a large population of 2B levels, the crystal is kept at an elevated temperature (60 °C or higher). Laser transitions occur around a wavelength of 760 nm within a width of about 100 nm.

The energy levels of Cr^{3+} in LiSAF and LiCaF are similar to the energy levels of Cr^{3+} in alexandrite. There is, however, an important difference: the energy minimum of the 2B band lies below the minimum of the 2 A band. Therefore, heating of the crystals is not necessary.

Alexandrite, Cr^{3+}:LiSAF, and Cr^{3+}:LiCaF lasers can be used for the same tasks as the titanium–sapphire laser. Titanium–sapphire has the advantage that the crystalline material has a larger hardness and a higher heat conductivity.

15.4 YAG Lasers

A Nd:YAG laser (YAG = $Y_3Al_5O_{12}$ = yttrium aluminum garnet) can have a high beam quality and can be operated as a cw or as a pulsed laser. Applications:

- Material processing: drilling, point welding, marking.
- Medicine: surgery, (Nd:YAG laser radiation can be guided with a glass fiber into the interior of a body and focused by a lens); eye surgery; applications in dermatology, *see*, for instance [127–129].

A neodymium YAG laser (Fig. 15.5a) makes use of energy levels of Nd^{3+}. The Nd atom has the electron configuration $4f^3 5s^2 5p^6 6s^2$. The free Nd^{3+} ion has the configuration $4f^2 5s^2 5p^6$; the two lowest energy levels are $^4I_{9/2}$ and $^4I_{11/2}$. The crystalline

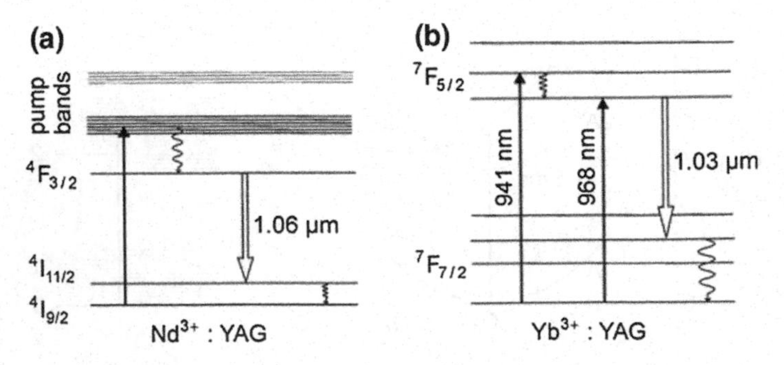

Fig. 15.5 YAG lasers. **a** Neodymium-doped YAG laser. **b** Ytterbium-doped YAG laser

Table 15.2 YAG lasers

Laser	λ	λ_{pump} (nm)	τ_{sp} (μs)	$\Delta\nu_g$	σ_{21} (m^2)
Nd:YAG	1.06 μm;	808	230	140 GHz	3×10^{-22}
Yb:YAG	1.03 μm;	941 968	960	1.7 THz	2.1×10^{-24}
Pr:YAG	1.03 μm	941			
Er:YAG	2.94 μm	800 970			

electric field causes a splitting of these levels (not shown in the figure). Energetically higher lying levels serve for optical pumping. Optical pumping and fast relaxation leads to population of the long-lived $^4F_{3/2}$ level (spontaneous lifetime 230 μs). The laser transition $^4F_{3/2} \rightarrow {}^4I_{11/2}$ corresponds to a wavelength of 1.064 μm (frequency ~300 THz).

YAG crystals can be prepared in a very high crystal quality. Nd^{3+} ions can replace about 1% of the Y^{3+} ions. Optical pumping of a neodymium YAG laser is possible with a lamp or with a semiconductor laser. Depending on the size of a laser crystal, a neodymium YAG laser can produce laser radiation at power levels of 1–10 W or more. Operated as a giant pulse laser, a Nd:YAG laser can generate pulses of an energy of 1 J.

Other laser frequencies of the Nd^{3+}:YAG laser lie at 0.914 μm. ($^4F_{3/2} \rightarrow {}^4I_{9/2}$ transitions) and at 1.35 μm ($^4F_{3/2} \rightarrow {}^4I_{13/2}$ transitions).

Table 15.2 shows a list of various other YAG lasers.

- *Ytterbium-doped YAG laser* (Fig. 15.5b). The ytterbium-doped YAG laser (Yb:YAG laser) emits at 1.03 μm, it is pumped with radiation (at 940 nm) of a semiconductor laser (InGaAs laser) by transitions between $^7F_{3/2}$ sublevels and $^7F_{5/2}$ sublevels. The Yb^{3+} ions can replace 6% of the Y^{3+} ions in YAG. The ytterbium-doped YAG laser is becoming a competitor of the neodymium-doped YAG laser. Due to a high concentration of impurity ions, ytterbium-doped YAG crystals are especially suited as active media of lasers of small length (namely disk lasers, Sect. 15.6).
- *Praseodymium-doped YAG laser*. Pr^{3+} ions replace Y^{3+} ions. The doping can be extraordinarily high; it is possible to replace about 26% of the Y^{3+} ions by Pr^{3+} ions. The Pr:YAG laser, pumped with radiation of a semiconductor laser, emits infrared radiation.
- *Erbium-doped YAG lasers* (Fig. 15.6). The free erbium (Er) atom has the configuration [Xe]$4f^{12}5s^25p^66s^2$. Removing three electrons leads to Er^{3+} with the electronic configuration [Xe]$4f^{11}5s^25p^6$. The Er^{3+} ion, doped into a solid, has a long-lived excited state $^4I_{11/2}$. Laser transitions, $^4I_{11/2} \rightarrow {}^4I_{13/2}$, generate infrared radiation (wavelength 2.94 μm). Pumping of the erbium-doped YAG laser, via the narrow $^4I_{11/2}$ or $^4I_{9/2}$ levels, is possible with a semiconductor laser (at 980 nm or 800 nm); it is possible to operate an erbium-doped YAG laser at 1.54 μm as a three-level laser (of ruby laser type).

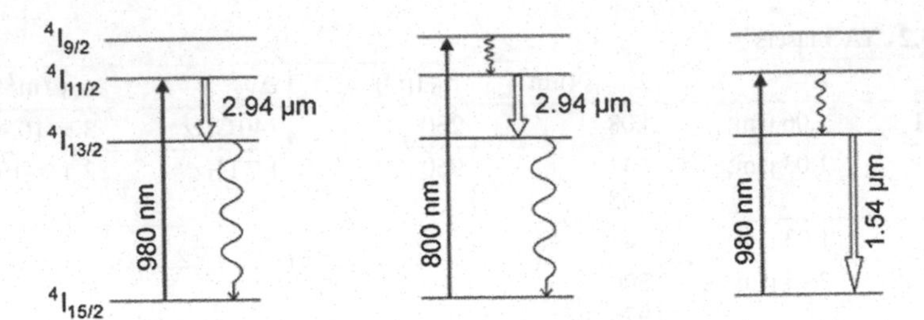

Fig. 15.6 Erbium-doped YAG lasers

The erbium-doped YAG laser with emission of radiation at 2.94 μm is of interest for biomedical applications.

15.5 Different Neodymium Lasers

Various other solids doped with Nd^{3+} are suitable as active media (Table 15.3).

- The Nd:YVO_4 laser emits at the same wavelength (1.064 μm) as the Nd^{3+}:YAG laser. With respect to applications, the Nd^{3+}:YVO_4 laser competes with the Nd^{3+}:YAG laser.
- Nd:YLF laser (= Nd:LiYF_4 = neodymium-doped lithium yttrium fluoride laser). The laser emits at 1.047 μm and 1.053 μm. Pumping is possible with a semiconductor laser (pump band at 804 nm, halfwidth 4 nm). The laser is also an alternative to the Nd:YAG laser.
- Neodymium-doped glass laser. In glass, Nd^{3+} ions occupy sites with different surroundings and different crystal fields. This leads to an inhomogeneous broadening of the excited-state levels. The halfwidth of the corresponding line (~6 THz) allows for generation of picosecond pulses. There are broad pump bands around 750 and 810 nm.

The lasers can generate radiation at power levels in the 10–100 W range.

Table 15.3 Neodymium-doped solid state lasers

Laser	λ (μm)	λ_{pump} (nm)	τ_{sp} (μs)	$\Delta\nu_g$	σ_{21} (m^2)
Nd:YAG	1.064	808	230	140 GHz	2.8×10^{-22}
Nd:YVO_4	1.06	809	90	210 GHz	$1.1 \times 10^{-22} (\pi)$ $4.4 \times 10^{-23} (\sigma)$
Nd:YLF	1.047 1.053	804	480	200 GHz	1.8×10^{-23} 1.2×10^{-23}
Nd:glass	1.05		300	6 THz	3×10^{-24}

15.6 Disk Lasers

A disk laser is a compact, highly efficient laser. It produces radiation of high power (1 kW or more). Applications lie in fields (cutting, welding, labeling), in competition with the Nd:YAG laser.

A disk laser (Fig. 15.7) consists of a disk (thickness 100–200 μm), pumped with a semiconductor laser. Because of the large diameter, the disk laser has a high beam quality. A large concentration of Yb^{3+} ions in Yb^{3+}:YAG allows for a compact design of the laser. We described the laser principle in the preceding section.

In comparison with a laser medium with a rod shape, the disk laser has a larger ratio of cooling area and active volume. The temperature distribution within the active medium has a nearly homogeneous radial distribution. This leads to a high beam quality.

A Nd:YVO_4 laser can be operated as miniature picosecond laser (Fig. 15.8). A semiconductor laser (808 nm) pumps the laser, which emits radiation at 1,064 nm. The laser can generate picosecond pulses (duration 10 ps) at a high repetition rate (e.g., 30 GHz). Mode locking is possible by use of a mirror with a reflectivity that depends on the radiation intensity. The reflectivity is small at small intensity and large at high intensity. The mirror is a semiconductor saturable absorber mirror (SESAM). A Nd:$GdVO_4$ laser has similar properties as the Nd:YVO_4 laser.

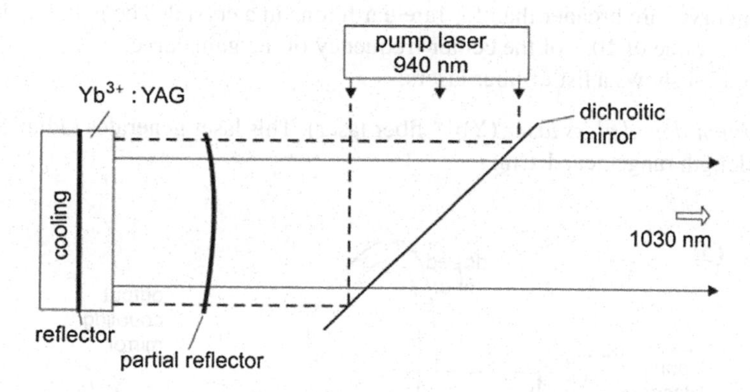

Fig. 15.7 Disk laser

Fig. 15.8 Picosecond disk laser

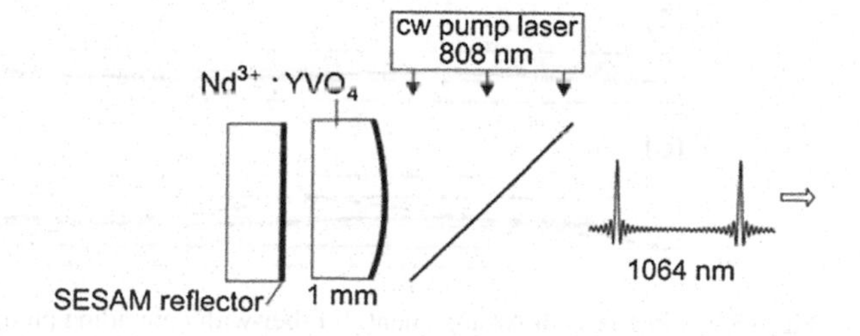

15.7 Fiber Lasers

Fiber lasers are important glass lasers. Fiber lasers have many applications in fields of material processing, chemistry, medicine, biology. The second harmonic radiation of glass fiber lasers serves for pumping of other lasers, e.g., of disk lasers. In comparison with other solid state lasers, fiber lasers are flexible and simple with respect to adjustment (or may not need adjustment at all).

The active medium of a fiber laser is a glass that is doped with rare earth ions. We describe here main features of a fiber laser (Fig. 15.9a):

- *Glass fiber* (length 1–10 m, or longer; diameter 5 μm), doped with ions.
- *Dichroitic end mirror*. It is highly reflecting for the laser radiation and transparent for the pump radiation.
- *Output coupling mirror*. In order to reach optimum efficiency, the reflectivity of the output coupling mirror is chosen appropriately.
- A fiber laser can be pumped with a semiconductor laser.

The pump waveguide either coincides with the laser waveguide (Fig. 15.9b) or has a larger diameter (Fig. 15.9c).

Fiber lasers are available in the 0.7–3 μm wavelength range. Rare earth ions in a glass occupy sites of different strength of the crystalline electric field. Therefore, the energy levels of the electronic states of ions in a glass are energetically distributed and the gain curves are broader than for rare earth ions in a crystal. The gain bandwidth can have a value of 10% of the center frequency of the gain curve.

Table 15.4 shows a list of fiber lasers:

- *Ytterbium-doped glass laser* (Yb^{3+} fiber laser). This laser generates radiation in a wavelength range near 1 μm.

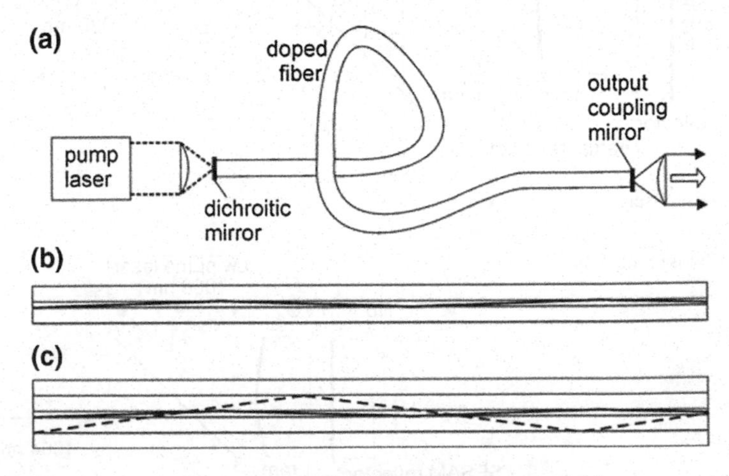

Fig. 15.9 Fiber laser. **a** Arrangement. **b** Fiber with coinciding pump and laser waveguide. **c** Fiber with pump and laser waveguide of different diameters

Table 15.4 Fiber lasers

Laser	λ (μm)	Doping % per weight
Yb^{3+} fiber	1.02–1.2	4
Yb^{3+}/Er^{3+} fiber	1.5–1.6	8/1
Yb^{3+}/Er^{3+} fiber	2.7–2.8	8/8
Tm^{3+} fiber	1.85–2.1	4
Ho^{3+} fiber	2.1 and 2.9	3

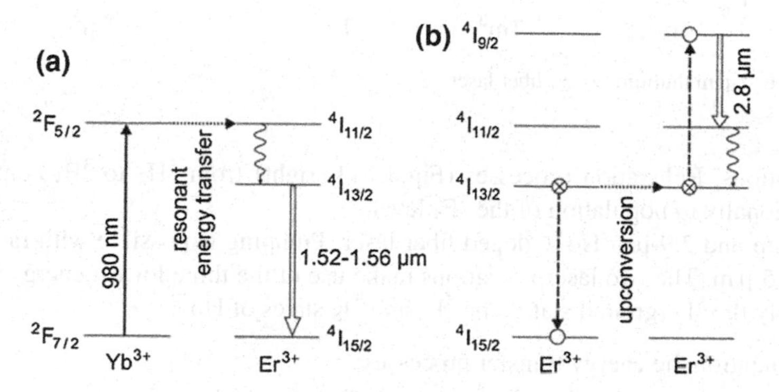

Fig. 15.10 Fiber lasers. **a** 1.5-μm erbium-doped fiber laser. **b** 2.8-μm erbium-doped fiber laser

- 1.5-μm *erbium-doped fiber laser* (Pr^{3+}/Er^{3+} fiber laser). The erbium-doped fiber laser makes use of the three energy levels $^4I_{15/2}$ (ground state), $^4I_{13/2}$ and $^4I_{11/2}$ of Er^{3+} (Fig. 15.10a). The 1.5-μm erbium-doped fiber laser is based on stimulated $^4I_{13/2} \rightarrow {}^4I_{15/2}$ transitions. The absorption coefficient for pump radiation is much larger if a glass contains, in addition to Er^{3+} ions, a large concentration of Yb^{3+} ions; the concentration can be ten times larger than the Er^{3+} concentration. The $^2F_4$ level of Pr^{3+} coincides with the energy level $^4I_{11/2}$ of Er^{3+}. Optical pumping and *resonant energy transfer* from Pr^{3+} ions to Er^{3+} ions leads to population inversion in the Er^{3+} ion ensemble. Co-doping with ytterbium enhances the absorptivity and allows for a more efficient optical pumping. The additional doping with ytterbium has only a small influence on the energy levels of the Er^{3+} ions.
- 2.8-μm *erbium-doped fiber laser* (Pr^{3+}/Er^{3+} fiber laser). The laser is pumped via Pr^{3+} ions. Above a concentration of about 1.5%, an excited Er^{3+} ion can transfer the excitation energy to a neighboring excited Er^{3+} ion by an *upconversion process*, leading to population of $^4I_{9/2}$ states (Fig. 15.10b). Laser radiation is due to $^4I_{9/2} \rightarrow {}^4I_{11/2}$ transitions.
- 2-μm *thulium-doped fiber laser* (Tm^{3+} fiber laser). Pumping results in population of $^3H_4$ levels (Fig. 15.11, left).*Cross relaxation* leads to population of $^3F_4$ levels; in a cross relaxation process, excitation energy is transferred to a neighboring unexcited ion (Fig. 15.11, center). Stimulated emission occurs by $^3F_4 \rightarrow {}^3H_6$

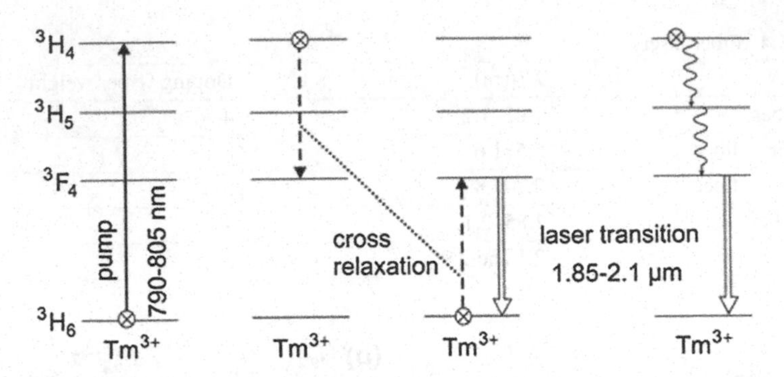

Fig. 15.11 2-μm thulium-doped fiber laser

transitions. Relaxation processes (Fig. 15.11, right) from $^3H_4$ to $^3F_4$ contribute additionally to population of the $^3F_4$ level.

- 2.1-μm and 2.9-μm Ho^{3+} doped fiber laser. Pumping is possible with radiation at 1.15 μm. The two laser transitions make use of the three lowest energy levels, namely the $^5I_8$ (ground state), the $^5I_7$, and $^5I_8$ states of Ho^{3+}.

We mention the energy transfer processes:

- Resonant energy transfer.
- Upconversion.
- Cross relaxation.
- Phonon-assisted energy transfer (Chap. 18).

Transfer of the excitation energy from an ion to another ion determines the microscopic dynamics of fiber laser media. We will treat the basis of energy transfer and the role of energy transfer processes in the microscopic dynamics of fiber media in Chap. 18. The treatment of the dynamics will provide the gain coefficient.

It is possible to use optically pumped fibers as amplifiers of radiation (Sect. 16.9 and Chap. 18). A special amplifier is the erbium-doped fiber amplifier—used in the optical communications. It is also possible to pump the active medium of an erbium-doped fiber amplifier with radiation at a wavelength (1.48 μm) that is only slightly smaller than the wavelengths (1.52–1.56 μm) of the range of gain (Chaps. 18 and 33.4).

15.8 A Short Survey of Solid State Lasers and Impurity Ions in Solids

The basic solid (a crystal or a glass) of a solid state laser (Fig. 15.12) is transparent for pump and laser radiation. The solid acts as a host of impurity ions. The electric field (crystal field) at the site of an impurity ion in a crystal or in glass mainly determines the energy levels of an impurity ion.

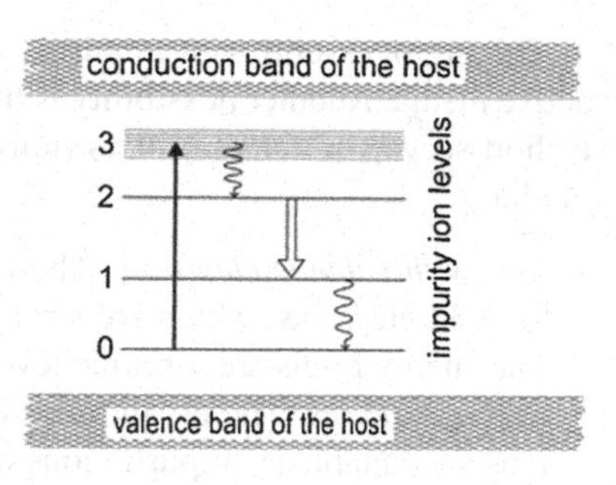

Fig. 15.12 Solid state laser: energy levels

Table 15.5 Maximum doping concentrations

Laser	λ (μm)	wt.%	N_0 (m^{-3})
Nd:YAG	1.06	1	1.4×10^{26}
Nd:YVO_4	1.06	1	1.5×10^{26}
Nd:YLF	1.05	1	1.3×10^{26}
Nd:glass	1.05	3.8	3.2×10^{26}
Yb:YAG	1.06	6.5	9×10^{26}
Pr:YAG	1.03	26	2.7×10^{28}
Er:YAG	2.8	0.7	1×10^{26}
Er:glass	1.5	3	2×10^{26}
Cr:Al_2O_3	0.69	0.05	1.6×10^{25}
Cr:LiSAF	0.9	15	1.5×10^{27}
Cr:LiCAF	0.8	15	1.5×10^{27}
Ti:Al_2O_3	0.83	0.1	3.3×10^{25}

Table 15.5 shows a list of few host crystals and impurity ions. The maximum concentration of impurity ions in a solid depends on the properties of the both solid and the impurity ions. Maximum doping concentrations lie between 0.1% by weight (Ti^{3+} in Al_2O_3) and 26% by weight (Pr^{3+} in YAG):

- Sapphire (Al_2O_3). The crystal field splitting of the 3d state of Ti^{3+} and the interaction of the electronic states with the lattice vibrations (phonons) are the basis of the titanium–sapphire laser.
- YAG (yttrium aluminum garnet = $Y_3Al_5O_{12}$). This material grows in a very high crystal quality. Doping with all three-valid rare earth ions is possible. Doping ions replace Y^{3+} ions. The doping concentration has a value of about 1% for all but two rare earths: the doping with Pr^{3+} can be exceptionally high (25%) and also the doping with Yb^{3+} can be very high (6%).
- YVO_4 (yttrium vanadium oxide). This host material became available in the last years as a high-quality crystalline material.
- $CaWO_4$, CaF_2, LaF_3 doped with rare earths can also be used as laser media.
- $LiYF_4$, LiSAF, LiCaF (Sect. 15.5).
- Alkali halides in color center lasers.

Doping of a solid with impurity ions creates the basic electronic states used in active media. Another possibility is the use of color centers. The following list gives a short survey of defect centers (ions and color centers) contained in various laser media:

- *Ions with valence electrons*. The ions of the transition metals (Ti^{3+}, Cr^{3+}, V^{3+}) have 3d electrons, which are strongly influenced by the crystalline electric field. The energy levels are vibronic levels.
- *Rare earth ions*. Three-valid ions (Nd^{3+}, Er^{3+}, Ho^{3+}, Pr^{3+}) as well as two-valid ions are suitable as impurity ions of laser media. The ions of the rare earths have closed external shells ($5s^25p^6$) and (internal) 4f states. A crystal field leads to a splitting of the 4f levels. The gain bandwidth of rare earth doped crystals at room temperature is of the order of 100 GHz (Sect. 15.9). The rare earth ions are excited via 4f levels or other energy levels. The crystal field splitting depends on the symmetry of the site of an impurity ion in a host material and on the lattice parameters of the host. Therefore, the wavelength of laser radiation, which is due to transitions between two particular energy levels of impurity ions, depends on the host material.
- *Color centers*. There are many different color centers. A color center can be an F center, which is an electron on an empty halide ion site in an alkali halide crystal, replacing the negative ion in an ionic crystal (LiF, NaF, KF, NaCl, KCl, CsCl). However, F centers that can be produced by irradiating a crystal with X-rays are not suitable as defect centers of active media. Suitable as laser media are alkali halides that contain F_2^+ centers. An F_2^+ center consists of two adjacent empty halide ion sites occupied with one electron. An F_2^+ center may be compared with an H_2^+ molecule. The electronic energy levels of an F_2^+ center are strongly influenced by the crystal surrounding: the electronic states of an F_2^+ center are vibronic states; for a discussion of vibronic lasers, *see* Chap. 17. The color center lasers are tunable. Different host crystals lead to different emission bands in the near infrared (from 0.8 to 4 μm). Most of the color center lasers require cooling to liquid nitrogen temperature. Today, color center lasers cannot compete with semiconductor lasers.

The laser medium can have, as we already mentioned, various geometrical shapes:

- *Circular cylindric rod*. There is a temperature gradient perpendicular to the rod axis; the rod is cooled mainly via the cylindric surface. The gain factor can be large.
- *Disk*. There is a temperature gradient perpendicular to the disk axis; the disk is cooled mainly via one of the plane surfaces.
- *Fiber*; Sect. 15.7 and Chap. 18.

Table 15.6 shows a selection of doping ions. The energy levels of the transition metals have electrons in 3d states. These are strongly influenced by the crystalline electric field giving rise to strong splitting of the energy levels and to strong vibronic sidebands (Chap. 17). A crystalline electric field splits the energy levels of a rare earth ion, too. However, the splitting energy is much smaller and vibronic sidebands are weak.

Table 15.6 A selection of ions

Atomic number	Element	Ion	Configuration
22	Titanium	Ti^{3+}	$3d\ ^2D_{3/2}$
24	Chromium	Cr^{3+}	$3d^3\ ^4F_{3/2}$
59	Praseodymium	Pr^{3+}	$4f^2\ ^3H_4$
60	Neodymium	Nd^{3+}	$4f^3\ ^4I_{9/2}$
64	Gadolinium	Gd^{3+}	$4f^3\ ^8S_{7/2}$
68	Erbium	Er^{3+}	$4f^{11}\ ^4I_{15/2}$
69	Thulium	Tm^{3+}	$4f^{12}\ ^3H_6$
70	Ytterbium	Yb^{3+}	$4f^{13}\ ^2F_{7/2}$

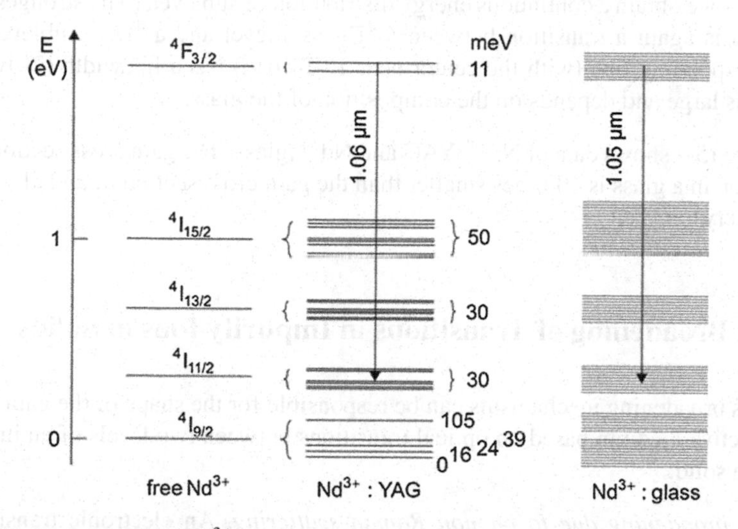

Fig. 15.13 Energy levels of Nd^{3+}

As an example of crystal field splitting of energy levels of a rare earth ion, we show energy levels of Nd^{3+} (Fig. 15.13):

- Free Nd^{3+} ion. The $4f^3$ state splits into states with different total angular momentum (quantum number J), due to spin–orbit interaction. The ground state is $^4I_{9/2}$. Optical transitions are forbidden.
- Nd^{3+}:YAG. The crystalline electric field splits a level with the quantum number J into $(2J + 1)/2$ sublevels (*Stark splitting*); a rare earth ion with an odd number of 4f electrons shows a twofold degeneracy (*Kramers degeneracy*)—*see*, for instance, [121]. The splitting of the sublevels has values in the range of several meV to about 100 meV [122–126]. Optical transitions are allowed due to spin–orbit interaction or due to the interplay of spin–orbit interaction and crystalline electric field. The strongest transition is a transition between a $^4F_{3/2}$ sublevel and a $^4I_{11/2}$ sublevel. The corresponding line (at 1.064 µm) has a linewidth of 12 GHz

Table 15.7 Gain data of Nd^{3+}:YAG and Nd^{3+}:glass

Laser	λ (μm)	$\Delta\nu_g$	τ_{sp} (μs)	σ_{21} (m^2)
Nd^{3+}:YAG	1.064	140 GHz	230	3×10^{-22}
Nd^{3+}:glass	1.054	7 THz	300	7×10^{-24}

at room temperature. The values of the crystal field splitting of levels of Nd^{3+} in other crystals are of the same order.

- Nd^{3+}:glass. Because of the great variety of the crystal field acting on ions at different sites in a glass, the energy of a sublevel differs strongly for ions at different sites—we obtain a continuous energy distribution of sublevels. The strongest transition is again a transition between a $^4F_{3/2}$ sublevel and a $^4I_{11/2}$ sublevel. The corresponding line (with the center near 1.054 μm) has a linewidth (7–10 THz) that is large and depends on the composition of the glass.

Table 15.7 shows data of Nd^{3+}:YAG and Nd^{3+}:glass. The gain cross section of an Nd^{3+} ion in a glass is 40 times smaller than the gain cross section of an Nd^{3+} ion in a YAG crystal.

15.9 Broadening of Transitions in Impurity Ions in Solids

Various broadening mechanisms can be responsible for the shape of the gain profile of an active medium based on optical transitions between two levels of an impurity ion in a solid:

- *Line broadening due to phonon Raman scattering*. An electronic transition is accompanied by a phonon Raman scattering process, i.e., by inelastic scattering of a phonon during the emission of a photon (Fig. 15.14). This process is frozen out at low crystal temperature (e.g., at 4 K or 77 K) but is the main broadening mechanism of many transitions in impurities in crystals at room temperature. Phonon Raman scattering leads to homogeneous line broadening. Fluorescence as well as absorption line have Lorentzian shape. We can interpret the mechanism as line broadening due to elastic collisions of an atom with phonons.

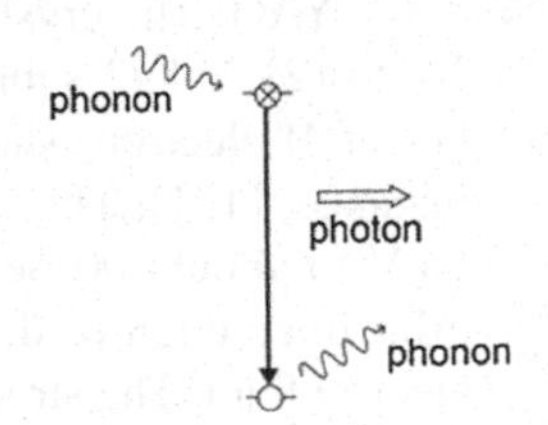

Fig. 15.14 Optical transition accompanied with phonon Raman scattering

Examples 1.06-μm line of Nd^{3+}:YAG at room temperature; the R_1 and R_2 lines of ruby at room temperature.

- *Line broadening due to the Stark effect*. Due to the Stark effect, the energy levels of ions at different sites of impurity ions in a glass have an energy distribution—and the transition energies too. Phonon-assisted energy transfer processes create a quasiband of excited ions (Chap 18).

At low temperature, the Stark effect of impurity ions in a solid (crystal or glass) leads to inhomogeneous broadening of absorption lines. An impurity ion occupies not exactly the position of an ion that it replaces. The strength of the crystal field at different impurity ion sites is slightly different. Due to the Stark effect, the energy levels and transition energies of the ions at different sites are different. The lineshape can be Gaussian.

Examples almost all lines that are due to transitions between 4f states of ions in crystals at low temperature (e.g., at 4 K); the R lines of ruby at low temperature; all lines of impurity ions in glasses at low temperature.

References [1–4, 6, 11, 31, 120–126].

Problems

15.1 Ruby laser.

(a) The crystal of a Q-switched ruby laser is optically pumped by the use of a flash lamp so that almost all Cr^{3+} ions are excited. Estimate the energy and the power of a laser pulse of 100 ns duration. [*Hint*: ignore oscillations that could cause a temporal structure in the pulse shape].

(b) Laser oscillation is possible with a ruby crystal cooled to low temperature (4 K) with two plane parallel surfaces as reflectors (refractive index of ruby $n = 1.76$). Estimate the threshold pump power of a laser with a ruby crystal (length 1 cm) pumped in a volume of 0.2 mm diameter by another laser (pump wavelength 530 nm); at low crystal temperature, the R_1 and R_2 lines are 100 times narrower than at room temperature.

15.2 Gain cross sections. Determine, by use of the data of linewidths and spontaneous lifetimes, the ratio of the gain cross section of the 1.06 μm line of Nd^{3+}:YAG and of the gain cross section at the line center of Ti^{3+}:Al_2O_3

15.3 Titanium–sapphire laser. Why is the energy distribution of vibronic energy levels of Ti^{3+} in Al_2O_3 continuous while the vibronic energy levels of N_2 are discrete?

15.4 Laser tandem pumping. A femtosecond titanium–sapphire laser can be pumped with the frequency-doubled radiation of a Nd^{3+}:YVO_4 laser, which itself is pumped by use of a semiconductor laser.

(a) Estimate the quantum efficiency of such an arrangement if the frequency doubling has a power conversion of 50%.
(b) What is the advantage of the tandem pumping in comparison with the direct pumping of the titanium–sapphire laser with a semiconductor laser?

15.5 Fiber laser. Estimate the efficiency of an erbium-doped fiber laser pumped with a pump power twice the threshold pump power.

15.6 Explain the nomenclature ($^4I_{9/2}$, $^4I_{11/2}$, $^7F_{7/2}$ etc.) used for characterization of atomic states.

Chapter 16
Some Other Lasers and Laser Amplifiers

We present further types of lasers: dye laser; chemical laser; X-ray laser; organic laser. And we discuss the principle of laser amplifiers. Another topic concerns optical damage.

16.1 Dye Laser

The dye laser was the first laser with a broad gain profile. The dye laser operates as a tunable cw laser or as picosecond laser (pulse duration ~1 ps). The tuning range of a dye laser is about 5% relative to the laser frequency. By the use of different dyes, the entire visible spectral range can be covered with laser radiation.

The dye laser is a vibronic laser (Fig. 16.1a). Transitions involve vibronic energy levels of the ground state (S_0) and of the first excited singlet state (S_1). The spatial extension of a molecule in the S_1 state is larger than in the S_0 state. Spontaneous emission of radiation determines the lifetime (2–5 ns). The vibronic levels are due to interaction of the electronic states with molecular vibrations. Optical pumping and fast nonradiative relaxation leads to population of the S_1 state. Laser radiation is generated by stimulated transitions from the lowest S_1 state of excited molecules to vibronic S_0 states. An optical transition is governed by the Franck–Condon principle.

In a dye laser (Fig. 16.1b), the solvent (water or an alcohol) that contains the dye molecules can continuously be pressed through a nozzle leading to a jet. The laser radiation passes the jet under the Brewster angle. The laser can be optically pumped with another laser (e.g., an argon ion laser) or with a lamp.

As an example of a dye molecule, we mention 7-hydroxycoumarin. The molecule has a benzene-like molecular structure (Fig. 16.1c). A corresponding laser contains coumarin solved in water (0.1 molar solution).

The $S_0 \rightarrow S_1$ absorption band of 7-hydroxycoumarin (Fig. 16.2) lies in the blue (450–470 nm) and the emission band in the green (580–600 nm). The fluorescence band (= fluorescence line) has a Gaussian-like shape. We attribute the line broadening to homogeneous broadening (Sect. 17.4). The linewidth of the

K.F. Renk, *Basics of Laser Physics*, Graduate Texts in Physics,
DOI 10.1007/978-3-319-50651-7_16

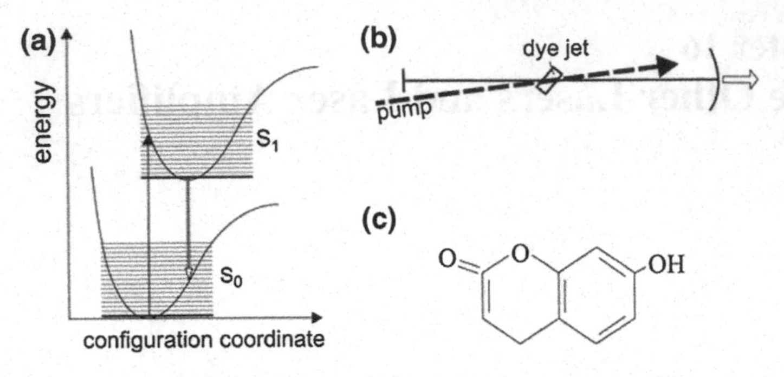

Fig. 16.1 Dye laser. a Principle. b Arrangement. c A dye molecule (*7-hydroxycoumarin*)

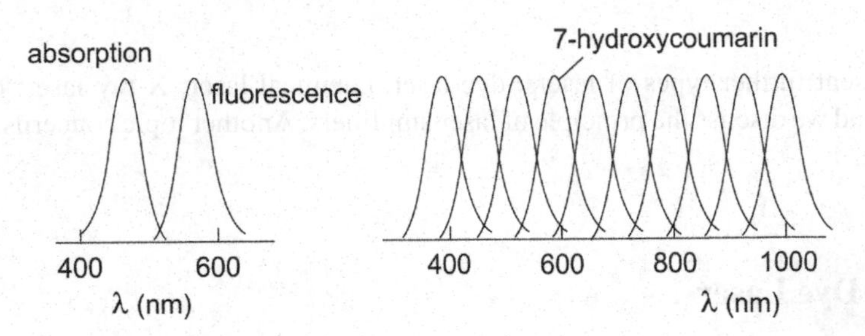

Fig. 16.2 Absorption and fluorescence of 7-hydroxycoumarin (*left*) and fluorescence bands of different dyes (*right*)

fluorescence line is about 20 THz. Dyes suitable as active media of dye lasers are available for the whole visible spectral range and also for the near UV and the near IR. The following list shows characteristic data.

- $\lambda_0 = 0.3$–$1.5\ \mu$m = wavelength of the line center of the gain curve; depending on the dye.
- $\tau_{sp} = 2$–5 ns.
- $\Delta\nu_g = 10$–20 THz.
- $\sigma_{21} = 5 \times 10^{-21}$ to $5 \times 10^{-19}\ m^2$.
- Concentration 10^{-4}–10^{-3} molar ($N_0 = 0.1$–$1 \times 10^{25}\ m^{-3}$).

Active media with dye molecules can have high gain coefficients (Problem 16.1).

Other applications of dyes. Dye molecules solved in water or in alcohol are saturable absorbers suitable for Q-switching of lasers (Sect. 12.6). Dye molecules find applications in medicine: dye molecules are suitable as markers in the photodynamic diagnosis and as active species in the photodynamic therapy of cancer [127–129].

16.2 Solid State and Thin-Film Dye Laser

The active medium of solid state dye laser can consist of a solid matrix, for instance polymethylmethacrylate, containing dye molecules. Suitable as pump sources are semiconductor lasers or diodes.

A thin-film dye laser consists of a thin film of dye molecules (embedded in a solid matrix) on a plane solid surface. A grating on the surface of the thin film can act as distributed feedback reflector (Sects. 25.4 and 34.4).

16.3 Chemical Laser

The basis of a chemical laser is a chemical reaction. In an HF laser, a gas discharge drives the reactions

$$F + H_2 \rightarrow H + (HF)^*, \tag{16.1}$$

$$F_2 + H \rightarrow F + (HF)^*. \tag{16.2}$$

The $(HF)^*$ molecules are in excited vibrational-rotational states and emit radiation in the 3-μm range by transitions between vibrational-rotational states.

A chemical reaction changes the enthalpy H of a system. The two reactions described by (16.1) and (16.2) are exothermic reactions—producing reaction energy (= reaction heat ΔH). In the first reaction, (16.1), the reaction heat is $\Delta H =$ 1.3 eV/molecule (132 kJ/mole). A portion of the reaction heat is transferred to energy of excitation of vibrational-rotational states of the v = 0, 1, 2 vibrational levels (Fig. 16.3). The second reaction, (16.2), has a larger reaction heat ($\Delta H =$ 4.0 eV/molecule) and results in excitation of vibrational-rotational energy levels up to the v = 6 vibrational level. The population of the different vibrational-rotational levels is a nonequilibrium population. Therefore, many laser transitions between different vibrational-rotational states can occur. The laser wavelengths are in the range between 2.7 and 3.3 μm.

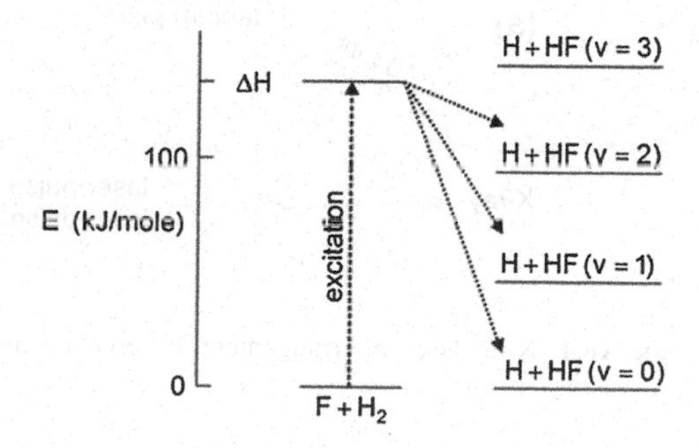

Fig. 16.3 Chemical laser: principle of pumping

The HF laser operates as a continuous wave laser (driven by a gas discharge) or as TEA laser. The continuous wave laser can generate radiation of a power of 10 kW. The TEA laser pulses (of about 100 ns duration) have pulse energies of several kJ. The energy of a pulse corresponds to the energy of 100 J of laser radiation that can be generated per liter of the active material.

Other chemical lasers operating with other gases produce laser radiation in slightly different wavelength regions (DF, 3.5–4.5 μm; HCl, 3.5–4.1 μm; HBr, 4.0–4.7 μm).

16.4 X-Ray Laser

There are first steps toward a table-top X-ray laser. Figure 16.4a shows the principle of an X-ray laser [131–133]. Two strong visible laser pulses, focused onto a titanium plate, pump an X-ray laser in a two-step excitation.

- A laser pulse 1 (wavelength around 600 nm; pulse energy 20 J; duration 1 ns) produces a plasma with a large concentration of Ti^{12+} ions; the configuration of a Ti^{12+} ion corresponds to a [Ne] configuration ($2p^6$).
- A laser pulse 2 (frequency around 600 nm; 4J; 1 ps) excites the plasma further. Then hot electrons in the plasma produce, by electron collisions, a population inversion, giving rise to stimulated emission of X-ray pulses.

In the second step, electrons collide with Ti^{12+} leading to excited Ti^{12+} ions in $2p^5 3s$ states (Fig. 16.4b). Transitions 3s → 2p result in laser radiation at 18.2 nm (pulse energy 30 μJ, repetition rate 1 s^{-1}). The 2p states decays by fast radiative transitions.

X-ray lasers with other solids (Ge, Pd, Ag, etc.) generate radiation pulses at other wavelengths (6–40 nm) in the soft X-ray region.

The X-ray laser presented here is a mirrorless laser, there is no feedback with a resonator. Laser radiation is generated by amplified spontaneous emission (ASE). During propagating through the plasma, spontaneously generated radiation is amplified by stimulated emission of radiation.

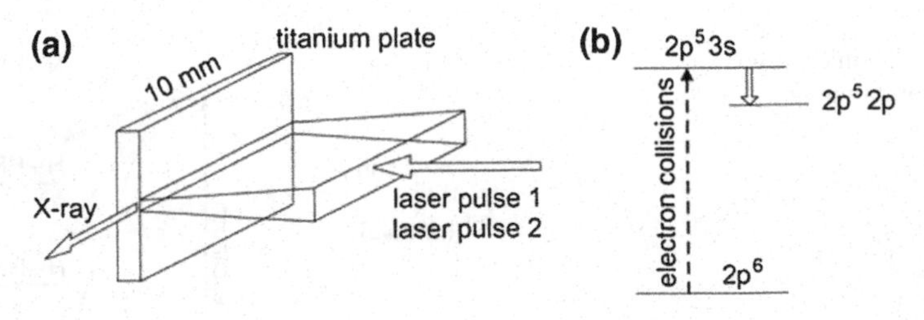

Fig. 16.4 X-ray laser. **a** Arrangement. **b** Laser transition in Ti^{12+}

16.5 Random Laser

A random laser can consist of an optical powder, for example a powder of Nd^{3+}:YAG crystallites. Due to light scattering at the powder particles, the light emitted spontaneously is amplified by stimulated emission; for information about solid state random lasers, *see* [134].

16.6 Optically Pumped Organic Lasers

We will treat optically pumped organic lasers in a later chapter (Sect. 34.4); then we will have available concepts, described in Chap. 18 and in chapters on semiconductor lasers, that are useful to explain how gain of radiation in an organic medium can occur.

16.7 Laser Tandem

A laser tandem is suitable for generation of laser radiation of high beam quality. A semiconductor laser, with a high efficiency of conversion of electric power to laser radiation, pumps a solid state laser. The frequency doubled radiation of this laser pumps a third laser. A semiconductor laser has a low beam quality. The combination of both type of lasers is most favorable: the use of a semiconductor laser as pump laser of a solid state laser allows for an efficient conversion of electric energy to high quality laser radiation.

Example A semiconductor laser pumps a Nd:YAG laser, then the radiation is frequency-doubled. The frequency doubled radiation finally pumps a titanium–sapphire laser.

16.8 High-Power Laser Amplifier

In a high-power laser system (Fig. 16.5) consisting of a laser and a laser amplifier, the laser beam is expanded by the use of a telescope in order to avoid optical damage of the active medium of the amplifier. The (single pass) gain factor G_1 of a laser amplifier can have a value of the order of 10. By the use of laser amplifiers in series, very large power levels can be obtained. Table 16.1 shows data of three high-power laser systems (t_p = pulse duration; W_p = pulse energy; P = pulse power; ν_{rep} = repetition rate).

- *Femtosecond titanium–sapphire laser amplifier*. The radiation of a femtosecond titanium–sapphire laser can be amplified with a laser amplifier containing optically

Fig. 16.5 Laser amplifier

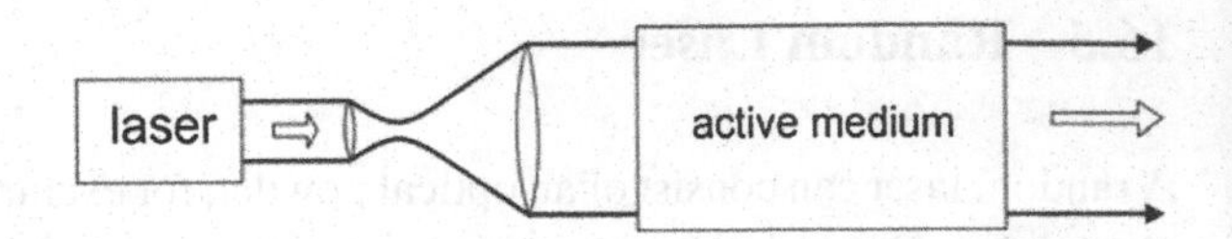

Table 16.1 High-power laser systems

Laser	λ	t_p	W_p	$P(W)$	ν_{rep}
TiS	780 nm	100 fs	1 mJ	10 GW	1 kHz
Nd:glass	1.06 μm	1 ns	10 kJ	10 TW	1 h
TEA CO_2	10.6 μm	100 ns	1 J	10 MW	10 Hz

pumped titanium–sapphire as the active material. An optical switch can reduce the pulse repetition rate (which is of the order of 100 MHz) of the radiation of a femtosecond titanium-sapphire laser to a value of, for example, 1 kHz.

- *Neodymium glass laser amplifier*. By amplification of a light pulse of a neodymium-doped glass laser with glass laser amplifiers, a pulse of extremely high pulse energy can be generated. The beam emitted by a glass laser is widened and amplified by a first amplifier, then widened and amplified by a second amplifier and so on. (It is possible to produce glass in cylinders of large diameter.) A radiation pulse generated by a laser amplifier system (or pulses generated by systems in parallel), focused on a target containing deuterium and tritium can heat up the target to a temperature at which nuclear fusion processes can occur (laser fusion); a laser pulse can produce a plasma of a temperature of the order of 100 million degrees.

16.9 Fiber Amplifier

Fiber amplifiers consisting of glass doped with rare earth ions are suitable for amplification of radiation in the 1–3 μm range; fiber amplifiers make use of the same rare earth-doped glasses as fiber lasers (Sect. 15.7 and Chap. 18). By the use of amplifiers, radiation at kW power levels can be generated.

The erbium-doped fiber amplifier—that is of great importance for long-distance optical communications—will be treated in Chap. 18.

16.10 Optical Damage

A strong radiation field in a transparent solid material can lead to optical damage. Different materials have different damage thresholds. The damage threshold of a material depends strongly on the wavelength of the radiation. The damage threshold is orders of magnitude larger for pulses of 10 fs duration than for pulses of 1 ns duration. Accordingly, the optical-damage threshold can have values between 10 kW per cm (or smaller) and 20 MW per cm.

Optical damage can be caused by interband transitions of electrons and subsequent impact ionization processes [135, 136]. An interband transition in a crystal in a strong electromagnetic field can be due to a multiphoton transition. Interband transitions excite electrons into the conduction band. Subsequently, the electrons in the conduction band gain energy by absorption processes, i.e., due to acceleration of the conduction electrons by the optical field. Highly excited conduction electrons excite, by impact ionization, further electrons from the valence band to the conduction band. The impact ionization is an avalanche process that can lead to optical breakdown associated with crystal damage.

16.11 Gain Units

The power of a light beam that traverses an amplifier increases from P_0 to P. We can characterize the increase in different ways, assuming that the gain does not change along the path of the beam:

- $G = P/P_0$ = gain factor.
- $G = e^{\alpha L}$, where α is the gain coefficient (in m^{-1}) and L (in m) the length of the gain medium.
- 1 dB (= 1 dB) $= 10 \times \log(P/P_0) = 10 \times 0.43 \times \alpha L = 4.3 \times \alpha L$.
- (1 B = 1 Bel = 10 dB).
- 1 dB/m $= L^{-1} \times 10 \times \log(P/P_0) = 4.3\alpha$.
- 1 dB m = 1 dB mW = 1 dB $\times$ 1 mW = a unit of gain of an amplifier.

Example erbium fiber amplifier; $\alpha = 0.5\ m^{-1}$ and $L = 14$ m; gain = 2.15 dB/m; $G = 10^3$.

References [127–136].

Problems

16.1 Dye laser (length of the active medium 1 mm; beam diameter 0.2 mm; reflectivity of the output coupling mirror $R = 0.7$; frequency 500 THz).

(a) Determine the threshold condition.
(b) Determine the output power at pumping 10 times above threshold.

16.2 Laser amplifier. To amplify femtosecond pulses emitted by a titanium-sapphire laser, an optical switch reduces the pulse repetition rate to 1 kHz. By passing through two amplifier stages (optically pumped titanium-sapphire crystals), each with a single path gain of 10, intense laser pulses are generated. Determine (by use of the data of Problem 13.2) the pulse power and the average power after amplification.

16.3 Momentum of a photon and radiation pressure.

(a) When an atom at rest emits a photon, then the atom experiences a recoil. Estimate the velocity of a neon atom that was originally at rest and emitted a photon (wavelength 632 nm).
(b) Estimate the average velocity of a spherical target (diameter 0.2 mm) consisting of frozen deuterium that absorbed an intense light pulse (energy 100 J, wavelength 1.05 μm).

16.4 Radiation of a titanium sapphire laser amplifier system (pulse power 1 GW, wavelength 780 nm, pulse duration 100 fs) is focused to an area of diameter 10 μm^2. Determine the intensity, the photon density, the energy density, the amplitude of the electric field in the focus.

16.5 Magnetic field of a light wave.
Determine for the example of the preceding problem the amplitude of the magnetic field of the electromagnetic wave in the focus. Compare the amplitude with Earth's magnetic field. The amplitude of the magnetic field of a plane wave is $B_0 = (1/c)A$, where A is the amplitude of the electric field. The magnitude of Earth's magnetic field on the surface ranges from 25 to 65 μT.

16.6 Material processing.
The radiation of a high power laser with amplifier (pulse duration 1 ps, pulse power 1 MW, diameter 0.02 mm) is used to drill a hole in a metal foil (thickness 0.2 mm). How many shots are necessary?

Chapter 17
Vibronic Medium

We study the origin of gain of radiation in a vibronic medium. We find that the gain coefficient of a vibronic medium like optically pumped titanium-sapphire has a Gaussian-like shape.

We introduce a one-dimensional model of a vibronic medium that illustrates the occurrence of vibronic transitions and we describe the results of theoretical investigations. The energy of an atomic state of a vibronic medium, i.e., the energy of a vibronic state, is composed of electronic energy of an impurity ion and of vibrational energy of the host crystal. The broad frequency distribution of lattice vibrations (phonons) of a crystal together with the possibility that many phonons can be involved in an optical transition lead to two broad vibronic sidebands of the zero-phonon line. One of the bands is observable as absorption band and the other as fluorescence band. In a laser, the absorption band is used for optical pumping and the other band for stimulated transitions.

In a classical description of vibronic transitions, we make use of the classical oscillator model of an atom to describe the electronic transition in an impurity ion and attribute a vibronic transition to an atomic oscillation experiencing frequency modulation by a vibration of the host crystal.

A vibronic laser like the titanium-sapphire laser is based on a homogeneous broadening mechanism, which determines the optical transitions.

17.1 Model of a Vibronic System

We first illustrate, by the use of a simple model, the origin of vibronic coupling. We consider a TiO_2 molecule (Fig. 17.1a); x_0 is the TiO distance. We describe the potential of the 3d electron of the Ti^{3+} ion in a TiO_2 molecule by a one-dimensional square well potential of infinite height (Fig. 17.1b). We make use of the Schrödinger equation

$$\left[-\frac{\hbar^2}{2m_0}\frac{\mathrm{d}^2}{\mathrm{d}\zeta^2} + E_{\mathrm{pot}}(\zeta)\right] \chi(\zeta) = E\,\chi(\zeta), \quad (17.1)$$

K.F. Renk, *Basics of Laser Physics*, Graduate Texts in Physics,
DOI 10.1007/978-3-319-50651-7_17

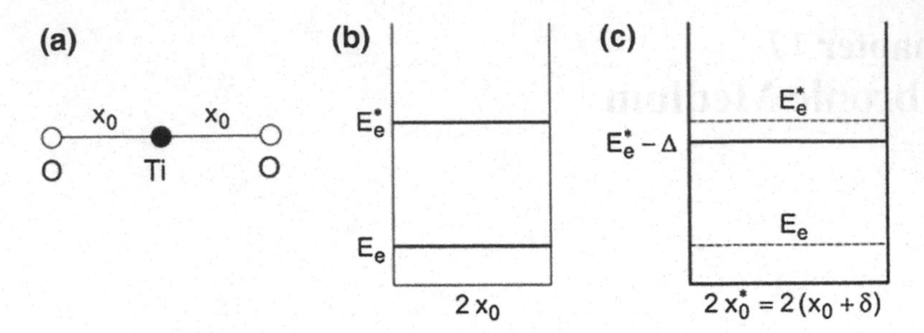

Fig. 17.1 Electronic excitation of a TiO_2 molecule. **a** TiO_2 molecule. **b** Electron in a one-dimensional potential well. **c** Excited electron in the potential well

where m_0 is the electron mass, ζ the spatial coordinate, and χ the wave function. The energy eigenvalue of the electronic ground state is equal to (Sect. 30.1)

$$E_e = \frac{\pi^2 \hbar^2}{8 m_0 x_0^2}, \tag{17.2}$$

where $2x_0$ is the width of the well. The first excited state has the energy

$$E_e^* = \frac{4\pi^2 \hbar^2}{8 m_0 x_0^2}. \tag{17.3}$$

We assume that an excitation of the 3d electron extends the TiO_2 molecule, i.e., that the TiO distance increases. The width $2x_0^*$ of the potential well corresponding to the excited state is larger than $2x_0$ and the energy of the excited state is smaller than E_e^* (Fig. 17.1c); δ is the increase of TiO distance and Δ is the decrease of energy.

The symmetric valence vibration of the TiO molecule causes a variation of the width of the potential well. The energy of the electronic ground state depends on the displacement $x - x_0$ according to the relation

$$E(x) = E_e + \frac{1}{2} f (x - x_0)^2 = \frac{\pi^2 \hbar^2}{8 m_0 x^2} + \frac{1}{2} f (x - x_0)^2 = E(x - x_0), \tag{17.4}$$

where f is a spring constant and $(1/2) f (x - x_0)^2$ is the elastic energy.

The energy of the excited state also depends on the TiO distance and therefore on the displacement $x - x_0^*$,

$$E^*(x) = E_e^* + \frac{1}{2} f (x - x_0^*)^2 = \frac{4\pi^2 \hbar^2}{8 m_0 s^2} + \frac{1}{2} f (x - x_0^*)^2 = E^*(x - x_0^*), \tag{17.5}$$

where x_0^* is the TiO distance in the excited state. We assume, for simplicity, that f has the same value as in the electronic ground state. But we take into account that the TiO distance is larger than in the case that the Ti ion is in the ground state.

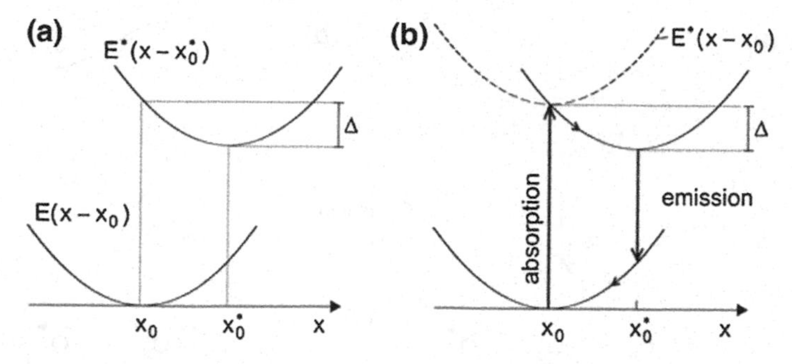

Fig. 17.2 Model of a vibronic system. **a** Vibronic energy levels. **b** Transitions between vibronic levels

Figure 17.2a (solid lines) shows parabolas describing the energy of the electronic ground state and of the excited state. The parabola describing the excited state is shifted toward larger x. According to the Franck–Condon principle, an electronic transition from the ground state to the excited state takes place without a change of the TiO distance—indicated in Fig. 17.2b as a "vertical" transition. A transition occurs to an energy that corresponds to the minimum of a parabola (dashed) that shifted in energy by the excitation energy of an electron in a rigid potential well. This energy is equal to

$$E^*(x - x_0) = \frac{4\pi^2\hbar^2}{8m_0 x_0^2} + \frac{f}{2}(x - x_0)^2. \tag{17.6}$$

We assume that the change of the distance between the oxygen ions is small, $x_0^* - x_0 \ll x_0$. Then we can write (Problem 17.1)

$$E^*(x - x_0) = E_0^* - \Delta + \frac{f}{2}(x - [x_0 + \delta])^2. \tag{17.7}$$

Δ is a relaxation energy and δ is the increase of the TiO distance. After an absorption process, the TiO molecule relaxes to the equilibrium position of the excited state. Emission of a photon occurs to a vibronic state of the ground state. Another relaxation process takes the system back to the ground state.

17.2 Gain Coefficient of a Vibronic Medium

Vibronic systems have been studied in detail by the use of appropriate quantum mechanical methods [137]. The configuration diagram (Fig. 17.3a) illustrates the role of lattice vibrations. The configuration coordinate Q replaces x of our one-dimensional model. Q_0 describes, in principal, the TiO distance of the oxygen

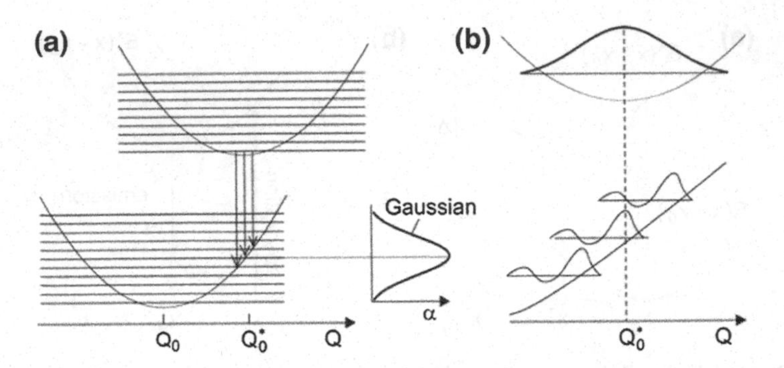

Fig. 17.3 Vibronic transitions in a crystal at low temperature. **a** Vibronic transitions and gain curve. **b** Wave functions

ions if Ti^{3+} is in its electronic ground state. Q_0^* describes the distance if Ti^{3+} is excited. $Q - Q_0$ is a measure of the displacement due to lattice vibrations; the configuration coordinate takes into account that a large number of lattice vibrations (phonons) of a crystal can couple to an electronic transition.

We discuss the gain coefficient of Ti^{3+}:Al_2O_3 at zero temperature. We make use of results obtained for vibronic transitions in Cr^{3+}:Al_2O_3 [137]. We assume that Ti^{3+}:Al_2O_3 shows a similar vibronic coupling strength as Cr^{3+}:Al_2O_3 as a comparison of the widths of the vibronic absorption bands suggests. Both Ti^{3+} and Cr^{3+} are transition metal ions; Ti^{3+} has one electron and Cr^{3+} two electrons in the 3d shell.

If the $E(Q - Q_0)$ and $E^*(Q - Q_0^*)$ curves have parabolic slopes (Fig. 17.3a), the gain coefficient $\alpha(\nu)$ is expected to have a Gaussian shape. Optical pumping leads to a population of the lowest excited-state level. Gain is due to transitions to vibronic levels of the ground state. The maximum of the gain curve corresponds, with respect to a single emission process, to emission of a photon and creation of about ten phonons of an average frequency of 7.5 THz. The halfwidth of the gain curve is about 10 times the average phonon energy. The density of states $D_1(\nu_{vib})$ of the phonons of Al_2O_3 extends from $\nu_{vib} = 0$ to a maximum vibrational frequency $h\nu_{max}$ of about 15 THz. The Gaussian shape of the gain curve reflects the Gaussian shape of the wave function of the lowest excited state (Fig. 17.3b). According to the Franck–Condon principle, the distance does not change during an optical transition from the excited state to a vibronic state of the electronic ground state.

In an optically pumped crystal at room temperature, transitions occur also from thermally populated vibronic states (Fig. 17.4a). The population is determined by the crystal temperature. Because of additional transitions, in comparison with low temperature, the gain curve shifts to higher energy and the halfwidth is larger than at low temperature (by a factor of 1.2 according to theory [137]). Figure 17.4b shows the $E(Q - Q_0)$ and $E^*(Q - Q_0^*)$ curves together with the shape of the gain coefficient α. The gain coefficient has a Gaussian shape. Due to anharmonicity of the lattice vibrations, the $E(Q - Q_0)$ deviates from a parabolic shape at $Q > Q_0$ (dashed). This is most likely the main reason of the deviation of the gain curve from a Gaussian

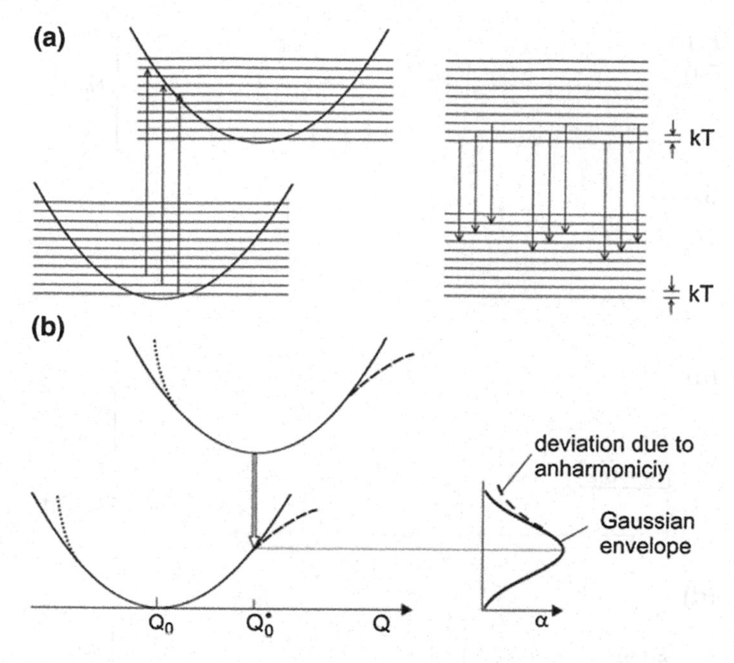

Fig. 17.4 Vibronic medium at room temperature. **a** Vibronic transitions. **b** $E(Q)$ curves and gain curves

curve at frequencies below the center frequency (*see* Fig. 7.7). Anharmonicity, occurring at room temperature, can lead to a broadening of the gain curve; the halfwidth (110 THz), derived from fluorescence data (Sect. 7.6), corresponds to about 15 times the average energy of a phonon in Al_2O_3.

While the gain curve deviates from a Gaussian shape at small frequencies, the experimental absorption curve deviates at large frequencies (*see* Fig. 7.8). This is in accordance with a deviation (Fig. 17.4, dotted) of the shape of the $E^*(Q - Q_0^*)$ curve from a parabolic shape due to anharmonicity for $Q < Q_0^*$. A structure in the absorption line, indicated in Fig. 7.8, is due to a splitting—caused by the Jahn-Teller effect [138]. The Jahn-Teller effect is most likely responsible that the absorption cross section of a Ti^{3+} ion in the ground state is by about a factor of 4 smaller than the gain cross section of an excited Ti^{3+} ion (*see* Fig. 5.3, upper part).

The maximum quantum efficiency of titanium-doped sapphire is about 80% [139].

17.3 Frequency Modulation of a Two-Level System

Instead of describing a vibronic state as a state with electronic and vibrational components, we can choose an alternative view.

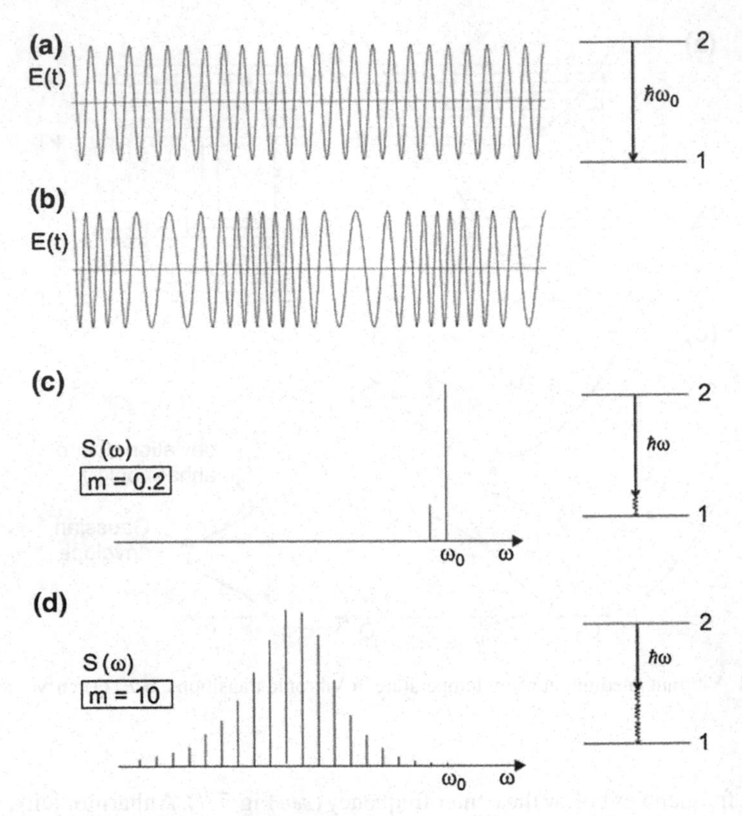

Fig. 17.5 Frequency modulation of an atomic oscillation by a lattice vibration. **a** Electric field without frequency modulation. **b** Frequency modulated electric field. **c** Luminescence spectrum at weak modulation. **d** Luminescence spectrum at strong modulation

We consider transitions between two discrete energy levels of a two-level system (with the upper level 2 and the lower level 1) of an impurity ion in a crystal at low temperature. We make use of the classical oscillator model of an atom (Sect. 4.9) and describe the electric field connected with an oscillating electron by (Fig. 17.5a)

$$E = A \cos \omega_0 t. \tag{17.8}$$

A is the amplitude of the field at the position of the electron and ω_0 the transition frequency; we assume, for simplicity, that the spontaneous lifetime is infinitely large. We now assume that a lattice vibration (frequency Ω) is present in the crystal. The vibrational wave modulates the crystalline field at the position of the impurity ion and therefore modulates the transition frequency of the electronic transition—corresponding to a *frequency modulation*. The instantaneous transition frequency is equal to

$$\omega_{\text{inst}} = \omega_0 + a \cos \Omega t, \tag{17.9}$$

where a is the maximum change of frequency toward larger and toward smaller frequency with respect to the "carrier" frequency ω_0. Due to frequency modulation, the instantaneous frequency varies periodically with time (Fig. 17.5b). It follows that the electric field is given by

$$E(t) = A \int_0^t \cos(\omega_0 + a \cos \Omega t')\mathrm{d}t' = A \cos(\omega_0 t + m \sin \Omega t), \qquad (17.10)$$

where

$$m = a/\Omega \qquad (17.11)$$

is the modulation degree; $m \ll 1$ corresponds to weak modulation and $m \gg 1$ to strong modulation. We make use of the relation [15]:

$$\cos(\alpha + m \sin \beta) = \sum_{n=-\infty}^{\infty} J_n(m) \cos(\alpha + n\beta). \qquad (17.12)$$

J_n is the Bessel function of nth order . A Fourier transformation leads to the spectrum of the frequency modulated field [15],

$$E(\omega) = \frac{A}{2} \sum_{n=-\infty}^{\infty} J_n(m) \left(\delta(\omega - \omega_0 - n\Omega) + \delta(\omega + \omega_0 + n\Omega)\right). \qquad (17.13)$$

The fluorescence spectrum is given by

$$S(\omega) = K E^2(\omega), \qquad (17.14)$$

where K is a constant that depends on the electronic properties of the impurity ion and on the experimental arrangement. If the modulation is weak, the spectrum (Fig. 17.5c) consists mainly of a strong zero-phonon line at ω_0 and a weak satellite line at $\omega_0 - \Omega$. If the coupling is strong (Fig. 17.5d), we obtain a vibronic spectrum with many lines (in principle an infinitely large number). The lines around $\omega_0 - m\Omega$ are strongest; now the spectral weight of the zero-phonon line is small. We omitted in Fig. 17.5c, d the sidebands at frequencies larger than ω_0, according to the situation of a crystal at low temperature that does not contain thermally excited phonons. At low temperature, the transition frequency is modulated due to the creation of phonon waves during emission of radiation. Taking into account that we have a continuous frequency distribution of phonons, we obtain a continuous multiphonon sideband (of Gaussian shape). For information on the zero-phonon line of Ti^{3+} in Al_2O_3, *see* [138].

17.4 Vibronic Sideband as a Homogeneously Broadened Line

In an active medium of a vibronic laser like the titanium–sapphire laser, all impurity ions contribute in the same way to stimulated emission of radiation. A titanium–sapphire laser can therefore be seen as a laser that operates on a homogeneously broadened transition.

References [31, 137–139].

Problem

17.1 Determine the dependencies of Δ and δ in (17.7) on x_0 and f.

Chapter 18
Amplification of Radiation in a Doped Glass Fiber

We study the dynamics of gain of fiber amplifiers and fiber lasers. We present a model—the quasiband model—that allows for derivation of an analytical expression for the gain coefficient of an optically pumped doped glass fiber. We concentrate the discussion mainly on the erbium-doped fiber amplifier. We will however discuss other fiber amplifiers and fiber lasers too.

A glass fiber of the worldwide optical fiber network contains, about every 50 km, an erbium-doped fiber amplifier. This amplifier operates in a frequency band (width ~5 THz) around 195 THz (1.54 μm). It is possible to pump the erbium-doped fiber amplifier with radiation of a semiconductor laser at a frequency (~202 THz) that lies just outside the range of gain. Alternatively, pumping with radiation at a much larger frequency is possible.

While an excited atom in a gas, a liquid, or a crystal keeps its excitation until a stimulated emission process takes place, the situation is completely different for a fiber glass medium. In a glass, an excited ion loses its excitation to another ion and this to a third ion and so on—the excitation migrates within the glass. The origin of the migration of excitation are phonon-assisted energy transfer processes. An excitation travels over a very large number of ions before a stimulated emission process takes place. The migration of excitation plays an essential role in the dynamics of gain of radiation in fiber amplifiers. We will introduce a model (quasiband model) that takes account of the migration of energy and that enables us to calculate the gain coefficient of a fiber.

We will begin this chapter with a short survey of the erbium-doped fiber amplifier: we first describe the gain coefficient and the quasiband model. Later in the chapter, we will justify the model and derive the gain coefficient. In the last section of the chapter, we will show that the quasiband model is in accord with experimental results of absorption, fluorescence, and gain measurements. We will also discuss three-level laser models often used for description of fiber lasers and amplifiers.

K.F. Renk, *Basics of Laser Physics*, Graduate Texts in Physics,
DOI 10.1007/978-3-319-50651-7_18

18.1 Survey of the Erbium-Doped Fiber Amplifier

Figure 18.1 shows the gain coefficient of an *erbium-doped fiber amplifier*; the gain coefficient was calculated (Sect. 18.6) under the assumption that ~60% of the erbium ions are in the excited state. The amplifier operates in a wavelength range near 1.54 μm (frequency 195 THz); *see* also Table 18.1. The gain bandwidth of about 40 nm (~5 THz) corresponds to ~2.5% of the center frequency. Radiation of a semiconductor laser serves as pump laser (pump wavelength 1.48 μm, frequency ~203 THz). According to the small difference between pump and laser wavelength, it is possible to reach a high quantum efficiency of conversion of pump radiation to laser radiation. The model of a glass fiber amplifier, presented in this chapter, has been published in 2010 [145].

We begin with mentioning few data of an erbium-doped fiber amplifier:

- Density of SiO_2 glass = $2.3 \times 10^3\,\mathrm{kg\,m^{-3}}$, corresponding to an SiO_2 number density of $2.3 \times 10^{28}\,\mathrm{m^{-3}}$.
- $N_0 = 7 \times 10^{25}\,\mathrm{m^{-3}}$ = density of Er^{3+} ions, corresponding to one Er^{3+} ion per 330 SiO_2 units; molar concentration of Er_2O_3 in quartz glass = 1,500 ppm = 1% by weight Er_2O_3 in SiO_2 glass.
- N = density of excited Er^{3+} ions.
- $\Delta\nu_g$ = 5 THz = gain bandwidth.
- Wavelength of maximum gain; λ = 1.54 μm (frequency 195 THz).
- Refractive index of quartz glass; n = 1.5.

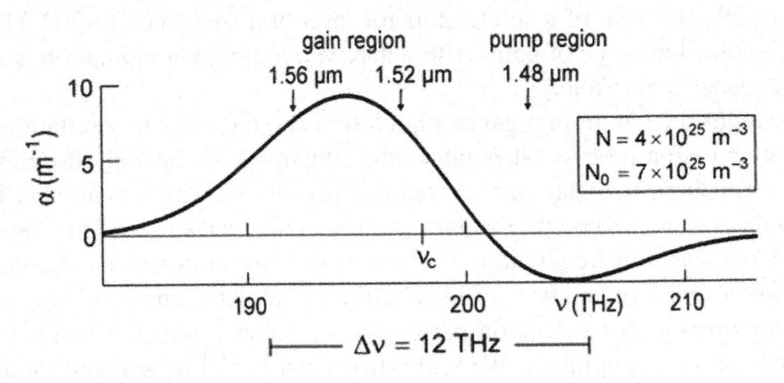

Fig. 18.1 Gain coefficient of an erbium-doped fiber

Table 18.1 Erbium-doped fiber amplifier

	λ	ν (THz)	Energy (meV)
Gain region	1.52–1.56 μm	193–197	799–815
Center of gain region	1.54 μm	195	807
Gain bandwidth	40 nm	5	20
Pump	1.48 μm	203	840

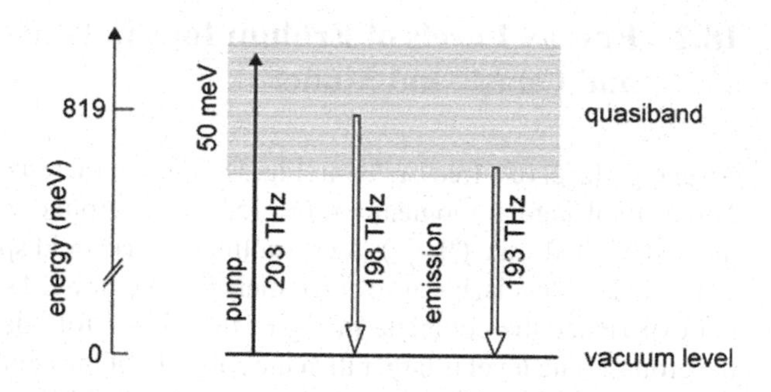

Fig. 18.2 Principle of the erbium-doped fiber amplifier

- $\tau_{sp} \sim 10\,\mathrm{ms}$ = spontaneous lifetime of the upper laser levels.
- $\alpha = 9\,\mathrm{m}^{-1}$ = maximum gain coefficient at a density of excited Er^{3+} ions of $N = 4 \times 10^{25}\,\mathrm{m}^{-3}$.
- Effective gain cross section at the wavelength of maximum gain $\sigma_{21} = 3 \times 10^{-25}\,\mathrm{m}^2$; *see* Sect. 18.7.
- Gain factor for a fiber of 10 m in length $G = 3 \times 10^3$; this correspond to a gain of $3.5\,\mathrm{dB\,m}^{-1}$.
- Pump rate

$$r = N/\tau_{sp} \sim 4 \times 10^{27}\,\mathrm{m}^{-3}\,\mathrm{s}^{-1}; \tag{18.1}$$

this corresponds to a pump power $P = r \times V \times h\nu \sim 20\,\mathrm{mW}$ if a fiber of 10 μm diameter and 0.5 m length, volume V is pumped with radiation at 1.48 μm.

Figure 18.2 illustrates the principle of the erbium-doped fiber amplifier (=Er^{3+}: glass amplifier) as we will explain in the following sections. Pump radiation creates excited-impurity quasiparticles in a quasiband via optical transitions from a vacuum level to the upper part of the quasiband. Stimulated emission of radiation by transitions from the lower part of the quasiband to the vacuum level gives rise to amplification of radiation. The width (~50 meV) of the quasiband is small compared to the center energy (819 meV); the width of the quasiband depends on the type of glass and differs by a factor 2–3 for glasses of different composition. Pumping via higher levels has already been discussed (*see* Fig. 15.10a).

The erbium-doped fiber laser at room temperature is a quasiband laser (Sect. 4.3)—the intraband relaxation time (10^{-13} s) is much smaller than the relaxation time of energy levels with respect to relaxation to the ground state ($\tau^*_{rel} = 10^{-2}$ s). Intraband relaxation is a nonradiative relaxation. Population inversion occurs if the occupation number difference is larger than zero, $f_2 - f_1 > 0$. Because the halfwidth of the quasiband is comparable to kT, population inversion requires that about half of the erbium ions are in the excited state (Sect. 18.6).

18.2 Energy Levels of Erbium Ions in Glass and Quasiband Model

An energy level of a free Er^{3+} ion (Fig. 18.3) is characterized by the quantum number J of the total angular momentum; J = 15/2 in the ground state ($^4I_{15/2}$) and J = 13/2 in the first excited state ($^4I_{13/2}$). A crystalline electric field splits a level into a multiplet of $J + 1/2$ sublevels; because of Kramers degeneracy [121], a state with an odd J does not experience the complete lifting of the $2J + 1$ fold degeneracy. The splitting of the ground state level is larger than the splitting of the excited-state level as indicated in the figure for Er^{3+} ions in a LaF_3 crystal [146, 147].

In a glass, the Er^{3+} ions are randomly distributed on sites of different crystalline electric field. Boltzmann's statistics determines the occupancy of the sublevels of an ion; thermal equilibrium of the sublevel population of an ion is established via spin-lattice relaxation; at room temperature, nonradiative relaxation by spin-lattice relaxation processes lead to a fast establishment of thermal equilibrium in an erbium-doped glass as long as pump radiation is absent.

Multiplet splitting and crystal field variations suggest widths of energy distributions (~50 meV for the ground state levels and ~25 meV for the excited-state levels), which are of the order of kT at room temperature (T = temperature; k = Boltzmann's constant). In a laser medium consisting of a doped crystal, an excited ion loses its excitation energy mainly via a stimulated optical transition. But in a doped-glass medium, an ion excited via a pump process transfers its excitation to another ion, this again to another ion and so on. On average, a laser transition process occurs only after 10^{11} transfer processes.

As an example of a glass, we discuss a quartz glass (=SiO_2 glass). A two-dimensional structural model (Fig. 18.4) illustrates the structure of glass. The SiO_2 glass consists of silicon ion and oxygen ions. The silicon and oxygen ions do not form

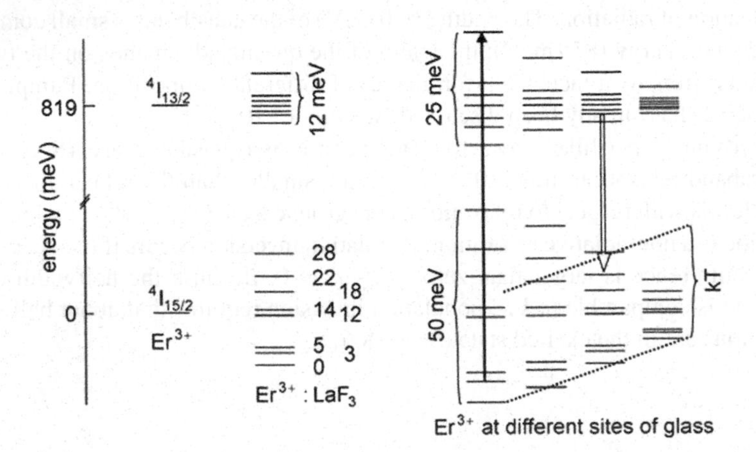

Fig. 18.3 Energy levels of free Er^{3+} ions, Er^{3+} ions in LaF_3, Er^{3+} ions in a glass and a pump and a laser transition (*arrows*)

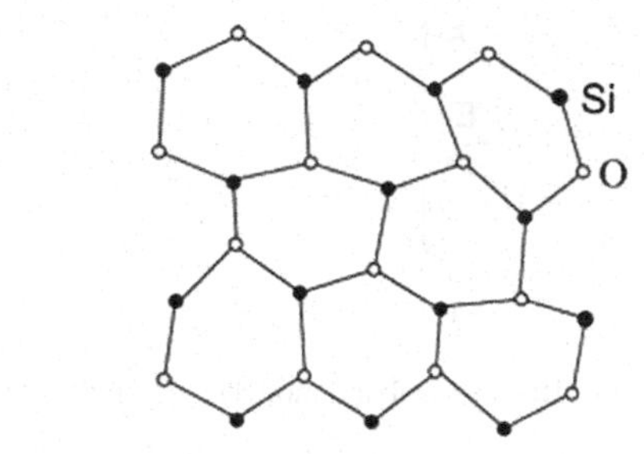

Fig. 18.4 Microscopic structure of glass

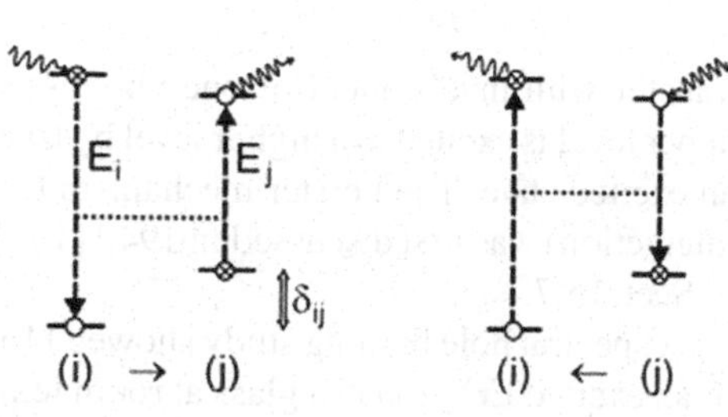

Fig. 18.5 A phonon-assisted energy transfer process and its reverse process

a periodic structure: the Si-O distance varies within the glass and the atomic arrangement shows no symmetry. Er^{3+} ions occupy different sites and therefore experience different crystal line fields. The energy of the ground state level, the energy of the excited state level, as well as the transition energy are different for erbium ions at different sites.

To discuss the role of energy transfer, we describe, for simplicity, an Er^{3+} ion as a two-level atomic system consisting of a ground state level and an excited-state level. An excited two-level system at site (i) with a transition energy E_i can transfer its excitation to a neighboring unexcited two-level system at site (j) that has a transition energy E_j (Fig. 18.5, left). The energies of the ground state levels differ by δ_{ij}. Energy conservation in a phonon assisted energy transfer process requires that

$$E_i + h\nu_{p1} = E_j + h\nu_{p2} + \delta_{ij}, \tag{18.2}$$

where ν_{p1} is the frequency of a phonon and ν_{p2} is the frequency of another phonon. The energy transfer rate depends on the concentration of impurity ions and the temperature of the glass. In the reverse process (Fig. 18.5, right), the sum of the energy of excitation of the ion at site (j), the energy of a phonon, and the energy of position is transferred to energy of excitation of ion (i) and of another phonon.

Energy transfer processes between rare earth ions in a glass at room temperature, with involvement of two phonons, first discussed in theoretical investigations [148, 149], are very efficient as experimental studies indicated [150, 151]. The microscopic process of energy transfer can be due to Coulomb interaction between two impurity ions.

Besides the two-phonon-assisted energy transfer, there are other energy transfer processes: resonant energy transfer (without the involvement of a phonon); energy

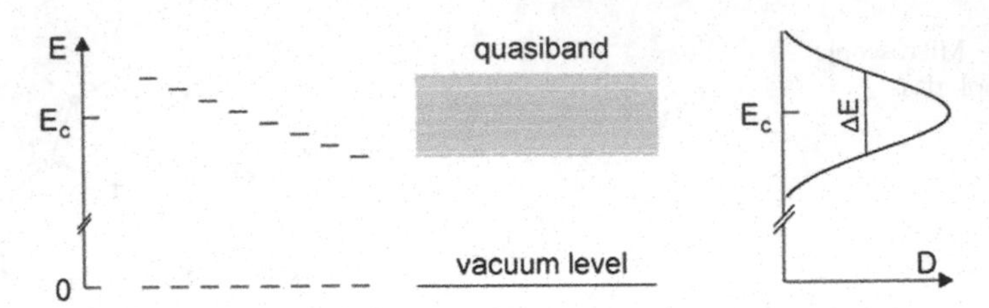

Fig. 18.6 Quasiband model of excited-impurity quasiparticles

transfer with involvement of one phonon; cross relaxation—an impurity ion in an upper level is excited to a higher level by transfer of energy from another ion that is in an excited state. The Förster mechanism (resonant energy transfer by dipole-dipole interaction) was first discussed in 1949 [152]. Some transfer processes are illustrated in Sect. 15.7.

A spectral hole burning study showed [153] that the broadening of an energy level of an excited Er^{3+} ion in a glass at room temperature corresponds to a lifetime of the order of 10^{-13} s (at a concentration of 1% Er_2O_3 by weight in glass). We associate the broadening to phonon-assisted energy transfer.

Now, we attribute the transition energies, sorted according to their values, to a quasiband (Fig. 18.6).

- $D(E)$ = density of states of the levels in the quasiband = number of levels per unit of volume and energy.
- $D(E)\mathrm{d}E$ = number of levels within the energy interval E, $E + \mathrm{d}E$ per unit of volume.
- N_0 = density of impurity ions = density of two-level atomic systems = number of impurity ions per unit of volume = number of lower levels per unit of volume (=number of upper levels per unit of volume).
- $N = N_2$ = density of excited ions.
- $N_0 - N$ = number of empty lower levels per unit of volume.
- N/N_0 = band filling factor.
- $f_2(E)$ = relative occupation number of level E = probability that the level with the energy E is occupied.
- $f_1 = 1 - f_2(E)$ = relative occupation number of the lower level = probability that the lower level, which belongs to the upper level of energy E, is occupied.
- $f_2 - f_1 = 2f(E) - 1$ = occupation number difference.
- $f_2(E)D(E)\mathrm{d}E$ = density of occupied levels in the energy interval E, $E + \mathrm{d}E$.

The integral over the density of states is equal to the density of impurity ions,

$$\int_0^\infty D(E)\mathrm{d}E = N_0. \tag{18.3}$$

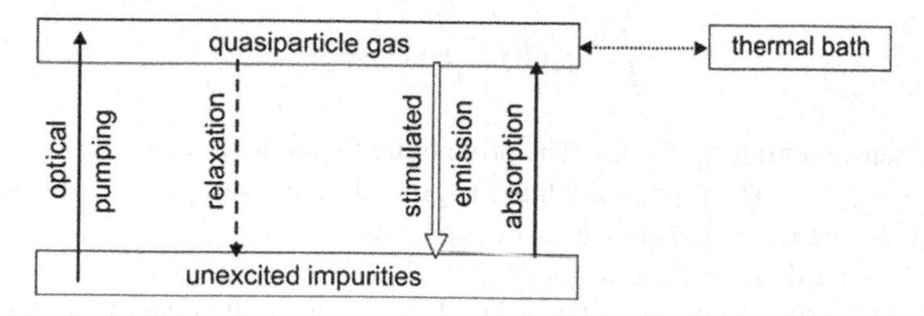

Fig. 18.7 Steady state of a gas of impurity-quasiparticles

We describe, for simplicity, the density of states by a Gaussian distribution,

$$D(E) = N_0 \times \sqrt{\frac{4\ln 2}{\pi}} \frac{1}{\Delta E} \exp\left[-\frac{\ln 2 \times (E - E_c)^2}{\Delta E^2/4}\right]. \tag{18.4}$$

E_c is the center and ΔE the halfwidth of the distribution.

Because of the phonon-assisted energy transfer processes, the quasiparticles interact with each other and couple to the thermal bath. The coupling to the thermal bath gives rise to the formation of a thermal equilibrium of the population in the quasiband (Fig. 18.7). At steady state, the average number of quasiparticles is constant. Continuous optical pumping compensates the loss of quasiparticles that is due to relaxation (mainly caused by spontaneous emission of radiation) and due to the net effect of stimulated emission and absorption of radiation.

Our model does not take into account that the ground state as well as the excited state of a single Er^{3+} ion are multiplets (due to crystal field splitting). Without pumping, the population in a multiplet of the ground state is in thermal equilibrium. This equilibrium is established via spin-lattice relaxation processes. During optical pumping, the ensemble of occupied ground state levels is not in a thermal equilibrium.

18.3 Quasi-Fermi Energy of a Gas of Excited-Impurity Quasiparticles

An energy level of the quasiband is, according to the Pauli principle, either empty or occupied with one quasiparticle. We apply to the ensemble of quasiparticles Fermi's statistics and describe the average occupation number of an energy level by the Fermi–Dirac distribution function

$$f_2(E) = \frac{1}{\exp\left[(E - E_F)/kT\right] + 1}. \tag{18.5}$$

E_F is the quasi-Fermi energy of the quasiparticle gas and T the temperature of the glass. The quasi-Fermi energy follows from the condition that

$$\int_0^\infty f_2(E)D(E)\mathrm{d}E = N. \tag{18.6}$$

N is the density of quasiparticles. The probability to find a quasiparticle in a level of energy E is $f_2(E)$. The probability that the ground state level, which corresponds to the excited-state level of energy E, is occupied, is $f_1 = 1 - f_2$.

We introduce dimensionless variables $x = E/kT, a = E_F/kT, b = E_c/kT, w = \Delta E/kT$ and write (assuming a Gaussian distribution of the density of states) the condition (18.6) in the form

$$\int_0^\infty \frac{\exp\left[-4\ln 2(x-b)^2/w^2\right]\mathrm{d}x}{\exp(x-a)+1} = 1.06\, w\, \frac{N}{N_0}. \tag{18.7}$$

If $b \gg 1$ ($E_c \gg kT$), which is the case for glass amplifiers and lasers, a numerical analysis of (18.7) yields a quasi-Fermi energy that does not depend on w; the integral is finite only in a small range of x around b and zero otherwise. The quasi-Fermi energy E_F (Fig. 18.8) increases with increasing filling factor N/N_0 and is equal to E_c at half filling; E_F is $-\infty$ at zero quasiband filling ($N = 0$) and $+\infty$ at complete filling ($N = N_0$). The quasi-Fermi energy E_F depends linearly on N/N_0 in a large range of the filling factor,

$$E_F = E_c + 4.44 \times (N/N_0 - 0.5)\, kT. \tag{18.8}$$

Figure 18.9 shows the occupation number difference $f_2 - f_1$ for quasiparticles at the center of the quasiband ($E = E_c$). The occupation number difference is -1 for $N = 0$. With increasing N, $f_2 - f_1$ increases, becomes zero at half filling of the quasiband and increases further. At complete filling ($N = N_0$), the occupation number difference is unity. The occupation number difference shows a linear dependence on the filling factor,

$$f_2 - f_1 \approx 2.22\,(N/N_0 - 1/2), \tag{18.9}$$

with small deviations near $N/N_0 = 0$ and $N/N_0 = 1$.

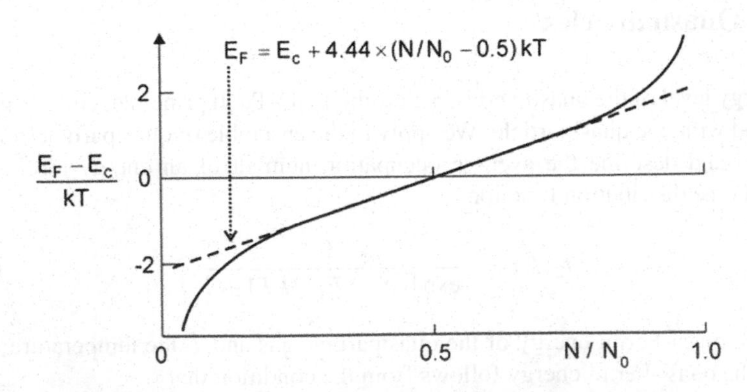

Fig. 18.8 Dependence of the quasi-Fermi energy on the filling factor of a quasiband

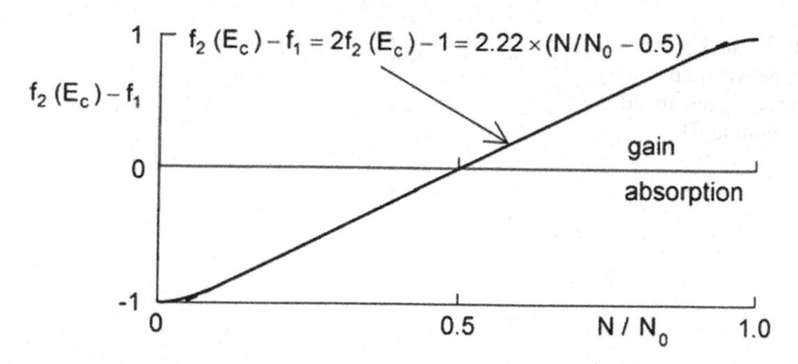

Fig. 18.9 Occupation number difference for quasiparticles in the center of the quasiband

At the center of the quasiband, the occupation number difference (*see* Fig. 18.9) increases linearly over almost the whole range of the filling factor,

$$f_2(E_c) - f_1 = 2f_2(E_c) - 1 = 2.22 \times (N/N_0 - 0.5). \tag{18.10}$$

The linear dependence of the occupation number difference on the filling factor, for $E = E_c$, appears to be characteristic for a Gaussian shape of the density of states of the quasiband; there are only small deviations from the linear dependence, occurring at N/N_0 near 0 and 1. The linear slope is slightly (11%) larger than for an ensemble of two-level systems that all have the same transition energy. We will show that $f_2 - f_1 \geq 0$ corresponds to gain and $f_2 - f_1 \leq 0$ to absorption.

A Fourier expansion of $f_2(E)$ around E_c indicates that the linear dependence of the occupation number difference $f_2(E_c) - f_1$ on N/N_0 follows directly from the linear dependence of the quasi-Fermi energy E_F on N/N_0 (Problem 18.3). The linear dependence of $f_2(E_c) - f_1$ extends, however, over a much larger range of the filling factor than the linear dependence of the quasi-Fermi energy.

18.4 Condition of Gain of Light Propagating in a Fiber

Electromagnetic radiation (frequency ν) that has a continuous energy distribution around the photon energy $h\nu = E$ interacts with a quasiparticle in a level of energy E (Fig 18.10) by absorption, stimulated and spontaneous emission. The transition rate (=number of transitions per s and m^3) of stimulated emission is given by

$$r_{\mathrm{em}}(h\nu) = \bar{B}_{21} f_2 \rho(h\nu) \tag{18.11}$$

and the rate of absorption by

$$r_{\mathrm{abs}}(\nu) = r_{\mathrm{abs}}(h\nu) = \bar{B}_{12} f_1 \rho(h\nu). \tag{18.12}$$

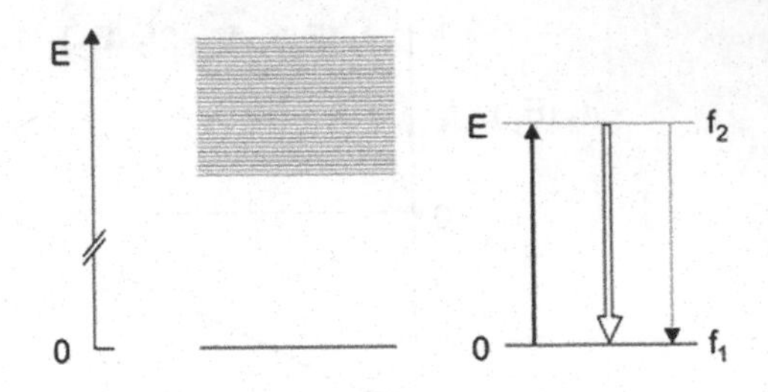

Fig. 18.10 Radiative transitions between an energy level of a quasiband and the vacuum level

We use the quantities:

- E = transition energy.
- $\bar{B}_{21}$ = Einstein coefficient of stimulated emission (in units of $\mathrm{m^3\,s^{-1}}$); $\bar{B}_{21} = hB_{21}$ (Sect. 6.6).
- $\bar{B}_{12}$ = Einstein coefficient of absorption; $\bar{B}_{12} = \bar{B}_{21}$.
- f_2 = probability that the upper level is occupied.
- $f_1 = 1 - f_2$ probability that the lower level is occupied.
- $\rho(h\nu)$ = spectral energy density of the radiation on the energy scale.

It is convenient to express the energy density on the energy scale.

The difference between the rates of stimulated emission and absorption is

$$r(h\nu) = \bar{B}_{21}(f_2 - f_1)\rho(h\nu). \tag{18.13}$$

Stimulated emission prevails if $f_2 - f_1 > 0$ or $f_2 > 1/2$. This is the condition for gain of light propagating in a fiber. The spontaneous emission rate is

$$r_{\mathrm{sp}} = A_{21} f_2. \tag{18.14}$$

A_{21} is the Einstein coefficient of spontaneous emission.

18.5 Energy Level Broadening

The phonon-assisted energy transfer processes cause a broadening of the levels of the quasiband (Fig. 18.11). We describe the broadening of a level of energy E by a lineshape function $g(h\nu - E)$ that has a halfwidth δE and is normalized,

$$\int g(h\nu - E)\,\mathrm{d}(h\nu) = 1. \tag{18.15}$$

The integral over all contributions $g(h\nu - E)\mathrm{d}(h\nu)$ in the photon energy interval $h\nu$, $h\nu + \mathrm{d}(h\nu)$ is unity. The net transition rate of monochromatic radiation, i.e., of radiation with $\rho(h\nu) \neq 0$ in the energy interval $h\nu$, $h\nu + \mathrm{d}(h\nu)$, where $\mathrm{d}(h\nu) \ll \delta E$, is given by

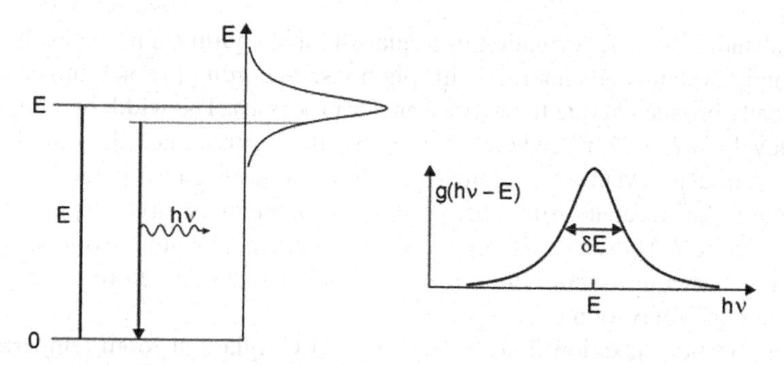

Fig. 18.11 Energy level broadening

$$(r_{\text{em}} - r_{\text{abs}})_{h\nu}\mathrm{d}(h\nu) = \bar{B}_{21}g(h\nu - E)(f_2 - f_1)\rho(h\nu)\mathrm{d}(h\nu). \tag{18.16}$$

Supposing that the lineshape function is a Lorentzian, we can write

$$g(h\nu - E) = \frac{\delta E}{2\pi}\frac{1}{(h\nu - E)^2 + (\delta E/2)^2}, \tag{18.17}$$

where δE is the halfwidth of the level broadening. The net transition rate is

$$r(\nu) = r(h\nu) = r_{\text{em}} - r_{\text{abs}} = \int \bar{B}_{21}g(h\nu - E)(f_2 - f_1)\rho(h\nu)\mathrm{d}(h\nu). \tag{18.18}$$

Introducing the energy density $u = \int \rho(h\nu)\mathrm{d}(h\nu) = Zh\nu$, where Z is the density of photons, leads to the net transition rate

$$\mathrm{r}(\nu) = h\nu\bar{B}_{21}g(h\nu - E)(f_2 - f_1)Z. \tag{18.19}$$

The net transition rate is proportional to the occupation number difference $f_2 - f_1 = 2f_2 - 1$ and to the photon density Z. The condition of gain is the same as derived for the case of neglected energy level broadening,

$$f_2 - f_1 = 2f_2(E) - 1 > 0 \quad \text{or} \quad f_2(E) > 1/2; \tag{18.20}$$

gain occurs for radiation of quantum energies

$$h\nu < E_{\mathrm{F}}. \tag{18.21}$$

Optical pumping is possible by using radiation of a quantum energy $h\nu$ that is larger than the quasi-Fermi energy E_{F}. The mechanism leading to the quasi-Fermi distribution is the intraband relaxation. Due to phonon-assisted energy transfer, the excited two-level atomic systems lose a portion of their excitation energy to phonons. This leads, at room temperature, to the formation of the quasi-Fermi distribution in the quasiband. After the formation of a quasithermal equilibrium, the excited two-level atomic systems still interact with phonons. Accordingly, each upper level is energetically broadened due to energy transfer processes. The width of a broadened energy level is $\delta E \approx \hbar/\tau_{in}$, where τ_{in} is the scattering time, i.e., the time between two energy transfer events. The scattering time τ_{in} depends on temperature. At room temperature, the occupation number of thermal phonons is large at phonon energies $kT \sim 25$ meV. Thus, a few energy transfer events (per excited two-level system) establish a quasiequilibrium after a few scattering events. Therefore, we regard τ_{in} as the intraband relaxation time.

The intraband relaxation time ($\sim 10^{-13}$ s) of Er^{3+}:glass at room temperature is much shorter than the lifetime (of the order of 10 ms) of an upper level with respect to spontaneous emission of a photon. The width of the broadening of an upper level, $\hbar/\tau_{in} \sim 4$ meV, is small compared to the range ($\sim kT$) of populated levels.

18.6 Calculation of the Gain Coefficient of a Doped Fiber

The temporal change of the density of excited ions due to stimulated transitions is

$$\frac{dN}{dt} = -h\nu \int \bar{B}_{21} g(h\nu - E)(f_2 - f_1) D(E)\, dE \times Z, \tag{18.22}$$

It follows that the temporal change of the photon density Z is given by the relation

$$dZ/dt = -dN/dt = \gamma Z, \tag{18.23}$$

where

$$\gamma = h\nu \int \bar{B}_{21} D(E)(f_2 - f_1)\, g(h\nu - E)\, dE \tag{18.24}$$

is the growth coefficient of radiation of frequency ν. With $dt = n dz/c$, where z is the direction of propagation of the radiation (along the fiber axis), c is the speed of light in vacuum, n (~ 1.5) the refractive index of the fiber glass, we find

$$dZ/dz = \alpha Z, \tag{18.25}$$

where

$$\alpha = \frac{n}{c} h\nu \int \bar{B}_{21} D(E)(f_2 - f_1)\, g(h\nu - E)\, dE \tag{18.26}$$

is the gain coefficient. The level broadening due to energy transfer is small compared to kT. Therefore, we can replace $g(h\nu - E)$ by a delta-function, $\delta(h\nu - E)$, and find

$$\alpha(\nu) = (n/c) h\nu \bar{B}_{21} D(E)(f_2 - f_1), \tag{18.27}$$

where $(f_2 - f_1) = 2 f_2(E) - 1$ and $E = h\nu$.

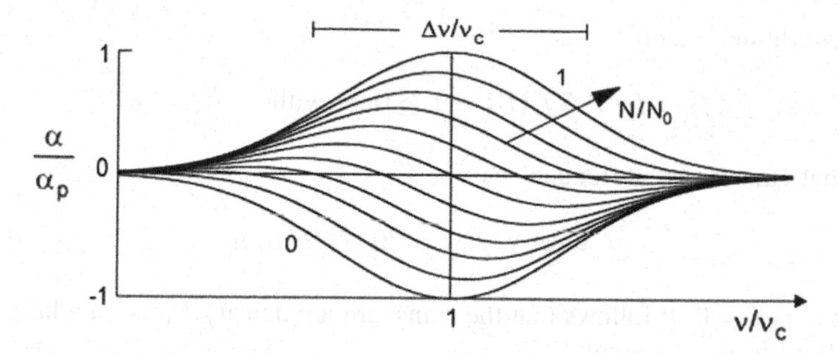

Fig. 18.12 Gain coefficient at different band filling factors N/N_0 *(in steps of 0.1)*

Under the assumption that B_{21} is the same for all two-level systems and does therefore not depend on E, the peak gain coefficient is

$$\alpha_p = (n/c)h\nu_c \bar{B}_{21} D(E_c). \tag{18.28}$$

Then the slope of the gain coefficient curves is given by the simple expression

$$\alpha(\nu)/\alpha_p = f_2 - f_1 = 2f_2(E) - 1; \quad E = h\nu. \tag{18.29}$$

Gain coefficient curves (Fig. 18.12) show that the range of gain increases with increasing quasiband filling according to the increase of the frequency

$$\nu_F = E_F/h. \tag{18.30}$$

The frequency ν_F is a *transparency frequency*. At complete filling, the gain coefficient has the same slope as the density of states and reaches the peak gain coefficient α_p at the center frequency $\nu_c = E_c/h$. If the quasiband is empty, the absorption coefficient has the same profile as the density of states.

From the peak gain coefficient, we obtain a peak gain cross section according to the relation

$$\alpha_p = N_0\, \sigma_p. \tag{18.31}$$

We find

$$\sigma_p = \frac{n}{c} h\nu \bar{B}_{21} D(E_c)/N_0 = 1.48 \frac{\Delta\nu_{hom}}{\Delta\nu} \frac{(\lambda/n)^2}{2\pi}, \tag{18.32}$$

with the quantities: $\Delta\nu = \Delta E/h$; $A_{21} = 8\pi\nu^3(n/c)^{-3}\bar{B}_{21}$ = Einstein coefficient of spontaneous emission; $\lambda = c/\nu_c$. Thus, σ_p has the same value as the peak gain cross section of an ensemble of noninteracting two-level systems with transitions cline of Gaussian shape.

The condition of gain

$$(f_2 - f_1) = 2f_2(E) - 1 \geq 0 \quad \text{with} \quad E = h\nu \tag{18.33}$$

means that gain occurs at frequencies

$$\nu < \nu_F(N/N_0) = E_F(N/N_0)/h \tag{18.34}$$

and that $\alpha(\nu_F) = 0$. It follows that the transparency density N_{tr} is, in a large range of the filling factor, given by

$$\frac{N_{tr}}{N_0} = 0.5 + \frac{E_F - E_c}{4.44\,kT} = 0.5 + \frac{\nu_F - \nu_c}{4.44\,kT/h}. \tag{18.35}$$

Example Gain coefficient of a fiber doped with 1% Er_2O_3 by weight ($N_0 = 7 \times 10^{25}$ m^{-3}) at a filling factor $N/N_0 = 0.6$ (*see* Fig. 18.1).

- $A_{21} = 100\,\mathrm{s}^{-1}$; $\Delta\nu_{hom} = A_{21}/(2\pi) \sim 16\,\mathrm{s}^{-1}$.
- $\bar{B}_{21} = 4.0 \times 10^{-18}\,\mathrm{m}^3\,\mathrm{s}^{-1}$.
- $E_c = 819\,\mathrm{meV}$; $\nu_c \sim 198\,\mathrm{THz}$.
- $c/n = 2 \times 10^8\,\mathrm{m\,s}^{-1}$.
- $\Delta E = 50\,\mathrm{meV} = 8 \times 10^{-21}\,\mathrm{J}$; $\Delta\nu \sim 12\,\mathrm{THz}$.
- $D(E_c) = 8.3 \times 10^{45}\,\mathrm{m}^{-3}\,\mathrm{J}^{-1}$.
- $\alpha_p = 22\,\mathrm{m}^{-1}$; $\sigma_p = 3.2 \times 10^{-25}\,\mathrm{m}^2$.

The gain coefficient (at a filling factor of 0.6) is positive below a frequency that is slightly larger than ν_c while a range of absorption follows at higher frequency; the maximum gain coefficient (~9 m^{-1}) is slightly smaller than half the peak gain coefficient α_p.

Figure 18.13 illustrates our result. Gain occurs up to the transparency frequency $\nu_F = E_F/h$. The quasi-Fermi energy E_F and thus ν_F increase with increasing band filling.

According to the linear dependence of the occupation number difference at the center of the quasiband on the filling factor (*see* Fig. 18.9), we find

$$\alpha(\nu_c)/\alpha_p \approx 2.22\,(N/N_0 - 1/2); \tag{18.36}$$

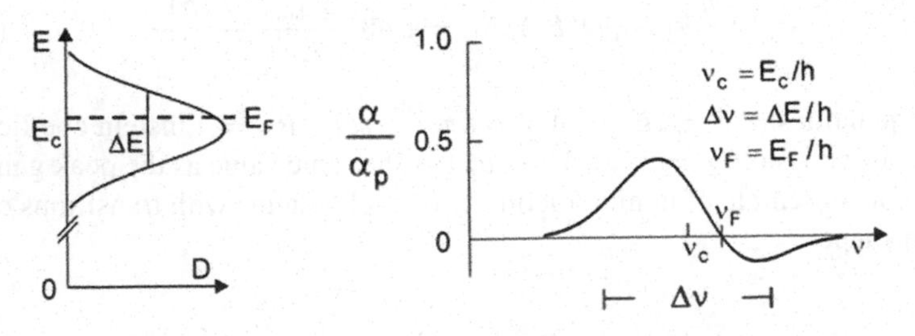

Fig. 18.13 Quasi-Fermi energy and transparency frequency

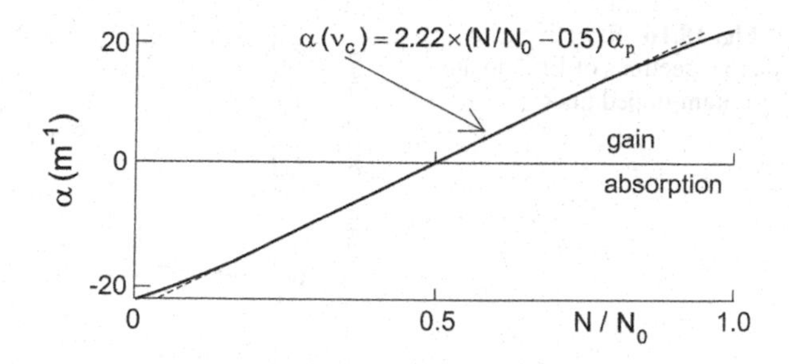

Fig. 18.14 Gain coefficient of an erbium-doped fiber at the center frequency $\nu_c = E_c/h$; $\alpha_p = N_0\,\sigma_p$, $N_0 = 7 \times 10^{25}$ m^{-3}, and $\sigma_p = 3.2 \times 10^{-25}$ m^2

the gain coefficient at the frequency $\nu_c = E_c/h$ increases linearly with the filling factor (Fig. 18.14).

18.7 Different Effective Gain Cross Sections

Here, we introduce three different effective gain cross sections: effective gain cross section σ; effective gain cross section $\bar{\sigma}_{\text{eff}}$; effective gain cross section σ_{eff}.

Figure 18.15 shows gain coefficients (solid lines) at different quasiband filling factors. The maximum gain coefficient α_{max} (dashed) depends on the filling factor. We can relate the maximum gain coefficient and the density of quasiparticles (i.e., the density of excited ions),

$$\alpha_{\text{max}} = N\,\sigma; \tag{18.37}$$

the effective gain cross section σ (Fig. 18.6, dashed) increases with increasing filling factor, from zero for the empty quasiband to σ_p at complete quasiband filling.

The effective gain cross section

$$\bar{\sigma}_{\text{eff}} = (\mathrm{d}\alpha_{\text{max}}/\mathrm{d}N)_{N/N_0} \tag{18.38}$$

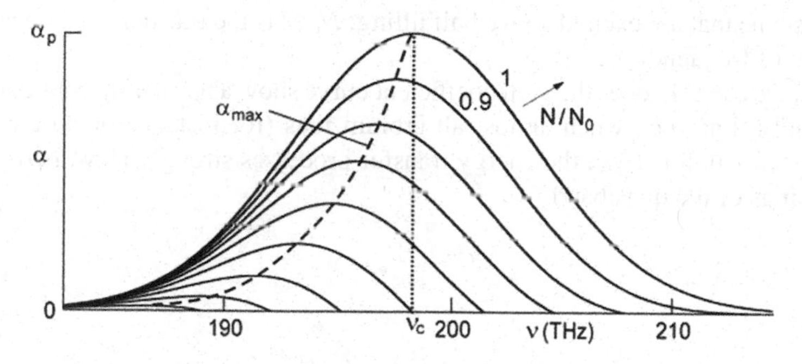

Fig. 18.15 Gain coefficient at different filling factors

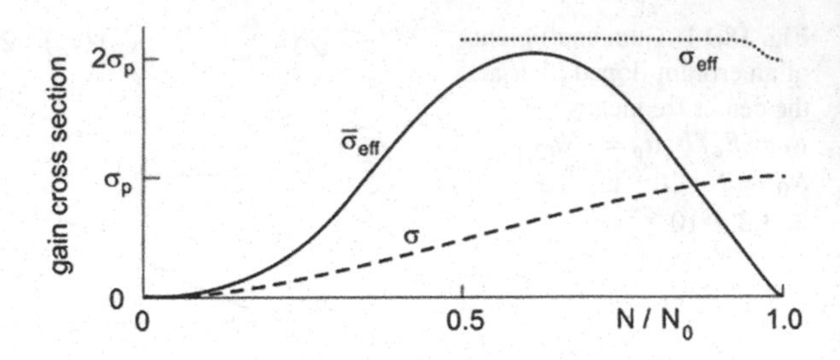

Fig. 18.16 Effective gain cross sections of Er^{3+} in an erbium-doped fiber

describes the change of α_{max} with N. With increasing N/N_0, the effective gain cross section $\bar{\sigma}_{eff}$ (Fig. 18.16, solid line) increases from zero near $N/N_0 = 0$, shows a maximum ($\sim 2\sigma_p$) for $N/N_0 \sim 0.6$, then decreases and approaches zero near $N/N_0 = 1$.

We can introduce another effective gain cross section, σ_{eff}, writing

$$\alpha(\nu_c) \approx 2.22(N/N_0 - 1/2)(n/c)h\nu \bar{B}_{21} D(E_c) = (N - N_0/2)\,\sigma_{eff} \tag{18.39}$$

and find

$$\sigma_{eff} \approx 2.22\,(n/c)h\nu \bar{B}_{21} D(E_c) = 2.22\,\sigma_p. \tag{18.40}$$

The effective gain cross section σ_{eff} (Fig. 18.16, dotted) has a constant value (2.22 σ_p) over a large range of the filling factor, from $N/N_0 = 0.5$ to nearly $N/N_0 = 1$, where it decreases to $2\sigma_p$. At complete filling, $\sigma_{eff} = 2\sigma_p$; the factor 2 is due to the different reference values, $N/N_0 - 0.5$ an N, respectively. At smaller band filling, σ_{eff} exceeds $2\sigma_p$ because band filling in the center of the quasiband leads to stronger gain than filling in the wing of the band for $N/N_0 \rightarrow 1$.

The effective gain cross section

$$\sigma_{eff} = (\mathrm{d}\alpha/\mathrm{d}N) \tag{18.41}$$

describes, for radiation of frequency ν_c, the gain cross section related to the two-level systems that are excited above half filling; $N_0/2$ is the transparency density for radiation of frequency ν_c.

Here, we can ask: does the gain coefficient curve show a narrowing near complete quasiband filling, i.e., when almost all erbium ions (for instance 90%) are in the excited state? In this case, the energy transfer processes strongly slow, particularly in the wings of the quasiband.

18.8 Absorption and Fluorescence Spectra of an Erbium-Doped Fiber

The shape of an absorption curve is given by

$$\bar{\alpha}_{\mathrm{abs}} = \alpha_{\mathrm{abs}}(\nu)/\alpha_{\mathrm{p}} = f_1 - f_2 = 1 - 2f_2(E), \tag{18.42}$$

where $E = h\nu$ and where α_{p} is the absorption coefficient at the frequency ν_{c} at zero quasiband filling. The shape of a fluorescence curve is given by

$$\bar{S}_\nu(\nu) = S_\nu(\nu)/S_{\nu,\mathrm{p}} = f_2(E); \quad E = h\nu. \tag{18.43}$$

$S_\nu(\nu)$ is the spectral distribution of the fluorescence radiation. $S_{\nu,\mathrm{p}}$ is the peak intensity, namely the intensity at the frequency ν_{c} in the case of complete quasiband filling. We neglect the frequency dependence of the Einstein coefficient of spontaneous emission.

Figure 18.17 shows an absorption curve and a fluorescence curve both for week quasiband filling ($N/N_0 = 0.1$). The absorption curve is slightly blue-shifted with respect to the absorption curve for zero quasiband filling. The fluorescence curve is red-shifted. The absorption and the fluorescence curves have different shapes. The shapes of the curves as well as the frequencies of their maxima depend on the filling factor.

The filling factor relates the absorption coefficient and the shape of the fluorescence curves according to the expression

$$\bar{S} = \frac{f_2}{1 - 2f_2} \times \bar{\alpha}_{\mathrm{abs}}. \tag{18.44}$$

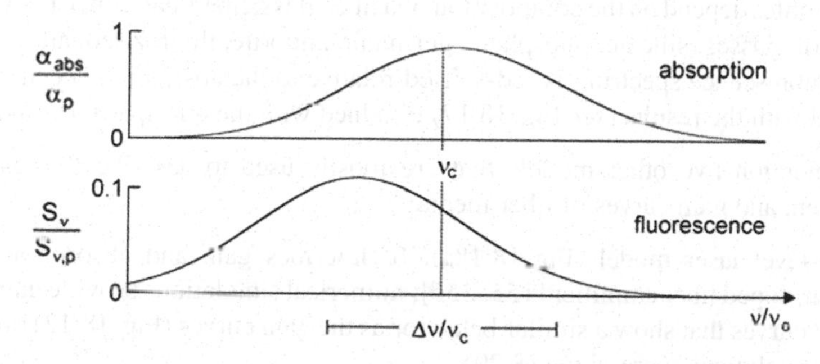

Fig. 18.17 Shapes of an absorption and a fluorescence curve of a fiber medium at the filling factor $N/N_0 = 0.1$

At week quasiband filling ($N/N_0 \ll 1$), the relation is

$$\bar{S} = f_2 \times \bar{\alpha}_{\text{abs}} = \frac{1}{\exp[(E - E_F)/kT] + 1} \times \bar{\alpha}_{\text{abs}}. \tag{18.45}$$

In this case, the shape of the fluorescence curve is determined by the product of the Fermi–Dirac distribution function and the absorption coefficient.

It is possible to modify the quasiparticle model taking into account that B_{21} can have different values for ions at different sites in a glass and that the density of states does not have a Gaussian shape. An analysis of absorption and fluorescence spectra measured for different filling factors N/N_0 (and different sample temperatures) may provide detailed information on $B_{21}(\nu)$ and $D(\nu)$.

There remains the question how to take account of the multiplet splitting, especially of the occupied sublevels of the ground state. During optical pumping, the population of the ensemble of sublevels of the nonexcited ions is not in thermal equilibrium as already mentioned.

18.9 Experimental Studies and Models of Doped Fiber Media

The gain coefficient curves at different filling factors (*see* Fig. 18.12) and the absolute values of the gain coefficients (*see* Fig. 18.1) are, in principle, in accord with experimental results. However, experimental studies of the shape of absorption curves, gain curves, and fluorescence curves of erbium-doped fibers indicate the following:

- The profiles of absorption spectra and of fluorescence spectra are non-Gaussian [154]; this shows that the densities of states of quasiparticles have non-Gaussian profiles and that—most likely—B_{21} does not have a constant value.
- The profiles depend on the composition of a fiber glass; fibers can consist of various types of glasses (silicate, phosphate, germanite, fluorite, fluorozirconate glass).
- The fluorescence spectrum is red-shifted relative to the absorption spectrum—in accord with the results (*see* Fig. 18.17) obtained with the quasiparticle model.

We mention two other models that are mostly used to describe fluorescence, absorption, and gain curves of fiber media:

- Three-level laser model (Fig. 18.18a). It describes gain and absorption of an erbium-doped fiber amplifier [155, 156]; numerical simulations provide gain coefficient curves that show a similar behavior as the gain curves (Fig. 18.12) obtained by the analytical expression (18.29).
- Three-level laser medium of the ruby laser type (Fig. 18.18b). It describes amplifiers and lasers, which are strongly pumped via high-lying pump levels [157, 158].

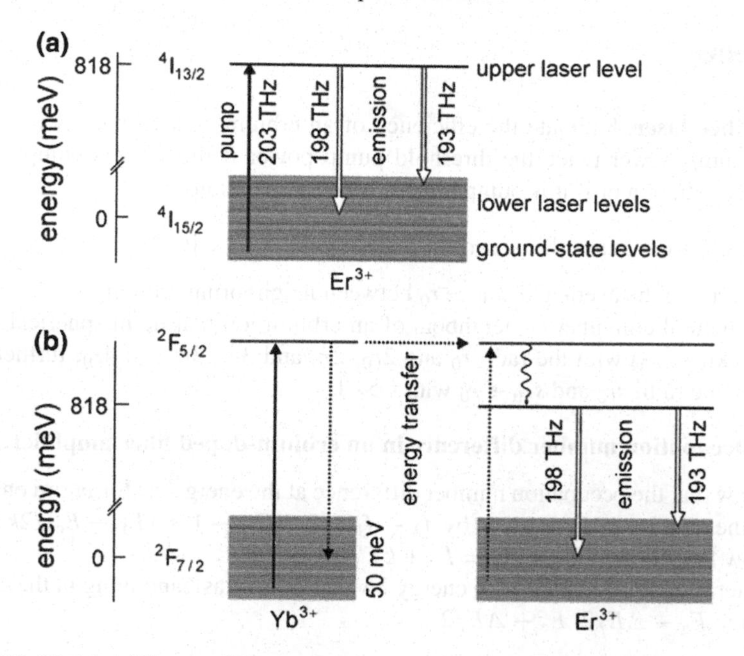

Fig. 18.18 Models of fiber amplifiers and lasers. **a** Three-level laser model of the erbium-doped fiber amplifier. **b** Ruby laser type model of an erbium-doped fiber laser

These models assume a quasithermal equilibrium of the population of the ground state levels. Although the assumption is not fully justified, the models provide a basis of the description of optical properties of a fiber.

The quasiband model should also be applicable to analyze active media of fiber lasers and amplifiers strongly pumped via high-lying energy levels mentioned in Sect. 15.7:

- 1.05-μm ytterbium-doped fiber laser [159].
- 1.5-μm ytterbium/erbium-doped fiber amplifiers [160].
- Thulium-doped fiber laser [161, 162].
- 2.1-μm holmium-doped fiber laser [163].
- 3-μm ytterbium/erbium-doped fiber laser [164, 165].

The quasiparticle model predicts a narrowing of the gain curve at nearly complete population inversion. When nearly all impurity ions are in the excited state, the rate of phonon-assisted energy transfer processes slows down; at complete population inversion, energy transfer processes are no longer possible.

References [140–165].

Problems

18.1 Fiber laser. Estimate the efficiency of an erbium-doped fiber laser pumped with a pump power twice the threshold pump power if the laser is pumped with 1480-nm radiation or if it is pumped with 980-nm radiation.

18.2 A glass contains erbium ions of a density $N_0 = 7 \times 10^{25}\,\mathrm{m}^{-3}$.

(a) Determine the average distance r_0 between neighboring erbium ions.
(b) Estimate the number of neighbors of an erbium ion that lie in spherical shells (thickness r_0) with the radii r_0 and $2r_0$; $2r_0$ and $3r_0$; $3r_0$ and $4r_0$; furthermore with the radii sr_0 and $sr_0 + r_0$ with $s \gg 1$.

18.3 Occupation number difference in an erbium-doped fiber amplifier.

(a) Show that the occupation number difference at the energy E_c for Fermi energies in the vicinity of E_c is given by $f_2 - f_1 = 2f(E_c) - 1 \approx (E_F - E_c)/2kT$.
(b) How large is $f_2 - f_1$ if $E_F = E_c + kT$?
(c) Determine the percentage of energy levels of the quasiband lying in the energy range $E_c - \Delta E/2$, $E_c + \Delta E/2$.

18.4 Discuss why the following lasers do not belong to the type "quasiband laser":

(a) Titanium–sapphire laser, alexandrite laser; and, generally, vibronic lasers.
(b) Helium–neon laser.
(c) Continuous wave CO_2 laser and TEA CO_2.

18.5 Density of states. We consider the following case: the density of states of quasiparticles in an erbium-doped fiber is the sum of two densities of state, $D = D_1 + D_2$; the center frequencies have a frequency distance of $4kT$ ($T = 300\,\mathrm{K}$); the halfwidth of both densities of states is $2kT$.

(a) Estimate the maximum gain coefficient α_{max}.
(b) Estimate the maximum gain coefficient in the case that the center frequencies have a frequency distance of kT.

18.6 Present arguments that show that it is most likely that the spontaneous lifetime τ_{sp} of the $^4I_{13/2}$ level of erbium ions in a glass fiber depend on the quasiband filling factor.

18.7 Einstein coefficients. Consider an impurity-doped fiber with a Gaussian shape of the density of states of quasiparticles.

(a) Design a dependence $B_{21}(E)$ that leads to a double peak in the gain curve.
(b) Then discuss the dependence of τ_{sp} on the filling factor.

18.8 Temperature coefficient. Make use of the quasiparticle model to estimate the temperature coefficient (in units of dB/°C) of an erbium-doped fiber amplifier of 10 m length for the temperature ranges 10–20, −50 to −40 and 50–60 °C:

(a) If the frequency of the radiation is equal to the center frequency.
(b) If the frequency of maximum gain occurs at a filling factor of 0.6.

18.9 Fiber laser and fiber amplifier. Determine the gain of radiation passing through an erbium-doped fiber (length 16 m) pumped at twice the transparency density; for data, *see* Sect. 18.6.

18.10 Why is the population of the multiplet levels of the ground state of Er^{3+} not in thermal equilibrium during optical pumping?

18.11 Spectral diffusion and quasiband model.
Describe diffusion of excitation energy in an infinitely long rectangular slab of a glass containing a large concentration of Er^{3+} ions. At time $t = 0$, excitation energy E_c at the center of the Gaussian quasiband is homogenously deposited over the slab, with the quasiparticle density N_0. [*Hint*: apply the one-dimensional diffusion equation $\partial f/\partial t = D_E \partial^2 f/\partial E^2$, where $f(E - E_c, t)$ is the distribution function and D_E the spectral diffusion constant, and replace the Gaussian shape of the density of states by a constant.]

(a) Show that $f(E - E_c, t) = N_0/(2\sqrt{2D_E t}) \exp(-(E - E_c)^2/4D_E t)$.
(b) Determine the variance and the halfwidth (FWHM) of the distribution at time t.
(c) Determine the average frequency range over which excitation energy of a two-level system traveled in a random walk after z ($\gg$1) inelastic scattering processes.
(d) Estimate the spectral diffusion constant, assuming that the average energy transferred in a spectral diffusion process according to (18.2) is $\delta = 0.1$ meV. [*Hint*: the spectral diffusion constant is $D_E = (1/3)\, v_E^2/\tau$, where $v_E = \delta/\tau$ is the velocity in the energy space and τ ($\approx 10^{-13}$s) the lifetime of an excited state level with respect to an energy transfer process.]
(e) How many scattering events are necessary to distribute the energy over the whole width of a quasiband of a width of 50 meV? Show that the corresponding time is still much shorter than the spontaneous lifetime of an excited state. (This is an essential condition for the applicability of the quasiband model.) [*Hint*: neglect the influence of thermal effects.]
(f) Determine, for the given numbers, the value of the maximum of the distribution function for the case that the energy is distributed over the whole quasiband.

18.12 Range of validity of the quasiband model.
The quasiband model is applicable for glass laser materials at room temperature. Cooling of the material leads to a slowing down of the energy transfer processes. Determine the temperature at which the quasiband model is no longer applicable if the lifetime of an excited state level with respect to an energy transfer process is inversely proportional to temperature. [*Hint:* make use of data of the preceding problem.]

18.13 Spectral-spatial diffusion in an active glass medium.
In a fiber laser, the laser field as well as the pump field is non-uniform over the cross section. Redistribution occurs by both spatial and spectral diffusion. It is the purpose of this problem to study the speed of redistribution by spectral-spatial diffusion.

(a) Describe spectral-spatial diffusion by a one-dimensional differential equation for the distribution $f(x, E, t)$.
(b) Solve the equation for the following case: The fiber has a quadratic cross section. At time $t = 0$, excitation energy is deposited at the center of a Gaussian quasi-band with a homogenous distribution over the cross section at $x = 0$. The two-dimensional quasiparticle density for $t = 0$ and $x = 0$ is N_0; x is the direction of the fiber.

18.14 Spectral-spatial diffusion in an erbium-doped glass fiber laser.
Apply the results of the preceding problem to a cw erbium-doped fiber laser assuming that the pump radiation is homogeneously distributed in the fiber. Assume that the laser field has a nearly Gaussian distribution within the fiber.

(a) How broad is the spatial hole?
(b) Estimate the time it takes to fill the spatial hole.
(c) Estimate the time it takes to fill the spectral hole.
(d) How deep is the spectral hole?

18.15 Formulate the formulas describing spectral-spatial diffusion (a) in dimensionless units and (b) in the frequency space $\nu = E/h$ instead of the energy space.

Chapter 19
Free-Electron Laser

In a free-electron laser (FEL), free-electrons of a velocity near the speed of light are passing through a periodic transverse magnetic field. Due to the Lorentz force, the electrons perform oscillations (free-electron oscillations) with displacements transverse to the propagation direction. Stimulated emission of radiation by the oscillating electrons is the origin of free-electron laser radiation. The frequency of the radiation increases quadratically with the electron energy. Frequency tuning over a large frequency range is possible by changing the electron energy; the range of the electron energy is determined by the particle accelerator that produces an external electron beam used to operate a free-electron laser.

Infrared and far infrared free-electron lasers generate pulses of radiation of high power; one type of the presently operating far infrared free-electron lasers generates quasi-continuous radiation.

Single-pass free-electron lasers, i.e., mirrorless lasers, are able to generate optical pulses of extremely large pulse power. An important single-pass free-electron laser is the SASE (self-amplified spontaneous emission) free-electron laser. It generates optical pulses by amplification of spontaneously emitted radiation. X-ray SASE free-electron lasers are successfully applied in many fields.

The equation of motion of an electron traversing a spatially periodic magnetic field provides the oscillation frequency (resonance frequency) ω_0 of the free-electron oscillation; ω_0 depends on the energy of the electron as well as on the period and the strength of the magnetic field.

Classical theory describes the dynamics of a free-electron laser by using Maxwell's equations together with classical laws for the generation of radiation by moving charges. Translational energy of electrons is converted to energy of optical radiation. Gain for radiation occurs at frequencies slightly below ω_0. Theory shows that the amplitude of the high frequency electric field in a free-electron laser medium is limited; even if the high frequency field in a laser resonator has no loss, the field cannot exceed a saturation field—conventional lasers do not have such a limitation. Numer-

K.F. Renk, *Basics of Laser Physics*, Graduate Texts in Physics,
DOI 10.1007/978-3-319-50651-7_19

ical simulations provide solutions of Maxwell's equations that allow for complete characterization of a free-electron laser.

This chapter contains an introduction to free-electron lasers. Instead of applying Maxwell's theory, we present a model of the free-electron laser, which we call *modulation model of the free-electron laser*. The transverse oscillation of an electron at the frequency ω_0 is joined with a transverse current at ω_0. The modulation model describes the interaction of an electron with a high frequency electric field of frequency ω as a phase modulation of the oscillation of the electron. Phase modulation results in another current, the *modulation current* at the frequency ω. The modulation leads, in an ensemble of electrons, to a modulation current density that mediates gain for the high frequency electric field; energy of translation of the electrons is transferred to energy of the high frequency electric field. An ensemble of electrons represent the active medium in a free-electron laser. The modulation model contains a numerical parameter, which is a measure of the strength of interaction of the electron oscillation and the high frequency electric field. We choose the parameter by comparison with experimental data of free-electron lasers. The modulation model provides analytical expressions for the saturation field amplitude and for the gain coefficient of the free-electron laser medium. The model allows, furthermore, for an analysis of the onset of oscillation, with the onset being initiated by radiation spontaneously emitted by oscillating electrons. The modulation model illustrates main properties of a free-electron laser. According to the modulation model, saturation of the amplitude of high frequency electric field is due to a transition of the active medium from a state of gain to a state in which absorption compensates stimulated emission of radiation.

We attribute, additionally, discrete energy levels, namely an energy-ladder system, to an oscillating electron; the energy levels are equidistant and have a next near energy level distance of $\hbar\omega_0$. In this description, radiation is generated by stimulated electronic transitions between discrete energy levels.

In the energy-ladder description of the elementary excitations in a free-electron laser medium, the free-electron medium is an active medium *with* a population inversion. In the description by classical Maxwell theory and also in the modulation model, the free-electron laser is a laser with an active medium *without* population inversion.

We will not discuss a radiation source—also called free-electron laser—that is operating at very large electron currents. In this type of free-electron laser, electron–electron interaction gives rise to charge density domains. Electromagnetic radiation interacts with a collective of electrons. These free-electron lasers, which are single-pass free-electron lasers, are also able to produce radiation of very large power; however, the radiation is not monochromatic.

19.1 Principle of the Free-Electron Laser

In a free-electron laser (Fig. 19.1), a beam of electrons (energy $E_{\mathrm{el},0}$) traverses a periodic magnetic field (period λ_{w}). Due to the Lorentz force, the electrons execute trans-

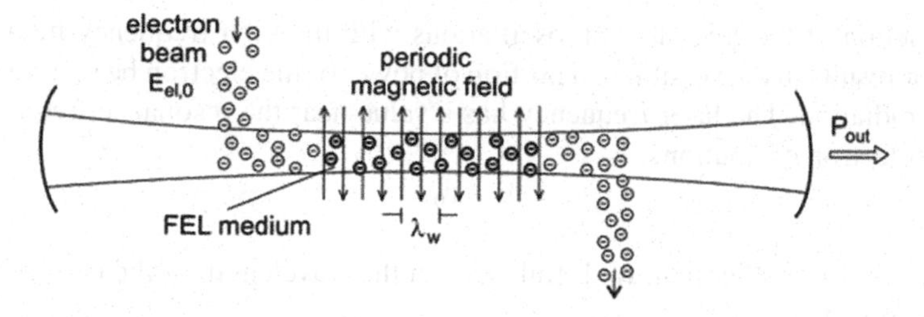

Fig. 19.1 Free-electron laser

verse oscillations. The oscillating electrons form an active medium (*free-electron laser medium* = FEL medium). Stimulated emission of radiation by the free-electron oscillators gives rise to buildup and maintenance of a high frequency field in a laser resonator. Laser radiation (output power P_{out}) is coupled out via a partial mirror.

The relativistic energy of an electron, which enters a periodic magnetic field with the velocity $v_{z,0}$, is given by

$$E_{el,0} = \gamma m_0 c^2, \tag{19.1}$$

where m_0 is the electron mass and

$$\gamma = \frac{1}{\sqrt{1 - v_{z,0}^2/c^2}} \tag{19.2}$$

is the Lorentz factor; γ measures the relativistic energy of an electron in units of m_0c^2; $\gamma = E_{MeV}/0.51\,MeV$. The oscillation frequency of a free-electron oscillation is equal to

$$\nu_0 = \frac{1}{1 + K_w^2/2} \frac{2c\gamma^2}{\lambda_w}. \tag{19.3}$$

K_w (= *wiggler parameter*) is the dimensionless wiggler strength. We will show (in Sect. 19.3) that it is given by

$$K_w = \frac{eB_w\lambda_w}{2\pi m_0 c}. \tag{19.4}$$

B_w is the maximum strength of a magnetic field assumed to vary sinusoidally along the wiggler axis; a value $K_w^2 = 1$, for instance, characterizes strong transverse oscillations. On the ω scale, the electron oscillation frequency is given by

$$\omega_0 = \frac{1}{1 + K_w^2/2} \frac{4\pi c\gamma^2}{\lambda_w}. \tag{19.5}$$

Interaction of the free-electron oscillations with the high frequency field in the resonator results in conversion of a portion of power of the electron beam into power of laser radiation. The laser frequency has a value near the resonance frequency of the free-electron oscillations,

$$\nu \sim \nu_0. \tag{19.6}$$

But ν is slightly smaller than ν_0. It follows that the wavelength of the laser radiation is about equal to

$$\lambda \sim \frac{\lambda_w}{2\gamma^2}\,(1 + K_w^2/2). \tag{19.7}$$

We describe the free-electron laser in more detail. A beam of relativistic electrons, produced by the use of an accelerator (Fig. 19.2a), traverses a spatially periodic magnetic field and excites a radiation field in the optical resonator. The electron beam, guided by a bending magnet into the resonator, passes the periodic magnetic field, which is produced by use of a periodic magnet structure, the *wiggler* (= undulator). The electron beam then leaves the resonator by means of a second bending magnet. Along the resonator axis (z axis), the magnetic field direction assumes the $+y$ direction and the $-y$ direction in turn. It varies in the simplest case sinusoidally,

$$B_y = B_w \sin \frac{2\pi}{\lambda_w} z. \tag{19.8}$$

The length of the wiggler is $L_w = N_w \lambda_w$ and N_w is the number of wiggler periods.

Due to the Lorentz force, the electrons execute oscillations perpendicular to the magnetic field direction (y direction) and perpendicular to the z direction; i.e., the electrons oscillate with elongations in $\pm x$ direction.

The wiggler can consist of two rows of equal magnets, with north poles N and south poles S arranged periodically (Fig. 19.2b). The magnetization of a magnet and the distance d between the rows determine the field strength B_w. Magnets prepared from a samarium-cobalt alloy, which have a high magnetization, are suitable as wiggler magnets. Alternatively, the wiggler is a superconducting magnet with a helical winding of the superconducting wires, leading to a circular sinusoidally varying transverse magnetic field.

The electron beam in the range between the wiggler magnets constitutes the free-electron laser medium (Fig. 19.2c). A radiation field propagating in $+z$ direction is amplified.

The length L of the optical resonator is larger than the length L_w of the wiggler; z_0 is the center of both the optical resonator and the wiggler. We consider the optical beam as a parallel beam within the active medium.

There are various comprehensive books and articles about free-electron lasers; *see*, for instance, [166, 167, 312].

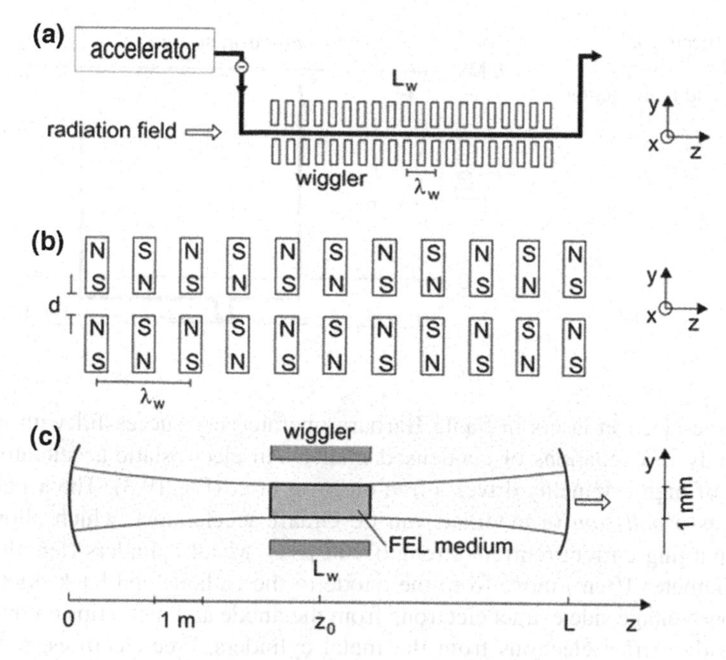

Fig. 19.2 Characterization of the free-electron laser. **a** Electron beam and wiggler. **b** Section of a wiggler. **c** Optical resonator with free-electron laser (FEL) medium

19.2 Free-Electron Laser Arrangements

Operation of a free-electron laser places great demands on the accelerator.

- *High current density*. The current density of a quasi-continuous free-electron laser should lie in the range 1–10 A. At smaller current, the gain is too small to reach laser threshold. At larger current, electron interaction (Coulomb repulsion) destroys the quality of the electron beam. The current density can be much larger in the case that the pulses are very short—1 kA for instance or even more for electron pulses of 0.1 ps duration.
- *High quality of the electron beam*. The energy distribution in the electron beam should be narrow, for instance 0.1–1% of $E_{el,0}$. The divergence of the electron beam should be small.

Almost all free-electron lasers make use of electron pulses, produced by linear accelerators, and therefore generate radiation pulses. There is one exception: free-electron lasers at the University of California, Santa Barbara (United States) produce quasi-continuous radiation.

The first free-electron lasers was operated at Stanford University [168, 169] and the second at the research center in Santa Barbara. After the first demonstrations of free-electron lasers, many research centers began to develop free-electron lasers.

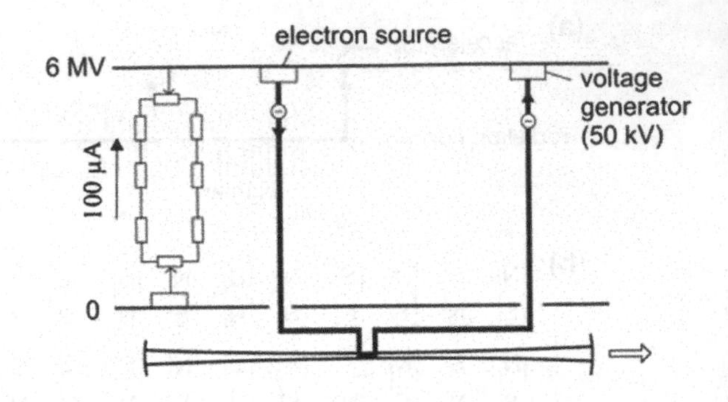

Fig. 19.3 Electrostatic accelerator with energy recovering and free-electron laser

The free-electron lasers in Santa Barbara operate very successful with respect to the study of excitations of condensed matter. An electrostatic accelerator with a design of high originality drives a free-electron laser (Fig. 19.3). The accelerator (6 MeV) is a *pelletron* (a modified van de Graaff accelerator), which allows for a high charging current (current strength 100 μA). Metal cylinders (length about 10 cm, diameter 10 cm) move from the anode to the cathode and back. Metal tips on the low-voltage side extract electrons from the anode and metal tips on the high-voltage side extract electrons from the metal cylinders. Free-electrons, produced with an electron gun at the cathode, are accelerated, pass through laser resonator and are then recovered by a deceleration system. The recovery of the electrons makes it possible that, during a certain time (∼30 μs), a constant current of high strength (∼2 A) is flowing through the free-electron laser. A voltage (∼50 kV), produced with a generator and a power supply at the cathode, accelerates the electrons to compensate energy loss due to generation of radiation in the free-electron laser. The wiggler consists of samarium-cobalt magnets. Radiation pulses of 30 μs duration are generated at a repetition rate of few Hz. The radiation has a very high degree of monochromaticity. The free-electron lasers generate millimeter wave and far infrared radiation of a power up to 10 kW. The free-electron lasers are continuously tunable. The free-electron laser laboratory in Santa Barbara is available as user facility.

Another user facility, available in a laboratory in Dresden (Germany), produces pulsed far infrared and infrared radiation also by using several free-electron lasers. These free-electron lasers are driven by linear accelerators (LINACs) with 15–45 MeV electron energy and peak currents up to 100 A. The lasers generate radiation pulses (duration 0.1 ps or longer) with pulse powers of 100 kW to 1 MW.

The SASE free-electron laser principle has been successfully demonstrated [170] for visible (wavelength 530 nm) and near ultraviolet radiation (385 nm); operation of SASE free-electron lasers generating radiation of GW power of VUV radiation in the 100 nm range (tuned in the range 95–105 nm) [171] and of radiation at 32 nm [172–174] has been reported. A SASE free-electron laser is a mirrorless laser; SASE = self-amplified spontaneous emission.

Presently (2016), X-ray SASE free-electron lasers experience an exciting development. X-ray SASE free-electron lasers are operating at wavelengths from 100 nm to less than 0.1 nm; for surveys, *see* [312, 313]. The lasers produce pulses of a few to 100 fs duration and peak powers of 10–100 GW. There is a wide field of applications in physics, chemistry, and biology. An example is the study of nonlinear X-ray excitation of atoms [314, 315]. X-ray SASE free-electron lasers generate radiation with transverse coherence. Longitudinal coherence is absent according to the initial process of spontaneously generated radiation; in comparison, in a free-electron laser, spontaneously generated radiation also initiates oscillation, however, feedback of radiation during the onset of laser oscillation by the laser resonator results in complete coherence of the free-electron laser radiation.

Seeded free-electron lasers, another type of single-pass free-electron lasers, provide fully coherent radiation; in a seeded free-electron laser, a coherent radiation pulse is injected, together with an electron pulse, into a first wiggler of a series of many wigglers in turn. Operation of seeded free-electron lasers has been demonstrated. High harmonic radiation pulses of mode locked titanium-sapphire lasers have been used to generate radiation at discrete frequencies in the 20–100 nm range [313, 317–321].

Extension of the wavelength of free-electron lasers by using harmonics of the fundamental frequency of a free-electron laser is, in principle, possible by operation of a free-electron laser at a harmonic of the fundamental oscillation frequency of the electrons in a laser; for recent studies, *see* [322, 323].

During the flight through a wiggler, an electron permanently loses energy due to stimulated emission of radiation. Therefore, the oscillation frequency decreases along the wiggler. In a *tapered wiggler*, the oscillation frequency is kept constant. A constant oscillation frequency can be achieved either by reducing, along the $+z$ direction, the wiggler period or by increasing the distance d between the wiggler magnets. Light guiding in the active medium of a SASE free-electron laser avoids spreading of the radiation beam.

The radiation of free-electron lasers is suitable for research (e.g., in solid state physics and biophysics), for technical applications (e.g., structuring).

19.3 Free-Electron Oscillation: Resonance Frequency and Spontaneously Emitted Radiation

In this section, we relate the resonance frequency of the free-electron oscillation and the electron energy. We will obtain, additionally, the spectrum of spontanously emitted radiation. We will proceed in three steps.

- First step: we study generation of spontaneously emitted radiation and take the relativistic Doppler effect into account. We will obtain the oscillation frequency of the free-electron oscillations,

$$\nu_0 = \frac{2c\gamma^2}{\lambda_w}, \tag{19.9}$$

and a spectrum of spontaneously emitted dipole radiation, centered at ν_0.
- Second step: we interpret the emission of radiation as relativistic Compton scattering. We will find that the analysis also leads to the frequency ν_0.
- Third step: we determine the modified resonance frequency ν_0 as it occurs in free-electron lasers.

An oscillating free-electron moving at a relativistic velocity emits radiation at a frequency according to the relativistic Doppler effect. The radiation observed in the laboratory frame has an extreme forward direction.

The motion of an electron through a wiggler takes the time (Fig. 19.4)

$$t = N_{\mathrm{w}}\lambda_{\mathrm{w}}/\mathrm{v}. \tag{19.10}$$

N_{w} is the number of wiggler periods, λ_{w} the wiggler period, and v the electron velocity along the wiggler. In the same time in which an electron traverses the wiggler, the electron emits an electromagnetic wave packet with N_{w} oscillation cycles. A wave packet of radiation emitted in z direction has the spatial length

$$(c - \mathrm{v})t = N_{\mathrm{w}}\lambda_{\mathrm{w}}(1 - \beta)/\beta, \tag{19.11}$$

where $\beta = \mathrm{v}/c$. Since $(c - \mathrm{v})t = N_{\mathrm{w}}\lambda$, where λ is the wavelength of the radiation, it follows that

$$N_{\mathrm{w}}\lambda = N_{\mathrm{w}}\lambda_{\mathrm{w}}(1 - \beta)/\beta \tag{19.12}$$

or, with $\beta \approx 1$,

$$\lambda \approx (1 - \beta)\lambda_{\mathrm{w}}. \tag{19.13}$$

With $1/\sqrt{1 - \beta^2} = \gamma$ and

$$\frac{1}{1 - \beta^2} = \frac{1}{(1 - \beta)(1 + \beta)} \approx \frac{1}{2(1 - \beta)} = \gamma^2, \tag{19.14}$$

we find

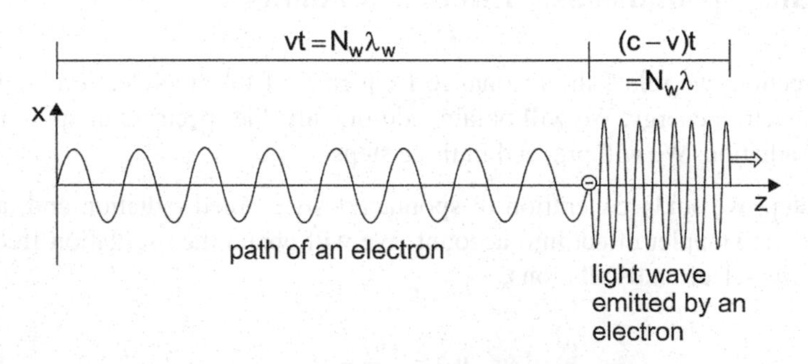

Fig. 19.4 Path of an electron through a wiggler field and dipole radiation

$$\lambda = \frac{\lambda_W}{2\gamma^2}. \tag{19.15}$$

The number of oscillation cycles in a wave packet is N_W. The rectangular envelope of the field has the temporal length

$$\Delta t = N_W \lambda / c. \tag{19.16}$$

It follows that the power spectrum of the spontaneously emitted radiation has the shape

$$\begin{aligned} S_\omega(\omega)/S_{\max} &= \left| \int_0^{\Delta t} e^{-i(\omega-\omega_0)t} \mathrm{d}t \right|^2 \\ &= \left| \frac{\sin[2\pi N_W(\omega-\omega_0)/(2\omega_o)]}{2\pi N_W(\omega-\omega_0)/(2\omega_0)} \right|^2 = \left| \frac{\sin X/2}{X/2} \right|^2. \end{aligned} \tag{19.17}$$

$S_\omega(\omega)$ is the spectral distribution of the radiation, $S_{\max}$ the maximum of the spectral distribution, $\omega_0 = 2\pi c/\lambda$ the resonance frequency, and

$$X = 2\pi N_W \left(\frac{\omega}{\omega_0} - 1 \right). \tag{19.18}$$

The halfwidth of the $[\sin(X/2)/(X/2)]^2$ curve (Fig. 19.5) is equal to $\delta X_0 = 5.7$. This yields an expression for the halfwidth $\Delta\omega_0$ of the spectrum of spontaneously emitted radiation,

$$\frac{\Delta\omega_0}{\omega_0} \approx \frac{1}{N_W}. \tag{19.19}$$

The halfwidth corresponds, for $N_W = 100$, to 1% of ω_0. The oscillations of the electrons are not synchronized to each other. Therefore, the phases of the wave packets, emitted by different electrons, are statistically distributed. A pulse containing N_p electrons leads to an average amplitude that is proportional to $\sqrt{N_p}$. The power is proportional to N_p, which itself is proportional to the strength I of the electron current. Thus, the power of spontaneously emitted radiation is proportional to the current strength. Since the electrons are moving at a relativistic velocity near the speed of light, the emission occurs into a narrow cone directed along the direction of the electron beam.

In the first step, we obtained an important result with respect to a beam of electrons, which have a narrow energy distribution. A beam of electrons traversing a periodic transverse magnetic field generates radiation by spontaneous emission:

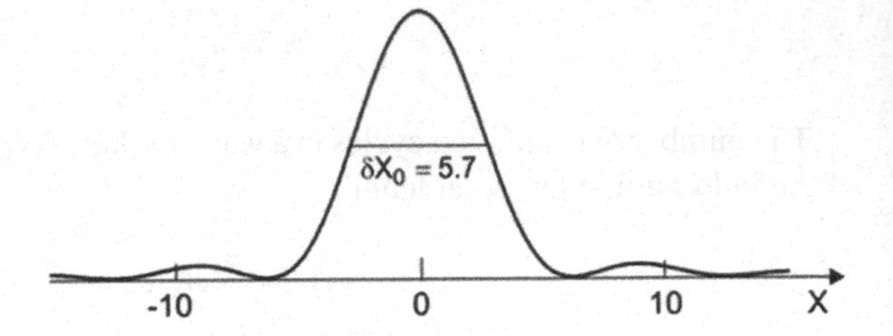

Fig. 19.5 Spectral distribution of radiation spontaneously emitted by relativistic electrons in a periodic magnetic field of small strength ($K_{el} = 1$)

- The spectrum of spontaneously emitted radiation has a maximum at the frequency $\nu_0 = 2c\gamma^2/\lambda_w$; this is the oscillation frequency (resonance frequency) of the free-electron oscillations if $K_w << 1$.
- The spectrum shows a frequency distribution varying as $(\sin X/X)^2$, where $X = N_w(\nu - \nu_0)/\nu_0$.
- The relative width of the spectrum is $1/N_w$.

In the second step, we attribute the origin of free-electron laser radiation to *Compton scattering*. We perform a relativistic transformation from the laboratory frame to the electron frame. We then describe the Compton scattering in the electron frame. The relativistic transformation back into the laboratory frame yields the wavelength of the radiation (Fig. 19.6);

- *Laboratory frame* (a). An electron has the velocity v and moves toward the wiggler field of wavelength λ_w.
- *Electron frame, before scattering* (b). The electron experiences an electromagnetic field of the wavelength $\lambda' = \lambda_w/(2\gamma)$. The wavelength λ' is smaller than λ_w because of the relativistic length contraction. The factor 1/2 occurs since the wiggler field in the laboratory frame is a static field (Weizsäcker-Williams theorem [59]).
- *Electron frame, scattering process* (c). Compton scattering of the electromagnetic radiation at the electrons reverses the direction of the electromagnetic field. Therefore, the wavelength λ' is the same before and after scattering (supposed that the recoil energy of an electron can be neglected).
- *Laboratory frame* (d). The transformation back into the laboratory frame leads, due to the relativistic contraction of length, to the wavelength $\lambda = \lambda_w/(2\gamma^2)$.

In the first two steps, we assumed that an electron has almost no energy of transverse motion and that an electron in the wiggler has therefore the γ value of a free-electron. We come to the third step. We assume that the average energy of longitudinal motion of an electron in the wiggler is notably reduced, in comparison with the free flight, because of the occurrence of transverse oscillations. The equation of motion in x direction is given by:

$$\gamma m_0 \frac{\mathrm{d}v_x}{\mathrm{d}t} = q v_z B_y \approx q c B_w \sin\left(\frac{2\pi}{\lambda_w} z\right). \tag{19.20}$$

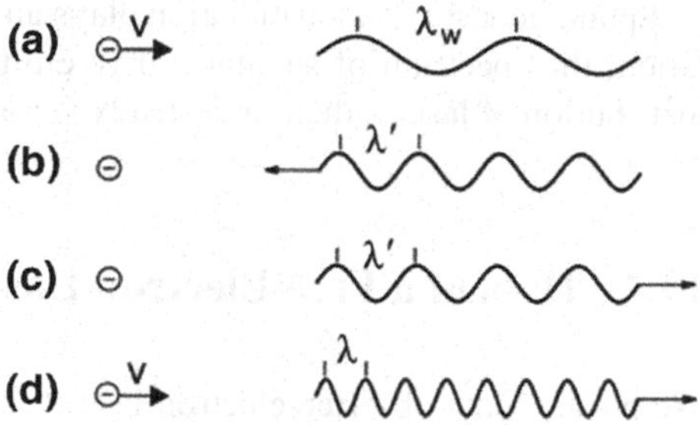

Fig. 19.6 Stimulated Compton scattering

The value of γ is a constant for electrons in the magnetic field; it is slightly smaller than for the electrons outside the magnetic field. Integration leads to

$$\mathrm{v}_x = \frac{qcB_w\lambda_w}{2\pi\gamma m_0}\cos\left(\frac{2\pi}{\lambda_w}z\right) = \frac{-K_w c}{\gamma}\cos\left(\frac{2\pi}{\lambda_w}z\right), \quad (19.21)$$

$$\mathrm{v}_z = c\sqrt{1-\frac{1}{\gamma^2}-\frac{\mathrm{v}_x^2}{c^2}} \approx \bar{\mathrm{v}}_z - \frac{K_w^2 c}{4\gamma^2}\cos\left(\frac{4\pi}{\lambda_w}z\right), \quad (19.22)$$

where

$$\bar{\mathrm{v}}_z = c\left(1-\frac{1+K_w^2/2}{2\gamma^2}\right) \quad (19.23)$$

is an average velocity in z direction and $K_w = e\lambda_w B_w/2\pi m_0 c$; the equations (19.20) through (19.23), are treated in Problems to this chapter. The average velocity in z direction decreases with increasing wiggler parameter. The average velocity determines the effective Lorentz parameter,

$$\bar{\gamma} = \frac{1}{\sqrt{1-\bar{\mathrm{v}}_z^2/c^2}} = \frac{\gamma}{\sqrt{1+K_w^2/2}}. \quad (19.24)$$

The transverse velocity is equal to

$$\mathrm{v}_x = \mathrm{v}_0 \cos\omega_0 t. \quad (19.25)$$

The peak velocity of the transverse motion,

$$\mathrm{v}_0 = cK_w/\gamma, \quad (19.26)$$

is small compared with the speed of light. We will discuss the velocities in more detail in Sect. 19.5.

Spontaneously emitted radiation plays an important role at the onset of laser oscillation; the spectrum of spontaneously emitted radiation determines the frequency distribution of laser radiation at steady state oscillation (Sect. 19.12).

19.4 Data of a Free-Electron Laser

We present data of a free-electron laser (Table 19.1). The data, oriented at infrared free-electron lasers, are concerning the laser resonator, the high frequency field (= optical field) in a resonator, the electron beam and the free-electron laser medium.

We choose the laser wavelength $\lambda = 5\,\mu\text{m}$ (laser frequency $\nu = 6 \times 10^{13}$ Hz; photon energy $h\nu = 0.25$ eV). A resonator of length $L = 10$ m is suitable as laser resonator; the beam waist has a radius $r_{\text{u},0} = 3.5$ mm. The lifetime of a photon in the resonator is $\tau_p = (2L/c)/(1 - R)$. We assume that the reflectivity of the output coupling mirror is $R = 0.9$.

We assume that the wiggler has a length of $L_\text{w} = 1.2$ m and a period of $\lambda_\text{w} = 2.4$ cm and that the number of wiggler periods is $N_\text{w} = 50$. The time it takes the electrons and the radiation to pass through the wiggler field is the transit time $t_\text{tr} = L_\text{w}/c$ (= 4 ns). The resonance frequency ν_0 of the free-electron oscillations is only slightly larger than the laser frequency ν. The gain bandwidth is 6×10^{11} Hz. A current $I_\text{el} = 100$ A corresponding to a current density $j = I_\text{el}/(\pi r_{\text{u},0}^2 ec) = 6.6 \times 10^4\,\text{A m}^{-2}$, an electron density $N_0 = j_\text{dc}/(ec) = 8 \times 10^{16}\,\text{m}^{-3}$, and a rate of electrons traversing the active medium $r_\text{el} = I/e\,(= 1.2 \times 10^{19}\,\text{s}^{-1})$. We assume that the small-signal gain is $G - 1 = 0.5$.

We assume that the output power P_out has a value of ~1 MW and that this output power corresponds to laser oscillation at saturation of the high frequency field within the laser resonator. Accordingly, the saturation of the high frequency field within the laser resonator. Accordingly, the saturation field amplitude is $A_\text{sat} = A_\infty \sim 4 \times 10^7\,\text{V m}^{-1}$ and the density of photons in the resonator is $Z_\infty = \varepsilon_0 A_\infty^2/2h\nu\,(\sim 2 \times 10^{22}\,\text{m}^{-3})$. The total photon output coupling rate is $r_\text{ph} = P_\text{out}/h\nu = (2.5 \times 10^{25}\,\text{s}^{-1})$. The output power,

$$P_\text{out} = \frac{Z_\infty}{\tau_\text{p}} \pi r_{\text{u},0}^2 L h\nu = (1/2)\pi r_{\text{u},0}^2 (1 - R) A_\infty^2, \tag{19.27}$$

of a free-electron laser is independent of the length of the active medium—because of field saturation. This is a main difference between a free-electron laser and a conventional laser: the density Z_∞ of photons in the resonator of a conventional laser and P_out increase if we increase the length of the active medium (and double the total pump strength).

Each electron traversing the active medium generates a large number of photons: the number of photons generated (per electron) by stimulated emission is s_stim. The time between two stimulated emission processes is $\tau_\text{stim} = t_\text{tr}/s_\text{stim}$ $(= 2 \times 10^{-14}$ s).

Table 19.1 Data of a free-electron laser

Quantity	Value	Notation
λ	5 μm	Laser wavelength
ν	6×10^{13} Hz	Laser frequency
$h\nu$	0.25 eV	Photon energy
L	10 m	Resonator length
$\tau_p = (2L/c)/(1-R)$	7×10^{-7} s	Lifetime of a photon in the resonator
$r_{u,0}$	3.5 mm	Radius of the optical beam waist = Radius of the electron beam
L_w	1.2 m	Length of wiggler
λ_w	2.4 cm	Wiggler period
N_w	50	Number of wiggler periods
ν_0	6×10^{13} Hz	Resonance frequency
$\Delta\nu_g = \nu_0/2N_w$	6×10^{11} Hz	Gain bandwidth
$t_{tr} = L_w/c$	4×10^{-9} s	Transit time
K_w	0.7	Wiggler strength
I_{el}	100 A	Current strength
$j_{el} = I/\pi r_{u,0}^2$	2.5×10^6 A m^{-2}	Current density
$N_0 = j_{el}/ec$	5×10^{16} m^{-3}	Electron density
$r_{el} = I/e$	6×10^{20} s^{-1}	Electron rate
$G - 1$	0.5	Small-signal gain
R	0.9	Reflectivity of output mirror
$G_\infty - 1 = 1/R - 1$	0.1	Steady state gain
$A_\infty = A_{sat}$	4×10^7 V m^{-1}	Amplitude of saturation field in the resonator
$P_{out} = (1/2)(1-R)c\epsilon_0 A_\infty^2 \pi r_{u,0}^2$	1 MW	Output power
$r_{ph} = P_{out}/h\nu$	2.5×10^{24} s^{-1}	Photon output coupling rate
s_{stim}	2.1×10^5	Emission processes per electron
$\tau_{stim} = t_{tr}/s_{stim}$	2×10^{-14} s	Time between two stimulated emission processes
$E_{el,0}$	50 MeV	Initial electron energy
$P_{el} = r_{el} E_{el,0}$	5×10^9 W	Power of electron beam
$\eta_P = P_{out}/P_{el}$	2×10^{-4}	Power conversion efficiency
$E_{el,0} - E_{el,\infty} = s_{stim} h\nu$	10 keV	Loss of energy of an electron

A free-electron laser for generation of radiation in the range around 5 μm can be realized by using a beam of electrons of an initial energy $E_{\rm el,0}$ near 50 MeV (at a wiggler period $\lambda_{\rm w} = 2.4\,{\rm cm}$). The power efficiency, $\eta_{\rm P} = P_{\rm out}/P_{\rm el}$ on the order of 0.1%. The loss of energy of an electron, $E_{\rm el,0} - E_{\rm el,\infty} = s_{\rm stim} h\nu$ (= 500 keV) due to stimulated emission corresponds to 10^{-2} $E_{\rm el,0}$. [The loss cannot be larger, because an electron cannot contribute to gain if its energy loss relative to the initial energy is larger than $1/(2N_{\rm w})$. There is another limitation: the saturation field is limited, *see* Sect. 19.8.]

The data of Table 19.1 characterize a pulsed laser and correspond to peak values. We obtain an average output $P_{\rm out,av} = P_{\rm out} t_{\rm p} r_{\rm p}$, where $t_{\rm p}$ is the pulse duration and $r_{\rm p}$ the pulse repetition rate; for example, $t_{\rm p} = 0.1\,{\rm ps}$, $r_{\rm p} = 10^7\ {\rm s}^{-1}$ and $P_{\rm out,av} = 2\ {\rm W}$.

Laser operation requires that the amplitude of the high frequency field in the resonator is smaller than the damage field of the laser mirrors. A short pulse duration is therefore necessary to reach the amplitude A_∞.

19.5 Rigid Coupling of Transverse and Longitudinal Oscillation of an Electron

Entering the wiggler, an electron is submitted to the Lorentz force of the transverse magnetic field. The electron starts a transverse oscillation. The transverse oscillation is rigidly coupled to a longitudinal oscillation (Fig. 19.7a). The orbit of the electron has the shape of an "eight" (*see* Problem 19.14). The x component, v_x, of the velocity of the electron is rigidly coupled to the z component, v_z. The coupling is mediated by the Lorentz force. The x component is given by (19.21) and the z component by (19.22). The longitudinal velocity oscillates twice as fast as the transverse velocity (Fig. 19.8b).

The transverse velocity of an electron is small compared with the speed of light. We can therefore describe the transverse motion nonrelativistically. An observer in

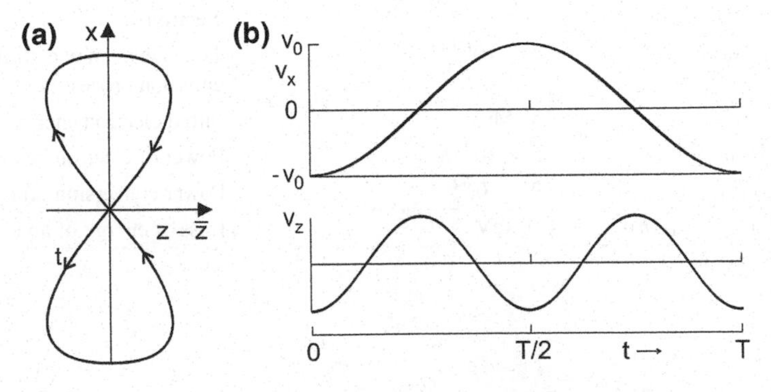

Fig. 19.7 Rigidly coupled transverse and longitudinal motion of an electron. **a** Displacements in the x, z plane *see* Problems to this chapter. **b** Temporal variations of transverse and longitudinal velocities

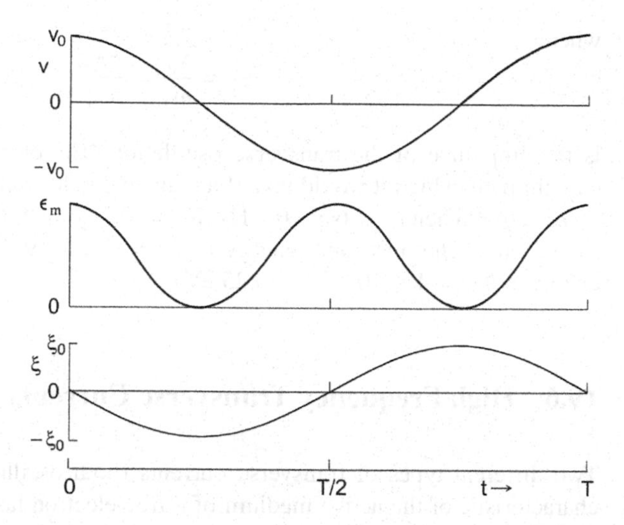

Fig. 19.8 Transverse oscillation of an electron; velocity, energy, and displacement

the laboratory system registers the following (Fig. 19.8). During the flight through the wiggler, an electron performs a transverse oscillation with the electron oscillation frequency

$$\omega_0 = \frac{1}{1+K_\mathrm{w}^2}\frac{4\pi\, c\gamma^2}{\lambda_\mathrm{w}}. \tag{19.28}$$

The transverse velocity is equal to

$$\mathrm{v} = \mathrm{v}_0 \cos \omega_0\, t;\ \mathrm{v}_0 = c\, K_\mathrm{w}/\gamma. \tag{19.29}$$

(We omit the subscript x.) The energy of transverse motion is equal to

$$\epsilon = \frac{m}{2}\mathrm{v}^2 = \epsilon_\mathrm{m} \cos \omega_0^2 t, \tag{19.30}$$

where

$$\epsilon_\mathrm{m} = \frac{m_0}{2}\mathrm{v}_0^2 = \frac{m_0 c^2 K_\mathrm{w}^2}{\gamma^2} \tag{19.31}$$

is the maximum kinetic energy of oscillation. It is proportional to the square of the wiggler parameter and inversely proportional to the square of the Lorentz parameter. It follows from the velocity that the trajectory for the transverse displacement is given by

$$\xi = -\xi_0 \sin \omega_0\, t, \tag{19.32}$$

where

$$\xi_0 = \frac{\mathrm{v_p}}{\omega_0} = \frac{cK_\mathrm{w}}{\omega_0 \gamma} \tag{19.33}$$

is the amplitude of the transverse oscillation. The electron oscillates around the straight path, which it would take if a magnetic field were absent.

Example When $\nu_0 = 6 \times 10^{13}$ Hz, $K_\mathrm{w} = 0.7$, $\gamma = 100$, then $\mathrm{v_p} = 7 \times 10^{-3}c$ and $\xi_0 = 11$ nm. The maximum energy is $\epsilon_\mathrm{m} = 2.2 \times 19^{-18}$ J (~14 eV). The photon energy is $\hbar\omega = 4 \times 10^{-20}$ J (~0.25 eV).

19.6 High Frequency Transverse Currents

Two different types of transverse currents (both oscillating in $\pm x$ direction) are characteristic of the active medium of a free-electron laser. We now introduce two types of currents.

- *Transverse electron current* $I_\mathrm{e}(\omega_0)$. The transverse electron current is joined with the transverse oscillation of a single electron in the wiggler field, $I_\mathrm{e} = -e\mathrm{v} = -e\mathrm{v}_0 \cos \omega_0 t$, where $\mathrm{v}_0 = cK_\mathrm{W}/\gamma$ is the amplitude of the transverse current and ω_0 the oscillation frequency.
- *Transverse modulation current* $I_\mathrm{mod}(\omega)$. The modulation current is a consequence of the interaction of the electron oscillation with a high frequency electric field of frequency ω. The interaction is due to phase modulation of the electron oscillation.

The modulation current of an electron leads to a transverse dynamical conductivity of an ensemble of electrons in a free-electron laser medium.

A high frequency electric field (oriented along x) of frequency ω and amplitude A,

$$\tilde{E}(\omega) = A\,\mathrm{e}^{\mathrm{i}\omega t}, \tag{19.34}$$

exerts an electric force on an electron. This causes a change of the trajectory of the electron in the wiggler. We will describe the action of the field as a phase modulation of the transverse velocity v (and accordingly of the transverse current I_e). The phase modulated oscillation of the electron leads to a transverse *modulation velocity*

$$\mathrm{v_{mod}}(\omega) \equiv \tilde{\mathrm{v}}(\omega) = \mathrm{v}_1(\omega) - \mathrm{i}\mathrm{v}_2(\omega) \tag{19.35}$$

at the frequency ω of the high frequency electric field. The modulation velocity depends linearly on the electric field:

$$\tilde{\mathrm{v}}(\omega) = \tilde{\eta}(\omega)\,\tilde{E}(\omega), \tag{19.36}$$

where $\tilde{\eta}(\omega)$ is the complex dynamical mobility of an electron for transverse motion under the action of a high frequency electric field. We can characterize an ensemble of electrons performing modulated oscillations at the frequency ω by a *transverse current density* $\tilde{j}(\omega)$. Assuming that electrons of an ensemble are oscillating synchronously, we obtain a transverse current density at a location of a free-electron laser medium that is given by

$$\tilde{j}(\omega) = N_0\, q\, \tilde{\mathrm{v}}(\omega) = N_0\, q\, \tilde{\eta}(\omega)\, \tilde{E}(\omega) \tag{19.37}$$

or

$$\tilde{j}(\omega) = \tilde{\sigma}(\omega)\, \tilde{E}(\omega). \tag{19.38}$$

N_0 is the electron density, $\tilde{\sigma}(\omega) = N_0\, q\, \tilde{\eta}(\omega)$ is the *transverse dynamical conductivity* of the free-electron laser medium at the frequency ω, and $q = (-e)$ is the electron charge.

We thus describe the linear response of the free-electron laser medium to a high frequency electric field by the transverse modulation current density in the free-electron laser medium (Fig. 19.9b),

The linear response is characterized by the complex high frequency conductivity

$$\tilde{\sigma}(\omega) = \sigma_1(\omega) - \mathrm{i}\, \sigma_2(\omega). \tag{19.39}$$

The current density,

$$\tilde{j}(\omega) = j_1(\omega) - \mathrm{i}\ j_2(\omega) = \sigma_1(\omega) A \cos \omega t - \mathrm{i}\sigma_2(\omega) A \sin \omega t, \tag{19.40}$$

has a real part

$$j_1(\omega) = \sigma_1(\omega) A \cos \omega t \tag{19.41}$$

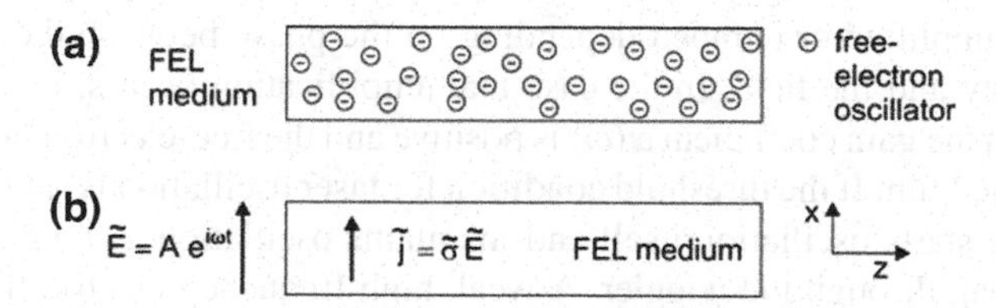

Fig. 19.9 Response of a free-electron laser medium to a high frequency field. **a** Free-electron laser medium: ensemble of free-electron oscillators. **b** Field and current density

that has the same phase as the field. Gain occurs if $\sigma_1(\omega) < 0$. The (negative) imaginary part

$$j_2(\omega) = \sigma_2(\omega) A \sin \omega t \tag{19.42}$$

has a phase of 90° relative to the field and corresponds to an inductive current. The current is phase shifted relative to the field,

$$j(\omega) = \sigma(\omega) A \cos[\omega t + \varphi(\omega)], \tag{19.43}$$

where

$$\sigma = \sqrt{\sigma_1^2 + \sigma_2^2} \tag{19.44}$$

is the absolute value of the conductivity and where φ is the phase between current density and field. The phase follows from the relation

$$\tan \varphi = \sigma_2/\sigma_1. \tag{19.45}$$

We assume that the optical beam has the same lateral extension as the electron beam and that the electron density N_0 and the amplitude A of the high frequency field do not vary over the cross section. Our goal is to derive $\sigma_1(\omega)$ and $\sigma_2(\omega)$. gain coefficient. From the real part of the high frequency transverse conductivity, we obtain the (small-signal) gain coefficient:

$$\alpha(\omega) = -\frac{1}{\varepsilon_0 c}\sigma_1(\omega). \tag{19.46}$$

Here, we give a short description, how a free-electron laser works according to the modulation model presented in the next section. Single electrons in the wiggler perform oscillations at the resonance frequency ω_0. A high frequency electric field of frequency ω modulates the oscillations. The modulation results in a transverse modulation current density $\tilde{j}(\omega)$ at the frequency of the electric field. The electric field can be amplified or damped-depending on the phase between the modulation current density and the field. In the case that amplification occurs, i.e., if $\sigma_1(\omega)$ is negative, then the gain coefficient $\alpha(\omega)$ is positive and the free-electron laser medium is an active medium. If the threshold condition for laser oscillation is satisfied, a free-electron laser starts oscillation itself and maintains oscillation as long as electrons are propagating through the wiggler. A weak high frequency electric field initiates oscillation of a free-electron laser. Mutual interaction of the field and the transverse modulation current leads to growth of both the field and the modulation current. The initial high frequency electric field is created by spontaneous emission of radiation by the transversely oscillating electrons.

Since the transverse dynamical conductivity is a local quantity, the gain coefficient $\alpha(\omega)$ is also a local quantity. It is a goal to determine the dependence of the gain coefficient α on frequency, location, and on time.

Introduction of a modulation current density for characterization of an active medium of a free-electron laser medium that is based on monopole oscillations of electrons is equivalent to the introduction of a polarization (Chap. 9) for characterization of an active medium of a laser that is based on dipole oscillations in an active medium (containing two-level atomic systems).

19.7 Modulation Model of the Free-Electron Laser

An electron that entered the wiggler performs a transverse oscillation, in x direction, with the transverse velocity

$$\mathrm{v} = \mathrm{v}_0 \cos \omega_0 t, \tag{19.47}$$

where

$$\mathrm{v}_0 = \frac{cK_\mathrm{w}}{\gamma} \tag{19.48}$$

is the amplitude of the transverse velocity and ω_0 the frequency of the oscillation.

We now assume that an oscillating electron travels along z together with the high frequency field E, which is oriented along x,

$$E(\omega) = A \cos \omega t. \tag{19.49}$$

A is the amplitude and ω the frequency of the field. The high frequency field causes a phase modulation of the electron oscillation. The electron has an instantaneous frequency

$$\omega_\mathrm{inst} = \omega_0 + \kappa A \cos \omega t, \tag{19.50}$$

where κ is the *coupling strength*. We will show that κ is proportional to the wiggler parameter K_w. We relate the instantaneous frequency to a phase according to the relationship $\omega = \partial\varphi/\partial t$. Integration leads to the instantaneous phase

$$\varphi_\mathrm{inst} = \omega_0(t - t_0) + \mu(\sin \omega t - \sin \omega t_0), \tag{19.51}$$

where t_0 is a starting time of the oscillation. The instantaneous transverse velocity is given by

$$\mathrm{v}_\mathrm{inst}(t, t_0) = \mathrm{v}_0 \cos\left[\omega_0(t - t_0) + \mu(\sin \omega t - \sin \omega t_0)\right], \tag{19.52}$$

where

$$\mu = \frac{\kappa A}{\omega} \tag{19.53}$$

is the *modulation index*. It is proportional to the coupling strength and to the amplitude of the high frequency electric field, and inversely proportional to the frequency. The modulation index is a measure for the deviation of the instantaneous phase from the phase $\omega_0 t$.

In the case that $\omega < \omega_0$, the electron oscillates faster than the field (Fig. 19.10). Accordingly, the phase between the electron oscillation and the field increases continuously with time—we denote this as *dephasing*. The continuous increase of the phase difference can be interrupted by a stimulated emission process. A stimulated emission process occurs preferentially when the phase between velocity and electric field is zero. A stimulated emission process changes the phase of the electron oscillation. After a stimulated emission process, the phase difference grows until a new stimulated emission process takes place. The time between two subsequent stimulated emission processes is the *dephasing time* τ; we suppose that the electron follows the Lorentz force by the static magnetic field adiabatically. Stimulated emission of a photon by an oscillating electron does not change the frequency ω of the high frequency field and the frequency ω_0 of the electron oscillation is the same before and after stimulated emission of a photon. The amplitude v_0 of the transverse velocity remains also unchanged. Stimulated emission processes in turn lead to synchronization of the electron oscillation to the electric field. We thus find that the transverse oscillation of the electron mediates gain for the field. Energy of transverse motion is converted to energy of radiation.

The free-electron oscillation of a free-electron in a periodic magnetic field represents a monopole oscillation (Sect. 4.12). Interaction with an external high frequency field changes the phase of the oscillation but not the amplitude.

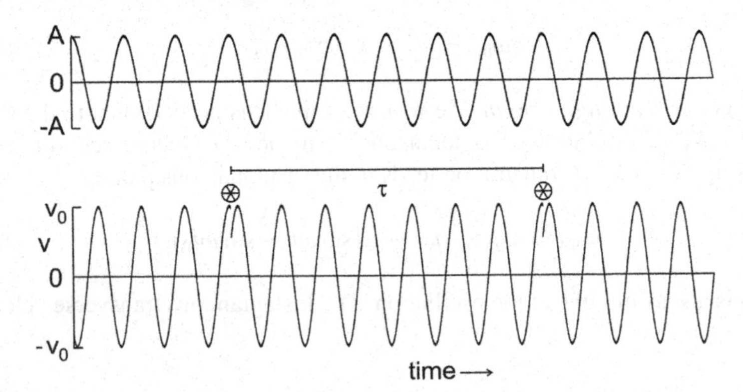

Fig. 19.10 Electric field and transverse velocity of an electron; star, stimulated emission process; τ, dephasing time

The probability that an oscillating electron does not undergo a stimulated emission process in the time interval $t - t_0$, t is given by

$$p(t, t_0) = \mathrm{e}^{-(t-t_0)/\tau}, \tag{19.54}$$

where τ is the dephasing time. The temporal average over all starting times yields the *transverse modulation velocity*

$$\mathrm{v}_{\mathrm{mod}}(t) = \mathrm{v}_0 \frac{1}{\tau}\int_{-\infty}^{t} p(t, t_0)\mathrm{v}(t, t_0)\mathrm{d}t_0. \tag{19.55}$$

The modulation-velocity varies periodically with the period $T = 2\pi/\omega$ of the high frequency field. A Fourier transformation [15] yields the amplitude of the real part of the modulation-velocity:

$$\mathrm{v}_1 = \mathrm{v}_0 \sum_{\mathrm{n}=-\infty}^{+\infty} J_n(\mu)\frac{(\omega_0 + n\omega)\tau}{(\omega_0 + n\omega)^2\tau^2 + 1}. \tag{19.56}$$

J_n is the Bessel function of nth order.

The modulation velocity shows resonances for $n = -1, -2, -3, \ldots$, i.e., if

$$\omega = \omega_0/|n|. \tag{19.57}$$

In the following, we will study free-electron lasers that operate on the fundamental resonance ($n = -1$). We will not treat higher order resonance free-electron lasers; these are working on higher order resonances ($n = -2, -3, \ldots$).

Using the relationship $J_{-1} = J_1$, we obtain, for $n = 1$, the real part of the modulation velocity

$$\mathrm{v}_1 = -2\mathrm{v}_{1,m}\bar{g}_{\mathrm{L,disp}}(\omega), \tag{19.58}$$

where

$$\mathrm{v}_{1,\mathrm{m}} = (\mathrm{v}_0/2)J_1(\mu) \tag{19.59}$$

is the maximum modulation velocity, $J_1(\mu)$ is the Bessel function of first order with the argument μ, and

$$\bar{g}_{\mathrm{L,disp}} = \frac{(\omega_0 - \omega)\Delta\omega_0/4}{(\omega_0 - \omega)^2 + \Delta\omega_0^2/4} \tag{19.60}$$

is the "normalized" Lorentz dispersion function; normalized is the corresponding normalized Lorentz resonance function:

$$\bar{g}_{L,res}(\omega) = \frac{\Delta\omega_0^2/4}{(\omega_0 - \omega)^2 + \Delta\omega_0^2/4}. \tag{19.61}$$

The halfwidth of the resonance function is determined by the dephasing time according to the relationship

$$\Delta\omega_0 = \frac{2}{\tau}. \tag{19.62}$$

The real part of the modulation velocity provides the real part of the mobility, $\eta_1 = v_1/E$. It follows that the modulation current density of an ensemble of electrons is equal to

$$\sigma_1(\omega) = 2\sigma_{1,m}\bar{g}_{L,disp}(\omega), \tag{19.63}$$

$$\sigma_{1,m} = \frac{N_0 q v_0 J_1(\mu)}{2A}. \tag{19.64}$$

The gain coefficient for radiation propagating through an ensemble of oscillating electrons is equal to

$$\alpha'(\omega) = -\frac{1}{c\,\varepsilon_0}\sigma_1(\omega) = 2\alpha'_m \frac{(\omega_0 - \omega)\,\Delta\omega_0/4}{(\omega_0 - \omega)^2 + \Delta\omega_0^2/4}, \tag{19.65}$$

where (with $v_0 = c\,K_w/\gamma$ and $q = -e$)

$$\alpha'_m = \frac{\sigma_{1,m}}{c\,\varepsilon_0} = \frac{N_0\,e\,K_w}{2\,\varepsilon_0\,\gamma} \times \frac{J_1(\mu)}{A} \tag{19.66}$$

is the maximum gain coefficient. The dash indicates that α' is the gain coefficient of the electron ensemble in the electron frame. According to the dimension (m^{-1}) of α', we find for the gain coefficient α in the laboratory frame the relationship $\alpha = \gamma\,\alpha'$ or

$$\alpha(\omega) = 2\alpha_m \frac{(\omega_0 - \omega)\,\Delta\omega_0/4}{(\omega_0 - \omega)^2 + \Delta\omega_0^2/4}, \tag{19.67}$$

where

$$\alpha_m \equiv \alpha_{max} = \frac{N_0\,e\,K_w}{2\,\varepsilon_0} \times \frac{J_1(\mu)}{A}. \tag{19.68}$$

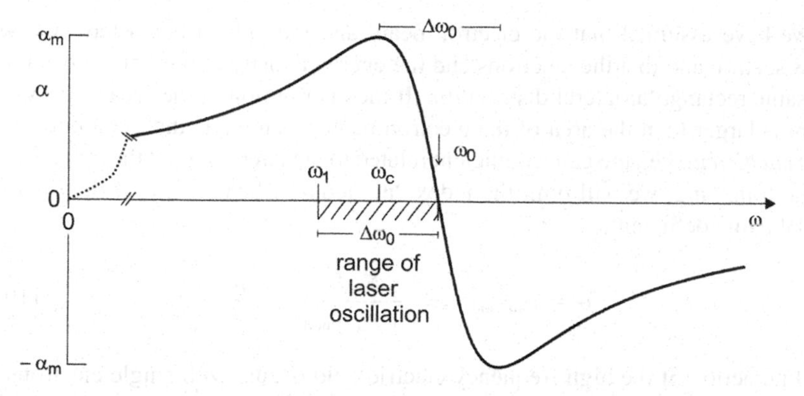

Fig. 19.11 Gain coefficient of a free-electron laser medium

We can expand the Bessel function, $J_1(\mu) = \mu/2$ for $\mu \leq 1$ and find

$$\alpha_{\mathrm{m}} = \frac{N_0 \, e \, K_{\mathrm{w}}}{4 \, \varepsilon_0} \times \frac{\mu}{A}. \tag{19.69}$$

The maximum gain coefficient is proportional to the ratio μ/A. The maximum gain coefficient is independent of $\Delta\omega_0$.

The gain coefficient curve (Fig. 19.11) is, in the vicinity of the resonance frequency, antisymmetric with respect to the resonance frequency ω_0; we suppose that $\Delta\omega_0 << \omega_0$. Gain occurs if $\omega < \omega_0$ and loss if $\omega > \omega_0$. The frequency distance between the extrema of the gain curve is equal to $\Delta\omega_0$. We will later show that laser oscillation occurs in a range (dashed) between ω_1 and ω_0 that has the width $\Delta\omega_0$. The center frequency of the range of laser oscillation is equal to ω_c. In a continuous wave free-electron laser, the laser frequency lies in the range $< \omega_1, \omega_0 >$. In a pulsed laser, all modes in the frequency range $< \omega_1, \omega_0 >$ are excited. (At small and large frequencies, the Lorentz dispersion function has to be replaced by the general Lorentz dispersion function. The dotted curve indicates the shape near $\omega = 0$.)

We can replace N_0 by the electron current density j_{el} by using the relationship

$$j_{\mathrm{el}} = N_0 \, e \, c \tag{19.70}$$

and find

$$\alpha_{\mathrm{m}} = \frac{e K_{\mathrm{w}} j_{\mathrm{el}}}{4 \, \varepsilon_0 \, e \, c} \times \frac{\mu}{A}. \tag{19.71}$$

With the current strength

$$I_{\mathrm{el}} = j_{\mathrm{el}} \, a_{\mathrm{mode}}, \tag{19.72}$$

we can write

$$\alpha_{\mathrm{m}} = \frac{K_{\mathrm{w}} I_{\mathrm{el}}}{4 \, c \, \varepsilon_0 \, a_{\mathrm{mode}}} \times \frac{\mu}{A}. \tag{19.73}$$

We have assumed that the electron beam and the optical beam have the same cross section and that the electrons and the energy density in the optical beam have the same rectangular lateral distribution. If the distribution of the area of the optical beam is larger than the area of the electron beam, then $\alpha_{\rm m}$ is the maximum *modal gain coefficient*, i.e., the gain coefficient related to the lateral size of the optical mode. In the following, we will omit the index "m,modal" when we are dealing with the modal gain coefficient,

$$\alpha \equiv \alpha_{\rm m,modal} = \frac{K_{\rm w} I_{\rm el}}{4\, c\, \varepsilon_0\, a_{\rm mode}} \times \frac{\mu}{A}. \tag{19.74}$$

(Interaction of the high frequency electric field occurs with single electrons and the response is linear. If the mode area of the optical mode is larger than the area of the electron beam, then the modal gain coefficient is almost independent of the shape of the distribution of the electrons in the electron beam; we suppose that the center of the electron beam coincides with the center of the optical beam.)

The Bessel function (Fig. 19.12, upper part) has a maximum (for $\mu = 1.8$). In the range of the modulation degree between $\mu \approx 1$ and $\mu = 1.8$, the gain coefficient has to be replaced by the differential gain coefficient

$$\alpha_{\rm d}(\omega) = \alpha_{\rm d,m} g_{\rm L,disp}, \tag{19.75}$$

where

$$\alpha_{\rm d,m} \equiv \alpha_{\rm d,max} = \frac{N_0\, e\, K_{\rm w}}{2\, \varepsilon_0\, A} \times \frac{{\rm d}J}{{\rm d}\mu} \tag{19.76}$$

is the maximum differential gain coefficient (Fig. 19.12, lower part). The maximum differential gain decreases strongly above $\mu \approx 1$, becomes zero for $\mu = 1.8$ and negative for $\mu > 1.8$. The amplitude of the high frequency electric field at which the gain coefficient is zero is given by the relationship $A(\mu = 1.8) = 1.8\, \omega_0/\kappa$.

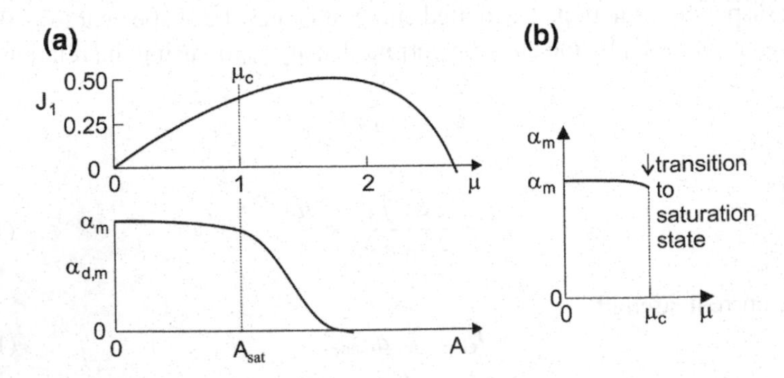

Fig. 19.12 Modulation index and maximum gain coefficient of a free-electron laser medium. **a** Bessel function and maximum differential gain coefficient; $\mu_{\rm c}$, critical modulation index and $A_{\rm sat}$, saturation field amplitude. **b** Transition to the saturation state

We will introduce (in the next two sections) a *saturation field amplitude* A_{sat} and attribute the saturation field amplitude to a *critical modulation index* μ_c. We will choose $\mu_c = 1$. If, at the onset of oscillation of a free-electron laser, the saturation field amplitude and thus the critical modulation index is reached, then the active medium goes over into the state of saturation (Fig. 19.12b). The maximum gain coefficient α_m is almost constant up to μ_c. Therefore, α_m is the small-signal gain coefficient as well as the large-signal gain coefficient. (However, we will differ between small-signal gain and large-signal gain; *see*, Sect. 19.12.)

19.8 Saturation Field and Energy of Distortion

In this section, we present a criterion that allows for an estimate of the saturation field amplitude A_{sat}. We study a free-electron laser at steady state oscillation. The field at a location in the laser resonator has the amplitude A_{sat},

$$E = A_{\text{sat}} \cos \omega t. \tag{19.77}$$

We consider the case that the phase between the velocity of an electron and the electric field is equal to zero,

$$\text{v} = \text{v}_0 \cos \omega_0 t; \; \text{v}_0 = c \, K_w/\gamma. \tag{19.78}$$

We ask the question: which is the maximum rate at which energy can be exchanged between an oscillating electron and an optical field? We introduce the criterion: *The work done by an electron during an oscillation cycle has an upper limit. At the upper limit, the work results in a strong distortion of the electron oscillation.* The work done by an electron under the action of the high frequency field has a maximum value if the phase between field and velocity of the electron oscillation is zero. Then, the work is given by

$$W_{1,\phi=0} = -e\text{v}_0 A_{\text{sat}} \int_0^T \cos(\omega_0 t) \, \cos(\omega t) \, \mathrm{d}t \tag{19.79}$$

The solution is, with $\omega_0 \approx \omega$, equal to

$$W_{1,\phi=0} = -\pi \, e \, \text{v}_0 A_{\text{sat}}/\omega. \tag{19.80}$$

We now assume that distortion occurs if the work is equal to the *distortion energy* E^*, i.e., if

$$W_{1,\phi=0} = -E^*. \tag{19.81}$$

Comparison of (1980) and (1981) provides the saturation field amplitude

$$A_{\text{sat}} = \frac{E^* \omega}{\pi \; e \; v_0} = \frac{(E^*/e) \; \gamma \; \omega}{\pi \; c \; K_{\text{w}}}. \tag{19.82}$$

The saturation field amplitude A_{sat} is proportional to the distortion energy, to the Lorentz factor, and to the frequency. It is inversely proportional to the wiggler parameter K_{w}. The frequency dependence corresponds to an $\omega^{3/2}$ dependence.

A process leading to distortion changes the phase of the oscillation by π. After a change of the phase, the reversed process occurs: the field transfers energy to the electron. The work is given

$$W_{1,\phi=\pi} = +ev_0 A_{\text{sat}} \int_0^T \cos(\omega_0 t) \; \cos(\omega t) \; \mathrm{d}t = + \; \pi \; ev_0 A_{\text{sat}}/\omega, \tag{19.83}$$

or,

$$W_{1,\phi=\pi} = +E^*. \tag{19.84}$$

Our analysis implies that a process that causes distortion is accompanied by stimulated emission of radiation and that the reversed process is accompanied by absorption of radiation. Absorption of radiation during a cycle of the electron oscillation follows on stimulated emission during the preceding cycle of the electron oscillation. At steady state oscillation of a free-electron laser, there is, in the saturation region of the wiggler, no net energy transfer of energy from the electrons to the field and vice versa, from the field to the electrons: absorption compensates stimulated emission of radiation. Analysis of free-electron laser data (Sect. 19.10) on basis of the ideas presented in this section suggest that a process causing distortion involves more than one photon per electron oscillation cycle. This corresponds to a fast cascade of stimulated emission of photons during one cycle of the electron oscillation. The reversed process involves the same number of photons and corresponds to a fast cascade of absorption processes during one cycle of the electron oscillation. The processes are also discussed in Sects. 19.13, 19.17, and 19.19.

We will treat the distortion energy as a parameter that can be determined from experimental data of free-electron lasers. We expect that the distortion energy depends on the wiggler parameter. We will show that experimental data for existing free-electron lasers are consistent with a distortion energy on the order of 1 eV ($E^*/e \approx 2$ V); *see* also Problem 19.16 suggesting a derivation of the distortion energy.

Example 5μm FEL (*see* Sect. 19.10). $\nu = 60$ THz, $E_{\text{el}} = 37$ MeV, $\gamma = 74$, $K_{\text{w}} = 0.7$, and $E^*/e = 2$ V. We find $A_{\text{sat}} = 4 \times 10^7$ V m^{-1}.

The power, P_{res}, of electromagnetic radiation in the resonator of a free-electron laser at steady state operation is given by

$$P_{\text{res}} = \frac{1}{2} c \; \varepsilon_0 \; A_{\text{sat}}^2 \; a_{\text{mode}}, \tag{19.85}$$

where a_{mode} is the cross sectional area of the area of the waist of the optical beam. If the beam is a Gaussian beam, then $a_{\text{mode}} = \pi\, r_{\text{u}}^2$, where r_{u} is the radius of the beam waist with respect to the distribution of the energy density of the electromagnetic radiation. The output power of a free-electron laser is given by

$$P_{\text{out}} = (1 - R)P_{\text{res}} = \frac{1}{2} c\, \varepsilon_0\, (1 - R)\, a_{\text{mode}}\, A_{\text{sat}}^2, \tag{19.86}$$

where $1 - R$ is the portion of power of radiation coupled out from the laser resonator.

19.9 Critical Modulation Index

The modulation index depends, according to its definition, on the coupling strength,

$$\mu = \frac{\kappa\, A}{\omega}. \tag{19.87}$$

We introduce the *critical modulation index* μ_{c} by the relationship

$$\frac{\mu}{\mu_{\text{c}}} = \frac{A}{A_{\text{sat}}} \quad \text{or} \quad \frac{\mu}{A} = \frac{\mu_{\text{c}}}{A_{\text{sat}}}. \tag{19.88}$$

The critical modulation index, a dimensionless number, corresponds to the modulation index that is reached if the amplitude of the high frequency electric field is equal to the saturation field amplitude A_{sat}.

In order to find out the value of μ_{c}, we consider the phase difference between the transverse oscillation of an electron and the high frequency electric field,

$$\phi(x) = x - \mu \sin ax, \tag{19.89}$$

where $x = \omega_0 t$ is the phase for $\mu = 0$ and $a = (\omega_0 - \omega)/\omega_0$ is the difference of the frequencies of the electron oscillation and the high frequency electric field, divided by ω_0. Without modulation ($\mu = 0$), the phase ϕ increases linearly with time. In the case that a high frequency electric field modulates the electron oscillation, the phase difference oscillates around the $\phi = x$ line. The amplitude of this oscillation increases with increasing μ. The phase difference ϕ increases continuously with time as long as $\mu < 1$. However, for $\mu > 1$, the same phase difference can be obtained for two different times. The change from the continuous increase to the more complicated behavior occurs at the critical modulation index μ_{c}. We find the value of μ_{c} from the condition that the derivative $\text{d}\phi/\text{d}x$ is zero, that is, $\text{d}\phi/\text{d}x = 1 - \mu\, a = 0$, which leads to

$$\mu_{\text{c}} = \frac{1}{a} \approx 1 \tag{19.90}$$

for $1 - a << 1$, i.e., if $\omega_0 - \omega << \omega_0$.

The critical modulation index, a dimensionless number, corresponds to the modulation index that is reached if the amplitude of the high frequency electric field is equal to the saturation field amplitude

$$A_{\mathrm{sat}} = \frac{(E^*/e)\gamma\omega}{\pi c K_{\mathrm{w}}}. \tag{19.91}$$

We find the coupling strength

$$\kappa = \frac{\omega\ \mu_{\mathrm{c}}}{A_{\mathrm{sat}}} = \frac{\pi\ c\ K_{\mathrm{w}}}{(E^*/e)\ \gamma}. \tag{19.92}$$

The modal gain coefficient is equal to

$$\alpha = \frac{K_{\mathrm{w}} I_{\mathrm{el}}}{4\ c\ \varepsilon_0\ a_{\mathrm{mode}}} \times \frac{1}{A_{\mathrm{sat}}} = \frac{\pi\ K_{\mathrm{w}}^2\ I_{\mathrm{el}}}{4\varepsilon_0\ (E^*/e)\ a_{\mathrm{mode}}\ \gamma\ \omega}. \tag{19.93}$$

The modulation model thus leads to the result (Fig. 19.13) that, at fixed frequency, the coupling strength κ increases as proportional to the wiggler parameter K_{w}, the saturation field amplitude A_{sat} is inversely proportional to K_{w}, and the gain coefficient α is proportional to the square of the wiggler parameter.

The coupling strength $\kappa\ \omega^{-1/2}$, the saturation field increases, with increasing frequency, as $\omega^{3/2}$, and the maximum gain coefficient decreases as $\omega^{-3/2}$. The coupling strength and the maximum gain coefficient are inversely proportional to the distortion energy E^*, while the saturation field amplitude is proportional to the distortion energy.

We find the relationship

$$\alpha\ A_{\mathrm{sat}} = \frac{K_{\mathrm{w}}\ I_{el}}{4\ c\ \varepsilon_0\ a_{\mathrm{mode}}}. \tag{19.94}$$

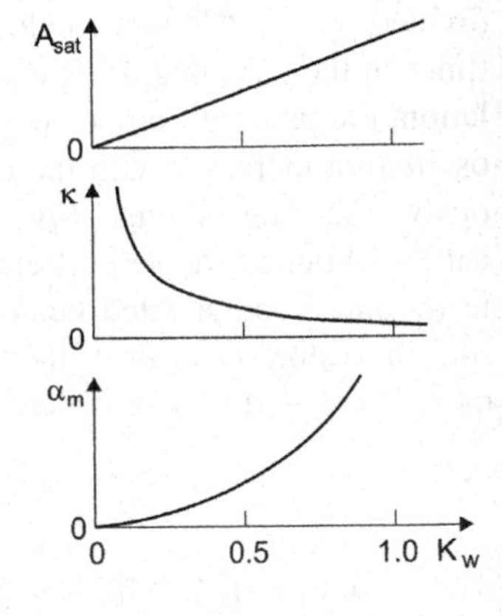

Fig. 19.13 Dependencies of coupling strength, saturation field, and maximum gain coefficient on the wiggler parameter

The product of the modal gain coefficient and the saturation field amplitude is proportional to the wiggler parameter K_w and to the modal current density. It is independent of frequency and of the distortion energy E^*.

In the case that the diameter of the electron beam is equal to the diameter of the optical beam, both supposed to be parallel beams, we obtain the saturation field (that is independent of the electron density)

$$A_{sat}(\omega) = \frac{(E^*/e)\,\gamma\,\omega}{\pi\,c\,K_w} \tag{19.95}$$

and the gain coefficient

$$\alpha(\omega) = 2\alpha_m(\omega)\,\bar{g}_{L,disp}(\omega); \qquad \text{with} \qquad \alpha_m(\omega) = \frac{\pi\,c\,e\,K_w^2\,N_0}{4\varepsilon_0(E^*/e)\,\gamma\,\omega}. \tag{19.96}$$

The expressions on the frequency scale are:

$$A_{sat}(\nu) = \frac{2(E^*/e)\,\gamma\,\nu}{c\,K_w} \tag{19.97}$$

and

$$\alpha(\nu) = 2\alpha_m(\nu)\,\bar{g}_{L,disp}(\nu); \qquad \text{with} \qquad \alpha_m(\nu) = \frac{c\,e\,K_w^2\,N_0}{8\,\varepsilon_0(E^*/e)\,\gamma\,\nu}. \tag{19.98}$$

19.10 Modulation Model and Data of Free-Electron Lasers

In this section, we compare results of the modulation model with free-electron laser data. The data concern free-electron lasers operating in the infrared, the far infrared and millimeter wave regions covering a wavelength range of 9 octaves (from 1.6 μm to 3 mm); tuning over more than one octave is possible with one device.

The saturation field is given by

$$A_{sat} = \frac{(E^*/e)\,\gamma\,\omega}{\pi\,c\,K_w} \tag{19.99}$$

and the (maximum) modal gain coefficient by

$$\alpha = \frac{\pi\,K_w^2\,I_{el}}{4\varepsilon_0(E^*/e)\,a_{mode}\,\gamma\,\omega}. \tag{19.100}$$

In order to obtain a survey of the saturation field amplitude and the modal gain coefficient, we chose the following parameters.

$K_w = 0.7$, wiggler parameter.
$\lambda_w = 4$ cm, wiggler wavelength.
$I_{el} = 100$ A, electron current strength.
$a_{mode} = 0.3\,cm^2$, area of beam waist; corresponding to a beam waist of $r_u = 3$ mm for a Gaussian beam; the quantities relate to the lateral distribution of the power.
$E^*/e = 2$ V, energy of distortion/e.

Further quantities are:

A_{sat}, saturation field amplitude.
α, modal gain coefficient; we omit the index m (for maximum).
t_p, pulse duration.
P_{res}, power of optical radiation in the laser resonator; peak power for a pulsed laser.
P_{out}, output power; $P_{out} = (1 - R)\, P_{res}$.
L_{res}, resonator length.

With $\omega = 2\pi\nu$ and $\nu = 10^{12}\, \nu_{THz}$, we find A_{sat}, in units of V m^{-1},

$$A_{sat} = 2 \times 10^5\, (\nu_{THz})^{3/2} \tag{19.101}$$

and α, in units of m^{-1},

$$\alpha = \frac{1.7 \times 10^3}{(\nu_{THz})^{3/2}}. \tag{19.102}$$

The solid lines of Fig. 19.14 show the saturation field amplitude and the maximum gain coefficient in a large frequency range, including millimeter waves and far infrared and infrared wavelengths.

Which are practical operation conditions?

- $G - 1 \approx 1$. Free-electron lasers operate preferably at a gain of about 100% per transit of radiation through the resonator. This corresponds, at a wiggler length of 1 m, to $\alpha_m = 0.7\,m^{-1}$. The gain coefficient is adjustable to a large extent, mainly by changing the current strength I_{el} and the mode area a_{mode}, i.e., by changing the modal current density. The gain coefficient can be varied, in comparison to the values of the solid curve for constant current density, given in the figure, by orders of magnitude.
- A_{sat}. Variation of the saturation field is only possible to a small extent.
- P_{res}. The power of optical radiation in the laser resonator can be increased, especially in the range of large gain coefficient, by choosing a large area of the cross section of the optical mode. The output power increases accordingly.

Changing the wiggler parameter allows for a further modification of the gain coefficient (and of the saturation field amplitude).

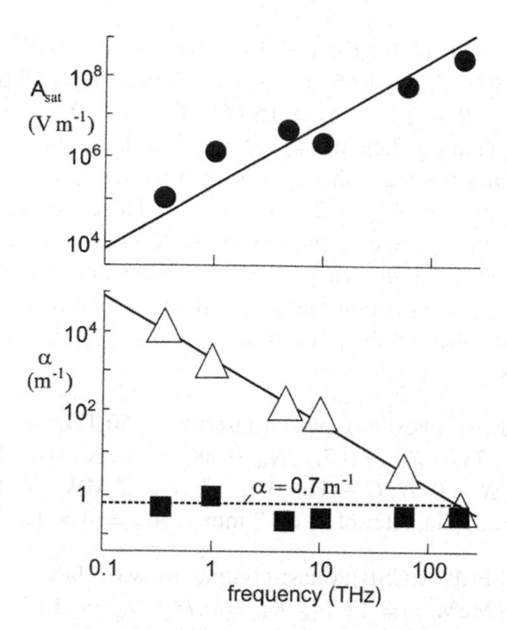

Fig. 19.14 Modulation model and laser data; *upper part* saturation field amplitude according to the modulation model for $K_w = 0.7$ (*solid line*) and laser data (*dots*); *lower part* gain coefficient according to the modulation model for $K_w = 0.7$, $I_{el} = 100$ A, and $r_u = 3$ mm (*solid line*) and laser data (*square dots*) for different lasers, operated at different current densities, and the corresponding values (*triangles*) supposing $K_w = 0.7$, $I_{el} = 100$ A, and $r_u = 3$ mm; *dotted line* corresponding to $\alpha = 0.7\,\mathrm{m}^{-1}$ or a gain of 100% per transit of radiation through the wiggler

Points in Fig. 19.14 represent data extracted from information about free-electron lasers in three laboratories. We will characterize a free-electron laser by one of the operation wavelengths. For some of the examples, the magnitude of the mode area is a guess.

- 1.6 μm FEL (JNFAF); pulsed laser. Free-electron laser at the Thomas Jefferson National Acceleration Facility, Newport News, USA. A systematic experimental study (published in 2007 [326]) was accompanied by a theoretical analysis via simulation studies based on Maxwell's equations.
- 5 μm FEL and further FELs (Dresden); pulsed lasers. Tunable free-electron lasers at the Helmholtz Zentrum Dresden (https://www.hzdr.de/).
- 30 μm FEL and further FELs (UCSB); quasi-continuous working lasers. Tunable free-electron lasers at the University of California, Santa Barbara (http://sbfel3.ucsb.edu/).

We mention here only very few experimental data. The data concern:
in *Example 1* experimental gain and power, but a_{mode} fitted;
in *Example 2* experimental gain and power for operation at maximum output power, but a_{mode} fitted;
in *Examples 2–6* design parameters (with known a_{mode}).

Example 1 1.6 μm FEL (JNFAF); $\nu = 188$ THz, $h\nu = 0.8$ eV, $E_{\mathrm{el}} = 115$ MeV, $\gamma = 230$, $N_{\mathrm{w}} = 30$, $L_{\mathrm{w}} = 1.65$ m, $\lambda_{\mathrm{w}} = 5.5$ cm, $K_{\mathrm{w}} = 0.64$, $I_{\mathrm{el}} = 100$ A, $L_{\mathrm{res}} = 30$ m, $1 - R = 0.1$, $t_{\mathrm{p}} = 150$ fs, $G - 1 = 0.7$, $P_{\mathrm{out}} = 2 \times 10^8$ W. We find the modal gain coefficient $\alpha_{\mathrm{m}} = 0.3$ m^{-1} and, assuming that the radius of the waist of the beam (in a near concentric resonator) was $r_{\mathrm{u}} = 2$ mm, the saturation field amplitude is given by $A_{\mathrm{sat}} = 2 \times 10^8$ V m^{-1}. The experimental value for the gain $G - 1$ and for the average output power (12 kW) agreed (within few percent) with theoretical values determined by a three-dimensional simulation [326]. The FEL emitted pulses at a repetition rate of 37 MHz (corresponding to the resonator length). For determination of the pulse power, we took into account that the pulses had Gaussian shape.

Example 2 5 μm FEL (Dresden), pulsed laser; $\nu = 60$ THz, $h\nu = 0.25$ eV, $E_{\mathrm{el}} = 37$ MeV, $\gamma = 74$, $K_{\mathrm{w}} = 0.7$, $N_{\mathrm{w}} = 68$, $t_{\mathrm{p}} = N_{\mathrm{w}}/\nu = 1$ ps, $L_{\mathrm{w}} = 2$ m, $I_{\mathrm{el}} = 100$ A, $1 - R = 0.07$, $G - 1 = 0.7$; $P_{\mathrm{out}} = 2$ MW. We find $\alpha = 0.4$ m^{-1} and, assuming a mode diameter of $r_{\mathrm{u}} = 2$ mm, $A_{\mathrm{sat}} = 4 \times 10^7$ V m^{-1}.

Example 3 30 μm FEL (UCSB); quasi-continuous wave laser. $\nu = 10$ THz, $h\nu = 40$ meV, $E_{\mathrm{el}} = 6$ MeV, $\gamma = 12.74$, $K_{\mathrm{w}} = 0.78$, $\lambda_{\mathrm{w}} = 1.85$ cm, $L_{\mathrm{w}} = 2.3$ m, $N_{\mathrm{w}} = 122$, $I_0 = 2$ A, $a_{\mathrm{mode}} = 0.4$ cm^2, $G - 1 \approx 1$, $P_{\mathrm{res}} = 89$ kW, $1 - R = 0.07$, $P_{\mathrm{out}} = 6$ kW. We find $\alpha = 0.3$ m^{-1} and $A_{\mathrm{sat}} = 1.2 \times 10^6$ V m^{-1}.

Example 4 63 μm FEL (UCSB); quasi continuous wave laser. $\nu = 4.7$ THz, $h\nu = 19$ meV, $E_{\mathrm{el}} = 6$ MeV, $\gamma = 12.7$, $K_{\mathrm{w}} = 0.13$, $\lambda_{\mathrm{w}} = 2$ cm, $L_{\mathrm{w}} = 3$ m, $N_{\mathrm{w}} = 150$, $I_{\mathrm{el}} = 2$ A, $a_{\mathrm{mode}} = 1$ cm $\times$ 1 cm, $G - 1 = 0.7$, $P_{\mathrm{res}} = 370$ kW, $1 - R = 0.03$, $P_{\mathrm{out}} = 11$ kW. We find $\alpha = 0.2$ m^{-1} and $A_{\mathrm{sat}} = 5 \times 10^6$ V m^{-1}.

Example 5 300 μm FEL (UCSB). $\nu = 1$ THz, $h\nu = 4$ meV, $E_{\mathrm{el}} = 6$ MeV, $\gamma = 12.7$, $K_{\mathrm{w}} = 0.7$, $\lambda_{\mathrm{w}} = 2$ cm, $L_{\mathrm{w}} = 3$ m, $N_{\mathrm{w}} = 150$, $I_{\mathrm{el}} = 2$ A, $a_{\mathrm{mode}} = 0.8$ cm^2, $1 - R = 0.01$, $G - 1 = 1.7$, $P_{\mathrm{res}} = 163$ kW, $1 - R = 0.04$, $P_{\mathrm{out}} = 7$ kW. We find $\alpha = 1$ m^{-1} and $A_{\mathrm{sat}} = 1.1 \times 10^6$ V m^{-1}.

Example 6 1 mm FEL (UCSB). $\nu = 0.33$ THz, $h\nu = 1.4$ meV, $E_{\mathrm{el}} = 2$ MeV, $\gamma = 4$, $K_{\mathrm{w}} = 0.73$, $\lambda_{\mathrm{w}} = 3$ cm, $L_{\mathrm{w}} = 0.48$ m, $N_{\mathrm{w}} = 16$, $I_{\mathrm{el}} = 2$ A, $a = 2$ cm^2, $L_{\mathrm{res}} = 0.58$ m, $1 - R = 0.07$, $G - 1 = 0.8$, $P_{\mathrm{out}} = 3$ kW. We find $\alpha_{\mathrm{m}} = 2.5$ m^{-1} and $A_{\mathrm{sat}} = 1.0 \times 10^5$ V m^{-1}.

The efficiency $\eta = P_{\mathrm{out}}/\left[(E_{\mathrm{el}}/e)\, I_{\mathrm{el}}\right]$ of the free-electron lasers for conversion of translational energy of the electrons to radiation energy can reach values on the order of 0.1%.

The far infrared lasers (*Examples 3–5*) are using hybrid resonators. A hybrid resonator consists of a parallel plate resonator combined with an optical resonator. The arrangement results in a mode volume that is larger than for a purely Gaussian beam resonator. Radiation is coupled out from a free-electron resonator via a hole in one of the optical mirrors. (Broadband dielectric mirrors are not available for radiation at wavelengths below a few micrometer. Metal film mirrors are not suitable as partially transparent mirrors because of a strong absorptivity in comparison to the transmissivity.) Diffraction of radiation in a resonator at a hole in a mirror causes internal loss. This loss is frequency dependent. Accordingly, the net gain and the output power of free-electron lasers operating at wavelengths below a few micrometer can strongly depend on frequency. In our discussion of the laser power, we consider maximum values reported, since these occur when internal loss has a small influence on gain and power.

We conclude from the analysis that an electron generates 10^4–10^7 photons during the flight through the wiggler at steady state oscillation of a free-electron laser. Accordingly, in the last part of the gain region of the wiggler, more than one photon per period of electron oscillation is generated.

Thus, the modulation model provides data for the saturation field amplitude that are, in principle, in accord with laser data known from experimental and design parameters. Experimental data for gain coefficients are in accord with the modulation model if we take account of the appropriate values of the current strength and the mode area of different lasers. We summarize results obtained by applying the modulation model.

- The gain coefficient varies as $\omega^{-3/2}$ (for fixed K_{w} and λ_{w} and fixed current density).
- The saturation field amplitude varies as $\omega^{3/2}$ (for fixed K_{w} and λ_{w}).
- The modulation model contains a parameter, the distortion energy E^*, which we adjust to experimental data or design data of free-electron lasers; for a derivation of the distortion energy, *see* Problem 19.16.
- The product of the gain coefficient and the saturation field amplitude is a constant that contains only the wiggler parameter and the modal current density; the constant is independent of frequency and of the distortion energy.
- At steady state oscillation of a free-electron laser, the orbit of an electron undergoes a transition into a distorted oscillation state if the electron traverses the saturation range of the wiggler.
- The modulation model seems to be applicable for characterization of lasers as well as of SASE free-electron lasers, that is, from millimeter waves to X-rays—*see* next section.

19.11 Modulation Model and SASE Free-Electron Lasers

In this section, we ask the question whether the modulation model is suitable to describe SASE free-electron lasers. There are worldwide many SASE free-electron lasers in operation or in planning. We will compare data following from the modulation model with experimental or design data of two SASE free-electron lasers:

SASE free-electron laser FLASH, operating at DESY, Hamburg, *see* https://flash.desy.de/ and [331].

SASE free-electron laser, simulation data, published in 2007 [173], *see* also https://portal.slac.stanford.edu for information about operating free-electron lasers at SLAC National Accelerator Laboratory, Stanford.

According to the modulation model, the gain coefficient is given by

$$\alpha = \frac{\pi\, K_{\mathrm{w}}^2\, I_{\mathrm{el}}}{4\varepsilon_0\, (E^*/e)\, a_{\mathrm{mode}}\, \gamma\, \omega}, \tag{19.103}$$

and the saturation field amplitude by

$$A_{\mathrm{sat}} = \frac{(E^*/e)\, \gamma\, \omega}{\pi\, c\, K_{\mathrm{w}}}. \tag{19.104}$$

We choose the same value for the distortion energy E^* as for the free-electron lasers discussed in the preceding section; the saturation field amplitude is most likely varying less strongly with frequency as discussed in Problem 19.17.

Example 1 5.8 nm SASE FEL (Flash, Hamburg). $\nu = 5 \times 10^{16}$ Hz, $h\nu = 200$ eV, $E_{\mathrm{el}} = 1.1$ GeV, $\gamma = 2.1 \times 10^3$, $L_{\mathrm{w}} = 27$ m, $\lambda_{\mathrm{w}} = 2.7$ cm, $K_{\mathrm{w}} = 1.1$, $I_{\mathrm{el}} = 80$ A, $t_{\mathrm{p}} = 20$ fs, $P_{\mathrm{out}} = 1 \times 10^9$ W. The Flash FEL produces radiation of a relative spectral width of 1%. The wiggler consists of periodically arranged magnets at constant period. During the transit through the wiggler, an electron loses 1% of its initial energy, corresponding to 11 MeV. We conclude that an electron generates 5×10^4 photons by stimulated emission, i.e., that the gain of radiation is equal to $G = 5 \times 10^4$. The number of photons in a pulse is equal to $Z = P_{\mathrm{out}}/h\nu = 3 \times 10^{11}$. Thus, the initial number of spontaneously emitted photons is of the order of 10^7. It follows from the value of the gain that the modal gain coefficient is equal to $\alpha = 0.4\ \mathrm{m}^{-1}$. This value follows from (19.103) for $E^*/e = 2$ V if we assume a radiation beam diameter of $r_{\mathrm{u}} = 0.06$ mm. The power generated by the SASE FEL is much smaller than the saturation field ($A_{\mathrm{sat}} = 10^{12}\ \mathrm{V\,m^{-1}}$), which follows from (19.104).

Example 2 0.15 nm SASE-FEL [173]; $\nu = 2 \times 10^{18}$ Hz, $h\nu = 8$ keV, $E_{\mathrm{el}} = 13.6$ GeV, $\gamma = 2.7 \times 10^4$, $L_{\mathrm{w}} = 110$ m, $\lambda_{\mathrm{w}} = 5.5$ cm, $N_{\mathrm{w}} = 3.7 \times 10^3$, $K_{\mathrm{w}} = 3.5$, $I_{\mathrm{el}} = 3.4$ kA, $\alpha = 0.23\ \mathrm{m}^{-1}$. The modulation model provides the same gain coefficient if we assume the values $E^*/e = 2$ V and $r_{\mathrm{u}} = 0.9$ mm. However, a gain of $G = \exp(\alpha L_{\mathrm{w}})$ cannot be realized because the output power of radiation would exceed the input power contained in the electron beam. We estimate the power in a different way. We assume that 1% of the energy of an electron is converted to radiation by stimulated emission of radiation. This corresponds to an optical power of $P = 460$ GW. An electron pulse, assumed to have a duration of $\delta t = 100$ fs, contains $I_{\mathrm{el}}\delta t/e$ $(= 2 \times 10^9)$ electrons. The number of photons in an optical pulse is given by the relationship $Pt_{\mathrm{p}} = Zh\nu$. We assume that the optical pulse duration is equal to the electron pulse duration, $t_{\mathrm{p}} = \delta t$, and find $Z = 4 \times 10^{13}$. It follows that the gain is equal to $G = 2 \times 10^4$. Each electron produces, by stimulated emission of radiation, 2×10^4 photons. The saturation field that follows from (19.104) cannot be reached because the corresponding power would exceed the initial power of an electron pulse.

An X-ray pulse in a SASE free-electron laser is guided by the electron beam according to the enhanced refractive index (Sect. 19.14) in the range of the electron beam.

In conclusion, the modulation model may, in principle, be applicable for an illustration of properties of a SASE free-electron laser. However, the modulation model cannot replace an adequate simulation based on Maxwell's equations.

19.12 Onset of Oscillation of a Free-Electron Laser

In this section, we discuss the onset of oscillation of a free-electron laser. We will find that the spectrum of radiation generated by a free-electron laser at steady state oscillation is determined by the dynamics during the onset of oscillation. We consider a continuous wave laser.

The gain coefficient of the free-electron laser medium is time dependent:

$$\alpha\,(\omega,t) = 2\alpha_{\rm m}\,\bar{g}_{\rm L,disp}(\omega,t) = 2\alpha_{\rm m}\,\frac{(\omega_0-\omega)\,\Delta\omega_0(t)/2}{(\omega_0-\omega)^2+\Delta\omega_0^2(t)/4}\,. \tag{19.105}$$

The maximum gain coefficient, $\alpha_{\rm m}$, is independent of time. However, $\Delta\omega_0$ depends on time.

The first electron pulses propagating through the wiggler at time $t=0$ excite, by spontaneous emission of radiation, modes of the laser resonator according to the spectrum of spontaneously emitted radiation (Fig. 19.15). Stimulated emission of radiation sets in at time $t\approx 0$. The initial rate of stimulated emission processes is very small. Therefore, at $t\approx 0$, the width $\Delta\omega_0$ of the gain coefficient curve is very narrow. Accordingly, initial stimulated emission of radiation occurs in a narrow frequency range near ω_0; the gain coefficient has the shape of the normalized general Lorentz dispersion function. With increasing number of transits of the radiation through the wiggler, the rate τ^{-1} $(=\tau_{\rm stim}^{-1})$ of stimulated emission processes increases. An increasing stimulated emission rate leads to broadening of the gain curve and a shift of the maximum of the gain coefficient curve (and of the maximum of the frequency distribution of radiation generated by stimulated emission) toward smaller frequency. At the oscillation onset time, $t_{\rm on}$, the amplitude of the high frequency electric field is equal to the saturation field amplitude $A_{\rm sat}$. During the onset of laser oscillation, the center frequency of the radiation in the resonator shifts from a value near ω_0 to the frequency of maximum gain, assumed to be near the center frequency $\omega_{\rm c}$ of the laser oscillation range $<\omega_1,\omega_0>$. The lower limit of the laser oscillation range occurs because spontaneous emission at ω_1 is absent. In the vicinity of this frequency, the spontaneous emission is too week to initiate strong laser oscillation, i.e. there is a bottleneck in the spontaneous emission of radiation at frequencies around ω_1 and at smaller frequencies.

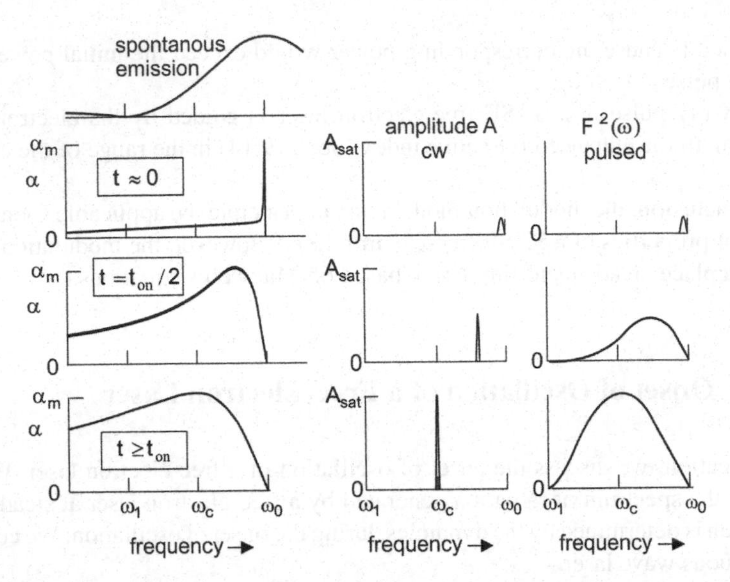

Fig. 19.15 Onset of oscillation of free-electron lasers. *Left part*, spectrum of spontaneously emitted radiation and gain coefficient. *Center*, onset of oscillation of a single line free-electron laser. *Right*, onset of oscillation of a pulsed (mode locked) free-electron laser

In a pulsed laser, the spectral width of the frequency distribution grows during the buildup of laser oscillation and corresponds to about $\Delta\omega_0/2 \approx \omega_0/2N_w$. The width $\Delta\omega_0$ of the gain coefficient curve characterizing an ensemble of electrons increases exponentially with time (Fig. 19.16). If the critical modulation index is reached, the ensemble loses its conductivity. Then stimulated emission and absorption of radiation determine the behavior of the electron ensemble. Accordingly, during the initial propagation of electron pulses through the wiggler, the width $\Delta\omega_0$ of the radiation field increases with time until it reaches a value that is slightly smaller than ω_0/N_w.

The gain per transit of radiation through the active medium is equal to

$$G - 1 = \exp(\alpha\, L_{eff}) - 1. \tag{19.106}$$

G is the gain factor, $G - 1$ the gain, α the modal gain coefficient, and L_{eff} is an effective length. The effective length changes during the onset of oscillation (Fig. 19.17):

$G - 1 = \exp(\alpha\, L_w) - 1$, small-signal gain; $L_{eff} = L_w$.
$G - 1 = \exp(\alpha L_w/2) - 1$, intermediate gain (at $t = t_{on}$); $L_{eff} = L_w/2$.
$G_\infty = \exp(\alpha L_w/3)$, large-signal gain at steady state oscillation of the laser; $L_{eff} = L_w/3$.

At begin of the onset of laser oscillation (at $t \approx 0$), radiation is amplified along the whole wiggler length. In the intermediate range, the modulation current density is

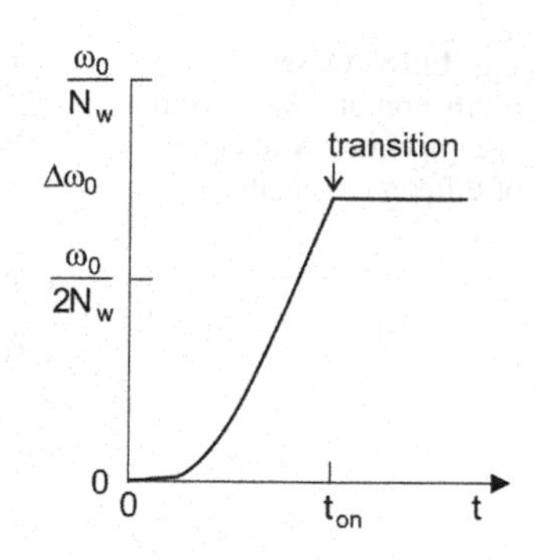

Fig. 19.16 Width $\Delta\omega_0$ of the gain coefficient curve for electron ensembles propagating, in turn, through the wiggler during the onset of oscillation of a free-electron laser

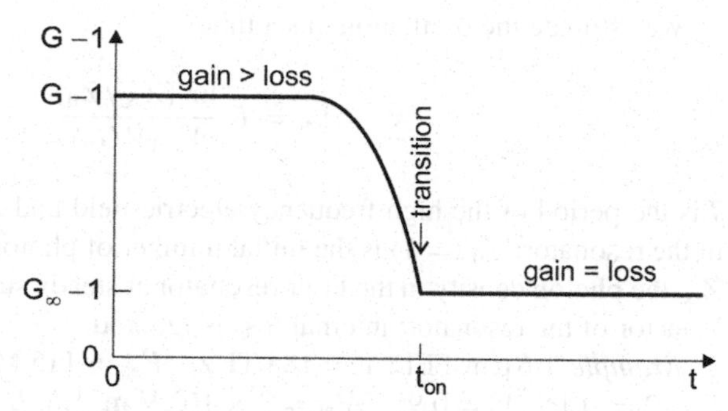

Fig. 19.17 Gain during onset of oscillation of a free-electron laser

built up in the first half of the wiggler and mediates gain mainly in the second half. At steady state oscillation, the modulation current density is built up in the first third of the wiggler and mediates gain mainly in the second third of the wiggler. In the third part of the wiggler, the saturation field amplitude is reached.

The threshold condition,

$$GV \geq 1 \quad \text{or} \quad G - 1 \geq \frac{1}{V} - 1, \tag{19.107}$$

is, of cause, a necessary condition for laser oscillation (V, V factor and $V^{-1} - 1$, loss per transit of radiation through the resonator).

The saturation field amplitude,

$$A_{\text{sat}} = \frac{(E^*/e)\,\gamma\,\omega}{\pi\,c\,K_{\text{w}}}, \tag{19.108}$$

does not depend on the length of the wiggler. Doubling of the current strength results in shortening of the oscillation onset time (Fig. 19.18).

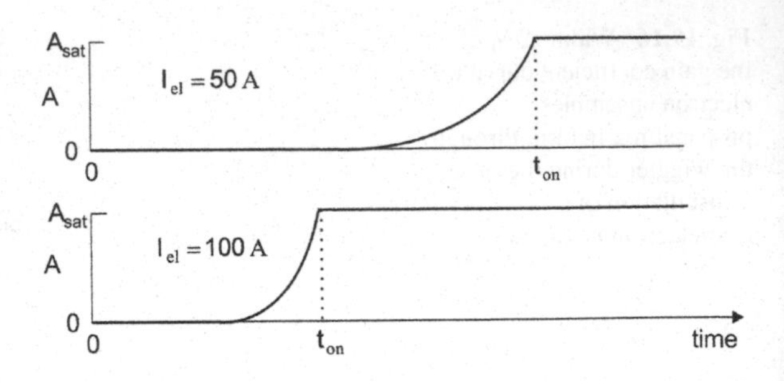

Fig. 19.18 Onset of oscillation of a free-electron laser operated with currents of different strength

We estimate the oscillation onset time,

$$t_{on} = T\,\frac{\ln\,(Z_\infty/Z_0)}{\ln\,(VG_0)}. \tag{19.109}$$

T is the period of the high frequency electric field and Z is the number of photons in the resonator; Z_0 (= 1) is the initial number of photons in the laser resonator and Z_∞ the photon density in the laser resonator at steady state oscillation. $V = R$ is the V factor of the resonator; internal loss is ignored.

Example 1.6 μm FEL; $\nu = 188$ THz, $E_{el} = 115$ MeV, $\gamma = 230$, $N_w = 30$, $1 - R = 0.1$, $V = 0.9$, $A_{sat} = 3 \times 10^8$ V m^{-1}, $L = 30$ m; $T = 2L/c = 0.2\,\mu$s, $G_m = 2$, $E_p = P_{res}\,t_p = 6 \times 10^{-4}$ J, $Z_\infty = E_p/h\nu = 5 \times 10^{15}$, $Z_0 = 1$; $I_0 =$ 100 A, $t_{on} = 35\,T = 7\,\mu$s; $I_0 = 50$ A, $t_{on} = 17\,T = 14\,\mu$s.

19.13 Phase Between Electron Oscillation and Optical Field

Dephasing between an electron oscillation in a free-electron medium and a high frequency electric field occurs because the frequency ω_0 of the electron oscillation is different from the frequency ω of the field.

Instead of the velocity v, we consider, by reason of convenience, the transverse current $I = -e\mathrm{v}$ connected with the transverse oscillation of an electron. The phase difference

$$\phi_I = \varphi_0 - \varphi_E = \omega_0 t - \omega t = \frac{\Delta\omega}{\omega_0}\,\omega_0 t \tag{19.110}$$

between the phase $\varphi_0 = \omega_0 t$ of the current and the phase $\varphi_E = \omega t$ of the electric field increases linearly with time (Fig. 19.19a) and is proportional to the relative frequency difference $\Delta\omega/\omega_0$. A stimulated emission process, which occurs preferably if $\phi \approx \pi$, causes a change of the phase (Fig. 19.19b). With increasing amplitude of the high frequency electric field, stimulated emission processes follow each other rapidly (Fig. 19.19c). In the case that the amplitude of the high frequency electric field is equal to the saturation field amplitude, a transition to a distorted oscillation state occurs by emission of a fast cascade of a large number of photons (Fig. 19.19d). Immediately

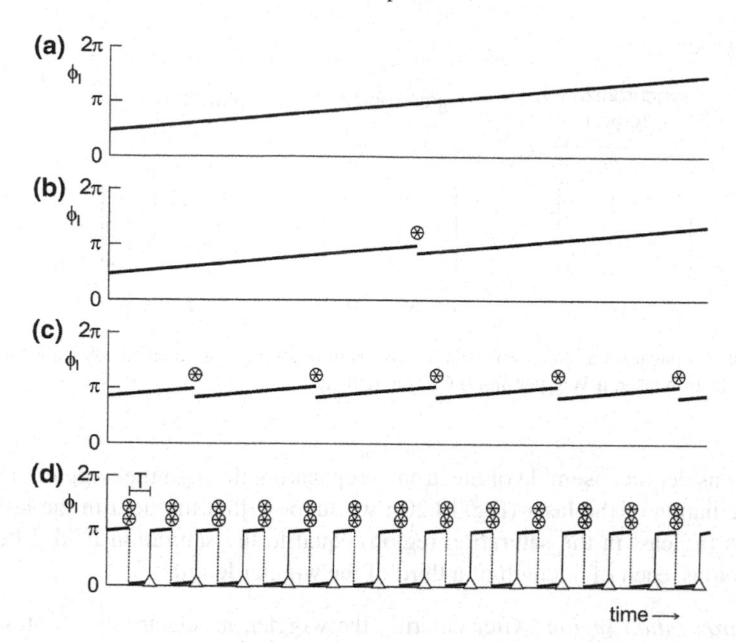

Fig. 19.19 Phase $\phi_{\rm I}$ between the transverse current $I_{\rm el}$ of an oscillating electron and the high frequency electric field. **a** Continuous dephasing due to continuous increase of $\phi_{\rm I}$. **b** Dephasing interrupted by a stimulated emission process (*star*). **c** Dephasing interrupted by stimulated emission processes, occurring in turn. **d** Saturation: stimulated emission of a cascade of photons (*double star*) followed by a cascade of absorption processes (*triangle*), in turn; $T = 2\pi/\omega$, period of the optical field

after generation of photons, a fast cascade of photon absorption processes follows. Fast cascades of stimulated emission and absorption of photons occur in turn. On time average, the work done by an electron during one period of oscillation is equal to E^* and the work done by the high frequency field during the following period is also equal to E^*. The phase difference jumps between 0 and π: the phase between the current and the field is zero after a fast cascade of stimulated emission processes and π after a fast cascade of absorption processes. The temporal average of the phase difference is $< \phi > \ = \pi/2$. There is, on average, no energy transfer from the electron to the field or, vice versa, from the field to the electron. Stimulated emission and absorption compensate each other. That is, if absorption compensates stimulated emission in the saturation region of the wiggler, then the modulation model is not applicable for describing the active medium in the saturation range, but, the active medium assumes a new type of state. This will be discussed in Sects. 19.17 and 19.18.

Example For $\Delta\omega = \Delta\omega_0/2 = \omega_0/N_{\rm w}$, the phase difference grows by π during a transit of an electron through the wiggler; the phase grows by $\pi/N_{\rm w}$ during the period $T = 2\pi/\omega$ of oscillation, if a high frequency field is absent.

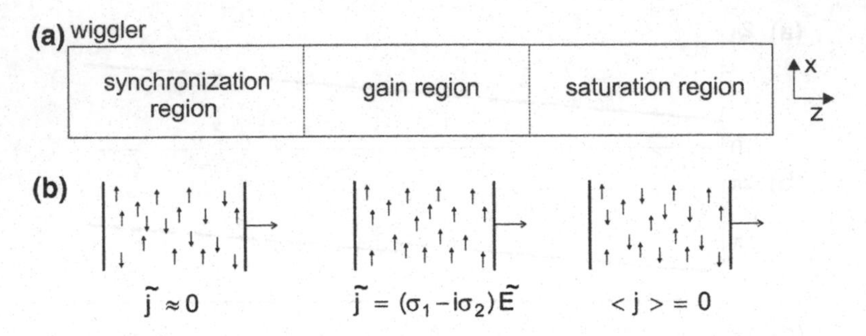

Fig. 19.20 Propagation of an ensemble of electrons through the wiggler at steady state oscillation of a free-electron laser. **a** Wiggler and **b** Current density

We consider an ensemble of electrons propagating through the wiggler at steady state oscillation of the laser (Fig. 19.20); we suppose that the field in the laser resonator is (at least in the saturation region) equal to the saturation field. There are three regions, each of a length of a third of the wiggler length:

- *Synchronization region.* After entering the wiggler, an electron oscillates at the frequency ω_0. Interaction of the electron with the high frequency electric field (frequency ω) results in a modulation of the electron oscillation. In an ensemble of electrons, interaction with the high frequency electric field results in the buildup of a modulation current density. The electron oscillations become synchronized to the high frequency electric field.
- *Gain region.* Interaction of the modulation current density with the high frequency electric field leads to growth of the modulation current density itself and to amplification of the optical radiation field; amplification during a transit of a radiation pulse through the laser resonator compensates internal loss and loss due to output coupling of radiation, occurring in the preceding transit of the radiation pulse.
- *Saturation region.* A modulation current is absent. The electrons are in a state in which stimulated emission of radiation during one cycle of the high frequency electric field is compensated by absorption of radiation during the following cycle. There is no net transfer of energy from the electrons to the field or, vice versa, from the field to the electrons.

The transverse modulation conductivity of an ensemble of electrons is a local quantity and changes during the propagation of the ensemble through the wiggler. Correspondingly, the gain coefficient of an ensemble of electrons is a local quantity and changes during the flight of the ensemble through the wiggler according to the change of the modulation conductivity.

19.14 Optical Constants of a Free-Electron Laser Medium

The modulation model provides the real part of the dynamical conductivity of a free-electron laser medium. In order to determine the imaginary part, we apply the Kramers-Kronig relations (Sect. 9.11). If the shape of the real part of a physical response function is a general Lorentz dispersion function, then the shape of the imaginary part is a general Lorentz resonance function. In the case that $\omega\tau << 1$, we can approximate, in the vicinity of a resonance, the general complex Lorentz function $\tilde{G}_L(\omega)$ by the complex Lorentz function $\tilde{g}_L(\omega)$. Accordingly, we find

$$\sigma_2(\omega) = -2\sigma_{1,m}\frac{\Delta\omega_0^2/4}{(\omega_0-\omega)^2+\Delta\omega_0^2/4}. \tag{19.111}$$

Real and imaginary part of the dynamical conductivity are shown in Fig. 19.21. The real part of the high frequency conductivity shows a dispersive behavior as derived in Sect. 19.7. The imaginary part shows a "negative" peak with the peak value $-2\sigma_m$ at the resonance frequency and with the halfwidth $\Delta\omega_0$.

It follows that the components of the complex refractive index $\tilde{n} = n_1 - in_2$ are given by

$$n_1 = 1 + \delta n\ \bar{g}_{L,disp}(\omega), \tag{19.112}$$

$$n_2 = \delta n\ \bar{g}_{L,res}(\omega), \tag{19.113}$$

$$\delta n \approx \frac{c\alpha_m}{\pi\ \omega_0}. \tag{19.114}$$

The real part of the refractive index, n_1, shows a resonance at ω_0 and is slightly larger than unity (Fig. 19.22). The free-electron laser medium can act as light guide for radiation propagating in $+z$ direction; for the radiation propagating in $-z$ direction, the refractive index is unity and the free-electron laser medium does not influence the radiation. The imaginary part of the refractive index has the shape of a Lorentz dispersion curve, with negative values below the resonance frequency.

Example 1.6 μm FEL. $\nu = 188$ THz, $\alpha_m = 0.6\ \mathrm{m}^{-1}$, $\delta n = 1 \times 10^{-7}$.

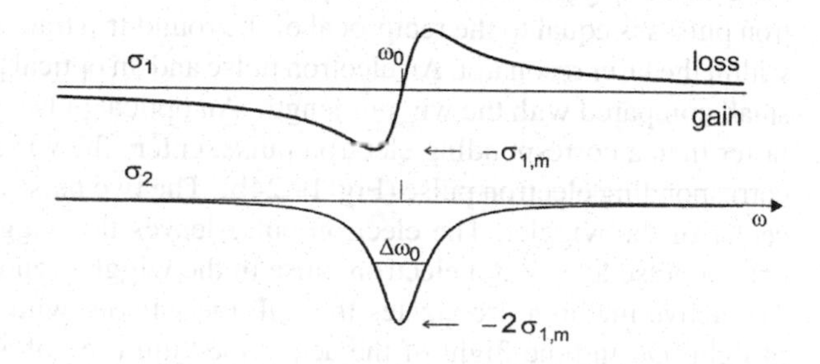

Fig. 19.21 Real and imaginary part of the transverse dynamical conductivity of a free-electron laser medium

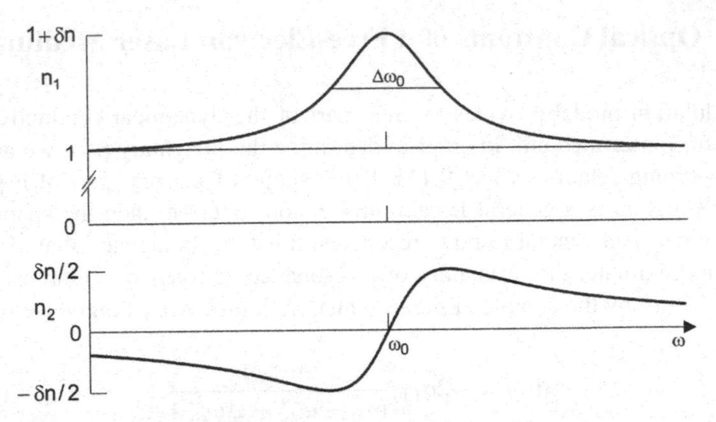

Fig. 19.22 Real and imaginary part of the refractive index of a free-electron laser medium

19.15 Mode Locked Free-Electron Laser

Figure 19.23 shows the result of our study of the onset of laser oscillation. At steady state oscillation, the spectral intensity profile $F^2(\omega)$ of the laser radiation extends from $\omega_1 = \omega_0 - \Delta\omega_0$ to ω_0. The center frequency is ω_c. The spectral intensity profile shows a maximum near ω_c. The halfwidth of the spectral distribution is equal to $\Delta\omega \approx \Delta\omega_0/2$. All resonator modes with frequencies in the laser oscillation range (of width ω_0/N_w) oscillate synchronously.

A sequence of electron pulses is driving a laser oscillation. Each pulse of the sequence propagates only once through the wiggler. Interaction of an electron pulse with the high frequency electric field in the laser resonator modifies the electron pulse. We now will specify the modifications. We suppose that the electron beam and the optical beam have the same Gaussian shape in time and, furthermore, the same Gaussian shape in the plane perpendicular to the beam axis (=resonator axis). We assume, for simplicity, that the beams are parallel beams. For a treatment of the dynamics, we simplify further: we replace the Gaussian shape by a rectangular shape.

An *optical pulse* circulates in the resonator and propagates at the same time through the wiggler as an electron pulse (Fig. 19.24a). The repetition rate of the electron pulses is equal to the reciprocal of the round trip transit time of the optical pulse within the laser resonator. An electron pulse and an optical pulse have lengths that are small compared with the wiggler length. An optical pulse, which propagates slightly faster than a corresponding electron pulse, enters the wiggler immediately after the corresponding electron pulse (Fig. 19.24b). The two pulses overlap completely at the center of the wiggler. The electron pulse leaves the wiggler immediately after the optical pulse leaves. An electron pulse in the wiggler represents the active medium. The active medium propagates through the wiggler with a velocity near the speed of light. During the flight of the active medium (i.e., of an electron pulse) through the wiggler, the optical pulse propagates through the active medium. Interaction of

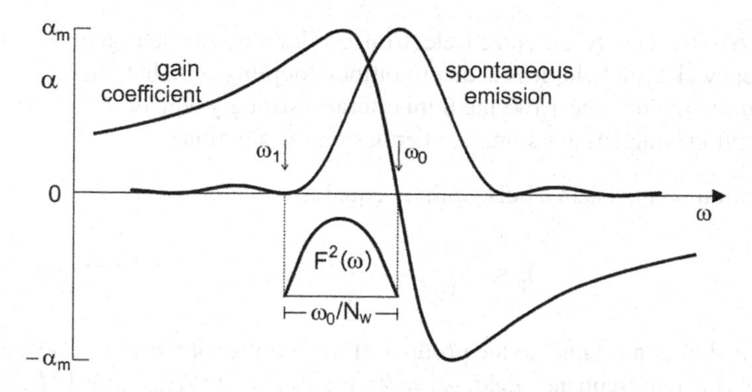

Fig. 19.23 Mode locked free-electron laser at steady state oscillation

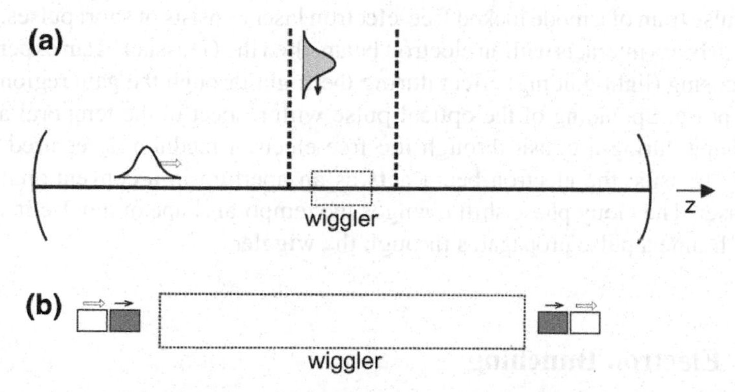

Fig. 19.24 Mode locked free-electron laser. **a** Principle of arrangement. **b** Electron and optical pulse before entering and after leaving the wiggler

the high frequency electric field of the optical pulse and the electrons of an electron pulse occurs during the whole flight of the two pulses through the wiggler. At the end of the wiggler, the interaction finishes, because of absence of a spatially periodic magnetic field.

A portion of radiation is coupled out from the laser resonator by means of an output coupling mirror. The radiation generated by the mode locked free-electron laser is a coherent pulse train. Each time an electron pulse travels through the wiggler, the optical pulse is amplified: at steady state oscillation of the mode locked free-electron laser, the optical gain per single transit is equal to the loss per round trip transit through the resonator.

At steady state oscillation of a mode locked free-electron laser, different modifications of the pulses occur in three regions of the wiggler, each with a length of about one third of the wiggler length:

- *Synchronization region*. The high frequency electric field synchronizes the transverse oscillations of the electrons.

- *Gain region.* The synchronized electron oscillations mediate gain for the high frequency electric field; loss is due to output coupling of radiation.
- *Saturation region.* The active medium interacts strongly with the optical radiation. Absorption compensates stimulated emission of radiation.

The duration of a pulse in a pulse train is equal to

$$t_{\mathrm{p}} \approx \frac{1}{\Delta\nu_0} \approx \frac{N_{\mathrm{w}}}{\nu_0} = N_{\mathrm{w}} T. \tag{19.115}$$

The pulse duration is equal to the product of the number of wiggler periods and the period of the high frequency field; $\omega_0 = 2\pi\nu_0$; $\Delta\omega_0 = 2\pi\Delta\nu_0$; $\nu_0 = 1/T$.
Example For $\nu = 60$ THz, $N_{\mathrm{w}} = 70$, we find $t_{\mathrm{p}} \approx 1$ ps.

The pulse train of a mode locked free-electron laser consists of short pulses. When a Gaussian beam interacts with an electron beam, then the Gaussian beam experiences a self-focusing (light guiding) effect during the flight through the gain region of the electron pulse. Spreading of the optical pulse with respect to the temporal and the lateral shape during a transit through the free-electron medium is repaired by the following transits; the electron beam acts as an aperture in a conventional phase locked laser. The Gouy phase shift changes the temporal shape of a pulse in a pulse train, each time a pulse propagates through the wiggler.

19.16 Electron Bunching

Synchronization of the transverse oscillations to the high frequency electric field leads also to synchronization of the longitudinal oscillations according to the rigid coupling of the transverse and the longitudinal oscillation of an electron. It follows that the transverse modulation current of an electron is joined with a longitudinal modulation current of the electron. In an ensemble of electrons, a longitudinal modulation current density is built up in the synchronization region of the wiggler. Accordingly, the electrons form bunches (*microbunches*) along the z direction. At a fixed location in the wiggler, the electron distribution, $N(t)$, is enhanced in temporal distances of half the period of the optical field and reduced in the regions in between (Fig. 19.25). This corresponds to bunching along the z axis, with two electron bunches every wavelength of the optical field.

At steady state oscillation of a free-electron laser, electrons enter the wiggler at different times. The high frequency electric field, which propagates through the electron beam, leads to buildup of bunches within the first third of the wiggler. In the second third of the wiggler, the electron bunches become more pronounced and mediate gain for the optical field. In the third part of the wiggler, electron bunches persist. Then, stimulated emission and absorption of radiation compensate each other.

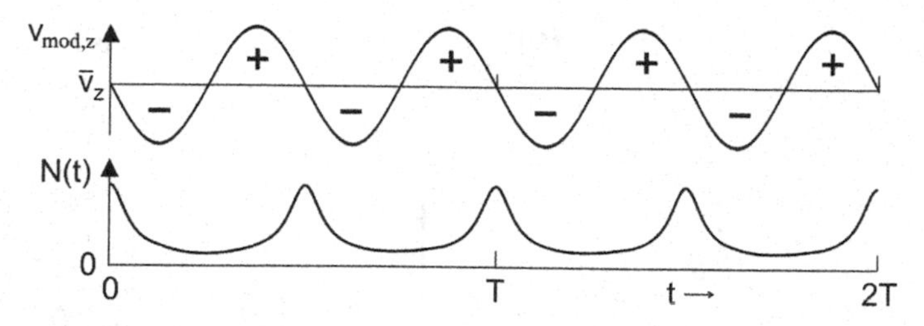

Fig. 19.25 Electron microbunches

19.17 Energy-Level Description of a Free-Electron Laser Medium

At begin of this section, we introduce the *hypothesis*: A transverse oscillation of a free-electron propagating through a spatially periodic magnetic field can be described as a quantum system, namely as an energy-ladder system. Characteristic of an energy-ladder system is the equidistance of the energy levels. An electron occupies one level of the corresponding energy-ladder system. (A motivation for the thesis will be given at the end of this section.)

Characteristic of an energy-ladder system (Fig. 19.26a), are the energy levels

$$E_l = l\, E_0, \tag{19.116}$$

where l is an integer and

$$E_0 = h\nu_0 \tag{19.117}$$

is the transition energy, i.e., the energy distance between two next-near energy levels. Electromagnetic radiation interacts via spontaneous emission, absorption, and stimulated emission according to the Einstein coefficients. However, absorption and stimulated emission processes have the same transition probability (Fig. 19.26b). Therefore, the average rate of absorption processes is the same as the average rate of stimulated emission processes if the phase of the radiation is equal to the resonance frequency, i.e., if $\nu = \nu_0$. The description as a phase modulation (Sect. 19.7) indicates that stimulated emission prevails if $\nu < \nu_0$ and absorption if $\nu > \nu_0$. Accordingly, the gain coefficient curve is not a Lorentz resonance curve but a Lorentz dispersion curve.

In a strong electromagnetic field, transitions between next near levels are also allowed as multiphoton transitions (Fig. 19.26c) corresponding to the condition

$$nh\nu = h\nu_0; \quad n = 1, 2, \ldots . \tag{19.118}$$

This corresponds to transverse velocity components of higher order according to (19.53).

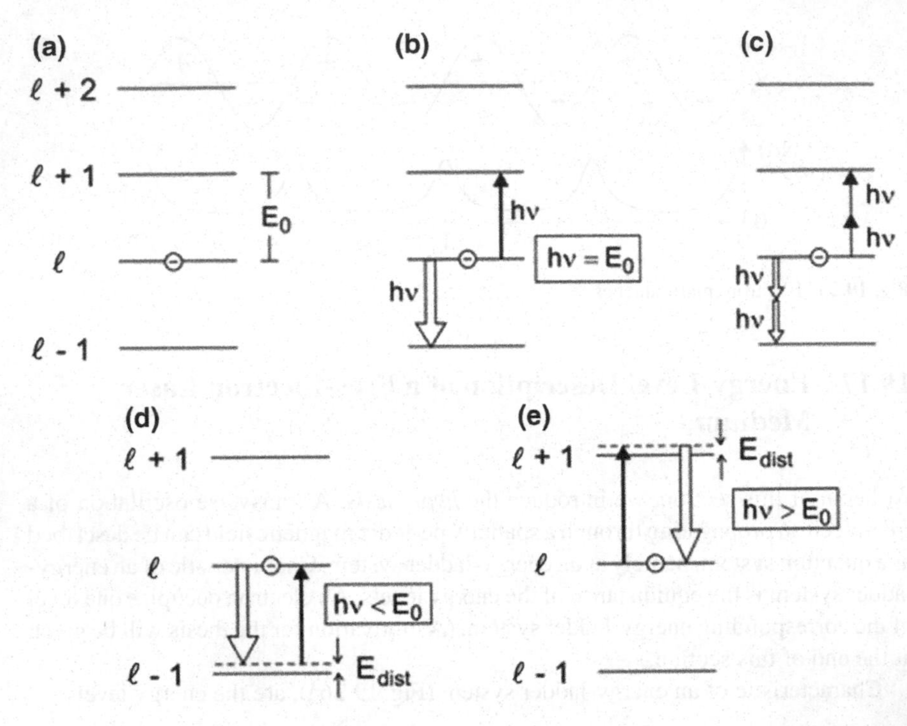

Fig. 19.26 Energy levels of an electron in a periodic magnetic field and transitions. **a** Energy ladder. **b** Absorption and stimulated emission. **c** Two-photon transitions. **d** Stimulated emission and absorption for $h\nu < E_0$. **e** Absorption and stimulated emission for $h\nu > E_0$

We have the following possibilities.

- A radiation field experiences a population inversion if $h\nu < E_0$ (*see* Fig. 19.26d). In a stimulated emission process by radiation at the frequency ν by an $l \rightarrow l - 1$ transition, the transition energy E_0 is converted to photon energy $h\nu$ and distortion energy, E_{dist}, connected with a single-photon transition:

$$E_0 = h\nu + E_{\text{dist}}. \tag{19.119}$$

The distortion energy E_{dist} for a transition is much smaller than the distortion energy E^* occurring if the amplitude of the high frequency electric field is equal to the saturation field amplitude,

$$E_{\text{dist}} <<< E^*. \tag{19.120}$$

Absorption does not occur since the states of distortion are almost unpopulated, i.e., the upper laser level has an occupation number of nearly unity, $f_2 \approx 1$, and the lower level of nearly zero, $f_1 \approx 0$. After a stimulated emission process, the weak distortion is repaired under the action of the Lorentz force of the wiggler.

- A radiation field does not experience population inversion if $h\nu > E_0$ (Fig. 19.26e). In an absorption process, a photon is converted into excitation energy E_0 and energy of distortion,

$$h\nu = E_0 + E_{\text{dist}}. \tag{19.121}$$

  The reverse process, namely stimulated emission by an $l+1 \rightarrow l$ process, does not occur since the states of distortion are almost unpopulated, i.e., the upper level has the occupation number of nearly zero, $f_2 \approx 0$, and the lower laser level of nearly unity, $f_1 \approx 1$.
- If $h\nu = E_0$, upward and downward transitions are equally strong and there is no net energy transfer from the field to the electrons and vice versa.

The states belonging to an energy-ladder system are transient states according to the finite time of flight of an electron through the wiggler. However, the time of flight of an electron through the wiggler is by many orders of magnitude larger than the period of a free-electron oscillation. A description of a transversely oscillating electron by an energy-ladder system may therefore be justified.

We are using of the following quantities:

- E_0 = transition energy = resonance energy.
- $\nu_0 = E_0/h$ = transition frequency = resonance frequency.
- ν = laser frequency (slightly smaller than the resonance frequency).
- τ_{stim} = time between two subsequent stimulated emission processes.

Figure 19.27 illustrates, in the energy-level description, the principle of the free-electron laser at steady state oscillation. An electron of energy $E_{\text{el},0}$ injected into the wiggler forms an energy ladder system. A stimulated emission process leads to a weakly distorted state that becomes an undistorted state. A cascade of stimulated

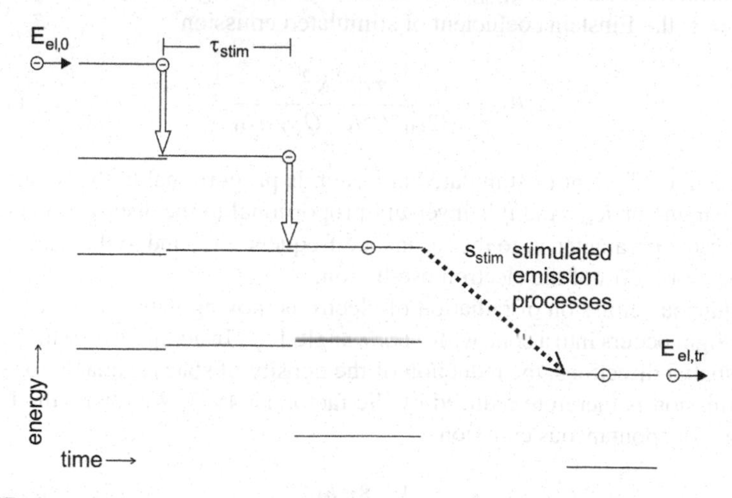

Fig. 19.27 Cascade of stimulated emission processes in an energy-ladder system (in the gain region of the wiggler); τ_{sim}, time between two subsequent stimulated emission processes

transitions in the energy-ladder system contributes to amplification of radiation. The electron leaves the gain region of the wiggler (and enters the saturation region) at an energy $E_{\mathrm{el,tr}}$. The energy difference $E_{\mathrm{el,0}} - E_{\mathrm{el,tr}}$ corresponds to the energy of the number s_{stim} of photons generated by stimulated emission. During the flight of an electron through the gain region of the wiggler, energy of longitudinal motion is converted to energy of the high frequency field; amplification occurs only for radiation propagating in $+z$ direction. A stimulated transition from a level l of an energy-ladder system occurs to the high-energy wing of the level $l - 1$ of the same energy-ladder system.

The energy of an electron after transit through the gain region is equal to

$$E_{\mathrm{el,tr}} = E_{\mathrm{el,0}} - s_{\mathrm{stim}} h\nu. \tag{19.122}$$

In an infrared free-electron laser, the number s_{stim} of stimulated emission processes is very large (on the order of 10^5).

We estimate the Einstein coefficients of stimulated emission and of absorption from the expression of the gain coefficient

$$\alpha(\nu) = 2\alpha_m \, \bar{g}_{\mathrm{L,disp}}(\nu) \tag{19.123}$$

by comparison with an expression, (7.31), derived earlier for a two-level atomic system,

$$\alpha(\nu) = (1/c) h\nu B_{21} \frac{2}{\pi \Delta\nu_0} \bar{g}_{\mathrm{L,res}}(\nu)(N_1 + N_2)(f_2 - f_1). \tag{19.124}$$

We replace the normalized Lorentz resonance function $\bar{g}_{\mathrm{L,res}}$ by the (normalized) Lorentz dispersion function $\bar{g}_{\mathrm{L,disp}}$ and obtain, by replacing $N_1 + N_2$ by N_0 and with $f_2 - f_1 = 1$, the Einstein coefficient of stimulated emission

$$B_{21} = \frac{\pi e c^2 K_{\mathrm{w}}^2}{32\varepsilon_0 (E^*/e)\, Q_0 \gamma h\nu_0}. \tag{19.125}$$

The Einstein coefficient of stimulated emission is proportional to the square of the wiggler parameter K_{w}. And it is inversely proportional to the distortion energy E^*, to the Lorentz parameter γ, to the resonance frequency ν_0, and to the quality factor $Q_0 = N_{\mathrm{w}} = \nu_0/\Delta\nu_0$ of the electron oscillation.

Spontaneous emission of radiation of electrons moving with a velocity near the speed of light occurs into a cone with a cone angle $1/\gamma$. In comparison with emission into all spatial directions, the reduction of the density of states available for spontaneous emission is therefore reduced by the factor $1/(4\gamma^2)$. We obtain the Einstein coefficient of spontaneous emission

$$A_{21} = \frac{1}{4\gamma^2} \frac{8\pi h\nu_0^3}{c^3} B_{21}. \tag{19.126}$$

Table 19.2 Einstein coefficients of a free-electron laser medium

	Value	
ν_0	6×10^{13} Hz	Resonance frequency
$\Delta\nu_0 = \nu_0/N_w$	1.2×10^{12} Hz	Width of resonance
γ	100	Lorentz parameter
K_w	1	Wiggler parameter
E^*/e	2 V	Distortion energy
$Q_0 = N_w = \nu_0/\Delta\nu_0$	50	Quality factor
B_{21}	2×10^{23} m^3 J^{-1} s^{-2}	Einstein coefficient
A_{21}	100 s^{-1}	Einstein coefficient
τ_{sp}	10 ms	Spontaneous lifetime

The spontaneous lifetime is given by

$$\tau_{sp} = 1/A_{21}. \tag{19.127}$$

Table 19.2 shows values of Einstein coefficients characterizing transitions between energy-ladder levels of a free-electron laser medium. The data are obtained for a free-electron laser presented in Sect. 19.6 (Table 19.1). The Einstein coefficient of stimulated emission is larger than that of active media of conventional lasers (for a comparison, *see* Table 6.1 in Sect. 6.5).

The gain cross section $\sigma_{21}(\omega)$ has the same frequency dependence as the gain coefficient. The maximum gain cross section is $\sigma_{21,m} = \alpha_m/N_0$ ($= 2 \times 10^{-18}$ m^2). In comparison, a naturally broadened two-level system propagating with a velocity corresponding to a Lorentz factor $\bar{\gamma}$ would have a gain cross section $4\bar{\gamma}^2\lambda^2/2\pi$ ($\sim 10^{-7}$ m^2).

The states belonging to an energy-ladder system are transient states according to the finite time of flight of an electron through the wiggler. However, the time of flight is by many orders of magnitude larger than the period of a free-electron oscillation. An illustration of a free-electron medium as an ensemble of energy-ladder sytems may therefore be justified.

Figure 19.28 is a modified version of the preceding figure. A beam of electrons of energy $E_{el,0}$ enters the wiggler. In the synchronization region, each electron loses, on average, a small amount of energy due to synchronization by stimulate emission of radiation. The electrons then enter the gain region with nearly the original energy. In the gain region, an electron loses energy of longitudinal motion due to a cascade of stimulated emission processes. In the saturation region, the electron energy remains almost unchanged although the transition rate of electronic transitions is large: a fast cascade of absorption processes follows on a fast cascade of stimulated emission processes. Absorption compensates stimulated emission. A fast cascade of stimulated emission processes occurs during a period of the electron oscillation and a fast cascade of absorption processes during the following period. A fast cascade of stimulated emission processes is joined with a distorted state of an electron. The reversed process leads back to the undistorted state.

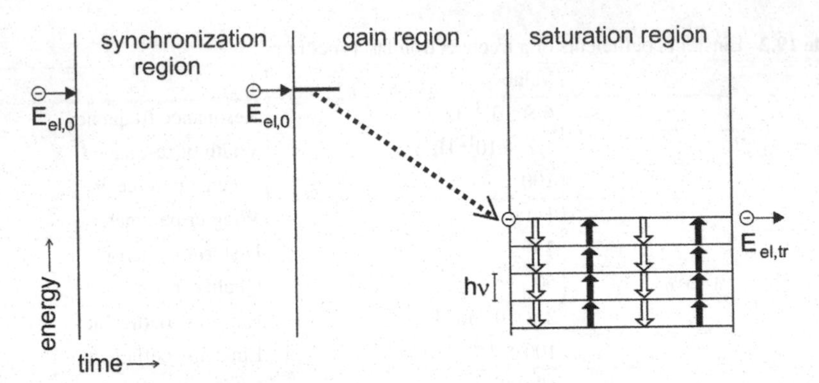

Fig. 19.28 Propagation of an electron beam through the synchronization region, the gain region, and the saturation region of the wiggler (with a fast cascade of absorption processes following on a fast cascade of stimulated emission processes, in turn)

In order to discuss the saturation behavior, we estimate the time between two stimulated emission processes. The electron transit rate is equal to $r_{el} = I_{el}/e$. The power of an electron is given by $P_{el} = U\ I_{el}$, where U is the voltage that corresponds to the electron energy. The power of the radiation emitted by the laser is equal to $P_{out} = \eta P_{el}$, where η is the efficiency of the laser. The rate of photon emission is given by $r_{ph} = P_{out}/(h\nu)$. The average time between two stimulated emission processes is equal to

$$\tau_{stim} = \frac{r_{el}}{r_{ph}} t_{tr} = \frac{(h\nu/e)\, t_{tr}}{\eta U}, \tag{19.128}$$

where t_{tr} is the transit time, i.e., the time an electron takes for a transit through the wiggler.

Example 1.6 μm FEL, pulsed laser; $\nu =$ 188 THz, $h\nu = 0.8$ eV, $U =$ 115 MV, $L_w = 2$ m, $t_{tr} = 6$ ns, $I_{el} = 100$ A, $\eta = 2 \times 10^{-2}$, $P_{out} =$ 200 MW. We find $\tau_{stim} = 2 \times 10^{-15} s$.

The time given in the example is about equal to half the period of the electron oscillation. Since the time τ_{stim} is an average taken over the time of flight of an electron through the whole wiggler, the time between two stimulated emission processes is much shorter in the final part of the gain region and, accordingly, in the saturation region. Thus, the assumption of a fast cascade of stimulated emission processes within one period of an electron oscillation cycle appears to be justified. The work E^* done by an electron in a cascade corresponds to stimulated emission of a number of photons. We can estimate this number n of photons involved in a cascade process within the saturation region of the wiggler assuming that the distortion energy is about equal to the energy of the phonons, $n\ h\nu \approx E^*$. The number n $(\approx E^*/h\nu \approx 3$ for the 1.6 μm FEL and for $E^* = 2$ V) increases with decreasing frequency. If the energy of a photon is larger than E^*, then saturation is expected if τ_{stim} is smaller than the period of the electron oscillation. According to (19.127), the time τ_{stim} increases proportional to U. Therefore, saturation of a SASE X-ray free-electron laser does not occur: τ_{stim} is larger than the period of the electron oscillation. This result is in accord with the conclusion that the saturation field cannot be reached in a SASE X-ray free-electron laser (Sect. 19.11).

Here, we discuss the motivation for the introduction of an energy-ladder system. It is known that a free-electron in a crystal propagating through a periodic electric potential can be described as a quantum system. The energy levels of the electron form an energy-ladder system, a Wannier-Stark ladder [274]. The levels are energetically equidistant.

There is a formal connection between a theoretically well studied superlattice Bloch laser and a free-electron laser:

- The active medium of a superlattice Bloch laser (Chap. 32) consists of an ensemble of free-electrons in a spatially periodic electric potential. The electrons execute, under the action of a static electric field and the periodic potential, free-electron oscillations. An electromagnetic field modulates the free-electron oscillations, which leads to a synchronization of the free-electron oscillations to the field and to gain for the field. The gain coefficient curve is a Lorentz dispersion curve. The states of an electron subject to both a periodic potential and a static field are quantum mechanical describable as Wannier–Stark states. The energy levels of an electron form a Wannier–Stark ladder—i.e., an energy-ladder with equidistant energy levels [175, 274]. An electron occupies one of the levels. A stimulated transition occurs from the occupied Wannier–Stark level to an intermediate level that corresponds to the distorted level of the energetically next near level of lower energy (Sect. 32.7 and [272]).
- The active medium of a free-electron laser consists of an ensemble of Free-electrons that execute, under the action of a periodic magnetic field, free-electron oscillations. An electromagnetic field modulates the free-electron oscillations, which leads to a synchronization of the free-electron oscillations to the field and to gain of the field. The gain coefficient curve is a Lorentz dispersion curve. We introduced—on the basis of the similarity of the formal description of the free-electron oscillations in a free-electron laser and the free-electron oscillations in a Bloch laser as monopole oscillations—an energy-ladder description.

The transverse oscillation of an electron in a free electron laser and a Bloch oscillation of an electron in a semiconductor superlattice have in common that the oscillations are monopole oscillations. An external perturbation can change the phase of the oscillation of a monopole oscillator, but not the amplitude. In comparison, an external perturbation can change the amplitude and the phase of a dipole oscillator. We used a dipole oscillator as a classical model of a two-level system. In an analogous way, a monopole oscillator may be seen as a classical model of an energy-ladder system. It is an open question, whether it is possible to develop a theory of transient energy states for relativistic electrons in a spatially periodic magnetic field.

19.18 Aspects of Free-Electron Laser Theory

Quantum effects taking account of the recoil of an electron due to emission of a photon are expected to be important for free-electron lasers generating radiation at short X-ray wavelengths. Theoretical aspects have been treated in various studies, for instance in [175, 176, 325, 326].

Analysis of a free-electron laser is possible on basis of Maxwell's equations together with laws of classical equations describing interaction of an accelerated or decelerated relativistic electron with an electromagnetic field. The equations are solvable by using methods of numerical simulation; for a treatment of a free-electron laser by a three-dimensional simulation, *see*, for instance, [326].

The equations of motion of electrons in the limit of negligibly small electromagnetic fields can approximately solved in a one-dimensional description and provide the gain [5, 166, 312]:

$$G^0 - 1 = (G_{\rm m}^0 - 1)g(X), \tag{19.129}$$

where

$$X = 2\pi N_{\rm w}\left(\frac{\omega}{\omega_0} - 1\right) \tag{19.130}$$

and

$$g(X) = -\frac{\rm d}{{\rm d}X}\left(\frac{\sin(X/2)}{X}\right)^2 = -\frac{1}{X^3}\left(1 - \cos X - \frac{X}{2}\sin X\right). \tag{19.131}$$

The gain curve (Fig. 19.29) is antisymmetric with respect to the resonance frequency $X = 0$; there is no gain at the resonance frequency ω_0. The gain curve shows a modulation according to the finite time of flight of the electrons through the wiggler.

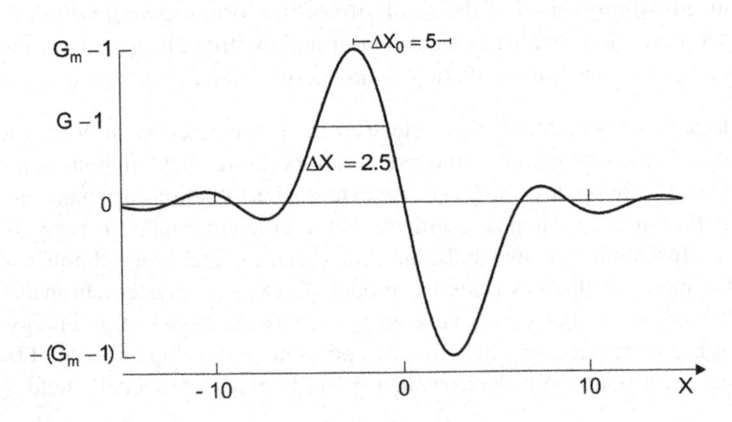

Fig. 19.29 Small-signal gain derived from equations of motion

The lineshape function $g(X)$ is equal to the derivative of the function that describes the spectral distribution of spontaneously emitted radiation. (This is sometimes called "Madey theorem".) Maximum gain, $G_\mathrm{m}^0 - 1$, occurs at a frequency ($X_\mathrm{m} \approx -2.6$) slightly smaller than the resonance frequency. The frequency distance ($\Delta X_0 \approx 5.2$) between the largest maximum and the corresponding minimum is given by

$$\frac{\Delta\omega_0}{\omega_0} \approx \frac{1}{N_\mathrm{w}}. \tag{19.132}$$

The halfwidth of the largest peak ($\Delta X \approx -2.5$) corresponds to a bandwidth

$$\frac{\Delta\omega}{\omega_0} \approx \frac{1}{2N_\mathrm{w}}. \tag{19.133}$$

$G_\mathrm{m}^0 - 1$ is given by [166, 312]

$$G_\mathrm{m}^0 - 1 = \frac{0.84\, j_\mathrm{el}\, L_\mathrm{w}\, N_\mathrm{w}^2}{\varepsilon_0\, (m_0\, c^2/e)} \frac{K_\mathrm{w}^2 \lambda_\mathrm{w}}{c\, \gamma^3}. \tag{19.134}$$

In comparision, the modulation model provides the maximum gain

$$G_\mathrm{m}^\mathrm{MM} - 1 = \frac{0.79\, j_\mathrm{el}\, L_\mathrm{w}}{\varepsilon_0\, (E^*/e)} \frac{K_\mathrm{w}^2}{\gamma\, \omega}. \tag{19.135}$$

The comparison shows that the frequency dependence is the same. Replacing the Lorentz factor, we find the expressions

$$G_\mathrm{m}^0 - 1 = \frac{37\, N_\mathrm{w}^2}{\varepsilon_0\, (m_0\, c^2/e)} L_\mathrm{w} \frac{I_\mathrm{el}}{a_\mathrm{mode}} \frac{c^{1/2}}{\lambda_\mathrm{w}^{1/2}} \frac{K_\mathrm{w}^2}{(1 + K_\mathrm{w}^2/2)^{3/2}}\, \omega^{-3/2} \tag{19.136}$$

and

$$G_\mathrm{m}^\mathrm{MM} - 1 = \alpha_\mathrm{m} L_\mathrm{w} = \frac{2.8}{\varepsilon_0\, (E^*/e)} L_\mathrm{w} \frac{I_\mathrm{el}}{a_\mathrm{mode}} \frac{c^{1/2}}{\lambda_\mathrm{w}^{1/2}} \frac{K_\mathrm{w}^2}{(1 + K_\mathrm{w}^2/2)^{1/2}}\, \omega^{-3/2}. \tag{19.137}$$

In comparison, the small-signal gain derived on basis of the equations of motion is proportional to N_w^2 while the small-signal gain following from the modulation model is inversely proportional to the distortion energy E^*. (There is also a difference in the dependence on the wiggler parameter.) The two expressions lead to the same magnitude of the gain for $N_\mathrm{w} = 140$. Thus, the gain determined by the two expressions is, for wiggler periods of the order of 100, on the same order of magnitude.

Example 1.6 μm FEL, *see*, Example 1, Sect. 19.10. $\nu = 188$ THz, $E_\mathrm{el} = 115$ MeV, $\gamma = 230$, $N_\mathrm{w} = 30$, $L_\mathrm{w} = 1.65$ m, $\lambda_\mathrm{w} = 5.5$ cm, $K_\mathrm{w} = 0.64$, $I_\mathrm{el} = 100$ A, $L_\mathrm{res} = 30$ m, $1 - R = 0.1$, $L_\mathrm{res} = 30$ m, $r_\mathrm{u} = 2$ mm. We find $G_\mathrm{m}^\mathrm{MT} - 1 = 0.012$ and $G_\mathrm{m}^\mathrm{MM} - 1 = 0.7$.

The gain $G_{\mathrm{m}}^{0} - 1$ is proportional to the length of the wiggler L_{w} and, additionally, proportional to the square of the number N_{w} of wiggler periods. This dependence corresponds, for a free-electron laser driven by short electron pulses, to a gain that is proportional to L_{w} and additionally, to N_{w} since the width of the laser oscillation interval decreases as $1/N_{\mathrm{w}}$. The gain $G_{\mathrm{m}}^{\mathrm{MM}} - 1$ is also proportional to the length of the wiggler. But it shows no additional dependence on N_{w}. This corresponds to a gain coefficient that is a local quantity. During the onset of laser oscillation $G_{\mathrm{m}}^{\mathrm{MM}} - 1$ remains constant while the spectrum of radiation broadens, joined with a shift of the maximum.

An experimental study [168] (performed in the High Energy Physics Laboratory, Stanford University) provided a gain curve that is in accordance with the theoretical curve (Fig. 19.31). In the experiment, CO_2 laser radiation (wavelength 10.6 μm) propagated through a spatially periodic magnet field (strength 0.24 T, period 3.2 cm, wiggler length 5.2 m) and interacted with an electron beam (24 MeV) that consisted of pulses (peak current 0.1 A). The electron energy was swept through a range in the vicinity of 24 MeV. The gain (during an electron pulse) was 7%. The power of the transmitted radiation increased by a value (4×10^3 W), which was 10^9 times larger than the power of spontaneously emitted radiation.

Shortly after the observation of amplification, the same laboratory published the first successful operation of a free-electron laser [169]. The arrangement was the same as for the amplification experiment, with two differences: the electrons had a larger energy (43 MeV) and radiation was stored in a laser resonator. Infrared radiation (wavelength 3.4 μm, relative spectral width 10^{-2}) had a power of 5×10^5 W within the resonator (output mirror transmissivity 1.5%) that was much larger than the power of the CO_2 laser radiation used for the amplification experiment. [John Madey once explained me (K.F.R.) in his laboratory in Stanford: a laser experiment is much easier to perform than an amplification experiment.]

Now, we mention the pendulum model of the free-electron laser. The model follows as a one-dimensional approximation of the equations of motion of an electron in a periodic magnetic field and a high frequency electric field: the equations of motion are given in the Problems to this chapter. The pendulum model (not treated in this book) provides a saturation field amplitude $A_{\mathrm{sat}}^{\mathrm{PM}}$ that corresponds to the criterion: in a passage through the wiggler, an electron can lose the maximum energy leading to a relative shift of the resonance frequency by $(\Delta\omega_0/2)/\omega_0 = 1/2N_{\mathrm{w}}$. This corresponds to an efficiency of conversion of electron energy to radiation power of $\eta = 1/2N_{\mathrm{w}}$. Accordingly, the maximum power generated by a free-electron laser is equal to $P = (1/2)c\varepsilon_0(A_{\mathrm{sat}}^{\mathrm{PM}})^2\pi r_u^2(1-R)$. With $P = \eta U l$, we find $A_{\mathrm{sat}}^{\mathrm{PM}} = (2Ul)^{1/2}(\eta c\varepsilon_0\pi r_u^2(1-R)^{-1/2}$. This saturation field amplitude increases with the current strength.

Here, we summarize aspects treated by different models.

- Maxwell's theory together with theorems treating interaction of relativistic electrons with electromagnetic fields; provides a complete description of a free-electron laser.
- Gain $G_m^0 - 1$ (often called small-signal gain of a free-electron laser) applies for negligibly small fields; it corresponds to gain following from the equations of motion at weak high frequency field.
- The pendulum model provides an ultimate upper limit of the saturation field; however, the field cannot be reached because the model ignores strong changes of the electron orbits in connection with saturation.
- The modulation model provides small-signal gain and saturation field and describes onset of laser oscillation; the model assumes that stimulated emission of a photon results in a change of the phase of the electron oscillation and that saturation is due to strong distortions of the electron orbits. A determination of the distortion energy E^* is performed by an analysis of free-electron laser data (Sect. 19.10). Problem 19.16 presents a study that provides the magnitude of E^* and the dependence on the strength of the wiggler field and on the wiggler wavelength.

19.19 Comparison of a Free-Electron Laser with a Conventional Laser

We illustrate dynamical processes occurring in a conventional laser and a free-electron laser. We will use the two different aspects, one of them based on a quantum mechanical description and the other based on classical oscillator models.

We consider, in the quantum mechanical description of a conventional laser, a two-level system (Fig. 19.30a). A pump process leads to population of the upper level. After a stimulated emission process and subsequent fast relaxation, the two-level system is unpopulated. A new pumping process leads again to population of the upper level. The processes occur repeatedly. Population inversion in an ensemble of two-level systems, which constitute the active medium of a laser, is a necessary condition of laser oscillation.

In the quantum mechanical description of the free-electron laser (Fig. 19.30b), an electron occupies a level of an energy-ladder system. A stimulated emission process leads to occupation of a distorted state at an energy slightly larger (by the distortion energy E_{dist}) than the energy of the next lower energy level. Stimulated emission processes occur repeatedly. Population inversion in an ensemble of energy-ladder systems, which constitute an active medium of a free-electron laser, is a necessary condition of free-electron laser oscillation.

We now discuss classical oscillator models of the active medium in a laser and of the active medium in a free-electron laser. In a conventional laser, the polarization current of a dipole oscillator oscillates, at a phase of π, synchronously to the high frequency electric field (Fig. 19.31a). After a stimulated emission process (indicated

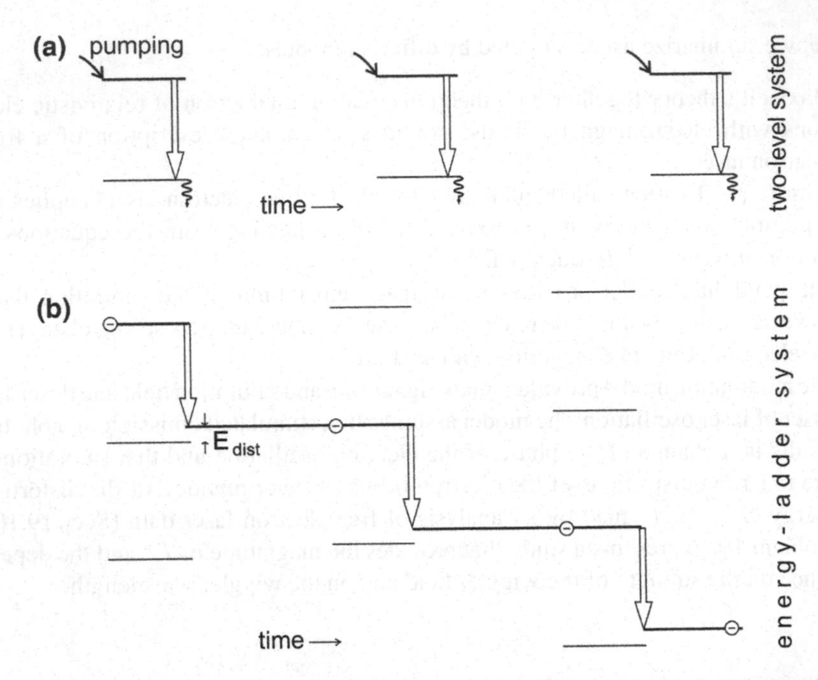

Fig. 19.30 Energy level based description of a conventional laser and of a free-electron laser. **a** Two-level system, with repeated processes of pumping, stimulated emission of a photon, and relaxation of the population of the lower level. **b** Energy-ladder system,

by a star) and subsequent relaxation of the lower level of the corresponding two-level system, the dipole oscillator disappears. A pumping excitation of the upper level results again in a dipole oscillation with the polarization current oscillating, at the phase of π, synchronously to the field. In a free-electron laser, an electron performs a monopole oscillation at the resonance frequency that is slightly larger than the frequency of the high frequency field in the laser resonator (Fig. 19.31b). Accordingly, the phase between the electric current of an electron oscillator and the high frequency field increases with time. This dephasing is an inherent property of a free-electron laser medium.

In a conventional laser, gain is mediated by a dielectric polarization of the active medium; during onset of laser oscillation the high frequency polarization grows together with the field. In a free-electron laser, gain is mediated by a high frequency electric current density; during onset of laser oscillation the high frequency current density grows together with the field.

In our classical treatment of a conventional laser, we related quantum mechanical dipole transitions and classical dipole oscillators, octroying the effect of population inversion by an appropriate adjustment of the phase between a field and a dipole oscillator (Chap. 9). In our treatment of the free-electron laser, we reversed the procedure. We introduced electron-monopole oscillators as elementary oscillators in a free-electron laser medium and, then, the energy-ladder system as a quantum mechanical description.

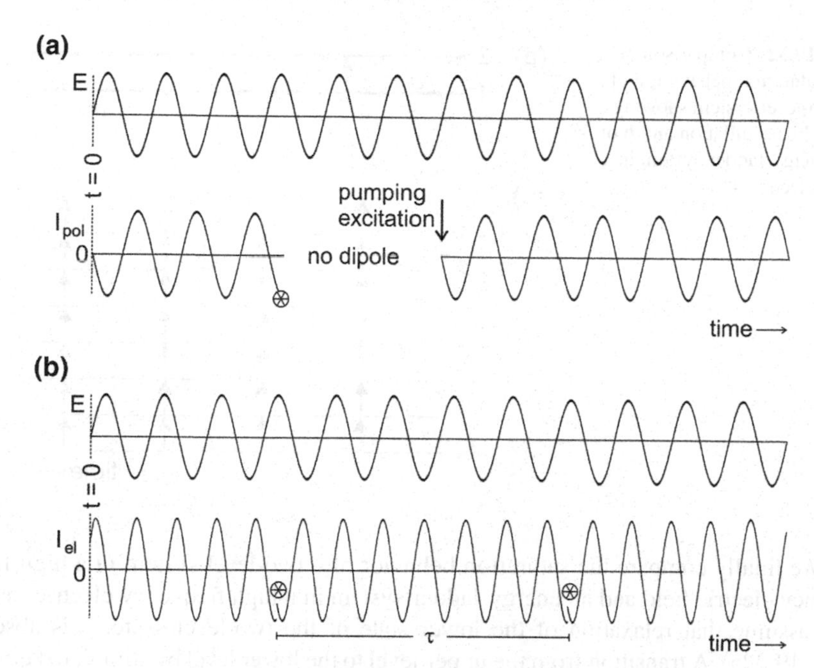

Fig. 19.31 Conventional laser and free-electron laser in classical model descriptions (with the star indicating stimulated emission of a photon). **a** Electric field and polarization current of a dipole oscillator in an active medium of a laser. **b** Electric field and transverse electric current of an electron-monopole oscillator in an active medium of a free-electron laser

Table 19.3 summarizes the comparison of a conventional laser with a free-electron laser.

Table 19.3 Conventional laser and free-electron laser: a comparison

	Conventional laser	Free-electron laser
Elementary system	Two-level system	Energy-ladder system
Classical model	Dipole oscillator	Monopole oscillator
Effect of an optical field	Induced dipole moment	Frequency modulation
Energy source	Pumping energy	Translational energy
Gain mediator	Dielectric polarization	Electric current density
Shape of gain coefficient	Lorentz resonance function	Normalized Lorentz dispersion function
Saturation field	No	Yes
Population inversion	Yes	Yes (quantum mechanical picture) No (modulation model) No (Maxwell's theory)

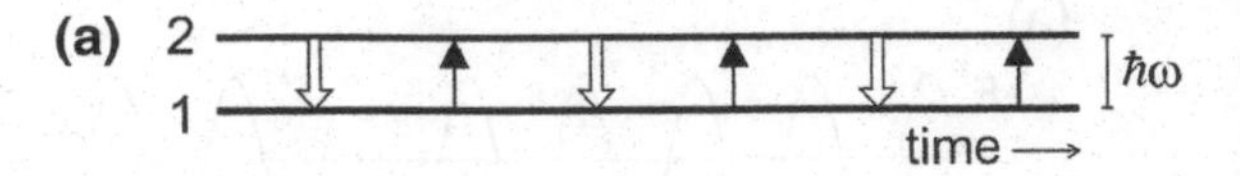

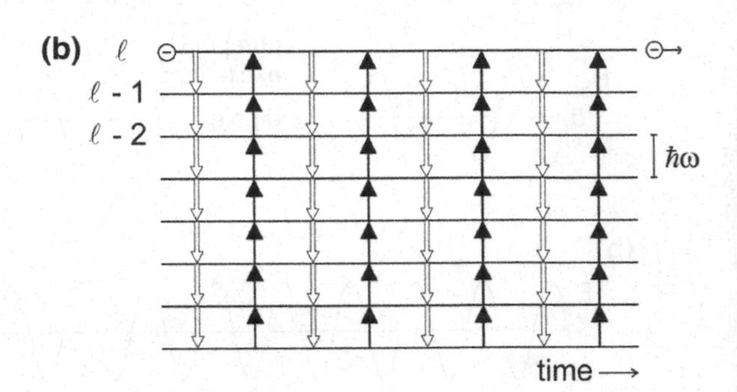

Fig. 19.32 Comparison of the saturation behavior, **a** of a two-level system showing a Rabi like oscillation and **b** of an energy-ladder system in a free-electron laser

We finally compare the saturation behavior of a two-level system in a high frequency electric field and an energy-ladder system in a high frequency electric field. We assume that relaxation of the lower state of the two-level systems is absent (Fig. 19.32a). A transition from the upper level to the lower level by stimulated emission of a photon follows a transition from the lower to the upper level by absorption of a photon. Stimulated emission and absorption follow each other. This behavior is consistent with the gain coefficient that is proportional to the population difference $N_2 - N_1$ of an ensemble of two-level systems (Chap. 9). It is therefore also consistent that the phase between polarization current and field is equal to π if a two-level system is in the upper state and that it is zero if the two-level system is in the lower state. In the case that a strong high frequency coherent field saturates an ensemble of two-level systems, the polarization and thus the polarization-current density follow the field coherently (*see* Problems to Chap. 9). Figure 19.32b illustrates saturation of a free-electron laser. A fast cascade of stimulated emission processes promotes an electron from a state *with* population inversion to a distorted state *without* population inversion. In the reversed process, a fast cascade of absorption processes promotes the electron back to the original level. Emission followed by absorption of radiation at a large amplitude of the high frequency electric field resembles the Rabi oscillation that is due to interaction of two-level systems with a strong high frequency electric field. (An analogy of the saturation behavior of a free-electron laser with a Rabi type of oscillation has already been suggested in connection with an analysis of a free-electron laser on basis of the pendulum model [312].)

Table 19.4 indicates the main difference between a conventional laser and a free-electron laser: the elementary excitation of a conventional laser medium, on one hand, can be characterized as a two-level system and the elementary system of a free excitation of a free-electron laser medium, on the other hand, as an energy-ladder system.

Table 19.4 Conventional laser and free-electron laser: a comparison

	Conventional laser	Free-electron laser
Elementary excitation	Two-level system	Energy-ladder system
Susceptibility, imaginary part	$\bar{g}_{\mathrm{L,res}}$	$\bar{g}_{\mathrm{L,disp}}$
Susceptibility, real part	$\bar{g}_{\mathrm{L,disp}}$	$\bar{g}_{\mathrm{L,res}}$
Saturation field	No	Yes

Accordingly, the response functions of the polarization—the susceptibilities—are different:

- *Conventional laser medium.* The shape of the imaginary part of the response function is a Lorentz resonance function and the real part is a Lorentz dispersion function. The amplitude of the high frequency field is not limited (besides limitations by the pump strength and due to optical damage).
- *Free-electron laser medium.* The shape of the imaginary part of the response function is a Lorentz dispertion function and the shape of the real part is a Lorentz resonance function. The amplitude of the high frequency field shows an intrinsic limitation.

The lasers have in common that the laser field synchronizes the elementary oscillations to the laser field, and that a population inversion between energy levels of a quantum system occurs.

19.20 Remark About the History of the Free-Electron Laser

We mention a few data concerning the history of the free-electron laser.

1933	P. Kapiza and P. Dirac discussed the possibility of stimulated Compton scattering.
1951	H. Motz (Oxford) proposed to use a wiggler configuration for generation of incoherent radiation.
1971/76	J. Madey (Stanford University) proposed a free-electron laser and realized the first free-electron laser (50 MeV LINAC; superconducting helical coil as wiggler magnet; wavelength of the radiation 3.4 μm).
1983	Petroff (Orsay) used radiation of a storage ring (150 MeV; the laser generated visible radiation (wavelength 650 nm).
1983	Stanford; 1 GeV LINAC.
1983	Los Alamos; 1 GeV LINAC.
1985	L. Elias et al. developed the free-electron laser in Santa Barbara, with a 6 MeV static accelerator; the laser was the first tunable far infrared laser.

1992 FELIX (Rijnhuizen, The Netherlands); pulsed infrared and far infrared laser.
2006 Rossendorf (near Dresden, Germany); pulsed infrared laser.

Presently, free-electron lasers in more than 30 laboratories worldwide are operating or are in planing.

References [5, 12, 166–176, 312–331].

Problems

19.1 Acceleration energies. Given is wiggler ($\lambda_w = 2.4\,\mathrm{cm}$ and $K = 1$). [*Hint*: make use of data of Table 19.1 to solve this and the following problems.]

(a) Determine the electron energy necessary to drive a terahertz FEL at 1 THz and determine the change of energy necessary to change the frequency by 1%
(b) Determine the electron energy necessary to drive an X-ray at a wavelength of 10 nm and determine the change of energy necessary to change the frequency by 1%

19.2 Frequency tuning. Relate a small change of energy relative to the energy E to the relative change of frequency and to the relative change of wavelength of a free-electron laser.

19.3 Show that the inhomogeneous broadening of the gain profile of a free-electron laser due to energy smearing is negligibly small if the condition $\Delta\gamma \ll \gamma/(2N_w)$ is fulfilled.

19.4 Refractive index of a free-electron laser medium.

(a) Estimate the frequency shift of a mode of a free-electron laser resonator that occurs when a free-electron laser is switched on.
(b) Determine the speed of light in a free-electron medium.
(c) Estimate the difference of the time it takes light and the time it takes an electron to propagate through the wiggler.

19.5 Determine the absolute number of electrons present in an active medium of a free-electron laser.

19.6 Estimate the time of onset of laser oscillation in a free-electron laser.

19.7 Determine characteristic quantities of the spectrum of spontaneously emitted radiation on the angular frequency scale: resonance frequency, halfwidth, frequency of the first minima, distance between this frequency and the resonance frequency.

19.8 Show that, on the time average, an electromagnetic field cannot exchange energy with an electron in free space.

19.9 Continuous wave free-electron laser. How is it possible to operate a continuous wave free-electron laser at the frequency of maximum gain although the initial gain at the frequency of maximum gain at steady state oscillation is negligibly small? [*Hint:* Consider the onset of oscillation.]

19.10 Energy spread of the electrons in an electron beam that enters a free-electron laser. (a) Estimate the inhomogeneous broadening of the frequency distribution of the radiation emitted by a mode locked free-electron laser due to a finite energy distribution of the electrons in the electron beam. (b) Estimate the broadening of the pulse duration. (c) If you would plan a mode locked free-electron laser for a wavelength optimized at a wavelength of 3 μm, what tolerance would you allow for the energy spread of the electrons? [*Hint:* the accelerator is the most expensive part of a free-electron laser; an answer like energy spread should be small compared with a value that you find by analyzing laser operation is not sufficient.]

19.11 X-ray SASE FEL. Show that an electron pulse (duration 100 fs) and a pulse of radiation do not separate in an X-ray SASE FEL (wavelength 0.1 nm) at a wiggler length of 100 m.

19.12 Relativistic electron in a periodic magnetic field. We describe, in the laboratory frame, the motion of an electron moving at a relativistic velocity (along the z axis).

(a) Which is the Lorentz force, assuming that an electric field is absent?
Answer: The Lorentz force is equal to $\mathbf{F} = q\mathbf{v} \times \mathbf{B}$, where $q(= -e)$ is the electron charge.
(b) Determine the equation of motion of an electron (mass m_0) moving at a relativistic velocity in a periodic magnetic field; assume that an optical field is absent and that loss of energy due to spontaneous emission of radiation is negligibly small.
Answer:

$$\frac{\mathrm{d}}{\mathrm{d}t}(\gamma m_0 \mathbf{v}) = q\mathbf{v} \times \mathbf{B}. \tag{19.138}$$

Since the electron does not lose energy, the Lorentz factor is a constant and we can write $\gamma \mathrm{d}\mathbf{v}/\mathrm{d}t = q\mathbf{v} \times \mathbf{B}$.
(c) Write the equation of motion for the x component and the z component of the velocity for the case that the magnetic field is oriented along the y direction, namely $\mathbf{B} = (0, B, 0)$ and $B = B_\mathrm{w} \sin k_\mathrm{w} z$, where $k_\mathrm{w} = 2\pi/\lambda_\mathrm{w}$ and λ_w is the period of the wiggler. *Answer*: $\gamma m_0 \mathrm{dv}_x/\mathrm{d}t = q\mathrm{v}_z B$ and $\gamma m_0 \mathrm{dv}_z/\mathrm{d}t = q\mathrm{v}_x B$.
(d) Replace in the equations of motion the term $k_\mathrm{w} z$ by the corresponding time dependent term.
Answer. We suppose that $\mathrm{v}_x << \mathrm{v}_z$ so that $z \approx \mathrm{v}_z t$ and define $\Omega_0 = k_\mathrm{w} \mathrm{v}_z$. We find $\gamma m_0 \mathrm{dv}_x/\mathrm{d}t = q\mathrm{v}_z B_\mathrm{w} \sin \Omega_0 t$ and $\gamma m_0 \mathrm{dv}_z/\mathrm{d}t = q\mathrm{v}_x B_\mathrm{w} \sin \Omega_0 t$.

(e) Determine the solutions.
Answer:

$$\mathrm{v}_x = \frac{qcB_\mathrm{w}\lambda_\mathrm{w}}{2\pi\gamma\mathrm{m}_0}\cos(\Omega_0 t) = -\frac{K_\mathrm{w}c}{\gamma}\cos(\Omega_0 t). \tag{19.139}$$

Using this expression, we find the differential equation

$$\frac{\mathrm{dv}_z}{\mathrm{d}t} = -\frac{qK_\mathrm{w}cB_\mathrm{w}}{\gamma^2 m_0}\sin\Omega_0 t\cos\Omega_0 t = -\frac{qK_\mathrm{w}cB_\mathrm{w}}{2\gamma^2 m_0}\sin 2\Omega_0 t. \tag{19.140}$$

The solution is

$$\mathrm{v}_z = \bar{\mathrm{v}}_z - \frac{K_\mathrm{w}^2 c}{4\gamma^2}\cos 2\Omega_0 t, \tag{19.141}$$

where $\bar{\mathrm{v}}_z = c\left(1 - \frac{1+K_\mathrm{w}^2/2}{\gamma^2}\right)$ is an average velocity for propagation along z and where $K_\mathrm{w} = \frac{e\lambda_\mathrm{w}B_\mathrm{w}}{2\pi m_0 c}$ is the wiggler parameter.
Instead of using the differential equation for v_z, we can make use of the Lorentz factor, $\gamma = \frac{1}{\sqrt{1-(\mathrm{v}_x^2+\mathrm{v}_z^2)}}$. From this relation we find the same expression for v_z.

(f) Determine the effective Lorentz factor (that is the Lorentz factor related to $\bar{\mathrm{v}}_z$) for $K_\mathrm{w} = 1$ and $\gamma = 100$.

(g) Show that the wiggler parameter can be written as the ratio of two energies. One energy term is equal to $ec\lambda_\mathrm{w}B_\mathrm{w}/2\pi$ and the other is the rest energy m_0c^2 of the electron; determine the value of $\lambda_\mathrm{w}B_\mathrm{w}$ for $K_\mathrm{w} = 1$.

(h) Determine the orbit of the electron.
Answer. Integration of the velocity components leads to

$$x(t) = -\frac{K_\mathrm{w}\lambda_\mathrm{w}}{2\pi\gamma}\sin\Omega_0 t, \tag{19.142}$$

$$z(t) = \bar{z}(t) - \frac{K_\mathrm{w}^2\lambda_\mathrm{w}}{8\pi\gamma^2}\sin 2\Omega_0 t, \tag{19.143}$$

with $\bar{z}(t) = \bar{\mathrm{v}}z$. The vector $(z-\bar{z}, x)$ describes the form of an "eight" in the z, x plane (*see* Fig. 19.7). The x component oscillatees with the frequency Ω_0 and the y component with $2\Omega_0$. The frequency $\Omega_0 \approx k_\mathrm{w}c$ is determined by the wiggler period. It is independent of γ and thus of the kinetic energy of the electron; $\Omega_0 = 2\times 10^7\,\mathrm{s}^{-1}$ for $\lambda_\mathrm{w} = 1\mathrm{cm}$.

19.13 Lorentz factor. Relate the relativistic momentum and the relativistic energy with the Lorentz factor.

Answer. The Lorentz factor is given by

$$\gamma = \frac{1}{\sqrt{1 - \mathbf{v}^2/c^2}}. \tag{19.144}$$

The relativistic momentum is equal to

$$\mathbf{p} = \gamma m_0 \mathbf{v} \tag{19.145}$$

and the relativistic energy

$$E = \sqrt{m_0^2 c^2 + c^2 \mathbf{p}^2}. \tag{19.146}$$

19.14 Relativistic electron submit to both a periodic magnetic field and a high frequency electric field. We describe, in the laboratory frame, the motion of an electron that propagates at a relativistic velocity through a periodic static magnetic field and is submit to a high frequency electric field. Which is the Lorentz force?
Answer: The Lorentz force is equal to

$$\mathbf{F} = q(\mathbf{E} + \mathbf{v} \times \mathbf{B}). \tag{19.147}$$

The electric field, which is oriented along the x direction, is given by

$$E = A \cos \omega t \tag{19.148}$$

and the magnetic field, which is oriented along y, is equal to

$$B = B_{\rm w} \cos \Omega_0 t. \tag{19.149}$$

The Lorentz force is oriented along x and is approximately given by

$$F = q(E + cB). \tag{19.150}$$

The frequency ω is much larger than Ω_0.

19.15 Critical field—a speculation. We speculate that a distortion of the electron orbit can occur when the amplitude of the electric force is equal to the amplitude of the magnetic force. We denote the corresponding electric field amplitude as critical field amplitude A^*.
(a) Determine the critical field amplitude A^*.
Answer.

$$A^* = c B_{\rm w}. \tag{19.151}$$

The critical field amplitude is determined by the amplitude of the wiggler field.
In the case that $B_{\rm w} = 1\,\mathrm{T}$, the critical field amplitude is equal to $A^* = 3 \times 10^8\,\mathrm{V\,m^{-1}}$.

19.16 Modulation model: an estimate of the distortion energy.
We assume that the critical field amplitude A^* plays a role for distortion of an electron orbit that we described in connection with the modulation model.

Characteristic of the modulation model is the distortion energy E^*. We now attribute the distortion energy a *critical frequency* ω^* by the relationship

$$\hbar\omega^* = E^*. \quad (19.152)$$

If an electron performs the work E^*, then, stimulated emission of radiation can result in a strong distortion of the electron orbit. We will differ between two cases:

- $\omega_0 \leq \omega^*$; the electron oscillation frequency $\omega_0 (\approx \omega)$ is smaller or equal to the critical frequency.
- $\omega_0 >> \omega^*$; the electron oscillation frequency is large compared with the critical frequency (*see* Problem 19.17).

Here, we consider the case that the electron oscillation frequency is equal or smaller than the critical frequency, $\omega_0 \leq \omega^*$. We ask for the work performed by an electron submitted to the critical field using the relation

$$q\int_0^T E v_x \mathrm{d}t = -\frac{eA^* c K_w}{\gamma^* \omega^*} = -E^*, \quad (19.153)$$

where γ^* is the Lorentz factor that corresponds to the energy of the electrons driving a free-electron laser at the frequency ω^*. We find, with $A^* = cB_w$, the distortion energy

$$E^* = \frac{ec^2 B_w K_w}{\gamma^* \omega^*}. \quad (19.154)$$

We find, from (19.152) and (19.154), using the relation between frequency and Lorentz factor,

$$\omega_0^* = \frac{1}{1+K_w^2/2}\,\frac{4\pi c(\gamma^*)^2}{\lambda_w}, \quad (19.155)$$

the critical frequency

$$\omega^* = (4\pi)^{1/5}(e/\hbar)^{2/5} c B_w^{2/5} \lambda_w^{-1/5} K_w^{-1/5} (1 + K_w/2)^{1/5}. \quad (19.156)$$

The distortion energy is equal to

$$E^* = \left[4\pi\hbar^3 e^2 c^5 \frac{B_w^2(1+K_w/2)}{\lambda_w K_w}\right]^{1/5}, \quad (19.157)$$

where $K_w = \dfrac{ecB_w\lambda_w}{4\pi m_0 c^2}$. Accordingly, the distortion energy depends on the strength of the wiggler field B_w and the wiggler wavelength λ_w.

(a) Determine E^* and ω^* for $B_w = 1$ T and $\lambda_w = 1$ cm; $K_w \approx 1$.

Answer.

$E^* = 1$eV. The calculated value of the distortion energy is comparable with the value (2 eV) that we extracted, using the modulation model, from experimental data of infrared and far infrared free-electron lasers. The critical frequency is equal to $\omega^*/2\pi = 2.5 \times 10^{14}$ Hz.

(b) Determine the distortion energy and the critical frequency for the limits $K_w << 1$ and $K_w >> 1$.

(c) Determine the coupling strength κ that describes, in the modulation model, the coupling between the electron oscillation and the high frequency electric field. *Answer.* The coupling strength is equal to

$$\kappa = \frac{\pi^{4/5} e^{3/5}}{4^{1/5} \hbar^{3/5}} \frac{\lambda^{1/5} K_w^{6/5}}{B_w^{2/5} (1 + K_w/2)^{1/5}}. \tag{19.158}$$

It depends on the strength of the wiggler field and on the wiggler wavelength.

(d) Determine the distortion energy, the critical frequency, and the coupling strength in the limits $K_w << 1$ and $K_w >> 1$.

(e) What is the reason that many expressions imply complicated dependences on the parameters; *see*, for instance the dependence of the distortion energy on the wiggler field and the wiggler wavelength.
Answer. Complicated dependences stem mainly from the dependence of the laser frequency on the Lorentz factor, $\omega_0(\gamma)$.

(f) What is the process that leads, according to the modulation model, to strong distortion of the electron orbit at saturation of the laser field?
Answer. At saturated laser field, the work done by an electron during one period of the electron oscillation is equal to the distortion energy. During a period of the electron oscillation, one photon is generated if the electron oscillation frequency is about equal to the critical frequency, $\omega_0 \approx \omega^*$. Many photons are generated by a cascade process if the electron oscillation frequency is small compared with the critical frequency. Then, the number n of photons generated in a cascade process is given by the relationship $n\omega_0 \approx \omega^*$. At saturated laser field, an absorption process in a period of the electron oscillation follows a stimulated emission process occurring in the preceding period.
Thus, the condition of saturation implies that stimulated emission and absorption processes follow each other at a period that is twice the period of the electron oscillation (= period of the laser field).

19.17 Saturation of the X-ray SASE free-electron laser. The laser frequency and thus the electron oscillation frequency of an X-ray SASE free-electron laser is much larger than the critical frequency, $\omega_0 >> \omega^*$. Stimulated emission of a photon is joined with a large distortion of the electron orbit. However, this distortion is repaired as long as an absorption process does not occur. Saturation does occur if the amplitude of the laser field is so strong that stimulated emission and absorption processes follow each other at a period that is twice the period of the electron oscillation.

(a) Determine the saturation field amplitude A_{sat} for an X-ray SASE free-electron laser. Show that A_{sat} shows an ω dependence; the amplitude increases, for frequencies $\omega >> \omega^*$, less strong than assumed in Sect. 19.11, equation (19.97), showing approximately a $\omega^{3/2}$ dependence.
(b) Which is the gain coefficient for radiation in an X-ray SASE free-electron laser? This is a yet open question. It may decrease, with increasing frequency, less strongly than discussed in Sect. 19.11, equation (19.96). If we assume that the product of the saturation field amplitude and the gain coefficient α is independent of frequency as at small frequencies, we would expect that α is proportional to ω^{-1} rather than proportional to $\omega^{-3/2}$.

19.18 Discuss the general equation of motion. The generals equation of motion, (19.146), takes into account that the Lorentz factor depends on time. Due to stimulated emission of radiation by an electron, the Lorentz factor decreases. We can write

$$m_0\mathbf{v}\frac{d\gamma}{dt} + \gamma m_0\frac{d\mathbf{v}}{dt} = q(\mathbf{E} + \mathbf{v} \times \mathbf{B}). \tag{19.159}$$

This differential equation is treated in many books and is used to derive the *pendulum model* of the electron motion that leads to the expression (19.127) for the small-signal gain; *see*, for instance, [5, 166, 312].

19.19 One-electron FEL. We consider a hypothetical *one-electron free-electron laser*. The FEL is driven by electron pulses, each pulse containing one electron. The pulse repetition rate is equal to the round trip transit rate of the optical pulses in the laser resonator. We assume that the laser resonator shows no loss. We choose the following data. Laser wavelength 1.6 μm; $\nu = 188$ THz, $E_{\text{el}} = 115$ MeV, $\gamma = 230$, $N_{\text{w}} = 100$, $L_{\text{w}} = 1$ m, $\lambda_{\text{w}} = 1$ cm, $B_{\text{w}} = 1$T, $K_{\text{w}} = 1$, $E^* = 2$ eV, $L_{\text{res}} = 30$ m, mode radius of the near concentric resonator $r = 2$ mm. Describe the dynamics by using the modulation model; instead of Gaussian distributions of the optical pulse we describe the pulse by a rectangular distributions of the high frequency field perpendicular to the beam and along the beam.

(a) Estimate the power of a radiation pulse in the laser resonator.
(b) Determine the oscillation onset time of the laser.
(c) Discuss the dynamics of the FEL during the oscillation onset and at steady state oscillation.
(d) Make clear that the one-electron FEL shows main features of a free-electron laser.

Part V
Semiconductor Lasers

Chapter 20
An Introduction to Semiconductor Lasers

Beginning with this chapter, we will treat semiconductor lasers—that are solid state lasers with active media based on semiconductor materials. We will concentrate our discussion on diode lasers (=laser diodes), that is, of current pumped semiconductor lasers and, in particular, on diode lasers operating at room temperature. These are presently the most important semiconductor lasers with respect to applications.

There are two families of semiconductor lasers: the bipolar semiconductor lasers and the unipolar semiconductor lasers.

The bipolar semiconductor lasers are two-band lasers. In a bipolar semiconductor laser, stimulated transitions between occupied electron levels in the conduction band and empty electron levels in the valence band generate the laser radiation. The photon energy of the laser radiation is about equal to the gap energy of the semiconductor.

The unipolar semiconductor lasers, realized as quantum cascade lasers, belong to the three-level laser type. Electrons are performing transitions between three energy levels in three different subbands of the conduction band. The photon energy of the laser radiation is much smaller than the gap energy.

We describe an electron in the conduction band of a semiconductor as a free electron, i.e., as an electron that can freely move (between two collisions) within a semiconductor. Accordingly, a hole in the valence band is a free hole. The free motion of an electron or a hole is either three-dimensional (in a bulk crystal), two-dimensional (in a quantum film), one-dimensional (in a quantum wire) or zero-dimensional (in a quantum dot); in a quantum dot, free motion of an electron is not possible, an electron is imprisoned in the quantum dot. The free motion of an electron in a direction is restricted if the extension of a semiconductor in the direction is, for a semiconductor at room temperature, about 10 nm or smaller.

The waveguide Fabry–Perot resonator, with reflectors formed by semiconductor surfaces only, is suitable as laser resonator for all types of bipolar lasers. Later (Chap. 25) we will discuss other types of resonators that can be used in semiconductor lasers.

K.F. Renk, *Basics of Laser Physics*, Graduate Texts in Physics,
DOI 10.1007/978-3-319-50651-7_20

We present the further program concerning semiconductor lasers. We will treat bipolar semiconductor lasers of different types: junction lasers, double heterostructure lasers, quantum well lasers, quantum wire lasers and quantum dot lasers. The active media of the junction and double heterostructure lasers are three-dimensional semiconductors while the active media of the other types of bipolar lasers are semiconductors of lower dimensions.

We also give a short survey of the frequency ranges of the different semiconductor lasers and mention the energy band engineering as the basis of the great variety of semiconductor lasers.

In comparison with other lasers, semiconductor lasers are unique with respect to their small sizes and, particularly, with respect to the possibility to design a semiconductor laser for a specific frequency—for any frequency in the near UV, visible and infrared. Realization of small-size semiconductor lasers is possible because both the Einstein coefficient B_{21} of stimulated emission and the density of two-level systems in an active semiconductor can have large values at the same time; in active media of other lasers, only one of the two quantities, either B_{21} or the density of two-level systems, has a large value.

20.1 Energy Bands of Semiconductors

We characterize (Fig. 20.1) a semiconductor by a conduction band c and a valence band v separated from each other by an energy gap:

- E_c = energy minimum of the conduction band.
- E_v = energy maximum of the valence band.
- $E_g = E_c - E_v$ = gap energy.

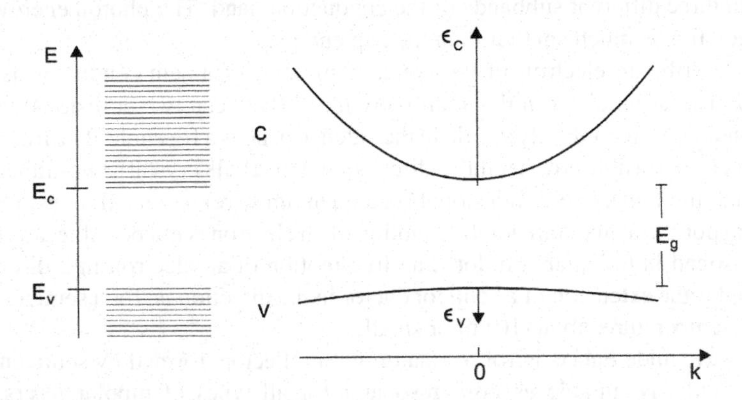

Fig. 20.1 Conduction band and valence band

We describe the electron states of the conduction band by free-electron waves

$$\psi(\boldsymbol{k}, \boldsymbol{r}) = A\, \mathrm{e}^{\mathrm{i}[\boldsymbol{kr} - (E/\hbar)t)]}, \tag{20.1}$$

where the quantities are:

- ψ = wave function.
- $\psi^*(\boldsymbol{r})\psi(\boldsymbol{r})\mathrm{d}V$ = probability to find an electron in the state ψ at the location $\boldsymbol{r}$ in the volume element $\mathrm{d}V$.
- A = amplitude of the electron wave.
- $\boldsymbol{k}$ = wave vector of the electron wave.
- $\boldsymbol{r}$ = spatial coordinate.
- E = energy.
- t = time.

We characterize the relation between the energy and the k vector by a parabolic dispersion relation

$$E = E_{\mathrm{c}} + \epsilon_{\mathrm{c}}, \tag{20.2}$$

where

$$\epsilon_{\mathrm{c}} = \frac{\hbar^2 k^2}{2m_{\mathrm{e}}} \tag{20.3}$$

is the energy within the conduction band, m_{e} the effective mass of a conduction band electron, and $1/m_{\mathrm{c}}$ the curvature of the dispersion curve in the energy minimum at $k = 0$. We can interpret ϵ_{c} as the kinetic energy of a conduction band electron.

With respect to the valence band, we are only interested in the range near the band maximum. We describe the wave function of a valence band electron also as a free-electron wave, however with another dispersion relation,

$$E = E_{\mathrm{v}} - \epsilon_{\mathrm{v}}, \tag{20.4}$$

where

$$\epsilon_{\mathrm{v}} = \frac{\hbar^2 k^2}{2m_{\mathrm{h}}} \tag{20.5}$$

is the energy of a valence band electron state measured from the top of the valence band, k is the wave vector, and m_{h} the effective mass of a valence band electron at the top of the valence band.

We denote an empty level in the valence band as a hole. The energy of a hole is equal to the energy of an electron in the valence band in the case that the electron occupies the empty level. The effective mass of a hole is equal to the effective mass of a valence band electron.

Example Effective masses of electrons and holes in GaAs at room temperature.

- $m_e = 0.07m_0$ = effective mass of a conduction band electron.
- $m_0 = 0.92 \times 10^{-30}$ kg = electron mass.
- $m_h = 0.43\, m_0$ = effective mass of a hole (in the valence band) = mass of a valence band electron (on top of the valence band).
- $m_h \sim 6\, m_e$; the effective mass of a hole is about six times the effective mass of a conduction band electron.

We utilize notations that are in use in semiconductor physics. A "free-electron" in a crystal is a conduction band electron (=electron in the conduction band) or a valence band electron (=electron in the valence band). We denote an ensemble of electrons in the conduction band as electron gas in the conduction band and, accordingly, an ensemble of electrons in the valence band as electron gas in the valence band.

20.2 Low-Dimensional Semiconductors

In a bulk semiconductor crystal, electrons move freely in space; we have a three-dimensional semiconductor (Fig. 20.2). The restriction of the free motion in one direction leads to a two-dimensional semiconductor realized as *quantum well* (= *quantum film*). The further restriction results in the one-dimensional semiconductor (*quantum wire*) and, finally, to the zero-dimensional semiconductor (*quantum dot*). Electrons in a quantum dot cannot move freely at all. The density of states (=level density) of electrons depends on the dimensionality of a semiconductor.

- Three-dimensional density of states = number of states per unit of energy and unit of volume:

$$D^{3D}(\epsilon) = \frac{1}{2\pi^2}\left(\frac{2m}{\hbar^2}\right)^{3/2} \epsilon^{1/2}; \tag{20.6}$$

relevant to the junction laser and the double-heterostructure laser.

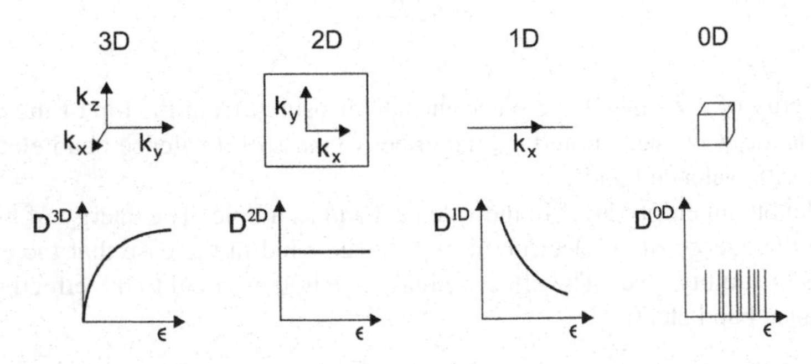

Fig. 20.2 Three-dimensional and low-dimensional semiconductors

- Two-dimensional density of states (=number of states per unit of energy and unit of area):

$$D^{2D}(\epsilon) = \frac{m}{\pi \hbar^2}; \qquad (20.7)$$

relevant to quantum well lasers.
- One-dimensional density of states (=number of states per unit of energy and unit of length):

$$D^{1D}(\epsilon) = \frac{1}{\pi \hbar}\sqrt{\frac{2m}{\epsilon}}; \qquad (20.8)$$

relevant to quantum wire lasers.
- Density of states of a zero-dimensional system $= D^{0D}(\epsilon)$; the energy levels are discrete; relevant to quantum dot lasers.

The spin degeneracy (allowing each k state to be occupied with two electrons of opposite spin) is taken into account. The density of states concern:

- Electrons in the conduction band; then $\epsilon = \epsilon_c$ is the energy within the conduction band and $m = m_e$ is the effective mass of a conduction band electron.
- Electrons or holes in the valence band near the top of the band; then $\epsilon = \epsilon_v$ is the energy within the valence band and $m = m_h$ is the effective mass of a valence band electron (=effective mass of a hole in the valence band).

20.3 An Estimate of the Transparency Density

The effective mass of an electron in the conduction band and the effective mass of an electron in the valence band are quite different. Therefore, the density (=level density) for the of states in the conduction band differs strongly from the density of states for the valence band. We expect, according to the criterion used to determine the transparency density (Sect. 2.3), that the transparency density has a value between

$$N^*_{\mathrm{tr,c}} = \frac{1}{2}\int_0^{kT} D_c(\epsilon)\mathrm{d}\epsilon \qquad (20.9)$$

and

$$N^*_{\mathrm{tr,v}} = \frac{1}{2}\int_0^{kT} D_v(\epsilon)\mathrm{d}\epsilon. \qquad (20.10)$$

Table 20.1 shows the lower limit ($N^*_{\mathrm{tr,c}}$) of the transparency density and the upper limit ($N^*_{\mathrm{tr,v}}$) together with the transparency density N_{tr} calculated by taking into account the appropriate occupation numbers. We will present a method for calculation of N_{tr} in the next two chapters. Our estimated limiting values indicate the orders of magnitudes of the values of the transparency density of GaAs semiconductors in

Table 20.1 Transparency density

	$N^*_{\mathrm{tr,c}}$	$N^*_{\mathrm{tr,v}}$	$\left(N^*_{\mathrm{tr,c}}N^*_{\mathrm{tr,v}}\right)^{1/2}$	N_{tr}
3D	$0.39\times 10^{26}\,\mathrm{m}^{-3}$	$5.9\times 10^{26}\,\mathrm{m}^{-3}$	$1.5\times 10^{26}\,\mathrm{m}^{-3}$	$1.2\times 10^{26}\,\mathrm{m}^{-3}$
2D	$0.82\times 10^{16}\,\mathrm{m}^{-2}$	$4.9\times 10^{16}\,\mathrm{m}^{-2}$	$2\times 10^{16}\,\mathrm{m}^{-2}$	$1.4\times 10^{16}\,\mathrm{m}^{-2}$
1D	$0.69\times 10^{8}\,\mathrm{m}^{-1}$	$1.7\times 10^{8}\,\mathrm{m}^{-1}$	$1.1\times 10^{8}\,\mathrm{m}^{-1}$	

different dimensions. The geometric averages $(N^*_{\mathrm{tr,c}}N^*_{\mathrm{tr,v}})^{1/2}$ are not far from the calculated values. (However, we cannot provide a real justification for taking the geometric average.)

20.4 Bipolar and Unipolar Semiconductor Lasers

In a *bipolar semiconductor laser* (Fig. 20.3a), the active medium contains nonequilibrium electrons in the conduction band and empty electron levels in the valence band. A current leads to injection of electrons into the conduction band and, at the same time, to extraction of electrons from the valence band. The electrons injected into the conduction band occupy mainly energy levels near the bottom of the conduction band—while empty electron levels that occur due to extraction of electrons from the valence band accumulate at the top of the valence band. The electrons in the conduction band are in a quasithermal equilibrium at the temperature of the semiconductor, and the electrons in the valence band are in a quasithermal equilibrium at the temperature of the semiconductor too. But the population of the conduction band is far out of equilibrium with respect to the population of the valence band. Stimulated transitions of electrons in the conduction band to empty electron levels in the valence band lead to generation of laser radiation. There is a distribution of transition energies E_{21}. The smallest transition energy is the gap energy:

$$E_{21} = E_2 - E_1 \geq E_{\mathrm{g}}. \tag{20.11}$$

The range of transition energies is small compared to the gap energy. Accordingly, the photon energy of laser radiation generated by a bipolar semiconductor laser is comparable to the gap energy,

$$h\nu \sim E_{\mathrm{g}}. \tag{20.12}$$

To cover the spectral ranges from the UV to the infrared with radiation of semiconductor lasers, semiconductors with quite different values of E_{g} are necessary. Thus, almost all semiconductors are candidates as basic materials of bipolar semiconductor lasers.

But to be suitable as an active semiconductor medium, a semiconductor has to fulfill an important condition: the semiconductor must be a direct gap semiconductor.

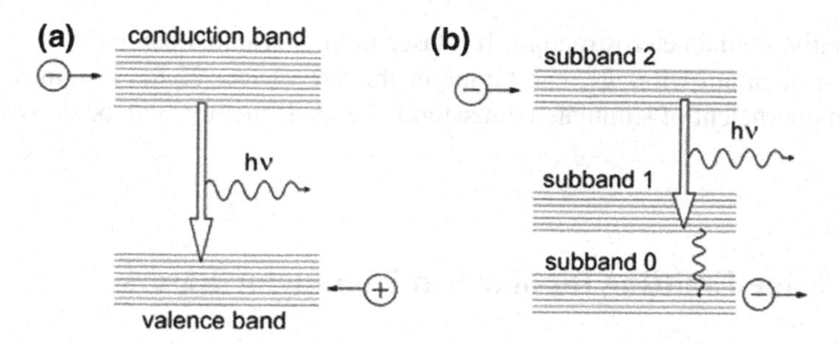

Fig. 20.3 Semiconductor lasers. **a** Bipolar semiconductor laser. **b** Quantum cascade laser

Table 20.2 Einstein coefficients of electronic transitions in GaAs-based semiconductor heterostructures used in lasers

Laser	λ (μm)	A_{21} (s^{-1})	B_{21} (m^3 J^{-1} s^{-2})	$B_{21}^{\parallel}$ (m^3 J^{-1} s^{-2})
Quantum well	0.8	3×10^9	2.2×10^{21}	1.5 B_{21}
Quantum wire	0.8	3×10^9	2.2×10^{21}	3 B_{21}
Quantum dot	0.8	3×10^9	2.2×10^{21}	
QCL	100		2×10^{21}	

Silicon and germanium have indirect gaps and are therefore not suitable as active media of bipolar semiconductor lasers. Most of the group III–V and group II–VI semiconductors have direct gaps and are usable as laser media. In a direct gap semiconductor, the conduction band minimum occurs at the same k value as the valence band maximum.

In a *unipolar semiconductor laser* (Fig. 20.3b), electronic transitions between subbands of the conduction band of a semiconductor heterostructure generate the laser radiation. Now, the design of the heterostructure mainly determines the transition energy. There is again a distribution of transition energies. The transition energies are smaller than the gap energy,

$$E_{21} = E_2 - E_1 \ll E_g, \tag{20.13}$$

and consequently, the laser frequencies are small compared to the gap frequency,

$$h\nu \ll E_{gap}. \tag{20.14}$$

The unipolar laser realized as quantum cascade laser is available in frequency ranges of the infrared and far infrared.

Table 20.2 shows data of Einstein coefficients for GaAs-based semiconductor heterostructures. The values of B_{21} for bipolar semiconductor structure follow from A_{21} (known from fluorescence studies). Due to anisotropy of a heterostructure, the Einstein coefficient B_{21} is different for different orientations of the electromagnetic field relative to the heterostructure. If a laser field is oriented parallel to a quantum well or a quantum wire, $B_{21}^{||}$ can be larger than B_{21} (Sect. 22.8). The value of the Einstein coefficient of stimulated emission for a QCL medium can be derived from theory.

20.5 Edge-Emitting Bipolar Semiconductor Lasers

A edge-emitting bipolar semiconductor laser emits radiation from one of the edges of a device (or from two edges), *see* Fig. 20.4. The resonator corresponds (in the simplest case) to a Fabry–Perot resonator with two uncovered semiconductor surfaces as reflectors. A central layer sandwiched between two other layers contains the active media. The central layer has a slightly larger refractive index (n_0) than the adjacent layers (refractive index n_1). The field is confined to the central layer, the laser light is *index guided* (=refractive index guided). The central layer has a thickness (a_2) that is equal to half a wavelength or larger than half a wavelength of the laser radiation in the resonator. The reflectivity of a single surface is

$$R_1 = R_2 = \frac{(n_0 - 1)^2}{(n_0 + 1)^2}. \tag{20.15}$$

The refractive index of a GaAs-based semiconductor is ~3.6 (or smaller) and the reflectivity of the surfaces $R_1 = R_2 \sim 0.3$. The V factor characterizing the loss per single transit through the resonator is $V_1 \sim 0.3$. The laser threshold condition, $G_1 V_1 \geq 1$, requires that the gain factor at a single transit of radiation through a laser resonator must have a value $G_1 \geq V_1^{-1}$ (~3) that is noticeably larger than unity. Laser radiation leaves the resonator via both ends of the resonator.

Example Resonator consisting of a semiconductor (GaAlAs) with $n_0 = 3.5$.

- $L = 1\,\text{mm}$.
- $T/2 = nL/c = 1.2 \times 10^{-11}$ s = half of the round trip transit time.
- $V_1 = 0.3$ = V factor related to a single transit of radiation through the resonator.
- $l_p = 0.8\,\text{mm}$; an average path length of a photon in the resonator.
- $\tau_p = 1 \times 10^{-11}$ s; lifetime of a photon in the resonator.

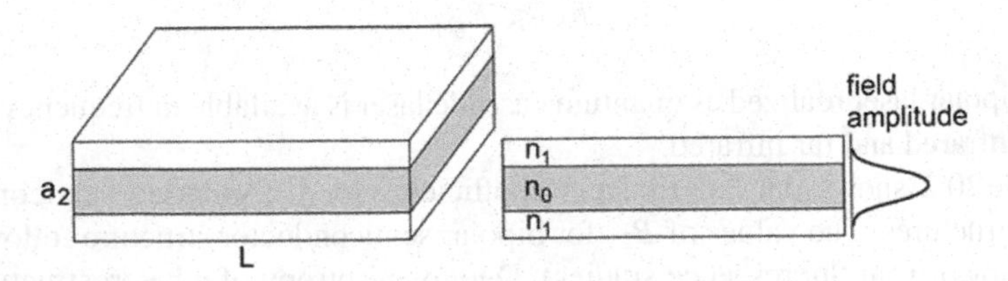

Fig. 20.4 Edge-emitting bipolar laser with a waveguide Fabry–Perot resonator

The lifetime of a photon is slightly smaller than the time of flight ($T/2$) of the light through the resonator. We will show later that waveguide Fabry–Perot resonators are nevertheless suitable for operation of bipolar lasers.

20.6 Survey of Topics Concerning Semiconductor Lasers

We will introduce (Fig. 20.5), on the basis of the properties of semiconductors and semiconductor heterostructures, the bipolar semiconductor lasers and the unipolar semiconductor lasers. We characterize the different types of the two families as types as follows.

- *Junction laser* (=homojunction laser = homostructure junction laser). The active medium is the junction region between an n-doped and a p-doped part of a crystal. This was the first semiconductor laser type.

  Example GaAs junction laser, containing an n GaAs/p GaAs junction.
- *Double heterostructure laser.* The active medium is an undoped film embedded in n- and p-doped materials. The undoped film is a well for electrons and holes. The thickness of the film is so large that the electrons move as free-electrons in all spatial directions. With this type, the laser design making use of heterostructures began.

  Example GaAs/GaAlAs laser, containing the layers n GaAlAs/GaAs/p GaAlAs.
- *Quantum well laser.* The active medium is an undoped quantum film. Adjacent to the quantum film, there is on one side n-doped material and on the other side p-doped material. The materials act as injectors of electrons and holes, respectively. Quantum well lasers are available for the visible, near infrared, and near UV spectral ranges and dominate presently the semiconductor laser field with respect to applications.

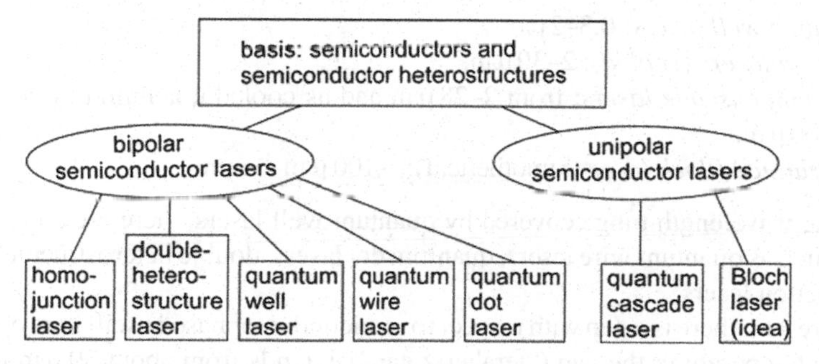

Fig. 20.5 Survey of topics concerning semiconductor lasers

Example GaAs quantum well laser, containing layers of n GaAlAs-2/n GaAlAs-1/GaAs/p GaAlAs-1/p GaAlAs-2; the numbers 1 and 2 refer to different compositions.

- *Quantum wire laser*. Quantum wire lasers, with quantum wires embedded in injector material, are in the first stage of realization.
- *Quantum dot laser.* This type of a bipolar semiconductor laser is being developed.
- *Quantum cascade laser* (QCL). This laser type is presently in a very active state of development. The radiation of quantum cascade lasers covers large wavelength ranges of the infrared and the far infrared (2–28 μm) and, as cooled QCL, wavelength ranges of the far infrared (70–300 μm).
- *Superlattice Bloch laser* (=Bloch laser = Bloch oscillator). This type of laser exists only as an idea on the basis of theoretical studies. The active element is a doped semiconductor superlattice. The superlattice is composed of two different semiconductor materials, for instance, GaAs and AlAs. An electron that propagates along the superlattice axis experiences a periodic potential. The energy is confined to a miniband of a width that is much smaller than the gap energy of the two semiconductors. In a strong static electric field (E_s), the states of an electron in a miniband form an energy-ladder system. An electron occupies one of the levels in an energy-ladder system. Stimulated transitions between energetically next-year levels in the energy-ladder systems give rise to gain. Bloch lasers—if realizable—should operate at room temperature and cover a frequency range beginning at a frequency below 1 THz up to several THz.

We will begin in the next section with a discussion of a bipolar laser and then concentrate the discussion during several chapters—because of its great importance—on the quantum well laser.

20.7 Frequency Ranges of Semiconductor Lasers

The frequency range covered by different semiconductor lasers extend from the near UV to the far infrared (Fig. 20.6).

- *Quantum well lasers*; 0.3–2 μm.
- *Junction lasers (cooled)*; 2–30 μm.
- *Quantum cascade lasers*; from 2–28 μm and as cooled quantum cascade lasers, 70–300 μm.
- *Superlattice Bloch laser* (hypothetical); ∼100 μm–1 mm.

In the wavelength range covered by quantum well lasers, there are other bipolar lasers such as quantum wire lasers, quantum dot lasers, double heterostructure lasers, and junction lasers.

At present, there is a gap with respect to semiconductor based oscillators operating at room temperature; the gap ("terahertz gap") extends from about 30 μm to 3 mm (Sect. 28.7). The cooled quantum cascade lasers cover a part of the gap.

20.8 Energy Band Engineering

An important tool used to design semiconductor lasers is the energy band engineering (tailoring of semiconductors), based on two principles.

- *The use of mixed crystals.* The energy gap of a mixed crystal is different from the gaps of the single components.
- *The preparation of heterostructures.* Heterostructures are spatial structures consisting of different semiconductor materials.

20.9 Differences Between Semiconductor Lasers and Other Lasers

Semiconductor lasers (=diode lasers = laser diodes) have a series of extraordinary properties.

- The transition energies can have, for different materials, very different values. Accordingly, lasers can be designed for the UV, visible, infrared, and far infrared spectral regions.
- The use of alloys of semiconductors makes it possible to design semiconductor lasers for each wavelength in the range from the near UV to the infrared.
- The use of heterostructures leads to an extraordinary extension of the possibility of designing lasers.

A laser diode

- Converts electric current directly to light.
- Has a high gain coefficient (e.g., $10\,\mathrm{cm}^{-1}$ or more), in comparison with the helium–neon laser ($10^{-3}\,\mathrm{cm}^{-1}$) or the CO_2 laser ($0.05\,\mathrm{cm}^{-1}$), *see* Fig. 7.6.
- Can reach a high efficiency (50% or more).
- Operates as cw or pulsed laser.
- Can be tailored for a given wavelength.

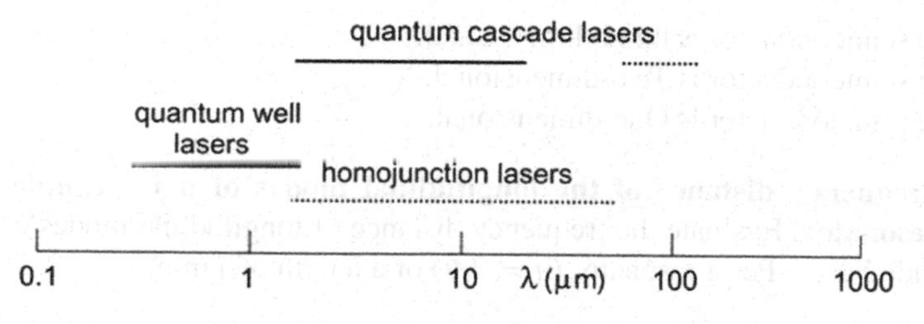

Fig. 20.6 Frequency ranges of semiconductor lasers; *solid lines*, lasers operated at room temperature and *dotted lines*, lasers operated at liquid nitrogen temperature

- Has a small dimension (typical sizes of a quantum well laser: 200 μm × 1 μm × 500 μm (edge-emitting laser) and 10 μm × 100 μm × 100 μm down to 10 μm × 10 μm × 10 μm or a corresponding circular area (vertical-surface emitting laser).
- Can be manufactured by mass production.

The combination of the radiation of a large number of semiconductor lasers leads to a high-power semiconductor laser; the radiation is monochromatic but not coherent.

The other lasers have, in comparison with semiconductors, other advantages: they are suitable for generation of radiation of high directionality and of high monochromaticity; some of the lasers allow for generation of ultrashort pulses; some are suitable as giant pulse lasers.

The Einstein coefficient B_{21} of transitions used in semiconductors is large and density of two-level atomic systems, which contribute to a laser oscillation, is large too. This has different reasons. A three-dimensional semiconductor can carry a large density of electrons contributing to the gain. A Low-dimensional semiconductor, with a large two-dimensional density of electrons contributing to gain, can be integrated in a integrated in a resonator of small size so that the so that the average density of two-level atomic systems in a photon mode has a large value too. In comparison with active media of semiconductor lasers, only one of the quantities of an active medium of other laser has a large value—either the Einstein coefficient B_{21} or the density of two-level atomic systems (Sects. 7.3 and 7.5).

References [177–186].

Problems

20.1 De Broglie wavelength. Estimate the ratio of the de Broglie wavelength of a conduction band electron in GaAs and a free-electron in vacuum that move at the same velocity.

20.2 Number of states. Evaluate the number of states in the conduction band of GaAs ($m_e = 0.07\,m_0$) that are available at the energy $E_g + 26\,\text{meV}$ in an energy interval of 1 meV for different cases.

(a) The semiconductor is three-dimensional.
(b) The semiconductor is Two-dimensional.
(c) The semiconductor is One-dimensional.

20.3 Frequency distance of the longitudinal modes of a waveguide Fabry–Perot resonator. Evaluate the frequency distance of longitudinal modes of a GaAs waveguide Fabry–Perot resonator ($n = 3.6$) of a length of 1 mm.

Chapter 21
Basis of a Bipolar Semiconductor Laser

We treat the basis of bipolar semiconductor lasers. We discuss: condition of gain; joint density of states; gain coefficient; laser equations; bipolar character of the active medium. And we derive, by use of Planck's radiation law, the Einstein coefficients for an ensemble of two-level systems that is governed by Fermi's statistics.

The first part of this chapter is dealing with three-dimensional semiconductors. Another part, Sects. 21.8 and 21.9, concerns quantum well lasers. Instead of following through these two sections, a reader may solve the Problems 21.5 and 21.6, or jump to the next chapter that contains, in a short form, the main conclusions with respect to quantum well lasers.

The active medium of a bipolar semiconductor laser is a semiconductor containing electrons in the conduction band and empty electron levels (holes) in the valence band. Permanent pumping (via a current delivered by a voltage source) leads to injection of electrons into the conduction band and to extraction of electrons from the valence band. The active medium carries no net charge. The density N of electrons in the conduction band is equal to the density of empty electron levels in the valence band. Laser radiation occurs due to stimulated transitions of electrons in the conduction band to empty levels in the valence band. The electrons in the conduction band have a Fermi distribution, f_2, corresponding to a quasi-Fermi energy E_{Fc}. The electrons in the valence band have another Fermi distribution, f_1, corresponding to a quasi-Fermi energy E_{Fv}.

We derive the condition of gain: gain occurs if the occupation number difference is larger than zero, $f_2 - f_1 > 0$. This is equivalent to the condition that the density of conduction band electrons is larger than the transparency density ($N > N_{tr}$). And it is also equivalent to the condition that the difference of the quasi-Fermi energies is larger than the gap energy, $E_{Fc} - E_{Fv} > E_g$. Gain occurs for photons of a quantum energy that is smaller than the difference of the quasi-Fermi energies, $h\nu < E_{Fc} - E_{Fv}$. The range of gain increases with increasing density of nonequilibrium electrons in the conduction band (and a corresponding increasing density of empty levels in the valence band).

K.F. Renk, *Basics of Laser Physics*, Graduate Texts in Physics,
DOI 10.1007/978-3-319-50651-7_21

We determine the reduced density of states (=joint density of states), taking into account that energy and momentum conservation laws have to be obeyed in a radiative transition. We derive expressions describing stimulated and spontaneous emission. Furthermore, we formulate laser equations. Their solutions provide the threshold condition, particularly the threshold current. The solutions reveal clamping of occupation number difference $f_2 - f_1$ and accordingly, clamping of the quasi-Fermi energies.

To combine, for calculation of gain, a quantum well that is a two-dimensional semiconductor with a radiation field that is three-dimensional, we use appropriate average densities. We introduced the method earlier (in Sects. 7.8 and 7.9 about gain mediated by a two-dimensional active medium). The topics we will treat with respect to the quantum well laser concern: condition of gain; quasi-Fermi energies; reduced mass; transparency density; gain characteristic; gain mediated by a quantum well oriented along the direction of a light beam; gain of radiation traversing a quantum well; laser equations and their solutions.

21.1 Principle of a Bipolar Semiconductor Laser

A bipolar semiconductor laser contains an electron gas in the conduction band and another electron gas in the valence band (Fig. 21.1a). The electron gas in the conduction band is in a quasithermal equilibrium with the thermal bath. The quasithermal

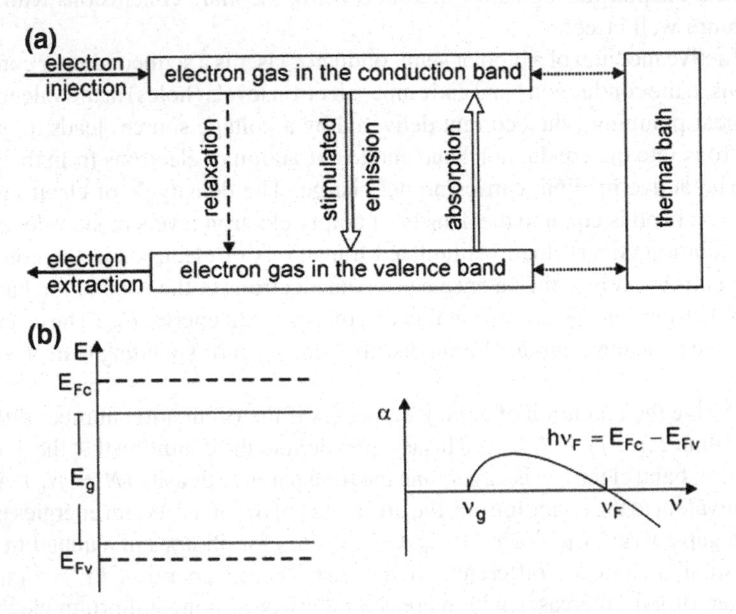

Fig. 21.1 Principle of a bipolar semiconductor laser. **a** Dynamics. **b** Quasi-Fermi energies and transparency frequency

equilibrium with the thermal bath, i.e., with the crystal lattice, is established via electron–phonon scattering. The electron gas in the valence band is also in a quasithermal equilibrium with the thermal bath. The quasithermal equilibrium with the thermal bath is also established via electron–phonon interaction. However, the two electron gases are far out of equilibrium with each other. Electron injection into the conduction band and electron extraction from the valence band maintain the nonequilibrium state.

We will characterize the electron gas in the conduction band by the quasi-Fermi energy E_{Fc} and the electron gas in the valence band by the quasi-Fermi energy E_{Fv} (Fig. 21.1b, left). The gain coefficient α of an active medium (Fig. 21.1b, right) is positive for radiation in the frequency range between ν_g, and ν_F, where $\nu_g = E_g/h$ is the gap frequency, E_g the gap energy, and where ν_F corresponds to the difference of the quasi-Fermi energies according to the relation

$$\nu_{\mathrm{F}} = (E_{\mathrm{Fc}} - E_{\mathrm{Fv}})/h. \tag{21.1}$$

This will be shown in the next sections.

21.2 Condition of Gain of Radiation in a Bipolar Semiconductor

We consider (Fig. 21.2) two discrete energy levels, a level 2 (energy E_2) in the conduction band and a level 1 (energy E_1) in the valence band. Radiative transitions between the two levels can occur by the three processes: absorption, stimulated and spontaneous emission. The transition rate of absorption (=number of transitions per m^3 and s) is equal to

$$r_{12}(\nu) = r_{12}(h\nu) = \tilde{B}_{12} f_1(1 - f_2)\rho(h\nu). \tag{21.2}$$

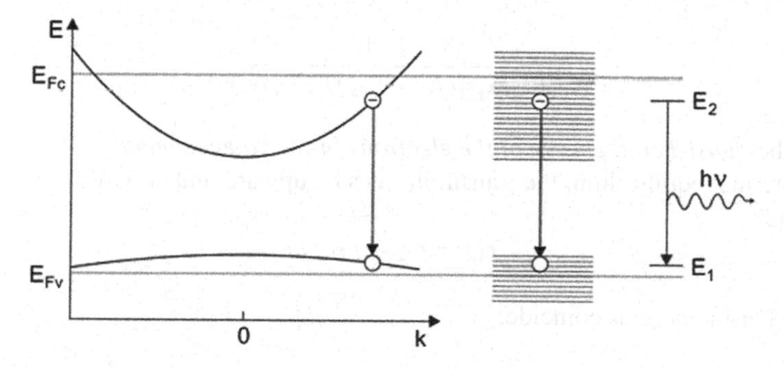

Fig. 21.2 Radiative transition in a bipolar semiconductor medium

We use the following quantities:

- E_c = energy of the bottom of the conduction band.
- E_v = energy of the top of the valence band.
- E_2 = energy of the upper laser level.
- E_1 = energy of the lower laser level.
- $E_{21} = E_2 - E_1$ = transition energy.
- $\bar{B}_{12}$ = Einstein coefficient of absorption (in units of $m^3\,s^{-1}$); $\bar{B}_{12} = hB_{12}$ (Sect. 6.6).
- $\bar{B}_{21}$ = Einstein coefficient of stimulated emission.
- A_{21} = Einstein coefficient of spontaneous emission.
- $f_1 = f_1(E_1)$ = probability that level 1 is occupied; $1 - f_1$ = probability that level 1 is empty.
- $f_2 = f_2(E_2)$ = probability that level 2 is occupied.
- $(1 - f_2)$ = probability that level 2 is empty.
- $\rho(h\nu)$ = spectral energy density of the radiation on the energy scale.

It is convenient to choose the energy scale. Consequently, the Einstein coefficients of absorption and stimulated emission, $\bar{B}_{12}$ and $\bar{B}_{21}$, differ from B_{12} and B_{21}.

The rate of stimulated emission processes is equal to

$$r_{21}(h\nu) = \bar{B}_{21} f_2(1 - f_1)\rho(h\nu). \tag{21.3}$$

The spontaneous emission rate is equal to

$$r_{21,sp}(h\nu) = A_{21} f_2(1 - f_1). \tag{21.4}$$

The occupation probability of level 2 is given by the Fermi–Dirac distribution

$$f_2 = \frac{1}{\exp[(E - E_{Fc})/kT] + 1}. \tag{21.5}$$

E_{Fc} is the *quasi-Fermi energy of the electrons in the conduction band* and T is the lattice temperature. The occupation probability of level 1 is

$$f_1 = \frac{1}{\exp[(E - E_{Fv})/kT] + 1}. \tag{21.6}$$

E_{Fv} is the *quasi-Fermi energy of the electrons in the valence band.*

At thermal equilibrium, the transition rates of upward and downward transitions are equal,

$$r_{12} = r_{21} + r_{21,sp}, \tag{21.7}$$

and the Fermi energies coincide,

$$E_{Fc} = E_{Fv} = E_F. \tag{21.8}$$

It follows that

$$\rho(h\nu) = \frac{A_{21} f_2(1 - f_1)}{\bar{B}_{12} f_1(1 - f_2) - \bar{B}_{21} f_2(1 - f_1)} \tag{21.9}$$

must be equal to the expression given by Planck's radiation law (now with the energy density given on the energy scale),

$$\rho(h\nu) = \frac{8\pi n^3 \nu^3}{c^3} \frac{1}{e^{h\nu/kT} - 1}. \tag{21.10}$$

The comparison yields

$$A_{21} = 8\pi n^3 \nu^3 c^{-3} \bar{B}_{21} \tag{21.11}$$

and

$$\bar{B}_{12} = \bar{B}_{21}. \tag{21.12}$$

We find again the Einstein relations. If a semiconductor is optically anisotropic, the value of $\bar{B}_{12}$ ($=\bar{B}_{21}$) depends on the direction of the electromagnetic field relative to the orientation of the semiconductor. Then a modification of the Einstein relations is necessary.

At nonequilibrium, the quasi-Fermi energies are different, $E_{Fc} \neq E_{Fv}$, i.e., the electrons in the conduction band are not in an equilibrium with respect to the electrons in the valence band. However, the electrons within the conduction band form an electron gas that is in a quasiequilibrium with the lattice at the temperature T—and the electrons within the valence band form another electron gas that is in a quasiequilibrium with the lattice at the temperature T. In a bibpolar semiconductor laser nonequilibrium state consists of two electron gases that are far out of equilibrium relative to each other.

The net rate of stimulated emission and absorption by transitions between the two energy levels of energy E_1 and E_2 is equal to

$$r_{21} - r_{12} = \bar{B}_{21}(f_2 - f_1)\rho(h\nu). \tag{21.13}$$

Stimulated emission prevails if the *occupation number difference* is larger than zero,

$$f_2 - f_1 > 0. \tag{21.14}$$

This condition corresponds to

$$E_{Fc} - E_{Fv} > E_g. \tag{21.15}$$

A bipolar medium is an active medium if the difference of the quasi-Fermi energies is larger than the gap energy. The injection of electrons leads to a density N of electrons in the conduction band. The extraction of electrons from the valence band leads to a density P of empty levels in the valence band (=density of holes in the valence band). Because of neutrality, the two densities are equal, $N = P$.

The quasi-Fermi energy of the electrons in the conduction band follows from the condition that the density of occupied levels in the conduction band is equal to the density N of nonequilibrium electrons in the conduction band,

$$\int_{-\infty}^{\infty} f_2(E) D_c(E) \mathrm{d}E = N. \tag{21.16}$$

The quantities are:

- $D_c(E)$ = density of states in the conduction band (in units of $\mathrm{m^{-3}\,J^{-1}}$).
- $f_2(E)D_c(E)\mathrm{d}E$ = density of occupied levels in the conduction band in the energy interval E, $E + \mathrm{d}E$.
- N = density of electrons = density of electrons injected into the conduction band (in units of $\mathrm{m^{-3}}$).

The density of unoccupied electron levels in the valence band is

$$\int_{-\infty}^{\infty} (1 - f_1) D_v(E) \mathrm{d}E = P \ (=N), \tag{21.17}$$

where the quantities are:

- $D_v(E)$ = density of states in the valence band.
- $f_1(E)D_v(E)\mathrm{d}E$ = density of occupied levels in the valence band within the energy interval E, $E + \mathrm{d}E$.
- $(1 - f_1)D_v(E)\mathrm{d}E$ = density of empty levels in the valence band within the energy interval E, $E + \mathrm{d}E$.
- P = density of empty levels in the valence band (=density of holes).

We can use the last two equations to determine, for a given electron density N, the quasi-Fermi energies E_{Fc} and E_{Fv}.

The description of the electronic states takes into account that the electrons in the conduction band obey the Pauli principle. Each of the states can be occupied with two electrons (of opposite spin). A Fermi distribution function describes the filling of the conduction band. With increasing electron density N, the quasi-Fermi energy E_{Fc} increases. Correspondingly, the extraction of electrons from the valence band leads, with increasing density N of empty levels, to a decrease of the quasi-Fermi energy E_{Fv}. The difference of the quasi-Fermi energies, $E_{Fc} - E_{Fv}$, increases with increasing N. The difference becomes equal to the gap energy,

$$E_{Fc} - E_{Fv} = E_g \ \text{ if } \ N = N_{tr}. \tag{21.18}$$

N_{tr} is the transparency density. At this electron density, the Fermi functions have, for $E_2 - E_1 = E_g$, the same values, $f_2(E_2) = f_1(E_1)$. Furthermore, the rates of stimulated emission and absorption are equal. Accordingly, the semiconductor is transparent for radiation of the photon energy $h\nu = E_2 - E_1 = E_g$. Gain occurs if the electron density exceeds the transparency density. The range of gain increases with increasing $N - N_{tr}$. We can express the result in other words: the range of gain increases with increasing filling of the conduction band with electrons and the simultaneous extraction of electrons from the valence band (i.e., with the filling of the valence band with holes).

21.3 Energy Level Broadening

We study transitions involving monochromatic radiation in the energy interval $h\nu$, $h\nu + d(h\nu)$ taking into account energy level broadening (Fig. 21.3). The net transition rate is equal to

$$(r_{21} - r_{12})_{\mathrm{hv}}\, \mathrm{d}(h\nu) = \bar{B}_{21} g(h\nu - E_{21})\, (f_2 - f_1)\, \rho(h\nu)\mathrm{d}(h\nu), \tag{21.19}$$

where $g(h\nu - E_{21})$ is the lineshape function that corresponds to the $1 \rightarrow 2$ absorption line. The lineshape function is normalized,

$$\int g(h\nu - E_{21})\, \mathrm{d}(h\nu) = 1; \tag{21.20}$$

the integral over all contributions $g(h\nu - E_{21})\mathrm{d}(h\nu)$ is unity. If the lineshape function is a Lorentzian, we can write

$$g(h\nu - E_{21}) = \frac{\delta E_{21}}{2\pi} \frac{1}{(h\nu - E_{21})^2 + \delta E_{21}^2/4}. \tag{21.21}$$

δE_{21} is the linewidth of the transition.

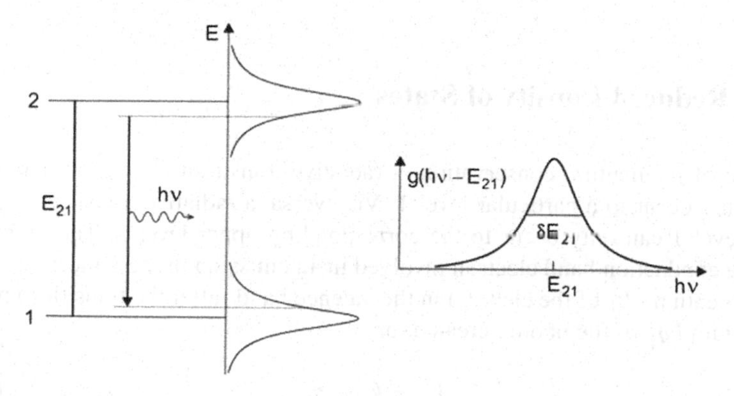

Fig. 21.3 Energy level broadening

If the radiation is monochromatic, $d(h\nu) \ll \delta E_{21}$, then the net transition rate of transitions between level 2 and level 1 is given by

$$r_{21} - r_{12} = \int (r_{21} - r_{12})_{h\nu} \mathrm{d}(h\nu) = \int \bar{B}_{21} g(h\nu - E_{21})(f_2 - f_1)\rho(h\nu)\mathrm{d}(h\nu). \tag{21.22}$$

With $u = \int \rho(h\nu)\mathrm{d}(h\nu)$ and the energy density $u = Zh\nu$, we can write

$$r_{21} - r_{12} = h\nu \bar{B}_{21} g(h\nu - E_{21}) \left[f_2(E_2) - f_1(E_1)\right] Z. \tag{21.23}$$

The transition rate is proportional to the occupation number difference $f_2 - f_1$ and to the photon density Z. The condition of gain remains the same, $f_2(E_2) - f_1(E_1) > 0$, as derived without taking account of energy level broadening. At the transparency density, where $f_2(E_2) - f_1(E_1) = 0$, there is no contribution to gain of radiation of the quantum energy $h\nu$ by $2 \to 1$ transitions—whether $h\nu$ lies in the line center or in the wing of the line. We thus have obtained the condition of gain:

$$f_2(E_2) - f_1(E_1) > 0 \tag{21.24}$$

or

$$E_{\mathrm{Fc}} - E_{\mathrm{Fv}} > E_{\mathrm{g}}, \tag{21.25}$$

which corresponds to

$$h\nu < E_{\mathrm{Fc}} - E_{\mathrm{Fv}}. \tag{21.26}$$

The photon energy can have a value that is smaller than the gap energy E_{g} because of the energy level broadening. An electron level (in the conduction band as well in the valence band) has a finite lifetime due to inelastic scattering of electrons at phonons (electron–phonon scattering).

The condition of gain, $h\nu < E_{\mathrm{Fc}} - E_{\mathrm{Fv}}$, is sometimes called *Bernard–Duraffourg relation* according to the authors of a corresponding publication [201].

21.4 Reduced Density of States

Because of momentum conservation, a radiative transition from a particular level 2 can only occur to a particular level 1. Vice versa, a radiative transition from the lower level 1 can only occur to the corresponding upper level 2. The momentum $\hbar \boldsymbol{k}_2$ of a conduction band electron involved in an emission process must be equal to the momentum $\hbar \boldsymbol{k}_1$ of the electron in the valence band (after the transition) plus the momentum $\hbar \boldsymbol{q}_{\mathrm{p}}$ of the photon created, or

$$\boldsymbol{k}_2 = \boldsymbol{k}_1 + \boldsymbol{q}_{\mathrm{p}}. \tag{21.27}$$

We assume, for simplicity, that

$$|\boldsymbol{q}_\mathrm{p}| \ll |\boldsymbol{k}_1|, |\boldsymbol{k}_2|. \tag{21.28}$$

It follows that

$$\boldsymbol{k}_1 = \boldsymbol{k}_2. \tag{21.29}$$

A radiative transition corresponds in the energy-wave vector diagram (*see* Fig. 21.2) to a "vertical" transition. The radiative transitions between the conduction and the valence band occur between states that have the same wave vector, i.e., radiation interacts with electrons in *radiative pair levels*. We consider a radiative transition from a level 2 of energy

$$E_2 = E_\mathrm{c} + \frac{\hbar^2}{2m_\mathrm{e}}k^2, \tag{21.30}$$

to a level 1 of energy

$$E_1 = E_\mathrm{v} - \frac{\hbar^2}{2m_\mathrm{h}}k^2. \tag{21.31}$$

A conduction band level and a valence band level belonging to states with the same wave vector have the energy difference

$$E_{21} = E_2 - E_1 = E_\mathrm{g} + \frac{\hbar^2}{2m_\mathrm{r}}k^2, \tag{21.32}$$

where

$$\frac{1}{m_\mathrm{r}} = \frac{1}{m_\mathrm{e}} + \frac{1}{m_\mathrm{h}} \tag{21.33}$$

is the reciprocal of the reduced mass m_r. Using the expressions of the energy of an electron in the conduction band (Fig. 21.4),

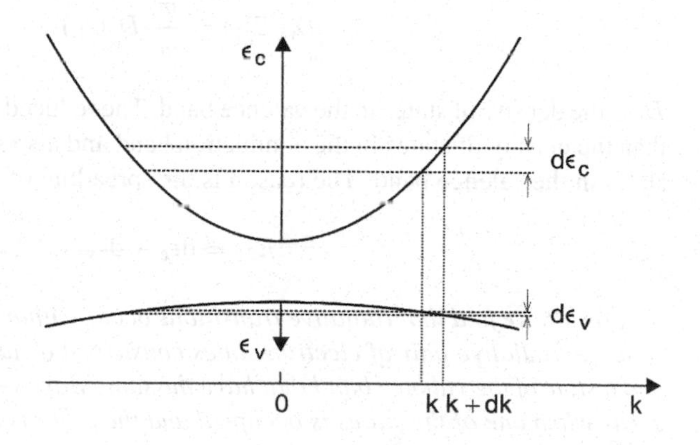

Fig. 21.4 Reduced density of states

$$\epsilon_c = E_2 - E_c = \frac{\hbar^2}{2m_e}k^2, \tag{21.34}$$

and of the energy of the corresponding level in the valence band,

$$\epsilon_v = E_v - E_1 = \frac{\hbar^2}{2m_h}k^2. \tag{21.35}$$

We can write

$$E_{21} = E_g + \epsilon_c + \epsilon_v. \tag{21.36}$$

By elimination of k^2, we obtain the relations

$$\epsilon_c = \frac{m_r}{m_e}(E_{21} - E_g) \tag{21.37}$$

and

$$\epsilon_v = \frac{m_r}{m_h}(E_{21} - E_g). \tag{21.38}$$

How many radiative pair levels are available in the energy interval E_{21}, $E_{21} + \mathrm{d}E_{21}$? The number of states, $D_r(E_{21})\mathrm{d}E_{21}$, is equal to the corresponding number of levels in the conduction band,

$$D_r(E_{21})\mathrm{d}E_{21} = D_c(\epsilon_c)\mathrm{d}\epsilon_c. \tag{21.39}$$

Thus, the *reduced density of states* (=joint density of states = density of states of radiative pair levels) is given by

$$D_r(E_{21}) = \frac{D_c(\epsilon_c)}{\mathrm{d}E_{21}/\mathrm{d}\epsilon_c} = \frac{m_r}{m_e}D_c(\epsilon_c). \tag{21.40}$$

D_c is the density of states in the conduction band. Correspondingly, we can write

$$D_r(E_{21}) = \frac{m_r}{m_h}D_v(\epsilon_h). \tag{21.41}$$

D_v is the density of states in the valence band. The reduced density of states is smaller than the density of states in the conduction band and also smaller than the density of states in the valence band. The reason is the spreading of the energy scale:

$$\mathrm{d}E_{21} = \mathrm{d}\epsilon_c + \mathrm{d}\epsilon_v. \tag{21.42}$$

As a result, we find that radiative transitions occur within radiative pairs of electron states. A radiative pair of electron states consists of a state of the conduction band and a state of the valence band that have the same wave vector. A transition can only occur when one of the states is occupied and the other is unoccupied.

21.5 Growth Coefficient and Gain Coefficient of a Bipolar Medium

The temporal change of the density N of conduction band electrons due to stimulated transitions is equal to

$$\frac{\mathrm{d}N}{\mathrm{d}t} = -h\nu \int \bar{B}_{21} g(h\nu - E_{21})(f_2 - f_1) D_{\mathrm{r}}(E_{21}) \mathrm{d}E_{21} Z. \tag{21.43}$$

It follows that the temporal change of the photon density is

$$\mathrm{d}Z/\mathrm{d}t = -\mathrm{d}N/\mathrm{d}t = \gamma Z, \tag{21.44}$$

where

$$\gamma = h\nu \int \bar{B}_{21} D_{\mathrm{r}}(E_{21})(f_2 - f_1) g(h\nu - E_{21}) \mathrm{d}E_{21} \tag{21.45}$$

is the growth coefficient of the semiconductor and, with $\mathrm{d}t = (n/c)\mathrm{d}z$, that

$$\mathrm{d}Z/\mathrm{d}z = \alpha Z, \tag{21.46}$$

where n is the refractive index, c the speed of light, and

$$\alpha = \frac{n}{c} h\nu \int \bar{B}_{21} D_{\mathrm{r}}(E_{21})(f_2 - f_1) g(h\nu - E_{21})\, \mathrm{d}E_{21} \tag{21.47}$$

is the gain coefficient of the semiconductor.

Figure 21.5 shows electron distributions for $T = 0$ and for a high temperature. At $T = 0$, all conduction band levels between E_{c} and E_{Fc} are occupied and all valence band levels between E_{v} and E_{Fv} are empty. Gain occurs, for $T = 0$, in the range

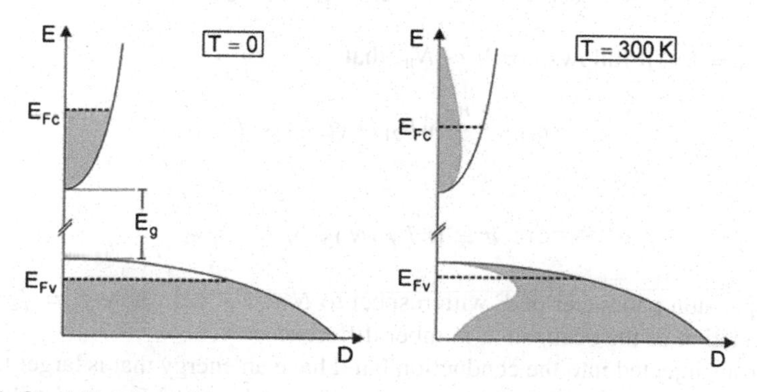

Fig. 21.5 Quasi-Fermi energies

$$E_g \leq h\nu < E_{Fc} - E_{Fv}. \tag{21.48}$$

At finite temperature, the electrons in the conduction band are distributed over a larger energy range and the quasi-Fermi energy E_{Fc} is smaller than in the case that $T = 0$. The empty levels in the valence band are distributed over a larger range too and the quasi-Fermi energy F_{Fv} has a larger value than for $T = 0$. At high temperatures, the energy levels broaden due to electron–phonon scattering. Therefore, the photon energy can be smaller than E_g and the condition of gain is

$$h\nu < E_{Fc} - E_{Fv}. \tag{21.49}$$

For $h\nu < E_g$, the gain coefficient decreases with decreasing quantum energy according to the lineshape function $g(h\nu - E_{21})$.

If $N \sim N_{tr}$, the maximum gain coefficient is equal to (Sect. 7.4)

$$\alpha_{max} = \sigma_{eff} \times (N - N_{tr}), \tag{21.50}$$

where

$$\sigma_{eff} = (\partial \alpha_{max}/\partial N)_{N=N_{tr}} \tag{21.51}$$

is the effective gain cross section. It follows that the growth coefficient is

$$\gamma_{max} = b_{eff} \times (N - N_{tr}) \tag{21.52}$$

and that

$$b_{eff} = (c/n)\sigma_{eff} \tag{21.53}$$

is the effective growth rate constant.

If thermal broadening of the energy levels is negligible, the gain coefficient is

$$\alpha = (n/c)h\nu \bar{B}_{21} D_r(E_{21})[f_2(E_2) - f_1(E_1)], \tag{21.54}$$

where $E_{21} = h\nu$. It follows, for $N \sim N_{tr}$, that

$$\sigma_{eff} = \frac{n}{c} h\nu \bar{B}_{21} D_r(E_{21}) \times d, \tag{21.55}$$

where

$$d = (\partial F/\partial N)_{N=N_{tr}} \tag{21.56}$$

is the expansion parameter of F with respect to $N - N_{tr}$ and where $F = f_2 - f_1$ is an abbreviation of the occupation number difference.

Electrons injected into the conduction band have an energy that is larger than the quasi-Fermi energy E_{Fc}. The mechanism leading to the quasi-Fermi distribution is the intraband relaxation of the electrons. The electrons lose energy by the emission of phonons. At finite temperature, emission and absorption of phonons leads to the establishment of the quasi-Fermi distribution of the electrons in the conduction band. After the establishment of a quasithermal equilibrium, the conduction band electrons still scatter permanently at phonons. Accordingly, each electron level is broadened. The width of a broadened energy level in the conduction band is $\Delta E_2 \approx \hbar/\tau_{\text{in}}$, where τ_{in} is the inelastic scattering time of an electron, i.e., the time between two inelastic scattering events. The scattering time τ_{in} depends on temperature.

The main process of electron–phonon scattering is the interaction with polar optic phonons; the energy of polar optic phonons of GaAs is about 40 meV. The inelastic scattering time ($\sim 10^{-13}$ s) of a conduction electron in GaAs at room temperature is much shorter than the lifetime (of the order of 1 ns) with respect to a radiative transition to the valence band by spontaneous emission of a photon. The width of broadening of a level in the conduction band is $\hbar/\tau_{\text{in}} \sim 6\,\text{meV}$.

The extraction of valence band electrons from the active region leads to the establishment of a quasi-Fermi distribution of the valence band electrons. Due to the non-radiative relaxation, the valence band, nearly filled with electrons, has empty states (holes) near the maximum of the band, as characterized by the quasi-Fermi energy E_{Fv}. The electrons in the valence band scatter also at phonons and the strength of the scattering is about the same as for the electrons in the conduction band. Accordingly, an electron level in the valence band has approximately the same lifetime with respect to inelastic scattering at phonons as an electron state in the conduction band—and the width of an energy level in the valence band is $\Delta E_1 \approx \hbar/\tau_{\text{in}}$ too.

Taking into account broadening of both the energy level in the conduction band and energy level in the valence band, which are involved in a radiative transition, we attribute to the transition a Lorentzian function $g(h\nu - E_{21})$ of a width that is by a factor of $\sqrt{2}$ larger than the value of the width of a level in a single band. The halfwidth of radiative transitions in GaAs is $\delta E_{21} \sim 10\,\text{meV}$.

21.6 Spontaneous Emission

Spontaneous emission of radiation is the origin of luminescence radiation. The rate of spontaneous emission of photons by transitions in the energy range $h\nu$, $h\nu + h\text{d}\nu$ is

$$R_{\text{sp},h\nu} h\text{d}\nu = \int g(h\nu - E_{21}) h\text{d}\nu \times A_{21} D_{\text{r}}(E_{21}) f_2(1 - f_1)\text{d}E_{21}. \qquad (21.57)$$

$R_{\text{sp},h\nu}$ is the spontaneous emission rate per unit of photon energy (and per unit of volume). The integration takes account of the contributions of all electrons in the conduction band and of the corresponding empty levels in the valence band. It follows that

$$R_{\text{sp},h\nu} = \int A_{21} D_{\text{r}}(E_{21}) f_2(1 - f_1) g(h\nu - E_{21})\text{d}E_{21}. \qquad (21.58)$$

Spontaneous emission can also occur at photon energies $h\nu < E_g$. This is the consequence of the broadening of the energy levels due to the finite lifetimes of the conduction band and valence band states with respect to inelastic scattering at phonons. The total spontaneous emission rate is

$$R_{sp} = \int\int A_{21} D_r(E_{21}) f_2(1 - f_1) g(h\nu - E_{21}) dE_{21} d(h\nu). \tag{21.59}$$

The lifetime of an electron in the conduction band with respect to spontaneous emission is

$$\tau_{sp} = \frac{1}{R_{sp}/N}. \tag{21.60}$$

N is the density of electrons in the conduction band. If $g(h\nu - E_{21})$ is a narrow function, we obtain

$$R_{sp,h\nu} = A_{21} D_r(E_{21}) f_2(1 - f_1). \tag{21.61}$$

The total spontaneous emission rate is

$$R_{sp} = \int A_{21} D_r(E_{21}) f_2(1 - f_1) d(h\nu) \tag{21.62}$$

and the decay constant is equal to

$$\frac{1}{\tau_{sp}} = A_{21} \frac{\int D_r(E_{21}) f_2(1 - f_1) d(h\nu)}{N}. \tag{21.63}$$

The occupation numbers of a continuously pumped crystal at zero temperature are $f_2 = 1$ in the region of populated energy levels and $f_1 = 0$ in the region of empty levels; the integral is equal to N. In this case, $\tau_{sp}^{-1} = A_{21}$.

At finite temperatures, transitions from an occupied electron level in the conduction band to an occupied electron level in the valence band cannot occur. Therefore, the decay constant is smaller than the Einstein coefficient of spontaneous emission, $\tau_{sp}^{-1} < A_{21}$. The value of τ_{sp}^{-1} depends on the electron density N and on the temperature.

21.7 Laser Equations of a Bipolar Semiconductor Laser

The laser equations (in the form of rate equations) of a continuously pumped single-mode bipolar laser are two coupled differential equations:

$$\frac{dN}{dt} = r - \frac{N}{\tau_{sp}} - \gamma Z, \tag{21.64}$$

$$\frac{dZ}{dt} = \gamma Z - \frac{Z}{\tau_p}. \tag{21.65}$$

The quantities are:

- dN/dt = temporal change of the electron density (=temporal change of the density of electrons in the conduction band = temporal change of the density of holes in the valence band).
- dZ/dt = temporal change of the density of photons.
- r = pump rate = number of electrons injected into the conduction band per m^3 and s (=number of electrons extracted from the valence band).
- N/τ_{sp} = loss of conduction band electrons due to spontaneous transitions to the valence band.
- τ_p = lifetime of a photon in the resonator.
- Z/τ_p = loss of photons from the resonator (e.g., due to output coupling of radiation).
- $-\gamma Z$ = rate of change of the density of electrons in the conduction band due to the net effect of stimulated emission and absorption of radiation.
- γZ = rate of change of the photon density in the resonator due to the net effect of stimulated emission and absorption of radiation.
- $\gamma = \int_0^\infty h\nu \bar{B}_{21} g(h\nu - E_{21})(f_2 - f_1) D_r(E_{21}) dE_{21}$ = growth coefficient.
- $f_2(E_2) - f_1(E_1)$ = occupation number difference.
- $E_{21} = E_2 - E_1$ = energy difference of radiative pair levels (=transition energy).
- $D_r(E_{21})$ = reduced density of states = density of states of radiative pair levels.
- $g(h\nu - E_{21})$ = lineshape function describing level broadening due to inelastic scattering of electrons at phonons.

At steady state, $dN/dt = 0$ and $dZ/dt = 0$, the second equation yields the threshold condition

$$\gamma_{th} = 1/\tau_p; \tag{21.66}$$

the photon generation rate is equal to the photon loss rate. This condition allows for determination of the threshold density N_{th}. We can write the threshold condition also in the form

$$\alpha_{th} L = \frac{nL}{c}\gamma_{th} = \frac{nL}{c\tau_p} \tag{21.67}$$

where L is the resonator length. The first laser equation leads to the photon density Z_∞ in the laser resonator at steady state,

$$Z_\infty = (r - r_{th})\tau_p, \tag{21.68}$$

where

$$r_{th} = N_{th}/\tau_{sp} \tag{21.69}$$

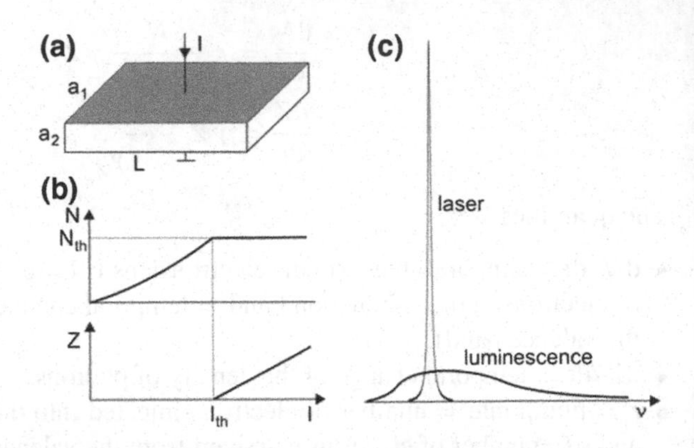

Fig. 21.6 Bipolar laser diode. **a** Device. **b** Dependence of the electron density and the photon density on the current. **c** Laser and luminescence radiation

is the threshold loss rate (in units of $m^{-3}\,s^{-1}$). The loss is due to spontaneous transitions of electrons from the conduction band to the valence band; we ignore other loss processes like the nonradiative recombination of electrons and holes.

In a bipolar semiconductor laser diode (Fig. 21.6a), the current I is flowing via the large area ($a_1 L$) through the active volume (height a_2). Below threshold, the electron concentration N increases (Fig. 21.6b) with increasing current strength until the current reaches the threshold current

$$I_{th} = r_{th} e a_1 a_2 L = N_{th} e a_1 a_2 L/\tau_{sp}. \tag{21.70}$$

The threshold current density is

$$j_{th} = \frac{I_{th}}{a_2 L} = \frac{N_{th} a_1 e}{\tau_{sp}}. \tag{21.71}$$

At stronger pumping, the carrier density remains at the value N_{th}. This means clamping of the following quantities: populations in the conduction and valence band; quasi-Fermi energy of the electrons in the conduction band; quasi-Fermi energy of the electrons in the valence band. Pumping above threshold leads to conversion of the additional pump power into photons and energy of relaxation. Above threshold, the rate of photon generation is equal to the additional rate of electron injection. The photon density in the laser resonator increases linearly with $I - I_{th}$.

The luminescence spectrum (Fig. 21.6c) is broad while the spectrum of the laser radiation is narrow.

At weak pumping (below threshold), luminescence radiation becomes stronger with increasing pump strength. Above threshold, clamping of luminescence occurs together with the clamping of the quasi-Fermi energies.

We can describe operation of a laser near the transparency density by the laser equations

$$dN/dt = r - N/\tau_{sp} - b_{eff}(N - N_{tr})Z, \tag{21.72}$$
$$dZ/dt = b_{eff}(N - N_{tr})Z - Z/\tau_p. \tag{21.73}$$

It follows that the threshold density is given by

$$N_{th} - N_{tr} = \frac{1}{b_{eff}\tau_p} = \frac{1}{\sigma_{eff} l_p} \tag{21.74}$$

and that the photon density is again $Z = (r - r_{th})\tau_p$, with $r_{th} = N_{th}/\tau_{sp}$.

If g is a narrow function, we obtain:

$$dN/dt = r - N/\tau_{sp} - h\nu \bar{B}_{21} D_r(E_{21})(f_2 - f_1)Z, \tag{21.75}$$
$$dZ/dt = h\nu \bar{B}_{21} D_r(E_{21})(f_2 - f_1)Z - Z/\tau_p. \tag{21.76}$$

Then, the threshold occupation number difference is then given by

$$(f_2 - f_1)_{th} = \frac{1}{h\nu \bar{B}_{21} D_r(E_{21})\tau_p} \tag{21.77}$$

and the photon density by

$$Z = (r - r_{th})\tau_p. \tag{21.78}$$

21.8 Gain Mediated by a Quantum Well

In two earlier sections (Sects. 7.8 and 7.9), we treated the question how we can combine a two-dimensional active medium with a light beam, which is three-dimensional. We introduced the two-dimensional gain characteristic H^{2D} and showed how we can determine the modal gain coefficient of radiation propagating along a two-dimensional gain medium and how we can determine the gain, $G_1 - 1$, of radiation crossing a quantum well. The topic of this section concerns the following questions.

- How can we determine semiconductor properties of an active quantum well (quasi-Fermi energies; strength of spontaneous emission of radiation; two-dimensional transparency density; two-dimensional gain characteristic)?
- How can we determine gain of radiation interacting with an active quantum well (according to the concepts presented in Sects. 7.8 and 7.9)?

Instead of proceeding with this section, a reader may jump to Sect. 21.10 *and then work out Problem* 21.5 *(gain mediated by a quantum well) and Problem* 21.6 *(quantum well laser).*

The quasi-Fermi energy of a two-dimensional gas of conduction electrons in a two-dimensional semiconductor follows from the condition

$$\int f_2 D_c^{2D} dE = N^{2D} \tag{21.79}$$

and the quasi-Fermi energy of the electrons in the valence band from

$$\int (1 - f_1) D_v^{2D} dE = P^{2D} = N^{2D}, \tag{21.80}$$

where we have the quantities:

- D_c^{2D} = two-dimensional density of levels in the conduction band.
- D_v^{2D} = two-dimensional density of levels in the valence band.
- N^{2D} = two-dimensional density of electrons in the conduction band.
- P^{2D} = two-dimensional density of empty states in the valence band = two-dimensional density of holes.
- $P^{2D} = N^{2D}$, due to neutrality.

The k vector of an electron is a vector in the plane of the two-dimensional semiconductor. The requirement of energy and momentum conservation leads to the two-dimensional reduced density of states

$$D_r^{2D}(E_{21}) = \frac{m_r}{m_e} D_c^{2D}(\epsilon_c). \tag{21.81}$$

The two-dimensional density of upper laser levels that contribute to stimulated radiative transitions in the energy interval E_{21}, $E_{21} + dE_{21}$ is

$$dN_2^{2D} = f_2(1 - f_1) D_r^{2D}(E_{21}) dE_{21} \tag{21.82}$$

and the corresponding density of electrons in the lower laser levels, which contribute to absorption, is

$$dN_1^{2D} = f_1(1 - f_2) D_r^{2D}(E_{21}) dE_{21}, \tag{21.83}$$

where $f_2 = f_2(E_2)$ and $f_1 = f_1(E_1)$ and $E_{21} = E_2 - E_1$. The spontaneous emission rate per unit of photon energy is given by

$$R_{sp,h\nu}^{2D} = \int A_{21} D_r^{2D}(E_{21}) f_2 (1 - f_1) g(h\nu - E_{21})\, dE_{21}. \tag{21.84}$$

The spontaneous emission rate per unit of volume is equal to

$$R_{sp}^{2D} = \int \int A_{21} D_r^{2D}(E_{21}) f_2 (1 - f_1) g(h\nu - E_{21}) dE_{21} d(h\nu). \tag{21.85}$$

The spontaneous lifetime of an electron in the condition band is

$$\tau_{sp} = N^{2D} / R_{sp}^{2D}. \tag{21.86}$$

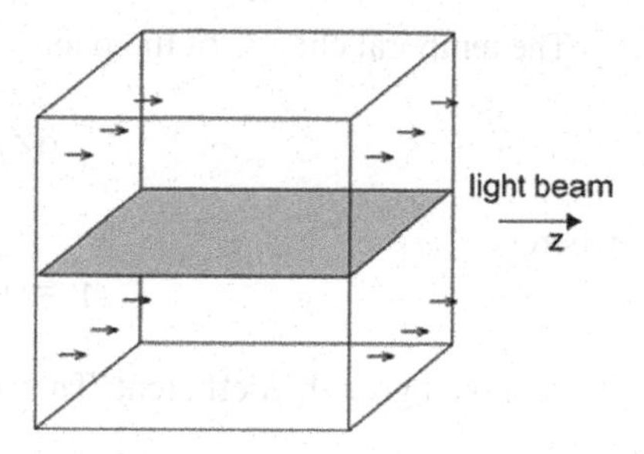

Fig. 21.7 Light beam propagating along a quantum well

If g is a narrow function, then

$$R^{2D}_{sp,h\nu} = A_{21} D^{2D}_{r}(E_{21}) f_2(1 - f_1) \tag{21.87}$$

and

$$R^{2D}_{sp} = \int A_{21} D^{2D}_{r}(E_{21}) f_2(1 - f_1) d(h\nu). \tag{21.88}$$

The two-dimensional densities D^{2D}_2 and D^{2D}_1 are completely determine the reduced density of states $D^{2D}_r(E_{21})$. The occupation numbers f_2 and f_1 depend on the two-dimensional density N^{2D} of electrons and the temperature.

To describe the interaction of the two-dimensional active medium with the three-dimensional radiation field, we consider two cases, namely that the propagation direction of the light is parallel to the plane of the two-dimensional active medium and that the propagation direction is perpendicular to the plane.

If the propagation direction is parallel to the plane of the active medium (Fig. 21.7), the average electron density in a photon mode is

$$N_{av} = N^{2D}/a_2, \tag{21.89}$$

where a_2 is the height of the mode. The temporal change of the average density is

$$dN_{av}/dt = -N_{av} Z = -(c/na_2) H^{2D} Z \tag{21.90}$$

and

$$H^{2D} = (n/c) \int h\nu \bar{B}_{21} g(h\nu - E_{21})(f_2 - f_1) D^{2D}_r(E_{21}) dE_{21} \tag{21.91}$$

is the two-dimensional gain characteristic. The expression of H^{2D} indicates: a two-dimensional semiconductor is a gain medium if $f_2 > f_1$. The condition is satisfied if the difference of the quasi-Fermi energies is larger than the gap energy, $E_{Fc} - E_{Fv} > E_g$. There is no net gain ($H^{2D} = 0$) if, at the two-dimensional transparency density N^{2D}_{tr}, the occupation number difference is zero, $f_2 - f_1 = 0$. This corresponds to the condition $E_{Fc} - E_{Fv} = E_g$. Thus, the condition of gain, $f_2 - f_1 > 0$, is the same as in the three-dimensional case.

The temporal change of the photon density in a photon mode is

$$\mathrm{d}Z/\mathrm{d}t = \gamma Z, \tag{21.92}$$

where

$$\gamma = (c/n)H^{2D}/a_2 \tag{21.93}$$

is the modal growth coefficient. The spatial change of the photon density is

$$\mathrm{d}Z/\mathrm{d}z = \alpha Z, \tag{21.94}$$

where

$$\alpha = \frac{H^{2D}}{a_2} \tag{21.95}$$

is the modal gain coefficient. The modal growth coefficient and the modal gain coefficient are inversely proportional to the lateral extension of the laser resonator mode.

If $N \sim N_{tr}$, we can write

$$\gamma_{max} = b_{eff} \frac{N^{2D} - N^{2D}_{tr}}{a_2}, \tag{21.96}$$

where

$$b_{eff} = \frac{(c/n)}{a_2} \left(\frac{\partial H^{2D}}{\partial N^{2D}} \right)_{N^{2D}=N^{2D}_{tr}} \tag{21.97}$$

is the effective growth rate constant and, furthermore,

$$\alpha_{max} = \sigma_{eff} \frac{N^{2D} - N^{2D}_{tr}}{a_2}, \tag{21.98}$$

where

$$\sigma_{eff} = \frac{1}{a_2} \left(\frac{\partial H^{2D}}{\partial N^{2D}} \right)_{N^{2D}=N^{2D}_{tr}} = \left(\frac{\partial \alpha_{max}}{\partial N^{2D}} \right)_{N^{2D}=N^{2D}_{tr}} \tag{21.99}$$

is an effective gain cross section (Sect. 7.4).

If g is a narrow function, we obtain

$$H^{2D} = h\nu \bar{B}_{21} D^{2D}_{r}(E_{21})(f_2 - f_1), \tag{21.100}$$

with $E_{21} = h\nu$. It follows that the modal growth coefficient is given by

$$\gamma = \frac{(c/n)}{a_2} h\nu \bar{B}_{21} D^{2D}_{r}(E_{21})(f_2 - f_1) \tag{21.101}$$

and the modal gain coefficient by

$$\alpha = \frac{1}{a_2} h\nu \bar{B}_{21} D_r^{2D}(E_{21})(f_2 - f_1). \tag{21.102}$$

Growth coefficient and gain coefficient are proportional to the occupation number difference.

In the case that $f_2 - f_1 \ll 1$, we can expand F with respect to N^{2D},

$$F = f_2 - f_1 = d^{2D} \times (N^{2D} - N_{tr}^{2D}), \tag{21.103}$$

where

$$d^{2D} = \left(\frac{\partial F}{\partial N^{2D}}\right)_{N^{2D}=N_{tr}^{2D}} \tag{21.104}$$

is the expansion coefficient of the occupation number difference with respect to $N^{2D} - N_{tr}^{2D}$. The expansion leads to

$$\gamma = b_{eff} \frac{N^{2D} - N_{tr}^{2D}}{a_2} = b_{eff}(N_{av} - N_{tr,av}) \tag{21.105}$$

and

$$b_{eff} = h\nu \bar{B}_{21} D_r^{2D}(E_{21}) d^{2D}. \tag{21.106}$$

The modal growth coefficient is proportional to the difference of the density of excited electrons and inversely proportional to the height of the photon mode. The unit of b_{eff} is the same as in the three-dimensional case since the product $D^{2D}d^{2D}$ has the same unit as the corresponding product in the three-dimensional case.

It follows that the modal gain coefficient is equal to

$$\alpha = \sigma_{eff} \frac{N^{2D} - N_{tr}^{2D}}{a_2}, \tag{21.107}$$

where

$$\sigma_{eff} = h\nu \bar{B}_{21} D_r^{2D}(E_{21}) d^{2D}. \tag{21.108}$$

In a disk of light traversing a two-dimensional bipolar medium (Fig. 21.8), the temporal change of the photon density is given by (*see* also Sect. 7.9):

$$\frac{\delta Z}{\delta t} = -\frac{\delta N_{av}}{\delta t} = \frac{c}{na_2} H^{2D} Z, \tag{21.109}$$

where $\delta t = n\delta z/c$ is the time it takes the disk of length δz to propagate over the medium (that has zero thickness). It follows that the gain of light traversing a two-dimensional bipolar medium is

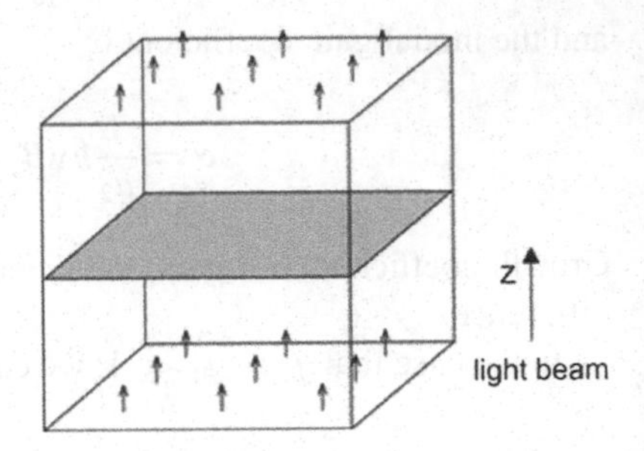

Fig. 21.8 Light beam traversing a quantum well

$$G_1 - 1 = \frac{\delta Z}{Z} = H^{2D}. \tag{21.110}$$

If g is a narrow function, we obtain

$$G_1 - 1 = h\nu \tilde{B}_{21} D_r^{2D}(E_{21})(f_2 - f_1). \tag{21.111}$$

Now, the *gain*, $G_1 - 1$, is proportional to $(f_2 - f_1)$.

21.9 Laser Equations of a Quantum Well Laser

The laser equations of a quantum well laser, with light propagating along the quantum well, are given by:

$$\frac{dN_{av}}{dt} = r_{av} - \frac{N_{av}}{\tau_{sp}} - (c/na_2)H_{av}Z, \tag{21.112}$$

$$\frac{dZ}{dt} = (c/na_2)H_{av}Z - \frac{Z}{\tau_p}. \tag{21.113}$$

$N_{av} = N^{2D}/a_2$ is the average electron density in the laser mode, $H_{av} = H^{2D}/a_2$ is the average gain characteristic—averaged over the laser mode volume. Furthermore,

$$r_{av} = r^{2D}/a_2 \tag{21.114}$$

is the pump rate averaged over the volume of the resonator, r^{2D} the two-dimensional pump rate (=pump rate per m^2) and a_2 the height of the resonator mode.

The solution describing steady state oscillation provides the threshold condition

$$H_{th}^{2D} = na_2/\tau_p \tag{21.115}$$

or, with $\tau_p = nl_p/c$,

$$H_{\rm th}^{\rm 2D} = \frac{a_2}{l_{\rm p}}. \tag{21.116}$$

$H_{\rm th}^{\rm 2D}$ is inversely proportional to the ratio of photon path length and extension of the resonator mode perpendicular to the plane of the active medium. A small value of $H_{\rm th}^{\rm 2d}$ corresponds to a small occupation number difference ($f_2 - f_1 \ll 1$) and to an electron density that is only slightly larger than the transparency density ($N^{\rm 2D} - N_{\rm tr}^{\rm 2D} \ll N_{\rm tr}^{\rm 2D}$).

Equation (21.112) yields the photon density Z_∞ in the laser resonator at steady state oscillation,

$$Z_\infty = \left(\frac{r^{\rm 2D}}{a_2} - \frac{r_{\rm th}^{\rm 2D}}{a_2}\right)\tau_{\rm p}, \tag{21.117}$$

where

$$r_{\rm th}^{\rm 2D} = \frac{N_{\rm th}^{\rm 2D}}{\tau_{\rm sp}} \tag{21.118}$$

is the two-dimensional threshold loss rate (=loss per s and m^2). The threshold current is

$$I_{\rm th} = \frac{eN_{\rm th}^{\rm 2D} a_1 a_2 L}{a_2} \times \frac{1}{\tau_{\rm sp}} = \frac{eN_{\rm th}^{\rm 2D} a_1 L}{\tau_{\rm sp}} \tag{21.119}$$

and the threshold current density

$$j_{\rm th} = \frac{eN_{\rm th}^{\rm 2D}}{\tau_{\rm sp}}. \tag{21.120}$$

In the case that g is a narrow function, the laser equations are

$$\frac{{\rm d}N_{\rm av}}{{\rm d}t} = r_{\rm av} - \frac{N_{\rm av}}{\tau_{\rm sp}} - \frac{1}{a_2} h\nu \bar{B}_{21} D_{\rm r}^{\rm 2D}(E_{21})(f_2 - f_1)Z, \tag{21.121}$$

$$\frac{{\rm d}Z}{{\rm d}t} = \frac{1}{a_2} h\nu \bar{B}_{21} D_{\rm r}^{\rm 2D}(E_{21})(f_2 - f_1)Z - \frac{Z}{\tau_{\rm p}}. \tag{21.122}$$

It follows that the threshold occupation number difference is equal to

$$F_{\rm th} = (f_2 - f_1)_{\rm th} = \frac{c/n}{h\nu \bar{B}_{21} D_{\rm r}^{\rm 2D} l_{\rm p}/a_2}. \tag{21.123}$$

If g is a narrow function and the threshold density is only slightly larger than the transparency density, $N_{\rm th} - N_{\rm tr} \ll N_{\rm tr}$, we can write

$$\frac{dN_{av}}{dt} = r_{av} - \frac{N_{av}}{\tau_{sp}} - b_{eff}\,(N_{av} - N_{tr,av})\,Z, \tag{21.124}$$

$$\frac{dZ}{dt} = b_{eff}\,(N_{av} - N_{tr,av})\,Z - \frac{Z}{\tau_p}, \tag{21.125}$$

where $r_{av} = r^{2D}/a_2$ is the average pump rate (averaged over the resonator volume), r^{2D} the pump rate per m^2, $b_{eff} = h\nu B_{21} D_r^{2D}(E_{21}) d^{2D}$ is the effective growth rate constant, and where

$$d^{2D} = \left(\frac{\partial F}{\partial N^{2D}}\right)_{N^{2D}=N_{tr}^{2D}}. \tag{21.126}$$

$F = f_2 - f_1$ is the occupation number difference. It follows that

$$N_{th}^{2D} - N_{tr}^{2D} = \frac{a_2}{b_{eff}\tau_p} = \frac{1}{\sigma_{eff} l_p / a_2}, \tag{21.127}$$

$$\frac{N_{th}^{2D} - N_{tr}^{2D}}{a_2} = \frac{1}{\sigma_{eff} l_p}. \tag{21.128}$$

The average density difference plays the same role as the density difference in a three-dimensional active semiconductor medium.

A laser containing a quantum well that is oriented perpendicular to the laser beam will be discussed in Sect. 22.7.

21.10 What Is Meant by "Bipolar"?

Instead of discussing empty electron states in the valence band of a bipolar laser medium, we can use the picture of holes: an empty level in the valence band is a hole in the valence band. Accordingly, a current leads to injection of electrons into the conduction band and to *injection of holes into the valence band*. In the electron–hole picture, the current is carried by electrons in the conduction band and by holes in the valence band—the current is carried by negatively charged quasiparticles (electrons) and positively charged quasiparticles (holes); the discussion that now follows can be in [236].

Involved in a radiative transition (Fig. 21.9) are an electron (in the conduction band) and a hole (in the valence band). In an absorption process, an electron and a hole annihilate recombine and create a photon. Conservation of momentum requires that the momentum before an emission process is equal to the momentum after the process,

$$\hbar \boldsymbol{k}_e + \hbar \boldsymbol{k}_h = \hbar \boldsymbol{q}_p, \tag{21.129}$$

where $\boldsymbol{k}_e$ is the wave vector of the electron, $\boldsymbol{k}_h$ the wave vector of the hole, and $\boldsymbol{q}_p$ the wave vector of the photon. If $q_p \ll k_e, k_h$, then

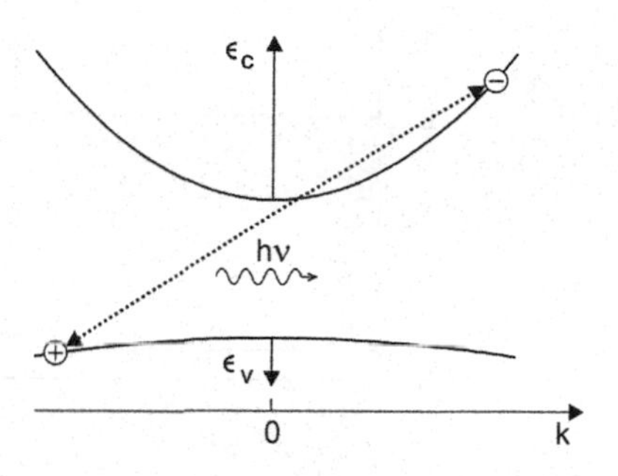

Fig. 21.9 Bipolar laser in an electron–hole picture

$$\boldsymbol{k}_{\mathrm{h}} = -\boldsymbol{k}_{\mathrm{e}}. \tag{21.130}$$

In the electron–hole picture, the wave vector conservation $\boldsymbol{k}_{\mathrm{e}} + \boldsymbol{k}_{\mathrm{h}} = 0$ corresponds to the wave vector conservation $\boldsymbol{k}_{\mathrm{ce}} = \boldsymbol{k}_{\mathrm{ve}}$ in the electron picture. The momentum of an empty electron state of the valence band is $-\hbar\boldsymbol{k}_{\mathrm{ve}}$ while $\hbar\boldsymbol{k}_{\mathrm{ve}}$ is the momentum of an electron that occupies this state. Accordingly, the wave vector $\boldsymbol{k}_{\mathrm{h}}$ of a hole (in the valence band) is

$$\boldsymbol{k}_{\mathrm{h}} = -\boldsymbol{k}_{\mathrm{ve}}. \tag{21.131}$$

Electron and hole have opposite wave vectors. A radiative pair—an electron–hole pair consisting of an electron and a hole of opposite wave vector—can annihilate (=recombine) by spontaneous or stimulated emission of a photon. Laser radiation in a bipolar laser is due to stimulated *electron–hole recombination*. The energy of a radiative pair is

$$E = E_{\mathrm{g}} + \frac{\hbar^2 k^2}{2m_{\mathrm{e}}} + \frac{\hbar^2 k^2}{2m_{\mathrm{h}}} = E_{\mathrm{g}} + \frac{\hbar^2 k^2}{2m_{\mathrm{r}}}, \tag{21.132}$$

where m_{r} is the reduced mass, m_{e} the electron mass, and m_{h} the hole mass.

The occupation number of a hole state is

$$f_{\mathrm{h}} = 1 - f_1. \tag{21.133}$$

It follows that the quasi-Fermi energy E_{Fh} of the holes is equal to the quasi-Fermi energy E_{Fv} of the valence band electrons,

$$E_{\mathrm{Fh}} = E_{\mathrm{Fv}} \tag{21.134}$$

and that

$$f_{\mathrm{h}} = \frac{1}{\exp{(E_{\mathrm{Fv}} - E)/kT + 1}}. \tag{21.135}$$

Figure 21.10 illustrates the connection between the electron picture and the electron–hole picture:

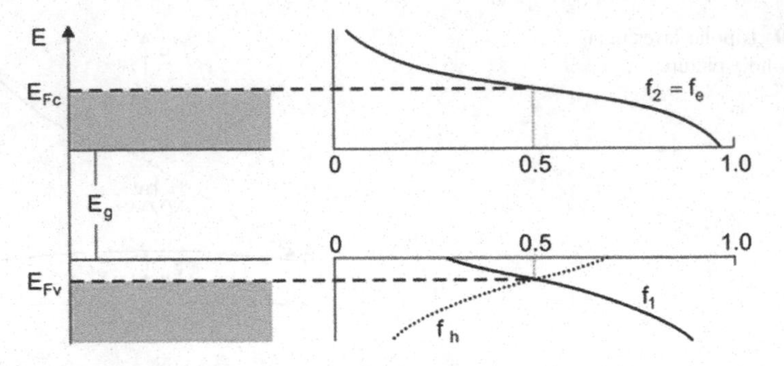

Fig. 21.10 Quasi-Fermi distributions of electrons and holes

- *Electron picture*. The conduction band contains an electron gas characterized by the quasi-Fermi energy $E_{\rm Fc}$ and the energy distribution $f_2(E)$. The valence band contains an electron gas characterized by the quasi-Fermi energy $E_{\rm Fv}$ and the distribution $f_1(E)$. The condition of gain requires that $f_2 - f_1 > 0$. Optical transitions occur between radiative pair levels.
- *Electron–hole picture*. The conduction band contains an electron gas characterized by $E_{\rm Fc}$ and the distribution $f_2 = f_{\rm e}$ (as in the electron picture). The valence band contains a hole gas characterized by the quasi-Fermi energy $E_{\rm Fv}$ and the distribution $f_{\rm h} = 1 - f_1$. The condition of gain now requires that

$$f_{\rm e} + f_{\rm h} - 1 > 0. \tag{21.136}$$

Optical transitions occur by recombination (annihilation) of radiative electron–hole pairs.

Because of bipolarity and charge neutrality of an active medium, the knowledge of the density N of electrons in the conduction band is sufficient for a complete characterization of a particular active medium (if the density of states of electrons and holes as well as the temperature are known).

The bipolarity of a medium manifests itself in the dependence of the spontaneous lifetime on the densities of positive and negative charge carriers. It turns out (analyzing $R_{\rm sp}$) that the rate $\tau_{\rm sp}^{-1}$ of spontaneous transitions of electrons in a semiconductor at room temperature is, for very small values of N, approximately proportional to the product of the density N of electrons and the density $P = N$ of holes,

$$\frac{1}{\tau_{\rm sp}} = KN^2. \tag{21.137}$$

K is a constant. At large values of N, the decay rate $\tau_{\rm sp}^{-1}$ is nearly independent of N. The behavior is characteristic of a bipolar system.

References [1–4, 6, 187–201].

Problems

21.1 Wave vector of nonequilibrium electrons in GaAs.

(a) Calculate the wave vector k of electrons in GaAs that have an energy of 100; 10; and 1 meV. Compare the values with the wave vector q_p of a photon with the energy $h\nu = E_g$ ($m_e = 0.07\,m_0$; $m_0 = 0.9 \times 10^{-30}$ kg; $E_g = 1.42$ eV; $n = 3.6$).

(b) Determine the energies ϵ_c and ϵ_v if $q_p = k$.

21.2 Wave vector of radiative pair levels. We supposed that the wave vector of a photon involved in a radiative transition is small compared to the wave vector of the electron and the hole that are involved in a radiative transition. Show that this is justified for electrons and holes of sufficient energies.

21.3 Electron and holes in an undoped GaAs quantum film in thermal equilibrium.

(a) What is the condition with respect to the quasi-Fermi energies that the electron gas and the hole gas are in thermal equilibrium?

(b) What is the corresponding condition with respect to ϵ_{Fc} and ϵ_{Fv}?

(c) Estimate the electron density $N^{2D}_{thermal}$ of subband electrons (=density of subband holes) in a quantum film at temperature T. Show that $N^{2D}_{thermal}$ is by many orders of magnitude smaller than the transparency density N^{2D}_{tr} of electrons in the quantum film.

21.4 Condition of gain. Show that the condition of gain, $E_{Fc} - E_{Fv} > E_{21} = E_2 - E_1$, follows from the condition $f_2 - f_1 > 0$.

21.5 Gain mediated by a quantum well. Given are the following quantities:

- D^{2D}_c = two-dimensional density of states of electrons in the conduction band.
- D^{2D}_v = two-dimensional density of states of electrons in the valence band (=two-dimensional density of states of holes).
- N^{2D} = two-dimensional density of nonequilibrium electrons in the conduction band (assumed to be equal to the two-dimensional density of nonequilibrium holes in the valence band).
- $g(h\nu - E_{21})$ = lineshape function.
- a_2 = height of a photon mode that contains the quantum well; the plane of the quantum well is oriented parallel to the propagation direction of the radiation.
- $F \equiv f_2 - f_1 = d^{2D} \times (N^{2D} - N^{2D}_{tr})$; this expansion implies that the quantum well is operated near the transparency density.

Formulate equations, which are suited to determine the following quantities:

(a) E_{Fc} = quasi-Fermi energy of electrons in the conduction band.

(b) E_{Fv} = quasi-Fermi energy of electrons in the valence band.

(c) N^{2D}_{tr} = two-dimensional transparency density.

(d) $R^{2D}_{\mathrm{sp},h\nu}$ = spontaneous emission rate per unit photon energy in the cases that the lineshape function is broad or narrow.
(e) R^{2D}_{sp} = total spontaneous emission rate (at a broad or a narrow lineshape function).
(f) τ_{sp} = lifetime of the nonequilibrium electrons with respect to spontaneous emission of radiation.
(g) H^{2D} = two-dimensional gain profile.
(h) γ = modal growth coefficient.
(i) α = modal gain coefficient.
(j) b_{eff} = effective growth rate constant.
(k) σ_{eff} = effective gain cross section.
(l) $G_1 - 1$ = gain of light traversing a quantum well.

The answers are found in Sect. 21.8.

21.6 Quantum well laser. Given are the quantities:

- H^{2D} = two-dimensional gain profile of a quantum well.
- a_2 = extension of the resonator perpendicular to the quantum well.
- a_1 = width of the resonator.
- $a_1 \times L$ = area of the quantum well.
- L = length of the resonator.
- N^{2D} = two-dimensional density of nonequilibrium electrons.
- r^{2D} = two-dimensional pump rate.
- $f_2 - f_1 = d^{2D} \times (N^{2D} - N^{2D}_{\mathrm{tr}})$; operation near the transparency density.
- b_{eff} = growth rate constant.
- $\sigma_{\mathrm{eff}} = nb_{\mathrm{eff}}/c$ = effective gain cross section.

(a) Formulate the laser equations (rate equations).
(b) Derive the threshold condition.
(c) Determine the threshold current and the threshold current density.
(d) Formulate the threshold condition of a quantum well laser operated at an electron density near the transparency density; neglect lineshape broadening. The answers can be found in Sect. 21.9.

21.7 Determine the de Broglie wavelength $\lambda_{\mathrm{dB}} = h/p$ of electrons of an energy of 10 meV that are propagating (a) in free space and (b) as conduction electrons in a GaAs crystal.

21.8 A three-dimensional GaAs semiconductor at zero temperature contains non-equilibrium electrons of a density that corresponds to a quasi-Fermi energy ϵ_{Fe} = 25 meV. Determine the following quantities.

(a) Density of electrons in the conduction band.
(b) Fermi momentum k_{F}, i.e., the momentum of the electrons at the Fermi surface.
(c) The de Broglie wavelength of the electrons that have Fermi momentum.
(d) Quasi-Fermi energy ϵ_{Fh} of the nonequilibrium holes in the valence band assuming crystal neutrality.

(e) Fermi momentum of the nonequilibrium holes.
(f) The de Broglie wavelength of the holes that have Fermi momentum.

21.9 Answer the questions of the preceding problem with respect to a *two-dimensional* GaAs semiconductor at zero temperature containing nonequilibrium electrons of a density that corresponds to a quasi-Fermi energy $\epsilon_{\mathrm{Fe}} = 25\,\mathrm{meV}$.

21.10 Answer the same questions with respect to a *one-dimensional* GaAs semiconductor at zero temperature containing nonequilibrium electrons of a density that corresponds to a quasi-Fermi energy $\epsilon_{\mathrm{Fe}} = 25\,\mathrm{meV}$.

Chapter 22
GaAs Quantum Well Laser

As an example of a bipolar semiconductor laser, we treat the GaAs quantum well laser (wavelength around 800 nm). In later chapters, we will study quantum well lasers consisting of other materials and bipolar lasers of other types.

We describe a quantum well by an electron subband and a hole subband (the heavy hole subband); we will, in a later chapter (Chap. 26), slightly modify the description of a quantum well laser by taking into account another hole subband (the light hole subband).

To characterize an active quantum well, we calculate the quasi-Fermi energies of electrons and holes; because the densities of states of electrons and holes have constant (energy-independent) values, we obtain analytic expressions of the quasi-Fermi energies. We consider a GaAs quantum well (at low temperature and at room temperature) carrying nonequilibrium electrons of different densities N^{2D}. We determine the quasi-Fermi energy, the occupation number difference $f_2 - f_1$ and the two-dimensional gain characteristic H^{2D}. We discuss modal growth and gain coefficients. We introduce the material gain coefficient; the material gain coefficient corresponds to a three-dimensional description of the quantum film, but with a two-dimensional density of states.

The quantum well laser consists of a heterostructure composed of at least five semiconductor layers. These have the tasks: to form a quantum well; to provide a light guide effect; to allow for injection of electrons and holes into the quantum well by means of a current. We describe the principle and the design of the edge emitting GaAs quantum well laser. We derive the laser threshold condition and determine the threshold current. The solutions to the rate equations of a quantum well laser indicate clamping of the quasi-Fermi energies of the electron and hole gases.

A multi-quantum well laser, containing several quantum wells in parallel, has a larger gain and a larger output power than a quantum well laser containing one quantum well only.

The arrangement of many laser diodes in a linear array or in a stack of arrays results in a high-power semiconductor laser. The radiation of a high-power semiconductor laser is not a single coherent wave but is composed of different coherent waves, which

K.F. Renk, *Basics of Laser Physics*, Graduate Texts in Physics,
DOI 10.1007/978-3-319-50651-7_22

permanently change the relative phase to each other. The radiation has, however, a high degree of monochromaticity.

Besides the edge emitting quantum well laser, we discuss the vertical-cavity surface-emitting laser (VCSEL). The importance with respect to applications—of both edge emitting quantum well laser and vertical-cavity surface-emitting laser—has already been discussed (in Sect. 1.4).

We finally point out that the laser radiation of an edge-emitting laser can be polarized, with the electric field vector of the laser radiation lies in the plane of the quantum well. As a last point, we determine the spectrum of luminescence radiation emitted by a quantum well laser in addition to laser radiation.

22.1 GaAs Quantum Well

A GaAs quantum well (Fig. 22.1) consists of a thin GaAs film embedded in GaAlAs. Electrons can assume lower energies in the GaAs layer than in GaAlAs and holes can assume higher energies. Because of the lateral restriction, free-electron motion is only possible along the film plane. The two-dimensional free-electron motion of a conduction band electron is characterized by the *electron subband*. Correspondingly, the two-dimensional free-electron motion of a valence band electron is characterized by a *hole subband*. The gap energy E_g^{2D} of the two-dimensional semiconductor is slightly larger than the gap energy E_g of the corresponding bulk semiconductor because of the zero point energy associated with the electron and the hole confinement.

Example GaAs quantum well (at room temperature).

- $E_g \sim 1.42\,\text{eV}$; gap energy.
- $Ga_{0.85}Al_{0.15}As$; $E_g \sim 1.51\,\text{eV}$ (that is 90 meV larger than for GaAs).
- $Ga_{0.75}Al_{0.25}$ As; $E_g \sim 1.60\,\text{eV}$ (that is 180 meV larger than for GaAs).
- $E_g^{2D} = 1.45\,\text{eV}$ (that is 30 meV larger than for bulk GaAs); two-dimensional gap energy of a quantum well of 10 nm thickness (Sect. 26.4).
- $\nu_g^{2D} = E_g^{2D}/h \sim 359\,\text{THz}$; gap frequency.
- $\lambda_g^{2D} = c/\nu_g^{2D} \sim 836\,\text{nm}$; vacuum wavelength corresponding to the gap frequency.
- $m_e \sim 0.07\, m_0$; $m_0 = 0.92 \times 10^{-30}\,\text{kg}$ = electron mass.

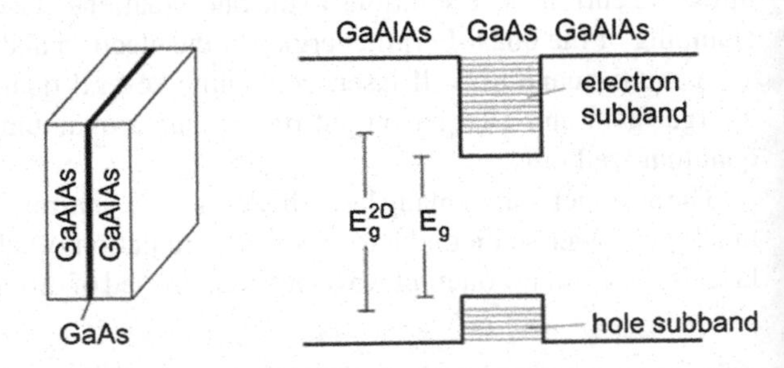

Fig. 22.1 GaAs quantum well

- $m_h = 0.43\, m_0$.
- $m_r \sim 0.06 m_0$.
- $D_c^{2D} = m_e/(\pi\hbar^2) = 2.0 \times 10^{36}\,\mathrm{J^{-1}\,m^{-2}}$; density of states of electrons in the conduction band.
- $D_v^{2D} = m_e/(\pi\hbar^2) = 12 \times 10^{36}\,\mathrm{J^{-1}\,m^{-2}}$; density of states of holes = density of states of electrons in the valence band.
- $D_r^{2D} = 1.7 \times 10^{36}\,\mathrm{J^{-1}\,m^{-2}}$ = reduced density of states for $E_{21} \geq E_g^{2D}$; *see* (21.81).

22.2 An Active Quantum Well

Injection of electrons into the conduction band (i.e., into the electron subband) and extraction of electrons from the valence band (i.e., injection of holes into the hole subband) results in an active quantum well (Fig. 22.2). An electron in the conduction band has the energy

$$E_c = E_c^{2D} + \epsilon_c, \tag{22.1}$$

where E_c^{2D} is the minimum of the electron subband and ϵ_c is the energy within the conduction band. An electron level in the valence band has the energy

$$E_v = E_v^{2D} - \epsilon_v, \tag{22.2}$$

where E_v^{2D} is the maximum of the hole subband and ϵ_v is the energy within the valence band. The conduction band electrons have a Fermi–Dirac distribution with the quasi-Fermi energy E_{Fc} and the electrons in the valence band have a Fermi–Dirac distribution with the quasi-Fermi energy E_{Fv}. We can write

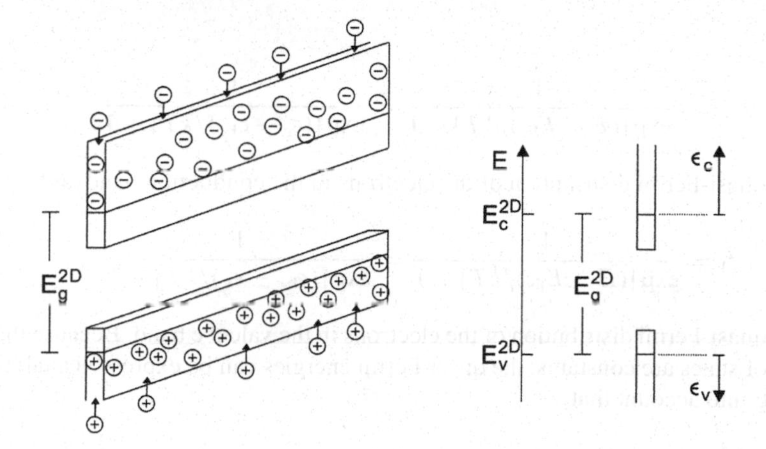

Fig. 22.2 An active quantum well and energy scales

$$E_{\text{Fc}} = E_{\text{c}}^{2\text{D}} + \epsilon_{\text{Fc}}, \tag{22.3}$$

$$E_{\text{Fv}} = E_{\text{v}}^{2\text{D}} - \epsilon_{\text{Fv}}, \tag{22.4}$$

where ϵ_{Fc} is the quasi-Fermi energy of the conduction band electrons relative to the energy $E_{\text{c}}^{2\text{D}}$ of the conduction band minimum and ϵ_{Fv} the quasi-Fermi energy of the valence band electrons relative to the energy $E_{\text{v}}^{2\text{D}}$ of the valence band maximum.

At zero temperature, all conduction band levels between $E_{\text{c}}^{2\text{D}}$ and the quasi-Fermi energy E_{Fc} are occupied while all valence band levels between $E_{\text{v}}^{2\text{D}}$ and E_{Fv} are completely empty. The quasi-Fermi energies at $T = 0$ are

$$E_{\text{Fc}} = E_{\text{c}}^{2\text{D}} + N^{2\text{D}}/D_{\text{c}}^{2\text{D}} = E_{\text{c}}^{2\text{D}} + \epsilon_{\text{Fc}}, \tag{22.5}$$

$$E_{\text{Fv}} = E_{\text{v}}^{2\text{D}} - N^{2\text{D}}/D_{\text{v}}^{2\text{D}} = E_{\text{v}}^{2\text{D}} - \epsilon_{\text{Fv}}. \tag{22.6}$$

E_{Fc} increases linearly with $N^{2\text{D}}$ and E_{Fv} decreases linearly with $N^{2\text{D}}$. Because of the larger density of states in the valence band, the energy range $E_{\text{Fc}} - E_{\text{c}}^{2\text{D}}$ of occupied energy levels in the conduction band is larger than the energy range $E_{\text{v}} - E_{\text{Fv}}$ of empty levels in the valence band.

At finite temperature, the Fermi distributions are broader. The quasi-Fermi energy of the electrons in the conduction band follows from the expression

$$\int f_2 D_{\text{c}}^{2\text{D}} \text{d}E = N^{2\text{D}} \tag{22.7}$$

and the quasi-Fermi energy of the electrons in the valence band from

$$\int (1 - f_1) D_{\text{v}}^{2\text{D}} \text{d}E = N^{2\text{D}}, \tag{22.8}$$

where

$$f_2 = \frac{1}{\exp[(E - E_{\text{Fc}})/kT] + 1} = \frac{1}{\exp[(\epsilon_{\text{c}} - \epsilon_{\text{Fc}})/kT] + 1} \tag{22.9}$$

is the quasi-Fermi distribution of the electrons in the conduction band and

$$f_1 = \frac{1}{\exp[(E - E_{\text{Fv}})/kT] + 1} = \frac{1}{\exp[(\epsilon_{\text{Fv}} - \epsilon_{\text{v}})/kT] + 1} \tag{22.10}$$

is the quasi-Fermi distribution of the electrons in the valence band. Because the densities of states are constants, the quasi-Fermi energies can be expressed analytically. Taking into account that

$$\int_0^\infty \frac{\text{d}x}{1 + \text{e}^{x-a}} = \ln \frac{\text{e}^x}{1 + \text{e}^x} = a + \ln(1 + \text{e}^{-a}), \tag{22.11}$$

we find the quasi-Fermi energies

$$\epsilon_{Fc} = kT \ln \left(-1 + \exp[N^{2D}/D_c^{2D} kT]\right), \tag{22.12}$$

$$\epsilon_{Fv} = kT \ln \left(-1 + \exp[N^{2D}/D_v^{2D} kT]\right). \tag{22.13}$$

The difference of the quasi-Fermi energies is given by

$$(E_c - E_{Fv})/kT = (\epsilon_{Fv} + \epsilon_{Fc})/kT = \ln \left(-1 + \exp[N^{2D}/D_c^{2D} kT]\right) + \ln \left(-1 + \exp[N^{2D}/D_v^{2D} kT]\right). \tag{22.14}$$

We now discuss a GaAs quantum well at room temperature (T = 300 K), which contains different electron densities N^{2D} (Fig. 22.3). The quasi-Fermi energy E_{Fc} of the electrons in the conduction band has a value of $-\infty$ at zero electron density. With increasing electron density the quasi-Fermi energy increases, reaches the minimum E_c^{2D} of the conduction band (where $E_{Fc} = E_c^{2D}$), and increases further. At large electron density the quasi-Fermi energy E_{Fc} increases proportionally to the electron density. The quasi-Fermi energy E_{Fv} of the electrons in the valence band is $+\infty$ at

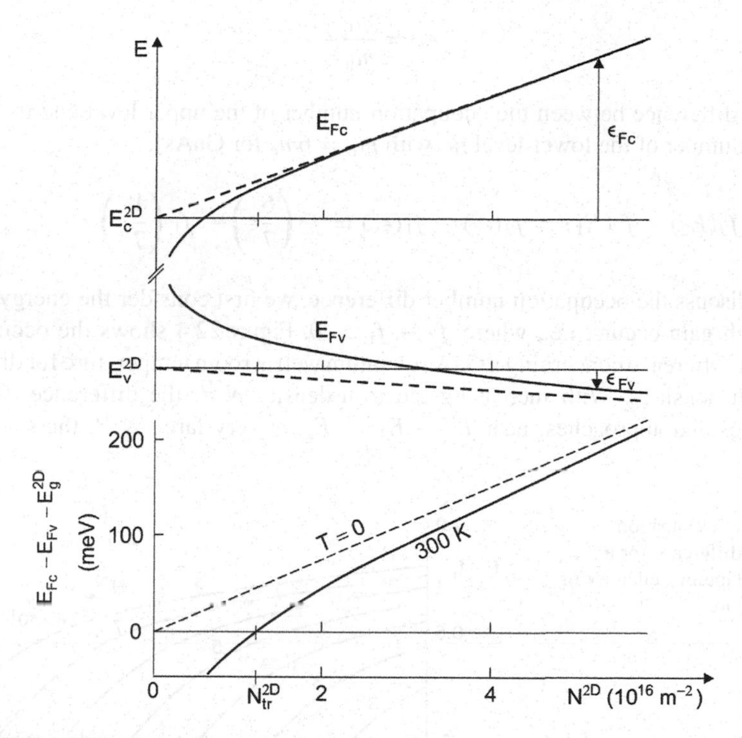

Fig. 22.3 Quasi-Fermi energies and their difference for a GaAs quantum well at room temperature (*solid lines*) and at zero temperature (*dashed*)

$N^{2D} = 0$, decreases with increasing N^{2D}, reaches the maximum of the valence band (where $E_{Fv} = E_v^{2D}$), and decreases further.

The difference of the quasi-Fermi energies (Fig. 22.3, lower part) increases with increasing electron density. The difference is 0 at thetransparency density N_{tr}^{2D} (=1.2 × 10^{16} m^{-2}). Gain occurs if $E_{Fc} - E_{Fv} > E_g^{2D}$. The difference of the quasi-Fermi energies reaches a value of 100 meV at an electron density of about 3×10^{16} m^{-2}. For a GaAs quantum well at zero temperature, $E_{Fc} - E_{Fv}$ is always larger than E_g^{2D} (dashed curves).

The energy difference between radiative pair levels is

$$E_2 - E_1 = E_g^{2D} + \epsilon, \tag{22.15}$$

where

$$\epsilon = E_2 - E_1 - E_g^{2D} = \epsilon_c + \epsilon_v \tag{22.16}$$

is the energy difference $E_2 - E_1$ minus the gap energy and where

$$\epsilon_c = \frac{m_r}{m_c}\,\epsilon, \tag{22.17}$$

$$\epsilon_v = \frac{m_r}{m_h}\,\epsilon. \tag{22.18}$$

The difference between the occupation number of the upper level and the occupation number of the lower level is (with $m_h = 6m_e$ for GaAs):

$$f_2(E_2) - f_1(E_1) = f_2(\epsilon_c) - f_1(\epsilon_v) = f_2\left(\frac{6}{7}\epsilon\right) - f_1\left(\frac{1}{7}\epsilon\right). \tag{22.19}$$

To discuss the occupation number difference, we first consider the energy range in which gain occurs, i.e., where $f_2 - f_1 > 0$. Figure 22.4 shows the occupation number difference concerning a GaAs quantum well at room temperature for different electron densities. With increasing electron density N^{2D}, the difference $f_2 - f_1$ increases and approaches, near $E_2 - E_1 = E_g$, at very large N^{2D}, the saturation

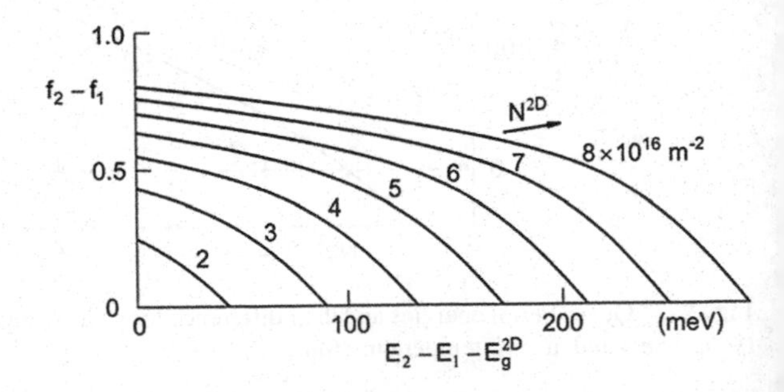

Fig. 22.4 Occupation number difference for a GaAs quantum well at room temperature

value $f_2 - f_1 = 1$. The range of gain increases with increasing electron density. When $f_2 - f_1$ is known, we can determine the different quantities describing gain.

- $H^{2D}(\nu) = (n/c)h\nu\bar{B}_{21}D_r^{2D}(E_{21})(f_2 - f_1)$ = two-dimensional gain characteristic.
- $\alpha(\nu) = (1/a_2)H^{2D}(\nu) = (n/c)h\nu\bar{B}_{21}(D_r^{2D}/a_2)(f_2 - f_1)$ = *modal gain coefficient* (=gain coefficient related to a mode); α depends on the extension of the radiation mode perpendicular to the propagation direction; D_r^{2D}/a_2 is the average density of states of radiative pairs levels within a mode of the radiation.
- $\gamma(\nu) = (c/na_2)h\nu\bar{B}_{21}D_r^{2D}(f_2 - f_1)$ = modal growth coefficient; it also depends on the extension a_2 of the mode perpendicular to the quantum well.

For completeness, we write the gain coefficient in the form

$$\alpha = \alpha_{\text{mat}} \times \Gamma, \tag{22.20}$$

where

$$\alpha_{\text{mat}}(\nu) = \frac{1}{s}H^{2D} = \frac{n}{c}h\nu\bar{B}_{21}\frac{D_r^{2D}}{s}(f_2 - f_1) \tag{22.21}$$

is the *material gain coefficient*, i.e., the gain coefficient of the quantum well that is now described as a three-dimensional system, with the two-dimensional density of states averaged over the quantum well thickness s, and where

$$\Gamma = a_2/s \tag{22.22}$$

is the ratio of the height of a photon mode and the quantum well thickness, sometimes called confinement factor. The material gain coefficient depends on the thickness of the quantum well while the modal gain coefficient is independent of the quantum well thickness but depends on the height of the photon mode.

Example Gain mediated by a GaAs quantum well, with a nonequilibrium electron density $N^{2D} = 2 \times 10^{16}\ \text{m}^{-2}$ corresponding to $(f_2 - f_1 = 0.25)$, for radiation of frequency $\nu = \nu_g^{2D} = E_g^{2D}/h$.

- $\nu_g^{2D} = 3.6 \times 10^{14}$ Hz; $n = 3.6$.
- $A_{21} = 3 \times 10^9\ \text{s}^{-1}$; $B_{21} = 2.2 \times 10^{21}\ \text{m}^3\,\text{J}^{-1}\,\text{s}^{-2}$; $\bar{B}_{21} = hB_{21}$.
- $H^{2D} = 1.5 \times 10^5$.
- $\gamma = 7.4 \times 10^{11}\ \text{s}^{-1}$ for $a_2 = 200$ nm.
- $\alpha = 8.9 \times 10^3\ \text{m}^{-1}$ for $a_2 = 200$ nm.
- $\alpha_{\text{mat}} = 1.8 \times 10^5\ \text{m}^{-1}$ for $s = 10$ nm.

We now determine the occupation number difference $f_2 - f_1$ for $\epsilon = 0$ ($E_2 - E_1 = E_g^{2D}$). The occupation number difference (Fig. 22.5) is -1 for $N^{2D} = 0$. With increasing electron (and hole) density, $f_2 - f_1$ increases, becomes zero at the transparency density and increases further. At very large electron density, it approaches $+1$. If the electron density has values near N_{tr}^{2D}, we can approximate the occupation number difference by

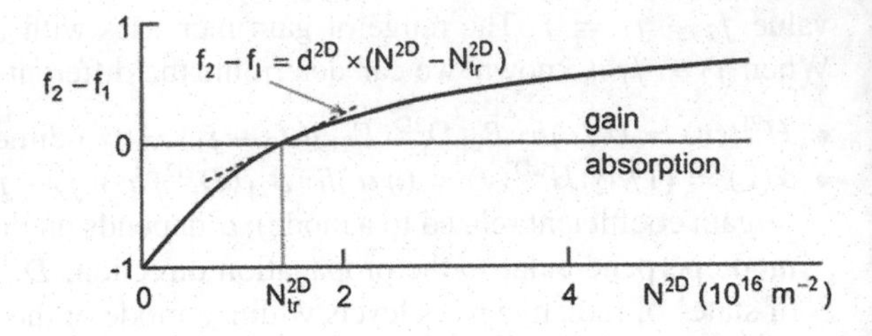

Fig. 22.5 Occupation number difference at the transition energy $E_2 - E_1 = E_{\mathrm{g}}^{\mathrm{2D}}$ for a GaAs quantum well at room temperature

$$f_2 - f_1 = d^{\mathrm{2D}} \times (N^{\mathrm{2D}} - N_{\mathrm{tr}}^{\mathrm{2D}}), \tag{22.23}$$

where $d^{\mathrm{2D}} = 3.8 \times 10^{-17}\,\mathrm{m}^2$. The gain characteristic is then given by

$$H^{\mathrm{2D}}(\nu_{\mathrm{g}}^{\mathrm{2D}}) = h\nu_{\mathrm{g}}^{\mathrm{2D}} \bar{B}_{21} D_{\mathrm{r}}^{\mathrm{2D}}(E_{\mathrm{g}}^{\mathrm{2D}}) d^{\mathrm{2D}} \times (N^{\mathrm{2D}} - N_{\mathrm{tr}}^{\mathrm{2D}}). \tag{22.24}$$

The modal gain coefficient is equal to

$$\alpha(\nu_{\mathrm{g}}^{\mathrm{2D}}) = \frac{H^{\mathrm{2D}}(\nu_{\mathrm{g}}^{\mathrm{2D}})}{a_2} = \sigma_{\mathrm{eff}}(N_{\mathrm{av}} - N_{\mathrm{tr,av}}) \tag{22.25}$$

and the modal growth coefficient

$$\gamma(\nu_{\mathrm{g}}^{\mathrm{2D}}) = \frac{cH^{\mathrm{2D}}(\nu_{\mathrm{g}}^{\mathrm{2D}})}{na_2} = b_{\mathrm{eff}}(N_{\mathrm{av}} - N_{\mathrm{tr,av}}), \tag{22.26}$$

where a_2 is the height of the photon mode, $b_{\mathrm{eff}} = h\nu \bar{B}_{21} D_{\mathrm{r}}^{\mathrm{2D}}(E_{21}) d^{\mathrm{2D}}$ is the effective growth constant, $\sigma_{\mathrm{eff}} = (n/c) b_{\mathrm{eff}}$ is the effective gain cross section, $N_{\mathrm{av}} = N^{\mathrm{2D}}/a_2$ the average electron density in the photon mode volume and $N_{\mathrm{tr,av}} = N_{\mathrm{tr}}^{\mathrm{2D}}/a_2$ the average transparency density.

Example Gain, mediated by a GaAs quantum well (at room temperature), for radiation of frequency $\nu_{\mathrm{g}}^{\mathrm{2D}}$.

- $N^{\mathrm{2D}} - N_{\mathrm{tr}}^{\mathrm{2D}} = 0.1\,\mathrm{N}_{\mathrm{tr}}^{\mathrm{2D}}$.
- $N_{\mathrm{tr}}^{\mathrm{2D}} = 1.4 \times 10^{16}\,\mathrm{m}^{-2}$.
- $N^{\mathrm{2D}} - N_{\mathrm{tr}}^{\mathrm{2D}} = 1.4 \times 10^{15}\,\mathrm{m}^{-2}$.
- $d^{\mathrm{2D}} = 3.8 \times 10^{-17}\,\mathrm{m}^2$; $\sigma_{\mathrm{eff}} = 2.7 \times 10^{-19}\,\mathrm{m}^2$.
- $H^{\mathrm{2D}} = 3.2 \times 10^{-4}$.
- $\gamma = 1.6 \times 10^{11}\,\mathrm{s}^{-1}$ for $a_2 = 200\,\mathrm{nm}$.
- $\alpha = 1.9 \times 10^3\,\mathrm{m}^{-1}$ for $a_2 = 200\,\mathrm{nm}$.
- $\alpha_{\mathrm{mat}} = 3.8 \times 10^4\,\mathrm{m}^{-1}$ for $s = 10\,\mathrm{nm}$.

We finally discuss the effect of thermal level broadening of energy levels (Fig. 22.6). Due to inelastic scattering of the electrons at phonons, the gain characteristic broadens and the maximum gain becomes smaller. The change of the gain curve is strongest in the range near the two-dimensional gap energy $E_{\mathrm{g}}^{\mathrm{2D}}$. The transparency

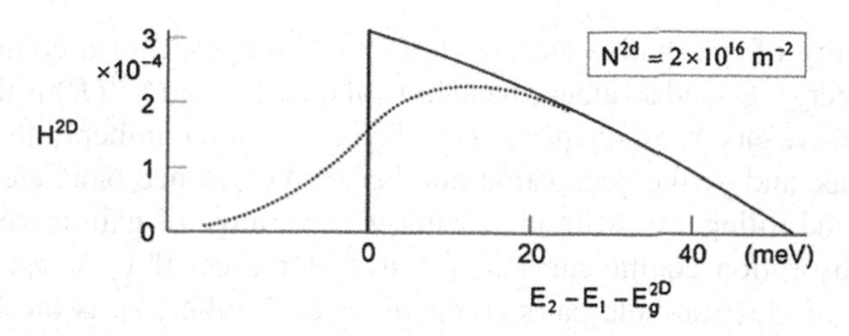

Fig. 22.6 Two-dimensional gain characteristic of a GaAs quantum well at room temperature without and with level broadening due to electron–phonon scattering

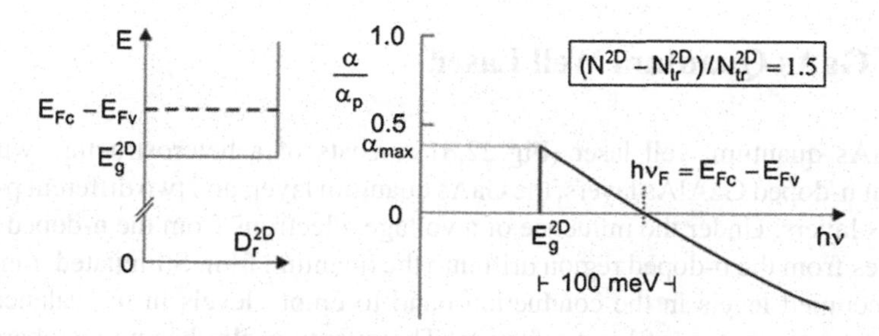

Fig. 22.7 Difference of the quasi-Fermi energies and transparency frequency for a GaAs quantum well at room temperature

density does not depend on the thermal broadening since the population difference $f_2(E_2) - f_1(E_1)$ is determined by the energy difference $E_{21} = E_2 - E_1$, i.e., by the energy difference between the center of level 2 and the center of level 1 rather than by the photon energy $h\nu$. Due to the thermal broadening, the maximum gain mediated by a quantum well at room temperature is reduced by a factor of about two, in comparison with the case that the thermal broadening is not taken into account. (This factor is compensated because of the anisotropy of a quantum well; *see* Sects. 22.8 and 26.8.)

Figure 22.7 shows the density of states of radiative electron-hole pairs and a modal gain coefficient of a GaAs quantum well at room temperature. Gain occurs at quantum energies up to

$$h\nu_F = E_{Fc} - E_{Fv}. \tag{22.27}$$

The gain coefficient (in the case that inelastic scattering is neglected) is given by

$$\alpha(h\nu)/\alpha_p = f_2(E_2) - f_1(E_1), \qquad \text{with} \qquad E = E_2 - E_1 = h\nu, \tag{22.28}$$

and with the peak gain coefficient

$$\alpha_p = n/(ca_2)h\nu \bar{B}_{21} D_r^{2D}(E). \tag{22.29}$$

E is the energy of a radiative electron-hole pair composed of a conduction band electron of energy E_2 and a valence band hole of energy E_1, $D_r^{2D}(E)$ is the density of states of radiative electron-hole pairs, f_2 is the occupation number of the conduction band electrons, and f_1 the occupation number of the valence band electrons. With increasing band filling, i.e., with increasing ν_F, the range of gain increases and the maximum absorption coefficient α_{max} ($< \alpha_p$) increases. If $\nu = \nu_F$, annihilation and creation of electron-hole pairs compensate each other; ν_F is the transparency frequency.

22.3 GaAs Quantum Well Laser

The GaAs quantum well laser (Fig. 22.8) consists of a heterostructure with two different n-doped GaAlAs layers, the GaAs quantum layer, and two different p-doped GaAlAs layers. Under the influence of a voltage, electrons from the n-doped region and holes from the p-doped region drift into the quantum film. Stimulated transitions from occupied levels in the conduction band to empty levels in the valence band give rise to generation of laser radiation. The quantum film has no net charge, the two-dimensional densities of nonequilibrium electrons and nonequilibrium holes are equal. The quantum well laser contains at least five different semiconductor layers. An n-doped GaAs substrate (n^+ GaAs substrate) supports the layers. The layer sequence can be, for instance, the following (beginning with the substrate).

- n^+ GaAs substrate.
- n $Ga_{0.75}Al_{0.25}As$.
- n $Ga_{0.9}Al_{0.1}As$.

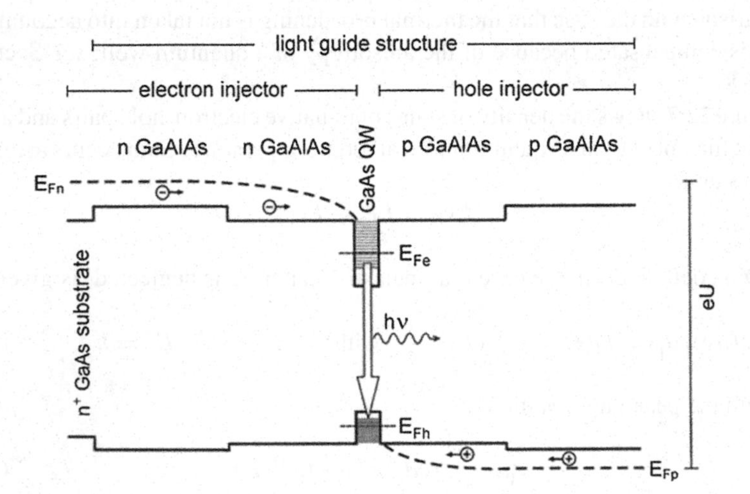

Fig. 22.8 GaAs quantum well laser (*principle*)

- GaAs quantum well (QW).
- p $Ga_{0.9}Al_{0.1}As$.
- p $Ga_{0.75}Al_{0.25}As$.

The layers fulfill the following different tasks.

- *Quantum well* (QW): GaAs.
- *Waveguide*: the $Ga_{0.9}Al_{0.1}As$ and $Ga_{0.75}Al_{0.25}As$ layers together.
- *Electron injector*: both n-doped GaAlAs layers together.
- *Hole injector*: both p-doped GaAlAs layers together.

The heavily doped substrate (n^+ GaAs) and an adjacent epitaxial n^+ GaAs layer are doped with silicon atoms and contain free-electrons of a concentration of (1–2) $\times 10^{19}$ m^{-3}; n^+ indicates a high n-doping concentration. The Fermi level E_{Fn} lies within the conduction band of GaAs. The concentration of excess electrons is smaller (by about two orders of magnitude) in the n GaAlAs layers. The Fermi level of the valence band electrons in the p-doped GaAs layers lies within the valence band.

The photon mode of a laser diode (Fig. 22.9) has submillimeter size (e.g., 100 μm × 0.2 μm × 500 μm). Metal films on top of the heterostructure and on the backside of the substrate serve as electrical contacts. Under the action of a voltage (U), a current (I) is flowing through the heterostructure. Electrons in the n-doped region and holes in the p-doped region carry the current. Electrons and holes recombine within the quantum well. Stimulated electron-hole pair recombination drives the laser oscillation. The laser is an *edge-emitting laser*.

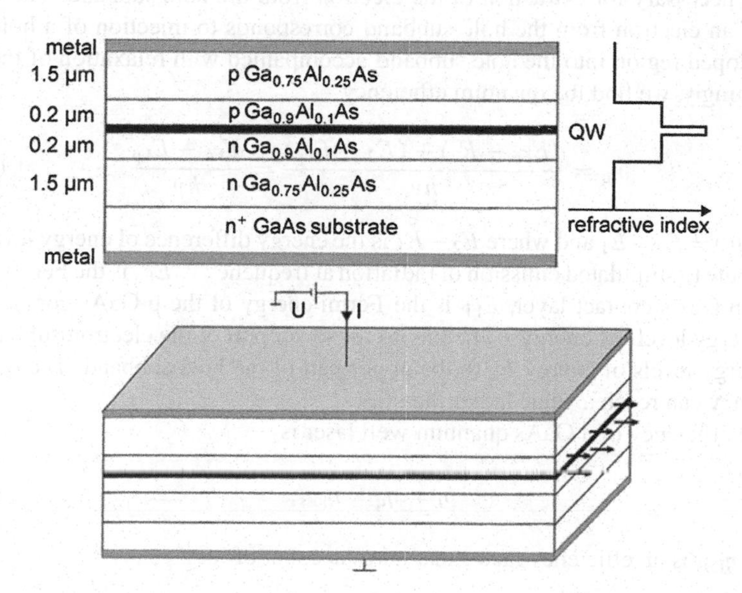

Fig. 22.9 GaAs quantum well laser; five semiconductor layers; refractive index profile; principle of the design

Example of a light guiding structure (at room temperature).

- GaAs; $n = 3.60$ for radiation with $h\nu = 1.42\,\text{eV}$.
- $Ga_{0.9}Al_{0.1}As$; refractive index $n = 3.52$.
- $Ga_{0.75}Al_{0.25}As$; $n = 3.41$.

The refractive is $n(Ga_{1-x}Al_xAs) \approx 3.60 - 0.71x$. Across the heterostructure, the refractive index has the largest value in the very thin GaAs quantum layer. The refractive index is larger in the $Ga_{0.75}Al_{0.25}As$ layers than in the $Ga_{0.9}Al_{0.1}As$ layers. This leads to a light guide effect. The light is concentrated in the $Ga_{0.9}Al_{0.1}As$ layers, which together have an optical thickness between one and two wavelengths. The GaAs film has a thickness (of the order of 10 nm) that is much smaller than the thickness of the adjacent $Ga_{0.9}Al_{0.1}As$ layers. Accordingly, only a small portion of the field overlaps with the quantum well.

The current through a quantum well corresponds to a migration of electrons from the n^+ substrate to the quantum well and, at the same time, to the migration of holes from the p-doped layers to the quantum well. The injection of electrons into the quantum well (*see* Fig. 22.8) leads to a nonequilibrium population of electrons in the electron subband, characterized by the quasi-Fermi energy E_{Fe}. The injection of holes into the quantum well leads to a nonequilibrium population of holes in the hole subband, characterized by the quasi-Fermi energy E_{Fv}. Under the action of a voltage U across the heterostructure, an electron migrating through the heterostructure loses its potential energy eU mainly due to the processes: relaxation within the electron subband; transition to the hole subband by stimulated emission of a photon; and energy necessary for extraction of the electron from the hole subband. The extraction of an electron from the hole subband corresponds to injection of a hole from the p-doped region into the hole subband accompanied with relaxation of the hole. Accordingly, we find the quantum efficiency

$$\eta_q = \frac{(E_{Fn} - E_2) + (E_2 - E_{Fp})}{h\nu} = \frac{F_{Fn} - E_{Fp}}{h\nu}, \tag{22.30}$$

where $h\nu = E_2 - E_1$ and where $E_2 - E_1$ is the energy difference of energy levels that contribute to stimulated emission of radiation at frequency ν, E_{Fn} is the Fermi energy of the n GaAs contact layer, E_{FP} is the Fermi energy of the p GaAs contact layer. The energy levels of energy E_2 belong to the lower part of the electron subband and the energy levels of energy E_1 to the upper part of the hole subband. The quantum efficiency can reach a value larger than 0.9.

The efficiency of a GaAs quantum well laser is

$$\eta = \eta_q \times \eta_{loss}, \tag{22.31}$$

where η_{loss} is an efficiency factor that takes account of loss.

22.4 Threshold Current of a GaAs Quantum Well Laser

To estimate the laser threshold condition of a GaAs quantum well laser at room temperature, we choose the simplest description.

- We ignore thermal broadening of the energy levels. The gain characteristic is then equal to

$$H^{2D}(\nu) = H^{2D}(h\nu) = (n/c)h\nu \bar{B}_{21} D_{\mathrm{r}}^{2D}(E_{21})\,[f_2(E_2) - f(E_1)], \tag{22.32}$$

  where $E_{21} = E_2 - E_1 = h\nu$.
- We choose $E_{21} = E_g^{2D}$.
- We assume that the threshold density has a value near the transparency density. Then we can write

$$f_2(E_2) - f_1(E_1) = d^{2D} \times (N^{2D} - N_{\mathrm{tr}}^{2D}), \tag{22.33}$$

  where $d^{2D} = 3.8 \times 10^{-17}$ m^2.
- The modal growth coefficient is

$$\gamma = \frac{(n/c)H^{2D}}{a_2} = b_{\mathrm{eff}} \times \frac{N^{2D} - N_{\mathrm{tr}}^{2D}}{a_2}, \tag{22.34}$$

  where a_2 is the height of the mode and

$$b_{\mathrm{eff}} = h\nu \bar{B}_{21} D_{\mathrm{r}}^{2D} d^{2D} \tag{22.35}$$

  is the effective growth rate constant.

The photon generation rate at the steady state osciallation of the laser is equal to the photon emission rate,

$$b_{\mathrm{eff}}(N_{\mathrm{av},\infty} - N_{\mathrm{tr,av}})Z = \frac{Z}{\tau_{\mathrm{p}}}. \tag{22.36}$$

$N_{\mathrm{av},\infty} = N_{\infty}^{2D}/a_2$ is the average threshold electron density and $N_{\mathrm{tr,av}} = N_{\mathrm{tr}}^{2D}/a_2$ the average transparency density in the laser resonator. This leads, with $N_{\infty}^{2D} = N_{\mathrm{th}}^{2D}$ (=threshold density) and $b_{\mathrm{eff}} = (c/n)\sigma_{\mathrm{eff}}$, to

$$N_{\mathrm{th}}^{2D} - N_{\mathrm{tr}}^{2D} = \frac{1}{\sigma_{\mathrm{eff}} l_{\mathrm{p}}/a_2}, \tag{22.37}$$

where σ_{eff} is the effective gain cross section and l_{p} is the photon mean free path in the resonator. The threshold current is, for $N_{\mathrm{th}}^{2D} - N_{\mathrm{tr}}^{2D} \ll N_{\mathrm{tr}}^{2D}$, equal to

$$I_{\mathrm{th}} = N_{\mathrm{tr}}^{2D} L a_2 e/\tau_{\mathrm{sp}} \tag{22.38}$$

and the threshold current density (with e = elementary charge) is

$$j_{th} = N_{tr}^{2D} e/\tau_{sp}. \tag{22.39}$$

Example GaAs quantum well laser.

- $h\nu = E_g^{2D}$.
- $L = 1\,\text{mm}$; $a_1 = 100\,\mu\text{m}$; $a_2 = 0.2\,\mu\text{m}$.
- $l_p = 1.2\,\text{mm}$; $l_p/a_2 \sim 7 \times 10^3$.
- $N_{tr}^{2D} = 1.4 \times 10^{16}\,\text{m}^{-2}$.
- $\sigma_{eff} = 2.7 \times 10^{-19}\,\text{m}^2$.
- $N_{th}^{2D} - N_{tr}^{2D} = 1 \times 10^{15}\,\text{m}^{-2}$.
- $\alpha_{th} = 700\,\text{m}^{-1}$.
- $\tau_{sp} = 2 \times 10^{-9}\,\text{s}$.
- $j_{th} = 1 \times 10^6\,\text{A}\,\text{m}^{-2}$; $I_{th} = 100\,\text{mA}$.

Because of the small height of the active volume, the ratio l_p/a_2 has a large value.

We can write the laser threshold condition in the form

$$\alpha_{th} l_p = 1, \tag{22.40}$$

where

$$\alpha_{th} = \sigma_{eff} \frac{N_{th}^{2D} - N_{tr}^{2D}}{a_2} \tag{22.41}$$

is the threshold gain coefficient.

Or we can write

$$\alpha_{th} = \sigma_{eff} \frac{N_{th}^{2D} - N_{tr}^{2D}}{s} \times \Gamma = \alpha_{mat,th} \times \Gamma, \tag{22.42}$$

where we have the quantities:

- s = thickness of a quantum well.
- N^{2D}/s = electron density within the quantum well described as a three-dimensional system.
- $\Gamma = s/a_2$ = confinement factor.
- $\alpha_{mat,th}$ = threshold material gain coefficient.

It follows for our example that $\alpha_{th} \sim 10^3\,\text{m}^{-1}$ and that $\Gamma = 1/20$ and $\alpha_{mat,th} = 10^4\,\text{m}^{-1}$ at a quantum well thickness of $s = 10\,\text{nm}$.

To determine the threshold current, we have taken into account the loss due to spontaneous emission of radiation. We ignored loss that is due to other processes such as nonradiative transitions of electrons from the conduction band to the valence band and loss of photons within the semiconductor materials.

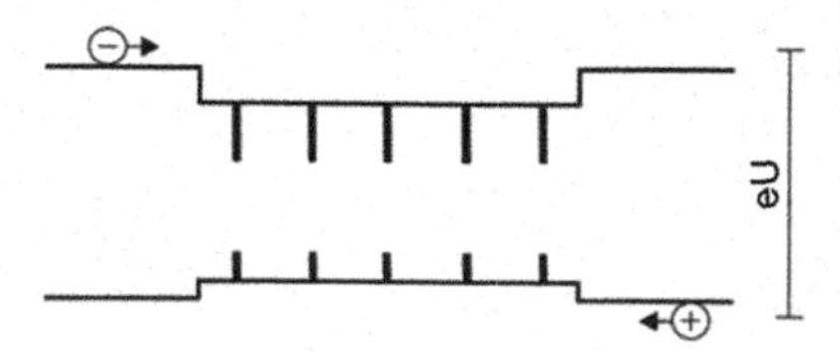

Fig. 22.10 Multi quantum well laser

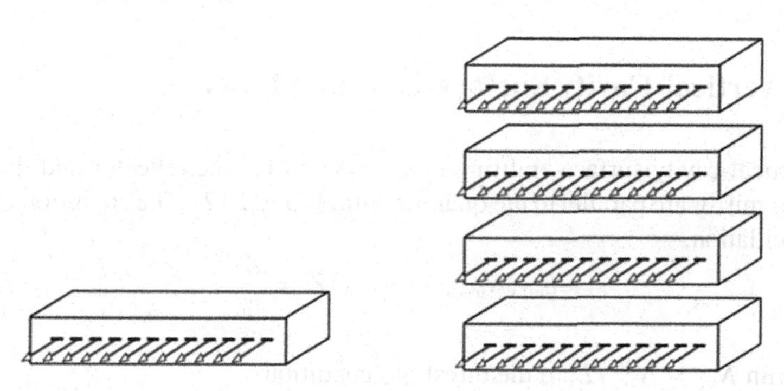

Fig. 22.11 Laser array and laser bar (*high-power semiconductor laser*)

22.5 Multi-Quantum Well Laser

A laser diode can contain (Fig. 22.10) more than one quantum well (e.g., five to ten quantum wells), arranged in parallel. This leads to a larger output power and a smaller threshold current. *The radiation of a multi-quantum well laser is coherent.*

22.6 High-Power Semiconductor Laser

A high-power semiconductor laser consists of laser diodes arranged in an array or as a bar of laser arrays (Fig. 22.11). A laser array contains 10–100 laser diodes. The laser diodes are kept at room temperature by the use of a cooler, which itself is cooled with air or via the mechanical support. Each single laser diode emits coherent radiation. However, the oscillations of different laser diodes are not in phase. Therefore, a high-power semiconductor laser generates a beam of incoherent monochromatic radiation. Depending on the number of arrays, a high-power semiconductor laser produces radiation with a power in the watt to kW range.

A diode array of laser diodes, each with a microlens collimating the radiation, emits radiation that has a divergence of $\sim 10°$ in the plane of the array and $1°$ perpendicular to the plane; without lenses, the divergence is $10°$ in the plane and $40°$ perpendicular to the plane.

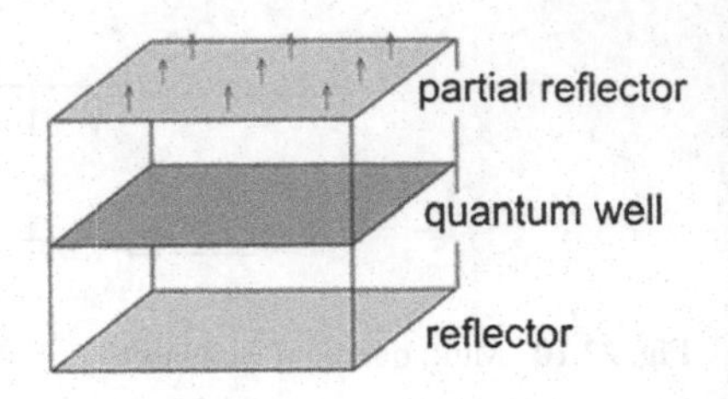

Fig. 22.12 Surface-emitting semiconductor laser

22.7 Vertical-Cavity Surface-Emitting Laser

In a vertical-cavity surface-emitting laser (=VCSEL), the reflector and the output coupling mirror are parallel to the quantum film (Fig. 22.12). The condition of steady state oscillation,

$$b_{\mathrm{eff}}(N_{\mathrm{av},\infty} - N_{\mathrm{tr,av}})Z = \frac{Z}{\tau_{\mathrm{p}}}, \tag{22.43}$$

leads, with $N_{\mathrm{av}} = N^{2\mathrm{D}}/L$, to the threshold condition

$$N_{\mathrm{th}}^{2\mathrm{D}} - N_{\mathrm{tr}}^{2\mathrm{D}} = \frac{1}{\sigma_{\mathrm{eff}} l_{\mathrm{p}}/L}, \tag{22.44}$$

where τ_{p} is the mean lifetime of a photon in the resonator and $l_{\mathrm{p}} = (c/n)\tau_{\mathrm{p}}$ the length of the path of a photon within the resonator. In order to obtain a large ratio l_{p}/L, the quality factor of the laser resonator has to be large. For a high-Q resonator, with a reflector (reflectivity = 1) and a partial reflector (reflectivity R), the threshold condition can be written, with $l_{\mathrm{p}}/L = 1/(1 - R)$, in the form

$$N_{\mathrm{th}}^{2\mathrm{D}} - N_{\mathrm{tr}}^{2\mathrm{D}} = \frac{1 - R}{\sigma_{\mathrm{eff}}}, \tag{22.45}$$

or

$$1 - R = \left(N_{\mathrm{th}}^{2\mathrm{D}} - N_{\mathrm{tr}}^{2\mathrm{D}}\right)\sigma_{\mathrm{eff}}. \tag{22.46}$$

Example Surface-emitting GaAs quantum well laser, with $N_{\mathrm{th}}^{2\mathrm{D}} \sim 2N_{\mathrm{tr}}^{2\mathrm{D}}$.

- $L = 10\,\mu\mathrm{m}$; $a_1 = 10\,\mu\mathrm{m}$; $a_2 = 10\,\mu\mathrm{m}$.
- $\sigma_{\mathrm{eff}} = 2.7 \times 10^{-19}\,\mathrm{m}^2$.
- $N_{\mathrm{th}}^{2\mathrm{d}} - N_{\mathrm{tr}}^{2\mathrm{D}} = 1.4 \times 10^{16}\,\mathrm{m}^{-2}$.
- $\tau_{\mathrm{sp}} = 4 \times 10^{-9}\,\mathrm{s}$.
- $1 - R = 1 \times 10^{-3}$.
- $j_{\mathrm{th}} = 5 \times 10^5\,\mathrm{A\,m}^{-2}$.
- $I_{\mathrm{th}} = 50\,\mu\mathrm{A}$.

To describe a case of stronger pumping, we use the laser equation involving the occupation number difference and find

$$(f_2 - f_1)_{\text{th}} = \frac{c/n}{h\nu \bar{B}_{21} d^{2\text{D}} l_{\text{p}}/L}. \tag{22.47}$$

As we have seen, the occupation number difference $f_2 - f_1$ saturates at large electron densities. Therefore, an increase of $N^{2\text{D}}$ to values much larger than a few times $N_{\text{tr}}^{2\text{D}}$ does not lead to noticeably larger values of $f_2 - f_1$ (Problem 22.4).

In comparison with the edge emitting laser, the vertical-cavity surface-emitting laser requires, as shown, a resonator with a high Q factor. The vertical-cavity surface-emitting laser has advantages:

- The radiation is less divergent.
- The size can be much smaller.
- The threshold current can be much smaller.

The lower threshold current results in a smaller heating effect.

22.8 Polarization of Radiation of a Quantum Well Laser

A quantum film is optically anisotropic. More detailed studies show that the Einstein coefficient $B_{21}^{\perp}$, i.e., with the electric field being perpendicular to the film plane, is zero and therefore $B_{21}^{||} = 1.5 B_{21}$. Accordingly, the radiation of an edge emitting bipolar laser—that generates radiation due to recombination of electrons and heavy holes—is polarized and the direction of the electric field of the electromagnetic wave is parallel to the plane of the quantum well. However, emission of radiation of well-defined polarization direction is limited to a narrow frequency range near the two-dimensional gap frequency. Toward higher frequency, light holes (Sect. 26.3) can give rise to generation of radiation of a less defined polarization direction.

It is a further consequence of the anisotropy that the Einstein relations have to be modified: A_{21} is related to an average value between $B_{21}^{||}$ and $B_{21}^{\perp}$.

22.9 Luminescence Radiation from a Quantum Well

Figure 22.13 (solid line) shows a luminescence spectrum calculated by the use of (21.61), modified corresponding to the two-dimensional density of states of electrons and holes in a quantum well, for a GaAs quantum well at room temperature containing electrons in the electron subband of a density of about twice the transparency density. S/S_{max} is equal to the ratio of the luminescence intensity at the frequency ν and the maximum luminescence intensity at the two-dimensional gap frequency

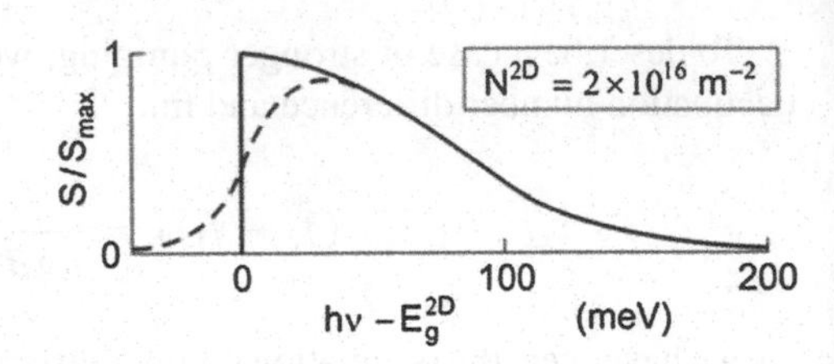

Fig. 22.13 Luminescence radiation from a quantum well

$\nu_g^{2D} = E_g^{2D}/h$, calculated without taking into account thermal level broadening. In comparison, a luminescence curve (dashed), which takes account of thermal level broadening, is wider and the maximum occurs at a larger frequency. The halfwidth of the luminescence curve is larger than kT. The luminescence radiation is emitted into the whole solid angle.

References [1–4, 6, 187–200].

Problems

22.1 Quasi-Fermi energies. A quantum film contains nonequilibrium electrons and holes. Determine the electron density at which the difference of the quasi-Fermi energies is $3kT$ (T = 300 K). [*Hint*: make use of the figure concerning the Fermi energies.]

22.2 Quantum well laser. A GaAs quantum well laser (length 0.5 mm, width 0.2 mm, resonator height 500 nm) contains 3 quantum wells and is operated at room temperature. Estimate the threshold electron density, threshold current density and threshold current. [*Hint*: neglect thermal broadening of the gain curve.]

22.3 Photons in a quantum well laser. A GaAs quantum well laser (length 0.5 mm, width 0.2 mm, resonator height 500 nm) contains 3 quantum wells, is operated at room temperature and emits, in two directions, laser radiation of a power of $P_{out} = 1$ mW into each of the directions.

(a) Determine the photon density in the resonator.
(b) Determine the total photon number Z_{tot} in the resonator.
(c) Compare Z_{tot} with the total number of nonequilibrium electrons in the quantum film.

22.4 Vertical-cavity surface-emitting laser. A vertical-cavity surface-emitting laser contains a GaAs quantum well and another laser contains five quantum wells; $(f_2 - f_1)_{th} = 0.5$ and $\tau_{sp} = 8 \times 10^{-9}$ s. Determine the following quantities:

(a) Threshold reflectivity of the output coupling mirror.
(b) Threshold current.
(c) Threshold current density.

Chapter 23
Semiconductor Materials and Heterostructures

We give a survey of semiconductor materials suitable for preparation of semiconductor lasers. The materials are compounds of elements of the third and fifth group of the periodic table or compounds of elements of the second and the sixth group. The compounds have energy gaps corresponding to gap frequencies ranging from the infrared to the near UV.

We describe the zinc blende crystal structure that is common to many of the semiconductor laser materials and introduce the monolayer as an important structural element of heterostructures. Heterostructures are suitable for the design of artificial materials with spatially varying energy bands.

We shortly mention the methods of preparation of semiconductor heterostructures.

After a survey of the different materials, we will concentrate the discussion on heterostructures composed of GaAs and AlAs.

We will present dispersion curves of electrons in GaAs and AlAs. And we will discuss absorption coefficients of GaAs and AlAs, characterizing absorption due to interband transitions in a direct semiconductor (GaAs) and an indirect semiconductor (AlAs).

23.1 Group III–V and Group II–VI Semiconductors

Group III–V semiconductors—materials composed of group III and group V elements of the periodic table—are well suitable for preparation of laser diodes. Figure 23.1 shows a section of the periodic table. The III–V semiconductors are the materials of laser diodes from the UV to the infrared. The III–V semiconductors consisting of atoms of small masses have large bandgaps and III–V semiconductors consisting of atoms of large masses have small bandgaps. AlN has a large bandgap and InSb a small one. The group IV semiconductors diamond, silicon, germanium and gray tin are indirect semiconductors. These are not suitable as active materials of bipolar semiconductor lasers. Group III–V semiconductors are known since 1952 [202].

K.F. Renk, *Basics of Laser Physics*, Graduate Texts in Physics,
DOI 10.1007/978-3-319-50651-7_23

Fig. 23.1 Section of the periodic table of the elements

II	III	IV	V	VI
	B	C	N	
Mg	Al	Si	P	S
Zn	Ga	Ge	As	Se
Cd	In	Sn	Sb	Te

Table 23.1 Energy gaps and gap wavelengths of III–V and II–VI semiconductors (at 300 K)

Semiconductor		E_g (eV)	λ_g
InSb	Indium antimonide	0.17	7.3 μm
InAs	Indium arsenide	0.36	3.4 μm
GaSb	Gallium antimonide	0.72	1.7 μm
GaAs	Gallium arsenide	1.42	873 nm
InN	Indium nitride	1.8	690 nm
AlAs	Aluminum arsenide	[2.2]	
GaP	Gallium phosphide	[2.3]	
GaN	Gallium nitride	3.4	370 nm
AlN	Aluminum nitride	6.2	200 nm
CdTe	Cadmium telluride	1.56	795 nm
CdSe	Cadmium selenide	1.8	690 nm
CdS	Cadmium sulfide	2.42	510 nm
ZnSe	Zinc selenide	2.7	460 nm
ZnS	Zinc sulfide	3.8	330 nm

The group II–VI semiconductors, composed of elements of the sixth group (S, Se, Te) and elements of the second main group (Mg) or of a side group of the second group (Zn, Cd), are direct semiconductors. Heterostructures of mixed crystals of II–VI semiconductors can be used to prepare green laser diodes.

Table 23.1 shows energy gaps and gap wavelengths $\lambda_g = h/E_g$ of semiconductors at room temperature. The semiconductors listed in the table are direct gap semiconductors, except AlAs and GaP that are indirect gap semiconductors. The values of E_g of AlAs and GaP correspond to the $k = 0$ gap (Sect. 23.8).

The crystal structure of most of the group III–V semiconductors used to prepare lasers is the zinc blende structure (Sect. 23.3). GaN can crystallize not only in the zinc blende structure but also in the wurtzite structure.

Table 23.2 shows effective masses of group III-V semiconductors; m_e = effective mass of a conduction band electron; m_h ($\equiv m_{hh}$) = effective mass of a heavy hole in the valence band; m_{lh} = effective mass of a light hole (Sect. 26.2).

Table 23.2 Effective masses

	m_e	m_h	m_{lh}
GaAs	0.067	0.43	0.09
InP	0.077	0.6	0.12
InAs	0.027	0.34	0.027
InSb	0.014	0.34	0.016
GaN	0.20	1.4	

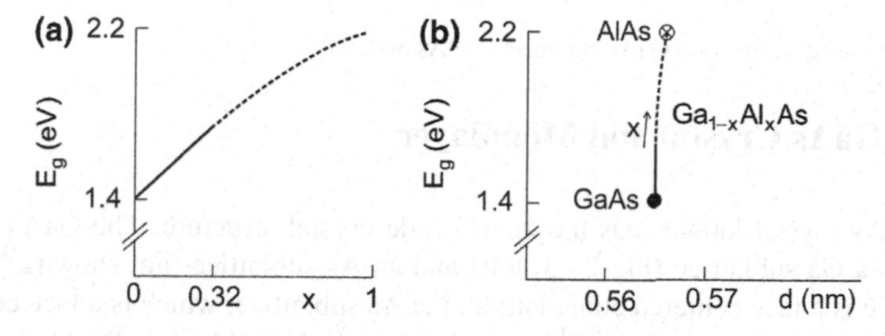

Fig. 23.2 Energy gap of $Ga_{1-x}Al_xAs$ mixed crystals. **a** Dependence of the gap energy on the composition. **b** Relation between gap energy and lattice constant

23.2 GaAlAs Mixed Crystal

It is possible to prepare $Ga_{1-x}Al_xAs$ mixed crystals of each mixing ratio x. In a GaAlAs crystal, Al atoms replace Ga atoms. The energy gap of $Ga_{1-x}Al_xAs$ (Fig. 23.2a) varies continuously with the mixing ratio, from the energy gap of GaAs ($x = 0$) up to the $k = 0$ gap of AlAs ($x = 1$); $Ga_{1-x}Al_xAs$ is a direct gap semiconductor for $x < 0.32$ and an indirect gap semiconductor for larger x. For the region of the direct gap, the energy gap (in units of eV) of a mixed GaAlAs crystal (at room temperature) follows from the relation $E_g\ (Ga_{1-x}Al_xAs) = 1.424 + 1.247\ x - 0.14\ x^2$.

A GaAs crystal and an AlAs crystal have a special property in common: they have nearly the same lattice constant. The cubic lattice constants (of crystals at room temperature) are:

- GaAs $d = 0.565326(2)$ nm.
- AlAs $d = 0.5660$ nm.

The difference between the lattice constants of GaAs and AlAs is only about a tenth of a percent and smaller for the $Ga_{1-x}Al_xAs$ ($x = 0 \ldots 1$) mixed crystals. Therefore, GaAs is an ideal substrate for deposition of $Ga_{1-x}Al_xAs$ layers, independently of the value of x. The lattice constant d increases, from the value of GaAs to the value of AlAs, linearly with x (Fig. 23.2b).

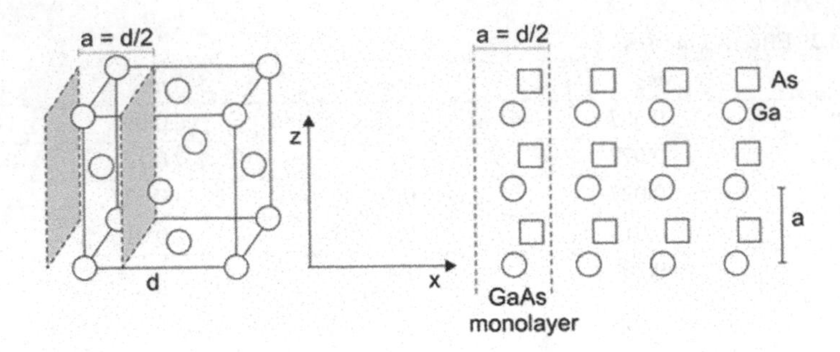

Fig. 23.3 Ga sublattice of a GaAs lattice and GaAs monolayer

23.3 GaAs Crystal and Monolayer

The GaAs crystal lattice has the zinc blende crystal structure. The GaAs crystal contains a Ga sublattice (Fig. 23.3, left) and an As sublattice (not shown). The Ga sublattice is a face centered cubic lattice; the As sublattice, which is a face centered cubic lattice too, is shifted by $(\frac{1}{4}, \frac{1}{4}, \frac{1}{4})d$ relative to the Ga sublattice. The GaAs crystal can be described as a sequence of Ga layers and of As layers in turn (Fig. 23.3, right). We introduce the *GaAs monolayer*, it has a lattice period a that is half the cubic lattice constant of GaAs,

$$a = \frac{1}{2}d. \tag{23.1}$$

23.4 GaAs/GaAlAs Heterostructure

A $Ga_{1-x}Al_xAs$/GaAs heterostructure (Fig. 23.4a) consists of a $Ga_{1-x}Al_xAs$ layer adjacent to a GaAs layer. A GaAs substrate (a GaAs crystal) supports the layers. The GaAs lattice structure of the substrate is continued in the heterostructure. The *lattice matching* between the $Ga_{1-x}Al_xAs$ layer and the GaAs substrate is nearly perfect because the lattice constants of GaAs and AlAs are only slightly different from each other. For $x = 1$, the *lattice mismatch*, measured relative to the monolayer thickness, is $\sim 10^{-3}$. It is smaller at smaller x (preceding section). Across an undoped GaAlAs/GaAs heterostructure (in 100 direction or x direction), there is a change of the values of various energies measured relative to a vacuum level of an electron, as illustrated in the figure:

- E_{vac} = vacuum level of an electron.
- E_{F} = Fermi energy.
- E_{c} = conduction band minimum.
- E_{v} = valence band maximum.
- δ_{c} = conduction band offset.
- δ_{v} = valence band offset.

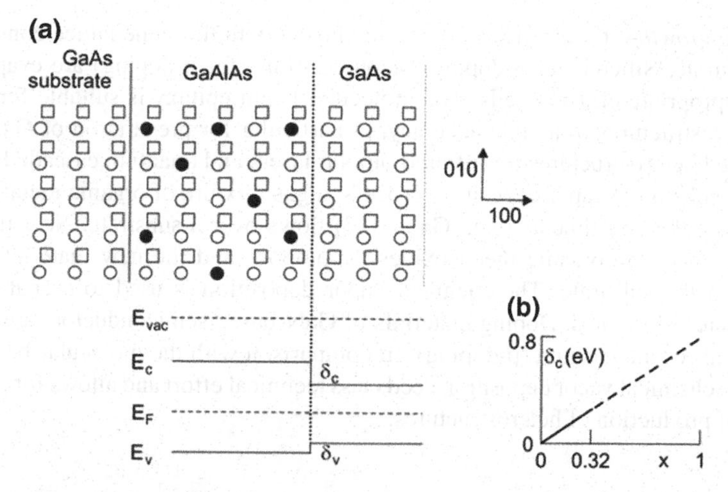

Fig. 23.4 GaAlAs/GaAs heterostructure. **a** Heterostructure and change of energy values across the heterostructure. **b** Conduction band offset

The Fermi-level of an undoped crystal has a value between E_c and E_v. We introduce:

- The conduction band profile $E_c(x) =$ minimum of the conduction band along the coordinate x across the heterostructure.
- The valence band profile $E_v(x) =$ maximum of the valence band across the heterostructure.

The gap energy $E_g = E_c - E_v$ of GaAs is smaller than that of GaAlAs. The change δE_g of the gap energy across the GaAlAs/GaAs interface is partly due to the conduction band offset and partly due to the valence band offset. Both vary linearly with the composition x. The mixed crystal is, as already mentioned, a direct gap semiconductor for $x < 0.32$ and an indirect gap semiconductor for larger x. The conduction band offset (Fig. 23.4b) is $\delta_c(x) = 0.67\delta E_g(x)$ and the valence band offset is $\delta_v(x) = 0.33\delta E_g(x)$.

23.5 Preparation of Heterostructures

There are two basic techniques of preparation of heterostructures, the *molecular beam epitaxy* (MBE) and the *metal oxide chemical vapor deposition* (MOCVD), which is a special method of *chemical vapor deposition* (CVD).

Molecular beam epitaxy is performed in a chamber with ultrahigh vacuum (pressure $<10^{-10}$ mbar). To grow a heterostructure containing GaAs and GaAlAs, the elements Ga and Al are evaporated from effusion cells. The chamber contains As at a very low pressure. The GaAs substrate has a temperature that is favorable for *epitactic growth*—for the growth of atomic layers with the same lattice constant as the substrate. Silicon (for n-doping) or phosphorus (for p-doping) are evaporated from appropriate effusion cells. The molecular beam epitaxy is suitable for growing heterostructures of atomic accuracy, in particular for preparation of GaAs and InP-based heterostructures for infrared bipolar lasers and quantum cascade lasers.

In the chemical vapor deposition process, a gas mixture of organic metal oxides containing the constituents (e.g., Ga and N) flows over a substrate. Near the substrate surface, the organic metal oxides decompose and the new material (GaN) grows on the substrate. The chemical vapor deposition is used to prepare GaN-based heterostructures. Doping materials of GaN-based semiconductors are silicon (n-doping) or magnesium (p-doping). In comparison with the molecular beam epitaxy, the chemical vapor deposition needs less technical effort and allows for a higher speed of production of heterostructures.

23.6 Preparation of Laser Diodes

To prepare laser diodes, a wafer covered with a heterostructure is laterally structured. Different steps of structuring include photolithography, chemical etching or plasma etching, and the preparation of ohmic contacts.

23.7 Material Limitations

Bipolar semiconductor lasers are realizable in a wide frequency range. There are limitations at large frequencies and at small frequencies.

The direct gap semiconductor with the widest gap (used to prepare semiconductor lasers) is AlN. The gap energy (6.2 eV) corresponds to a gap frequency of 1.5×10^{15} Hz (wavelength 200 nm).

At small frequencies, bipolar lasers operated at room temperature are limited to frequencies of about 1.3×10^{14} Hz (vacuum wavelength 2 μm), corresponding to an energy gap of 0.6 eV. At smaller frequencies (smaller gap energies) nonradiative transitions between conduction band and valence band become strong and population inversion is not possible. By cooling, the nonradiative transitions slow down and laser oscillation at smaller frequencies is possible. With cooled lead salts, bipolar semiconductor lasers up to frequencies of about 8 THz (wavelength 40 μm) can be produced (Sect. 28.4).

23.8 Energy Bands and Absorption Coefficients of GaAs and AlAs

Figure 23.5 shows dispersion curves of electrons in GaAs and AlAs crystals, with the k vector oriented along the 100 direction. The energy bands are periodic with the period $2\pi/(d/2)$, where d is the cubic lattice constant; the Brillouin zone extends in the 100 direction from $-2\pi/d$ to $2\pi/d$.

The energy maximum (E_v) of the valence band of GaAs occurs at the same k vector as the energy minimum (E_c) of the conduction band—GaAs is a direct gap semiconductor (Fig. 23.5, left). Accordingly, the electronic transitions between the two bands of GaAs are strong at frequencies $\nu > \nu_g$; GaAs is transparent for $\nu < \nu_g$.

The absorption coefficient α_{abs} of GaAs (Fig. 23.6, left) reaches a value of the order of $10^4\ \mathrm{cm}^{-1}$. Population inversion results in a gain coefficient that can, in principle, be of the same order of magnitude. That means that a crystal with an inverted population

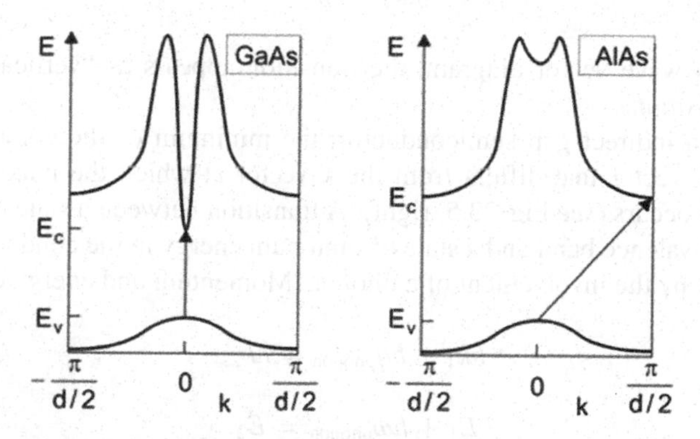

Fig. 23.5 Dispersion curves for electrons in GaAs and AlAs, with the wave vector oriented along the (100) direction; the arrows indicate direct transitions (in *GaAs*) and indirect transitions (in *AlAs*)

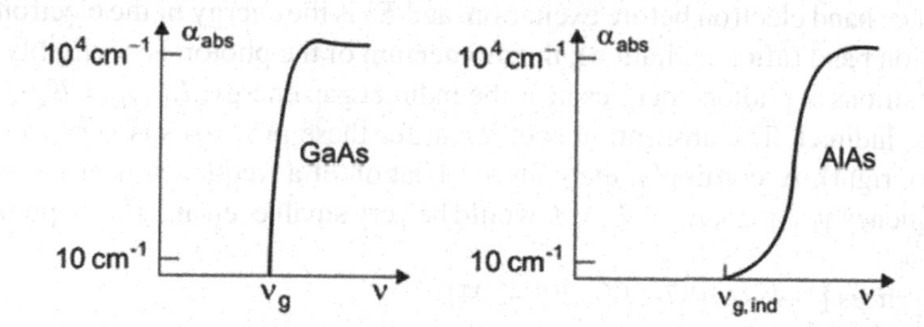

Fig. 23.6 Absorption coefficients of GaAs and AlAs

can have a gain factor $G = e^{\alpha L} = e = 2.6$ already at a length of 1 μm. The large coefficient of absorption due to interband transitions, with a corresponding large gain coefficient, in case of a population inversion is the basis of the bipolar semiconductor lasers.

In an absorption process, a photon (momentum $\hbar q_{\rm photon}$) is absorbed and an electron (momentum $\hbar k_1$) in the valence band is excited to the conduction band where it has the momentum $\hbar k_2$. Momentum conservation requires that

$$\hbar k_1 + \hbar q_{\rm photon} = \hbar k_2. \quad (23.2)$$

The sum of the momentum $\hbar k_1$ of a valence band electron and the momentum $\hbar q_{\rm photon}$ of a photon has to be equal to the momentum $\hbar k_2$ of the conduction band electron after the excitation. Since the momentum of a photon is small compared to the wave vector at the Brillouin zone boundary, $\hbar q_{\rm photon} \ll 2\pi/d$, the condition of an electronic transition is

$$\hbar k_1 \approx \hbar k_2. \quad (23.3)$$

In the energy-wave vector diagram, the transition appears as "vertical" transition (=*direct transition*).

AlAs is an indirect gap semiconductor: the minimum of the conduction band occurs at a k vector that differs from the k vector at which the maximum of the valence band occurs (*see* Fig. 23.5, right). A transition between a state of maximum energy in the valence band and a state of minimum energy in the conduction band is possible only by the involvement of a phonon. Momentum and energy conservation require that

$$\hbar k_1 + \hbar q_{\rm phonon} = \hbar k_2, \quad (23.4)$$

$$E_1 + \hbar\omega_{\rm phonon} = E_2, \quad (23.5)$$

where $\hbar q_{\rm phonon}$ is the momentum, $\hbar\omega_{\rm phonon}$ the energy of a phonon, E_1 is the energy of the valence band electron before excitation, and E_2 is the energy of the electron in the conduction band (after excitation); the momentum of the photon is negligibly small. The transitions at photon energies near the indirect gap energy, $E_{\rm g,ind} = E_{\rm c} - E_{\rm v}$, of AlAs are indirect. The absorption coefficient for these processes is very small (*see* Fig. 23.6, right). Accordingly, the gain of radiation of a frequency near the indirect gap frequency $\nu_{\rm g,ind} = (E_{\rm c} - E_{\rm v})/h$ would be very small even at a strong population inversion.

References [1–4, 6, 187–200, 202–204].

Problems

23.1 Wave vectors of light and of electrons. Compare the wave vector of visible light with the wave vector of electrons in GaAs at the Brillouin zone boundary.

23.2 Indirect gap semiconductor. An indirect gap semiconductor can absorb or emit light by the involvement of phonons; however, the processes are much weaker than the direct processes (processes without phonons). Formulate the energy and momentum conservation laws for the indirect processes:

(a) Absorption of a photon and simultaneous generation of a phonon.
(b) Emission of a photon and simultaneous generation of a phonon.

23.3 Determine the absorption coefficient in the vicinity of the gap frequency (a) of bulk GaN and (b) of a GaN quantum well.

23.4 Determine the transparency density of a GaN quantum well at room temperature.

23.5 A GaN VCSEL has a diameter of 10 μm and contains 10 quantum wells. Determine the reflectivity of the output coupling mirror that is necessary to reach laser threshold. Calculate the threshold current.

Chapter 24
Quantum Well Lasers from the UV to the Infrared

Quantum well lasers are available in a large wavelength range, extending from the near UV to the near infrared. Basic materials are: GaN for UV and blue lasers; GaAs for red lasers; InP for near infrared lasers; GaN, GaAs or ZnSe for green lasers. We will discuss the design of different lasers.

24.1 A Survey

Figure 24.1 shows a selection of quantum well materials, together with barrier and substrate materials. At each wavelength in the range of 0.3–2 μm, a laser diode is in principle available. The materials used for preparing a laser diode must have appropriate energy gaps. There are further requirements.

- *Red and infrared laser diodes*. The materials must have a very good lattice matching. This condition requires the use of binary, ternary and quaternary compounds. Suitable substrates are GaAs and InP. Heterostructures are prepared by molecular beam epitaxy.
- *Blue and UV laser diodes*. The material basis is GaN. The lattice matching is not critical. Sapphire has a large lattice mismatch to GaN, but it is nevertheless suitable as a suited. Heterostructures can be prepared by chemical vapor deposition (CVD).
- *Green laser diodes*. The basic materials are GaN, GaAs, or ZnSe.

24.2 Red and Infrared Laser Diodes

By mixing GaAs with the heavier InAs, all energy gaps between the gap of GaAs (1.4 eV) and the gap of InAs (0.4 eV) are available (Fig. 24.2a). All ternary $Ga_{1-x}In_xAs$ ($x = 0 \ldots 1$) compounds are direct semiconductors. InP is an appropriate substrate material. As a rule of determination of a property a (like gap energy or

K.F. Renk, *Basics of Laser Physics*, Graduate Texts in Physics,
DOI 10.1007/978-3-319-50651-7_24

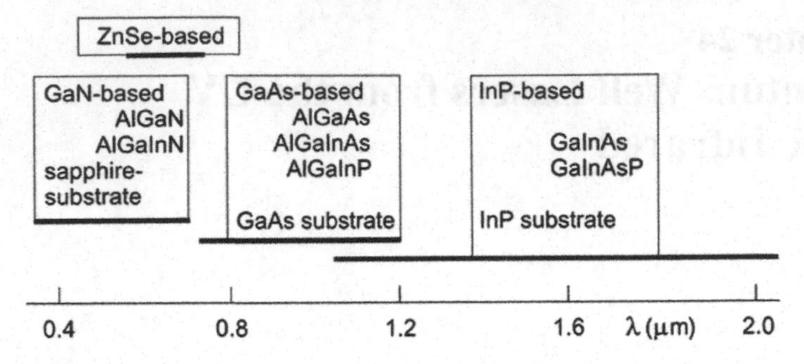

Fig. 24.1 Quantum well lasers: materials and wavelength regions

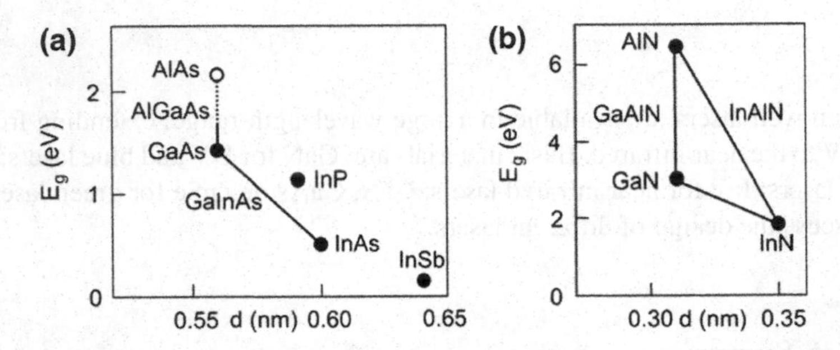

Fig. 24.2 Energy gaps **a** of GaAs-based and **b** of GaN-based semiconductors

lattice constant) of a semiconductor consisting of the compounds A, B, and C, we can use the relation $a(A_{1-x}B_xC) = (1-x) \times a(AC) + x \times a(BC)$.

The following materials are lattice matched to InP.

- InP and $Ga_{0.52}In_{0.48}As$ (a combination of a binary and a ternary semiconductor). The energy gap of $Ga_{0.52}In_{0.48}As$ has the value $E_g = 0.75$ eV and the refractive index is $n = 3.56$ while the refractive index of InP is $n = 3.16$ at the gap energy (1.2 eV) of InP.
- InP and $Ga_{1-x}In_xAs_{1-y}P_y$ (a combination of a binary and a quaternary III–V compound); Ga is partly replaced by the heavier In and As by the lighter P.

These materials, together with InP substrates, are suitable for the preparation of a variety of lasers.

- $Ga_{1-x}In_xAs$/GaAs laser; wavelength in the range 900–1100 nm; application: pump lasers.
- $Ga_{0.8}In_{0.2}As$/GaAs; 980 nm; application: pump laser of the Er^{3+}: glass fiber laser and amplifier.
- $Ga_{1-x}In_xAs_{1-y}P_y$/GaInAsP; 1.2–1.6 μm.
- GaInAs/GaInAlAs; 1.8–2.1 μm.

The following materials, lattice matched to InP, are suited to prepare lasers used in optical communications.

- $\lambda = 1.32\,\mu m$; $Ga_{0.27}In_{0.73}As_{0.58}P_{0.42}$.
- $\lambda = 1.55\,\mu m$; $Ga_{0.42}In_{0.58}As_{0.9}P_{0.1}$.

Lasers with GaAs substrates.

- GaInP (quantum layer)/AlGaInP; wavelength $\lambda \sim 630$–700 nm; pump laser of other lasers.
- $Ga_{1-x}Al_xAs/Ga_{1-y}Al_yAs$; 720–850 nm; pump lasers.

24.3 Blue and UV Laser Diodes

In 1997, S. Nakamura and coworkers at a small Japanese company (Nichia Chemicals) succeeded in preparing blue diode lasers [196, 203, 204]. In 2014, Nakamura received, together with H. Amano and I. Akasaki, the Nobel Price in Physics. The basic materials are nitrides (Fig. 24.2b), belonging to the group III–V semiconductors:

- GaN; $E_g = 3.4$ eV ($\lambda_g = 365$ nm).
- AlN; $E_g = 6.2$ eV.
- InN; $E_g = 1.8$ eV.

$Ga_{1-x}Al_xN$ and $Ga_{1-x}In_xN$ mixed materials are most suitable for preparation of blue and near UV laser diodes. Although sapphire (Al_2O_3) has a large mismatch (16%) to GaN, it serves as a substrate; SiC is suitable as substrate too.

An example of a GaN-based laser diode is shown in Fig. 24.3a. The laser diode (emitting at a wavelengths of 413 nm) consists of the following layers.

- InGaN quantum well layers (thicknesses 3 nm).
- GaN barrier layers.
- GaAlN (p type) electron blocking layer; it acts as a reflector of electrons.
- GaAlN layers, n-doped on one side and p-doped on the other side of the GaN layer.

At a wavelength of 400 nm, the refractive index of GaN is $n = 2.55$, while the refractive index of AlGaN is smaller. Doping with silicon leads to n-type conductivity and doping with magnesium to p-type conductivity.

The design of a blue laser diode is shown in Fig. 24.3b. The different layers are (beginning at the Al_2O_3 substrate): a very thin undoped GaN layer (e.g., of a thickness of 50 nm) as buffer layer; an n-doped AlGaN cladding layer; then the layers embedding the layers containing the multi-quantum wells (MQWs); finally, a p-doped AlGaN cladding layer. The pump current flows from the metallic anode through the heterostructure to the metallic cathode.

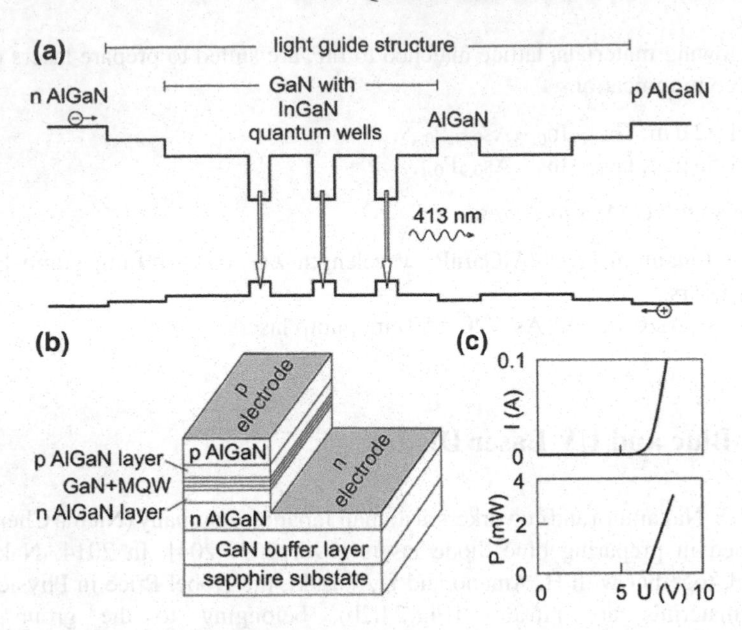

Fig. 24.3 Blue laser diode. **a** Principle. **b** Design. **c** Dependence of the current and the laser output power on the voltage across the diode

The current-voltage (I–V) curve (Fig. 24.3c) shows that current flow sets in at a voltage above 5 V. At a voltage of 7 V, the threshold current I_{th} is reached. In the range $I > I_{th}$, the current increases strongly due to generation of laser radiation. The laser output power P_{out} increases almost proportionally to $I - I_{th}$.

24.4 Group II–VI Materials of Green Lasers

As already mentioned, green laser diodes consist of GaAs-, GaN- or ZnSe-based materials. The ZnSe-based mixed materials have energy gaps between 2 and 4 eV (Fig. 24.4a). There are various possibilities to prepare mixed crystal materials composed of elements of group II and group VI in the periodic table.

- *Binary II–VI semiconductors*: ZnS, ZnSe, CdSe ....
- *Ternary II–VI semiconductors*: ZnSSe, ZnSeTe, CdSSe, CdZnSe with energy gaps between 3.8 eV (ZnS) and 1.8 eV (CdSe); lattice matched to GaAs (substrate).

The layer sequence of a ZnS-based laser diode [205, 206] is shown in Fig. 24.4b. On an n-doped GaAs substrate, first a GaAs buffer layer is grown in order to obtain a perfect crystal structure to which the further layers are added. The different tasks of the layers are as follows:

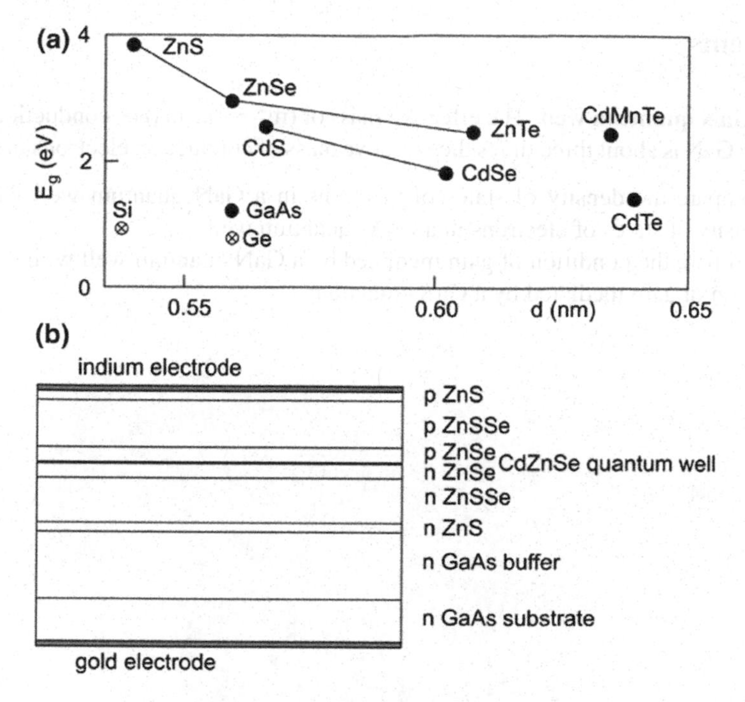

Fig. 24.4 Green semiconductor laser. **a** Energy gaps. **b** Device

- CdZnSe (quantum well).
- p ZnSSe/p ZnSe/CdZnSe/ n ZnSe/ n ZnSSe (light guide).
- p-doped layers (hole injector).
- n-doped layers; electron injector.
- An indium film on the heterostructure serves as anode and a gold film on the backside of the highly doped substrate as cathode.

Heterostructures of group II–VI semiconductors are not as stable as those of group III–V semiconductors.

24.5 Applications of Semiconductor Lasers

We mention a few applications of semiconductor lasers: optical storage (e.g., compact disc; blue ray disc); color projection; laser printer; sensor devices; micro controllers. In comparison with a red laser, a blue laser emits radiation of smaller wavelength. Therefore, a blue laser allows for a higher storage density.

References [1–4, 6, 187–201, 203–206].

Problems

24.1 GaN quantum well. The effective mass of ($m^* \sim 0.2\,m_0$) of conduction electrons in GaN is about three times the effective mass of conduction electrons in GaAs.

(a) Compare the density of states of electrons in a GaN quantum well with the density of states of electrons in a GaAs quantum well.
(b) Compare the condition of gain mediated by a GaN quantum well with the condition of gain mediated by a GaAs quantum.

Chapter 25
Reflectors of Quantum Well Lasers and of Other Lasers

We discuss different reflectors: distributed feedback reflector; Bragg reflector and photonic crystal reflector; total internal reflector leading to whispering gallery modes. The reflectors are suited as reflectors not only in quantum well lasers but also in quantum wire and quantum dot lasers (Chap. 27). Depending on the type of reflector, it is possible to design semiconductor lasers of submillimeter size down to $(10\ \mu\text{m})^3$. In connection with photonic crystals, we mention the photonic crystal fiber as a dielectric light guiding structure. The one-dimensional photonic crystal reflector (= Bragg reflector = multilayer reflector) is in use for almost all types of lasers.

We consider propagation of electromagnetic waves in layered materials (stratified media) in the special—but important—case that radiation is propagating in the direction perpendicular to the layers. We introduce the plane-wave transfer matrix that describes transfer of a wave from one side of an interface to the other side—and the propagation matrix, which characterizes propagation of a wave within a medium. The plane-wave matrix method is based on the boundary conditions for fields at an interface. We treat: thin film between two media; dielectric multilayer; one-dimensional photonic crystal. We will apply (Sect. 30.3) the plane-wave matrix method also to investigate electron waves passing through an interface of two semiconductor media.

25.1 Plane Surface

We have already discussed the edge-emitting quantum well laser with two uncovered crystal surfaces as reflectors. Cleaving a substrate (together with the layers on the substrate) results in a plane surface and cleaving along two parallel planes results in a resonator. Disadvantages and advantages of an edge-emitting laser with cleaved surfaces (Fig. 25.1a) are the following:

- The reflectivity, determined by the refractive index of the cleaved semiconductors material has a fixed value.

K.F. Renk, *Basics of Laser Physics*, Graduate Texts in Physics,
DOI 10.1007/978-3-319-50651-7_25

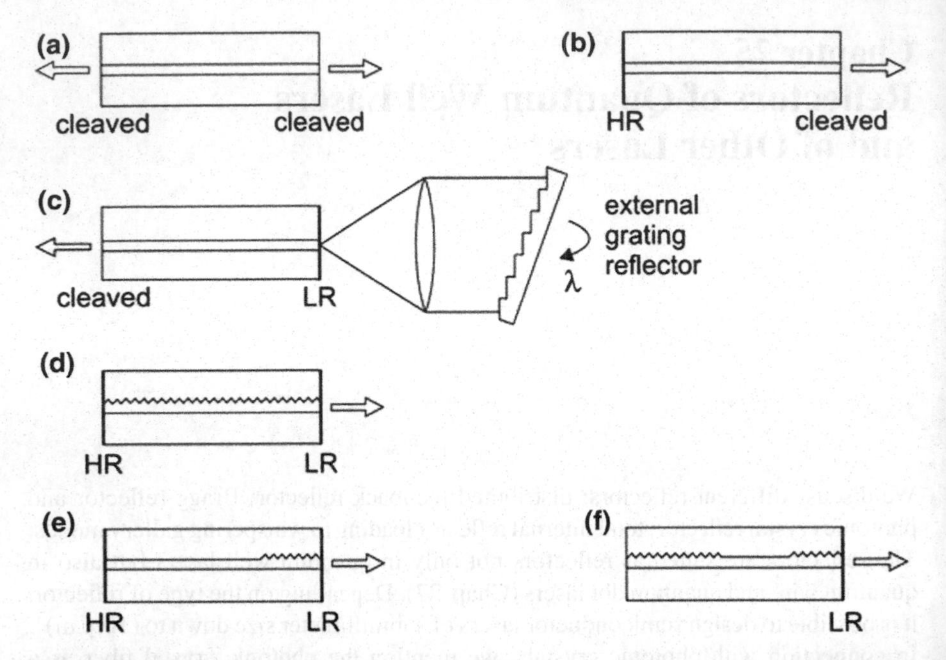

Fig. 25.1 Resonators of semiconductor lasers. **a** Resonator with cleaved surfaces. **b** Resonator with a coated and an uncoated surface. **c** Resonator with an external reflector. **d** Resonator with distributed feedback. **e** Resonator with a distributed Bragg reflector. **f** Resonator with two distributed Bragg reflectors

- Emission occurs into two directions.
- The laser beam has a large angle of aperture in the plane perpendicular to the active layer.
- It is easy to prepare a cleaving surface.

We now will discuss other possibilities to realize laser resonators of semiconductor lasers.

25.2 Coated Surface

By deposition of a dielectric coating on a surface (Fig. 25.1b), one of the reflectors has a high reflectivity (HR). The other surface can remain without coating. Thus, the laser emits radiation in one direction only.

25.3 External Reflector

An external reflector (Fig. 25.1c) makes it possible to realize a tunable semiconductor laser. The reflector is an echelette grating in Littrow arrangement. Rotation of the grating results in a change of the wavelength of the laser radiation. Resonances between the surface and the grating are avoided by the use of an antireflecting coating (a coating with a low reflectivity, LR) on one of the surfaces of the laser diode. The length of the external resonator limits the tuning range—up to about 50 GHz—for tuning on one mode.

25.4 Distributed Feedback Reflector

The integration of a grating into the light guiding structure (Fig. 25.1d) leads to *distributed feedback*. The wavelength of the laser radiation is mainly determined by the period of the distributed reflector. Distributed feedback together with coated cleaved surfaces makes it possible to optimize the laser output of distributed feedback edge-emitting semiconductor lasers.

Distributed feedback reflectors are also suitable as reflectors of solid state dye lasers and organic and polymer lasers (Sect. 34.4).

25.5 Distributed Bragg Reflector

A distributed reflection grating separated from the gain region is a distributed Bragg reflector (Fig. 25.1e). A distributed Bragg reflector acts as output coupler. The surface opposite to the Bragg reflector is highly reflecting.

By the use of two distributed Bragg reflectors (Fig. 25.1f), a high reflectivity at one end of the active region and an optimized output coupler at the other end can be realized.

Distributed Bragg reflectors are well suitable as reflectors of bipolar semiconductor lasers.

25.6 Total Reflector

Internal total reflection in a circularly shaped solid results in a resonator with a very high Q factor. Light is propagating in a whispering gallery mode (Fig. 25.2). Output coupling of radiation is possible by positioning a prism near the surface, resulting in frustrated total reflection in a small region of the resonator. The distance d between the prism and the surface of the resonator regulates the output power. The quality factor can be of the order of 10^6.

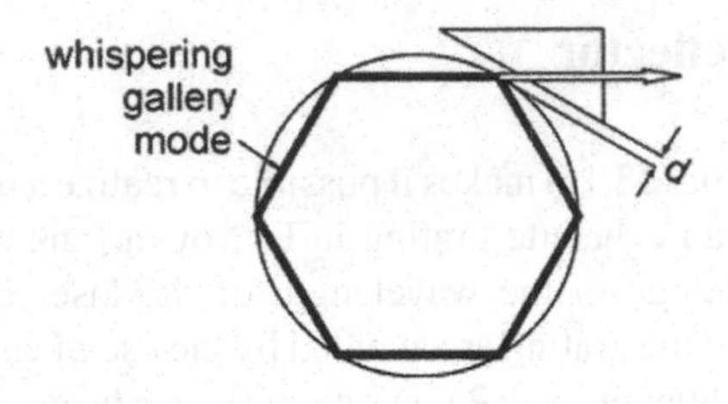

Fig. 25.2 Whispering gallery mode

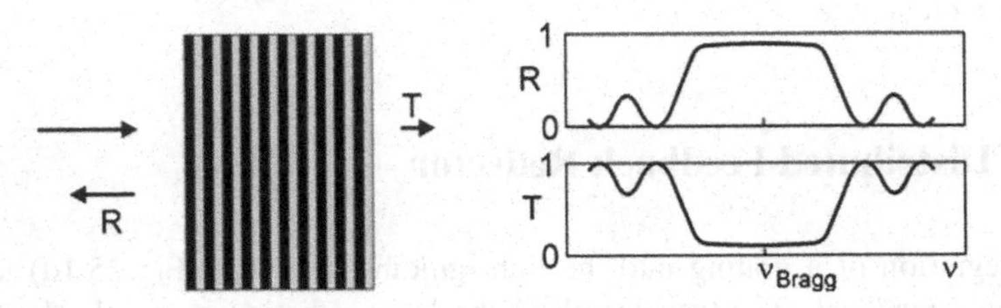

Fig. 25.3 Bragg reflector

25.7 Bragg Reflector

A very successful type of reflector is the Bragg reflector (=dielectric multilayer reflector = 1D photonic crystal reflector); it can consist of a multilayer coating on a transparent substrate.

A Bragg reflector (Fig. 25.3, left) consists of dielectric layers of two different materials of different refractive indices. Each layer has a thickness that is equal to a fourth of the wavelength of radiation in the corresponding material; a material (refractive index n_1) has the thickness $\lambda/(4n_1)$ and the other material (n_2) has the thickness $\lambda/(4n_2)$. With increasing number of quarter-wavelength layers (of two materials in turn) the reflectivity increases and can reach a value very near unity. A Bragg reflector can have a high reflectivity R over nearly one octave of the spectrum (Fig. 25.3, right) around a central frequency ν_{Bragg}. The transmissivity T of the Bragg reflector is $T = 1 - R$. Radiation incident on a Bragg reflector is either reflected or transmitted. Bragg reflectors are essential for operation of vertical-cavity surface-emitting lasers (VCSELs). A Bragg reflector can consist of quarter-wavelength layers of two semiconductors—for instance, of GaAs and AlAs for red and infrared lasers.

Bragg reflectors (=dielectric mirrors) consisting of other materials (e.g., layers of two glass types with different refractive indices) can be used as reflectors and as partial reflectors of almost all types of lasers (Chaps. 14–19).

25.8 Photonic Crystal

A medium that has a spatially periodic dielectric constant is a photonic crystal. A medium consisting of a periodic metal structure with holes or dielectric inclusions can be a photonic crystal too. All photonic crystals have in common that the light propagation is anisotropic and that there can be a photonic bandgap—radiation of a frequency that lies in the bandgap cannot propagate in a photonic crystal. There are three types of photonic crystals (Fig. 25.4, upper row):

- 1D (one-dimensional) photonic crystal. A 1D photonic crystal is, with respect to the optical properties, periodic in one direction and has no structure along the two other directions; the 1D photonic crystal is a three-dimensional medium. The frequency spectrum $\omega(\boldsymbol{k})$ can have a gap for electromagnetic waves propagating along the direction of periodicity (Sect. 25.14).
- 2D (two-dimensional) photonic crystal. The 2D photonic crystal is periodic in two directions and has no structure along the third direction; the 2D photonic crystal is a three-dimensional medium. The frequency spectrum $\omega(\boldsymbol{k})$ can show gaps for electromagnetic waves with wave vectors in the plane, which contains the two directions of periodicity.
- 3D (three-dimensional) photonic crystal. The 3D photonic crystal is periodic in three directions. The frequency spectrum $\omega(\boldsymbol{k})$ can show gaps for electromagnetic waves in all three spatial directions.

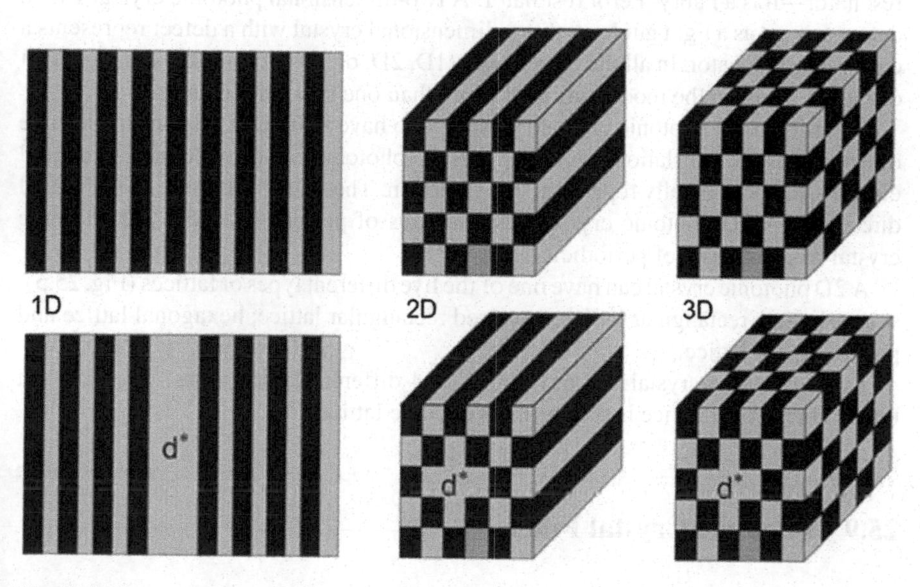

Fig. 25.4 Photonic crystals without and with a defect

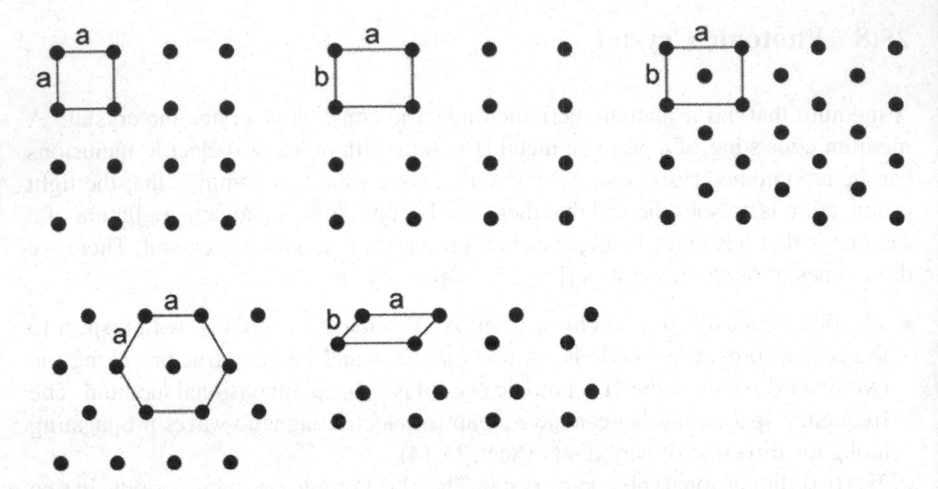

Fig. 25.5 Elementary cells of 2D photonic crystal lattices

A photonic crystal of finite length can act as a partial reflector for radiation of a photon energy in the photonic gap: radiation incident on a photonic crystal of finite thickness is partly reflected and partly transmitted.

If a single layer of a one-dimensional crystal is missing (Fig. 25.4, lower row), the photonic crystal contains a defect (d*). A photonic crystal with a defect represents a resonator—it is a Fabry–Perot resonator. A two-dimensional photonic crystal with a defect can act as a light guide. A three-dimensional crystal with a defect represents a cavity-like resonator. In all the three cases (1D, 2D, or 3D photonic crystal), a defect can also consist in the modification of more than one structural element.

Common to all photonic crystals (assumed to have infinite extensions in all three directions) is the translational symmetry. A 1D photonic crystal contains a structural element that periodically repeats in one direction. There is no structure in the lateral directions. A 2D photonic crystal has two axes of periodicity and a 3D photonic crystal has three axis of periodicity.

A 2D photonic crystal can have one of the five different types of lattices (Fig. 25.5): square lattice; rectangular lattice; centered rectangular lattice; hexagonal lattice and parallelogram lattice.

A 3D photonic crystal can have one of 14 different lattice types. The simplest three-dimensional lattice is the (primitive) cubic lattice.

25.9 Photonic Crystal Fiber

A photonic crystal fiber (Fig. 25.6) can consist of a fiber with an internal two-dimensional photonic crystal (e.g., a hexagonal two-dimensional lattice). The structure is composed of two different glass materials, one with a higher dielectric con-

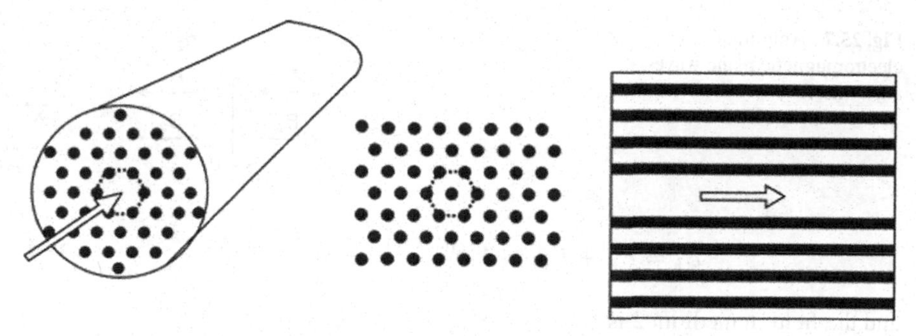

Fig. 25.6 Photonic crystal fiber

stant than the other. A defect allows for propagation of light (along the defect). A photonic bandgap for propagation of radiation along a direction perpendicular to the axis of the two-dimensional photonic crystal avoids spreading of the radiation.

25.10 Remark About Photonic Crystals

In 1887, Lord Rayleigh explained an experiment, which indicated that a periodic dielectric multilayer stack showed a spectral range of high reflectivity [217], corresponding to a stop-band of radiation. Such multilayer stacks (later called Bragg reflectors or photonic crystals) are widely studied and applied as reflectors or partial reflectors. The term "photonic crystal" describing inhomogeneous but periodic structures was introduced by E. Yablonovitch [218] in 1987.

Reflectors of the two-dimensional photonic crystal type were first used as reflectors of microwave Fabry–Perot interferometers in 1957 [219] and of far infrared Fabry–Perot interferometers up to frequencies of several THz in 1962 [220]; a far infrared reflector consists of a thin metal mesh and a Fabry–Perot interferometer of two meshes in parallel.

25.11 Plane-Wave Transfer Matrix Method Characterizing an Optical Interface

We consider the interface of two optically isotropic, nonabsorbing, and nonmagnetic materials (Fig. 25.7). The refractive indices of the two media are n_1 and n_2. We study the special case of monochromatic radiation (frequency ω) propagating along the x or $-x$ direction. The field in medium 1 is

Fig. 25.7 Amplitudes of electromagnetic plane waves at an interface

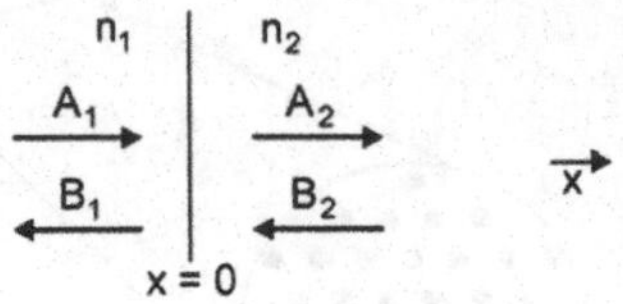

$$E_1 = E_1^+ + E_1^- = A_1 e^{i(\omega t - k_1 x)} + B_1 e^{i(\omega t + k_1 x)} \tag{25.1}$$

and the field in medium 2 is

$$E_2 = E_2^+ + E_2^- = A_2\, e^{i(\omega t - k_2 x)} + B_2\, e^{i(\omega t + k_2 x)}. \tag{25.2}$$

$A_1, B_1, k_1, -k_1$ and $A_2, B_2, k_2, -k_2$ are the amplitudes and the wave vectors of the waves in medium 1 and medium 2 in x and $-x$ direction, respectively.

We assume that the electric field is oriented along the y axis and that, according to Maxwell's equations, H is therefore oriented along the z direction. It follows from Maxwell's equation

$$\boldsymbol{H} = \frac{i}{\mu_0 \omega} \nabla \times \boldsymbol{E} \tag{25.3}$$

that the magnetic field strength is given by

$$H_n = \frac{i}{\mu_0 \omega} \frac{dE}{dx}. \tag{25.4}$$

The boundary conditions for the electromagnetic fields (continuity of E_t and H_n) require that E and dE/dx are continuous at the boundary ($x = 0$),

$$E_1 = E_2 \quad \text{at } x = 0, \tag{25.5}$$

$$dE_1/dx = dE_2/dx \quad \text{at } x = 0, \tag{25.6}$$

or

$$A_1 + B_1 = A_2 + B_2, \tag{25.7}$$

$$k_1 A_1 - k_1 B_1 = k_2 A_2 - k_2 B_2. \tag{25.8}$$

We can write

$$M_1 \begin{pmatrix} A_1 \\ B_1 \end{pmatrix} = M_2 \begin{pmatrix} A_2 \\ B_2 \end{pmatrix}, \tag{25.9}$$

where, with l =1, or 2,

$$M_l = \begin{pmatrix} 1 & 1 \\ k_l & -k_l \end{pmatrix}, \tag{25.10}$$

$$\begin{pmatrix} A_1 \\ B_1 \end{pmatrix} = M_1^{-1} M_2 \begin{pmatrix} A_2 \\ B_2 \end{pmatrix} = M_{12} \begin{pmatrix} A_2 \\ B_2 \end{pmatrix}, \quad (25.11)$$

and

$$M_{12} = \begin{pmatrix} \frac{1}{2}(1 + k_2/k_1) & \frac{1}{2}(1 - k_2/k_1) \\ \frac{1}{2}(1 - k_2/k_1) & \frac{1}{2}(1 + k_2/k_1) \end{pmatrix}. \quad (25.12)$$

The matrix M_{12} is the *plane-wave transfer matrix*. It relates the amplitudes of electromagnetic plane waves in medium 1 and the amplitudes of electromagnetic plane waves in medium 2.

25.12 Thin Film Between Two Media

A thin film (thickness a) located between two media (Fig. 25.8) has two boundaries. The boundary conditions lead to

$$\begin{pmatrix} A_1 \\ B_1 \end{pmatrix} = M_1^{-1} M_2 \begin{pmatrix} A_2 \\ B_2 \end{pmatrix} = M_{12} \begin{pmatrix} A_2 \\ B_2 \end{pmatrix}, \quad (25.13)$$

$$\begin{pmatrix} A_2' \\ B_2' \end{pmatrix} = P_2 \begin{pmatrix} A_2 \\ B_2 \end{pmatrix} = \begin{pmatrix} e^{i\varphi_2} & 0 \\ 0 & e^{-i\varphi_2} \end{pmatrix} \begin{pmatrix} A_2 \\ B_2 \end{pmatrix}, \quad (25.14)$$

$$\begin{pmatrix} A_2 \\ B_2 \end{pmatrix} = M_2^{-1} M_3 \begin{pmatrix} A_3 \\ B_3 \end{pmatrix} = M_{23} \begin{pmatrix} A_3 \\ B_3 \end{pmatrix}, \quad (25.15)$$

where $\varphi_2 = k_2 a$ and $K_2 = n_2\omega/c$. The matrix

$$P_2 = \begin{pmatrix} e^{i\varphi_2} & 0 \\ 0 & e^{i\varphi_2} \end{pmatrix} \quad (25.16)$$

is the propagation matrix taking into account the phase change due to propagation.

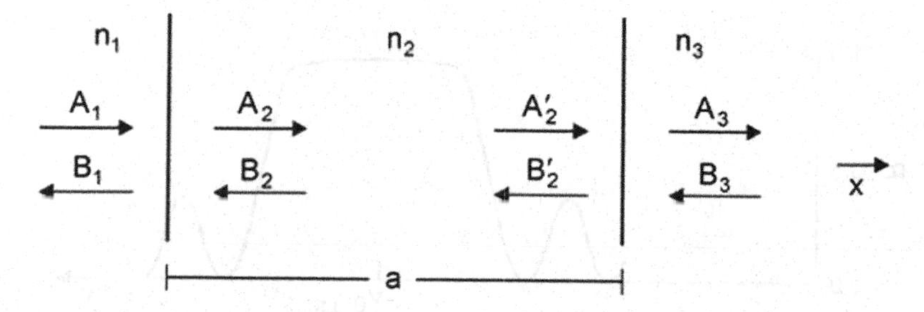

Fig. 25.8 Thin film (refractive index n_2, thickness a) between two extended media

25.13 Dielectric Multilayer

We study a dielectric multilayer system (Fig. 25.9, upper part) for radiation propagating along the axis (x axis) that is perpendicular to the layers. We consider a system of N layers (n_l = refractive index and d_l = thickness of the lth layer) on a substrate (refractive index n_s). The multilayer system is covered with a medium of refractive index n_0. We apply the matrix method and find:

$$\begin{pmatrix} A_l \\ B_l \end{pmatrix} = \begin{pmatrix} M_{11} & M_{12} \\ M_{21} & M_{22} \end{pmatrix} \begin{pmatrix} A_{l+1} \\ B_{l+1} \end{pmatrix}, \tag{25.17}$$

$$\begin{pmatrix} M_{11} & M_{12} \\ M_{21} & M_{22} \end{pmatrix} = M_0^{-1} \left[\prod_{l=1}^{N} M_l P_l M_l^{-1} \right] M_s, \tag{25.18}$$

$$P_l = \begin{pmatrix} e^{i\varphi_l} & 0 \\ 0 & e^{-i\varphi_l} \end{pmatrix}, \tag{25.19}$$

$$\varphi_l = k_l a_l. \tag{25.20}$$

Making use of the dispersion relations

$$k_l = n_l \omega / c, \tag{25.21}$$

and with $B_s = 0$, we can determine the reflectivity and the transmissivity of the multilayer system (A_0, B_0, A_s = amplitudes of incident, reflected, transmitted field):

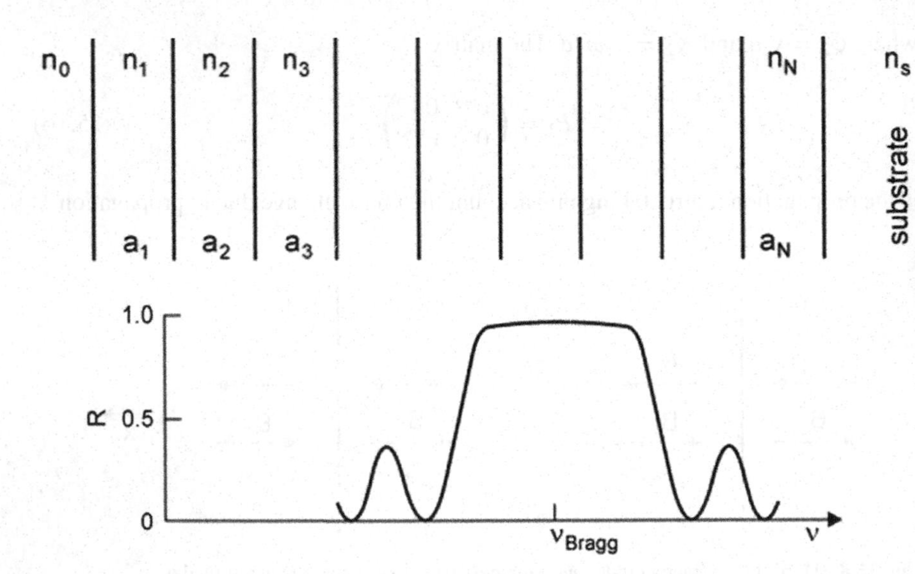

Fig. 25.9 Dielectric multilayer reflector (*Bragg reflector*) and reflectivity

$$R = (B_0/A_0)^2, \tag{25.22}$$

$$T = (A_s/A_0)^2. \tag{25.23}$$

In the special case that the multilayer system consists of a sequence of two layers (refractive indices n_1 and n_2 that have quarter-wavelength thicknesses $d_1 = \lambda_0/(4n_1)$ and $d_2 = \lambda_0/(4n_2)$ for radiation of wavelength λ_0, the reflectivity at λ_0 is given by

$$R = \left(\frac{1 - (n_s/n_0)(n_2/n_1)^{2N}}{1 + (n_s/n_0)(n_2/n_1)^{2N}}\right)^2, \tag{25.24}$$

where N is the number of double-layers. The reflectivity approaches unity if N becomes very large.

Figure 25.9 (lower part) shows the reflectivity of a GaAs/AlAs Bragg reflector.

25.14 One-Dimensional Photonic Crystal

We consider a stratified periodic medium consisting of a series of double layers (Fig. 25.10). A double layer consists of a layer 1 (refractive n_1, thickness a_1) and a layer 2 (refractive n_2, thickness a_2). The stratified medium is spatially periodic, the spatial period is

$$a = a_1 + a_2. \tag{25.25}$$

The unit cell consists of a double layer. We suppose that we have an infinite number of cells, numbered $l = \ldots -1, 0, 1, \ldots$.

We study the propagation of a monochromatic plane wave (frequency ω). Our goal is to find the dispersion relation $\omega(k)$. We use the ansatz (Boch theorem):

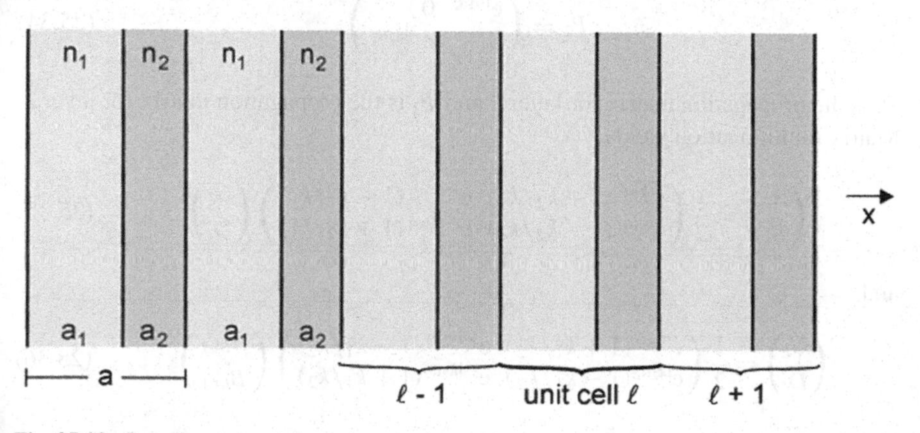

Fig. 25.10 One-dimensional photonic crystal

$$E(x) = A_k(x)\,\mathrm{e}^{\mathrm{i}(\omega t - kx)}, \tag{25.26}$$

where k is the wave vector and $A_k(x)$ an amplitude that is a periodic function,

$$A_k(x + a) = A_k(x). \tag{25.27}$$

The field $E(x)$ is a Bloch wave. The amplitude varies within a periodicity interval. However it is, for a particular k, lattice-periodic. The propagation of the plane wave over the distance x causes, as it is typical for plane waves, a change kx. The wave vector k of the plane wave depends on the frequency, $k = k(\omega)$, or

$$\omega = \omega(k). \tag{25.28}$$

The field in the lth cell is

$$E(x) = A_l \mathrm{e}^{\mathrm{i}k_1(x-la)} + B_l \mathrm{e}^{\mathrm{i}k_1(x-la)} \quad \text{in layer 1 of cell } l, \tag{25.29}$$

$$E(x) = C_l \mathrm{e}^{\mathrm{i}k_2(x-la)} + D_l \mathrm{e}^{\mathrm{i}k_2(x-la)} \quad \text{in layer 2 of cell } l, \tag{25.30}$$

with $k_1 = n_1\omega/c$ and $k_2 = n_2\omega/c$.

We relate, in a first step, the electric fields in three neighboring cells (*see* Fig. 25.10):

$$\begin{pmatrix} A_l \\ B_l \end{pmatrix} = M_1^{-1} M_2 P_1 \begin{pmatrix} C_l \\ D_l \end{pmatrix}, \tag{25.31}$$

$$\begin{pmatrix} C_l \\ D_l \end{pmatrix} = M_2^{-1} M_1 P_2 \begin{pmatrix} A_{l+1} \\ B_{l+1} \end{pmatrix}, \tag{25.32}$$

$$P_1 = \begin{pmatrix} \mathrm{e}^{\mathrm{i}k_1 a_1} & 0 \\ 0 & \mathrm{e}^{-\mathrm{i}k_1 a_1} \end{pmatrix}, \tag{25.33}$$

$$P_2 = \begin{pmatrix} \mathrm{e}^{\mathrm{i}k_2 a_2} & 0 \\ 0 & \mathrm{e}^{-\mathrm{i}k_2 a_2} \end{pmatrix}. \tag{25.34}$$

P_1 is the propagation matrix for layer 1 and P_2 is the propagation matrix for layer 2. Matrix multiplication yields

$$\begin{pmatrix} A_l \\ B_l \end{pmatrix} = \frac{1}{2} \begin{pmatrix} \mathrm{e}^{\mathrm{i}k_1 a_1}(1 + k_2/k_1) & \mathrm{e}^{-\mathrm{i}k_1 a_1}(1 - k_2/k_1) \\ \mathrm{e}^{\mathrm{i}k_1 a_1}(1 - k_2/k_1) & \mathrm{e}^{-\mathrm{i}k_1 a_1}(1 + k_2/k_1) \end{pmatrix} \begin{pmatrix} C_l \\ D_l \end{pmatrix} \tag{25.35}$$

and

$$\begin{pmatrix} C_l \\ D_l \end{pmatrix} = \frac{1}{2} \begin{pmatrix} \mathrm{e}^{\mathrm{i}k_2 a_2}(1 + k_1/k_2) & \mathrm{e}^{-\mathrm{i}k_2 a_2}(1 - k_1/k_2) \\ \mathrm{e}^{\mathrm{i}k_2 a_2}(1 - k_1/k_2) & \mathrm{e}^{-\mathrm{i}k_2 a_2}(1 + k_1/k_2) \end{pmatrix} \begin{pmatrix} A_{l+1} \\ B_{l+1} \end{pmatrix}. \tag{25.36}$$

We write

$$\begin{pmatrix} A_l \\ B_l \end{pmatrix} = \begin{pmatrix} A & B \\ C & D \end{pmatrix} \begin{pmatrix} A_{l+1} \\ B_{l+1} \end{pmatrix}, \tag{25.37}$$

where

$$A = \mathrm{e}^{\mathrm{i}k_1 a_1} \left[\cos k_1 a_1 + \frac{1}{2}\mathrm{i} \left(\frac{k_2}{k_1} + \frac{k_2}{k_2} \right) \sin k_1 a_1 \right], \tag{25.38}$$

$$B = \mathrm{e}^{-\mathrm{i}k_1 a_1} \left[\frac{1}{2}\mathrm{i} \left(\frac{k_2}{k_1} - \frac{k_1}{k_2} \right) \sin k_1 a_1 \right], \tag{25.39}$$

$$C = \mathrm{e}^{\mathrm{i}k_2 a_2} \left[-\frac{1}{2}\mathrm{i} \left(\frac{k_2}{k_1} - \frac{k_1}{k_2} \right) \sin k_2 a_2 \right], \tag{25.40}$$

$$D = \mathrm{e}^{-\mathrm{i}k_2 a_2} \left[\cos k_2 a_2 - \frac{1}{2}\mathrm{i} \left(\frac{k_2}{k_1} + \frac{k_1}{k_2} \right) \sin k_2 a_2 \right]. \tag{25.41}$$

We have the relation

$$AD - BC = 1. \tag{25.42}$$

It follows that the amplitude in the lth cell and the amplitude in the zeroth cell are related:

$$\begin{pmatrix} A_l \\ B_l \end{pmatrix} = \begin{pmatrix} A & B \\ C & D \end{pmatrix}^{-1} \begin{pmatrix} A_0 \\ B_0 \end{pmatrix} \tag{25.43}$$

or

$$\begin{pmatrix} A_l \\ B_l \end{pmatrix} = \begin{pmatrix} D & -B \\ -C & A \end{pmatrix} \begin{pmatrix} A_0 \\ B_0 \end{pmatrix}. \tag{25.44}$$

If we specify A_0 and B_0, all amplitudes can be calculated.

We make use of the periodicity of the multilayer system and write

$$\begin{pmatrix} A_{l+1} \\ B_{l+1} \end{pmatrix} = \mathrm{e}^{-\mathrm{i}ka} \begin{pmatrix} A_l \\ B_l \end{pmatrix}, \tag{25.45}$$

leading to

$$\begin{pmatrix} A & B \\ C & D \end{pmatrix} \begin{pmatrix} A_{l+1} \\ B_{l+1} \end{pmatrix} = \mathrm{e}^{\mathrm{i}ka} \begin{pmatrix} A_{l+1} \\ B_{l+1} \end{pmatrix}. \tag{25.46}$$

The phase factor $\exp(\mathrm{i}ka)$ is the eigenvalue of the matrix ABCD. We find

$$\mathrm{e}^{\mathrm{i}ka} = \frac{1}{2}(A + D) \pm \sqrt{\frac{1}{4}(A + D)^2 - 1}. \tag{25.47}$$

The sum A + D is real. It follows, with

$$\mathrm{e}^{\mathrm{i}ka} = \cos ka + \mathrm{i} \sin ka, \tag{25.48}$$

that

$$\cos ka = \frac{1}{2}(A + D) \tag{25.49}$$

and

$$\sin ka = \pm\sqrt{1 - \frac{1}{4}(A + D)^2}. \tag{25.50}$$

We obtain, with

$$\xi = k_1/k_2 = n_1/n_2, \tag{25.51}$$

the dispersion relation

$$\cos ka = \cos(k_1 a_1)\cos(k_2 a_2) - \frac{1}{2}\left(\xi + \frac{1}{\xi}\right)\sin(k_1 a_1)\sin(k_2 a_2) \tag{25.52}$$

or

$$\cos ka = \cos\left(n_1\frac{\omega}{c}a_1\right)\cos\left(n_2\frac{\omega}{c}a_2\right) - \frac{1}{2}\left(\frac{n_2}{n_1} + \frac{n_1}{n_2}\right)\sin\left(n_1\frac{\omega}{c}a_1\right)\sin\left(n_2\frac{\omega}{c}a_2\right). \tag{25.53}$$

We discuss the dispersion relation in the special case that the optical paths in layer 1 and layer 2 are equal,

$$n_1 a_1 = n_2 a_2. \tag{25.54}$$

The dispersion relation has the form

$$\cos ka = \cos^2\left(n_1\frac{\omega}{c}a_1\right) - \frac{1}{2}\left(\frac{n_2}{n_1} + \frac{n_1}{n_2}\right)\sin^2\left(n_1\frac{\omega}{c}a_1\right) \tag{25.55}$$

or

$$k = \frac{1}{a}\cos^{-1}\left[\cos^2(n_1\frac{\omega}{c}a_1) - \frac{1}{2}\left(\frac{n_2}{n_1} + \frac{n_1}{n_2}\right)\sin^2(n_1\frac{\omega}{c}a_1)\right]. \tag{25.56}$$

The frequency increases proportionally to the wave vector k at long waves ($ka \ll 1$):

$$k = n_{\text{eff}}\frac{\omega}{c}, \tag{25.57}$$

where

$$n_{\text{eff}} = \sqrt{n_1 n_2} \tag{25.58}$$

is an effective refractive index. The appearance of an effective refractive index with a value between n_1 and n_2 is a consequence of the reflection of the radiation at the interfaces. There are frequency gaps for

$$k = \pm\frac{\pi}{a}, \ \pm\frac{3\pi}{a}, \ldots . \tag{25.59}$$

Because of the periodicity,

$$\omega\left(k + \frac{2\pi}{a}\right) = \omega(k), \tag{25.60}$$

we can restrict the k values to the first Brillouin zone,

$$-\frac{\pi}{a} < k \leq \frac{\pi}{a}. \tag{25.61}$$

The values $-\pi/a$ and π/a are the Brillouin zone boundaries.

The curve $k = n_{\mathrm{eff}}\omega/c$ reaches the Brillouin zone boundary at the Bragg frequency

$$\omega_{\mathrm{Bragg}} = \frac{\pi c}{n_{\mathrm{eff}} a}, \tag{25.62}$$

which corresponds to the Bragg wavelength

$$\lambda_{\mathrm{Bragg}} = 2n_{\mathrm{eff}} a. \tag{25.63}$$

For radiation of this vacuum-wavelength, the multilayer system represents a stack of quarter-wavelength layers. Bragg reflection of the radiation at wavelengths around λ_{Bragg} is responsible for the occurrence of a frequency gap.

It follows that the field in layer 1 of the lth cell is given by

$$E(x) = \left[A_0\, \mathrm{e}^{\mathrm{i}n_1(\omega/c)(x-la)} + B_0\, \mathrm{e}^{\mathrm{i}n_1(\omega/c)(x-la)}\right] \mathrm{e}^{\mathrm{i}(\omega t - lka)}. \tag{25.64}$$

Example Dispersion relation of radiation in a GaAs/AlAs photonic crystal (Fig. 25.11); $n_1 = 3.3$; $n_2 = 2.9$; $a_1 = 152\,\mathrm{nm}$; $a_2 = 173\,\mathrm{nm}$; $a = 325\,\mathrm{nm}$; $n_1 a = n_2 b$; $n_{\mathrm{eff}} = 3.09$.

25.15 Bragg Reflection as Origin of Energy Gaps

The occurrence of energy gaps is a consequence of the ability of radiation to undergo Bragg reflection. Bragg reflection occurs for radiation with discrete values of the wave vector, namely for $k = k_{\mathrm{Bragg}}$. In the case of a one-dimensional crystal, $k_{\mathrm{Bragg}} = \pi/a$. A two-dimensional photonic crystal has Bragg vectors that lie on a plane. A two-dimensional photonic crystal can show a photonic bandgap (frequency gap) for radiation of all k vectors in a plane. A three-dimensional photonic crystal has Bragg vectors in the three-dimensional k space. A three-dimensional photonic crystal can have energy gaps for k vectors of all spatial directions.

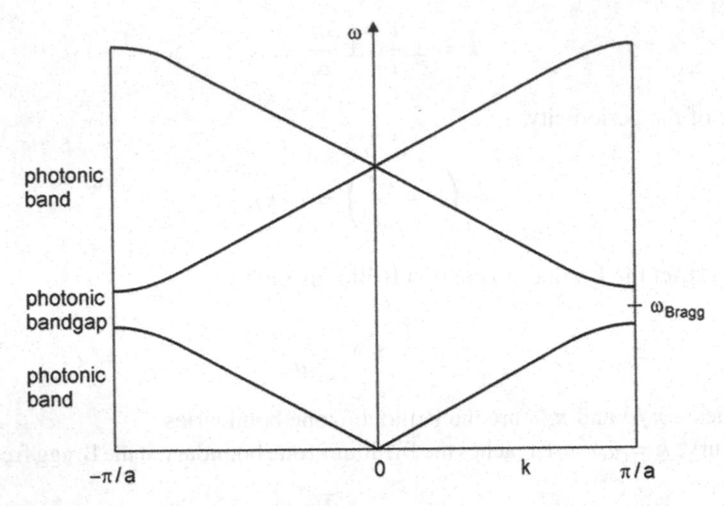

Fig. 25.11 One-dimensional photonic crystal: dispersion relation of radiation, with the propagation direction parallel to the axis of a periodic multilayer structure

We come back to Fig. 25.11. The speed of light is equal to c/n_{eff} for long wavelengths, i.e., for small frequencies. The speed of light is almost constant at small wave vectors, decreases at large wave vectors and becomes zero for $k = \pi/a$. A photonic crystal is thus able to slow down an electromagnetic wave.

References [26, 28, 177, 207–220].

Problems

25.1 Bragg reflection. Formulate the conditions for the occurrence of Bragg reflection of Bragg reflection of electromagnetic radiation in different systems. (a) A 1D photonic crystal, (b) 2D photonic crystal, (c) 3D photonic crystal.

25.2 Bragg reflection of X-rays.

(a) Formulate the conditions for the occurrence of Bragg reflection of X-rays.
(b) Why are Bragg peaks at X-rays extremely sharp?
(c) Estimate the width of an energy gap expected for X-rays. [*Hint*: estimate the refractive index of X-rays—it is slightly smaller than unity—and describe a crystal (e.g., with respect to the 100 direction) as a 1D photonic crystal with the electrons distributed in thin layers perpendicular to the propagation direction of the X-rays.]

25.3 One-dimensional photonic crystal.

(a) Estimate the widths of forbidden frequency bands in the case that $n_2 - n_1 \ll 1$.
(b) Estimate, for radiation of the vacuum wavelength 1 μm, the widths of forbidden frequency bands in the case that the photonic crystal consists of a stack of GaAs/AlAs quarter-wavelength films.

25.4 One-dimensional photonic crystal consisting of freestanding plates.

(a) Determine the effective refractive index, the Bragg frequency and the Bragg wavelength of thin freestanding silicon plates (thickness 1 μm, refractive index $n = 4$) separated by air under the assumption that the plates and the space between two plates have the same optical thickness.
(b) Calculate the dispersion relation of radiation in such a one-dimensional photonic crystal.

25.5 How many quarter-wavelength films of GaAs and AlAs films on a GaAs substrate are necessary to obtain reflectivities $R \sim 70, 80, 90, 95$ or, 99, 99.9%?

25.6 Antireflecting coating.

(a) Show, by use of the matrix method, that the reflectivity of the surface of an optical substrate (refractive index n_s) covered with a quarter-wavelength film, thickness $\lambda/(4n)$, is zero if the refractive index n of the film satisfies the condition $n_s - n^2 = 0$. [*Hint*: assume that the substrate has infinite thickness, so that no reflection from the end surface of the substrate occurs.]
(b) Show that the multiple beam method (introduced in Sect. 3.5) yields the same result. [*Hint:* add all beams reflected by the two surfaces of the film, taking multiple reflection into account.]

25.7 Determine the Airy formula (Sect. 3.5) by use of the matrix method.

25.8 Double-resonator. We consider a double-resonator (Fabry–Perot resonator) with three lossless mirrors of equal reflectivity R. The distance between mirror 1 and mirror 2 is L_1 and the distance between mirror 2 and mirror 3 is L_2. Derive, by the use of the matrix method, the transmission curve of a double resonator for (a) $L_1 = L_2$; (b) $L_2 \ll L_1$, (c) $L_1 = \lambda/2$. (d) Choose $R = 0.95$ and $\lambda = 1$ μm for a discussion of the results.

25.9 Boundary between two dielectric media.

(a) Show that the boundary conditions for normal incidence are consistent with the requirement that the energy flux density is the same in medium 2 as in medium 1. [*Hint*: describe the energy flux density by the Poynting vector $\boldsymbol{P} = \boldsymbol{E} \times \boldsymbol{H}$.]
(b) Derive the Fresnel coefficient of reflection for normal incidence by the use of the matrix method.

25.10 Bloch theorem.. Derive the Bloch theorem for the one-dimensional photonic crystal, i.e., justify the ansatz (25.26). [*Hint*: make use of periodic boundary conditions.]

25.11 Propagating of radiation in a one-dimensional crystal. Discuss the dependence of group and phase velocity on the wave vector of radiation belonging to the two lowest branches of the dispersion curves shown in Fig. 25.11.

25.12 Determine, by use of the matrix method, the halfwidth of the resonance curve of a Fabry–Perot resonator (Sect. 3.6) that has a reflector of a reflectivity of unity and a partial reflector.

25.13 Derive the Airy formula for a Fabry–Perot resonator containing an active medium (Sect. 3.7), by the use of the matrix method. [*Hint*: assume that one of the mirrors has a reflectivity of unity.]

25.14 Reflection of radiation by a perfect conductor.

(a) Show that the reflectivity of a perfect conductor is 1.
(b) The radiation penetrates into the conductor. Derive an expression of the penetration depth of the electric field and of the radiation energy.
(c) Calculate the penetration depth of radiation reflected by a perfect conductor, which contains electrons of a concentration $N = 10^{28}\ \mathrm{m}^3$, for radiation of of 1 mm and of 0.5 μm wavelength.

25.15 A perfect mirror. A thin film consisting of a perfectly conducting material can act as a partial mirror. [*Hint*: a perfect conductor for currents at microwave frequencies is superconducting lead at a temperature well below the superconducting transition temperature of 7 K.]

(a) Determine the complex transmission coefficient $\tilde{t}$, the complex reflection coefficient $\tilde{r}$, the phase φ of the reflected beam, the phase φ_t of the transmitted beam, transmissivity T and the reflectivity R (*see* Sect. 3.4). [*Hint*: make use of the matrix method; treat the film as a free-standing film surrounded by air].
(b) Design partial mirrors that have reflectivities $R \sim 70, 80, 90, 95, 99, 99.9\%$ for radiation of 1 mm wavelength, assuming that the mirror is perfectly conducting and contains electrons of a concentration $N = 10^{28}\ \mathrm{m}^3$.
(c) Calculate, for a Fabry–Perot resonator resonator formed by two (perfect) partial mirrors as reflectors, the change of phase per round trip transit of radiation of 1 mm wavelength in the case that the reflectivity of each mirror is $R = 0.9$.

25.16 Methods of describing the field in a resonator. Show that the three methods of describing a field in a resonator lead to the same result:

(a) The method of multiple reflection (Sect. 3.5).
(b) The method directly based on the boundary conditions (this chapter).
(c) A method directly based on the boundary conditions but that immediately introduces the complex transmission coefficient $\tilde{t} = B_1/A_1$ and the complex reflection coefficient $\tilde{r} = B_2/A_1$ of a mirror; use this method to derive the Airy formula.

25.17 Bulk metal. We study the optical properties of a metal like copper (free-electron concentration $N = 10^{28}\ \mathrm{m}^{-3}$, relaxation time $\tau = 10^{-13}$ s).

(a) Determine, by use of the complex optical constants, the frequency dependence of the reflectivity.
(b) Compare the reflectivity of the metal with the reflectivity of a perfect conductor that contains electrons of the same density.
(c) Determine the optical constants and the reflectivity of a metal for radiation of long wavelengths (i.e., for $\omega \ll \omega_p = \sqrt{Ne^2/\epsilon_0 m_0}$ = plasma frequency).

25.18 Metal film. Study optical properties of a metal film (e.g., a copper film). Restrict the discussion to long wavelengths.

(a) Determine the dependence of transmissivity T, reflectivity R and absorptivity A of a metal film on the thickness of the film by use of the matrix method.
(b) Show that there is a film thickness where $T = R = 0.25$ and $A = 0.5$, and that $T \ll A$ for thicker films.

Chapter 26
More About the Quantum Well Laser

We continue the discussion of subbands with a description of wave functions and energy bands of electrons in a quantum well. We also show how light holes modify the gain profile. Furthermore, we discuss the influence of inhomogeneous broadening on the properties of a quantum well laser.

26.1 Electron Subbands

Electrons can move freely in the GaAs plane (y, z plane) of a GaAs quantum well (Fig. 26.1, left). The motion perpendicular to the plane, along the x direction, is spatially limited. The potential energy $E_{\rm pot}(x, y, z)$ of electrons (Fig. 26.1, center) is equal to the conduction band profile $E_{\rm c}(x, y, z)$.

We treat the GaAs layer as an infinitely extended layer. The Schrödinger equation of an electron in the GaAs quantum layer (quantum film) has the form

$$\left[-\frac{\hbar^2}{2m_{\rm e}}\nabla^2 + E_{\rm pot}(x)\right]\Psi = {\rm i}\hbar\frac{\partial\Psi}{\partial t}. \tag{26.1}$$

We assume that the effective mass of an electron in a quantum well is the same as for bulk GaAs ($m_{\rm e} = 0.07\, m_0$). To determine the wave function Ψ, we use an ansatz of stationary states,

$$\Psi(x, y, z, t) = \psi(x, y, z){\rm e}^{-{\rm i}(E/\hbar)t}. \tag{26.2}$$

E is the energy of a stationary state. We obtain

$$\left[-\frac{\hbar^2}{2m_{\rm e}}\nabla^2 + E_{\rm pot}(x)\right]\psi = E\psi. \tag{26.3}$$

K.F. Renk, *Basics of Laser Physics*, Graduate Texts in Physics,
DOI 10.1007/978-3-319-50651-7_26

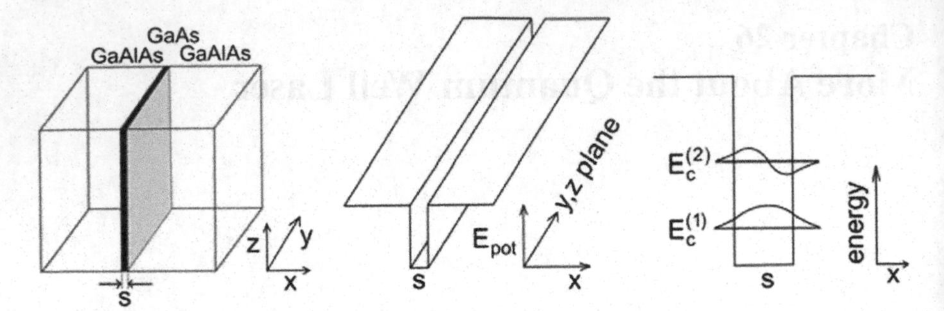

Fig. 26.1 Quantum well

The ansatz

$$\psi = \chi(x)\,\eta(y, z), \tag{26.4}$$

leads to the differential equation

$$\left[-\frac{\hbar^2}{2m_e}\frac{\partial^2}{\partial x^2} + E_{pot}(x)\right]\eta\chi - \frac{\hbar^2}{2m_e}\left(\frac{\partial^2}{\partial y^2} + \frac{\partial^2}{\partial z^2}\right)\eta\chi = E_{tot}\,\eta\chi\,. \tag{26.5}$$

By dividing by $\eta\chi$, we obtain

$$\frac{1}{\chi}\left(-\frac{\hbar^2}{2m_e}\frac{\partial^2}{\partial x^2} + E_{pot}(x)\right)\chi + \frac{1}{\eta}\left(-\frac{\hbar^2}{2m_e}\right)\left(\frac{\partial^2}{\partial y^2} + \frac{\partial^2}{\partial z^2}\right)\eta = E_{tot}. \tag{26.6}$$

The two terms on the left side must have constant values. The total energy is given by

$$E_{tot} = E_\perp + E_{||}, \tag{26.7}$$

where

- $E_\perp$ is the energy of electron motion perpendicular to the layer and
- $E_{||}$ is the energy of electron motion along the film plane.

The Schrödinger equation describing motion along the layer plane is

$$-\frac{\hbar^2}{2m_e}\left(\frac{\partial^2}{\partial y^2} + \frac{\partial^2}{\partial z^2}\right)\eta = E_{||}\eta. \tag{26.8}$$

As a solution, we obtain the wave function

$$\eta = C e^{i\boldsymbol{k}_{||}\boldsymbol{r}_{||}}. \tag{26.9}$$

C is a constant and

- $\boldsymbol{k}_{||} = (k_y, k_z) = \boldsymbol{k}$ is the wave vector parallel to the film plane and
- $\boldsymbol{r}_{||} = (y, z)$ is a location in the plane of the quantum layer.

The differential equation yields the energy

$$E_{||} = \frac{\hbar^2}{2m_e} k^2. \tag{26.10}$$

The Schrödinger equation of the electron motion perpendicular to the quantum layer,

$$\left[-\frac{\hbar^2}{2m_e} \frac{\partial^2}{\partial x^2} + E_{\rm pot}(x) \right] \chi(x) = E_{\perp} \chi(x), \tag{26.11}$$

is the equation of an electron in a one-dimensional square well potential. The energy eigenvalues of a quantum well with infinitely high walls are given by

$$E_c^{(n)} = \frac{\pi^2 \hbar^2}{2m_e s^2} n^2; \quad n = 1, 2, \ldots, \tag{26.12}$$

where s is the thickness of the quantum film. The quantized motion leads to discrete energy eigenvalues $E_c^1, E_c^2, \ldots$. The energy E_c^1 is the *zero point energy* of the electron in the quantum well. The wave functions $\chi_n(x)$ are cosine and sine functions within the film and are zero at the borders of the film and outside the film.

The energy values of wells with walls of finite height are smaller than in the case that the potential walls are infinitely high. The wave functions are cosine and sine shaped within the well and decrease exponentially outside the quantum well (Fig. 26.1, right).

The zero point energy of a quantum well with infinitely high walls varies as $1/s^2$ (Fig. 26.2, solid line). A quantum well, like a GaAs quantum well has walls of finite height. Therefore, the zero point energy is smaller (Fig. 26.2, dashed) but decreases also strongly with increasing layer thickness. A calculation of the zero point energy in the case that the potential walls have finite height is possible by applying appropriate boundary conditions taking into account the different effective masses of GaAs and AlAs (Sect. 30.5).

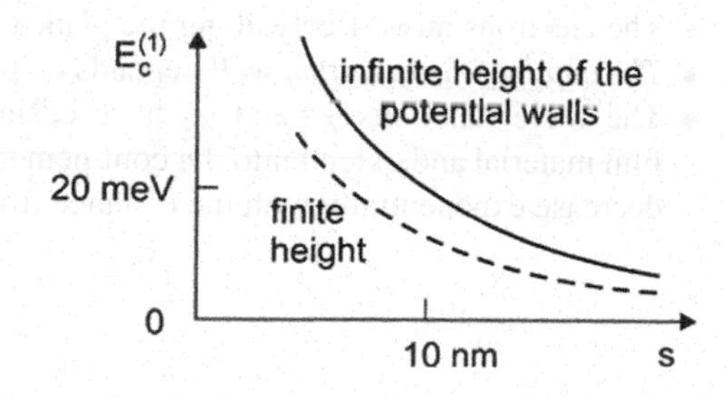

Fig. 26.2 Zero point energy of an electron ($m_e = 0.07\, m_0$) in a one-dimensional square well potential

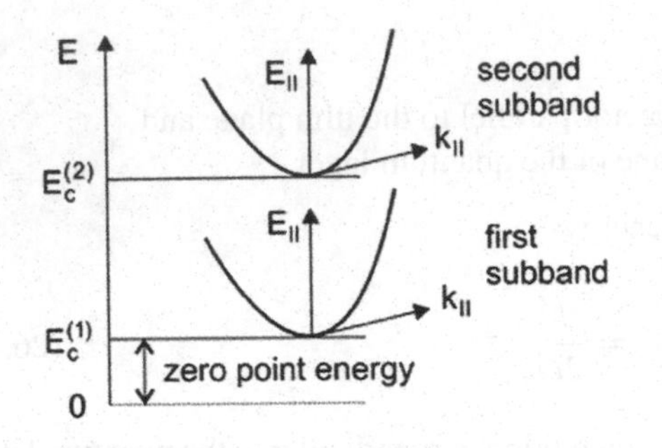

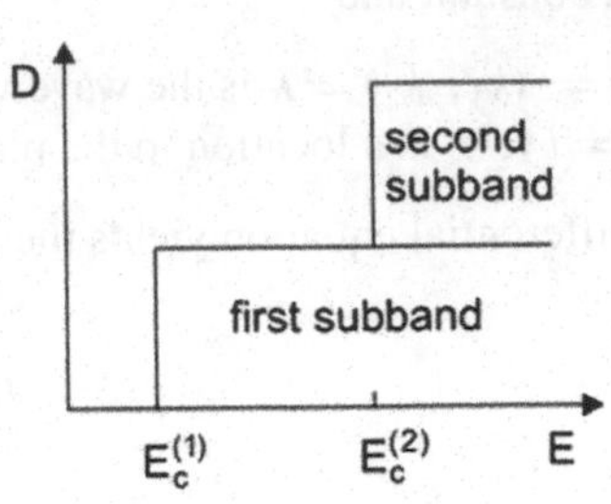

Fig. 26.3 Subbands of electrons in a quantum film

The energy of an electron is given by

$$E = E_c^{(n)} + E_{||}, \tag{26.13}$$

where n is the number of a subband. The conduction band of a quantum layer consists of *electron subbands*. Figure 26.3 (left) shows the zero point energy and the first and the second electron subband. The energy of the motion perpendicular to the layer is discrete while the motion within the quantum layer corresponds to the motion of a free-electron. The density of states in a subband (Fig. 26.3, right) is a constant (Problem 27.1):

$$D_c^{2D}(\epsilon) = \frac{m_e}{\pi \hbar^2}. \tag{26.14}$$

The total density of states is the sum of the densities of states in the different subbands.

That the density of states of electrons in a quantum film is independent of the thickness of the film is plausible: the film thickness determines the zero point energy, which is due to the lateral confinement of an electron, while the dispersion relation for electrons in a two-dimensional semiconductor determines the propagation along the plane.

The *quantum confinement* of an electron in a quantum film has consequences:

- *Subbands.*
- *Discrete energy values* for motion perpendicular to the quantum layer.
- *Zero point energy* ($E_c^{(1)}$). The value of $E_c^{(1)}$ depends on the thickness s of the quantum film.
- The electrons move freely along the plane of the quantum film.
- *The depth of the quantum well* depends on the composition of the GaAlAs layers.
- The wave functions χ_1 and χ_2 have cosine and sine shapes within the GaAs film material and extend into the confinement material GaAlAs. Their amplitudes decrease exponentially with the distance from the GaAs film.

26.2 Hole Subbands

GaAs and other group III–V semiconductors have three hole bands (Fig. 26.4): the heavy hole band; the light hole band; the split-off band. In a laser diode, the split-off band is completely populated and does not play any role. However, the light holes ($m_{\mathrm{lh}} \sim 0.08\, m_0$) influence the gain coefficient curves.

In a quantum well, the zero point energy of a heavy hole,

$$E_{\mathrm{v}}^{(1)} = \frac{\pi^2 \hbar^2}{2 m_{\mathrm{h}} s^2}, \tag{26.15}$$

is by the factor $m_{\mathrm{h}}/m_{\mathrm{e}}$ smaller than the zero point energy of an electron in the conduction band while the zero point energy $E_{\mathrm{v,lh}}^{(1)}$ of the light hole is comparable with the zero point energy of a conduction band electron since $m_{\mathrm{lh}} \sim m_{\mathrm{e}}$. There is an energy range, between $E_{\mathrm{v}}^{(1)}$ and $E_{\mathrm{v,lh}}^{(1)}$, without light hole energy levels. The density of states of light holes is much smaller than that of heavy holes.

The conduction band states of GaAs have their origin in s-like hybrid states composed of s-states of Ga and As atoms. The s-like hybrid states overlap spatially. This leads to electron waves extended over the whole crystal and to the conduction band (Sect. 30.2); the dispersion relation $E(\boldsymbol{k})$ characterizes the states of the conduction band. The valence band states stem from hybrid states composed of p-states (i.e., p_x, p_y, p_z states) of Ga and As atoms. The three p-state components give rise to three different energy bands and dispersion relations—i.e., a wave function of a valence band state can assume, for the same k vector, three different energy values. At the Brillouin zone center ($k = 0$), two of the dispersion curves have the same energy value indicating energy degeneracy; however, the two dispersion curves have completely different shapes described by the different effective masses,

$$m_{\mathrm{h}} \equiv m_{\mathrm{hh}} = (\mathrm{d}^2 E_{\mathrm{hh}}/\mathrm{d}^2 k)_{k=0}, \tag{26.16}$$

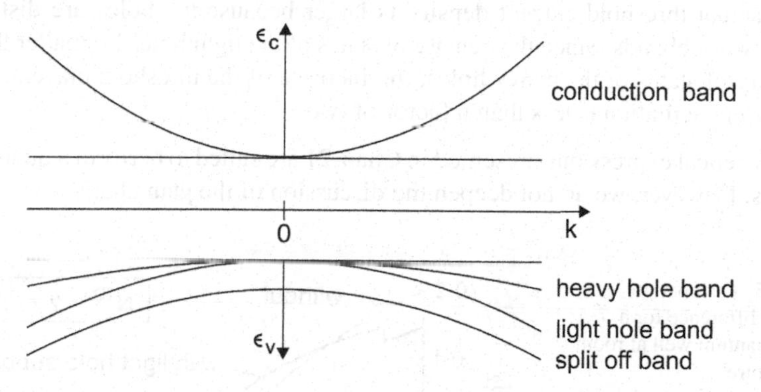

Fig. 26.4 Energy bands of electrons in GaAs: conduction band; heavy hole band; light hole band; split off band

$$m_{\mathrm{lh}} = (\mathrm{d}^2 E_{\mathrm{lh}}/\mathrm{d}^2 k)_{k=0}. \tag{26.17}$$

The third band, the split off band, is shifted to smaller energies relative to the two other valence bands due to spin-orbit interaction.

26.3 Modification of the Gain Characteristic by Light Holes

The quasi-Fermi energy of the electrons in the conduction band is determined by the density of nonequilibrium electrons in the electron subband. The quasi-Fermi energy of the electrons in the hole subbands follows from the condition

$$\int f_1(D_{\mathrm{v}}^{\mathrm{2D}} + D_{\mathrm{v,lh}}^{\mathrm{2D}})\mathrm{d}E = N^{\mathrm{2D}}, \tag{26.18}$$

where $D_{\mathrm{v}}^{\mathrm{2D}}$ is the density of states of heavy holes, $D_{\mathrm{v,lh}}^{\mathrm{2D}}$ is the density of states of light holes and f_1 is the Fermi function of the electrons in the heavy hole and light hole subbands. The equation yields the quasi-Fermi energy of the valence band electrons.

In our earlier treatment of quantum wells, we ignored the light hole band. Now, we discuss, qualitatively, the modification of a quantum well laser at room temperature that is due to the light hole band. Figure 26.5 shows (qualitatively) the occupation number difference $f_2 - f_1$ for a GaAs quantum well at room temperature, with $N^{\mathrm{2D}} = 2 \times 10^{16}\ \mathrm{m}^{-2}$. In comparison with the case of a single hole subband, we have a different situation:

- There are two peaks in the gain characteristic H^{2D}, which is proportional to $f_2 - f_1$.
- The gain curve has a larger width and the maximum of the gain characteristic has a smaller value.
- The actual threshold current density is larger because the holes are distributed over two subbands. Since the density of states of the light holes is smaller than the density of states of the heavy holes, the increase of the threshold current density for laser oscillation is less than a factor of two.

The theoretical expressions presented in Chap. 21 are suited to perform a quantitative analysis. However, we do not deepen the discussion of the gain characteristic.

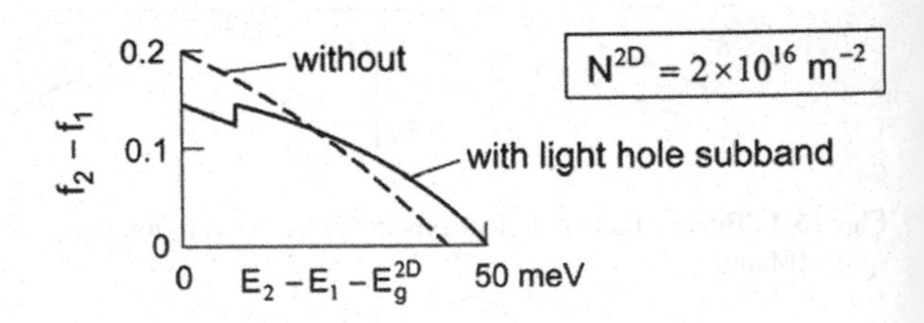

Fig. 26.5 Occupation number difference for a GaAs quantum well at room temperature

26.4 Gap Energy of a Quantum Well

Taking into account the zero point energy of electrons and holes, we find that the gap energy of a quantum well of infinitely high potential walls is equal to

$$E_{\mathrm{g}}^{2\mathrm{D}} = E_{\mathrm{g}} + \frac{\pi^2\hbar^2}{2m_r s^2}. \tag{26.19}$$

Example GaAs quantum film (thickness $s = 10\,\mathrm{nm}$) at room temperature ($\hbar = 1.04 \times 10^{-34}\,\mathrm{J\,s}$; $m_0 = 0.9 \times 10^{-30}\,\mathrm{kg}$).

- $E_{\mathrm{g}} = 1.42\,\mathrm{eV}$ = gap energy of bulk GaAs.
- $E_{\mathrm{g}}^{2\mathrm{D}} = E_{\mathrm{g}} + 61\,\mathrm{meV}$; energy gaps of a GaAs quantum well in the case of infinitely high walls.
- $\nu_g = 344\,\mathrm{THz}$; $\lambda_g = 872\,\mathrm{nm}$.
- $\nu_g^{2\mathrm{D}} = 358\,\mathrm{THz}$; $\lambda_g^{2\mathrm{D}} = 838\,\mathrm{nm}$.

The actual zero point energy is smaller because of the finite height of the walls. Through the choice of the composition of the quantum film and of the barrier material as well as of s, different values of the gap energy $E_{\mathrm{g}}^{2\mathrm{D}}$ ($> E_{\mathrm{g}}$) can be realized .

26.5 Temperature Dependence of the Threshold Current Density of a GaAs Quantum Well Laser

The electrons occupy energy levels in the electron subband in the range from $E_{\mathrm{c}}^{2\mathrm{D}}$ to E_{Fc}, with a spread of kT. The holes in the heavy hole subband occupy energy levels between $E_{\mathrm{v}}^{2\mathrm{D}}$ and E_{Fv}. With increasing temperature, the energy distribution of the electrons in the electron subband broadens. The energy distribution of the holes in the heavy hole subbands broadens too. It follows that the threshold current of a quantum well laser operating at room temperature is much larger than the threshold current of a quantum well laser operating at low temperature.

26.6 Gain Mediated by a Quantum Well with Inhomogeneous Well Thickness

When the thickness of a quantum film is different at different positions of the film, interband transitions are inhomogeneously broadened. Both the zero point energies $E_{\mathrm{c}}^{(1)}$ and $E_{\mathrm{v}}^{(1)}$ show variations. We obtain an energy gap distribution of a width $\Delta E_{\mathrm{g}}^{2\mathrm{D}}$ that is given by the relation

$$\frac{\Delta E_g^{2D}}{E_c^{(1)} + E_v^{(1)}} = -\frac{2\Delta s}{s}, \tag{26.20}$$

where Δs is an average variation of the thickness.
Example GaAs quantum well of thickness $s = 10\,\text{nm}$ and an inaccuracy of the well thickness of 0.1 nm. We obtain $E_c^{(1)} + E_v^{(1)} \sim 49\,\text{meV}$ and $\Delta E_g^{2D} \sim 1\,\text{meV}$. This is, for a GaAs quantum well at room temperature, smaller than the broadening (10 meV) that is due to the inelastic scattering of the electrons at phonons.

26.7 Tunability of a Quantum Well Laser

A single-mode quantum well laser operating at room temperature emits radiation at a laser frequency that is determined by the gain characteristic and the resonance frequency of the laser resonator. The position of the energy gap, the frequency of the maximum of the gain characteristic, and the refractive index of a semiconductor depend on temperature. Therefore, the frequency of laser oscillation depends on temperature. The temperature of a laser diode changes if the temperature of the surrounding or if the current through the diode is varied. A shift of several percent of the frequency of a quantum well laser can be achieved. Tuning over a small frequency range is possible by the use of an external resonator (Sect. 25.3).

26.8 Anisotropy of a Quantum Well

The quantum theory of the optical transitions in a quantum well shows that transitions in which heavy holes are involved are only allowed if the electric field vector of the electromagnetic field lies in the plane of the quantum well. If light holes are involved, transitions for the field vector parallel and perpendicular to the quantum well are allowed too.

References [187–192].

Problems

26.1 Two-dimensional density of states. Determine the density of states of a two-dimensional electron gas.

26.2 Subpicosecond quantum well laser. Is it possible to generate subpicosecond pulses with a quantum well laser? Divide the procedure of answering this question into three parts.

(a) Is it in principle possible to generate subpicosecond pulses with a quantum well laser?
(b) Is it possible to use a quantum well laser of 1 mm length, supposed that the reflector on one side is a SESAM reflector?
(c) Discuss a semiconductor laser that uses an external broadband reflector.

Chapter 27
Quantum Wire and Quantum Dot Laser

The next steps of spatial restriction of free motion of electrons in a semiconductor lead to quantum wire lasers and quantum dot lasers.

A quantum wire is a carrier of electrons, which can move freely in one dimension only. The electronic levels form subbands. The density of states of electrons and of holes are now characteristic of a one-dimensional semiconductor. We study gain mediated by quantum wires and the quantum wire laser.

In a quantum dot, electrons cannot perform a free motion. The motion of an electron is limited by a potential, which is formed by the boundary of a semiconductor material. (A comparison with electrons in an atom shows: the electrons in an atom cannot perform a free motion; the limitation of the motion is due to the atomic potential.) We discuss the energy levels of radiative electron-hole pairs in a quantum dot and their use in quantum dot lasers. A large number of quantum dots. A single quantum dot in a photonic crystal (at low temperature) could be the basis of a nanolaser.

27.1 Quantum Wire Laser

In the quantum wire laser (Fig. 27.1a), the active medium consists of a quantum wire, with a one-dimensional conduction band (the electron subband) and a one-dimensional valence band (the hole subband). Stimulated transitions of electrons in the conduction band to empty levels of the valence band are the origin of gain. The quantum wire (Fig. 27.1b) is embedded in an undoped semiconductor layer, which itself is sandwiched between an n-doped layer and a p-doped layer. The undoped layer serves as light guide. Under the action of a static voltage, electrons migrate from the n-doped layer through the undoped layer into the quantum wire. Correspondingly, holes migrate from the p-doped layer into the quantum wire. The direction of the quantum wire is perpendicular to the propagation direction of the electromagnetic field.

The use of a series of parallel quantum wires leads to an enhancement of gain.

K.F. Renk, *Basics of Laser Physics*, Graduate Texts in Physics,
DOI 10.1007/978-3-319-50651-7_27

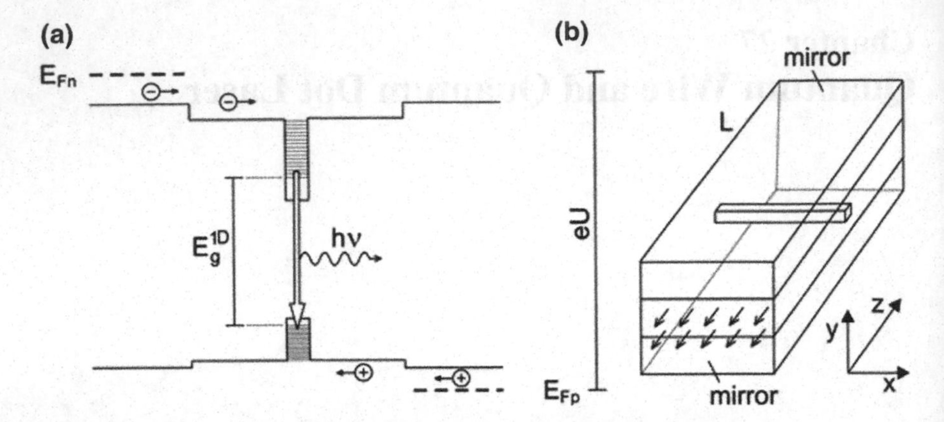

Fig. 27.1 Quantum wire laser. **a** Principle. **b** Arrangement

27.2 Quantum Wire

We write the wave function of an electron in a quantum wire of rectangular cross section in the form

$$\psi = \chi(x)\,\eta(y)\,\zeta(z). \tag{27.1}$$

The wave function

$$\chi(x) = C\exp\left(\mathrm{i}(\epsilon_{\mathrm{c}}/\hbar)t - \mathrm{i}kx\right) \tag{27.2}$$

characterizes the free-electron motion along the direction x (direction of the wire); k is the wave vector for propagation along the wire. The wave functions $\eta(y)$ and $\zeta(z)$ describe the electron states of the two other degrees of freedom of an electron. If infinitely high walls limit the quantum wire, the wave functions perpendicular to the wire are standing waves and the energy of an electron is

$$E_{\mathrm{c}}^{n_1 n_2} + \epsilon_{\mathrm{c}} = \frac{\pi^2\hbar^2}{2m_{\mathrm{e}}}\left(\frac{n_1^2}{s_1^2} + \frac{n_2^2}{s_2^2}\right) + \epsilon_{\mathrm{c}}. \tag{27.3}$$

$E_{\mathrm{c}}^{n_1 n_2}$ is the energy due to the confinement ($n_1 = 1, 2, \ldots; n_2 = 1, 2, \ldots$), s_1 and s_2 are the widths of the quantum wire and

$$\epsilon_{\mathrm{c}} = \frac{\hbar^2}{2m_{\mathrm{e}}}k^2 \tag{27.4}$$

is the energy of free motion of a conduction band electron. We assume, for simplicity, that the quantum wire is infinitely long and introduce periodic boundary conditions,

$$\chi(x + L_{\mathrm{p}}) = \chi(x). \tag{27.5}$$

L_p is the periodicity interval. We obtain discrete k values,

$$k = 0, \pm 2\pi/L_p, \pm 4\pi/L_p, \ldots. \tag{27.6}$$

The one-dimensional density of states (per unit of length) in k space is equal to

$$D_c^{1D}(k) = \frac{1}{2\pi}. \tag{27.7}$$

Taking into account that each orbital state can contain two electrons of opposite spin, we find, with $2D_c^{1D}(k)\mathrm{d}k = D_c^{1D}(\epsilon_c)\mathrm{d}\epsilon$,

$$D_c^{1D}(\epsilon_c) = \frac{1}{\pi\hbar}\sqrt{\frac{m_e}{2\epsilon_c}}. \tag{27.8}$$

The one-dimensional density of states is independent of the lateral extensions of the quantum wire. The density of states of electrons in a quantum wire varies as $1/\sqrt{\epsilon_c}$. It becomes infinitely large for $\epsilon_c = 0$.

The energy of a level in the valence band is given by

$$E_v^{n_1 n_2} - \epsilon_v = \frac{\pi\hbar^2}{2m_h}\left(\frac{n_1^2}{s_1^2} + \frac{n_2^2}{s_2^2}\right) - \epsilon_v, \tag{27.9}$$

where $\epsilon_v = \hbar^2 k^2/2m_h$ is the energy of free motion of a hole. The density of states in the valence band is

$$D_v^{1D}(\epsilon_v) = \frac{1}{\pi\hbar}\sqrt{\frac{m_h}{2\epsilon_v}}. \tag{27.10}$$

The reduced density of states is given by

$$D_r^{1D}(\epsilon) = \frac{1}{\pi\hbar}\sqrt{\frac{m_r}{2\epsilon}}, \tag{27.11}$$

where

$$\epsilon = \epsilon_c + \epsilon_v = \frac{\hbar^2 k^2}{2m_r} \tag{27.12}$$

is the sum of the energies of free motion of the electron and the hole belonging to a radiative electron-hole pair and m_r is the reduced mass. The energy separation between the levels belonging to the $k = 0$ states is

$$E^{n_1 n_2} = E_g + E^{n_1 n_2} + E_v^{n_1 n_2} = E_g + \frac{\pi^2\hbar^2}{2m_r}\left(\frac{n_1^2}{s_1^2} + \frac{n_2^2}{s_2^2}\right). \tag{27.13}$$

E_g is the gap energy of the bulk. The energy difference between the lowest level of the conduction band and the highest level of the valence band is the gap energy of the one-dimensional semiconductor,

$$E_g^{1D} = E_g + \frac{\pi^2 \hbar^2}{2m_r}\left(\frac{1}{s_1^2} + \frac{1}{s_2^2}\right). \tag{27.14}$$

It is the sum of the gap energy of the bulk, the zero point energy of the electron, and the zero point energy of the hole.

We now discuss radiative transitions. We assume that we have an ideal quantum wire (without a variation of the thicknesses s_1 and s_2). We consider a quantum wire cooled to low temperature and assume that the only level broadening is due to spontaneous radiative transitions. A radiative transition between a particular level of the electron subband and a level of the hole subband results in a broadened luminescence line. The linewidth is the natural linewidth ΔE_{nat} $(=\Delta\epsilon_{nat})$. The largest number of states within an energy interval $\Delta\epsilon_{nat}$ lies in the energy interval 0, $\Delta\epsilon_{nat}$. Making use of the integral $\int dx/\sqrt{x} = 2\sqrt{x}$, we obtain the density of radiative pair levels within the interval 0, $\Delta\epsilon_{nat}$,

$$n_{nat,0}^{1D} = \int_0^{\Delta\epsilon_{nat}} D_r^{1D}(\epsilon)d\epsilon = \frac{2}{\pi\hbar}\sqrt{2m_r\Delta\epsilon_{nat}}. \tag{27.15}$$

The density of radiative pair levels within the natural linewidth $\Delta\epsilon_{nat}$ has a maximum value for $\epsilon = 0$, i.e., for $h\nu = E_g^{1D}$. The maximum density is $n_{nat,0}^{1D}$. Toward higher energies, the density of levels within the natural linewidth decreases.

Example Subband of a GaAs quantum wire, in the limit of infinitely high potential walls.

- $m_r = 0.06\, m_0$.
- $s_1 = s_2 = 10\,\text{nm}$.
- $E^{11} = 122\,\text{meV}$ = zero point energy (of electrons and holes together).
- $E^{21} = 305\,\text{meV}$.
- $E_g^{1D} = E_g + E^{11} = (1.42 + 0.12)\,\text{eV} = 1.54\,\text{eV}$.
- $\nu_g = E_g^{1D}/h = 373\,\text{THz}$.
- $\lambda_g = c/\nu_g = 804\,\text{nm}$.
- $\tau_{sp} = 1/A_{21} = 3 \times 10^{-10}\,\text{s}$; spontaneous lifetime (at low temperature).
- $A_{21} = 3 \times 10^9\,\text{s}^{-1}$.
- $\Delta\nu_{nat} = (2\pi\tau_{sp})^{-1} = 6 \times 10^8\,\text{Hz}$; natural linewidth (due to spontaneous recombination of radiative pairs).
- $\Delta E_{nat} = h\Delta\nu_{nat} = 3.6 \times 10^{-25}\,\text{J} = 2\,\mu\text{eV}$; natural linewidth on the energy scale.
- $n_{nat,0}^{1D} = 1.3 \times 10^6\,\text{m}^{-1}$.

Because of the finite height of the walls, the actual zero point energy of electrons in a quantum wire and the value of E^{21} are smaller than we calculated for infinitely high potential walls.

27.3 Gain Mediated by a Quantum Wire

The average electron density in a disk of light (height a_1, length δz) crossing a quantum wire (*see* Fig. 27.1b) is equal to

$$N_{\mathrm{av}} = \frac{N^{1\mathrm{D}}}{a_1 \delta z}. \tag{27.16}$$

The temporal change of the photon density is given by

$$\frac{\delta Z}{\delta t} = -\frac{\delta N_{\mathrm{av}}}{\delta t} = \frac{(c/n)H^{1\mathrm{D}}}{a_2 \delta z} Z, \tag{27.17}$$

where

$$H^{1\mathrm{D}} = \frac{c}{n} h\nu \bar{B}_{21} D_{\mathrm{r}}^{1\mathrm{D}}(E_{21})(f_2 - f_1) \tag{27.18}$$

is the one-dimensional gain characteristic. The gain is, with $\delta t = \frac{n}{c}\delta z$, equal to

$$G - 1 = \frac{\delta Z}{Z} = \frac{H^{1\mathrm{D}}}{a_2} = \frac{n}{c} h\nu \bar{B}_{21} \frac{n_{\mathrm{nat},0}^{1\mathrm{D}}}{a_1 \Delta\epsilon_{\mathrm{nat}}} (f_2 - f_1). \tag{27.19}$$

The gain increases with decreasing height of the photon mode.
Example GaAs quantum wire embedded in GaAlAs with an ideal quantum wire at zero temperature.

- $\nu_{\mathrm{g}}^{1\mathrm{D}} = 373\,\mathrm{THz}$.
- $f_2 - f_1 = 0.1$.
- $a_1 = 0.2\,\mu\mathrm{m}$; $n = 3.5$.
- $B_{21} = 2.2 \times 10^{21}\,\mathrm{m}^3\,\mathrm{J}^{-1}\,\mathrm{s}^{-2}$; $\bar{B}_{21} = hB_{21}$.
- $G_1 - 1 = 7 \times 10^{-2}$.

Figure 27.2a shows the reduced density of states at energies near the bandgap ($\epsilon = 0$). Figure 27.2b exhibits the gain mediated by a weakly pumped quantum wire at zero temperature. The gain G − 1 is drawn versus the energy ϵ. The photon energy is equal to $h\nu = E_{\mathrm{g}}^{1\mathrm{D}} + \epsilon$. In the case that $\epsilon = 0$, the frequency is equal to the one-dimensional gap frequency $\nu_{\mathrm{g}}^{1\mathrm{D}}$. The gain bandwidth has a width (2 μeV) that is due to natural line broadening.

The electrons and holes in a quantum wire at room temperature have quasithermal energy distributions. The energy levels broaden due to inelastic scattering. We choose as inelastic scattering time a value ($\tau_{\mathrm{in}} = 10^{-12}$ s), which is by an order of magnitude smaller than that of a quantum well; a smaller value is expected because of the smaller density of states of the electrons, leading to a smaller probability of inelastic scattering of electrons at phonons. The corresponding halfwidth $\Delta E_{\mathrm{in}} \sim 1\,\mathrm{meV}$ is 500 times larger than $\Delta\epsilon_{\mathrm{nat}}$. The density of radiative pair levels, which contribute to transitions

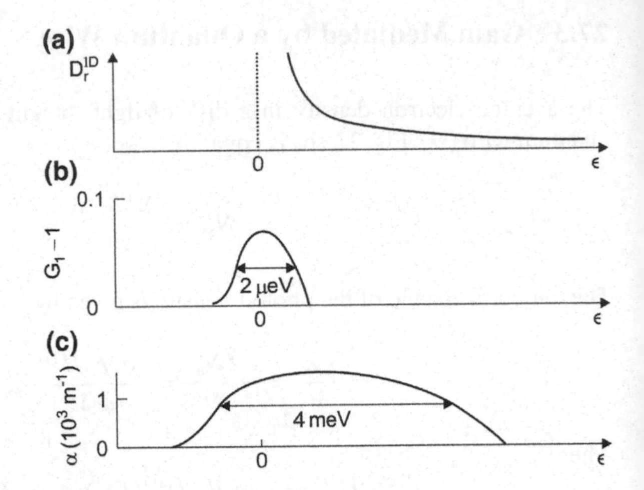

Fig. 27.2 GaAs quantum wire. **a** Density of states of radiative pair levels. **b** Gain mediated by a quantum wire at zero temperature. **c** Modal gain coefficient of a multi-quantum wire device at room temperature

within the energy range E_g^{1D}, $E_g^{1D} + \Delta E_{in}$, is equal to $n_{in,0}^{1D} = (\Delta E_{in}/\Delta\epsilon_{nat})^{1/2} n_{nat,0}^{1D}$. It follows that the gain in a quantum wire laser at room temperature, for $f_2 - f_1 = 0.1$, is $G_1 - 1 = 3 \times 10^{-3}$.

27.4 Multi Quantum Wire Laser

A much larger gain is obtainable for a series of quantum wires (Fig. 27.3), arranged in parallel (within a plane). The average electron density in a photon mode is

$$N_{av} = \frac{n_{nat,0}^{1D}}{pa_2}, \tag{27.20}$$

where p is the period of the quantum wires, i.e., the distance between neighboring quantum wires, arranged in parallel in a plane. The modal gain coefficient at the gap frequency (ν_g^{1D}) is, for $T = 0$, given by

$$\alpha_{nat}(\nu_g^{1D}) = \frac{n}{c} h\nu_g^{1D} \bar{B}_{21} \frac{n_{nat,0}^{1D}}{pa_2 \Delta\epsilon_{nat}} (f_2 - f_1). \tag{27.21}$$

Example For $p = 100$ nm, $a_2 = 200$ nm, and zero temperature, the modal gain coefficient is $\alpha(\nu_g^{1D}) = 7 \times 10^5\ \mathrm{m}^{-1}$. It follows, with $f_2 - f_1 = 0.1$, that the modal gain coefficient of the quantum wires at room temperature is $\alpha(\nu_g^{1D}) = 3.1 \times 10^3\ \mathrm{m}^{-1}$.

If the wire thickness varies along a quantum wire, the zero point energy varies too. This causes inhomogeneous broadening of the interband transitions. The relative broadening (for $s_1 = s_2 = s$) is equal to

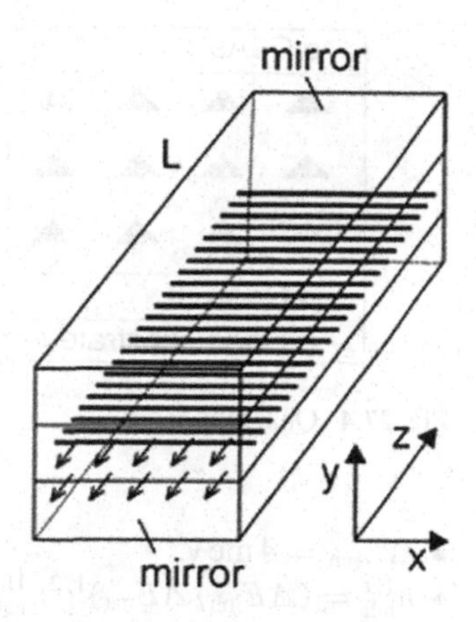

Fig. 27.3 Multi quantum wire laser

$$\Delta E_g^{1D}/E^{11} = 4\Delta s/s, \tag{27.22}$$

where $\Delta s/s$ is the relative variation of the wire thickness. Taking into account inhomogeneous broadening, we find the modal gain coefficient

$$\alpha_{inh}(\nu_g^{1D}) = \frac{n}{c} h\nu_g^{1D} B_{21} \frac{n_{inh}^{1D}}{pa_2 \Delta\epsilon_{inh}} (f_2 - f_1). \tag{27.23}$$

A variation $\Delta s/s = 5\%$ causes an energy variation of $\Delta E_g^{1D}/E_g^{11} \sim 20\%$, which corresponds to an inhomogeneous width $\Delta\epsilon_{inh} = \Delta E_g^{1D} \sim 4\,\text{meV}$. This value, which is larger than the broadening due to inelastic scattering at room temperature, leads (with the parameters we have chosen in the last example) to $\alpha(\nu_g^{1D}) = 1.5 \times 10^3\,\text{m}^{-1}$ (for $f_2 - f_1 = 0.1$).

If $f_2 - f_1 \ll 1$, then the threshold density is only slightly larger than the one-dimensional transparency density. Then, the laser threshold current is (approximately) equal to

$$I_{th} = \frac{L}{p} \frac{N_{th}^{1D} a_1 e}{\tau_{sp}}, \tag{27.24}$$

where a_2 is the width of the resonator (that is equal to the length of the quantum wires). The threshold current density is

$$j_{th} = \frac{N^{1D} e}{p \tau_{sp}}. \tag{27.25}$$

Example GaAs multi-quantum wire laser at room temperature.

- $\nu = 373\,\text{THz}$.
- $a_1 = 100\,\mu\text{m}$; $a_2 = 200\,\text{nm}$; $L = 1\,\text{mm}$.
- $p = 100\,\text{nm}$.

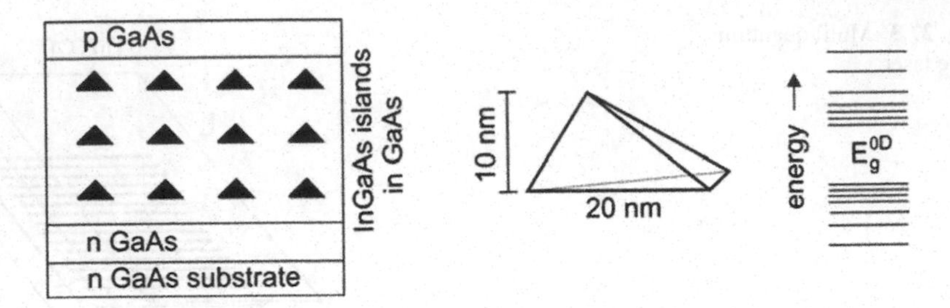

Fig. 27.4 Quantum dot

- $\Delta E_{\text{inh}} = 4\,\text{meV}$.
- $n_{\text{inh}}^{1\text{D}} = (\Delta E_{\text{inh}}/\Delta E_{\text{nat}})^{1/2} n_{\text{nat},0}^{1\text{D}} = 3 \times 10^7\,\text{m}^{-1}$.
- $f_2 - f_1 = 0.1$.
- $N_{\text{tr}}^{1\text{D}} \sim 10^8\,\text{m}^{-1}$.
- $B_{21} = 2.2 \times 10^{21}\,\text{m}^3\,\text{J}^{-1}\,\text{s}^{-2}$; $\bar{B}_{21} = hB_{21}$.
- $\alpha(\nu_g^{1\text{D}}) = 1.5 \times 10^3\,\text{m}^{-1}$.
- $G_1 = 4.5$.
- $\tau_{\text{sp}} = 10^{-9}\,\text{s}$.
- $I_{\text{th}} = 160\,\mu\text{A}$; $j_{\text{th}} = 1.6 \times 10^5\,\text{A}\,\text{m}^{-2}$.

Figure 27.2c shows the modal gain coefficient. According to our estimate, the gain coefficient is large enough for operation of a quantum wire laser at room temperature as an edge emitting laser (without special coatings on the surfaces).

Because of the anisotropy of a quantum wire, the orientation must be parallel to the field; the Einstein coefficient B_{21} and the gain coefficient is larger (by a factor of 3) than we estimated. (For the orientation of the quantum wire perpendicular to the field, transitions that involve higher subbands or light holes can occur.)

27.5 Quantum Dot

Quantum dot lasers are currently being developed and may become important for optical communications (by the use of laser radiation at wavelengths 1.32 and 1.55 μm) and for other applications.

To prepare quantum dots (Fig. 27.4, left), one can make use of the mismatch between two materials. The deposition of a small amount of GaInAs on a plane GaAs surface results in formation of GaInAs islands. Further deposition of GaAs fills up the space between the islands and the GaAs surface becomes plane again. The result is a layer with a large number of quantum dots. By further deposition of the two materials, it is possible to obtain further layers with quantum dots. A quantum dot can have a pyramidal shape (Fig. 27.4, center) of extensions of the order of 10 nm. The motion of electrons and holes is restricted with respect to all three spatial directions. Therefore, the energy levels are discrete.

To understand the main properties of a quantum dot laser, we make a few simplifications.

- We consider a quantum dot of rectangular shape; s_i is the side length of the ith side of the dot ($i = 1, 2, 3$).
- We treat the quantum dot as a three-dimensional quantum box with infinitely high walls.

The energy levels of an electron are

$$E_{\mathrm{c}}^{\mathrm{n_1 n_2 n_3}} = \frac{\pi^2 \hbar^2}{2m_{\mathrm{e}}} \left(\frac{n_1^2}{s_1^2} + \frac{n_2^2}{s_2^2} + \frac{n_3^2}{s_3^2} \right), \quad (27.26)$$

where $n_1 = 1, 2, \ldots; \;\; n_2 = 1, 2, \ldots; \;\; n_3 = 1, 2, \ldots$.

The lowest conduction band energy level is, for $s_1 = s_2 = s_3 = s$, equal to

$$E_c^{111} = \frac{3\pi^2 \hbar^2}{2m_{\mathrm{e}} s^2}. \quad (27.27)$$

The next higher level is $E_{\mathrm{c}}^{211} = 1.7 E_{\mathrm{c}}^{111}$. The energy levels of electron as well as of holes are discrete (Fig. 27.4, right).

The lowest energy of a radiative electron-hole pair (for $s_1 = s_2 = s_3 = \mathrm{s}$) is the gap energy of the zero-dimensional system,

$$E_{\mathrm{g}}^{\mathrm{0D}} = E_{\mathrm{g}} + \frac{3\pi^2 \hbar^2}{2m_{\mathrm{r}} s^2}, \quad (27.28)$$

where m_{r} is the reduced mass.

Example GaAs quantum dot with $s_1 = s_2 = s_3 = 20\,\mathrm{nm}$. $E_{\mathrm{c}}^{111} \sim 45\,\mathrm{meV}$; $E_{\mathrm{c}}^{211} \sim 90\,\mathrm{meV}$; $E_{\mathrm{v}}^{111} \sim 7\,\mathrm{meV}$; $E_{\mathrm{v}}^{211} \sim 14\,\mathrm{meV}$; $E_{\mathrm{g}}^{\mathrm{0D}} = E_{\mathrm{g}} + 52\,\mathrm{meV}$.

27.6 Quantum Dot Laser

In the quantum dot laser (Fig. 27.5), electrons migrate from an n-doped region via an undoped region into the dots. Holes migrate from a p-doped region via the undoped region into the dots. Stimulated recombination of radiative electron-hole pairs within the quantum dots is the source of laser radiation. It is possible to prepare a quantum dot laser as an edge emitting laser or as a surface emitting laser. In the following, we treat the edge emitting laser.

We consider an array of quantum dots. The number of dots per unit of area is $1/(p_1 p_2)$, where p_1 and p_2 are the periods in x and y direction. Transitions between

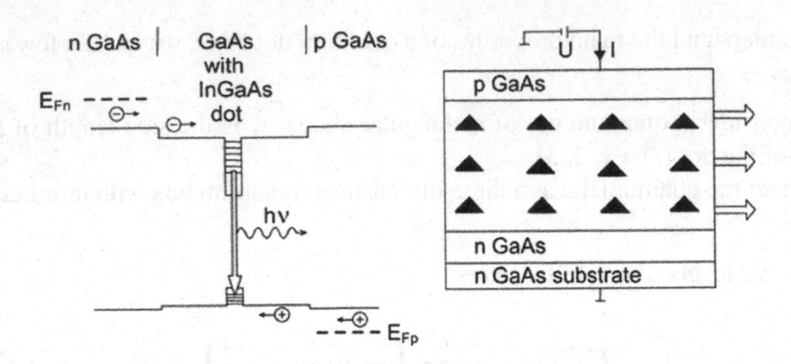

Fig. 27.5 Quantum dot laser

the electron sublevel 111 and the hole sublevel 111 lead to the modal gain coefficient

$$\alpha = \frac{2N_{\rm dot}^{\rm 2D}}{a_1 c/n} h\nu \bar{B}_{21} g(h\nu - E_{21})\,(f_2 - f_1). \tag{27.29}$$

$N_{\rm dot}^{\rm 2d} = 1/(p_1 p_2)$ is the two-dimensional density of quantum dots (=number of dots per m^2). The factor 2 accounts for the possibility that two electrons of opposite spin orientation occupy a level.

Example GaAs quantum dot laser at room temperature.

- $B_{21} = 2.2 \times 10^{21}\,\mathrm{m^3\,J^{-1}\,s^{-2}}$; $\bar{B}_{21} = hB_{21}$.
- $a_1 = 100\,\mu\mathrm{m}$; $a_2 = 200\,\mathrm{nm}$; $n = 3.5$.
- $\nu = 3.5 \times 10^{14}\,\mathrm{Hz}$.
- $p_1 = p_2 = 100\,\mathrm{nm}$; $N_{\rm dot}^{\rm 2D} = 10^{14}\,\mathrm{m^{-2}}$.
- $\Delta s_1/s_1 = \Delta s_2/s_2 = \Delta s_3/s_3 = 5\%$.
- $\Delta E_{\rm inh} = 8\,\mathrm{meV}$.
- $g(h\nu - E_{21}) = 2\pi/\Delta E_{\rm inh}$ for $h\nu = E_{\rm g}^{\rm 0D}$.
- $f_2 - f_1 = 0.5$.
- $\alpha = 1000\,\mathrm{m^{-1}}$.
- $L = 2\,\mathrm{mm}$.
- $\alpha L \sim 2$.
- $G_1 V_1 = 1.4$.

The threshold current is given by

$$I_{\rm th} = 2N_{\rm dot}^{\rm 2D} a_1 L e/\tau_{\rm sp}. \tag{27.30}$$

We find, with $\tau_{\rm sp} = 10^{-9}$ s, the threshold current $I_{\rm th} \sim 3\,\mathrm{mA}$.

27.7 One-Quantum Dot Laser

We deposit a single quantum dot in a defect of a photonic crystal (Fig. 27.6), which serves as high-Q resonator. The size of the resonator is $a_1 a_2 L$. What is the threshold condition?

The photon generation rate has to be equal to the photon loss rate,

$$\frac{2}{a_1 a_2 L}(f_2 - f_1) h\nu \bar{B}_{21} g(h\nu - E_{21}) Z = \frac{Z}{\tau_p}. \tag{27.31}$$

Laser oscillation at the line center ($h\nu = E_{21}$), where $g(h\nu - E_{21}) = 2\pi/\Delta E_{21}$, is possible if $f_2 - f_1 > (f_2 - f_1)_{th}$, where

$$(f_2 - f_1)_{th} = \frac{\pi a_1 a_2 L \Delta E_{21}}{4 h\nu \bar{B}_{21} \tau_p} \tag{27.32}$$

is the threshold population difference. If $(f_2 - f_1)_{th}$ is given, then the minimum lifetime of a photon necessary to reach the threshold condition is equal to

$$\tau_{p,th} = \frac{\pi a_1 a_2 L \Delta E_{21}}{4 h\nu \bar{B}_{21} (f_2 - f_1)_{th}}. \tag{27.33}$$

Example One-quantum dot laser at 4 K.

- $a_1 = a_2 = L = 0.5\,\mu\text{m}$.
- $\nu = 3.5 \times 10^{14}$ Hz; $n = 3.5$.
- $\tau_{sp} = 0.6\,\text{ns} = A_{21}^{-1}$.
- $\Delta E_{nat} = \hbar/\tau_{sp} = 2 \times 10^{-24}\,\text{J} \sim 1\,\mu\text{eV}$.
- $B_{21} = 1.1 \times 10^{21}\,\text{m}^3\,\text{J}^{-1}\,\text{s}^{-2}$; $\bar{B}_{21} = h B_{21}$.
- $(f_2 - f_1)_{th} = 0.2$.
- $\tau_{p,th} = 2 \times 10^{-12}$ s.
- $Q_{min} = 2\pi\nu\tau_{p,th} = 4{,}000$.

Operation of a one-quantum dot laser at 4 K requires a Q factor of the order of ten thousand.

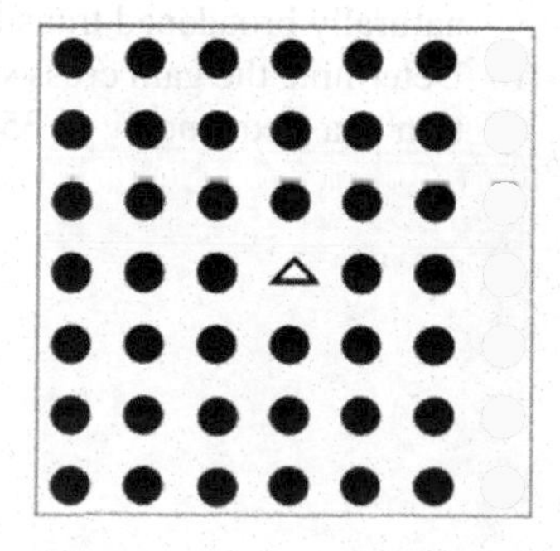

Fig. 27.6 One-quantum dot laser

A one-quantum dot laser at low temperature (2 K or 4 K) operates by the use of a naturally broadened transition line. The laser oscillation involves two levels only.

References [221, 222].

Problems

27.1 Density of states of electrons in a quantum wire. Derive the density of states of electrons in a quantum wire.

27.2 Absorption by a quantum wire. Estimate the maximum modal absorption coefficient in the case that monochromatic light propagates in an optical waveguide of 0.2 μm height and 10 μm width along a quantum wire ($n = 3.5$). [*Hint*: assume that the cross section of the GaAs quantum wire shows no variation and that the quantum wire has a low temperature (near 0 K); transitions occur if heavy holes are involved.]

27.3 Calculate the threshold current of a nanolaser that contains one quantum dot and operates (at low temperature) at an occupation number difference $f_2 - f_1 = 0.8$.

27.4 Bipolar semiconductor laser as a two-level laser. We can describe a bipolar semiconductor laser as a two-level laser. The ground state is the vacuum level $E_{\mathrm{vac,pair}}$ of pairs and the excited states are radiative pair states. The pair levels are populated by injection of electrons and holes into the active medium.

(a) Characterize the density of states of radiative pairs.
(b) Determine the gain.
(c) Formulate the condition of gain.
(d) Apply the bipolar picture to a single radiative pair level in a quantum dot (at low temperature) that is continuously pumped. Determine the lineshape of the radiation emitted by spontaneous electron-hole recombination.

27.5 Laser operating on a naturally broadened line.

(a) Why is the quantum dot laser (operated at low temperature) so special with respect to line broadening? Is there any other cw laser operating by use of a naturally broadened transition between two laser levels?
(b) Determine the gain cross section of a quantum dot at low temperature for radiation at a frequency $\nu = 350\,\mathrm{THz}$.

Chapter 28
A Comparison of Semiconductor Lasers

The simplest heterostructure laser is the double heterostructure laser. A GaAs double-heterostructure consists of three layers: n-doped GaAlAs; GaAs; p-doped GaAlAs. It corresponds to two heterostructures, a GaAlAs/GaAs and a GaAs/GaAlAs heterostructure that have in common the GaAs layer. The GaAs layer forms a well—not a quantum well. The well width is so large that the electrons and the holes can, in principle, move freely in all three dimensions. The double-heterostructure also acts as a light guide. The successful realization of the double heterostructure laser initiated the development of the semiconductor lasers with the more complex heterostructures that we discussed.

The junction laser (=homostructure laser = homojunction laser) was the first semiconductor laser type. A GaAs junction laser consists of an n-doped GaAs layer in direct contact to a p-doped GaAs layer. Without applied voltage, the contact (=junction) region is a depletion layer. The contact region does not contain free-electrons or free holes. A voltage applied across the junction causes a drift of electrons from the n-doped GaAs into the depletion layer and, at the same time, a drift of holes from the p-doped GaAs into the depletion layer. Recombination of electrons and holes in the depletion layer by stimulated optical transitions is the origin of laser radiation. The gain region provides a weak light guiding effect. Junction lasers, cooled to liquid nitrogen temperature, are available in the infrared spectral range up to wavelengths of about 30 μm. However, quantum cascade lasers are taking over the tasks of infrared junction lasers.

We show how the laser threshold decreased since the realization of the first semiconductor lasers.

Finally, we will present a comparison of different types of semiconductor lasers, including the quantum cascade laser (that we will discuss in the next chapter). In a spectral range—called the terahertz gap—semiconductor lasers (quantum cascade lasers) are presently in development.

Heterostructures made it possible to design artificial, spatially varying energy bands, which is the basis of the many different types of semiconductor lasers. In 1874,

K.F. Renk, *Basics of Laser Physics*, Graduate Texts in Physics,
DOI 10.1007/978-3-319-50651-7_28

Ferdinand Braun introduced the *contact* between two materials as an important physical object. He found that the strength of current flowing in one direction through a contact between a metal and a conducting crystal was different from the strength of current flowing in the other direction at opposite voltage across the contact. This effect led to the first device suitable for rectification of high frequency radiation. Later, a contact between n-doped germanium and p-doped germanium was the basis of semiconductor junction transistors discovered by Bardeen, Brattain and Shockley. Making use of heterostructures instead of contacts represented an essential change in semiconductor physics. This change began by 1960 and resulted in a miniaturization of electronic devices.

28.1 Gain of Radiation in a Bulk Semiconductor

In the double heterostructure laser and the junction laser, the extensions of the active medium are large in all three dimensions. The electrons form a three-dimensional electron gas and the holes a three-dimensional hole gas.

The density of states of a three-dimensional electron gas, including the spin degeneracy, is equal to

$$D_{\mathrm{c}}(\epsilon_{\mathrm{c}}) = \frac{1}{2\pi^2}\left(\frac{2m_{\mathrm{e}}}{\hbar^2}\right)^{3/2} \epsilon_{\mathrm{c}}^{1/2}. \tag{28.1}$$

The quasi-Fermi energy of a three-dimensional electron gas follows from the relation

$$\frac{1}{2\pi^2}\left(\frac{2m_{\mathrm{e}}}{\hbar^2}\right)^{3/2} \int_0^{\infty} \frac{\epsilon^{1/2}\mathrm{d}\epsilon}{\exp\left[(\epsilon - \epsilon_{\mathrm{Fc}})/kT + 1\right.} = N. \tag{28.2}$$

N is the density of electrons in the conduction band. For $T = 0$, we have the relation

$$\int_0^{\epsilon_{\mathrm{Fc}}} \epsilon_c^{1/2}\mathrm{d}\epsilon = \frac{2}{3}\epsilon_{\mathrm{Fc}}^{3/2} \tag{28.3}$$

and the quasi-Fermi energy is given by

$$\epsilon_{\mathrm{Fc}}(T = 0) = \frac{\hbar^2}{2m_{\mathrm{e}}}(3\pi^2 N)^{2/3}. \tag{28.4}$$

The reduced density of states,

$$D_{\mathrm{r}}(\epsilon) = \frac{1}{2\pi^2}\left(\frac{2m_{\mathrm{c}}}{\hbar^2}\right)^{3/2} \epsilon^{1/2}, \tag{28.5}$$

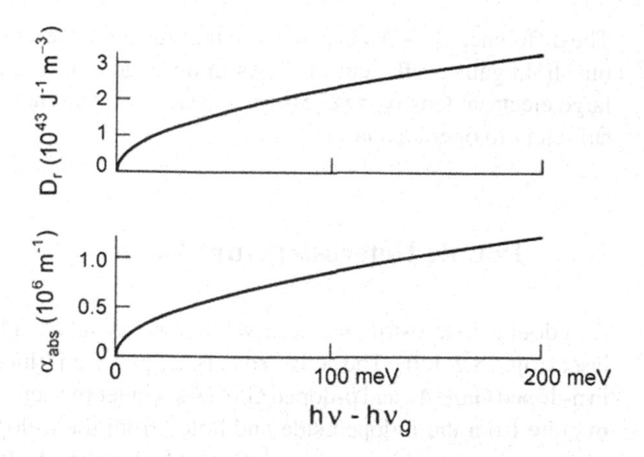

Fig. 28.1 Reduced density of states and absorption coefficient of bulk GaAs

is proportional to $\sqrt{E - E_g}$. The absorption coefficient for $T = 0$ is

$$\alpha_{abs} = \frac{n}{c} h\nu \bar{B}_{21} D_r(E_{21}). \tag{28.6}$$

Figure 28.1 shows the reduced density of states and the absorption coefficient for GaAs at $T = 0$ on the energy scale, with $h\nu = E_{21}$ and $h\nu_g = E_g$. The gain coefficient of a crystal containing nonequilibrium electrons and holes is equal to

$$\alpha = \frac{n}{c} h\nu \bar{B}_{21} D_r(E_{21})(f_2 - f_1), \tag{28.7}$$

where $f_2 = f(E_2)$, $f_1 = f_1(E_1)$, and $E_2 - E_1 = E_{21}$. Injection of electrons into the conduction band (and of holes into the valence band) of a crystal at room temperature has to be sufficiently strong to reach the transparency density. The maximum gain coefficient of GaAs at $T = 300$ K is

$$\alpha_{max} = \sigma_{eff}(N - N_{tr}), \tag{28.8}$$

where $N_{tr} \sim 2 \times 10^{24}$ m^{-3} is the transparency density and $\sigma_{eff} = 1.5 \times 10^{-20}$ m^2 the effective gain cross section [6]. Already a small increase of N above N_{tr} results in a large gain coefficient.

Example Gain coefficient of GaAs at 300 K.

- $N_{tr} = 2 \times 10^{24}$ m^{-3}.
- $\sigma_{eff} = 1.5 \times 10^{-20}$ m^2.
- $N - N_{tr} = 0.1 \times 10^{24}$ m^{-3}.
- $\alpha_{max} = 1.5 \times 10^3$ m^{-1}.

The difference $N - N_{tr}$ is chosen so that the gain coefficient α_{max} corresponds to the threshold gain coefficient of GaAs in an edge emitting laser. The data show that a large electron density is necessary to reach transparency. A slightly larger density is sufficient to operate a laser.

28.2 Double Heterostructure Laser

As a double heterostructure laser we discuss a GaAs/GaAlAs double heterostructure laser (Fig. 28.2, left). The active zone is a GaAs well (thickness 0.2–1 μm) embedded in n-doped GaAlAs and p-doped GaAlAs. Under the action of a voltage (U), electrons migrate from the n-doped side and holes from the p-doped side into the well. This results in nonequilibrium populations of electrons in the conduction band and of holes in the valence band. Stimulated recombination of radiative electron-hole pairs leads to laser radiation.

A double heterostructure laser diode (Fig. 28.2, right) consists of different layers forming an n GaAlAs/GaAs/p GaAlAs heterostructure grown on a highly doped GaAs substrate (doped with silicon; electron concentration $2 \times 10^{24}\,\mathrm{m}^{-3}$). The GaAs layer, with a larger refractive index than the neighboring GaAlAs material, acts as a light guide. The crystal surfaces perpendicular to the light guide serve as reflectors. The threshold current is, for $N \approx N_{tr}$,

$$I_{th} \approx N_{tr} a_1 a_2 L e / \tau_{sp}. \tag{28.9}$$

Example Double heterostructure GaAs laser.

- $a_1 = 100\,\mu\mathrm{m}$; $a_2 = 0.2\,\mu\mathrm{m}$; $L = 0.5\,\mathrm{mm}$.
- $G_1 V = 1$.
- $\tau_{sp} \sim 3\,\mathrm{ns}$.
- $I_{th} \sim 2.1\,\mathrm{A}$; $j_{th} \sim 2.1 \times 10^7\,\mathrm{A\,m^{-2}}$.

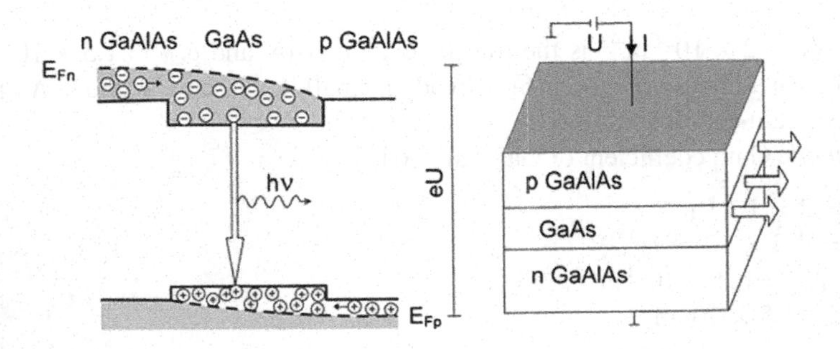

Fig. 28.2 Double heterostructure laser

The critical current is much larger than that of a quantum well laser. Below threshold, the double heterostructure emits luminescence radiation (electro-luminescence radiation). Above laser threshold, the quasi-Fermi energies of electrons and holes remain at their threshold values. Accordingly, the luminescence spectrum remains unchanged when the current exceeds the threshold current (Sect. 21.7).

28.3 GaAs Junction Laser

The junction laser (=homostructure junction laser = homojunction laser) contains nonequilibrium electrons and holes in a junction within a homogenous semiconductor material.

The GaAs junction laser (Fig. 28.3a, b) consists of a GaAs crystal, with n-doped GaAs adjacent to p-doped GaAs. A voltage (U) causes a current flow (strength I). Electrons move from one side and holes from the other side into the junction region where they recombine by emission of laser radiation. The emission wavelength of a GaAs junction laser lies in the near infrared (860 nm). Without applied voltage, the Fermi energy E_F has everywhere within the crystal the same value (Fig. 28.3c). In the junction region, there is a depletion zone (thickness ~1 μm for carrier densities of 10^{23} m^{-3} in the n-doped and the p-doped regions) in which no free carriers arc present. E_{Fc} lies above E_c in n GaAs and E_{Fv} lies below E_v in p GaAs.

Example GaAs junction laser at 300 K.

- $a_1 = 100$ μm; $a_2 = 1$ μm; L = 1 mm.
- $G_1 V = 1$ at $\alpha \sim 1.5 \times 10^3$ m^{-1}.
- $\tau_{sp} \sim 3$ ns.

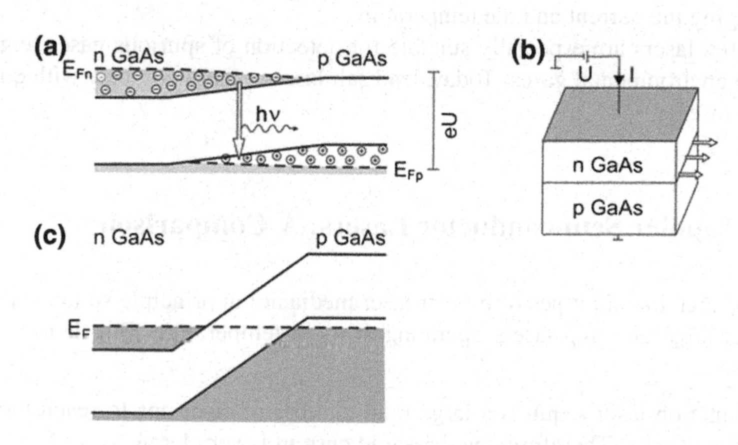

Fig. 28.3 GaAs junction laser. **a** Principle. **b** Device. **c** An n GaAs/p GaAs junction without applied voltage

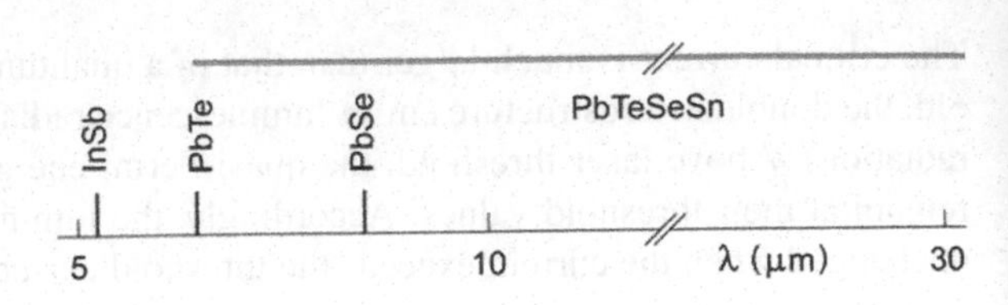

Fig. 28.4 Frequency regions of lead salt lasers

- $I_{th} \sim 10\,A$.
- $j_{th} \sim 10^8\,A\,m^{-2}$.

A junction laser operating at room temperature requires strong cooling.

In the junction laser, the active layer has a slightly larger refractive index than the surrounding n-doped GaAs and p-doped GaAs; the difference of the refractive indices is about 0.02. Therefore, there is a (weak) light guiding effect.

28.4 Junction Lasers in the Infrared

Infrared junction lasers (Fig. 28.4) consist of mixed crystals of lead salts. Lead salts have small energy gaps (PbS, $E_g \sim 270\,meV$; PbTe, 170 meV; PbSe, 130 meV). Mixed crystals of lead salts and tin salts have still smaller gap energies. The lead salt lasers operate at low temperature, at the temperature of liquid nitrogen or at lower temperature. In principle, it is possible to build a lead salt laser that generates radiation at a specific wavelength in a large range (4–30 μm). Nonradiative relaxation due to electron-hole recombination via phonons limits the wavelength range of lead salt lasers at large wavelengths.

A lead salt laser is tunable on a single mode over a very small frequency range by changing the current and the temperature.

Infrared lasers are especially suitable for detection of spurious gases (e.g., NO, NO_2) in environmental gases. Today, lead salt lasers cannot compete with quantum cascade lasers.

28.5 Bipolar Semiconductor Lasers: A Comparison

We have seen that all types of bipolar laser media are in principle suitable as active media of edge emitting lasers operating at room temperature. But there are great differences.

- The junction laser requires a large total number of electrons to reach the transparency density. Therefore, the threshold current is very large.
- The double heterojunction laser also requires a large total number of electrons. The light guiding effect is more favorable than for the junction laser.

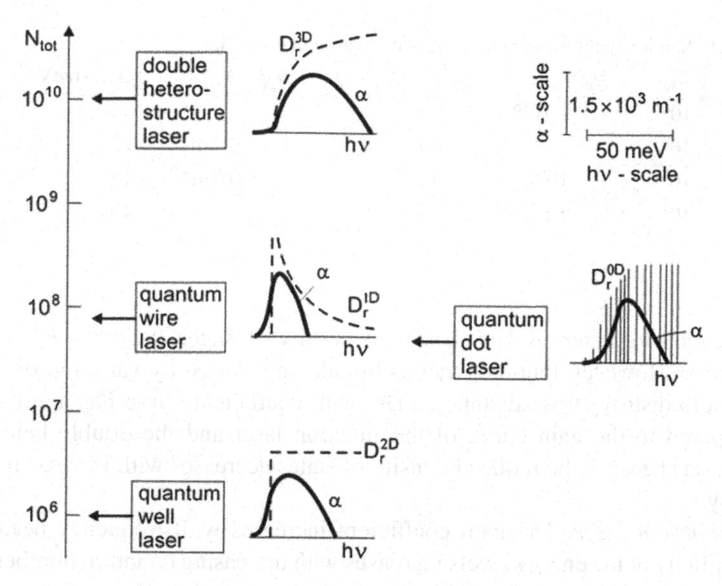

Fig. 28.5 A comparison of different bipolar semiconductor lasers at 300 K

- The quantum well laser requires a much lower total number of electrons. Preparation of quantum well lasers is possible by mass production: quantum well lasers of a high reliability are available.
- The quantum wire laser is in an early state of development.
- The quantum dot laser is being developed. The quantum dot lasers can at present not yet compete with the quantum well lasers.

We perform a quantitative comparison of different bipolar lasers operating at room temperature (Fig. 28.5). We ask how many excited electrons are necessary to drive a bipolar laser in an edge-emitting arrangement that is the same for each of the laser types ($a_1 a_2 L = 100\,\mu\text{m} \times 200\,\text{nm} \times 1\,\text{mm} = 2 \times 10^{-14}\,\text{m}^3$). The arrows indicate the total number $N_{\text{tot}} = N_{\text{tr}} a_1 a_2 L$ of electrons necessary to reach transparency density; the total number of electrons at laser threshold, $N_{\text{th,tot}}$, is only slightly (10%) larger than N_{tot}. We take into account the main broadening mechanisms. The figure shows, for each type of lasers, the shape of the reduced density of state curves $D_{\text{r}}(h\nu)$ and the $\alpha(h\nu)$ curves. All $\alpha(h\nu)$ curves have the same α scale and the same $h\nu$ scale. The maximum of the gain coefficient curve is equal to threshold gain coefficient. An increase of the number of excited electrons (for example, to ten times the total number at transparency) allows laser oscillation to occur at frequencies belonging to the whole gain bandwidths (=halfwidts of the gain coefficient curves). The survey indicates the following.

- *Double heterostructure laser*. Because of the three-dimensionality, a large number of electrons is necessary for the band filling.

Table 28.1 Semiconductor lasers (at a frequency around 400 THz)

Laser	N_{tot}	N_{tr}	$N_{th}/N_{tr}-1$	N_{dot}^{2D} [N_{wire}^{2D}]	$\Delta\nu_g$ (meV)	(THz)
Qu well	10^6	$1.4\times10^{16}\,m^{-2}$	0.1		30	7
Qu dot	10^8		0.1	$1\times10^{14}\,m^{-2}$	22	5
Qu wire	10^8	$1\times10^{8}\,m^{-1}$	0.1	[$5\times10^{6}\,m^{-2}$]	17	4
Bulk	10^{10}	$2\times10^{24}\,m^{-3}$	0.1		40	9

- *Quantum wire laser*. A large reduced density of states (at $h\nu = E_g^{1D}$) is very favorable. However, inhomogeneous broadening caused by variation of the wire thickness distroys this advantage. The gain coefficient curve has a small width (compared to the gain curve of the junction laser and the double heterostructure laser) because the reduced density of states decreases with increasing photon energy.
- *Quantum dot laser*. The gain coefficient increases with frequency because the multiplicity of the energy levels increases with increasing quantum numbers of the energy levels. The gain curve is continuous because of inhomogeneous broadening.
- *Quantum well laser*. The quantum well laser operates with the smallest number of electrons. Because of the constant two-dimensional density of states, inhomogeneous broadening is much less effective than for the quantum wire laser and the quantum dot laser.

Table 28.1 shows data used to compare different bipolar lasers (frequency 400 THz; length L = 1 mm; crystal surfaces as reflectors; width of the active medium 100 μm; height of the photon mode 200 nm); the data have been estimated in the preceding sections or chapters. The table lists the quantities:

- N_{tot} = total number of excited electrons necessary to fulfill the threshold condition.
- N_{tr} = transparency density.
- $(N_{th}-N_{tr})/N_{tr}$ = threshold density minus transparency density, divided by the transparency density.
- N_{dot}^{2D} = two-dimensional density of quantum dots in a layer of quantum dots.
- N_{wire}^{2D} = density of quantum wires in a layer of quantum wires.
- $\Delta\nu_g$ = gain bandwidth.

28.6 Development of Semiconductor Lasers

The threshold current of semiconductor lasers (Fig. 28.6) has been strongly reduced since the first operation of a semiconductor laser in 1970. The junction laser needs the largest threshold current and has to be cooled (with few exceptions) to low temperature (100 K or lower) in order to suppress relaxation via phonons. The double heterostructure laser (since 1980) operates at room temperature. The threshold

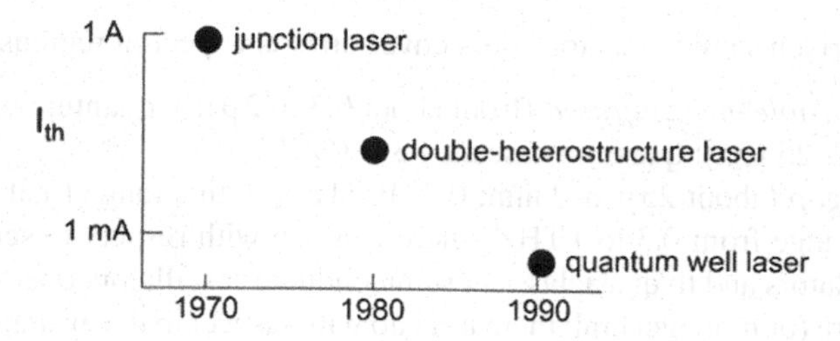

Fig. 28.6 Threshold current of bipolar semiconductor lasers operated at room temperature

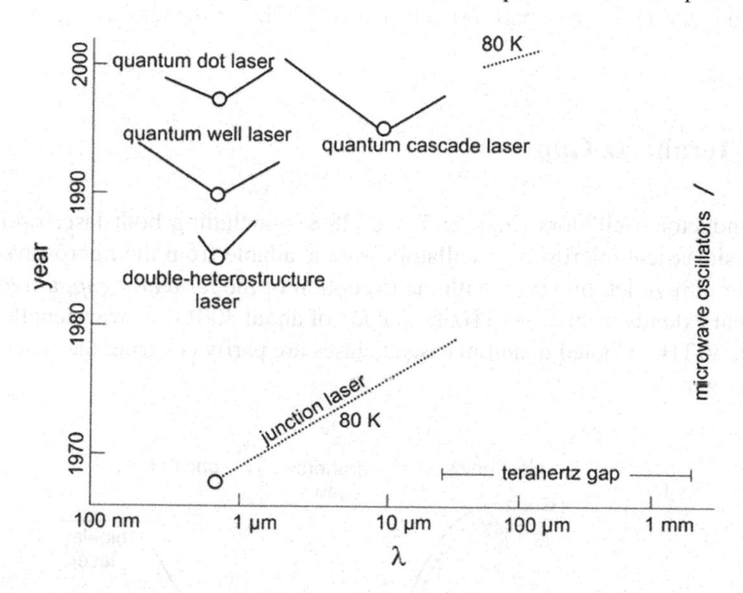

Fig. 28.7 Development of semiconductor lasers

current lies in the range of 10–100 mA. The quantum well laser reached a further remarkable decrease of the threshold current.

Together with the development of lasers of small threshold current, the reliability increased. Already in 1995, the monthly production rose to about one million laser diodes.

The development of the current-driven semiconductor lasers (Fig. 28.7) began with the junction laser. After the operation in the near infrared, the range was extended by the use of lead salt compounds into the far infrared up to a wavelength near 30 μm. In the near infrared, the junction laser, then the double heterostructure laser and the quantum well laser were introduced. The wavelength range of the quantum well laser was extended up to the near UV. In the far infrared, the junction lasers are in a large part of the spectrum replaced by the quantum cascade lasers, which generate radiation in the range from about 2–30 μm. Quantum cascade lasers, cooled to liquid nitrogen temperature, generate far infrared radiation (*see* next section).

Current-driven semiconductor lasers cover different spectral regions.

- *Near UV, visible, near infrared* (from about 0.3 to 2 μm): quantum well laser.
- *Infrared* (2–25 μm): quantum cascade laser (QCL).
- *Terahertz gap* (about 25 μm–1 mm; 0.3–10 THz): in this range (that includes the sub-THz range from 0.3 to 1 THz), there is a gap with respect to semiconductor laser oscillators and to quasiclassical semiconductor oscillators operating at room temperature (or more general: there is a gap with respect to the availability of solid state electronic devices and solid state photonic devices). Quantum cascade lasers working at 80 K cover a part (60–300 μm; 1–5 THz) of the terahertz gap.

28.7 Terahertz Gap

Semiconductor oscillators (Figs. 28.7 and 28.8)—including both laser oscillators and quasiclassical microwave oscillators—are available from the microwave range up to the ultraviolet, however, with the exception of the *terahertz gap*, a frequency range that extends from a sub-THz frequency of about 300 GHz (wavelength 1 mm) to about 30 THz. Cooled quantum cascade lases are partly covering the range of the terahertz gap.

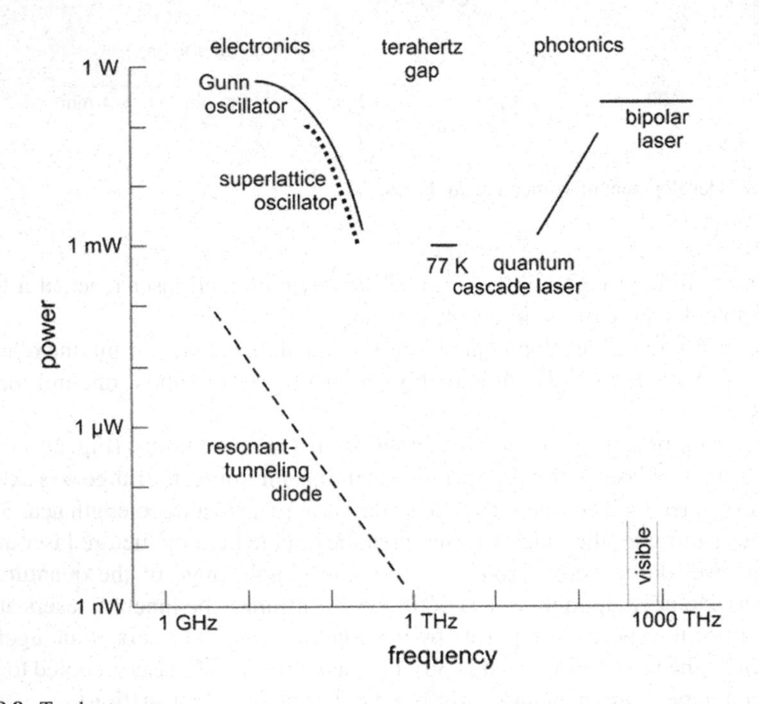

Fig. 28.8 Terahertz gap

Why are room-temperature quantum cascade lasers not available in the range between 4 and 30 THz? There are several reasons: The energy difference $E_2 - E_1$ is of the order of kT or smaller so that the population difference is smaller than at low temperature. Relaxation via phonons is stronger at high temperatures than at low temperatures. If the energy difference $E_2 - E_1$ coincides with the energy of polar optic phonons, relaxation by one-phonon processes occurs, resulting in very short lifetimes of electrons in the upper subband (subband 2); the energy of a polar optic phonon of GaAs is 36 meV (corresponding to a frequency of 8.6 THz).

Another type of unipolar semiconductor laser should be mentioned here, the p germanium laser. This is a unipolar semiconductor laser pumped by a current. It is operated at temperatures below liquid nitrogen temperature. A current pulse applied to a p germanium crystal in a magnetic field gives rise to a nonequilibrium hole population with a population inversion. By feedback with a resonator, laser oscillation occurs. The laser is tunable over a very wide frequency range (0.3–3 THz). Tuning is possible by varying the strength of the magnetic field.

There are, furthermore, CO_2 laser pumped semiconductor lasers emitting far infrared radiation; these have also to be cooled to temperatures below liquid nitrogen temperature. The laser transitions occur between discrete energy levels of impurity ions in semiconductor crystals.

We mention semiconductor oscillators of the range of electronics at frequencies above 100 GHz (*see* Fig. 28.8).

- *Gunn oscillator* (Sect. 31.2). Gunn oscillators are microwave oscillators, commercially available up to about 200 GHz.
- *Semiconductor superlattice oscillator* (Sect. 31.3). Semiconductor superlattice oscillators (up to 200 GHz) are being developed.
- *Resonant tunnel diode oscillator* (Sect. 31.7). Operation of resonant tunnel diode oscillators have been demonstrated up to 700 GHz; however, the output power was very small.

Radiation at frequencies above 100 GHz up to 10 THz can be generated by frequency multiplication of microwave radiation.

Backword wave oscillators are continuous wave oscillators operating in the range from 200 GHz to ~1.5 THz.

References [187–192].

Problems

28.1 Efficiency of bipolar lasers. Compare the efficiency η of different types of GaAs bipolar lasers (double heterostructure laser, quantum wire laser, quantum dot laser, quantum well laser) driven at a current that is ten times larger than the threshold current. Assume, for simplicity, that the quantum efficiency is unity.

28.2 Explain qualitatively why the refractive index in the active regions of a GaAs junction laser is smaller than in the adjacent n-doped GaAs and p-doped GaAs regions.

28.3 Estimate the intensity of luminescence radiation (emitted into the whole space) of the lasers mentioned in Fig. 28.5.

Chapter 29
Quantum Cascade Laser

A quantum cascade laser contains active regions and conducting regions in turn. An active region contains three electron subbands. An electron injected into the upper subband undergoes a stimulated transition to the lower subband and reaches, after a nonradiative relaxation process, the lowest subband. The electron leaves the lowest subband by spatial escape to the neighboring conducting region. Then, the electron is injected into another upper subband, undergoes another stimulated emission process, relaxes, escapes, and so on. Passing, for example, through a hundred gain regions, an electron can produce a hundred photons by stimulated transitions. A single gain region is in principle a three-level system. An electron performs a cascade of stimulated emission processes in subsequent three-level systems.

How can we obtain a gain region in a quantum cascade laser and how can we inject an electron into a gain region and extract an electron from a gain region?

We can realize a gain region by the use of coupled quantum wells. Tunnel splitting of energy levels leads to appropriate subbands. Superlattices connect next-near gain regions. Injection of electrons into a gain region and extraction of electrons from a gain region are due to tunneling processes under the action of a static electric field.

The quantum cascade laser operating at room temperature is a radiation source of the infrared; it is available at wavelengths just beyond the wavelengths of bipolar lasers, from about 2–28 μm (11–150 THz). Quantum cascade lasers cooled to liquid nitrogen temperature operate in the frequency range of about 1–5 THz.

It is expected that terahertz radiation may be of importance for applications in the areas of communications, the environment, medicine, and security. Pioneering work is done in infrared and millimeter wave astronomy through the use of oscillators as local oscillators of heterodyne detectors (that are most sensitive).

K.F. Renk, *Basics of Laser Physics*, Graduate Texts in Physics,
DOI 10.1007/978-3-319-50651-7_29

29.1 Principle of the Quantum Cascade Laser

The quantum cascade laser (QCL) is a three-level laser. The three levels (Fig. 29.1a) belong to subband 0, subband 1, and subband 2. Under the action of a static field E_s, electrons are injected into the subband 2 and perform stimulated transitions to subband 1. The subband 1 is depopulated by relaxation via the emission of phonons and the subband 0 is depopulated by the spatial extraction of electrons. In principle, an electron passing through a subband system with 100 periods can produce 100 photons! Injection occurs by means of a conducting superlattice. In the superlattice region, the energy of an electron is energetically constrained to a miniband, i.e., to an energy band that is much smaller than an energy band of a bulk semiconductor. Extraction occurs to another superlattice. A spatial period is repeated about 100 times or more (Fig. 29.1b).

An external voltage, leading to a voltage U_1 per period, drives the electron through the cascade system. The voltage per period is approximately

$$U_1 \approx E_2 - E_0, \tag{29.1}$$

where E_2, E_1, and E_0 are levels belonging to the three subbands. The quantum efficiency $\eta_q \approx (E_2 - E_1)/(E_2 - E_0)$ have a value near 1. The overall power efficiency can be larger than 0.5.

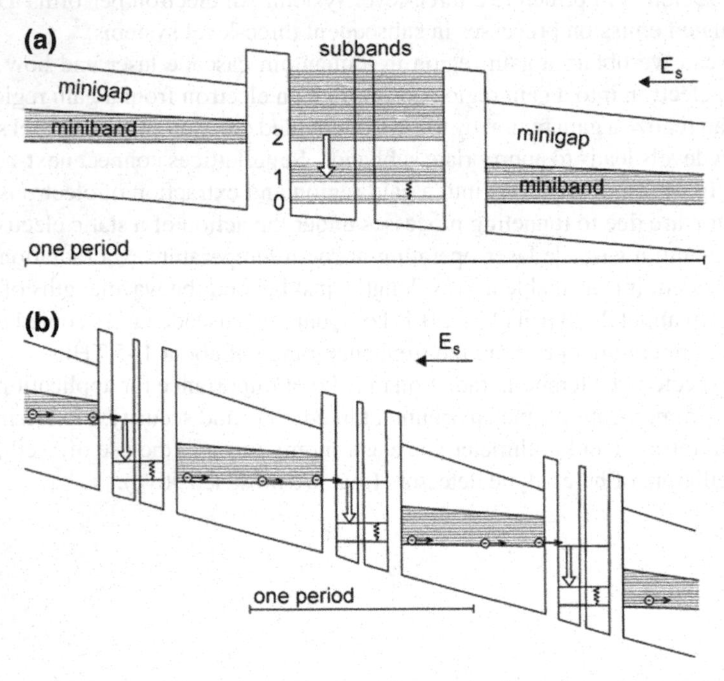

Fig. 29.1 Quantum cascade laser. **a** Single period and **b** three periods (*out of a hundred periods*)

Under the action of a static field, an electron propagates through a miniband, then through a tunnel barrier into the active region with the three subbands and, after an optical transition and relaxation, into another miniband. The active region contains two coupled quantum wells. A narrow tunnel barrier between the wells provides the coupling.

A minigap (a gap between two minibands) prevents the tunneling of electrons out of the subband 2. Perpendicular to the heterostructure, the electrons can move freely. The injector consists of a miniband, with a continuously decreasing miniband width, realized by quantum film layers and quantum well layers in turn (Sect. 29.3).

The heterostructure of a quantum cascade laser can be grown by molecular epitaxy.

29.2 Infrared Quantum Cascade Laser

The infrared quantum cascade laser (Fig. 29.2) consists of a quantum cascade heterostructure on a conducting substrate (GaP or GaAs). The heterostructure contains conduction electrons ($\sim 10^{22}$ m^{-3}), introduced into the heterostructure during its preparation by doping with silicon. An electric power of the order of 1 W (voltage 5 V; current 0.2 A) leads (in a structure of 1 mm length; 100 μm width; 10 μm thickness) to radiation of a power of several mW.

Room temperature QCLs and low temperature QCLs cover different wavelength regions.

- 2–28 μm; InGaAs/InAlAs heterostructure grown on GaP substrate; operation at room temperature; power 1–100 mW.
- 60–360 μm; GaAs/GaAlAs heterostructures grown on a GaAs substrate; operation at the temperature of liquid nitrogen; power 1–10 mW.

Before discussing the far infrared quantum cascade laser, we introduce superlattices and minibands.

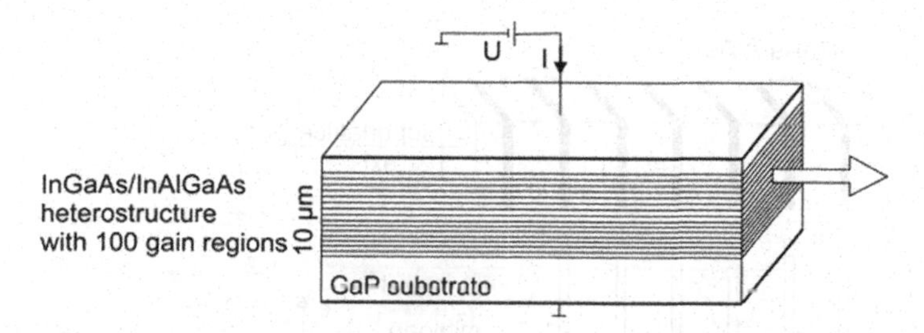

Fig. 29.2 Quantum cascade laser

29.3 Semiconductor Superlattice and Minibands

Semiconductor superlattices with minibands play an important role for quantum cascade lasers. We describe here properties of a GaAs/AlAs superlattice.

A GaAs/AlAs superlattice (Fig. 29.3) consists of a periodic sequence of GaAs layers and AlAs layers. An electron propagating along the superlattice axis experiences the AlAs layers as potential barriers. The periodic potential leads to minibands separated by minigaps. We characterize the dispersion relation of electrons in the lowest miniband by:

$$\epsilon = \epsilon_{\rm m}\left(\frac{1}{2} - \frac{1}{2}\cos k_x a\right), \tag{29.2}$$

where ϵ is the energy, $\epsilon_{\rm m}$ is the miniband width and k_x the wave vector along the superlattice axis. The dispersion curve is periodic in k_x. Therefore, we can restrict the wave vector k_x to the mini-Brillouin zone $-\pi/a < k_x \leq \pi/a$. The calculation of minibands is possible by the use of a Kronig–Penney model (Chap. 30).

The motion perpendicular to the superlattice axis, within the GaAs layers, corresponds to a free motion of a conduction electron. The energy of a miniband electron is given by

$$E = E_{\rm c} + E_{\rm zp} + \epsilon_{\rm m}\left(\frac{1}{2} - \frac{1}{2}\cos k_x a\right) + \frac{\hbar^2(k_y^2 + k_z^2)}{2m_{\rm c}}. \tag{29.3}$$

$E_{\rm c}$ is the energy of an electron at the minimum of the conduction band and $E_{\rm zp}$ the zero point energy of a miniband electron. The last term corresponds to the energy of motion perpendicular to the superlattice axis; $m_{\rm c}$ is the effective mass of an electron in the minimum of the conduction band of GaAs.

The value of $\epsilon_{\rm m}$ of a GaAs/AlAs superlattice is adjustable by the choice of the period a of the superlattice and of the width of the AlAs barrier layers. The largest

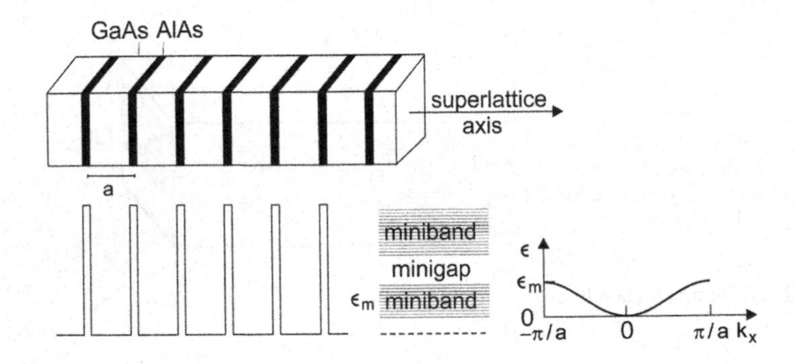

Fig. 29.3 GaAs/AlAs superlattice and minibands

value of the miniband width of a GaAs/AlAs superlattice is $\epsilon_m \sim 0.14$ eV, which corresponds to 10% of the gap energy of GaAs. InGaAs/InGaAlAs superlattices can have larger miniband widths (up to 0.3 eV) because the effective mass of a conduction electron in InAs is smaller than in GaAs.

29.4 Transport in a Superlattice

We discuss the electric transport in a superlattice (*miniband transport*). A static electric field E_s oriented along the superlattice axis accelerates the electrons. Relaxation gives rise to an ohmic conductivity (Fig. 29.4). The ohmic conductivity is given by

$$\sigma = \frac{N_0 e^2 \tau}{m^*}. \tag{29.4}$$

N_0 is the density of electrons in a superlattice, m^* the effective mass of an electron at the bottom of the miniband (at $k_x \approx 0$), and τ is the intraminiband relaxation time; $\tau \sim 10^{-13}$ s for an electron in a GaAs/AlAs superlattice at room temperature. The value of the effective mass m^* depends on the period of the superlattice and the barrier width. Ohmic conductivity is limited to not too strong static fields (Sect. 32.3).

In a superlattice used in a quantum cascade laser as injector (and as extractor), the layer thicknesses of GaAs and AlAs—or of InGaAs and InGaAlAs—are varying along the superlattice axis. Accordingly, the zero point energy, the widths of the minibands, and the widths of the minigaps are varying as well. An applied voltage leads to a static field along the superlattice axis. The upper boundary of the miniband limits the maximum energy an electron can reach in the static field. The first minigap prevents escape of excited electrons from the gain region as already mentioned.

Miniband transport will be treated in more detail in Sect. 32.3.

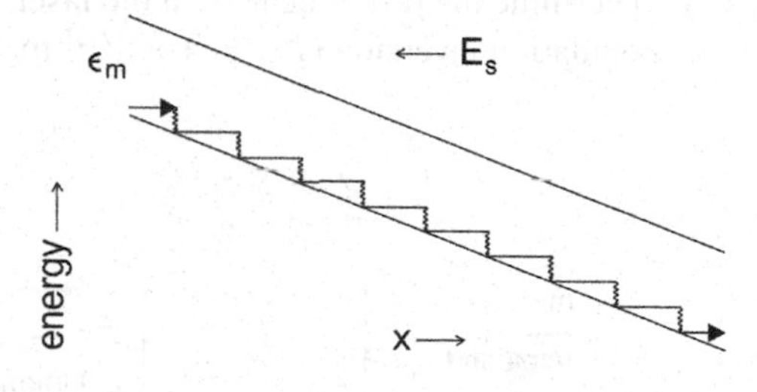

Fig. 29.4 Ohmic transport in a superlattice

29.5 Far Infrared Quantum Cascade Laser

In the far infrared quantum cascade laser (Fig. 29.5), the energy separation between sublevel 2 and sublevel 1 corresponds to a frequency in the far infrared. The energy difference between the laser levels is of the order of kT at room temperature. To reach the threshold population difference, a far infrared QCL has to be cooled, for instance to the temperature (77 K) of liquid nitrogen.

References [223–225].

Problems

29.1 Determine the frequency (and the wavelength) of laser radiation at which cooling of the active medium is favorable.

29.2 Estimate the gain (gain coefficient and gain factor per round trip) of radiation propagating in the active medium of a quantum cascade laser. [*Hint*: assume that the Einstein coefficient of stimulated emission is the same as for interband transitions.]

29.3 A quantum cascade laser cannot be realized if the laser transition frequency ν_1 coincides with the longitudinal optic frequency of the semiconductor material. Then, fast nonradiative relaxation of the upper laser level, by emission of a longitudinal optic phonon (frequency ν_{LO}) near the Brillouin zone center, makes population inversion almost impossible. Relaxation is still strong if ν_1 lies in the vicinity ν_{LO} (=8.7 THz for GaAs).

(a) Determine the frequency range for which the nonradiative lifetime of the upper laser level is shorter than 10^{-6} s, assuming that the nonradiative lifetime is 10^{-12} s at resonance ($\nu_1 = \nu_{LO}$) and that the relaxation rate decreases for laser frequencies around ν_{LO} according to a Lorentz resonance function.

(b) Determine the power density in the laser medium that is necessary for reaching population inversion ($B_{21} = 4 \times 10^{21}$ m^3 J^{-1} s^{-2}).

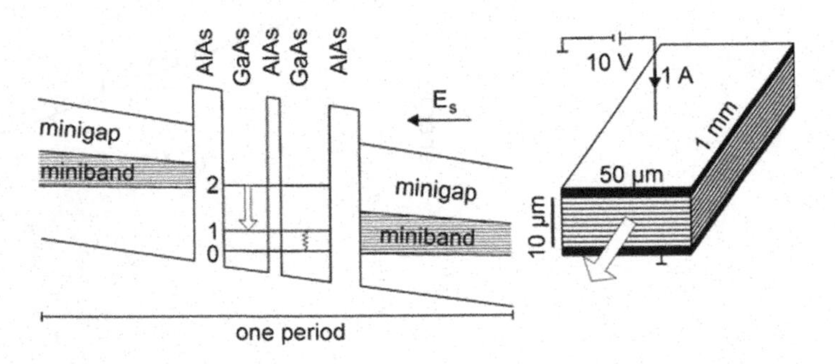

Fig. 29.5 Far infrared quantum cascade laser

(c) Discuss the influence due to thermal broadening of the electron distribution in the upper laser subband.

29.4 Strong absorption at the transversal optical frequency (=8.0 THz for GaAs) at the zone center results in damping of optical waves. In which frequency range is the optical thickness of the laser material (length 1 mm) of a quantum cascade laser larger than 0.02)? [*Hint*: assume that the absorption coefficient varies according to a Lorentz resonance function and that the absorption coefficient has a maximum value of 10^5 cm^{-1}.]

Chapter 30
Electron Waves in Semiconductor Heterostructures

We study electron waves in one-dimensional potentials and in semiconductor heterostructures.

We will begin with the discussion of the one-dimensional square well potential, then describe the origin of energy bands for electrons in a one-dimensional periodic potential. We make use of the tight binding method.

We will introduce the plane-wave transfer matrix method to describe, for an interface of two semiconductors, how a wave function of a semiconductor continues in the other semiconductor. The requirement that the energy flux through a boundary is steady provides the boundary conditions for electron waves at an interface of two different semiconductors. The plane-wave transfer matrix method allows for determination of the energy bands (minibands) of a superlattice. Finally, we will treat the quantum well and the double quantum well.

The plane-wave transfer matrix method is the same we used to describe electromagnetic plane waves in layered systems (Sect. 25.11). The difference of the results comes from the different dispersion relations: the wave vector of a free-electron wave in vacuum (or in a semiconductor) varies with the square root of energy while the wave vector of an electromagnetic wave in vacuum (or in a homogeneous medium) shows a linear dependence on frequency.

30.1 Electron in a One-Dimensional Square Well Potential

An electron wave with the wave vector k obeys the dispersion relation

$$E = \frac{\hbar^2 k^2}{2m_0}. \qquad (30.1)$$

E is the energy of an electron and m_0 the electron mass. The energy E increases quadratically with k. We describe free-electrons (in a bulk semiconductor) propagating in x direction by the use of the time-independent Schrödinger equation

K.F. Renk, *Basics of Laser Physics*, Graduate Texts in Physics,
DOI 10.1007/978-3-319-50651-7_30

$$-\frac{\hbar^2}{2m_0}\frac{\mathrm{d}^2\varphi}{\mathrm{d}x^2} = E\varphi, \tag{30.2}$$

where $\varphi(x)$ is the wave function. A solution is

$$\varphi(x) = A\mathrm{e}^{ikx}; \quad k = +\sqrt{2m_0E/\hbar^2}. \tag{30.3}$$

We consider an electron in a one-dimensional square well potential (width a) with rigid walls (Fig. 30.1a). The Schrödinger equation for $|x| < a/2$,

$$-\frac{\hbar^2}{2m_0}\frac{\mathrm{d}^2\varphi}{\mathrm{d}x^2} = E\varphi, \tag{30.4}$$

has the general solution

$$\varphi(x) = A\sin kx + B\cos kx; \quad k = +\sqrt{2m_0E/\hbar^2}. \tag{30.5}$$

The boundary conditions require that $\varphi(\pm a/2) = 0$ or

$$A\sin(ka/2) + B\cos(ka/2) = 0, \tag{30.6}$$

$$-A\sin(ka/2) + B\cos(ka/2) = 0. \tag{30.7}$$

Solutions are

- $A = 0$ and $\cos(ka/2) = 0$ leading to

$$\varphi(x) = B\cos\frac{n\pi x}{a}, \quad n = 1, 3, \ldots \text{ (even solution)}; \tag{30.8}$$

- $B = 0$ and $\sin(ka/2) = 0$ leading to

$$\varphi(x) = A\sin\frac{n\pi x}{a}, \quad n = 2, 4, \ldots \text{ (odd solution)}. \tag{30.9}$$

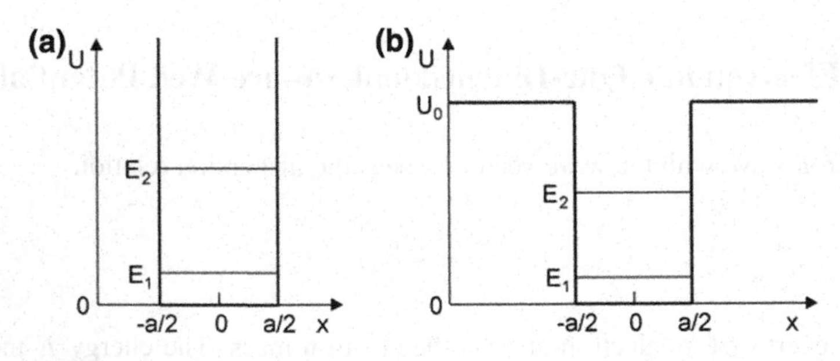

Fig. 30.1 One-dimensional square well potential **a** with infinitely high walls and **b** with walls of finite height

The energy eigenvalues are

$$E_{\mathrm{n}} = \frac{\pi^2 \hbar^2 n^2}{2m_0 a^2}. \tag{30.10}$$

For a square well potential with finite potential steps (Fig. 30.1b), the Schrödinger equation is unaltered for $|x| < a/2$. The wave equation for $|x| > a/2$,

$$-\frac{\hbar^2}{2m_0}\frac{\mathrm{d}^2\varphi}{\mathrm{d}x^2} + U_0\varphi = E\varphi, \tag{30.11}$$

has the solution

$$\varphi(x) = C\mathrm{e}^{-\kappa x} + D\mathrm{e}^{\kappa x}; \quad \kappa = +\sqrt{2m_0(U_0 - E)/\hbar^2}. \tag{30.12}$$

The boundary conditions require that $\varphi(x)$ and $\mathrm{d}\varphi/\mathrm{d}x$ are continuous for $x = \pm a/2$. The application of the boundary conditions would allow for determination of A, B, C, D and of the eigenvalues. Instead, we make use of the symmetry of the potential. The ansatz of the even solutions

$$\varphi(x) = B\cos kx \quad \text{for} \quad |x| < a/2, \tag{30.13}$$

$$\varphi(x) = C\mathrm{e}^{\ \kappa x} \quad \text{for} \quad |x| > a/2, \tag{30.14}$$

and the condition of continuity of φ and $\mathrm{d}\varphi/\mathrm{d}x$ at $|x| = a/2$ lead to

$$B\cos(ka/2) = C\mathrm{e}^{-\kappa a/2}, \tag{30.15}$$

$$kB\sin(ka/2) = \kappa C\mathrm{e}^{-\kappa a/2} \tag{30.16}$$

or

$$k\tan(ka/2) = \kappa. \tag{30.17}$$

The odd solutions are

$$\varphi(x) = A\sin kx \quad \text{for} \quad |x| < a/2, \tag{30.18}$$

$$\varphi(x) = C\mathrm{e}^{-\kappa x} \quad \text{for} \quad |x| > a/2. \tag{30.19}$$

The boundary conditions of the odd solutions require that

$$k\cot(ka/2) = \kappa. \tag{30.20}$$

The conditions $k\tan(ka/2) = \kappa$ and $k\cot(ka/2) = \kappa$ provide a finite number of discrete energy eigenvalues E_{n}.

30.2 Energy Bands of Electrons in a Periodic Square Well Potential

We describe a model (tight binding model) that illustrates the occurrence of energy bands and of dispersion for electron waves in a periodic potential.

A single isolated square well potential at position x_l (Fig. 30.2, upper part) is characterized by the wave equation

$$\left[-\frac{\hbar^2}{2m_0}\frac{\mathrm{d}^2}{\mathrm{d}x^2} + U_l(x - x_l)\right] \varphi_l(x - x_l) = E_0\, \varphi_l(x - x_l). \tag{30.21}$$

We regard E_0 as the energy E_1 of the lowest state in a square well potential (Sect. 30.1) and $\varphi_l = \varphi_l(x - x_l)$ as the corresponding wave function. The wave function is normalized,

$$\int_{-\infty}^{\infty} \varphi_l^*(x - x_l)\varphi_l(x - x_l)\mathrm{d}x = 1. \tag{30.22}$$

The wave equation of a periodic sequence of identical square well potentials (Fig. 30.2, center) is

$$\left(-\frac{\hbar^2}{2m_0}\frac{\mathrm{d}^2}{\mathrm{d}x^2} + U(x)\right) \psi(x) = E\psi(x). \tag{30.23}$$

The potential energy is a periodic function,

$$U(x + a) = U(x), \tag{30.24}$$

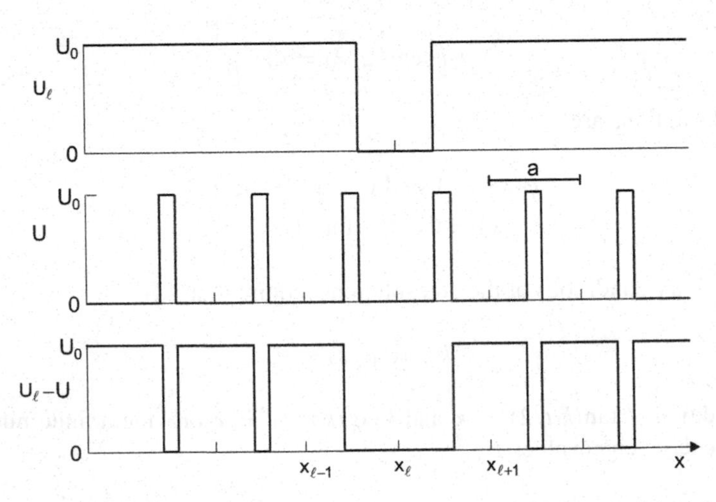

Fig. 30.2 A single square well potential, an infinite series of square well potentials and the difference between the two potentials

where a is the period. We describe the wave function of the periodic system by a linear combination of the wave functions of the single wells,

$$\psi(x) = \sum_l c_l\, \varphi_l(x - x_l). \tag{30.25}$$

Making use of the periodicity, we write

$$\psi(x) = \frac{1}{\sqrt{N}} \sum_{l=0}^{N-1} \mathrm{e}^{ikla}\, \varphi_l(x - la). \tag{30.26}$$

N is the number of quantum wells in a periodicity interval. We apply periodic boundary conditions, $\psi(x + Na) = \psi(a)$, and find

$$k = \frac{2\pi l}{Na}; \quad l = 0, 1, \ldots N - 1. \tag{30.27}$$

We restrict the k values to the first Brillouin zone

$$-\frac{\pi}{a} < k \le \frac{\pi}{a}. \tag{30.28}$$

Inserting $\psi(x)$ into the wave equation provides

$$\sum_l \mathrm{e}^{ikla}\, [U(x) - E]\, \varphi_l = -\sum_l \mathrm{e}^{ikla} \left(-\frac{\hbar^2}{2m_0} \frac{\mathrm{d}^2}{\mathrm{d}x^2} \right) \varphi_l. \tag{30.29}$$

We add on both sides the term $(-U_l(x - x_l) + E_0)\varphi_l$ and obtain

$$\sum_l \mathrm{e}^{ikla}\, [U(x) - U_l(x - x_l) - E + E_0]\, \varphi_l$$
$$= -\sum_l \mathrm{e}^{ikla} \left[-\frac{\hbar^2}{2m_0} \frac{\mathrm{d}^2}{\mathrm{d}x^2} - U_l(x - x_l) + E_0 \right] \varphi_l. \tag{30.30}$$

The right side is zero. We find

$$(E - E_0) \sum_l \mathrm{e}^{ikla} \varphi_l = \sum_l \mathrm{e}^{ikla}\, [U(x) - U_l(x - x_l)]\, \varphi_l. \tag{30.31}$$

We multiply the equation by

$$\psi^*(x) = \frac{1}{\sqrt{N}} \sum_{m=0}^{N-1} \mathrm{e}^{-ikma} \varphi_m^*(x - ma) \tag{30.32}$$

and integrate over the length L of the periodicity interval. We obtain

$$(E - E_0) \sum_{l,m} e^{ik(l-m)a} \int_0^L \varphi_m^* \varphi_l dx = \sum_{l,m} e^{ik(l-m)a} \int \varphi_m^* [U(x) - U_l(x - x_l)] \varphi_l dx. \tag{30.33}$$

Neglecting the weak overlap of φ_l^* and φ_l for $l \neq m$, we obtain for the term on the left side $N(E - E_0)$. Because of the large values of $U - U_l$ (Fig. 30.2, lower part) at positions of the cells $m \neq l$, we cannot neglect the terms with $m \neq l$ on the right side of the equation. We assume that $\varphi_l(x_l - la)$ decreases strongly at large distance $|x - x_l|$. Then we can restrict the double sum to terms that correspond to neighboring cells. We obtain

$$\begin{aligned} N \times (E - E_0) = N \times & \int \varphi_l^* [U(x) - U_l(x - x_l)] \varphi_l dx \\ & + N \times e^{-ika} \int \varphi_{l-1}^* [U(x) - U(x - x_l)] \varphi_l dx \\ & + N \times e^{ika} \int \varphi_{l+1}^* [U(x) - U(x - x_l)] \varphi_l dx. \end{aligned} \tag{30.34}$$

It follows, with

$$\alpha = \int \varphi_l^* [U_l(x - x_l) - U(x)] \varphi_l dx \tag{30.35}$$

and

$$\gamma = \int \varphi_{l-1}^* [U_l(x - x_l) - U(x)] \varphi_l dx = \int \varphi_{l+1}^* [U_l(x - x_n) - U(x)] \varphi_l dx, \tag{30.36}$$

that the energy is equal to

$$E = E(k) = E_0 - \alpha - \gamma \cos ka. \tag{30.37}$$

We introduce

$$\epsilon(k) = E_0 - E(k) - \alpha - \gamma. \tag{30.38}$$

With

$$\epsilon_{\rm m} = -2\gamma \tag{30.39}$$

we find

$$\epsilon(k) = \epsilon_{\rm m} \left(\frac{1}{2} - \frac{1}{2} \cos ka \right). \tag{30.40}$$

We obtain the lowest energy band (Fig. 30.3), with a minimum at $k = 0$ (width $\varepsilon_{\rm m}$), then an energy gap, and a second band (maximum at $k = 0$ since $\gamma < 0$). We can interpret $E_0 - \alpha - \gamma$ as zero point energy of an electron in the periodic potential.

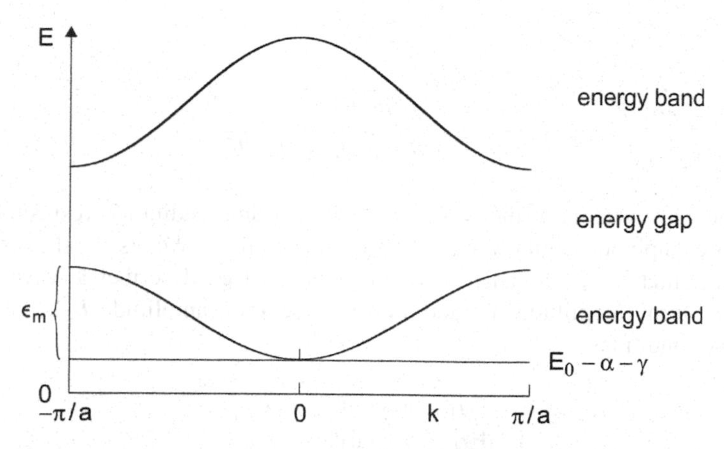

Fig. 30.3 Energy bands

Fig. 30.4 Electron wave at an interface of two semiconductors

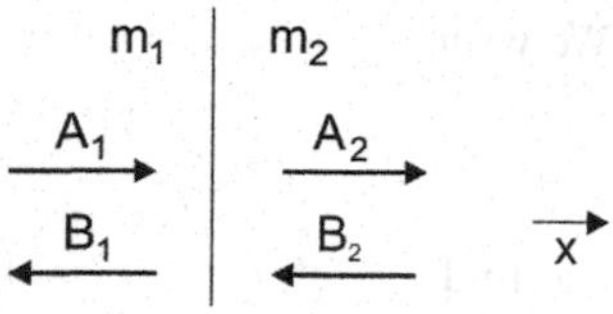

30.3 Plane-Wave Transfer Matrix Method of Characterizing a Semiconductor Interface

We consider (Fig. 30.4) propagation of an electron wave through an interface of two semiconductor materials (for instance GaAs and AlAs). At the interface, the potential energy and the effective mass of an electron change abruptly. We describe an electron wave propagating in x direction (perpendicular to the interface) by the time-independent Schrödinger equation

$$\left(-\frac{\hbar^2}{2m(x)}\frac{\mathrm{d}^2}{\mathrm{d}x^2} + U(x)\right)\psi(x) = E\,\psi(x), \tag{30.41}$$

where $m(x)$ is the effective mass and $U(x)$ the potential energy. We look for wave functions $\psi(x)$ and energies E that satisfy the equation. We describe the wave functions in medium 1 and medium 2 by the ansatz:

$$\psi_1 = A_1 \mathrm{e}^{\mathrm{i}k_1x} + B_1 \mathrm{e}^{-\mathrm{i}k_1x}, \tag{30.42}$$

$$\psi_2 = A_2\,\mathrm{e}^{\mathrm{i}k_2x} + B_2\,\mathrm{e}^{-\mathrm{i}k_2x}, \tag{30.43}$$

With

$$k_1 = +\sqrt{2m_1 E/\hbar^2}, \tag{30.44}$$
$$k_2 = +\sqrt{2m_2(E - U_0)/\hbar^2} \tag{30.45}$$

being the wave vectors of the waves in medium 1 and medium 2, respectively. A_1 and B_1 are amplitudes of the waves of opposite directions. We restrict the discussion to the case that $E < U_0$. Then k_2 is imaginary and ψ_2 describes a wave with an increasing term (amplitude A_2) and a decreasing term (amplitude B_2). We use the boundary conditions

$$\psi_1 = \psi_2 \quad \text{at} \quad x = 0, \tag{30.46}$$
$$\frac{1}{m_1}\frac{\mathrm{d}\psi_1}{\mathrm{d}x} = \frac{1}{m_2}\frac{\mathrm{d}\psi_2}{\mathrm{d}x} \quad \text{at} \quad x = 0. \tag{30.47}$$

We write

$$M_1\begin{pmatrix} A_1 \\ B_1 \end{pmatrix} = M_2\begin{pmatrix} A_2 \\ B_2 \end{pmatrix} \tag{30.48}$$

and find

$$M_l = \begin{pmatrix} 1 & 1 \\ k_l & -k_l \end{pmatrix}; \quad l = 1, 2. \tag{30.49}$$

It follows that

$$\begin{pmatrix} A_1 \\ B_1 \end{pmatrix} = M_1^{-1} M_2 \begin{pmatrix} A_2 \\ B_2 \end{pmatrix} = M_{12}\begin{pmatrix} A_2 \\ B_2 \end{pmatrix}, \tag{30.50}$$

where

$$M_{12} = \begin{pmatrix} \frac{1}{2}(1 + k_2/k_1) & \frac{1}{2}(1 - k_2/k_1) \\ \frac{1}{2}(1 - k_2/k_1) & \frac{1}{2}(1 + k_2/k_1) \end{pmatrix}. \tag{30.51}$$

is the transfer matrix. It has exactly the same form as the transfer matrix for a plane electromagnetic wave at an interface; *see* (25.12).

The continuity conditions we used follow from the requirements that the probability density $\rho(x) = \psi^*(x)\psi(x)$ and the probability current density $\partial\rho/\partial t$ are continuous. The first condition is fulfilled if $\psi(x)$ is continuous. To discuss the second condition, we replace in the time-dependent Schrödinger equation

$$-\frac{\hbar}{\mathrm{i}}\frac{\partial}{\partial t}\psi(x) = \left(-\frac{\hbar^2}{2m_0}\nabla^2 + U(x)\right)\psi(x, t) \tag{30.52}$$

the first term on the right side:

$$-\frac{\hbar^2}{2m_0}\frac{\partial^2}{\partial x^2}\psi(x) \to \frac{\hbar^2}{2}\frac{\partial}{\partial x}\left(\frac{1}{m(x)}\frac{\partial \psi(x)}{\partial x}\right). \tag{30.53}$$

Then

$$\begin{aligned}\frac{\partial}{\partial t}(\psi^*\Psi) &= \frac{i\hbar}{2}\frac{\partial}{\partial x}\left(\frac{1}{m(x)}\frac{\partial}{\partial x}\psi^*\right)\psi \\ &= \frac{i\hbar}{2}\left[\frac{\partial}{\partial x}\left(\psi^*\frac{1}{m(x)}\frac{\partial}{\partial x}\psi\right) - \psi\frac{1}{m}\frac{\partial}{\partial x}\Psi^*\right] = 0.\end{aligned} \tag{30.54}$$

This condition is satisfied at an interface (at $x = 0$) between two semiconductors if

$$\frac{1}{m_1}\frac{\partial \psi}{\partial x} = \frac{1}{m_2}\frac{\partial \psi}{\partial x} \quad \text{at} \quad x = 0. \tag{30.55}$$

30.4 Minibands

The potential energy of an electron in a superlattice is a periodic function,

$$U(x + a) = U(x), \tag{30.56}$$

where a is the period. We describe the wave function of an electron in the periodic system as a linear combination of the wave functions of the single wells,

$$\psi(x) = \sum_l c_l\, \varphi_l(x - x_l). \tag{30.57}$$

Making use of the periodicity, we write

$$\psi(x) = \frac{1}{\sqrt{N}}\sum_{l=0}^{N-1} e^{ikla}\varphi_l(x - la). \tag{30.58}$$

N is the number of quantum wells in a periodicity interval. We apply periodic boundary conditions, $\psi(x + Na) = \psi(x)$, and find

$$k = \frac{2\pi l}{Na}; \quad l = 0, 1, \ldots N - 1. \tag{30.59}$$

We restrict the k values to the first Brillouin zone

$$-\frac{\pi}{a} < k \le \frac{\pi}{a}. \tag{30.60}$$

We now use the transfer matrix method. We described in the preceding section the transfer matrix of an electron wave at a boundary. Taking account of propagation, we find the same equations, (25.31)–(25.34), as for electromagnetic plane waves in a one-dimensional photonic crystal. The equations lead, as shown in Sect. 25.14, to the dispersion relation

$$\cos ka = \cos k_1 a_1 \cos k_2 a_2 - \frac{1}{2}\left(\xi + \frac{1}{\xi}\right) \sin k_1 a_1 \sin k_2 a_2. \tag{30.61}$$

For an electron wave in a superlattice, the quantities are:

$$k_1 = +\sqrt{2m_1 E/\hbar^2}, \tag{30.62}$$

$$\xi = \frac{k_1}{k_2}\frac{m_2}{m_1} = -\mathrm{i}\frac{k_1}{\kappa}\frac{m_2}{m_1}, \quad \kappa = +\sqrt{2m_2(U_0 - E)/\hbar^2}. \tag{30.63}$$

We can write the dispersion relation of a miniband electron in the form

$$\cos ka = \cos k_1 a_1 \cosh \kappa a_2 - \frac{1}{2}\left(|\xi| + \frac{1}{|\xi|}\right) \sin k_1 a_1 \sinh \kappa a_2 = f(E). \tag{30.64}$$

This equation, $\cos ka = f(E)$, cannot be solved analytically. However, we can obtain an approximate solution. We expand $f(E)$ around the eigenvalue $E_{0,\mathrm{n}}$ of an isolated quantum well,

$$f(E) \approx f(E_{0,\mathrm{n}}) + \left(\frac{\mathrm{d}f}{\mathrm{d}E}\right)_{E=E_0} \times (E - E_0). \tag{30.65}$$

We find

$$E(k) = E_0 - \alpha - \gamma \cos ka, \tag{30.66}$$

$$\alpha = f(E_0)/(\mathrm{d}f/\mathrm{d}E)_{E=E_0}, \tag{30.67}$$

$$-\gamma = [(\mathrm{d}f/\mathrm{d}E)^{-1}]_{E=E_0}. \tag{30.68}$$

It follows, with $-2\gamma = \epsilon_{\mathrm{m}}$, that

$$\epsilon(k) = \epsilon_{\mathrm{m}}\left(\frac{1}{2} - \frac{1}{2}\cos ka\right). \tag{30.69}$$

Taking into account the free motion perpendicular to the superlattice axis, we obtain, with $k = k_x$, the total energy

$$\epsilon(\boldsymbol{k}) = \epsilon_{\mathrm{m}}\left(\frac{1}{2} - \frac{1}{2}\cos k_x a\right) + \frac{\hbar^2}{2m_{\mathrm{e}}}(k_y^2 + k_z^2), \tag{30.70}$$

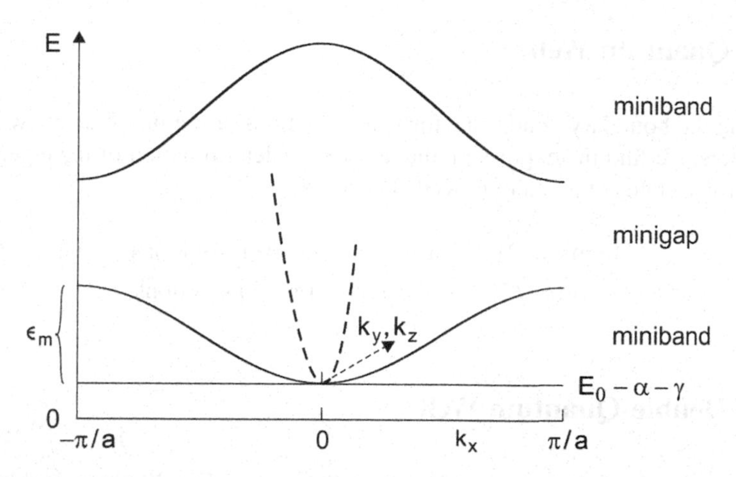

Fig. 30.5 Minibands

where m_e is the effective mass of an electron in GaAs in a GaAs/AlAs superlattice and $\boldsymbol{k} = (k_x, k_y, k_z)$. The energy dispersion curves (Fig. 30.5) indicate minibands (and minigaps) of electron wave vectors oriented along the superlattice axis. There is no gap for electrons that have wave vectors with components (k_y, k_z) perpendicular to the superlattice axis. The energy $E = 0$ is equal to the energy of the minimum of the conduction band of bulk GaAs.

The widths of the minibands and of the minigaps depend on the superlattice parameters:

- a_1 = thickness of a quantum well layer.
- a_2 = thickness of a barrier layer.
- $a = a_1 + a_2$ = superlattice period.

It is possible to design superlattices for a great range of values of ϵ_m, namely ϵ_m = 5–140 meV for GaAs superlattices and ϵ_m up to 300 meV for GaInAs/GaAlInAs superlattices.

If we neglect the difference of the effective masses of the superlattice materials, the matrix method yields the same result as obtained via the superposition of the wave functions of the single wells (Sect. 30.2).

A remark. The method of superposition of elementary wave functions (tight binding model) was introduced by Felix Bloch in 1928 [251]. Ralph Kronig and William Penney [231] introduced (in 1931) the square well potential (Kronig-Penney potential) and derived the dispersion relation (30.40). Gerard Bastard [232, 233] extended the model (extended Kronig–Penney model) to describe energy bands of semiconductor superlattices—with different effective masses of an electron in different layers of a superlattice.

30.5 Quantum Well

Knowing the boundary conditions for wave functions at the interface of two semiconductors, we find the expression that allows for determination of the eigenvalues of electronic states of a quantum well (Problems):

$$k \tan(ka/2) = -\alpha\, m_2/m_1 \quad \text{for even solutions,} \tag{30.71}$$

$$k \cot(ka/2) = -\alpha\, m_2/m_1 \quad \text{for odd solutions.} \tag{30.72}$$

30.6 Double-Quantum Well

The energy levels of electrons in a double-well potential (Fig. 30.6) are doublets. The energy level E_1 of the lowest state of isolated potential wells splits into two levels E_1^+ and E_1^-. Correspondingly, the level E_2 splits into two levels (E_2^- and E_2^+); *see* Problem 30.2.

References [31, 178, 186, 226–233, 251].

Problems

30.1 Quantum well.
Estimate the eigenvalues E_1 and E_2 of an electron in an AlAs/GaAs/AlAs quantum well (barrier height 2.2 eV; $m_{\mathrm{GaAs}} = 0.07\, m_0$; $m_{\mathrm{AlAs}} \sim 3\, m_{\mathrm{GaAs}}$) if the well consists of films of different thickness.

(a) Film thickness = 14 GaAs monolayers
(b) Film thickness = 2 GaAs monolayers.

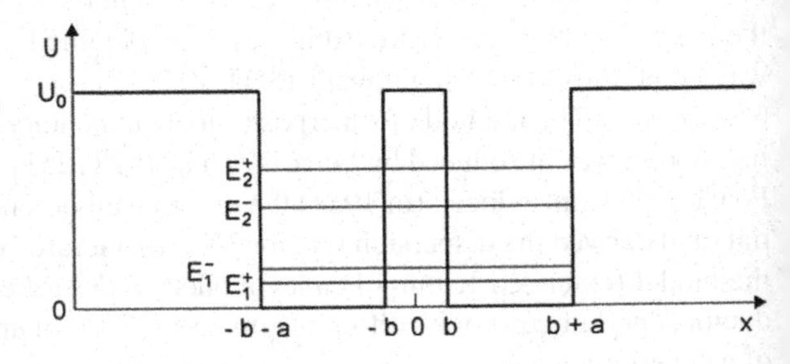

Fig. 30.6 Double-well potential

30.2 Double-quantum well.

(a) Determine the eigenvalues of a one-dimensional double well, which correspond to the two lowest energy levels ($s = 1, 2$) of a single one-dimensional double well. [*Hint*: make use of the symmetry.]
(b) Determine the energy level splitting $E_1^- - E_1^+$ for the two lowest levels.
(c) Sketch the wave functions that correspond to the four lowest levels.
(d) Calculate the level splitting occurring in an AlAs/GaAs/AlAs/GaAs/AlAs double quantum well (Fig. 30.6) for $a_1 = 10\,\text{nm}$ and $a_2 = 2\,\text{nm}$.

30.3 Dispersion of electrons in a periodic potential.
Derive the dispersion relation of electrons in a periodic potential by the use of the matrix method.

30.4 Interface.

(a) Electrons (energy ϵ) propagate toward a GaAs/AlAs interface and are reflected. Determine the average penetration depth of electrons. [*Hint*: take into account the difference between the penetration depth of the wave function and of the electrons.]
(b) Determine the penetration depth for $\epsilon = 10\,\text{meV}$ and $100\,\text{meV}$.
(c) Show that the reflectivity is $R = |k_1 - i\kappa|/|k_1 - i\kappa_1|$.
(d) Explain the electron total reflector used in a GaN quantum well laser (Sect. 24.3, Fig. 24.3a).

30.5 Tunneling.

(a) Determine the transmissivity of an AlAs barrier in a GaAs/AlAs/GaAs heterostructure for electrons of energy ϵ.
(b) Determine the transmissivity for electrons of energy $\epsilon = 10\,\text{meV}$ and $100\,\text{meV}$ at a barrier width of 2 monolayers of AlAs and for a barrier of 10 monolayers of AlAs.

30.6 Resonance state.

(a) Given is a GaAs/AlAs/GaAs/AlAs/GaAs heterostructure. Determine the energy dependence of the transmissivity for electron waves of different energies.
(b) Design a heterostructure that is transparent for electrons of $\epsilon = 10\,\text{meV}$.
(c) Design a heterostructure that is transparent for electrons of ϵ $100\,\text{meV}$.

30.7 Injector of a quantum cascade laser.

(a) Design a quantum cascade laser of AlAs/GaAs/AlAs/GaAs/AlAs heterostructures embedded in chirped GaAs/AlAs superlattices for a quantum cascade laser that may be able to generate radiation at a frequency of 4 THz.
(b) Estimate the thicknesses of the different layers.
(c) Discuss the role of the superlattice, especially in view of the result of the preceding problem.

30.8 Semiconductor superlattice.

(a) Determine the effective mass m^* of an electron in a superlattice (for propagation along the superlattice axis) for an electron with $k \sim 0$.
(b) Determine m^* of a GaAs/AlAs superlattice with 14 monolayers GaAs and 2 monolayers AlAs ($\epsilon_m \sim 140$ meV).
(c) Determine m^* of a GaAs/AlAs superlattice with 4 monolayers GaAs and 2 monolayers AlAs ($\epsilon_m \sim 40$ meV).
(d) Determine the effective mass m^* of an electron in a superlattice (for propagation along the superlattice axis) for arbitrary k and discuss the slope $m^*(k)$ and $m^*(\epsilon)$.
(e) Determine the group velocity $v_g(k)$ and the peak group velocity.
(f) Sketch the wave functions of the lowest miniband for $k \sim 0$ and $k = \pi/a$.
(g) Sketch the wave functions of the second miniband for $k \sim 0$ and $k = \pi/a$.

Chapter 31
A Comparison of Laser Oscillators and Quasiclassical Solid State Oscillators

We present three types of quasiclassical oscillators that are able to generate microwave radiation of high frequency: Gunn oscillator (used as source of radiation up to ~200 GHz); superlattice oscillator (in development, up to 200 GHz); resonant-tunnel diode oscillator (demonstrated up to 700 GHz). These oscillators are solid state oscillators, driven by active media. An active medium of a solid state oscillator makes use of the nonlinear transport in a semiconductor (Gunn oscillator) or a semiconductor heterostructure (superlattice oscillator and resonant-tunnel diode oscillator). The nonlinear transport is due to a negative mobility of conduction electrons. The origin of negative differential mobility is of quantum mechanical nature. However, the transport can be described classically.

A laser oscillator and a quasiclassical solid state oscillator have in common that gain is mediated by a high frequency polarization of an active medium and, additionally, that the active medium experiences a change during the buildup of an oscillation.

What makes the difference between a laser oscillator and a quasiclassical solid state oscillator? In a laser oscillator, polarization occurs via interaction of a high frequency field with single particles (atoms, molecules, free-electrons). In a quasiclassical solid state oscillator, polarization occurs via interaction of a high frequency field with charge density domains, i.e., with collectives of free-electrons. The formation of domains and thus of the polarization is due to nonlinear transport properties of the active medium—and not by a population inversion. A quasiclassical solid state oscillator shows an upper frequency limit that is determined by a relaxation time; this is the time it takes the electrons to establish a collective. Oscillation is only possible if the period of the high frequency field is larger than the relaxation time.

There is, beside the mechanism of interaction of radiation with a medium, a difference in the techniques used to couple radiation to a medium. The active medium of a laser fills a resonator partly or completely. A solid state diode that drives a quasiclassical solid state oscillator can have extensions that are small compared to the wavelength of the radiation. An antenna serves for coupling of the active medium to the radiation. It is possible to use an active medium of small volume because the gain of classical active media can be much larger than the gain of laser media.

K.F. Renk, *Basics of Laser Physics*, Graduate Texts in Physics,
DOI 10.1007/978-3-319-50651-7_31

According to Kroemer, there are two types of domains and therefore two modes of operation of a quasiclassical solid state oscillator—the pure charge accumulation mode and the propagating dipole domain mode. Here, we treat the pure charge accumulation mode, which is rarely described in textbooks, and we present an experiment that demonstrates the occurrence of the pure charge accumulation mode in a quasiclassical solid state oscillator.

We will, furthermore, discuss a classical oscillator model—the van der Pol oscillator. The model describes an equivalent circuit containing a nonlinear resistance that drives a self-excited oscillation in the circuit. The resistance of a van der Pol oscillator does not undergo a change during the buildup of an oscillation.

The chapter provides a connection to textbooks that treat microwave oscillators.

31.1 Interaction of Radiation with an Active Medium of a Laser or a Quasiclassical Oscillator

A comparison of a laser oscillator and a quasiclassical solid state oscillator shows the following.

- A laser oscillator and a quasiclassical oscillator have in common: interaction of an active medium with a high frequency field results in a high frequency polarization, which is synchronized to the field; mutual interaction of field and polarization leads to the buildup of both field and polarization.
- An active medium of a laser is, with respect to the charge distribution ρ, homogeneous (Fig. 31.1a). The corresponding material equation has the form $\nabla \cdot \boldsymbol{D} = 0$. An active medium of a laser contains high frequency dipole moments carried by atomic excitations. Interaction of these single-particle excitations with a high frequency electromagnetic field leads to gain for the high frequency field.
- In the active medium of a solid state oscillator, the charge distribution is inhomogeneous, $\nabla \cdot \boldsymbol{D} \neq 0$ (Fig. 31.1b). The periodic buildup and destruction of charge density domains gives rise to a high frequency polarization of the active medium. A quasiclassical solid state oscillator shows an upper frequency limit that is determined by the relaxation time of the electrons, which constitute a domain. The material properties responsible for the occurrence of charge density domains are based on quantum mechanical properties of a semiconductor (or a semiconductor heterostructure).

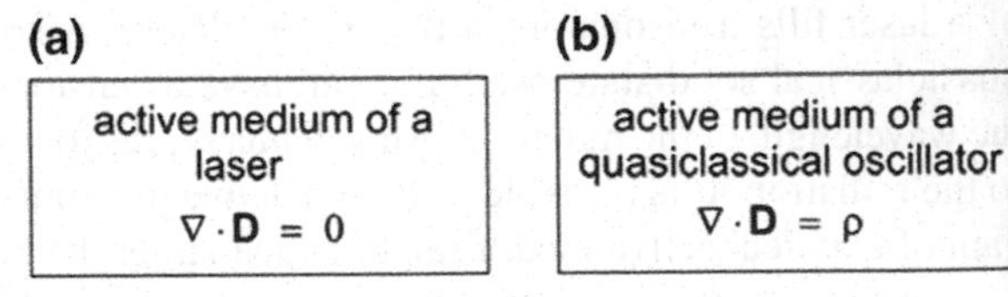

Fig. 31.1 Active media. **a** Laser medium. **b** Active medium of a solid state oscillator

31.2 Solid State Oscillators

There are various types of solid state oscillators for generation of microwave radiation. We mention three types.

- *Gunn oscillator*. The active device (active medium and electrodes together) of a Gunn oscillator is a Gunn diode. We describe a GaAs Gunn diode. The active device consists of a doped GaAs layer embedded in highly doped GaAs layers carrying metallic contacts. Nonlinearity is due to transfer of conduction electrons from a high-mobility state to a low-mobility state in GaAs. The electron transfer, which is of quantum mechanical nature, gives rise to a negative differential mobility for voltages larger than a critical voltage. The negative differential mobility causes formation of charge density domains. Gunn oscillators are available as microwave oscillators up to frequencies of ~200 GHz. Gunn oscillators are described in many textbooks and survey articles; *see*, for instance, [234–239].
- *Semiconductor superlattice oscillator*. The basis of the nonlinearity of a semiconductor superlattice oscillator is the miniband transport. At voltages across a superlattice that are larger than a critical voltage, miniband electrons show a negative differential mobility. The negative differential mobility causes formation of charge density domains.
- *Resonant tunnel diode oscillator*. The active medium is a resonant-tunneling diode (Sect. 31.7).

There are two modes of operation of a Gunn oscillator or of a superlattice oscillator.

- *Pure charge accumulation mode* [239] (Fig. 31.2a). Under the action of a static field, a negative differential mobility medium extracts electrons from the cathode. The excess electrons in the medium and the positive charges at the cathode represent a dipole domain connected with a quasistatic polarization of the medium. Under the action of both a static field and a high frequency field, the number of excess electrons within the medium (and thus the density of positive charge at the cathode) increases and decreases periodically at the frequency of the high frequency field. The corresponding high frequency polarization P mediates gain. In the pure charge accumulation mode of operation of an oscillator, negative charge flows periodically from the cathode into the negative differential mobility medium and back to the cathode while the positive charge is bound to the cathode.

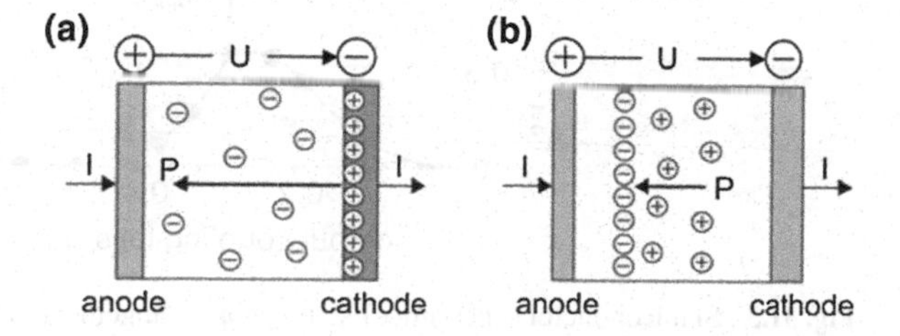

Fig. 31.2 Dipole domains in an active medium of a solid state oscillator. **a** Dipole domain caused by pure charge accumulation. **b** Propagating dipole domain

- *Propagating dipole domain mode* [239] (Fig. 31.2b). Under the action of a static field and of a high frequency field, negative and positive charges within the negative differential mobility medium separate giving rise to dipole domains. A dipole domain is formed near the cathode, travels through the medium, and disappears at the anode. The periodic formation and destruction of domains at the frequency of the high frequency field is joined with a high frequency polarization P of the active medium. The polarization mediates gain. The formation of propagating domains requires special boundary conditions for the field at the boundary between cathode and the negative differential mobility medium.

We will consider a particular solid state oscillator, namely a semiconductor superlattice oscillator, operating in a pure charge accumulation mode.

31.3 Semiconductor Superlattice Oscillator

In a semiconductor superlattice oscillator (Fig. 31.3a), a superlattice in a cavity resonator drives the oscillation. The superlattice is electromagnetically coupled to the field in the resonator via an antenna (a metal whisker). The antenna is also connected

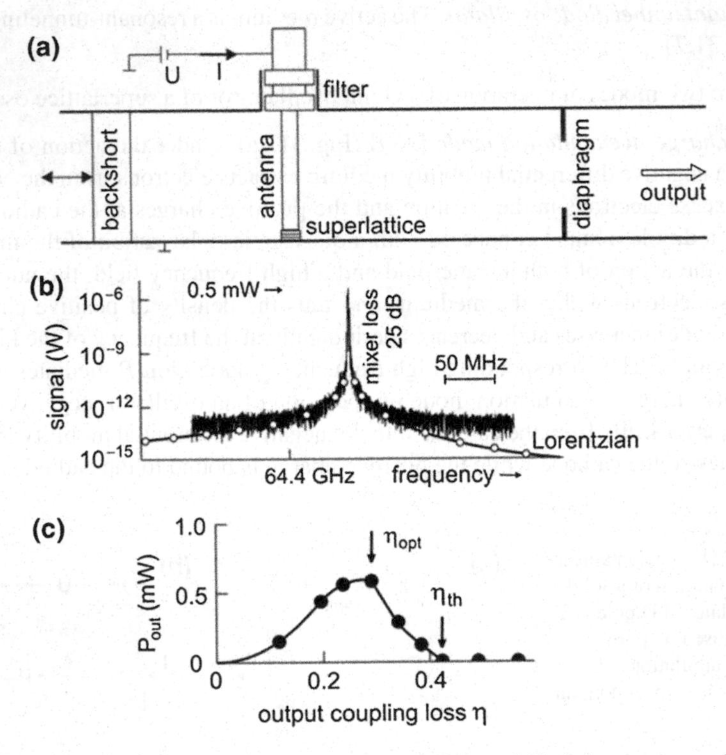

Fig. 31.3 Semiconductor superlattice oscillator. **a** Arrangement. **b** Emission spectrum. **c** Threshold behavior

to a bias circuit containing a voltage source (voltage U), which delivers a direct current I. A filter in the bias circuit avoids loss of radiation to the bias circuit. Radiation is coupled out from the resonator via the output port that contains a diaphragm. The oscillator is suited to generate microwave radiation. The emission spectrum (Fig. 31.3b), of an oscillator generating radiation near a frequency of 64 GHz, shows a bandwidth (200 kHz) that is determined by the spectrum analyzer used to register the spectrum. The emission line is, as indicated by the slope in the far wings, a Lorentzian line; a small deviation is due to background of the spectrum analyzer. For description of a superlattice oscillator, we follow [246].

Figure 31.3c (points and solid line) shows the output power P_{out} of the oscillator for different strengths η of output coupling loss; a measure of the output coupling loss η is the ratio of the aperture area and the area of the completely open output port. At small η, with radiation stored in the resonator, P_{out} is small. With increasing η, P_{out} increases, shows a maximum corresponding to optimum output coupling at η_{opt}, and then decreases to zero at the threshold loss η_{th}. A solid state oscillator shows an oscillation threshold behavior as a laser oscillator does.

To illustrate the principle of a superlattice oscillator, we consider the current-voltage (I–V) curve of a superlattice (Fig. 31.4a). With increasing voltage, the current increases linearly at small voltage, then less than linearly, reaches a peak value I_{p} at a critical voltage U_{c}, and remains constant for $U_{\text{s}} > U_{\text{c}}$. A static voltage $U_{\text{s}} > U_{\text{s}}$ causes the buildup of a high frequency current $I(t)$ and voltage $U(t)$. The active medium experiences feedback from the high frequency field stored in the resonator, which results in a reduction δI of the direct current. The current reduction is equal to the amplitude of the high frequency current (Fig. 31.4b). The current reduction occurs stepwise: at increasing U_{s}, the direct current shows plateau-like slopes. A current reduction, indicating oscillation, can occur already for $U < U_{\text{c}}$.

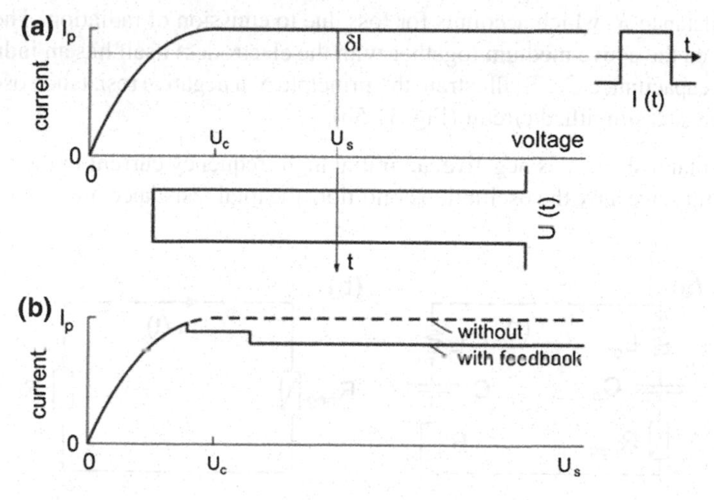

Fig. 31.4 Principle of the semiconductor superlattice oscillator. **a** I–V curve and time-dependent current and voltage. **b** I–V curve without and with feedback from radiation

We conclude from the occurrence of oscillations that the real high frequency current and high frequency voltage contain components of opposite phases, i.e., that the high frequency resistance of the superlattice is negative, which that is a condition of gain. The high frequency resistance is equal to

$$R_{\mathrm{neg}} = -\,\hat{U}/\hat{I}. \tag{31.1}$$

$\hat{U}$ is the amplitude of the high frequency voltage and $\hat{I}$ the amplitude of the high frequency current.

Example A particular GaAs superlattice (diameter 4 μm; length 0.6 μm; electron density $N_0 = 5 \times 10^{22}\ \mathrm{m}^{-3}$) has a critical voltage of 0.6 V and a peak current of 10 mA. Oscillation at 65 GHz results in a reduction of the current amplitude of $\delta I = \hat{I} = 2\,\mathrm{mA}$. The amplitude of the high frequency voltage is $\hat{U} = 0.9$ V (for $U_s = 2\ U_c$). Thus, the negative resistance is equal to $R_{\mathrm{neg}} = -450\ \Omega$. The experimental output power at optimum output coupling is ~0.5 mW corresponding to an efficiency of 4% for conversion of electric power to power of microwave radiation.

31.4 Model of a Solid State Oscillator

We follow [234]. We characterize a (quasiclassical) solid state oscillator by an equivalent resonance circuit. The resonance circuit can be a parallel or series resonance circuit. We choose a parallel resonance circuit.

The equivalent circuit (Fig. 31.5a) describes a high frequency circuit containing an active device with a negative resistance R_{neg}, a capacitance C, an inductance L and a resistance R, which accounts for loss due to emission of radiation. The active device (i.e., the active medium together with the electrodes) itself has an inductance L_d and a capacitance C_d. To illustrate the principle of a negative resistance oscillator, we make use a simplified circuit (Fig. 31.5b).

- If the total resistance is negative, an initial high frequency current in the loop will grow; thus, we have the oscillation condition: the total resistance must be negative.

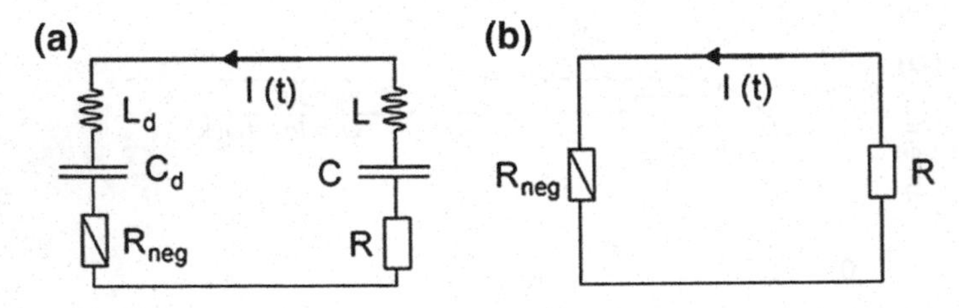

Fig. 31.5 Negative resistance oscillator. **a** Equivalent circuit and **b** simplified equivalent circuit

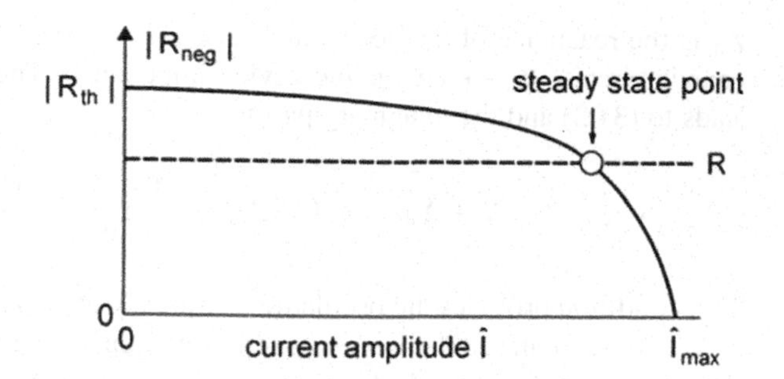

Fig. 31.6 Dependence of the magnitude of a negative resistance on the current amplitude

- At steady state oscillation, the sum of the resistances is zero; during onset of oscillation, the magnitude of the negative resistance decreases from a small-signal value to a large-signal value.
- If the total resistance is positive, an initial high frequency current will be damped and oscillation will not start.

The magnitude of the negative resistance depends on the amplitude of the high frequency current (Fig. 31.6). The absolute value of R_{neg} is largest for a small current amplitude $\hat{I}$ and is zero at maximum current amplitude $\hat{I}_{max}$ obtained for $R = 0$. The resistance R determines the point of steady state oscillation.

The output power of the oscillator is

$$P_{out} = (1/2)R\hat{I}^2 \tag{31.2}$$

if the condition

$$R_{neg} + R = 0 \tag{31.3}$$

is satisfied. An appropriate choice of the value of R—for instance, by an appropriate choice of the output coupling aperture of the resonator—leads to optimum output coupling. In the description of an equivalent parallel circuit, the threshold condition of a solid state oscillator is given by

$$R < |R_{th}|; \tag{31.4}$$

the loss resistance R must be smaller than the absolute value of the threshold resistance R_{th}.

To maintain a steady state oscillation, the high frequency voltage across the loop described by the complete equivalent circuit (*see* Fig. 31.5a) must be zero according to Kirchhoff's rules of voltages and currents in an electrical circuit,

$$I_0\,(R_{neg} + \mathrm{i}\,X_d) + I_0\,(R + \mathrm{i}\,X) = 0. \tag{31.5}$$

X_d is the reactance of the device and $X = \omega L - 1/(\omega C)$ the reactance of the resonance circuit. $R_{neg} + \mathrm{i}\, X_d$ is the device impedance. The real part of the equation leads to (31.3) and the imaginary part to

$$X + X_d = \omega(L + L_d) - \frac{1}{\omega}\left(\frac{1}{C} + \frac{1}{C_d}\right) = 0. \qquad (31.6)$$

This condition provides the oscillation frequency at steady state oscillation.

We consider an oscillator with an active element carrying a high frequency current (frequency ω) of amplitude $\hat{I}$,

$$I(t) = \hat{I} \cos \omega t. \qquad (31.7)$$

The voltage across the active device is given by

$$U(t) = R_{neg}\, \hat{I}\, \cos \omega t - X_d\, \hat{I}\, \sin \omega t; \qquad (31.8)$$

we neglect higher harmonics. Voltage and current have a phase shift of

$$\tan \varphi = -X_d/R_{neg}. \qquad (31.9)$$

Without loss ($R = R_{neg} = 0$), the phase shift between current and voltage is $\pi/2$.

We can write the oscillator equation in the form

$$L\frac{\mathrm{d}I}{\mathrm{d}t} + RI + \frac{1}{C}\int I \mathrm{d}t + U = 0. \qquad (31.10)$$

Inserting (31.7) and (31.8) in (31.10) leads to the conditions of steady state oscillation, $R_{neg} + R = 0$ and $\omega L - 1/(\omega C) + X_d = 0$.

A negative resistance device based on nonlinear properties of conduction electrons in a semiconductor has internal degrees of freedom: the charge density distribution in an active device (=active medium and electrodes together) can be inhomogeneous. The degree of inhomogeneity depends nonlinearly on the voltage across the device. The value of R_{neg} depends therefore on the internal dynamics.

We consider an oscillator operated at a fixed $R = |R_{neg}|$, i.e., at a fixed static voltage. In the case that the oscillator is submitted to a small additional time dependent voltage $U_1(t)$, the oscillator equation is given by

$$L\frac{\mathrm{d}I}{\mathrm{d}t} + RI + \frac{1}{C}\int I \mathrm{d}t + U = U_1(t). \qquad (31.11)$$

We solve the equation by using the ansatz:

$$I(t) = \hat{I}(t) \cos[\omega t + \varphi(t)], \qquad (31.12)$$

where higher harmonic currents are neglected. It follows that the high frequency voltage is equal to

$$U(t) = R_{\text{neg}}\, \hat{I}\, \cos[\omega t + \varphi(t)] - X_{\text{d}}\, \hat{I}\, \sin[\omega t + \varphi(t)]. \tag{31.13}$$

We assume that $\hat{I}(t)$ and $\varphi(t)$ do not vary appreciably over one cycle of the oscillation (slowly varying envelope approximation) and find the differential equations

$$\frac{\mathrm{d}I}{\mathrm{d}t} = -\hat{I}\left(\omega + \frac{\mathrm{d}\varphi}{\mathrm{d}t}\right)\sin(\omega t + \varphi) + \frac{\mathrm{d}\hat{I}}{\mathrm{d}t}\cos(\omega t + \varphi), \tag{31.14}$$

$$\int I \mathrm{d}t = \left(\frac{\hat{I}}{\omega} - \frac{\hat{I}}{\omega^2}\frac{\mathrm{d}\varphi}{\mathrm{d}t}\right)\sin(\omega t + \varphi) + \frac{1}{\omega^2}\frac{\mathrm{d}\hat{I}}{\mathrm{d}t}\cos(\omega t + \varphi). \tag{31.15}$$

Using (31.12) and (31.13), multiplying by $\cos(\omega t + \varphi)$ and $\sin(\omega t + \varphi)$ and integrating over a period $T = 2\pi/\omega$, we find from (31.14) and (31.15) *two oscillator equations*

$$\left(L + \frac{1}{\omega^2 C}\right)\frac{\mathrm{d}\hat{I}}{\mathrm{d}t} + (R_{\text{neg}} + R)\,\hat{I} = \frac{2}{T}\int_{t-T}^{t} U_1(t)\cos(\omega t + \varphi)\mathrm{d}t \tag{31.16}$$

and

$$\left(-\omega L + \frac{1}{\omega C} - \bar{X}\right) - \left(L + \frac{1}{\omega^2 C}\right)\frac{\mathrm{d}\varphi}{\mathrm{d}t} = \frac{2}{\hat{I}T}\int_{t-T}^{t} U_1(t)\sin(\omega t + \varphi)\mathrm{d}t. \tag{31.17}$$

If an external voltage is absent, these differential equations describe self-excited oscillation of a quasiclassical solid state oscillator. The description of onset of oscillation and steady state oscillation of a specific oscillator requires knowledge about the parameters R, L, C of the passive elements and the parameters $R_{\text{neg}}(\hat{I})$, $C_{\text{d}}(\hat{I})$, and $L_{\text{d}}(\hat{I})$ of the active device. If an external voltage is present, the equations describe phase locking of a classical oscillator to an external (weak) high frequency voltage (that is delivered, for instance, by a highly stabilized oscillator).

In comparison with a laser oscillator coupled to an external field—characterized by five differential equations of first order (Sect. 9.9)—the quasiclassical solid state oscillator coupled to an external field can be characterized by only two differential equations of first order, an equation for the amplitude of the current, and another equation for the phase between current and external field. The equations are coupled equations that have in common the parameters of the active device.

We can describe the superlattice oscillator as a regenerative amplifier with a resonator mediating feedback. Amplification of thermal radiation leads to phase and amplitude fluctuations and therefore to a noise bandwidth of the oscillator radiation. The spectral distribution of the radiation has a Lorentzian lineshape (Sect. 4.5).

The quality factor for radiation generated in a single mode oscillation is equal to (Sect. 8.9)

$$Q_{rad} = Q_{res} Z / Z_0. \tag{31.18}$$

Q_{res} is the quality factor of the resonator, Z the average occupation number of photons in the resonator mode at steady state oscillation, and Z_0 the average occupation number of thermal photons in the resonator mode without oscillation. Z follows from the relation $P_{out} = Zh\nu/\tau_p$, where $\tau_p = Q_{res}/\omega$ is an average lifetime of a photon in the resonator. The thermal occupation number is $Z_0 = kT/h\nu$; k is Boltzmann's constant and T the temperature.

Example Superlattice oscillator with a superlattice described in the preceding and the following example. Frequency $\nu = 6.5 \times 10^{10}$ Hz; $Q_{res} = 30$; output power $P_{out} = 0.5$ mW; $Z_0 \sim 100$; $Z = 2 \times 10^8$; $Q_{rad} \sim 10^8$.

A more detailed treatment of noise in solid state oscillators can be found, for example, in [244, 245].

31.5 Dynamics of Gain Mediated by a Semiconductor Superlattice

We describe a particular superlattice (Fig. 31.7a). It consists of layers of GaAs and of AlAs in turn. The superlattice is doped and contains free-electrons. Adjacent to the superlattice are, on both ends, highly doped GaAs layers (electron concentration 2×10^{24} m^{-3}). One of these layers connects the superlattice to a highly doped GaAs substrate and the other layer is covered with a metallic contact layer.

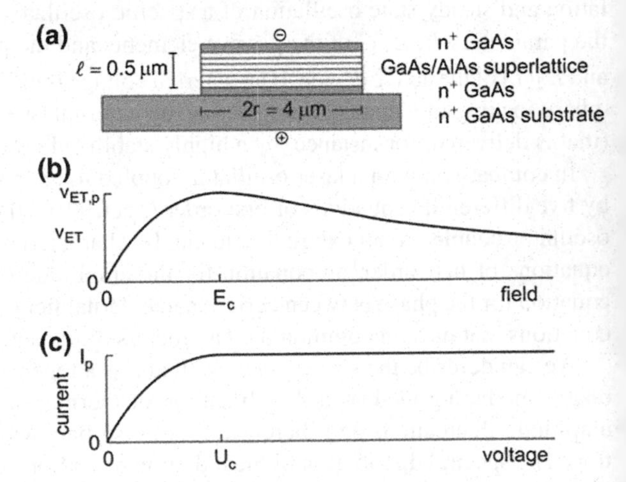

Fig. 31.7 Semiconductor superlattice. **a** Geometric structure. **b** Drift velocity-field characteristic for a homogeneous field along the superlattice axis. **c** Experimental I–V curve (*simplified*)

The drift velocity-field characteristic (Fig. 31.7b) is expected to have the form of an Esaki-Tsu characteristic:

$$v_{ET} = v_{ET,m} \frac{2\,E_s/E_c}{1+(E_s/E_c)^2} \quad (31.19)$$

E_s is the static field, $v_{ET,m}$ the maximum drift velocity reached at the critical field E_c. The differential mobility $\mu = dv_{ET}/dE_s$ is equal to the ohmic mobility $\mu_{ohm} = 2v_{ET,m}/E_c$ for fields around $E_s = 0$ and is negative for $E_s > E_c$. The negative differential mobility has the largest absolute value for $E = 1.7\,E_c$, where $\mu = -\mu_{ohm}/8$. We will derive the Esaki–Tsu characteristic in Sect. 32.3. We will show that the critical field is determined by the superlattice period a and a relaxation time τ according to $E_c = \hbar/ea\tau$; the relaxation time indicates how fast an equilibrium is established in an ensemble of free-electrons in a superlattice.

If the field along the superlattice axis is homogeneous even if the field exceeds the critical field, we obtain the Esaki–Tsu I–V characteristic, which is given by

$$I_{ET} = I_p \frac{2\,U_s/U_c}{1+(U_s/U_c)^2}. \quad (31.20)$$

$U_s = E_s l$ is the static voltage across the superlattice, $U_c = E_c l$ is the critical voltage, and

$$I_p = \pi r^2 N_0 e v_{ET,m} \quad (31.21)$$

is the peak current; r is the radius of the superlattice and N_0 the electron density. The ohmic resistance around $U_s = 0$ is equal to

$$R_{ohm} = \frac{U_c}{2I_p}. \quad (31.22)$$

Example of a superlattice (radius $r = 2\,\mu$m; $l = 0.6\,\mu$m; electron density $N_0 = 5.5 \times 10^{22}\,m^{-3}$); $v_{ET,m} = 10^5\,m\,s^{-1}$; $I_p = 11\,mA$; $E_c = 10^6\,V\,m^{-1}$; $U_c = 0.6\,V$; $\tau \sim 1.5 \times 10^{-13}\,s$; ohmic resistance $R_{ohm} = 27\,\Omega$ (around $U_s = 0$) and ohmic mobility $\mu_{ohm} = 0.20\,m^2\,V^{-1}\,s^{-1}$; ohmic conductivity $\sigma_{ohm} = 1.8 \times 10^3\,\Omega^{-1}\,m^{-1}$.

Now, the experimental I–V curve (Fig. 31.7c) shows a constant current (peak current $I_p = N_0 e v_{ET,m}$) at voltages above U_c. The origin of the excess current $I_{exc}(U) = I_p - N_0 e v_{ET}(E)$, with $U = El$, are excess electrons extracted from the cathode.

A constant current ($I = I_p$) corresponds to an excess electron density n (Fig. 31.8a) that increases with U_s ($>U_c$) according to

$$n(U_s) = N_0 \left(\frac{v_{ET,m}}{v_{ET}(U_s)} - 1 \right) = N_0 \frac{(1 - U_s/U_c)^2}{2U_s/U_c}. \quad (31.23)$$

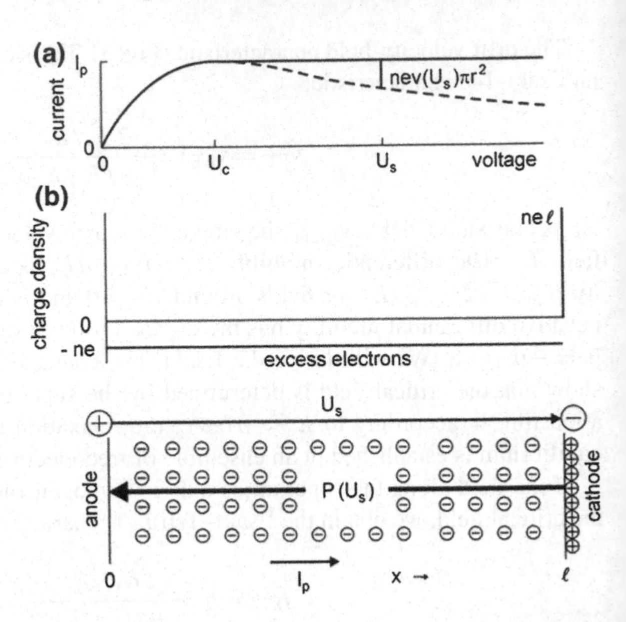

Fig. 31.8 Superlattice biased with a static voltage. **a** I–V curve and excess current at a static voltage U_s. **b** Charge density and polarization

The excess electron density is zero at $U_s = U_c$, equal to N_0 at $U_s = 3.7\ U_c$ and increases linearly with U_s at $U_s/U_c \gg 1$. For $U_s/U_c \gg 1$, the excess electron density increases linearly with the static voltage, $n = (1/2)N_0U_s/U_c$. We suppose that the positive charges are distributed on a plane (at the cathode) adjacent to the superlattice boundary. The excess charge within the superlattice is equal to the positive charge at the cathode (Fig. 31.8b, upper part). The density of the excess charge in the superlattice is *-ne* and the area density of the charge at the cathode is *nel*. The excess charge in the superlattice together with the positive charge at the cathode form a dipole domain (Fig. 31.8b, lower part). It consists of a charge density domain within the superlattice and a positive area charge bound to the cathode. In the absence of current oscillations, a dipole domain is associated with a quasistatic polarization $P(U_s) = -n(U_s)el/2$. The direction of polarization is opposite to the direction of the direct current I_p.

The density $n(U_s)$ increases with increasing U_s (Fig. 31.9a). If the voltage across the superlattice suddenly changes from U_s to $U_s + U_1$, additional excess electrons flow into the superlattice until the excess charge density is equal to $n(U_s + U_1)$ in the whole superlattice. If the voltage suddenly changes from $U_s + U_1$ to $U_s - U_1$, all excess electrons escape from the superlattice. For $U_s = 2\ U_c$ and $U_1 = 1.5\ U_c$, the characteristic time of a cycle of filling of the superlattice with excess electrons and their escape is $t_c \sim 2 \times l/(0.8 v_{\mathrm{ET,m}})$. The critical rate of generation of a full domain and its destruction is

$$\nu_c = 0.4\ v_{\mathrm{ET,m}}/l. \tag{31.24}$$

The critical rate is ~70 GHz at a superlattice of a length $l = 0.6\ \mu$m.

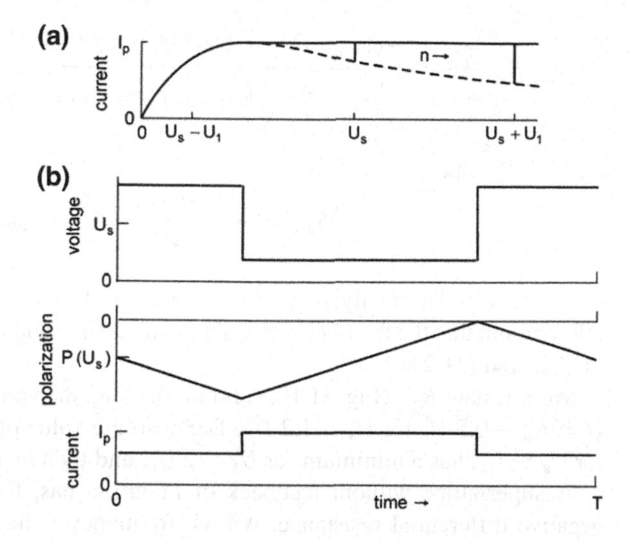

Fig. 31.9 Dynamics of gain. **a** I–V curve. **b** Voltage, polarization and current during an oscillation cycle

Under the influence of both a static and a high frequency voltage, the high frequency voltage causes a periodic change of polarization. The temporal change of polarization is equal to a polarization-current density. A high frequency voltage (frequency ν_c) of a rectangular shape (Fig. 31.9b) produces a polarization that has, in a simplified picture, a triangular shape and is phase-shifted by $\pi/2$. The current has a rectangular shape and is phase-shifted by π relative to the voltage. A Fourier transformation yields the amplitude $\hat{U} = (4/\pi)\, U_1$ of the high frequency voltage $U = \hat{U}\cos\omega t$ and the amplitude $\hat{P} = nel/2\pi$ of the high frequency polarization $P = \hat{P}\sin\omega t$. The high frequency polarization current is equal to $I = -\hat{I}\cos\omega t$, where $\hat{I} = \pi r^2 \omega \hat{P} = r^2 l \nu n e$ is the amplitude of the current; the high frequency polarization-current is continued outside the superlattice by a high frequency current (flowing through the antenna). For static voltages that are noticeably larger than U_c, the electrons are not fast enough to follow the high frequency voltage. Therefore, the effective length l_{eff} of an excess charge domain is shorter than l. We write $l_{\text{eff}} = 0.2 v_{\text{ET,m}}/\nu_c$. The product nl_{eff} $(= 1.2\, N_0 l)$ and thus $\hat{P}$ are independent of U_s. We obtain a constant current amplitude

$$\hat{I} = 1.2\, r^2 \nu l N_0 e. \tag{31.25}$$

A constant amplitude of the high frequency current results in a plateau in the $I - V$ curve for the superlattice in the oscillating state. This is in accord with the experimental result.

An analysis of the large-signal behavior of the amplitude of the high frequency voltage and the amplitude of the high frequency current leads to a negative differential resistance of the superlattice operating in the accumulation mode, $R_{\text{acc}} = -\hat{U}/\hat{I}$, which is equal to

$$R_{\mathrm{acc}} = -\frac{2\pi v_{\mathrm{ET,m}}}{\nu l} \frac{\hat{U}}{U_{\mathrm{c}}} \frac{2\left(U_{\mathrm{s}}/U_{\mathrm{c}} + (\pi/4)\hat{U}/U_{\mathrm{c}}\right)}{\left(|1 - U_{\mathrm{s}}/U_{\mathrm{c}} - (\pi/4)\hat{U}/U_{\mathrm{c}}|\right)^2} R_{\mathrm{ohm}} \tag{31.26}$$

for $U_{\mathrm{s}} \sim U_{\mathrm{c}}$ and

$$R_{\mathrm{acc}} = -\frac{1.6\pi v_{\mathrm{ET,m}}}{\nu l} \frac{\hat{U}}{U_{\mathrm{c}}} R_{\mathrm{ohm}}. \tag{31.27}$$

for $U_{\mathrm{s}}^2 \gg U_{\mathrm{c}}^2$. This analysis is oriented at the I–V curve (*see* Fig. 31.9a). It is taken into account that the flow of excess charge takes time and it was made use of (31.20)–(31.22) and (31.24).

We estimate R_{acc} (Fig. 31.10, solid line), using the values: $\hat{U} = 0.2\, U_{\mathrm{c}}$; $U_{\mathrm{s}} \sim U_{\mathrm{c}}$; $\hat{U} = U_{\mathrm{s}} - 0.5\, U_{\mathrm{c}}$ for $U_{\mathrm{s}} > 1.2\, U_{\mathrm{c}}$. The absolute value of R_{acc} has the largest value for $U_{\mathrm{s}} \sim U_{\mathrm{c}}$, has a minimum for $U_{\mathrm{s}} \sim 2\, U_{\mathrm{c}}$, and then increases with increasing U_{s}.

A superlattice without feedback of radiation has, for $U_{\mathrm{s}} > U_{\mathrm{c}}$, a small-signal negative differential resistance. A high frequency voltage $U = \hat{U} \cos \omega t$ of small amplitude $\hat{U}$ causes the high frequency polarization $P = -(el/2)n$. This leads, with $\mathrm{d}n/\mathrm{d}t = (\mathrm{d}n/\mathrm{d}U_{\mathrm{s}})\mathrm{d}U_{\mathrm{s}}/\mathrm{d}t$, to the current amplitude

$$\hat{I} = \pi r^2 (el/2)\omega \hat{U}\, \mathrm{d}n/\mathrm{d}U_{\mathrm{s}}, \tag{31.28}$$

where

$$\frac{\mathrm{d}n}{\mathrm{d}U_{\mathrm{s}}} = \frac{N_0}{U_{\mathrm{c}}} \frac{U_{\mathrm{s}}^2/U_{\mathrm{c}}^2 - 1}{2U_{\mathrm{s}}^2/U_{\mathrm{c}}^2}. \tag{31.29}$$

It follows that the small-signal differential resistance for the superlattice operating in a pure charge accumulation mode, $R_{\mathrm{acc,0}} = -\hat{U}/\hat{I}$, is given by

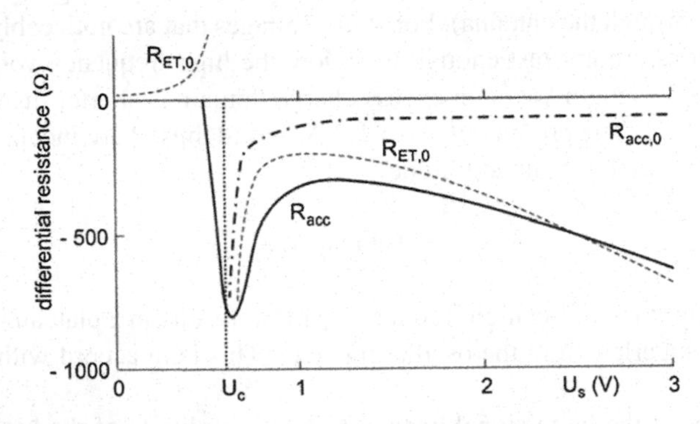

Fig. 31.10 Differential resistances of a superlattice; $R_{\mathrm{acc,0}}$, small-signal negative differential resistance of a superlattice operating in the pure charge accumulation mode; R_{acc}, large-signal negative differential resistance of a superlattice in an oscillator operating in the pure charge accumulation mode; $R_{\mathrm{ET,0}}$, Esaki–Tsu small-signal differential resistance

$$R_{\text{acc},0} = -\frac{2v_{\text{ET,m}}}{\pi \nu l}\frac{2U_s^2/U_c^2}{U_s^2/U_c^2 - 1}R_{\text{ohm}}. \tag{31.30}$$

$R_{\text{acc},0}$ (Fig. 31.10, dashed dotted) is equal to $-\infty$ for $U_s = U_c$ and assumes the constant value $-(2v_{\text{ET,m}}/\pi \nu l)(U_s^2/U_c^2)R_{\text{ohm}}$ for $U_s^2 \gg U_c^2$.

Thermal radiation in a resonator is amplified according to the small-signal negative resistance $R_{\text{acc},0}$. Due to fluctuations of amplified thermal radiation, a superlattice can be promoted into a state of larger negative resistance. If this resistance reaches R_{acc}, stable oscillation can occur. The resistance R_{acc} corresponds to the threshold resistance R_{th} since, for $R < |R_{\text{th}}|$, feedback is strong enough to start oscillation. Then at steady state oscillation, the superlattice resistance $|R_{\text{acc}}|$ assumes the value R. Because of fluctuations of the field, i.e., because of noise, oscillation can occur also for $U_s < U_c$.

It follows from (31.19) that the small-signal Esaki-Tsu differential resistance, $R_{\text{ET},0} = 1/(\mathrm{d}I/\mathrm{d}U_s)$ is equal to

$$R_{\text{ET},0} = \frac{(1 + U_s^2/U_c^2)^2}{1 - U_s^2/U_c^2}R_{\text{ohm}}. \tag{31.31}$$

$R_{\text{ET},0}$ (Fig. 31.10, dashed) is equal to the ohmic resistance R_{ohm} near $U_s = 0$, then increases and becomes infinitely large for $U \to U_c$. $R_{\text{ET},0}$ is negative for $U_s \geq U_c$, varies from $-\infty$ at $U_s = U_c$ to a value of $-8R_{\text{ohm}}$ for $U_s \sim 2U_c$ and is equal to $-(U_s^2/U_c^2)R_{\text{ohm}}$ for $U_s^2/U_c^2 \gg 1$.

In the voltage range of oscillation, the small-signal Esaki–Tsu resistance $R_{\text{ET},0}$ is comparable with the large-signal resistance R_{acc}. However, the absolute value of the large-signal Esaki–Tsu negative resistance R_{ET} is smaller than the absolute value of $R_{\text{ET},0}$ according to the slope of the Esaki–Tsu I–V curve. Therefore, $|R_{\text{acc}}|$ is larger than $|R_{\text{ET}}|$. This means that the interaction of the high frequency field with an electron collective of a pure charge accumulation mode is associated, with respect to the negative resistance, with a larger nonlinearity than the interaction of the high frequency field with single electrons in the case that the field in the superlattice is homogeneous.

31.6 Balance of Energy in a Superlattice Oscillator

The electric field associated with a domain has a triangular shape, with a low-field value E_1 at the anode and a high field value E_2 at the cathode; E_1 ($<E_c$) is also the field immediately after domain destruction. From the Poisson equation

$$\nabla \cdot \boldsymbol{D}\rho, \tag{31.32}$$

we obtain the relation $ne/2 = \epsilon\epsilon_0(E_2 - E_1)/l$. A fully developed dipole domain carries the field energy $\pi r^2 l \epsilon\epsilon_0(E_2 - E_1)^2/2$. The field energy of a domain stems from the high frequency field in the resonator. During a half cycle of the field, the domain transfers its field energy to the high frequency field and during the following half cycle, energy of the high frequency field is used to build up the field of the domain.

Energy balance requires that the power delivered by the voltage source is equal to the sum of the losses (we modify a discussion in [236]):

$$\pi r^2 l \nu N_0 e U_s = \pi r^2 l \nu N_0 e E_1 l + P_{out} + P_{dom}. \tag{31.33}$$

The loss terms concern:

- $\pi r^2 l \nu N_0 e E_1 l$ = loss due to the current carried by the electrons (of density N_0).
- P_{out} = loss due to output coupling of radiation.
- P_{dom} = loss due to dissipation caused by relaxation processes during domain formation and destruction.

We find

$$U_s = E_1 l + U_{rad} + U_{dom}, \tag{31.34}$$

where $U_{rad} = P_{out}/(\pi r^2 l \nu N_0 e)$ and $U_{dom} = P_{dom}/(\pi r^2 l \nu N_0 e)$. The static voltage across the superlattice is equal to the sum of three terms: the voltage necessary to drive the normal electrons by the field E_1; the voltage U_{rad} necessary to compensate loss of radiation and the voltage U_{dom} necessary for compensation of energy of dissipation associated with domains. The normal electrons drift with the average velocity $\mathrm{v}(E_1)$ through the superlattice. The domains, with the positive charges bound to the cathode, appear and disappear at the repetition rate ν.

Example (*for the superlattice already discussed*) For $U_s = 2\ U_c$ (=1.2 V) and $P_{rad} = P_{out}$ at optimum output, the analysis yields the data: $E_1 = 0.7\ E_c$; $E_2 = 4\ E_c$; $E_1 l = 0.5$ V; $U_{rad} = 0.25$ V; $U_{dom} = 0.5$ V. Accordingly, the dissipation energy is, for $U_s = 2\ U_c$, equal to half the field energy of a fully developed domain. The direct current strength is determined by the drift velocity at the lower field and is given by the expression $I_{dc}/I_p = N_0 e \mathrm{v}(E_1)/I_p$ (~0.8).

The upper limit frequency ν_{limit} is determined by the intraminiband relaxation of the electrons in a superlattice. It follows from the intraminiband relaxation time (1.5×10^{-13} s) that ν_{limit} is ~1 THz. The appropriate superlattice length, according to the relation $\nu_c = 0.4\ \nu_{ET,m}/l$ has a value of ~10 nm. This means that we are no longer dealing with a superlattice but with a resonant-tunneling diode like structure (next section).

A more detailed discussion of the pure charge accumulation mode observed for superlattice oscillators can be found in [246]. The study presents a method that is suited to investigate the mechanism of gain of a solid state oscillator. Such studies may contribute to an improvement of the efficiency of microwave oscillators, particularly in the range above 100 GHz.

31.7 Resonant-Tunneling Diode Oscillator

The resonant-tunnel diode oscillator [234, 247] is a quasiclassical solid state oscillator that reaches very high oscillation frequencies. Because of small radiation power, resonant-tunneling diode oscillators are not in use.

A resonant-tunneling diode (Fig. 31.11a) consists, for instance, of two AlAs layers separated by a GaAs layer, embedded in n GaAs. The layers form a quantum well with a discrete energy level for electron motion perpendicular to the layers. Under the action of a static voltage U_s, electrons tunnel through the quantum well from one n GaAs region to the other n GaAs region, which results in a current. If the energy of the tunneling electrons coincides with the energy of the discrete energy level (Fig. 31.11b), the tunnel current has a maximum as indicated in the I–V curve (Fig. 31.11c). The I–V curve has, for a voltage above a critical voltage U_c, a negative slope, which corresponds to a negative differential resistance. The I–V curve is a hypothetical curve: because of the negative differential resistance, the charge distribution is inhomogeneous.

The negative differential resistance gives rise to a self-excited oscillation if the active element is coupled to a resonance circuit; the oscillation frequency is determined by the resonator. Radiation generated in first order has been observed in frequency ranges from 10 GHz up to several hundred GHz. The power decreased strongly at frequencies above 100 GHz. The highest frequency of radiation emitted

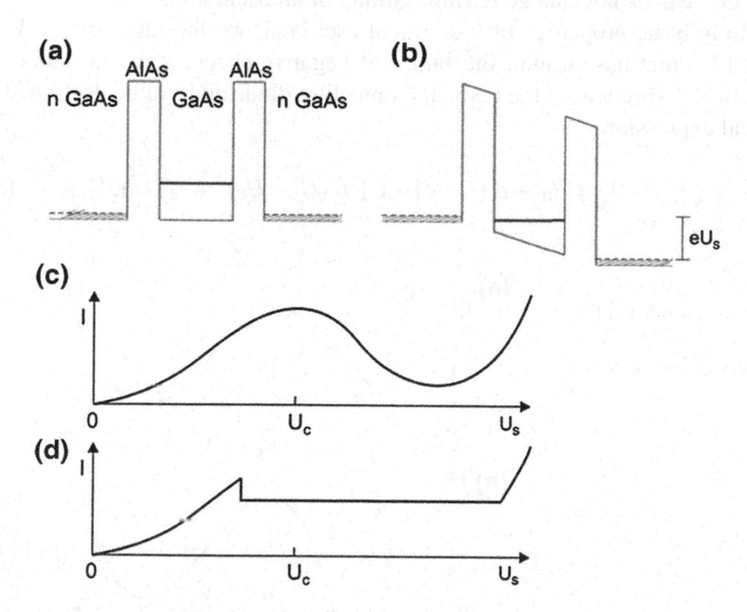

Fig. 31.11 Resonant tunnel diode. **a** Quantum well. **b** Voltage-biased quantum well. **c** Hypothetical I–V curve. **d** I–V curve in the case of occurrence of feedback from a high frequency field

by a GaAs/AlAs resonant-tunneling diode oscillator was near 400 GHz [247] and near 700 GHz for an InAs/AlSb resonant-tunneling diode oscillator [248].

Example InAs/AlAs resonant-tunneling diode [249]. Double barrier structure with 1.5 nm thick undoped barriers separated by a 6.4 nm thick undoped InAs quantum well; diameter 1.8 μm; current 5 mA; voltage 1.3 V; $R_{opt} \sim -50\ \Omega$; power 0.3 μW at 712 GHz.

The resonant-tunneling diode oscillators operated most likely in the pure charge accumulation mode.

31.8 Van der Pol Oscillator

We discuss a model of a classical electric oscillator, namely the van der Pol oscillator. The active device is a resistance that shows a negative differential resistance above a critical voltage U_c (Fig. 31.12a); the $I-V$ curve resembles the hypothetical $I-V$ curve of the resonant-tunneling diode. Under the action of a static voltage (bias voltage U_0), with the resistance coupled to a resonant circuit, a self-excited oscillation can occur. It is characteristic of this classical oscillator model that the current through the active device and the voltage across the device always follow the $I-V$ curve and that the curve does not change during buildup of an oscillation.

To study basic properties of a classical oscillator, we introduce an $I-V$ curve (Fig. 31.12b) that has, around the range of negative shape, a similar slope as the hypothetical $I-V$ curve of the resonant-tunneling diode and can be described by an analytical expression,

$$I(U) = I_0 - a\,(U - U_0) + b\,(U - U_0)^3 = f(U), \qquad (31.35)$$

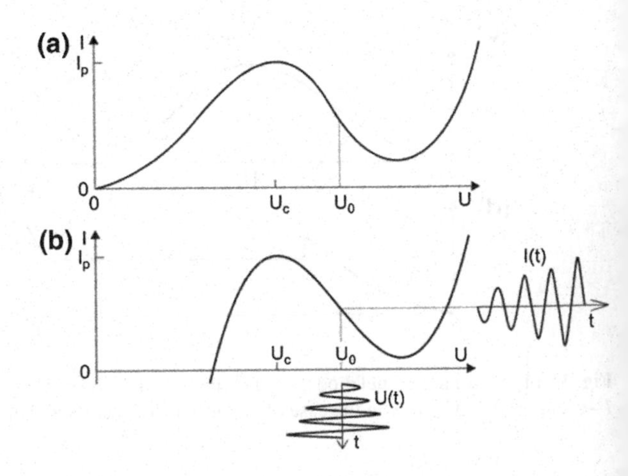

Fig. 31.12 Classical oscillator. **a** Hypothetical $I-V$ curve of a tunnel diode. **b** $I-V$ curve of a van der Pol oscillator

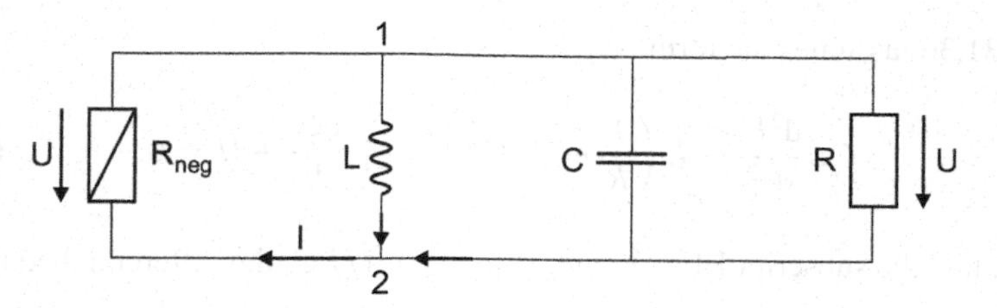

Fig. 31.13 Equivalent circuit of a negative resistance oscillator of the van der Pol type

where $a > 0$ and $b > 0$ are constants. The $I-V$ curve has the largest negative slope for $U = U_0$. The slope of the $I-V$ curve according to (31.35) is unrealistic for voltages $U \ll U_0$ and $U \gg U_0$. In the range around U_0, it shows for appropriate parameters a, b, U_0, and I_0, the hypothetical characteristic of a tunnel diode.

We consider a parallel equivalent circuit (Fig. 31.13) containing a negative resistance $R_{\rm neg}$, an inductance L, a capacitance C, and a loss resistance R, which describes loss due to emission of radiation. The high frequency currents through the capacitance ($I_{\rm c}$), through the inductance ($I_{\rm L}$), and through the resistance ($I_{\rm R}$) are related to the high frequency voltage $U_{\rm HF}$ between the points 1 and 2,

$$I_{\rm C} = C\,\frac{{\rm d}U_{\rm HF}}{{\rm d}t}; \quad I_{\rm L} = \frac{1}{L}\int U_{\rm HF}{\rm d}t; \quad I_{\rm R} = \frac{U_{\rm HF}}{R}. \tag{31.36}$$

The sum of the total current in point 1 of the circuit must be zero,

$$I_{\rm C} + I_{\rm L} + I_{\rm R} + I_{\rm d} = 0. \tag{31.37}$$

$I_{\rm d}$ is the current through the nonlinear device. The signs follow from Kirchhoff's rules for voltages and currents in a circuit taking into account that the instantaneous voltage $U_{\rm HF}(t)$ across the active element has a sign that is opposite to the sign of the high frequency current flowing through the resistance. The sum of all currents through a knot is zero and the sum of all voltages in a loop is zero according to Kirchhoff's rules. By differentiation, we obtain

$$CL\frac{{\rm d}^2 U_{\rm HF}}{{\rm d}t^2} + U_{\rm HF} + \frac{L}{R}\frac{{\rm d}U_{\rm HF}}{{\rm d}t} = -L\,\frac{{\rm d}I_{\rm d}}{{\rm d}t}. \tag{31.38}$$

The current, i.e., the derivative of the current with respect to time, is the source of the high frequency voltage.

We can write

$$\frac{{\rm d}I_{\rm d}}{{\rm d}t} = \frac{{\rm d}I_{\rm d}}{{\rm d}U}\frac{{\rm d}U}{{\rm d}t} = \frac{{\rm d}f}{{\rm d}U}\frac{{\rm d}U}{{\rm d}t}. \tag{31.39}$$

Then (31.36) assumes the form

$$CL\frac{d^2U}{dt^2} + L\left(\frac{1}{R} + f'(U_0 + U)\right)\frac{dU}{dt} + U = 0. \tag{31.40}$$

We omitted the subscript HF. We find, with $\omega_0^2 = 1/LC$, the differential equation

$$\frac{d^2U}{dt^2} + (-\gamma + \kappa)\frac{dU}{dt} + \omega_0^2 U = 0, \tag{31.41}$$

where

$$\gamma = \gamma(U) = -C^{-1}\left(\frac{\partial I_d}{\partial U}\right)_U \tag{31.42}$$

is the growth coefficient and

$$\kappa = 1/RC \tag{31.43}$$

the damping coefficient. In the active element of a classical oscillator, the growth coefficient γ depends on the instantaneous voltage $U(t)$ at time t.

Using the analytical form (31.20) of the I–V curve, we can write

$$I_{HF} = -aU_{HF} + bU_{HF}^3. \tag{31.44}$$

It follows that

$$\left(\frac{\partial I}{\partial U}\right)_U = -a + 3bU^2; \tag{31.45}$$

we again omit the subscript HF. We find the growth coefficient

$$\gamma = -a/C + 3b/C\ U^2 \tag{31.46}$$

and obtain the differential equation (*van der Pol equation*) for the high frequency voltage

$$\frac{d^2U}{dt^2} + \left(-\gamma_0 + \kappa + \frac{3b}{C}U^2\right)\frac{dU}{dt} + \omega_0^2 U = 0. \tag{31.47}$$

The differential equation describes a self-excited oscillator with the small-signal growth coefficient

$$\gamma_0 = \frac{a}{C} \tag{31.48}$$

and two damping terms. The first damping term, κ, characterizes output coupling of electromagnetic radiation and the second term intrinsic loss in the active element. This loss is zero for $U = 0$ and increases proportionally to the square of U. The van der Pol equation describes an oscillation that is strongly nonlinear, except in the case that the net gain is small,

$$\gamma_0 - \kappa \ll \omega_0. \tag{31.49}$$

In this case the van der Pol equation has a solution that corresponds to a nearly harmonic oscillation. With the ansatz

$$U = A(t) \cos \omega_0 t, \tag{31.50}$$

where $A(t)$ is a slowly varying function, $\gamma_0 - \kappa \ll \omega_0$, we obtain

$$\frac{\mathrm{d}U}{\mathrm{d}t} = \frac{\mathrm{d}A}{\mathrm{d}t} \cos \omega_0 t - \omega_0 A \sin \omega_0 t, \tag{31.51}$$

$$\frac{\mathrm{d}^2 U}{\mathrm{d}t^2} = -2\omega_0 \frac{\mathrm{d}A}{\mathrm{d}t} \sin \omega_0 t - \omega_0^2 A \cos \omega_0 t. \tag{31.52}$$

The differential equation leads, with $(\gamma_0 - \kappa) \left| \frac{\mathrm{d}A}{\mathrm{d}t} \right| \ll \omega_0 \left| \frac{\mathrm{d}A}{\mathrm{d}t} \right|$ (SVEA), to

$$-2\omega_0 \frac{\mathrm{d}A}{\mathrm{d}t} \sin \omega_0 T - (-\gamma_0 + \kappa)\omega_0 A \sin \omega_0 t - \frac{3b\omega_0}{C} A^3 \cos^2 \omega_0 t \sin \omega_0 t = 0. \tag{31.53}$$

Using the relation

$$\cos^2 \alpha \sin \alpha = \frac{1}{2}(1 + \cos 2\alpha) \sin \alpha = -\frac{1}{4} \sin \alpha + \frac{1}{4} \sin 3\alpha \tag{31.54}$$

and neglecting the higher order term $\sin 3\alpha$, we find

$$\frac{\mathrm{d}A}{\mathrm{d}t} + \frac{1}{2}(-\gamma_0 + \kappa)A + \frac{3b}{8C} A^3 = 0. \tag{31.55}$$

This differential equation has exactly the same form as the differential equation (9.144) derived for the amplitude of the field in a laser oscillator. The solution is

$$A(t) = \frac{A_\infty}{\sqrt{1 + (A_\infty / A_0)^2 \,\mathrm{e}^{-(\gamma_0 - \kappa)t}}}. \tag{31.56}$$

$A_0 = A(t = 0)$ is the initial amplitude of the voltage and

$$A_\infty = 2\sqrt{(\gamma_0 - \kappa)C/3b} \tag{31.57}$$

is the amplitude of the high frequency voltage at steady state oscillation. After a sudden turning on of the active element, a small high frequency voltage initiates the buildup of an oscillation. The initial high frequency voltage stems from noise in the resonance circuit.

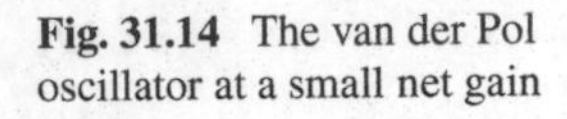

Fig. 31.14 The van der Pol oscillator at a small net gain

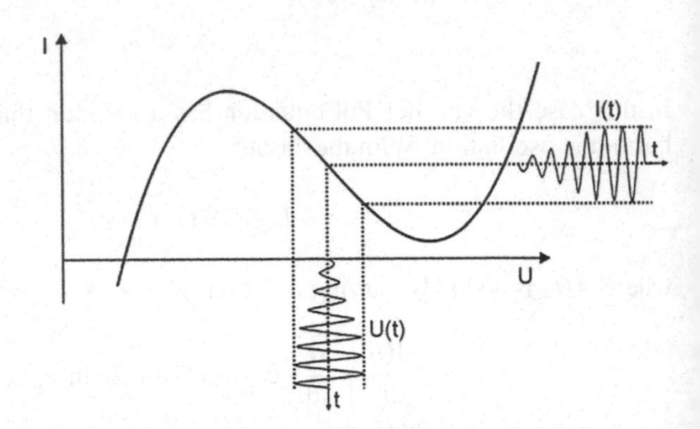

The van der Pol oscillator of small net gain $\gamma_0 - \kappa$ is driven in a range of the voltage amplitude that corresponds to the range of almost constant negative slope of the I–V curve (Fig. 31.14). At small amplitude of the voltage, the intrinsic damping is negligibly small. At large amplitude and steady state oscillation, the intrinsic damping becomes efficient during each cycle at instantaneous voltages in the ranges $U \approx \pm A$. This leads, as our analysis shows, to the same form of the first-order differential equation for the amplitude of the voltage in the classical oscillator as we found for the amplitude of the field in a laser oscillator, although the nonlinearities have completely different origins.

The van der Pol oscillator represents a model oscillator of a negative-resistance oscillator that is discussed in many textbooks; *see*, for instance, [250].

References [240–243].

Problems

31.1 Equivalent circuit.

(a) Replace the equivalent circuit of Fig. 31.5 by a parallel resonant circuit; the active device has the negative admittance $G_{\rm d}$ and the loss resistor the admittance G.
(b) Derive the differential equation for the high frequency voltage.
(c) Discuss the dependence of the negative admittance of the device on the voltage across the device.
(d) Show that the output power of the oscillator is $P_{\rm out} = (1/2)G\hat{U}^2$, where $G + G_{\rm neg} = 0$ is the condition of steady state oscillation.

31.2 Electric polarization.

(a) Determine the electric polarization of a dipole domain consisting of a positive area charge $\rho l e$ at $x = 0$ and a negative charge of density ρ in the range $0, l$.

(b) Determine the polarization of a dipole domain consisting of a negative charge of area density $\rho l e$ at $x = x_0$ and a positive charge of density ρ in the range $x_0, x_0 + l$.

31.3 Van der Pol oscillator.

(a) Evaluate for a van der Pol oscillator (with the data: $a = 10^{-2}\ \Omega^{-1}$; $b = 10^{-2}\ \Omega^{-1}\ \mathrm{V}^{-2}$; $\omega_0 = 2\pi \times 10^{10}$ Hz; $C = 1$ pF; and $G = 1\Omega^{-1}$) the small-signal net growth coefficient and show that it is small compared to ω_0.
(b) Determine the voltage amplitude for the steady state oscillation.
(c) Determine the current amplitude for the steady state oscillation. [*Hint*: make use the relation $\cos^3\alpha = \frac{3}{4}\cos\alpha + \frac{1}{3}\cos 3\alpha$ and neglect the term with 3α.]

31.4 Van der Pol equation.

(a) Show that the van der Pol equation can be written in dimensionless units,

$$\frac{\mathrm{d}^2 y}{\mathrm{d}\tau^2} + \epsilon(-1 + y^2)\frac{\mathrm{d}y}{\mathrm{d}\tau} + y = 0,$$

where y is the voltage in dimensionless units, $\tau = \omega_0 t$ the dimensionless time and ϵ the small-signal net gain coefficient in dimensionless units.
(b) Solve the van der Pol equation for $\epsilon \ll 1$ at steady state oscillation. [*Hint*: make use of the relation $\cos^3\tau = \frac{3}{4}\cos\tau + \frac{1}{3}\cos 3\tau$ and neglect the term with $\cos 3\tau$.]

31.5 Which of the following differential equations describe a self-sustained oscillation?

(a) $\dfrac{\mathrm{d}^2 y}{\mathrm{d}t^2} + \dfrac{\mathrm{d}y}{\mathrm{d}t} + y = 0.$

(b) $\dfrac{\mathrm{d}^2 y}{\mathrm{d}t^2} - \dfrac{\mathrm{d}y}{\mathrm{d}t} + y = 0.$

(c) $\dfrac{\mathrm{d}^2 y}{\mathrm{d}\tau^2} + \varepsilon(-1 + y^2)\dfrac{\mathrm{d}y}{\mathrm{d}\tau} + y = 0.$

31.6 Compare a classical oscillator and a laser oscillator. (A classical oscillator has a stable I–V characteristic, Fig. 31.12, while a laser oscillator has a current-density-field characteristic that varies during onset of oscillation, Fig. 9.7.)

31.7 Show that the impedance $Z(\omega)$ of a resonance electrical circuit has Lorentzian lineshape.

Chapter 32
Superlattice Bloch Laser: A Challenge

The superlattice Bloch laser (also called Bloch oscillator) exists only as an idea. We discuss this type of laser for two reasons. First, a superlattice Bloch laser would provide a semiconductor source of coherent radiation in the 1–10 THz range—with operation at room temperature. Second, there are, with respect to the formal description of a Bloch laser medium, many similarities to a free-electron laser medium, although the origin of gain is completely different.

In a superlattice Bloch laser, free-electrons in a semiconductor superlattice perform, under the action of a static electric field directed along the superlattice axis, Bloch oscillations. The oscillation frequency (=Bloch frequency = resonance frequency) is determined by the strength of the static field and the period of the superlattice. A high frequency electric field, also oriented along the superlattice axis, causes a phase modulation of the Bloch oscillations. This results in a high frequency drift current along the superlattice axis. Interaction of the high frequency drift current with the high frequency electric field mediates gain for the high frequency field; gain is due to transfer of energy of translation of the electrons to energy of the high frequency field. The Bloch frequency increases linearly with the strength of the static field. Frequency tuning over a large range is possible by changing the static field strength, that is, the voltage across a superlattice. The amplitude of the high frequency field in an active medium of a Bloch laser medium is limited; even if the high frequency field in the laser resonator has no loss, the field cannot exceed a saturation field—conventional lasers do not have such a limitation.

Suitable as an active medium is a semiconductor superlattice submitted to a homogeneous static electric field E_s of a strength that is larger than a critical field E_c. However, under this condition, the electrons tend to form charge density domains, which destroy the homogeneity of the field. We will mention methods that may be suited to avoid domains.

We study transport properties of a superlattice in a homogeneous electric field—assuming that the field remains homogeneous even if $E_s > E_c$. We characterize Bloch oscillations and derive the current-voltage characteristic of a superlattice.

K.F. Renk, *Basics of Laser Physics*, Graduate Texts in Physics,
DOI 10.1007/978-3-319-50651-7_32

We derive the small-signal gain coefficient and the saturation field amplitude. Gain can occur up to the Bloch frequency ν_B.

We also present an energy-level description of the superlattice Bloch laser. The energy levels of an electron, which executes Bloch oscillations, form an energy-ladder system (a Wannier–Stark ladder): the energy levels are equidistant and have a next-near energy distance of $h\nu_B$. In this description, radiation is generated by stimulated transitions of electrons. An electron, which occupies a level of the energy-ladder system, emits a photon by a transition to a distorted state at an energy slightly above the next near energy level at lower energy.

32.1 Principle of a Superlattice Bloch Laser

Figure 32.1 illustrates the principle of a superlattice Bloch laser. A voltage source (voltage U_s) produces a direct current (I) that flows as electron current through an n-doped semiconductor superlattice. In the superlattice, miniband electrons carry the current. The miniband electrons execute—under the action of the static field along the superlattice axis—free-electron oscillations (=Bloch oscillations). The resonance frequency of a free-electron oscillator is the Bloch frequency

$$\nu_B = \frac{eaE_s}{h}. \tag{32.1}$$

E_s is the strength of the static field within the superlattice and a the superlattice period. The Bloch frequency is proportional to E_s and to a.

We assume that a superlattice fills a resonator, which is formed by the metallic anode, the metallic cathode, and four free surfaces of the superlattice. Interaction of the electrons, performing Bloch oscillations, with a high frequency electric field can result in gain (*Bloch gain*) for the high frequency electric field. Radiation (power P_{out}) is coupled out via free surfaces of the superlattice.

The gain coefficient is, approximately, given by (Sect. 32.4):

$$\alpha(\nu) = \alpha_p \, g_{L,disp}(\nu), \tag{32.2}$$

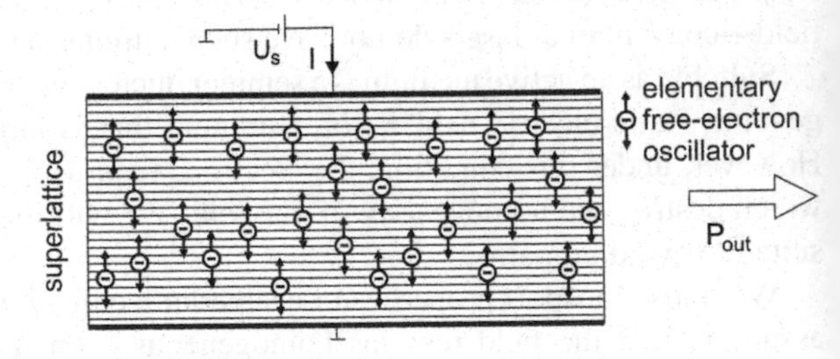

Fig. 32.1 Principle of a superlattice Bloch laser

where

$$\alpha_{\mathrm{p}} = f(T)\frac{N_0 e^2}{4\pi (c/n)\varepsilon_0 m^* \nu_{\mathrm{B}}} \tag{32.3}$$

is a peak gain coefficient, $f(T) \leq 1$ a temperature parameter, N_0 the electron density, c/n the speed of light in the superlattice, n the refractive index at the laser frequency, m^* the effective mass of an electron at the bottom of the miniband of the superlattice

$$\bar{g}_{\mathrm{L,disp}}(\nu) = \frac{\nu_0 - \nu}{\Delta\nu_{\mathrm{B}}/2}\, \bar{g}_{\mathrm{L,res}}(\nu) \tag{32.4}$$

is the normalized Lorentz dispersion function, and

$$\bar{g}_{\mathrm{L,res}}(\nu) = \frac{\Delta\nu_{\mathrm{B}}^2/4}{(\nu_{\mathrm{B}} - \nu)^2 + \Delta\nu_{\mathrm{B}}^2/4} \tag{32.5}$$

the corresponding Lorentz resonance function; it has the halfwidth $\Delta\nu_B = 1/(\pi\tau)$, where τ is the dephasing time of the Bloch oscillations. (Dephasing is mainly due to energy relaxation. Therefore, the dephasing time is equal to the energy relaxation time of an electron.)

The gain curve (Fig. 32.2) is antisymmetric with respect to the Bloch frequency ν_{B}, in the vicinity of the Bloch frequency. The gain coefficient is positive for $\nu < \nu_{\mathrm{B}}$. The maximum small-signal gain coefficient is equal to

$$\alpha_{\mathrm{m}} = \frac{\alpha_{\mathrm{p}}}{2} = f(T)\frac{N_0 e^2}{4\pi (c/n)\varepsilon_0 m^* \nu_{\mathrm{B}}}. \tag{32.6}$$

The distance between the frequency of the maximum and the frequency of the minimum of the gain coefficient curve is equal to $\Delta\nu_{\mathrm{B}}$. The maximum of the gain coefficient occurs at the frequency $\nu_{\mathrm{B}} - \Delta\nu_{\mathrm{B}} = 2$. The maximum gain coefficient is proportional to the electron density and inversely proportional to the Bloch frequency.

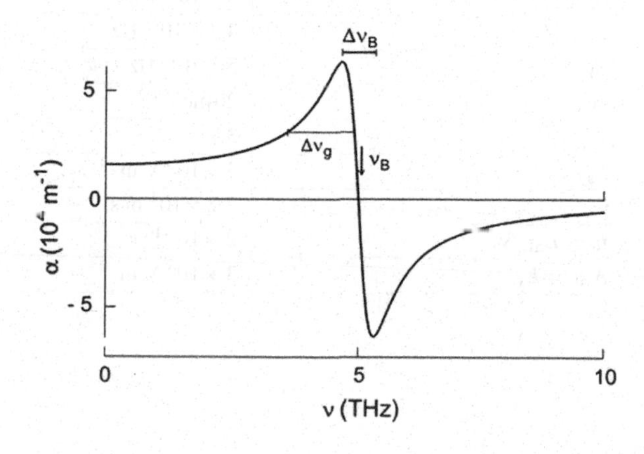

Fig. 32.2 Calculated gain coefficient of a semiconductor superlattice

The amplitude of the field in a Bloch laser is limited. The saturation field amplitude is equal to (Sect. 32.5)

$$A_{\mathrm{sat}} = \frac{h\nu_{\mathrm{B}}}{ea}. \tag{32.7}$$

It follows that the output power is given by

$$P_{\mathrm{out}} = \frac{\varepsilon_0\, A_{\mathrm{sat}}^2\, a_1 a_2 L}{2\tau_{\mathrm{p}}}, \tag{32.8}$$

where $a_1 a_2 L$ is the volume of the superlattice and τ_{p} the lifetime of a photon in the laser resonator; we assume that emission occurs, for instance, mainly via the two areas opposite to the long axis of a 101 rectangular resonator.

Table 32.1 shows data used for calculation of the gain coefficient of a particular GaAs/AlAs superlattice; a superlattice with 14 monolayers of GaAs and 2 monolayers of AlAs per period ($a = 4.2$ nm). A superlattice with these data can be prepared with high quality by the use of molecular beam epitaxy. It has the largest miniband width ϵ_{m} (=0.14 eV) that can be achieved for a GaAs/AlAs superlattice; the effective mass m^* of a superlattice with these data is almost the same as for bulk GaAs. The table lists data that concern different quantities.

Table 32.1 Data of a GaAs/AlAs superlattice Bloch laser

Quantity	Value	
a	4.2 nm	Superlattice period
ϵ_{m}	140 meV	Miniband width
$m^* = 2\hbar^2/\varepsilon_{\mathrm{m}} a^2$	6×10^{-32} kg	Effective mass
τ	5×10^{-13} s	Relaxation time
$\Delta\nu_{\mathrm{B}} = 1/\pi\tau$	6×10^{11} Hz	Width of resonance
$Q_{\mathrm{B}} = \nu_{\mathrm{B}}/\Delta\nu_{\mathrm{B}}$	8	Quality factor
$\Delta\nu_{\mathrm{g}} \sim \Delta\nu_{\mathrm{B}}/2$	6×10^{11} Hz	Gain bandwidth
E_{c}	3×10^5 V m^{-1}	Critical field
ν	4.7×10^{12} Hz	Laser frequency
ν_{B}	5×10^{12} Hz	Bloch frequency
$h\nu_{\mathrm{B}}$	20 meV	
n	3.7	Refractive index
E_{s}	3×10^6 V m^{-1}	Static field strength
v	2.7×10^4 m s^{-1}	Drift velocity
$\tau_{\mathrm{d}} = L_{\mathrm{SL}}/\mathrm{v}$	2×10^{-10} s	Drift time
$A_{\mathrm{sat}} = E_s$	3×10^6 V m^{-1}	Saturation field amplitude

(continued)

Table 32.1 (continued)

Quantity	Value	
a_1	9 μm	Width of superlattice
$a_2 = L_{SL} = 1.2 \times 10^3\ a$	5 μm	Height of resonator
$a_3 = L$	20 μm	Length of resonator
$P_{out} =$ $(2c/4n)(1-R)\ A_{sat}^2\ a_1 a_2$	20 mW	Output power (for R = 0.9)
$r_{ph} = P_{out}/h\nu$	6×10^{18} s^{-1}	Photon emission rate
τ_{stim}	4×10^{-12} s	Time between two stimulated emission processes
N_0	3×10^{22} m^{-3}	Electron density
$I = N_0 e v a_1 a_3$	23 mA	Current
$r_{el} = I/e$	1.4×10^{17} s^{-1}	Electron transit rate
r_{ph}/r_{el}	50	Photons per electron
$U = L_{SL} E_s$	15 V	Voltage across superlattice
$P_{el} = UI$	350 mW	Electric power
$\eta_P = P_{out}/P_{el}$	6%	Power efficiency
$f(300K)$	0.6	Temperature parameter
$\alpha_m =$ $f(300\,K)N_0 e^2/8\pi c_m \varepsilon_0 m^* \nu_B$	7×10^4 m^{-1}	Small-signal gain coefficient
$\sigma_{21} = \alpha_m/N_0$	2×10^{-18} m^2	Gain cross section
$G_1 = \exp(\alpha_m a_3)$	4	Small-signal gain factor

- Superlattice data: period; miniband width; effective mass; intraminiband relaxation time of an electron; resonance bandwidth $\Delta\nu_B$; quality factor of the Bloch oscillation; gain bandwidth; critical field.
- Data of a Bloch laser for a particular frequency ($\nu = 4.7$ THz): refractive index of the superlattice material; Bloch frequency leading to maximum gain and the corresponding static field strength E_s; drift velocity of an electron at the field E_s; time (τ_d) it takes an electron to drift through the superlattice; saturation field amplitude; L_{SL} = length; and a_1 = width of superlattice; L = length of resonator.
- Extensions of the superlattice; output power for an output coupling loss $1 - R$ for a reflectivity $R = 0.9$; photon emission rate; time between subsequent stimulated emission processes.
- Electron density; current; rate of electron transits through the superlattice; number of photons generated by an electron.
- Voltage across the superlattice; electric power; power efficiency.
- Temperature parameter (Sect. 32.10) $f(T) = J_0(\varepsilon_m/kT)/J_1(\varepsilon_m/kT)$; J_0, Bessel function of zeroth order and J_1, of first order; small-signal gain coefficient of the superlattice medium; gain cross section of an electron; small-signal gain factor at the maximum of the gain curve.

We will show:

- Gain is due to modulation of the Bloch oscillations by the high frequency field.
- A saturation field limits the gain at steady state oscillation.
- An oscillating electron is describable as an energy-ladder system and the active medium of a Bloch laser as an ensemble of energy-ladder systems.

32.2 Bloch Oscillation

Figure 32.3a shows the dispersion curve $\epsilon(k_x)$ of a miniband electron,

$$\epsilon = (1/2)\epsilon_{\mathrm{m}}(1 - \cos k_x a); \tag{32.9}$$

ϵ is the energy of propagation along the superlattice axis (=x axis), ϵ_{m} the maximum energy in the miniband and k_x the wave vector along the superlattice axis. Around the minimum, $\epsilon = 0$, where $k_x a \ll 1$, the dispersion relation is approximately given by

$$\epsilon = \frac{\hbar^2 k_x^2}{2m^*}, \tag{32.10}$$

where

$$m^* = \frac{2\hbar^2}{\epsilon_{\mathrm{m}} a^2} \tag{32.11}$$

is the effective mass of an electron at the bottom of the miniband.

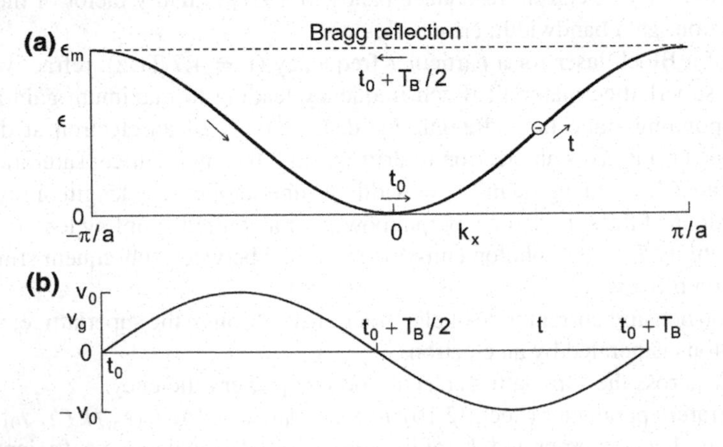

Fig. 32.3 Bloch oscillation of an electron. **a** Dispersion curve and Bragg reflection in the k space. **b** Group velocity

We describe an electron propagating in x direction as a wave packet composed of plane waves of different wave vectors k_x that has a central wave vector $k_{x,c}$. The central wave vector corresponds to a de Broglie wavelength $\lambda_{dB} = 2\pi/k_{x,c}$. In the following, we consider the temporal change of the central wave vector under the influence of a force. We omit, for convenience, the subscript "c".

We first study the motion of an electron under the action of a static electric field E_s along the x axis. The field leads to acceleration of an electron (charge $q = -e$) according to the equation of motion

$$\hbar dk_x/dt = qE_s. \tag{32.12}$$

In this semiclassical equation of motion, $\hbar k_x$ plays the role of the classical momentum; *see*, for instance, [179]. The equation (also called acceleration theorem) corresponds to Newton's equation of motion in classical physics. The solution is

$$k_x = (qE_s/\hbar)(t - t_0), \tag{32.13}$$

where t_0 is the time the electron starts with the wave vector $k_x = 0$. The wave vector increases linearly with time. Multiplying k_x by a, we find that the phase

$$k_x a = (qaE_s/\hbar)\ (t - t_0) \tag{32.14}$$

increases linearly with time. The group velocity is

$$v_g = \frac{1}{\hbar}\frac{\partial\epsilon}{\partial k}\ v_0\ \sin[\omega_B(t - t_0)], \tag{32.15}$$

where

$$v_0 = \frac{\epsilon_m a}{2\hbar} \tag{32.16}$$

is the maximum group velocity and ω_B the Bloch frequency, which is given by the relationship

$$\omega_B = eaE_s/\hbar. \tag{32.17}$$

The maximum group velocity increases proportional to ϵ_m. The Bloch frequency is proportional to the strength of the static field. The energy $\hbar\omega_B$ is the energy an electron can gain in the field E_s when it travels over of a superlattice period a. An electron executes Bloch oscillations with the period $T_B = 2\,\pi/\omega_B$.

An electron starting with $k(t_0) = 0$ is accelerated, reaches the mini-Brillouin zone boundary after $T_B/2$ (*see* Fig. 32.3a), experiences a Bragg reflection, is decelerated until it begins a new oscillation cycle. The group velocity of the electron wave packet (Fig. 32.3b) varies harmonically with the period T_B.

The spatial coordinate $\xi = \int_0^t v_g(t')dt'$ also varies periodically (Fig. 32.4a),

$$\xi = (1/2)\ \xi_m\ (1 - \cos[\omega_B(t - t_0)]), \tag{32.18}$$

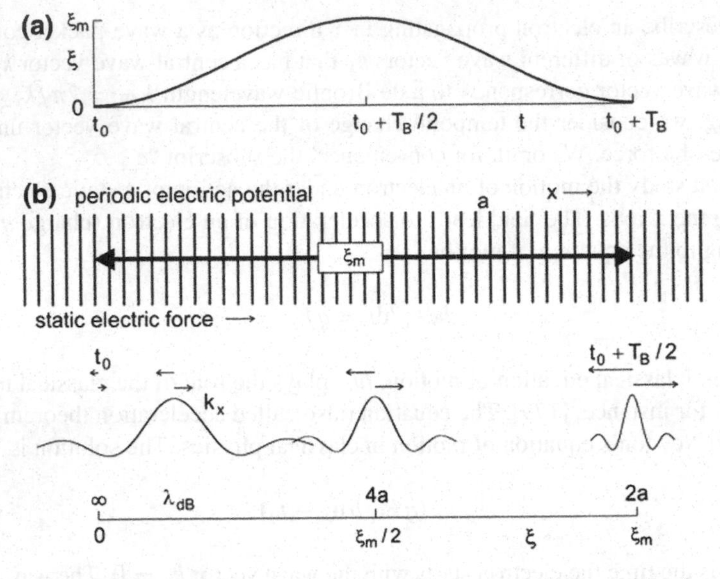

Fig. 32.4 Bloch oscillation of an electron in space. **a** Displacement. **b** Bragg reflection

where

$$\xi_m = \frac{\epsilon_\mathrm{m}}{eE_\mathrm{s}} = \frac{\epsilon_\mathrm{m}}{\hbar\omega_\mathrm{B}}\, a \tag{32.19}$$

is the length of the trajectory. The Bloch oscillation thus corresponds to a periodic motion of an electron wave packet in space around $\xi_m/2$. We suppose that $\varepsilon_\mathrm{m} \gg \hbar\omega_\mathrm{B}$, i.e., that the trajectory extends over many superlattice periods (Fig. 32.4b, upper part). The electron wave packet, periodically accelerated and decelerated, has a large spatial extension at small central wave vectors and a small extension at large central wave vectors. The de Broglie wavelength of the electron

$$\lambda_\mathrm{dB} = \frac{2\pi}{k_x} \tag{32.20}$$

is infinitely large at the bottom of the miniband and reaches the value $2a$ at the mini-Brillouin zone boundary (Fig. 32.4b, lower part). The electron wave undergoes a Bragg reflection when the de Broglie wavelength is equal to twice the spatial period of the superlattice.

An electron oscillates around a fixed position. A static field does therefore not lead to a direct current. Relaxation, however, gives rise to a direct current, as we will *see* in the next section.

Example of a Bloch oscillation. 14/2 GaAs/AlAs superlattice (14 monolayers of GaAs and 2 monolayers of AlAs); $a = 14$ monolayers ($\sim$4.2 nm); $\epsilon_\mathrm{m} = 140\,\mathrm{meV}$; $E_\mathrm{s} = 10\,\mathrm{kV/cm}$; $\nu_\mathrm{B} = \omega_\mathrm{B}/2\pi = 1\,\mathrm{THz}$; $\xi_m \sim 30\,a$ ($\sim$120 nm).

32.3 Esaki–Tsu Characteristic

A drift of an oscillating electron arises due to intraminiband relaxation. In a relaxation process, an electron loses energy and reaches another trajectory. The trajectory is, relative to the original trajectory, shifted along the direction of the electric force (Fig. 32.5a). The drift velocity is given by

$$\mathrm{v} = \frac{1}{\tau}\int_{-\infty}^{0} \mathrm{e}^{-t_0/\tau}\mathrm{v_g}\,\mathrm{d}t_0. \tag{32.21}$$

The exponential is equal to the probability that an electron starting a Bloch oscillation at time t_0 does not undergo a relaxation process in the time interval t_0,0. The integration takes into account that the starting time can have a value between $-\infty$ and 0. The contribution of starting times $t_0 \ll -\tau$ is small in comparison with the contributions for starting times $t_0 \sim -\tau$. This model supposes that an electron relaxes in an intraminiband relaxation process to the bottom of the miniband. Integration yields the Esaki–Tsu drift velocity

$$\mathrm{v_{ET}} = \mathrm{v_{ET,m}}\,\frac{2\omega_{\mathrm{B}}\tau}{1+\omega_{\mathrm{B}}^2\tau^2}, \tag{32.22}$$

where

$$\mathrm{v_{ET,m}} = \frac{\mathrm{v_0}}{2} = \frac{\epsilon_{\mathrm{m}}a}{4\hbar} \tag{32.23}$$

is the maximum drift velocity. The drift velocity (Fig. 32.5b) varies linearly with the electric field around zero field and shows a maximum at a critical field

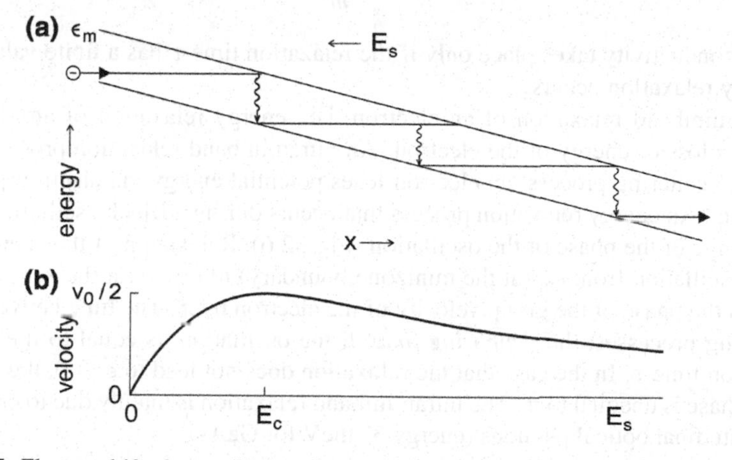

Fig. 32.5 Electron drift of an oscillating electron in a static electric field. **a** Electron drift. **b** Drift velocity-field curve

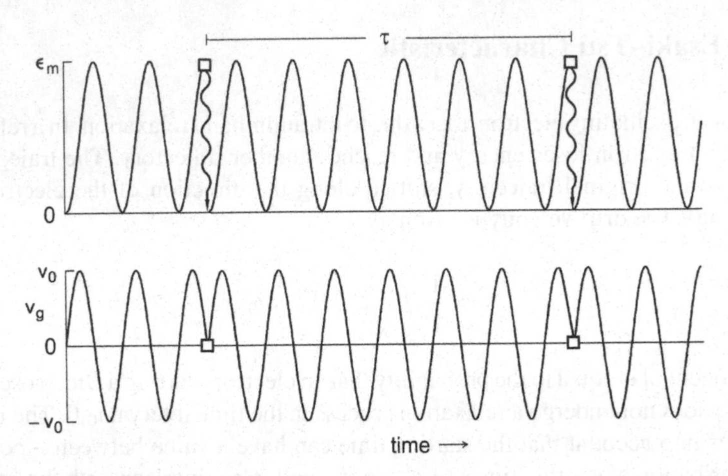

Fig. 32.6 Dephasing of a Bloch oscillation by energy relaxation: energy and group velocity of an oscillating electron; *square*, intraminiband energy relaxation process joined with a phase relaxation process

$$E_c = \frac{\hbar}{ea\tau}, \tag{32.24}$$

which corresponds to $\omega_B\tau = 1$. For $E \gg E_c$, the drift velocity decreases inversely proportional to E_s. With increasing E_s ($> E_c$), the number of Bragg reflections per unit of time increases and therefore the drift velocity decreases.

For $E_s \ll E_c$, we obtain the ohmic conductivity of a superlattice,

$$\sigma_{\text{ohm}} = \frac{N_0 e^2 \tau}{m^*}. \tag{32.25}$$

Ohmic conductivity takes place only if the relaxation time τ has a finite value, i.e., if energy relaxation occurs.

Intraminiband relaxation of an electron, i.e., energy relaxation of an electron, results in loss of energy of the electron. An intraminiband relaxation process is an inelastic scattering process: an electron loses potential energy via electron-phonon scattering. An energy relaxation process that occurs during a Bloch oscillation leads to a change of the phase of the oscillation (Fig. 32.6). Relaxation of the energy of a Bloch oscillation from ϵ_m (at the minizone boundary) to $\epsilon = 0$ (at the zone center) changes the phase of the group velocity of the electron by π. The time between two dephasing processes, the *dephasing time* of the oscillation, is equal to the energy relaxation time τ. In the case that the relaxation does not lead to $\epsilon = 0$, the change of the phase is unequal to π. The intraminiband relaxation is mainly due to emission of longitudinal optical phonons (energy 37 meV for GaAs).

We assumed that the superlattice has a temperature near $T = 0$. The thermal distribution of miniband electrons in a superlattice at temperature T reduces the drift velocity. The shape of the Esaki–Tsu curve remains unchanged (supposed that the relaxation time remains unchanged), but the maximum drift velocity is reduced,

$$\mathrm{v}_{\mathrm{ET,m}}(T) = f(T)\,\frac{\epsilon_{\mathrm{m}} a}{4\hbar}, \tag{32.26}$$

where $f(T)$ (<1) is a temperature parameter (Sects. 32.1 and 32.10).

Example GaAs superlattice (at room temperature) with the data of the last example; $\tau \sim 5 \times 10^{-13}$ s; $\omega_{\mathrm{B}}\tau = 1$ occurs for $\omega_{\mathrm{B}}/2\pi = 0.2$ THz. The Bloch oscillation of a free-electron in a semiconductor superlattice is a monople oscillation (Sect. 4.12). Intraminiband relaxation changes the phase of the oscillation, but not the amplitude.

32.4 Modulation Model of a Bloch Laser

Under the influence of both a static field E_{s} and a high frequency field (amplitude A, frequency ω),

$$E = A\cos\omega t, \tag{32.27}$$

oriented along the superlattice axis, a miniband electron is accelerated according to the equation of motion

$$\hbar\frac{\mathrm{d}k_x}{\mathrm{d}t} = qE_{\mathrm{s}} + qA\cos\omega t. \tag{32.28}$$

Integration yields the phase

$$k_x a = \omega_{\mathrm{B}}(t - t_0) + \frac{qaA}{\hbar\omega}(\sin\omega t - \sin\omega t_0), \tag{32.29}$$

where t_0 is the time at which a Bloch oscillation starts. It follows that the phase $k_x a$ is phase-modulated with the modulation degree

$$\mu = \frac{eAa}{\hbar\omega}. \tag{32.30}$$

The instantaneous group velocity is equal to

$$\mathrm{v}_{\mathrm{g}} = \mathrm{v}_0 \sin[\omega_{\mathrm{B}}(t - t_0) - \mu(\sin\omega t - \sin\omega_{\mathrm{B}} t_0)]. \tag{32.31}$$

We write $\mathrm{v}_{\mathrm{g}} = \mathrm{v}_0 \sin\phi(t)$ and find the instantaneous frequency

$$\omega_{\mathrm{inst}} = \frac{\mathrm{d}\phi}{\mathrm{d}t} = \omega_{\mathrm{B}} + \kappa A\cos\omega(t - t_0), \tag{32.32}$$

where

$$\kappa = ea/\hbar \tag{32.33}$$

is the *coupling strength* characterizing the coupling between the Bloch oscillation of an electron and the high frequency electric field. The coupling strength is proportional to the superlattice period. The high frequency field causes phase modulation of the Bloch oscillation.

We now take account of relaxation. The probability that an electron does not undergo an energy relaxation process in the time interval $t - t_0$, t is

$$p(t, t_0) = e^{-(t-t_0)/\tau}. \tag{32.34}$$

The average over all starting times yields a *modulation velocity*

$$v_{mod}(t) = \frac{1}{\tau}\int_{-\infty}^{t} p(t, t_0) v_g(t, t_0) dt_0 \tag{32.35}$$

that varies periodically with the period $T = 2\pi/\omega$ of the high frequency field. A Fourier transformation [15] yields the amplitude of the real part of the modulation velocity:

$$v_1 = v_0 \sum_{n=-\infty}^{+\infty} J_n(\mu) \frac{(\omega_B + n\omega)\tau}{(\omega_B + n\omega)^2\tau^2 + 1}. \tag{32.36}$$

J_n is the Bessel function of nth order. The terms for $n = -1, -2, -3, ...$ describe resonances at which $|n|\omega = \omega_B$.

In the following, we will discuss the case $n = \pm 1$. Making use of the relation $J_{-1} = -J_1$, we obtain

$$v_1 = -v_0 J_1(\mu)\left(\frac{(\omega_B - \omega)\tau}{(\omega_B - \omega)^2\tau^2 + 1} - \frac{(\omega_B + \omega)\tau}{(\omega_B + \omega)^2\tau^2 + 1}\right). \tag{32.37}$$

We can write, with $\Delta\omega_0 = 2/\tau$, the last expression in the form

$$v_1 = -v_0 J_1(\mu)\,[\bar{g}_{L,disp}(\omega) - K(\omega)], \tag{32.38}$$

where

$$\bar{g}_{L,disp}(\omega) = \frac{\omega_B - \omega}{\Delta\omega_B/2}\bar{g}_{L,res} \tag{32.39}$$

is the (normalized) Lorentz dispersion function and

$$\bar{g}_{L,res} = \frac{\Delta\omega_B^2/4}{(\omega_B - \omega)^2 + \Delta\omega_B^2/4} \tag{32.40}$$

the corresponding normalized Lorentz resonance function. The halfwidth of the Lorentz resonance function is equal to $\Delta\omega_{\rm B} = 2/\tau$. The quality factor of the Bloch resonance is

$$Q_{\rm B} = \omega_{\rm B}\tau/2. \tag{32.41}$$

The term

$$K(\omega) = \frac{\omega_{\rm B}+\omega}{\Delta\omega_{\rm B}/2}\,\frac{\Delta\omega_{\rm B}^2/4}{(\omega_{\rm B}+\omega)^2+\Delta\omega_{\rm B}^2/4} \tag{32.42}$$

contributes strongly to v_1 if $\omega_{\rm B}\tau$ has a value that is not much larger than 1.

We consider the limit of small modulation degree, where $J_1(\mu) = \mu/2$. We find, with $v_0 = \epsilon_{\rm m}a/2\hbar$, $m = eAa/\hbar\omega$ and $m^* = 2\hbar^2/\epsilon_{\rm m}a^2$, the amplitude of the velocity:

$$v_1 = -\frac{eA}{2m^*\omega}\,[\bar{g}_{\rm L,disp}(\omega) - K(\omega)]. \tag{32.43}$$

The real part of the high frequency mobility is equal to

$$\mu_1 = \frac{v_1}{A} = -\frac{e}{2m^*\omega}\,[\bar{g}_{\rm L,disp}(\omega) - K(\omega)] = -\frac{e}{2m^*\omega_{\rm B}}\,\frac{\omega_{\rm B}}{\omega}\,[\bar{g}_{\rm L,disp}(\omega) - K(\omega)]. \tag{32.44}$$

The factor $1/\omega$ reflects the dependence of the modulation degree on the frequency.

It follows that the high frequency conductivity is given by

$$\sigma_1(\omega) = -\sigma_{\rm p}\,\frac{2\omega_{\rm B}\tau(1+\omega^2\tau^2-\omega_{\rm B}^2\tau^2)}{1+2\omega^2\tau^2+\omega^4\tau^4+2\omega_{\rm B}^2\tau^2-2\omega^2\tau^2\omega_{\rm B}^2\tau^2+\omega_{\rm B}^4\tau^4}, \tag{32.45}$$

where

$$\sigma_{\rm p} = \frac{N_0e^2}{2m^*\omega_{\rm B}} \tag{32.46}$$

is a peak conductivity.

Figure 32.7 shows $\sigma_1/\sigma_{\rm p}$ for a particular Bloch frequency ($\nu_{\rm B} = 5$ THz) and two different values of the relaxation time. At the smaller relaxation time, the transparency frequency is slightly smaller than the Bloch frequency; the shift of the transparency frequency is due to the term $K(\omega)$.

The small-signal gain coefficient is equal to

$$\alpha(\omega) = \frac{-\sigma_1(\omega)}{(c/n)\varepsilon_0}. \tag{32.47}$$

In Sect. 32.1 (*see* Fig. 32.2) we already discussed the absorption coefficient $\alpha(\omega)$ for $\nu_{\rm B} = 5$ THz and $\tau = 5\times10^{-13}$ s.

For $\omega_{\rm B}\tau \gg 1$ and $\omega \sim \omega_{\rm B}$, we can neglect $K(\omega)$ and obtain the high frequency conductivity

$$\sigma_1(\omega) = -\sigma_{\rm p}\,\bar{g}_{\rm L,disp}(\omega). \tag{32.48}$$

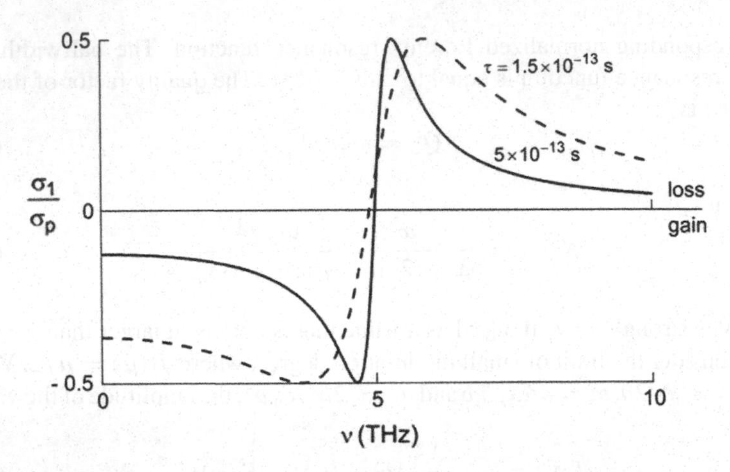

Fig. 32.7 High frequency conductivity of a superlattice in a static electric field corresponding to a Bloch frequency of 5 THz

In Sect. 9.11, we mention the Kramers–Kronig relations, which relate the real part of a physical response function and the imaginary part. If the shape of the imaginary part of a response function is given by a Lorentz resonance function, the shape of real part of the response function is a Lorentz dispersion function and vice versa. The Kramers–Kronig relations have been derived for systems in thermal equilibrium. We now assume that the Kramers–Kronig relations are also valid for a nonequilibrium system. Accordingly, we find that the imaginary part of the high frequency conductivity is given, for frequencies around the resonance frequency, by

$$\sigma_2(\omega) = -\sigma_p\, \bar{g}_{L,res}(\omega). \tag{32.49}$$

(The same result of the small-signal conductivities σ_1 and σ_2 has been obtained by a direct analysis of the response of miniband electrons to a high frequency electric field [258].)

The response function of the current density is the complex conductivity

$$\tilde{\sigma}(\omega) = -\,\sigma_p \frac{\Delta\omega_B/2}{i\,(\omega_B - \omega) + \Delta\omega_B/2}. \tag{32.50}$$

The shape of the real part of the dynamical conductivity corresponds to a Lorentz dispersion function and the shape of the imaginary part to a Lorentz resonance function (Fig. 32.8a). The maximum of $|\sigma_2|$ is determined by the peak conductivity σ_p. The extrema of σ_1 have the values $\mp\sigma_p/2$. The small-signal gain curve (Fig. 32.8b) is, around the resonance frequency, antisymmetric with respect to ω_B.

At finite temperature, the frequency dependences are the same as for the low temperature case. However, the peak conductivity is reduced by the temperature factor (Sects. 32.1 and 32.10),

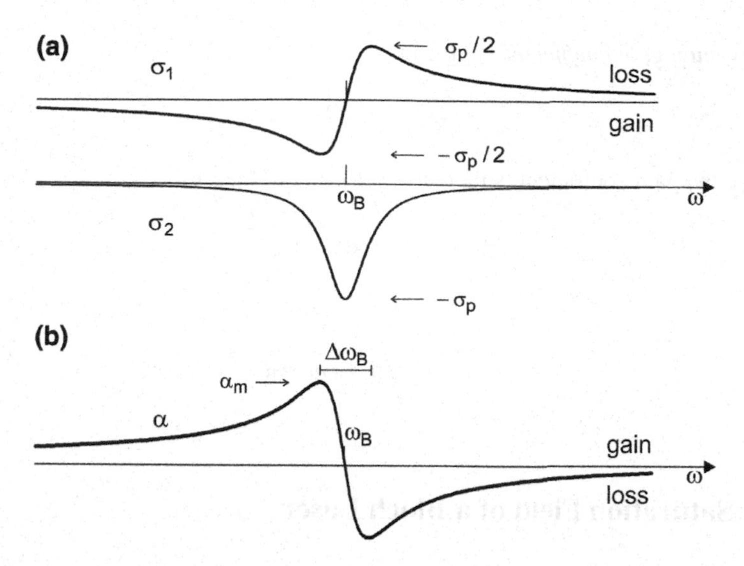

Fig. 32.8 Bloch gain for $\omega_B \tau \gg 1$. **a** High frequency conductivities. **b** Gain coefficient

$$\sigma_p(T) = f(T)\,\frac{N_0 e^2}{2m^* \omega_B}. \tag{32.51}$$

It follows that the small-signal gain coefficient for $\omega_B \tau \gg 1$, and therefore $K(\omega) \sim 0$, is given by

$$\alpha(\omega) = \alpha_p\, \bar{g}_{L,\mathrm{disp}}(\omega); \tag{32.52}$$

$$\alpha_p = f(T)\,\frac{N_0 e^2}{2(c/n)\varepsilon_0 m^* \omega_B}. \tag{32.53}$$

A comparison indicates the following. The real part of the polarization conductivity of an atomic system with population inversion shows a resonance and the imaginary part a dispersion like behavior; gain occurs at frequencies around the resonance frequency. But the real part of the conductivity of a system of oscillating free-electrons shows a dispersion-like behavior and the imaginary part a resonance; gain occurs at frequencies below the resonance frequency.

We summarize the result of the modulation theory with respect to the gain coefficient. The modulation model provides the following results.

- The gain coefficient has an inflection point at the frequency ω_B^*, which is slightly smaller than ω_B.
- $\omega = \omega_B^*$. Field and Bloch oscillation interact strongly. In the time average, there is no net energy exchange between field and Bloch oscillation.
- $\omega < \omega_B^*$. The gain coefficient curve is determined by the normalized Lorentz dispersion function, modified by $K(\omega)$.
- $\omega > \omega_B^*$. The gain coefficient is negative, radiation experiences absorption.

The *maximum gain coefficient* is given by

$$\alpha_m = \alpha_p/2. \tag{32.54}$$

We write the gain coefficient, supposing $f(T) = 1$, in the form

$$\alpha(\omega) = 2\alpha_m\, \bar{g}_{L,disp}(\omega), \tag{32.55}$$

$$\alpha_m = \frac{N_0 e^2}{4(c/n)\, \varepsilon_0\, m^* \omega_B}. \tag{32.56}$$

32.5 Saturation Field of a Bloch Laser

We discuss the saturation behavior. We assume, for simplicity, that $\omega_B \tau \gg 1$. Therefore, we can neglect $K(\omega)$. Then, the maximum gain coefficient is given by

$$\alpha_m(\mu) = \frac{N_0 e^2}{2(c/n\varepsilon_0) m^* \omega_B} J_1(\mu). \tag{32.57}$$

The Bessel function J_1 increases linearly with the modulation degree μ (Fig. 32.9) and shows a maximum at $\mu = 1.8$ (and then decreases to zero and becomes negative).

There are two possibilities to find a saturation field. The first possibility follows from the slope of the gain curve. In an oscillator, gain is, in principle, limited to the

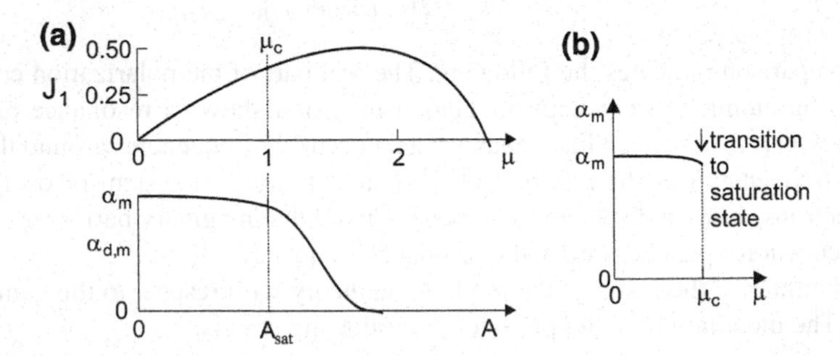

Fig. 32.9 Saturation field of a Bloch laser. **a** Bessel function and differential gain coefficient. **b** Transition to saturation

amplitude $A_{\mu=1.8}$. The maximum gain coefficient has to be replaced by the differential maximum gain coefficient,

$$\alpha_{\mathrm{d,m}}(\mu) = \frac{N_0 e^2}{(c/n\varepsilon_0)m^*\omega_{\mathrm{B}}} \frac{\mathrm{d}J_1(\mu)}{\mathrm{d}\mu}. \tag{32.58}$$

The differential gain coefficient is zero for $\mu = 1.8$. Accordingly, $A_{\mu=1.8} = 1.8h\nu_{\mathrm{B}}/(ea)$. We find a different value for the saturation field if we apply the criterion (*see* also Sect. 19.9):

Saturation of the high frequency electric field occurs if the modulation degree assumes the critical modulation degree $\mu_{\mathrm{c}} = 1$.

We consider the phase difference between the transverse oscillation of an electron and the high frequency electric field,

$$\phi(x) = x - \mu \sin ax, \tag{32.59}$$

where $x = \omega_{\mathrm{B}}t$ is the phase for $\mu = 0$ and $a = (\omega_B - \omega)/\omega_{\mathrm{B}}$ is the difference of the frequencies of the Bloch oscillation and the high frequency electric field, divided by ω_{B}. Without modulation ($\mu = 0$), the phase ϕ increases linearly with time (if intraminiband relaxation is ignored). In the case that a high frequency electric field modulates the electron oscillation, the phase difference oscillates around the $\phi(t) = x(t)$ line. The amplitude of this oscillation increases with increasing μ. The phase difference ϕ increases continuously with time as long as $\mu < 1$. However, for $\mu > 1$, the same phase difference can be obtained for two different times. The change from the continuous increase to the more complicated behavior occurs at the critical modulation index μ_{c}. We find the value of μ_{c} from the condition that the derivative $\mathrm{d}\phi/\mathrm{d}x$ is zero, $\mathrm{d}\phi/\mathrm{d}x = 1 - \mu\, a = 0$, which leads (for a $\ll 1$) to

$$\mu_{\mathrm{c}} = \frac{1}{a} \approx 1 \tag{32.60}$$

With this argument, we find the saturation field amplitude

$$A_{\mathrm{sat}} = \frac{\hbar\omega}{ea}. \tag{32.61}$$

We supposed that $\omega_{\mathrm{B}}\tau \gg 1$. Gain occurs for $\omega \approx \omega_{\mathrm{B}}$. We thus find that

$$A_{\mathrm{sat}} = E_{\mathrm{s}}.$$

The amplitude of the saturation field is equal to the strength of the static electric field E_{s} (Fig. 32.10).

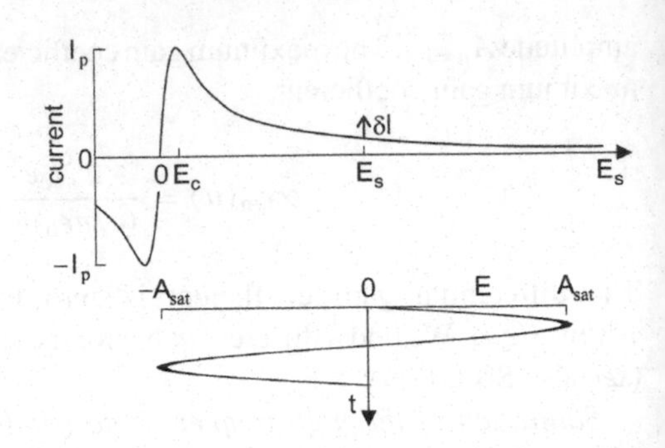

Fig. 32.10 Current-field characteristic of a superlattice and high frequency electric field in a Bloch laser

It is evident that the saturation field based on the criterion that the modulation degree is limited to unity, leads to a well understandable result. We discuss the current-field curve together with the high frequency electric field (Fig. 32.10). The instantaneous field, which is the sum of the static field E_s and the high frequency field E, causes the electrons to perform Bloch oscillations resulting in gain for the high frequency field. However, if the instantaneous field is smaller than the critical field E_c, the instantaneous field gives rise to loss of radiation. Therefore, maximum gain is expected for $|A| \leq E_s - E_c$. This corresponds, for $\omega_B \tau \gg 1$, to the condition $A_{sat} = E_s$. The quantities in the current-field curve are: I_p, peak current; δI, excess current due to laser oscillation; E_c, critical field; E_s, static field, corresponding to the bias voltage U_s; $E(t)$, high frequency field; and A_{sat}, saturation field amplitude.

The output power of a Bloch laser is equal to

$$P_{out} = \frac{\varepsilon_0 \, \varepsilon \, A_{sat}^2 a_1 a_2 L}{2\tau_{res.}}. \tag{32.62}$$

The electric power delivered by a voltage source (delivering a direct current at constant voltage) is equal to

$$U_s I = U_s I_0 + P_{out}. \tag{32.63}$$

where U_s is the static bias voltage, I the direct current at presence of radiation and I_0 the current at absence of radiation. Laser oscillation results in an enhancement of the direct current. The power that corresponds to the excess current

$$\delta I \equiv I - I_0 = P_{out}/U_s \tag{32.64}$$

is converted to radiation.

The efficiency for conversion of electric power to radiation is given by

$$\eta = \frac{P_{out}}{U\,I} = \frac{1}{1 + P_0/P_{out}}, \tag{32.65}$$

where $P_0 = U_s I_0$ is the electric power at absence of a high frequency field.

Example see Table 32.1.
There is a characteristic difference between a classical solid state oscillator operating with charge density domains and a Bloch laser. For a classical solid state oscillation, excitation of a high frequency field results in a reduction of the direct current (Sect. 31.5); this is a consequence of the domain dynamics. In a (yet hypothetical) Bloch laser, generation of high frequency radiation leads to an enhancement of the direct current. The excess current is converted to laser radiation. The excess current is expected to increase linearly with the power of the high frequency radiation.

32.6 Energy of Distortion in a Bloch Laser

In the state of saturation, a high frequency electric field is present and interacts strongly with the Bloch oscillations. Accordingly, absorption compensates stimulated emission of radiation. However, it is not clear, how the state of saturation is realized.

The modulation current may have a phase of $\pi/2$ relative to the current as in the case of resonance absorption. This appears to be unlikely because distortion of the oscillating state seems to be disregarded.

We rather suggest that the phase between the modulation current of the Bloch oscillation and the high frequency electric field is equal to π or 0. The two corresponding oscillation states allow for the strongest interaction between Bloch oscillation and high frequency field. In order to find a possible origin of saturation, we determine the work done by the electron during a cycle of a Bloch oscillation. The work is given by

$$W_{1,\phi=0} = -ev_0\, A_{\mathrm{sat}} \int_0^T \cos(\omega_{\mathrm{B}} t)\cos(\omega t)\mathrm{d}t. \tag{32.66}$$

The solution is, with $\omega_0 \approx \omega$ equal to

$$W_{1,\phi=0} = -\pi\, ev_0\, A_{\mathrm{sat}}/\omega. \tag{32.67}$$

We now assume that the work results in a distortion of the electron oscillation and that the distortion occurs if the work is equal to the *distortion energy* E^*, i.e., we assume that

$$W_{1,\phi=0} = -E^*. \tag{32.68}$$

Comparison of (32.67) and (32.68) provides the saturation field amplitude

$$A_{\mathrm{sat}} = \frac{E^*\omega_{\mathrm{B}}}{\pi e v_0}.$$

Since we know A_{sat}, we find the distortion energy,

$$E^* = \frac{\pi e v_0 A_{\mathrm{sat}}}{\omega_{\mathrm{B}}}. \tag{32.69}$$

It follows that

$$E^* = (\pi/2)\epsilon_{\mathrm{m}}.$$

The distortion energy is equal to the miniband width (multiplied by $\pi/2$). We can write the condition in the form

$$e\,a\,A_{\mathrm{sat}} = (2/\pi)\hbar\omega_{\mathrm{B}}. \tag{32.70}$$

Accordingly, the saturation energy is reached, when the maximum energy gain within a period a of the superlattice is equal to the quantum energy of a photon (multiplied by $\pi/2$).

The distortion changes the phase of the oscillation by π. After a change of the phase, the reversed process occurs: the field transfers energy to the electron. The work is given

$$W_{1,\phi=\pi} = e v_0 A_{\mathrm{sat}} \int_0^T \cos(\omega_0 t)\cos(\omega t)\mathrm{d}t = +\pi e v_0 A_{\mathrm{sat}}/\omega, \tag{32.71}$$

or,

$$W_{1,\phi=\pi} = +E^*. \tag{32.72}$$

A process of repairing a distortion follows on a process that leads to distortion of an electron oscillation. At saturation, absorption compensates stimulated emission of radiation. On average, there is no energy exchange between the electrons and the high frequency field.

The saturation field amplitude does not depend on the length of the Bloch laser medium. Doubling of the length of the Bloch laser medium results in a shortening of the oscillation onset time.

32.7 Synchronization of Bloch Oscillations to a High Frequency Field

The dynamics of synchronization of the high frequency drift current to a high frequency electric field can be analyzed by studying the temporal change of the wave vector of an electron. The method, leading *to phase space bunching* (*k-space bunching*), has been introduced by Kroemer [258]. We illustrate the result of the gain calculation. Electrons in a static field perform Bloch oscillations at the Bloch frequency. The oscillations of different electrons are uncorrelated to each other (Fig. 32.11, left).

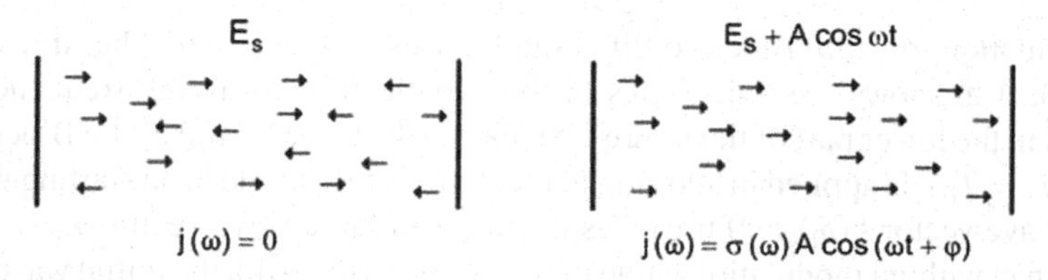

Fig. 32.11 Bloch oscillations of miniband electrons under the action of a static field (*left*), and of both a static and a high frequency field (*right*)

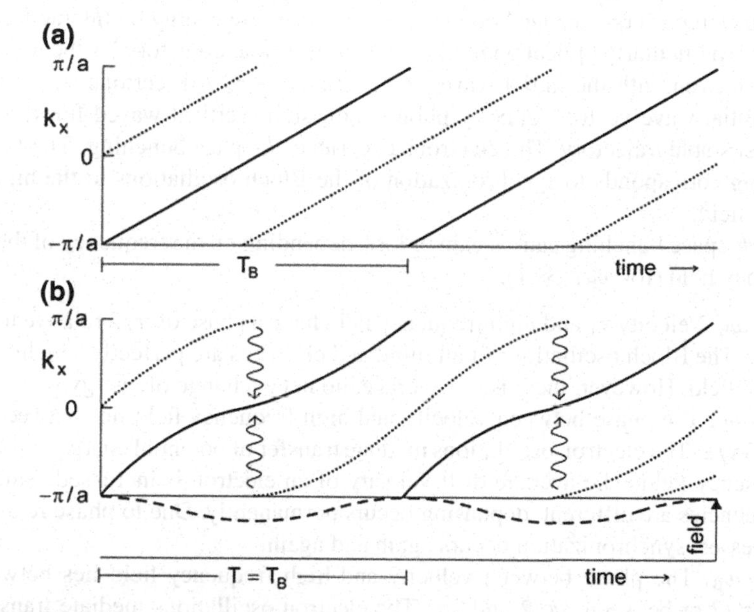

Fig. 32.12 *k*-space bunching. **a** *k*-space trajectories of two electrons in a static field. **b** *k*-space trajectories of two electrons submitted to both a static and a high frequency field

The average high frequency current at the Bloch frequency is zero. A high frequency field (frequency ω) forces the electrons to oscillate with the same average phase relative to the phase of the field (Fig. 32.11, right). For $\omega = \omega_B$, the phase between the current and the field is $\pi/2$. There is, in the time average, no net exchange of energy. For $\omega < \omega_B$, the phase is larger than $\pi/2$ and gain occurs while at $\omega > \omega_B$, the phase is smaller than $\pi/2$, which corresponds to absorption.

In a static field along the superlattice axis, the wave vector of an electron increases linearly with time. We restrict the wave vector to the mini-Brillouin zone (Fig. 32.12a). An electron (with a path represented by the dotted line) that starts with the initial wave vector $k(t_0) = 0$ experiences a Bragg reflection after half a period and then after each further period $T_B = 2\pi/\omega_B$ of the Bloch frequency. An electron with the initial wave vector $k(t_0) = -\pi/a$ (solid line) undergoes the first Bragg reflection after one temporal period, the second after two periods and so on. The k-space trajectories are straight lines.

Modulation of the Bloch oscillations causes a change of the trajectories. Figure 32.12b shows the trajectories of the two electrons if a high frequency field (dashed in the lower part of the figure), which has the same period as the Bloch oscillation ($T = T_B$), is applied in addition to the static field. The electron starting with the initial wave vector $k(t_0) = 0$ traverses the range of large wave vector ($k_x \sim \pm\pi/a$) slower than without modulation while the electron starting with the initial wave vector $k(t_0) = -\pi/a$ traverses the range of large wave vector faster. Relaxation processes occur preferably when an electron has a wave vector near the minizone boundary: due to its large energy, a zone boundary electron can lose energy by interaction with phonons (particularly optical phonons). The "most stable trajectory" is the trajectory of the electron with the initial wave vector $k(t_0) = -\pi/a$. Electrons starting with other initial wave vectors relax by phonon emission (vertical waved lines) toward the most stable trajectory. The electrons experience k-space bunching. The k-space bunching corresponds to synchronization of the Bloch oscillations to the high frequency field.

The k-space bunching causes gain or loss, depending on the frequency of the high frequency field (for $\omega_B\tau \gg 1$):

- $\omega = \omega_B$. Velocity v_1 and high frequency field have a phase of $\pi/2$ relative to each other. The Bloch oscillations of all miniband electrons are perfectly synchronized to the field. However, there is, on average, no net exchange of energy.
- $\omega < \omega_B$. The phase between velocity and high frequency field lies between $\pi/2$ and $3\pi/2$. The electron oscillations mediate transfer of potential energy to the high frequency field. The average drift velocity of an electron is increased. Since the frequencies are different, dephasing occurs permanently. Due to phase relaxation processes, synchronization occurs again and again.
- $\omega > \omega_B$. The phase between velocity and high frequency field lies between 0 and $\pi/2$ or between $3\pi/2$ and 2π. The electron oscillations mediate transfer of radiation energy to potential energy of the electrons. The average drift velocity of an electron is decreased. Since the frequencies are different, dephasing occurs permanently. Due to energy relaxation processes, synchronization occurs again and again.

In all three cases, modulation of the Bloch oscillations by the high frequency field results in a current at the frequency of the high frequency field.

32.8 Energy-Level Description of the Superlattice Bloch Laser

An electron in a superlattice submitted to a static field occupies a level in an energy-ladder system (Fig. 32.13a):

$$E_l = lE_0, \tag{32.73}$$

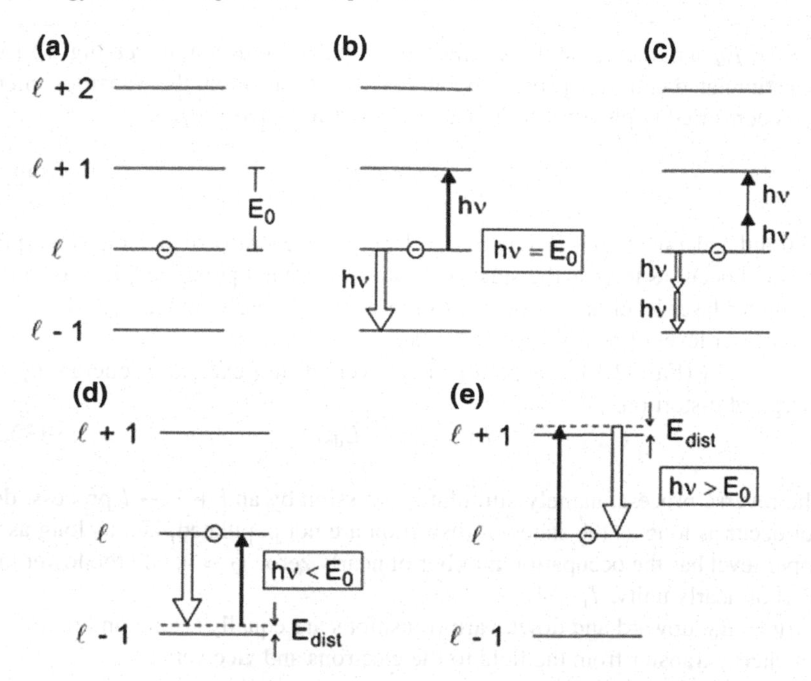

Fig. 32.13 Energy levels of an electron in a superlattice submitted to a static field and optical transitions. **a** Energy ladder. **b** Absorption and stimulated emission. **c** Two-photon transitions. **d** Stimulated emission. **e** Absorption

$E_0 = h\nu_B$ is the energy separation between next near energy levels (=transition energy). Electronic transitions between next near levels are allowed. Electromagnetic radiation interacts via spontaneous emission, absorption and stimulated emission according to the Einstein coefficients. However, absorption and stimulated emission processes have the same transition probability (Fig. 32.13b). Therefore, the average rate of absorption processes is the same as the average rate of stimulated emission processes if the frequency of the radiation is exactly equal to the resonance frequency, i.e., if $\nu = \nu_B$. The description as a frequency modulation (Sect. 32.4) indicates that stimulated emission prevails if $\nu < \nu_B$ and absorption if $\nu > \nu_B$. Accordingly, the gain curve is a Lorentz dispersion curve rather than a Lorentz resonance curve.

In a strong electromagnetic field, transitions between next-near levels are also allowed as multiphoton transitions (Fig. 32.13c) according to the condition

$$nh\nu = E_0 - h\nu_B; \quad n - 1, 2, \ldots \tag{32.74}$$

The cases n > 1 correspond to velocity components of higher order according to (32.37).

Whether a radiation field experiences a population inversion in the Bloch laser medium, depends on the frequency of the field.

- If $h\nu < E_0$, a radiation field experiences a population inversion (*see* Fig. 32.13d). In a stimulated emission process by an $l \rightarrow l - 1$ transition, the transition energy E_0 is converted to photon energy $h\nu$ and distortion energy E_{dist},

$$E_0 = h\nu + E_{\text{dist}}. \tag{32.75}$$

  A stimulated transition in an energy-ladder system leads to a distortion. Absorption does not occur as long as the states of distortion are not populated, i.e., as long as the upper laser level has an occupation number of nearly unity, $f_2 \approx 1$, and the lower laser level of nearly zero, $f_1 \approx 0$.
- If $h\nu > E_0$ (Fig. 32.13e), a photon is converted into excitation energy E_0 and energy of distortion,

$$h\nu = E_0 + E_{\text{dist}}. \tag{32.76}$$

  The reverse process, namely stimulated emission by an $l + 1 \rightarrow l$ process, does not occur as long as the states of distortion are not populated, i.e., as long as the upper level has the occupation number of nearly zero, $f_2 \approx 0$, and the lower laser level of nearly unity, $f_1 \approx 1$.
- If $h\nu = E_0$, upward and downward transitions are equally strong and there is no net energy transfer from the field to the electrons and vice versa.

The distortion energy, joined with a transition, is small compared with the transition energy.

Figure 32.14 illustrates, in the energy-level description, the principle of the superlattice Bloch laser; the figure illustrates the situation during the onset of oscillation

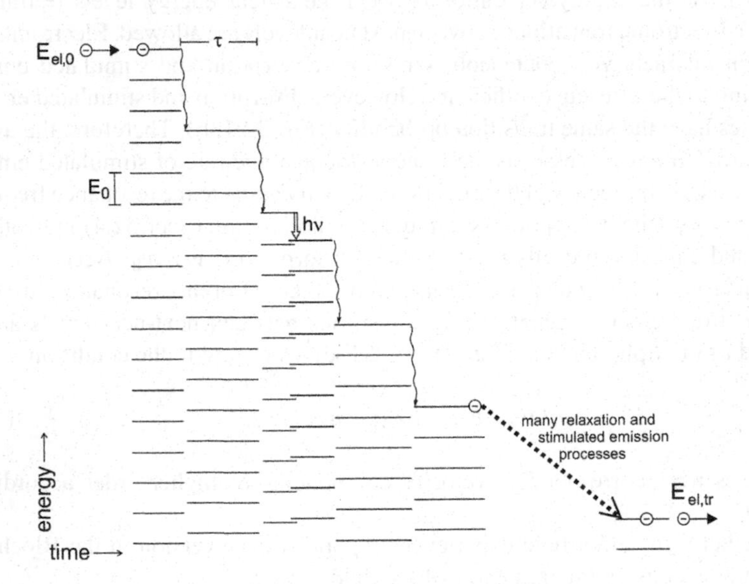

Fig. 32.14 Principle of the Bloch laser in an energy-level description

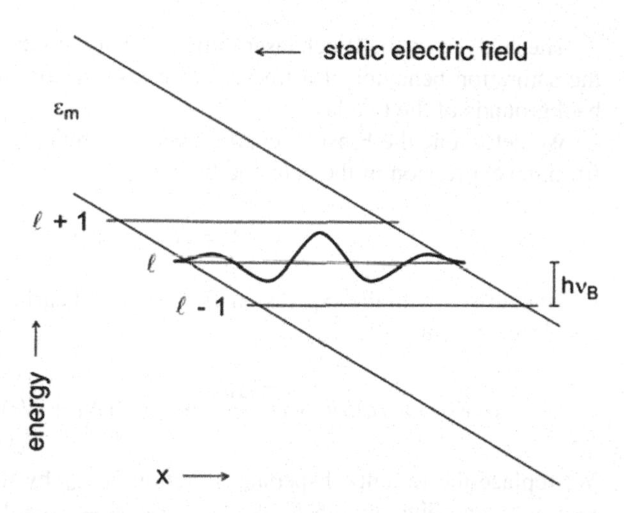

Fig. 32.15 Wannier state

of the laser. An electron of energy $E_{\mathrm{el},0}$ injected into a superlattice forms an energy-ladder system. A nonradiative relaxation process (intraminiband relaxation time τ) leads to formation of a new energy ladder system. A stimulated transition occurs to a disturbed state; then, the electron forms a new energy-ladder system. Relaxation processes and stimulated emission processes go on until the electron reaches the cathode. The electron leaves the superlattice with energy $E_{\mathrm{el,tr}}$. The relaxation processes are associated with a drift. During the drift of an electron through the superlattice, the potential energy $U_{\mathrm{s}} = (E_{\mathrm{el},0} - E_{\mathrm{el,tr}})L_{\mathrm{SL}}2$ is converted to energy of radiation plus energy of relaxation.

At saturation of the Bloch laser, an absorption process during a cycle of the Bloch oscillation follows on the stimulated emission of a photon. In an ensemble of electrons, the average occupation number of an undisturbed electron state is equal to one half, $f_2 = 1/2$, and the average occupation number of the distorted state is also one half, $f_1 = 1/2$. In this case, the time between two stimulated emission processes is equal to twice the period of the Bloch frequency, $\tau_{\mathrm{stim}} = 2/\nu_{\mathrm{B}}$. This analysis is consistent with data of Table 32.1. The time τ_{stim} given in the table is an average time between two stimulated emission processes for the case that radiation is coupled out from the laser resonator. In a closed resonator without any loss, the saturation field may be reached in the whole resonator volume and the time τ_{stim} may correspond to twice the period of the Bloch frequency.

(Another possibility of the occurrence of saturation may be connected with the limit of the quantum theory of the Wannier-Stark states. A strong high frequency electric field interacts with the Bloch oscillations giving rise to new dynamical coupled states, Floquet states. These are expected to occur for an amplitude of the electric field that is given by the relationship $eaA = 2.4\hbar\omega_{\mathrm{B}}$ [333]. The value of A that follows from this criterion is, however, slightly larger than the value of A_{sat} that follows from the modulation model.)

The analysis of the Bloch laser indicates that the saturation behavior differs from the saturation behavior of a free-electron laser according to the different physical backgrounds of the two lasers.

We determine the Einstein coefficients of stimulated emission and of absorption from the expression of the gain coefficient

$$\alpha(\nu) = 2\alpha_\mathrm{m}\, \bar{g}_{\mathrm{L,disp}}(\nu), \tag{32.77}$$

by comparison with the expression (7.31) derived earlier for an ensemble of atomic two-level systems,

$$\alpha(\nu) = (n/c)h\nu_0 B_{21} \frac{\pi\,\Delta\nu_0}{2}\, \bar{g}_{\mathrm{L,res}}(\nu)(N_1 + N_2)(f_2 - f_1). \tag{32.78}$$

We replace the Lorentz dispersion function $\bar{g}_{\mathrm{L,disp}}$ by the Lorentz resonance function $\bar{g}_{\mathrm{L,res}}$ and obtain, with $N_1 + N_2 = N_0$, $f_2 - f_1 = 1$, and $\nu_0 = \nu_\mathrm{B}$, the Einstein coefficient of stimulated emission

$$B_{21} = \frac{e^2}{8\varepsilon_0 h\nu_\mathrm{B} Q_\mathrm{B} m^*}. \tag{32.79}$$

The Einstein coefficient of stimulated emission is inversely proportional to the effective mass m^* of a miniband electron, to the photon energy $h\nu_\mathrm{B}$, and to the quality factor

$$Q_\mathrm{B} = \frac{\nu_\mathrm{B}}{\Delta\nu_\mathrm{B}} \tag{32.80}$$

of the Bloch oscillations. The Einstein coefficient of absorption is equal to the Einstein coefficient of stimulated emission, $B_{12} = B_{21}$.

Table 32.2 Einstein coefficients of transitions between Wannier–Stark levels of electrons in a superlattice

	Value	
ν_B	5×10^{12} Hz	Bloch frequency
$\Delta\nu_\mathrm{B}$	6×10^{11} Hz	Width of resonance
A_sat	$3 \times 10^{6}\ \mathrm{V\,m^{-1}}$	Amplitude of saturation
τ	5×10^{-13} Hz	Relaxation time
$Q_\mathrm{B} = \nu_\mathrm{B}/\Delta\nu_\mathrm{B}$	8	Quality factor
m^*	6×10^{-32} kg	Effective mass
$B_{21} = e^2/(8\varepsilon_0 h\nu_\mathrm{B} Q_\mathrm{B} m^*)$	$1 \times 10^{23}\ \mathrm{m^3\,J^{-1}\,s^{-2}}$	Einstein coefficient
$A_{21} = (8\pi h\nu_\mathrm{B}^3/c_m^3)\, B_{21}$	$2 \times 10^{9}\ \mathrm{s^{-1}}$	Einstein coefficient
τ_sp	4×10^{-10} s	Spontaneous lifetime

Table 32.2 shows values of Einstein coefficients. We determined the Einstein coefficient of spontaneous emission by using the corresponding Einstein relation:

$$A_{21} = \frac{8\pi h\nu^3}{(c/n)^3} B_{21} = \frac{\pi e^2 \nu_B^2}{(c/n)^3 \varepsilon_0 Q_B m^*}. \quad (32.81)$$

The Einstein coefficient of spontaneous emission is proportional to the square of the Bloch frequency. It is inversely proportional to the effective mass m^* and to the quality factor of the Bloch oscillation. The spontaneous lifetime is $\tau_{sp} = 1/A_{21}$. The Einstein coefficient of stimulated emission has a value that is of the same order of magnitude as for a quantum cascade lasers of the same frequency range.

The energy-ladder we introduced is a Wannier–Stark ladder. The energy distance between next-nearest levels is $h\nu_B$. The wave functions are Wannier functions. A Wannier function (Fig. 32.15) is spatially localized in the range ξ_m of a spatial trajectory of a Bloch oscillation. The wave function decreases exponentially outside this range.

32.9 Possible Arrangements of a Bloch Laser

A possible arrangement of a Bloch laser, already discussed in Sect. 32.1, is shown in Fig. 32.16a. A large-area superlattice fills out a resonator completely. The resonance frequency of the resonator is determined by the length L, the width a_1 of the resonator, and the optical properties of the superlattice material. Superlattices with extensions L_{SL} up to ~10 μm can be grown by molecular epitaxy. The laser frequency is expected to correspond to the lowest order resonance of the resonator. Bias

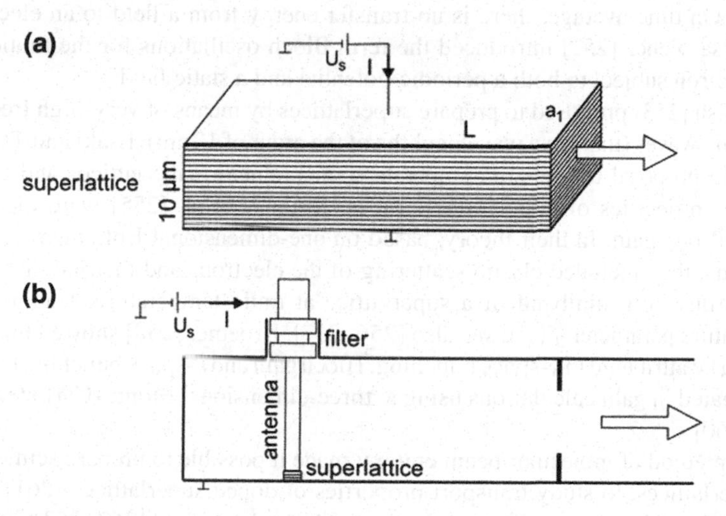

Fig. 32.16 Possible arrangements of a superlattice Bloch laser. **a** Large-area superlattice in a resonator. **b** Superlattice coupled to a resonator via an antenna

oscillations should be avoidable by the use of filters in the bias circuit. The output power is determined by the saturation field amplitude, the output coupling strength, and the area of two opposite output surfaces; the mismatch between the superlattice and air should provide sufficient feedback.

Another possible arrangement is shown in Fig. 32.16b. The superlattice is coupled to a resonator via an antenna. In this case, the resistance of the superlattice has to be matched to the resonator. The frequency is determined by the geometry of the resonator and the resistance of the superlattice in the active state. In this arrangement, superlattices with much smaller extensions and thus less heating would be appropriate. However, the preparation of the resonator and the filter would require more effort.

32.10 References to the Bloch Laser and Discussion

In 1928, Felix Bloch (then a Ph.D. student of Werner Heisenberg) described [251] a one-dimensional quantum theory of the electric conductivity of crystals and introduced a number of basic concepts that allow for a description of properties of conduction electrons in a crystal,

- The (one-dimensional) tight binding method.
- Energy bands.
- The acceleration theorem of conduction electrons.

Bloch also realized that energy relaxation via phonons is a necessary condition for the occurrence of electric conduction in a crystal. Without energy relaxation, a conduction electron interacting with a static electric field is accelerated and decelerated in turn. On time average, there is no transfer energy from a field to an electron or vice versa. Zener [252] introduced the term Bloch oscillations for the motion of a free-electron subject to both a periodic potential and a static field.

Keldish [253] proposed to prepare superlattices by means of very-high frequency ultrasonic waves (that have wavelenghths of the order of 10 nm). Esaki and Tsu [254] made the proposal to prepare composite semiconductor superlattices and to study transport properties of doped superlattices. Ktitorov et al. [255] were the first to predict Bloch gain. In their theory, based on one-dimensional Boltzmann transport equations, they included elastic scattering of the electrons and thermal distribution of electrons in a miniband of a superlattice at finite temperature, leading to the temperature parameter $f(T)$; *see* also [256, 257]. Kroemer [258] showed that Bloch gain can be attributed to k-space bunching. Bloch gain and k-space bunching have also been treated in gain calculations using a three-dimensional Monte–Carlo technique [259, 260].

The method of molecular beam epitaxy made it possible to prepare semiconductor superlattices, to study transport properties of doped superlattices [261], and to investigate response of oscillating electrons to a terahertz field [262, 263]. Evidence of Bloch gain of THz radiation has been reported by Allen et al. [264].

It is an open question whether it is possible to avoid formation of space charge domains in a superlattice in order to realize a THz Bloch laser. Proposals concern the use of weakly doped superlattices [264, 265] and the use of an additional THz radiation source [266]. A microwave-terahertz double oscillator is another proposal [246]. The idea is to operate a superlattice as a microwave oscillator (for instance, at a frequency near 65 GHz) based on domains as described in Sects. 31.3–31.6. The use of a lossless microwave resonator may lead, during a certain time during each microwave period, to a homogeneous field distribution that may be associated with Bloch gain at THz fields. If a superlattice is coupled, at the same time, to a microwave resonator and to a resonator for a THz field, a THz field may be amplified during a part of each cycle of a microwave oscillation.

Bloch oscillations of miniband electrons have been observed by the use of femtosecond optical techniques [267, 268]; for surveys, *see* [269, 270].

Wannier–Stark ladders and Bloch oscillations are discussed in [271]. Willenberg et al. [272] introduced intermediate states (distorted states) in Wannier-Stark ladders for calculation of the small-signal gain coefficient. Johannes Stark (1874–1957; born in Freihung, Bavaria) predicted and observed the Stark effect [273], namely the splitting of energy levels of atoms and molecules in electric fields. In 1919, he received the Nobel Prize in Physics. Gregory Wannier (1911–1983; born in Basel, Switzerland) developed the Wannier functions [274, 275].

Finally, we discuss the question of the upper limit frequency of a Bloch laser. The limit is determined by the condition that the length of the trajectory of a Bloch oscillation should be larger than a superlattice period. This leads to a limiting frequency, where $\xi = 2\,a$ of $\sim$15 THz for a GaAs/AlAs superlattice (with a maximum miniband width of 140 meV) and $\sim$30 THz of an InGaAs/InAlAs superlattice. Gain can also occur at larger frequencies via stimulated transitions between Wannier–Stark states that are localized within a few superlattice periods or within one period [270]. In the range of infrared active phonons (near 8 THz) intrinsic absorption caused by an infrared active lattice vibration can be stronger than gain.

References [177, 178, 251–275, 333]

Problems

32.1 High frequency limit frequency of a Bloch laser.

(a) Give a general condition of the high frequency limit of a Bloch laser.
(b) Determine the high frequency limit of a Bloch laser based on a GaAs/AlAs superlattice with a miniband of a width of 140 meV.

32.2 Large-signal amplitude.

Estimate the large-signal amplitude of a semiconductor superlattice Bloch laser and compare it with E_c,

(a) for $\nu = 1$ THz and (b) $\nu = 5$ THz. [*Hint*: choose data of the superlattice described in the text.]

32.3 Estimate the Bloch frequency of a conduction electron in an energy band of GaAs in the case that a static field is applied along the (100) crystal direction and compare it with the Bloch frequency of a miniband electron in a GaAs/AlAs superlattice. [Hint: Bloch oscillations are not observable for bulk GaAs by various reasons: intervalley scattering of electrons and impact ionization are extremely strong for electrons submitted to a strong electric field; furthermore, electrons can reach higher conduction bands.]

32.4 Determine the absolute number of electrons in an active medium of a superlattice Bloch laser described in Sect. 32.1.

32.5 Design a resonator for a superlattice Bloch laser oscillating at 6 THz.

32.6 Estimate the change of the refractive index at the Bloch frequency of the superlattice of a Bloch laser if the laser is switched on.

32.7 Determine the time of onset of laser oscillation of a Bloch laser.

32.8 Relate the high frequency peak-conductivity σ_p and the ohmic conductivity of a superlattice.

32.9 Electric conductivity.
Show the following: Without energy relaxation, a conduction electron (in a crystal) interacting with a static electric field is accelerated and decelerated in turn, but it is, on average, not possible to transfer energy from a field to an electron and vice versa. Electric conduction by materials like copper cables makes use of energy relaxation of the conduction electrons. [*Hint*: for demonstration, make use of a one-dimensional simple conduction band, separated by an energy gap from the next higher conduction band; *see* also Fig. 30.3. Assume that the field is not strong enough to cause electron tunneling between the lowest conduction band and the next-higher band or impact ionization.]

Part VI
Laser Related Topics

Chapter 33
Optical Communications

An important field of application of lasers is optical communications by means of glass fibers.

To transfer information over very large distances (e.g., around the world), radiation of a wavelength of 1.55 μm is most suitable by two reasons. First, glass fibers have the smallest loss of radiation in a wavelength band around 1.55 μm. Second, a glass fiber amplifier—the erbium-doped fiber amplifier—allows for amplification of laser radiation at 1.55 μm. A glass fiber with integrated light amplifier (installed every 100 km) can transport information over any distance on earth.

The transfer of optical waves over shorter distances (up to about 50 km) is possible with radiation at wavelengths around 1.32 μm. At this wavelength, the absorptivity of glass fibers is also small and the dispersion is zero (resulting in less distortion of optical pulses in comparison to pulses of 1.55 μm radiation). However, there is no efficient light amplifier available for 1.32 μm radiation.

In the past, the transfer rate increased more and more. The use of radiation at many wavelengths at the same time—corresponding to many frequency bands available for information transfer—and increase of the modulation bandwidth enhanced the transfer rate. For long-distance transfer of information via fiber-optic cable networks, a frequency band of about 5 THz is available; the width of the band corresponds to about 2.5% of an average frequency (about 200 THz) of radiation of the 1.55 μm band.

33.1 Principle of Optical Communications

The basis of optical communications is the guidance of light by means of optical fibers. The principle of optical communications is illustrated in Fig. 33.1. Light, coupled by means of a transmitter into a fiber, propagates through the fiber to a receiver. For long-distance transfer, the fiber contains laser amplifiers. The following components belong to an optical transfer system.

K.F. Renk, *Basics of Laser Physics*, Graduate Texts in Physics,
DOI 10.1007/978-3-319-50651-7_33

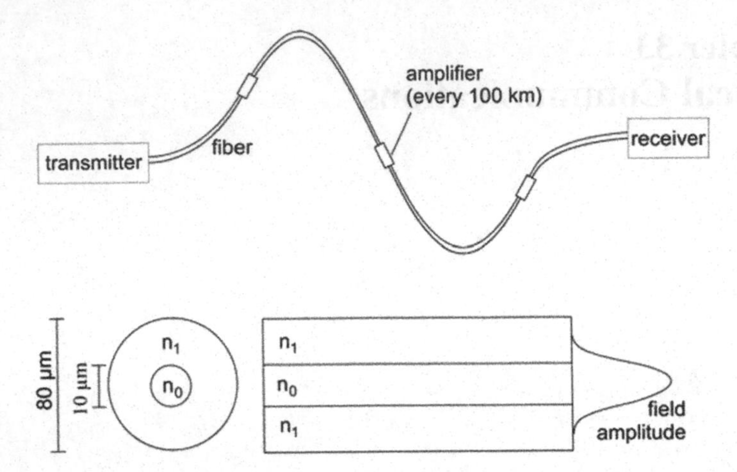

Fig. 33.1 Optical communications (*principle*) and glass fiber

- *Transmitter*. Laser + modulator + coupler.
- *Laser*. Quantum well laser, a heterostructure of $Ga_xIn_{1-x}As_yP_{1-y}$ material on n-type or p-type InP, lattice matched for $y = 1.2x$. The laser wavelengths are 1.55 μm ($x = 0.42$) for long distance transmittance and 1.32 μm ($x = 0.27$) for transmittance up to about 50 km.
- *Receiver*. Photodetector + demodulator.

Of course, communication with free light waves is also possible. On earth, it is restricted because of damping of light in the atmosphere and because of the effort that is necessary for changing the propagation direction of light.

33.2 Glass Fiber

Glass fibers can be prepared with high accuracy from quartz glass. A glass fiber (Fig. 33.1) consists of two parts.

- *Core*. Diameter about 10 μm; SiO_2 doped with germanium; refractive index $n_0 = 1.52$.
- *Cladding (mantle)*. Diameter about 80 μm; SiO_2; $n_1 = 1.48$.

The basis of the guidance of light is the total reflection. The light is propagating in the 00 mode of a fiber (monomode fiber). The field distribution is Gaussian like. The amplitude of the field is large within the core and decreases exponentially in the mantle. Accordingly, a portion of the light is propagating in the core, another portion in the mantle. While a Gaussian beam in free space is always divergent, a Gaussian like beam in a fiber remains confined.

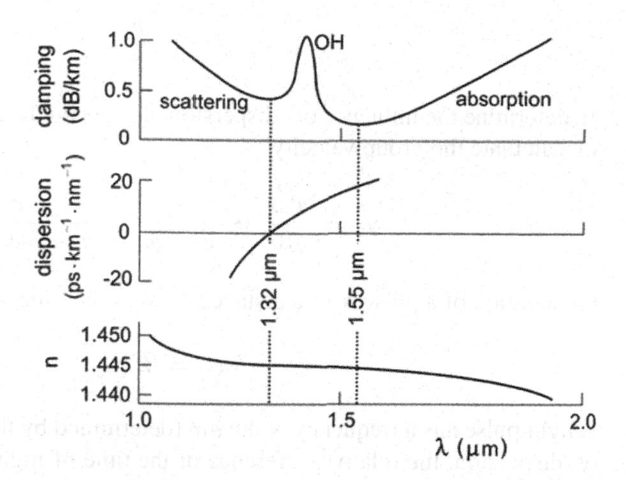

Fig. 33.2 Damping, dispersion and refractive index of light in a glass fiber

Light in a glass fiber experiences damping (Fig. 33.2, upper part). The damping has two minima at slightly different wavelengths.

- 1.32 μm; damping 0.4 dB/km.
- 1.55 μm; damping 0.2 dB/km.

Toward small wavelengths, the damping increases because of Rayleigh scattering. This is a consequence of small irregularities at the interface between core and mantle. Toward large wavelengths the damping increases due to absorption caused by lattice vibrations (phonons) in glass. OH impurities in glass are responsible for an absorption line near 1.4 μm. The OH concentration (5 ppm) in a glass fiber corresponds to 1 OH group per 2×10^8 SiO_2 molecules.

33.3 Pulse Distortion Due to Dispersion

A light pulse propagating in a fiber is damped and changes its shape. This is due to dispersion (Fig. 33.2, center). The dispersion is zero at 1.32 μm. It has a value of about 15 ps per km and nm at 1.55 μm. The dispersion relation of light in an isotropic medium (refractive index n) is

$$\omega = v_{ph} k = \frac{c}{n} k, \tag{33.1}$$

where ω is the angular frequency, v_{ph} the phase velocity, and k the wave vector. The refractive index of SiO_2 glass (Fig. 33.2, lower part) shows in the near infrared a weak decrease with increasing wavelength and a point of inflection at 1.32 μm. We can write the dispersion relation in the form

$$k = \frac{\omega}{c} n. \tag{33.2}$$

To determine the influence of dispersion of glass on the propagation of light pulses, we calculate the group velocity

$$\mathrm{v_g} = \frac{\mathrm{d}\omega}{\mathrm{d}k} = \frac{1}{\mathrm{d}k/\mathrm{d}\omega} = \frac{c}{n + \omega \mathrm{d}n/\mathrm{d}\omega}. \tag{33.3}$$

Propagation of a pulse over a distance L takes the time

$$\tau_g = L/\mathrm{v_g}. \tag{33.4}$$

A light pulse has a frequency width $d\omega$ (determined by the pulse duration). Because of dispersion, the relative difference of the time of flight of light of a frequency ω and a frequency $\omega + \mathrm{d}\omega$ is

$$\begin{aligned} \beta_2 &= \frac{1}{L}\frac{\mathrm{d}\tau_g}{\mathrm{d}\omega} = \frac{\mathrm{d}}{\mathrm{d}\omega}\left(\frac{1}{\mathrm{g_g}}\right) \\ &= \frac{\mathrm{d}}{\mathrm{d}\omega}\left(\frac{n}{c} + \frac{\omega}{c}\frac{\mathrm{d}n}{\mathrm{d}\omega}\right) = \frac{2}{c}\frac{\mathrm{d}n}{\mathrm{d}\omega} + \frac{\omega}{c}\frac{\mathrm{d}^2 n}{\mathrm{d}\omega^2}. \end{aligned} \tag{33.5}$$

The first term of β_2 describes a delay of a pulse and the second term a distortion of the shape of the pulse. The unit of β_2 is s $\mathrm{m}^{-1}\,\mathrm{Hz}^{-1}$.

33.4 Erbium-Doped Fiber Amplifier

In a long distance fiber cable, an erbium-doped fiber amplifier (EDFA) compensates damping of radiation.

Light is amplified (Fig. 33.3) by stimulated transitions in Er^{3+} ions that are pumped with radiation of a semiconductor laser (wavelength 1,480 nm or 980 nm). The erbium-doped glass amplifier can amplify radiation in the range of 1,520–1,560 nm. The relative bandwidth ($\delta\nu/\nu \sim 2.6\%$) corresponds to a bandwidth $\delta\nu \sim 5$ THz at the frequency $\nu \sim 2 \times 10^{14}$ Hz. Thus, the band available near 200 THz for optical communications has a width of about 5 THz. The mechanism of gain of radiation in an erbium fiber is discussed in Chap. 18; *see*, particularly, Fig. 18.1.

A fiber amplifier consists of an erbium-doped region of a long glass fiber cable. pump radiation of a semiconductor laser is coupled into a fiber via an optocoupler. An amplifier (length 30 m) has a gain factor of the order of 1,000 (gain 30 dB). In parallel to a a long-distance fiber cable, there is a current carrying cable delivering the electric energy necessary to operate the pump laser of the erbium-doped fiber amplifier.

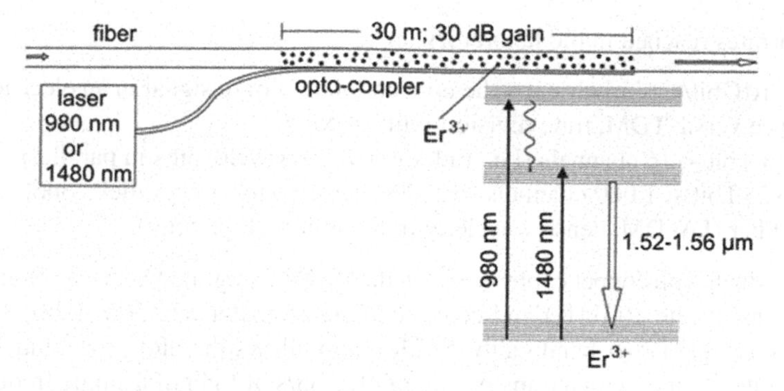

Fig. 33.3 Erbium-doped fiber amplifier (EDFA)

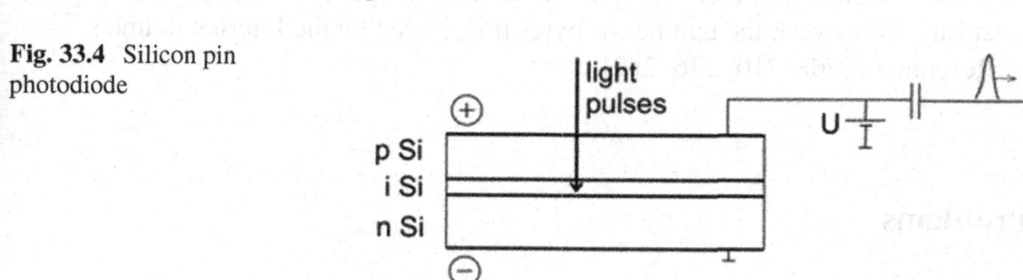

Fig. 33.4 Silicon pin photodiode

33.5 Detector

Photodiodes are suitable as detectors of radiation in optical communication systems. A silicon pin photodiode (Fig. 33.4) consists of a thin p-doped silicon layer, an intrinsic (1) silicon layer, and an n-doped silicon layer on an n-doped silicon substrate. A transparent metallic anode film on the p-doped silicon layer and a metal cathode on the backside of the substrate serve as metallic contacts. A light pulse traversing the anode film creates electron-hole pairs in the intrinsic layer. A static voltage across the photodiode accelerates electrons and holes, created by a light pulse, giving rise to an electric pulse that is registered electronically.

33.6 Transfer Rates

The transfer rates of large-distance communication systems increased permanently.

- Before 1996, a copper cable in the ocean reached a transfer rate of 280 Mbit/s (about 4,000 phone calls at the same time).
- 1996. An optical cable in the ocean (all-optical cable) reached 2.5 Gbit/s.
- Since 1999. Faster networks are in operation.

Transfer rates reached in the laboratory are:

- 1993. 10 Gbit/s. Limitation by the conversion of a light signal in an electric signal and vice versa. TDM, time division multiplexing.
- 2000. 1 Tbit/s; 16 channels (i.e., radiation at 16 wavelengths in parallel).
- 2001. 25 Tbit/s; 1,000 channels (=1,000 wavelengths in parallel); optical broadband fiber (DWDM, dense wavelength division multiplexing).

In comparison with copper cables (used in the ISDN, Integrated Services Digital Network; transfer rate 100 kHz) and coaxial cables (transfer rate 300 MHz), the glass fiber has a much larger bandwidth (5 THz). and allows therefore for a much larger transfer rate. A fiber with a transfer rate of 40 Gbit/s of light of a single frequency is presently the basis of the global network (i.e., the global system of mobile communication). Every year, the number of bytes transferred by the Internet doubles.

References [208–210, 276–282].

Problems

33.1 Estimate the electric power needed to maintain an optical fiber cable with 10 fibers that extends around the earth.

Chapter 34
Light Emitting Diode and Organic Laser

The development of semiconductor lasers is accompanied by the development of light emitting diodes (LEDs). An LED, based on spontaneous electron-hole recombination, does not require a cavity and can therefore have a simpler design than a laser. We discuss properties of diodes and mention various areas of applications. There is a growing market of LEDs as lighting sources. Superluminescent semiconductor diodes can reach high efficiencies. Stimulated emission just below laser threshold is favorable for a high efficiency. In 2014, S. Nakamura, H. Amano, and I. Akasaki received the Nobel Prize in physics for the invention of efficient blue light emitting diodes, which has enabled bright and energy-saving light sources.

Besides the LED, the organic LED (OLED) is being developed. The basis of the OLED is the spontaneous recombination of electrons and holes in molecular crystals or polymers. Production of OLEDs is possible at low costs as large-area films suitable for outdoor lighting of large areas.

We will also mention organic lasers.

34.1 LED Preparation and Market

An LED converts electric power into light. Most efficient are LEDs based on direct semiconductors.

In an LED, electrons and holes recombine giving rise to spontaneous emission of radiation at frequencies near the gap frequency of a semiconductor. An LED is easier to prepare than a laser diode. Antireflecting coatings ensure that the active medium of an LED experiences almost no feedback or only a weak feedback from radiation reflected at the crystal surfaces. Thus, the light is incoherent. An LED emits the light into a large solid angle.

Of great importance for the development of the LED market was the discovery of the blue LED (in 1992); *see* Sect. 1.9, Nobel Prizes. The lifetime of an industrial LED lies in the range of 10,000–100,000 h of operation. Already in 2006, about 25 billion LEDs (about 60% InGaN LEDs, 38% AlInGaP LEDs, and 2% LEDs of other

K.F. Renk, *Basics of Laser Physics*, Graduate Texts in Physics,
DOI 10.1007/978-3-319-50651-7_34

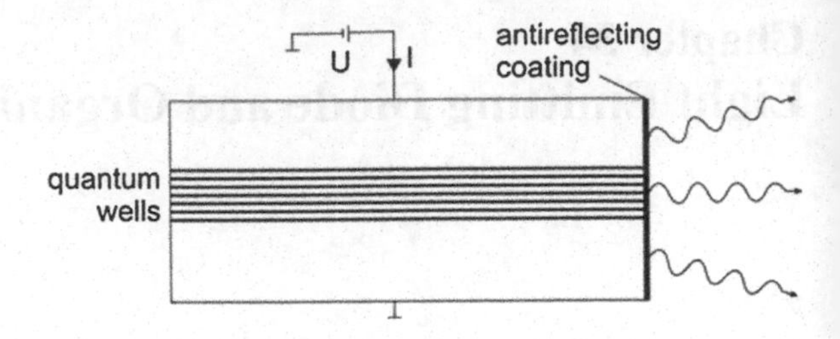

Fig. 34.1 Superluminescent LED

materials) have been prepared, mainly by metal oxide chemical vapor deposition (MOCVD). The size of a chip is typically 250 μm × 250 μm.

The luminous efficiency of LEDs was (by 2006) about 50 lm/W (lumen per watt electric power). The record in the laboratory was 100 lm/W. The goal of the development is a further increase of efficiency. The luminous flux (in units of lm) is a measure of power of light perceived by the human eye. The power of broadband visible radiation (e.g., from a blackbody source) of 1 mW corresponds to about 15 lm.

A superluminescent LED is based on stimulated emission below laser threshold. Figure 34.1 shows the principle of the *superluminescent LED (=high-power LED = laser diode operating below threshold)*. Under the action of a voltage U (current I), electrons migrate from the n-doped material into quantum wells and holes migrate from the p-doped material into the quantum wells. Radiation is generated by electron-hole recombination in the quantum wells. Antireflecting coatings on the surfaces avoid cavity resonances. The superluminescent LED has a broad output spectrum, like an LED. The power can be as large as that of a corresponding laser diode. The light has a higher directionality than the light of a normal diode.

The market is permanently growing. More than 100 firms are worldwide active in the LED industry. In 2005, the LED market had a turnover of 4 billion US $. The main portion concerned mobile phones. Although prices of LEDs steadily decrease, it is expected that the turnover would double every five years. The growth rate of the LED market with respect to the number of LEDs was (2010) ~50% per year. The largest growth is expected in various fields—illumination; full color displays; television and monitor screens; outdoor large-area screens; and projectors.

34.2 Illumination

Besides the use of LEDs for color applications (screens, traffic lights), the generation of white light is of great importance. We compare different illumination elements.

- *Light bulb.* Radiative efficiency about 20 lm/W (efficiency of conversion of electric power into light ~4%); 1,000 h of operation; for 1US $, one obtains 800 lm.
- *Energy-saving lamp.* Radiative efficiency ~50 lm/W. The energy-saving lamp contains mercury, which should be avoided in future. In the energy-saving lamp (Fig. 34.2, left), a gas discharge excites Hg atoms by collisions with electrons.

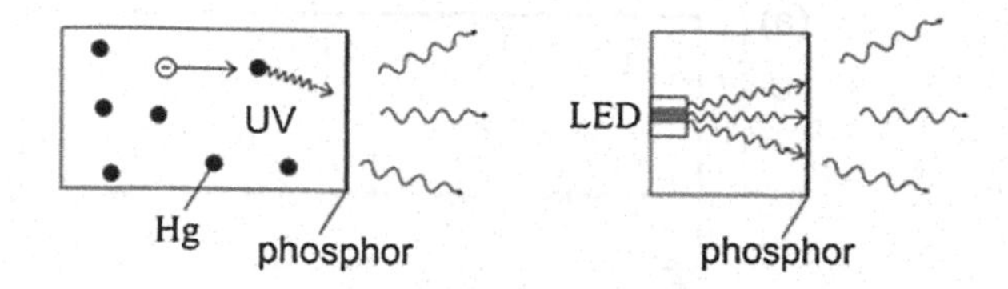

Fig. 34.2 Energy-saving lamp and LED based white light source

Excited Hg atoms emit UV radiation (at a wavelength of 254 nm). The UV radiation is absorbed by a phosphor giving rise to luminescence radiation in the visible.

- *LED*. 50 lm/W (future 200 lm/W). 10,000–100,000 h of operation; for 1 US $, one obtains 20 lm. [Problem: power supplies have short lifetimes.]
- *LED-based white light lamps*. A blue LED (Fig. 34.2, right) illuminates a phosphor on the inner surface of a glass plate.

LED-based illumination should lead to an energy saving of hundreds of billions US $ per year in the year 2025. Worldwide, the generation of light consumes about 20% of the total electric power. The replacement of bulbs is a worldwide task.

A problem concerns the quality of the white light produced with LEDs. Many people have the feeling that the light is cold. This feeling has a physical background. The light emitted by a phosphor, irradiated by an LED, does not have the same spectrum as the radiation of a light bulb. If white light is generated by three LEDs emitting radiation in the red, yellow and blue spectral regions, the radiation is, of course, also quite different from white light of a light bulb.

34.3 Organic LED

It is possible to produce organic LEDs of large areas at a low price per square meter. The luminous efficiency is about 30 lm/W. There are worldwide efforts to increase the efficiency and to improve the reproducibility.

The center of an organic LED (OLED) is a molecular layer or a polymer layer sandwiched between electric contact layers. The organic LED (Fig. 34.3a) consists in principle of 3 layers on a substrate:

- Metal contact.
- Organic layer.
- Transparent conductor (e.g., indium tin oxide) as a metallic contact.
- Transparent polymer substrate.

Under the action of a static field produced by a voltage (U) between the metal contacts, electrons migrate into the organic layer from one side and holes from the other side. Electron-hole recombination generates light.

In the organic LED, molecular levels (Fig. 34.3b) play an essential role. The organic layer, which is undoped and isolating, contains π-conjugated molecules or polymers. The electron states are molecular orbitals. An organic molecule of an

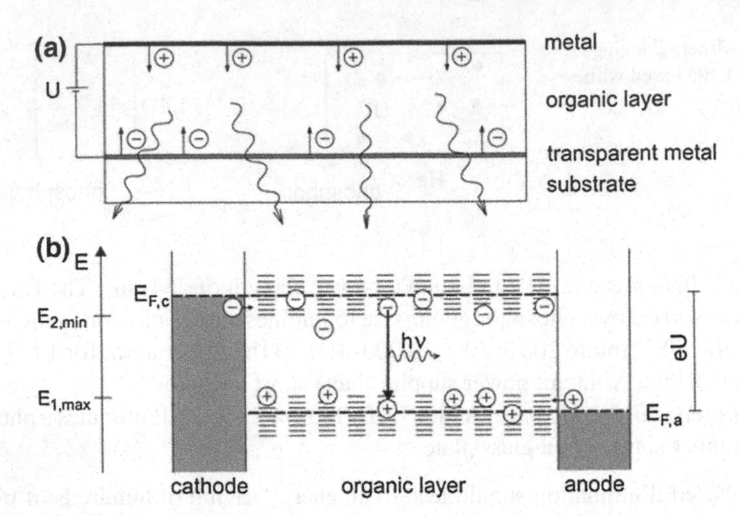

Fig. 34.3 Organic LED (OLED). **a** Arrangement. **b** Energy levels of an organic molecular film and principle of the organic LED

organic LED has occupied energy levels that have a broad energy distribution and, separated by a gap, empty levels that have also a broad energy distribution. An electron excited in a molecule migrates via a hopping process to another molecule. The holes migrate via hopping processes too. Characteristic energy levels are:

- $E_{1,max}$ = highest energy of occupied energy levels.
- $E_{2,min}$ = lowest energy of empty molecular energy levels.
- $E_{2,min} - E_{1,max}$ = energy gap.

A voltage U forces electrons to drift from the cathode and holes from the anode through the molecular film. Electron-hole recombination leads to generation of luminescence radiation. The voltage is equal to the energy difference between the Fermi energy $E_{F,c}$ of the cathode material and the Fermi energy $E_{F,a}$ of the anode material, $eU = E_{F,c} - E_{F,a}$.

Special emphasis lies in the choice of the cathode and anode materials. The cathode material has a work function $W_{cathode}$ that has a similar value as $E_{2,min}$ (related to the vacuum level E_{vac}). The anode material has a work function $W_{anode} \sim E_{1,max}$. Appropriate materials are the following:

- *Cathode materials.* Calcium, lithium, magnesium or alloys of these materials.
- *Anode materials.* Indium tin oxide ($InSnO_2$) film; this is transparent for visible radiation.

Without voltage, there is a built-in potential, establishing a constant Fermi level over the whole device. By application of a sufficiently large voltage, a nonequilibrium state with electrons and holes gives rise to the electron-hole recombination and to luminescence radiation.

34.4 Organic and Polymer Lasers

Operation of *optically pumped organic lasers* and *optically pumped polymer lasers* has been demonstrated for wavelengths in the whole visible spectral range [285–296]. Organic and polymer lasers are bipolar lasers (Fig. 34.4). We can describe a polymer laser (or an organic laser) as a two-quasiband laser, with an upper quasiband and a lower quasiband. The upper quasiband consists of energy levels of excited molecules. An excited molecule can transfer its excitation energy to a nonexcited, neutral molecule. Energy transfer is possible via the Förster mechanism and other energy transfer processes (Sect. 18.2). Optical pumping and intramolecular relaxation leads to quasithermal population in the upper quasiband. The density of states $D_2(E) = D_2(\epsilon_2)$, of the levels in the upper quasiband depends on the molecular properties of the polymer; $\epsilon_2 = E - E_{2,\text{min}}$ is the energy within the upper quasiband. Injection of electrons leads to band filling, characterized by the quasi-Fermi energy $E_{2,\text{F}}$ ($\epsilon_{2,\text{F}}$).

Extraction of electrons from nonexcited molecules results in a nonequilibrium distribution of empty levels in the lower quasiband. Energy transfer is again possible via energy transfer processes. Intramolecular relaxation processes are responsible that the occupied levels if the lower quasiband have also a quasithermal distribution. The density of states, $D_1(E) = D_1(\epsilon_1)$, of the levels of the lower quasiband depends on the molecular properties too; $\epsilon_1 = E_{1,\text{max}} - E$ is the energy within the lower quasiband. Extraction of electrons leads to empty levels in the lower quasiband, characterized by the quasi-Fermi energy $E_{1,\text{F}}$ ($\epsilon_{1,\text{F}}$).

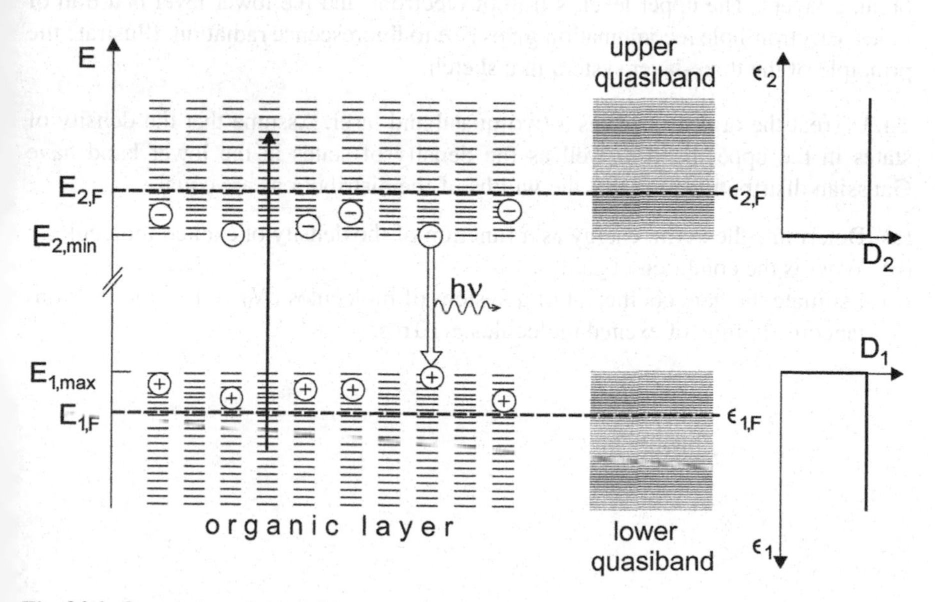

Fig. 34.4 Organic laser (*principle*)

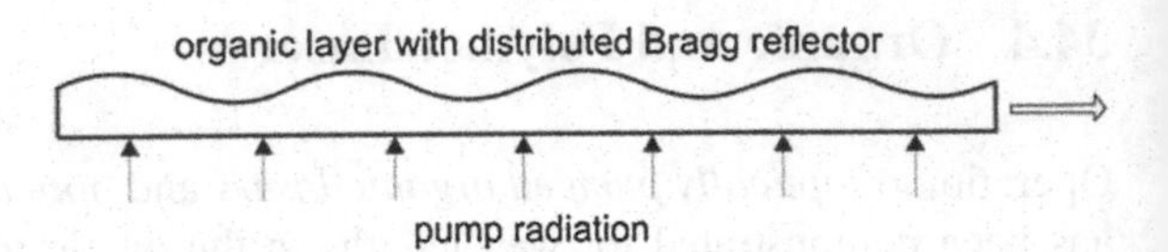

Fig. 34.5 Organic laser (*arrangement*)

Optical pumping creates electrons in the upper quasiband and holes in the lower quasiband. Stimulated electron-hole recombination is the origin of generation of laser radiation.

A polymer laser (Fig. 34.5) can consist of a polymer film (thickness of the order of a wavelength). A distributed Bragg reflector (Sect. 25.4) can be used to produce feedback.

A challenge is the development of an organic laser driven by a current.

References [31, 283–296].

Problems

34.1 Estimate the quantum efficiency of an OLED.

34.2 The actual efficiency of an OLED can be enhanced if three organic layers are used instead of one. In this case, the third layer is embedded in two organic layers. The third layer (e.g., a layer with a dye) has two energy levels in the gap of the organic layers. The upper level is trap of electrons and the lower level is a trap of holes. Electron-hole recombination gives rise to fluorescence radiation. Illustrate the principle of the three-layer system in a sketch.

34.3 Treat the organic laser as a two-quasiband laser. Assume that the density of states in the upper band as well as the density of states in the lower band have Gaussian distributions and that the widths of the distributions are equal.

(a) Determine the Fermi energy as a function of the density of excited molecules.
(b) What is the condition of gain?
(c) Estimate the gain coefficient of a system of molecules ($N_0 = 10^{24}\,\mathrm{m}^{-3}$); spontaneous lifetime of excited molecules = 10 ms.

Chapter 35
Nonlinear Optics

The polarization of a dielectric medium depends nonlinearly on the amplitude of the electromagnetic field. Nonlinear dielectric media are suitable for frequency conversion of radiation. Nonlinear media can be crystals, glasses, liquids or vapors. We discuss: frequency multiplication; difference frequency generation; parametric oscillation; four wave mixing; stimulated Raman scattering. In connection with four-wave mixing, we show how the frequencies of a frequency comb can be determined.

We will present only a very narrow view on the fascinating field of Nonlinear Optics. Our main aspect concerns the question: how can we convert coherent radiation of one frequency to coherent radiation of other frequencies?

35.1 Optics and Nonlinear Optics

In Maxwell's theory, the matter equations describe the electric properties of a dielectric medium are expressed by the relation between the dielectric polarization $\boldsymbol{P}$ of the medium and the electric field $\boldsymbol{E}$ in the medium,

$$\boldsymbol{P} = \boldsymbol{P}(\boldsymbol{E}). \tag{35.1}$$

In *Optics* (Linear Optics), the relation is

$$\boldsymbol{P} = \varepsilon_0 \chi^{(1)} \boldsymbol{E}, \tag{35.2}$$

where $\chi^{(1)}$ is the (complex) dielectric susceptibility. The susceptibility of an optically isotropic medium is a scalar. It is a tensor if a medium is anisotropic. The polarization has the same frequency as the electromagnetic field. The susceptibility $\chi^{(1)}(\omega)$ characterizes optical properties of a material. The study of $\chi^{(1)}(\omega)$ leads to an understanding of basic microscopic properties of matter.

K.F. Renk, *Basics of Laser Physics*, Graduate Texts in Physics,
DOI 10.1007/978-3-319-50651-7_35

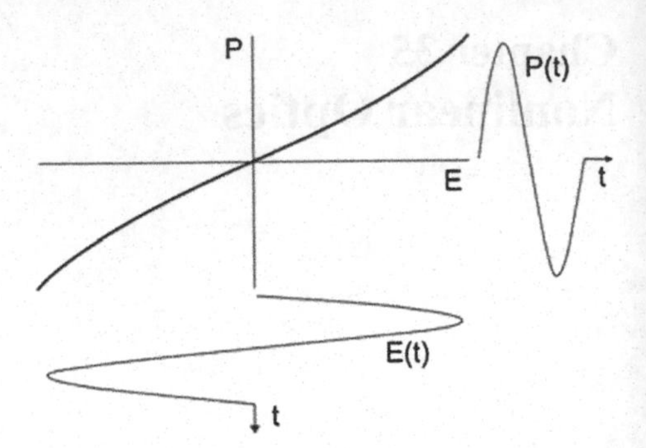

Fig. 35.1 Nonlinear polarization

The basis of *Nonlinear Optics* is the nonlinearity of the polarization at large amplitudes of the electric field (Fig. 35.1). We characterize the polarization by the relation

$$P = \varepsilon_0 \chi^{(1)} E + \varepsilon_0 \chi^{(2)} E^2 + \varepsilon_0 \chi^{(3)} E^3 + \ldots \tag{35.3}$$

We make use of two simplifications. We neglect the vector character of $\boldsymbol{E}$ and $\boldsymbol{P}$ as well as the tensor properties of $\chi^{(1)}$, $\chi^{(2)}$, $\chi^{(3)}$ etc. We assume that the field is spatially homogeneous in the direction of a light beam—we ignore that the field changes the phase during propagation. Thus, we neglect phase effects, which can be of great importance. Nevertheless, the simplified representation of the relation between polarization and field allows for developing an understanding of the principle of generation of radiation by means of the nonlinear polarization. Nonlinear polarization is applicable, for instance, to convert monochromatic radiation to radiation at other frequencies.

The electric field that causes a polarization can consist of fields of different frequencies. We can write, instead of (35.3),

$$P = \varepsilon_0 \chi^{(1)} E + \varepsilon_0 \chi^{(2)} E_1 E_2 + \varepsilon_0 \chi^{(3)} E_1 E_2 E_3 + \ldots, \tag{35.4}$$

where $E_1, E_2, \ldots$ are fields of different frequencies and where $E = E_1 + E_2 + \ldots$ is the sum of the fields.

35.2 Origin of Nonlinear Polarization

At which amplitude of an electromagnetic field do we expect a nonlinear polarization? We consider a hydrogen atom in a static electric field. We describe the H atom by Bohr's atomic model. In a distance of the Bohr radius ($a_0 = 0.053$ nm) to the nucleus (proton), an electron experiences the field strength

$$|E_{\mathrm{at}}| = \frac{e}{4\pi \varepsilon_0 a_0^2} \approx 10^{11}\ \mathrm{V/m}. \tag{35.5}$$

A static external field polarizes an H atom. The center of the positive charge and the center of the negative charge do not coincide with each other. Therefore, the H atom represents an electric dipole. The dipole moment increases with the field strength. At large field strength, the dipole moment depends nonlinearly on the field strength. The nonlinearity is extremely large when the field strength is of the order of the internal field produced by the proton at the site of the electron.

A strong high frequency field with a sinusoidal time dependence applied to a nonlinear medium leads to a time dependent polarization

$$P(t) = \varepsilon_0 \chi^{(1)} E(t) + \varepsilon_0 \chi^{(2)} E^2(t) + \ldots. \tag{35.6}$$

We describe the nonlinear polarization in the classical model of an atom (Sect. 4.8). Under the action of a strong electric field, an electron oscillates unharmonically—leading to a nonlinear dipole moment. Accordingly, the polarization of a medium depends nonlinearly on the amplitude of the electric field. The time-dependent polarization, which depends nonlinearly on the electric field, contains frequency components not only at the driving frequency but also at other frequencies. Therefore, the nonlinear polarization is the source of electromagnetic radiation at frequencies that differ from the driving frequency.

Atoms, ions, or molecules in gases, liquids and solids show nonlinear polarization. The strength of the nonlinearity strongly depends on the specific material. Especially large nonlinear susceptibilities are known for a variety of crystals (e.g., KDP, $LiNbO_3$).

We will now discuss applications that are based on the nonlinear polarization.

35.3 Optical Frequency Doubler

A frequency doubler (Fig. 35.2a) converts radiation of frequency ν to radiation at the doubled frequency (2ν). A filter blocks the radiation of frequency ν that is not converted. The frequency doubling makes use of the quadratic term of the polarization,

$$P = \varepsilon_0 \chi^{(1)} E(\omega) + \varepsilon_0 \chi^{(2)} E^2(\omega). \tag{35.7}$$

An electric field

$$E = A \cos \omega t \tag{35.8}$$

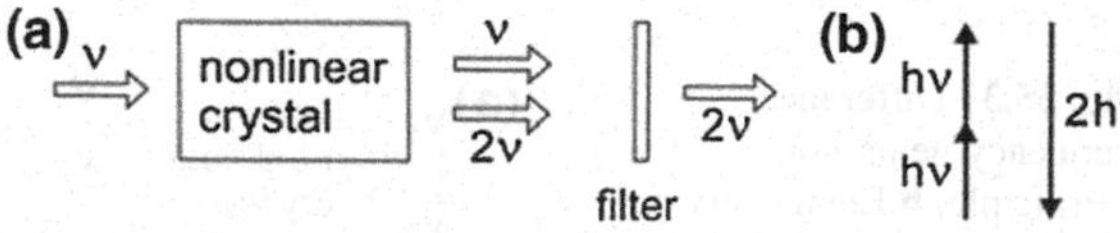

Fig. 35.2 Optical frequency doubler. **a** Principle. **b** Elementary process

causes a polarization

$$P = \varepsilon_0 \chi^{(1)} A \cos \omega t + \varepsilon_0 \chi^{(2)} A^2 \cos^2 \omega t. \tag{35.9}$$

It follows, with $\cos^2 \omega t = \frac{1}{2} + \frac{1}{2} \cos(2\omega t)$, that

$$P = \varepsilon_0 \chi^{(1)} A \cos \omega t + \frac{1}{2} \varepsilon_0 \chi^{(2)} A^2 + \frac{1}{2} \varepsilon_0 \chi^{(2)} A^2 \cos 2\omega t. \tag{35.10}$$

We obtain a polarization at the frequency 2ω that is the source of radiation at the frequency 2ω. The additional term corresponds to a static polarization (optical rectification). In an elementary process of frequency doubling (Fig. 30.2b), two photons of the quantum energy $h\nu$ are annihilated and a photon of energy $2h\nu$ is created.

Examples

- A frequency doubler converts infrared radiation to green radiation. As nonlinear crystals, $LiNbO_3$ or KDP are suitable: the conversion efficiency can reach 40%.
- In a titanium–sapphire laser, a frequency doubler located within the laser resonator produces frequency-doubled radiation in the violet and green (according to the tunability of the laser).
- Frequency doubling of green or blue radiation leads to UV radiation.

35.4 Difference Frequency Generator

In a difference frequency generator (Fig. 35.3a), two sinusoidal fields of different frequencies (ω_1 and ω_2; with $\omega_1 > \omega_2$) produce a nonlinear polarization in a nonlinear crystal. The nonlinear polarization is the source of an electromagnetic field at the difference frequency (beat frequency)

$$\omega_3 = \omega_1 - \omega_2. \tag{35.11}$$

An electric field of frequency ω_1

$$E_1 = A_1 \cos \omega_1 t \tag{35.12}$$

and another field

$$E_2 = A_2 \cos \omega_2 t \tag{35.13}$$

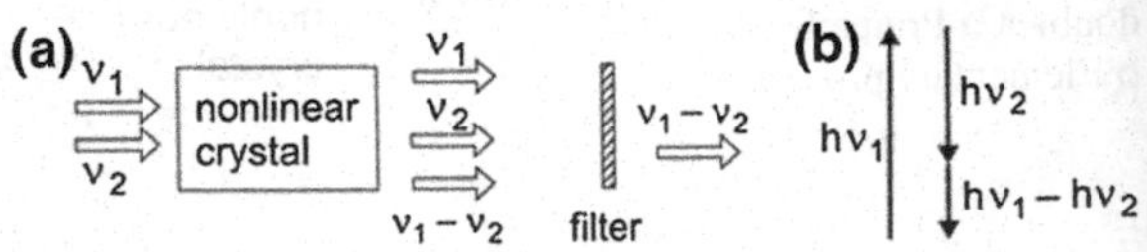

Fig. 35.3 Difference frequency generator. **a** Principle. **b** Elementary process

superimposed to each other lead to the field

$$E = E_1 + E_2. \tag{35.14}$$

This produces the polarization

$$\begin{aligned} P = &\varepsilon_0 \chi^{(1)} (A_1 \cos \omega_1 t + A_2 \cos \omega_2 t) \\ &+ \varepsilon_0 \chi^{(2)} (A_1 \cos \omega_1 t + A_2 \cos \omega_2 t)^2. \end{aligned} \tag{35.15}$$

The polarization contains the term

$$P_{\omega_1 - \omega_2} = \frac{1}{2} \varepsilon_0 \chi^{(2)} A_1 A_2 \cos(\omega_1 - \omega_2) t \tag{35.16}$$

that is the source of the field at the difference frequency. The polarization and, accordingly, the field at the difference frequency $\omega_1 - \omega_2$ are proportional to the product of the amplitudes A_1 and A_2. In an elementary process of difference frequency generation, a photon (energy $h\nu_1$) is annihilated, a photon at the energy $h\nu_2$ and another photon at the energy $h\nu_3 = h\nu_1 - h\nu_2$ are created (Fig. 35.3b).

The frequency difference generation obeys the *Manley-Rowe rule*. A photon of the quantum energy $h\nu_1$ can only produce one photon of energy $h\nu_3$. This corresponds to the energy conservation law of the elementary process. Thus, the efficiency of conversion of radiation at frequency ν_1 to radiation of frequency ν_3 is

$$\eta_{\text{diff}} = \nu_3 / \nu_1. \tag{35.17}$$

If the frequency ν_3 is much smaller than ν_1 (and ν_2), only a small portion of the power of radiation at the frequency ν_1 is converted to power of radiation at the difference frequency.

Application. The superposition of two visible or near infrared laser fields of different frequencies can lead to generation of far infrared radiation.

35.5 Optical Parametric Oscillator

An optical parametric oscillator (OPO) converts radiation of a pump frequency (ν_p) to tunable radiation at two other frequencies (ν_1 and ν_2). Radiation of one of the frequencies ν_1 or ν_2 (or of both frequencies) is stored in a resonator in order to produce a feedback to the nonlinear crystal (Fig. 35.4a). The OPO shows threshold behavior. Above a threshold amplitude of the pump field, optical parametric oscillation sets in. The oscillation frequency ν_1 depends on the resonator. Changing the eigenfrequency of the resonator leads to a variation of ν_2 and ν_3.

(Together with the change of the resonance frequency, phase matching of the fields of different frequencies is a necessary condition for operation of a parametric

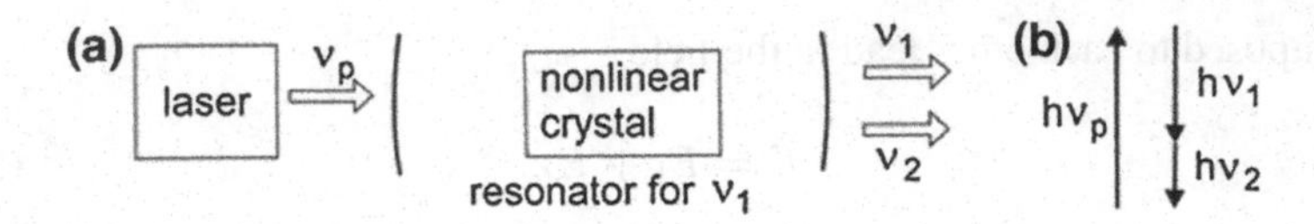

Fig. 35.4 Optical parametric oscillator; OPO. **a** Principle. **b** Elementary process

oscillator. Phase matching can be achieved by the choice of an appropriate orientation of the nonlinear crystal, e.g., of a $LiNbO_3$ crystal; changing the frequency of the signal wave then requires a rotation of the crystal. Another possibility is the change of the crystal temperature, suitable for KDP.)

In the photon picture (Fig. 35.4b), the elementary process in the OPO crystal corresponds to the decay of a photon into two photons of smaller quantum energy. The energy conservation law holds,

$$h\nu_p = h\nu_1 + h\nu_2. \tag{35.18}$$

An optical parametric oscillator, pumped with radiation near 1 μm, is suitable for generation of tunable infrared radiation with frequencies in the range from 1 THz to 15 THz; the threshold pump power of a parametric oscillator is of the order of 1 MW/cm^2 (for $LiNbO_3$ as the nonlinear crystal).

The notations—ν_p = pump frequency, $\nu_1 = \nu_s$ = signal frequency and $\nu_2 = \nu_i$ = idler frequency—have originally been introduced in the fields of high frequency technique and of microwave technique. "Parametric" means that the pump field modulates a parameter and that the parameter gives rise to a frequency converting process. In the OPO, the parameter is the refractive index of the nonlinear medium.

35.6 Third-Order Polarization

As an example of third-order polarization $P^{(3)} = \varepsilon_0 \chi^{(3)} E^3$, we consider the effect of a harmonic field $E = A \cos \omega t$ and find

$$P^{(3)} = \frac{1}{4}\varepsilon_0 \chi^{(3)} A^3 \cos 3\omega t + \frac{3}{4}\varepsilon_0 \chi^{(3)} A^3 \cos \omega t. \tag{35.19}$$

The third-order polarization due to a monochromatic field causes two different effects, frequency tripling and a change of the refractive at the frequency ω.

The first term is the source of an electric field at the frequency 3ν (Fig. 35.5a). Radiation of frequency ν is converted to radiation of frequency 3ν. In an elementary process, three photons of energy $h\nu$ are annihilated and a photon of energy $3h\nu$ is created (Fig. 35.5b).

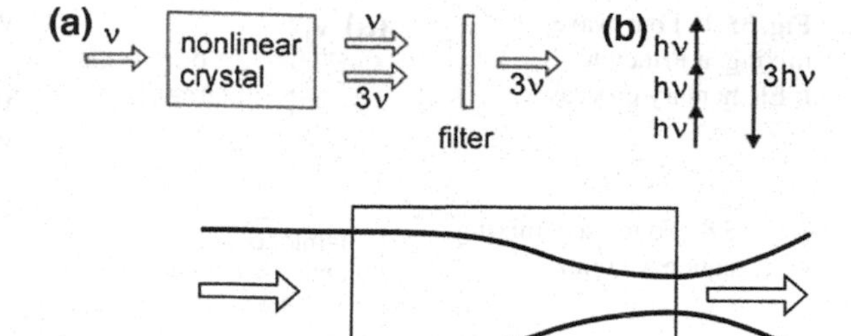

Fig. 35.5 Frequency tripler. **a** Principle. **b** Elementary process

Fig. 35.6 Self-focusing

The polarization at the frequency ω can be written in the form:

$$P = \varepsilon_0 \chi^1 A \cos \omega t + \frac{3}{4} \varepsilon \chi^{(3)} A^3 \cos \omega t = \varepsilon_0 \left(\chi^{(1)} + \frac{3}{4} \chi^{(3)} A^2 \right) A \cos \omega t$$
$$= \varepsilon_0 \chi^{(1)} \left(1 + \frac{3\chi^{(3)}}{4\chi^{(1)}} A^2 \right) A \cos \omega t. \tag{35.20}$$

It follows, with $n = \sqrt{1 + \chi^{(1)}}$, that $n = n_0 + n_2 I$, where n_0 is the refractive index and

$$n_2 = \frac{6\chi^{(3)}}{c\chi^{(1)}\varepsilon_0} \tag{35.21}$$

is a factor that accounts for the change of the refractive index due to third-order nonlinearity (Problems 35.2 and 35.3). The intensity-dependent refractive index gives rise to self-focusing (Fig. 35.6; the Kerr lens mode locking (Sect. 13.2) makes use of self-focusing).

35.7 Four-Wave Mixing and Optical Frequency Analyzer

In a four-wave mixing experiment (Fig. 35.7a) two fields (frequency ν_1 and ν_2) produce fields at two other frequencies (ν_3 and ν_4). The polarization

$$P = \varepsilon_0 \chi^{(3)} E_1 E_2 E_3, \tag{35.22}$$

with $E_1 = A_1 \cos \omega_1 t$, $E_2 = A_2 \cos \omega_2 t$ and $E_3 = A_3 \cos \omega_3 t$, is responsible for the mixing process. The polarization P is the source of a field $E_4 = A_4 \cos \omega_4 t$. The nonlinear medium can be a crystal or a glass. The elementary process of the four-wave mixing (Fig. 35.7b) corresponds to the conversion of two photons (energy $h\nu_1$ and $h\nu_2$) to two other photons (energy $h\nu_3$ and $h\nu_4$). Four-wave mixing is a stimulated process. It occurs at strengths above appropriate threshold fields of E_1 and E_2.

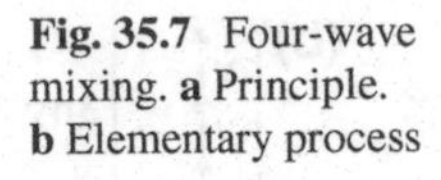

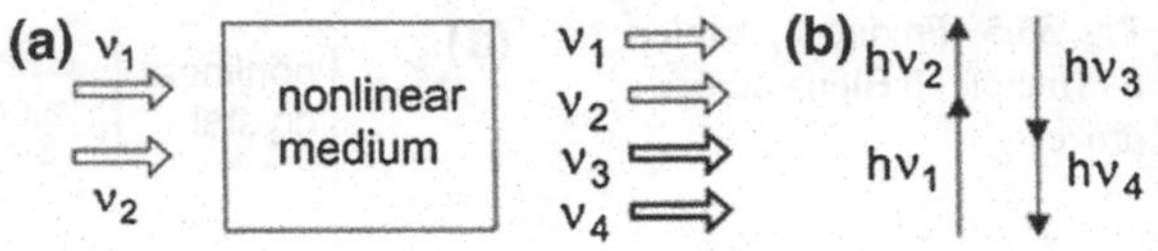

Fig. 35.7 Four-wave mixing. **a** Principle. **b** Elementary process

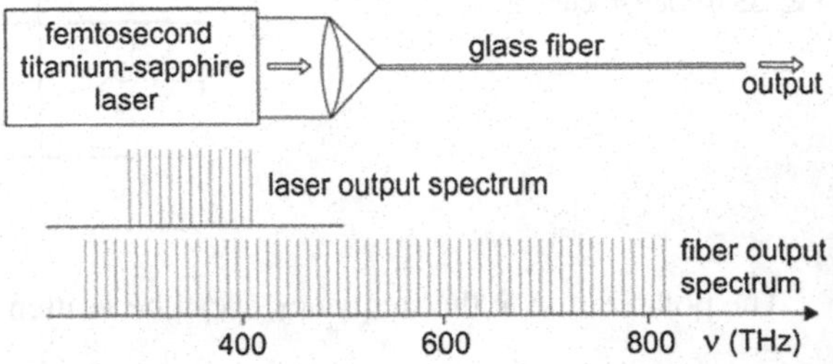

Fig. 35.8 Four-wave mixing of a frequency comb

Four-wave mixing has many applications (*see* books on nonlinear optics). Here, we discuss as an example the role of four-wave mixing for the optical frequency analyzer.

Example four-wave mixing of radiation consisting of an optical frequency comb. In our earlier treatment of the optical spectrum analyzer, we have seen that the pulses emitted by a femtosecond laser are not exactly multiples of the round trip frequency f_r of the radiation pulses in the laser, but that the frequencies are shifted by a frequency offset f_o. The frequency of the nth line of the optical frequency comb is equal to

$$f_n = nf_r + f_o. \tag{35.23}$$

We now discuss how the frequencies f_r and f_o can be determined (Fig. 35.8). A frequency comb generated by a titanium–sapphire laser is strongly focused to a glass fiber (inner diameter ~1 μm). In a four-wave mixing process, the frequency comb broadens and all lines show the same frequency shift f_o. This follows from an analysis of the polarization. The term $\cos\omega_1 t \cos\omega_2 t \cos\omega_3 t$ of the polarization contains the angular frequency $\omega_1 + \omega_2 - \omega_3$ and is the source of a field of frequency

$$f_4 = f_1 + f_2 - f_3 = n_4 f_r - f_o, \tag{35.24}$$

where $n_4 = n_1 + n_2 - n_3$ is an integer. By the mixing of radiation of different frequencies f_1, f_2, and f_3, the frequency comb becomes very broad. The optical frequency analyzer involves the following frequencies.

- f_r = repetition rate of the femtosecond pulses = number of pulses per second, measured by counting the pulses.
- f_o = offset frequency; measured by mixing of frequency-doubled radiation.
- $2f_n = 2nf_r + 2f_o$ = frequencies of frequency-doubled radiation. The mixing of the frequency-doubled radiation with radiation generated by four-wave mixing provides the difference frequencies

$$2f_n - f_m = (2n - m)f_r - f_o, \tag{35.25}$$

where n and m are integers. For $2n = m$, the difference frequency is f_o. By measuring different combinations of $(2n - m)$, also f_r can be determined by a mixing experiment: a photodiode serves as nonlinear device, which produces microwaves at the difference frequencies.

In an optical frequency analyzer, the frequency f_o is kept constant. For this purpose, the laser resonator of the titanium–sapphire laser is stabilized: the distance between the resonator mirrors is piezoelectrically controlled. After starting a femtosecond titanium–sapphire laser and reaching stable operation, the offset frequency f_o is kept constant

35.8 Stimulated Raman Scattering

An efficient way of converting monochromatic radiation into coherent radiation of another frequency is the *stimulated Raman scattering*.

In a Raman scattering process (Fig. 35.9), the energy $\hbar\omega_1$ of a photon is converted to energy $\hbar\omega_2$ of another photon and internal excitation energy E_{int} of a medium. The Raman scattered light is incoherent. Above a threshold pump field, stimulated Raman scattering results in the generation of a coherent field. Internal excitations can be phonons in crystals, phonons in glasses, vibrational-rotational excitations, or rotational excitations in molecular gases.

Example Stimulated Raman scattering of radiation of a CO_2 laser at molecules (e.g., CH_3F molecules in a gas) can lead to coherent far infrared radiation; the internal excitation is a vibrational-rotational state in a molecular gas (Sect. 14.9).

References [12, 297–307].

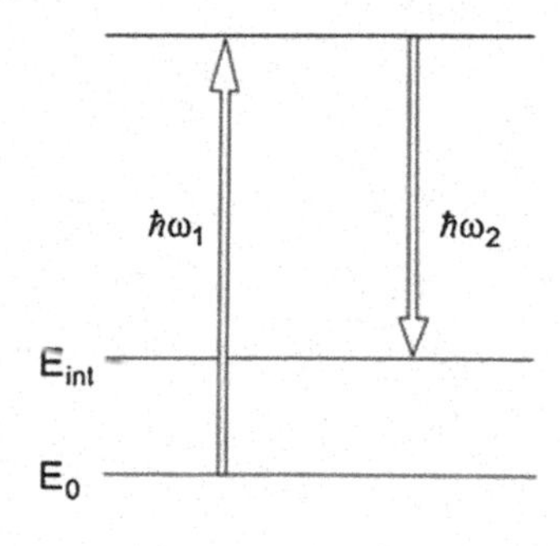

Fig. 35.9 Raman scattering

Problems

35.1 Two monochromatic optical waves (wavelengths near 600 nm) are focused on a photodetector. The photodetector generates a microwave signal at beat frequencies. The smallest beat frequency is 200 MHz. Calculate the wavelength difference of the two optical waves.

35.2 Nonlinear polarization.

(a) Show that a strong electric field applied to a hydrogen atom gives rise to nonlinear polarization of any order. [*Hint*: make use of the Taylor expansion of $(1+x)^{-2}$].
(b) Estimate the values of $\chi^{(2)}$, $\chi^{(3)}$ and $\chi^{(4)}$; $\chi^{(1)}$ is of the order of unity. [*Hint*: the lowest-order correction term $P^{(2)}$ would be comparable to $P^{(1)}$ if the amplitude A of the field is of the order of the strength of field acting on an electron in an H atom.]

35.3 Show that (35.21) follows from (35.20). Estimate the value of n_2. [*Hint*: use the estimate of $\chi^{(3)}$ in the preceding problem.]

Solutions to Selected Problems

Problems of Chap. 1

1.1 Physical constants

(a) $c = 2.99792458 \times 10^8\,\mathrm{m\,s^{-1}}$ (exact).
(b) $h = 6.6261 \times 10^{-34}\,\mathrm{J\,s}$.
(c) $\hbar = 1.0545 \times 10^{-34}\,\mathrm{J\,s}$.
(d) $e = 1.6022 \times 10^{-19}\,\mathrm{C}$.
(e) $m_0 = 0.9109 \times 10^{-30}\,\mathrm{kg}$.
(f) $\mu_0 = 4\pi \times 10^{-7}\,\mathrm{V\,s\,A^{-1}\,m^{-1}}$.
(g) $\varepsilon_0 = 1/(\mu_0 c^2) = 0.8854 \times 10^{-11}\,\mathrm{A\,s\,V^{-1}\,m^{-1}}$.
(h) $k = 1.3807 \times 10^{-23}\,\mathrm{J\,K^{-1}}$.
(i) $N_A = 6.022 \times 10^{26}$ molecules per Mole.
(j) $R = kN_A = 8.315 \times 10^3\,\mathrm{J\,K^{-1}}$ per Mole.
(k) $L_0 = 2.687 \times 10^{25}$ molecules per $\mathrm{m^3}$ at $0\,^\circ\mathrm{C}$ and normal pressure.

1.2 Frequency, wavelength, wavenumber and energy scale

(a) $1\,\mu\mathrm{m}$; $300\,\mathrm{THz}$; $10^4\,\mathrm{cm^{-1}}$; $1.9878 \times 10^{-20}\,\mathrm{J}$; $1.2407\,\mathrm{eV}$.
(b) $300\,\mu\mathrm{m}$; $1\,\mathrm{THz}$; $3300\,\mathrm{m^{-1}} = 33.33\,\mathrm{cm^{-1}}$; $6.626 \times 10^{-23}\,\mathrm{J}$; $4.136\,\mathrm{meV}$.
(c) $1\,\mathrm{nm}$; $300\,\mathrm{PHz}$; $2.0 \times 10^{-17}\,\mathrm{J}$; $1.240\,\mathrm{keV}$.
(d) $1\,\mathrm{m^{-1}}$; $1\,\mathrm{m}$; $300\,\mathrm{MHz}$; $1.24\,\mu\mathrm{eV}$.
(e) $1.2407\,\mu\mathrm{m}$; $241.8\,\mathrm{THz}$; $1.6022 \times 10^{-19}\,\mathrm{J}$; $1\,\mathrm{eV}$.

1.3

(a) $T = 300\,\mathrm{K} \equiv kT = 4.142 \times 10^{-21}\,\mathrm{J} \equiv kT/e = 25.85\,\mathrm{meV} \equiv \nu = kT/h = 6.625\,\mathrm{THz} \equiv kT/(hc) = 208\,\mathrm{cm^{-1}} \equiv hc/(kT) = 48\,\mu\mathrm{m}$.
(b) $1\,\mathrm{meV} \equiv 8.06\,\mathrm{cm^{-1}} \equiv 0.2418\,\mathrm{THz}$.
(c) $1\,\mathrm{cm^{-1}} \equiv 30\,\mathrm{GHz}$.
(c) $10\,\mathrm{cm^{-1}} \equiv 1.2408\,\mathrm{meV}$.

1.4 Power of the sun light and laser power

(a) $140\,\mathrm{mW}$. (b) $1.8\,\mathrm{kW/cm^2}$. (c) $1.3\,\mathrm{kW/cm^2}$.

K.F. Renk, *Basics of Laser Physics*, Graduate Texts in Physics,
DOI 10.1007/978-3-319-50651-7

Problems of Chap. 2

2.1 $5.6 \times 10^{22}\,\mathrm{m}^{-3}$; $5.6 \times 10^{25}\,\mathrm{m}^{-3}$; $5.6 \times 10^{28}\,\mathrm{m}^{-3}$.

2.2 Field amplitude

(a) $\varepsilon_0 A^2/2 = u$; $A = \sqrt{2u/\varepsilon_0}$; $\varepsilon_0 = 0.89 \times 10^{-11}\,\mathrm{A\,s\,V^{-1}\,m^{-1}}$; $A = 4.7 \times 10^5\,\mathrm{V\,m^{-1}}$.
(b) $Z = 10^6\,\mathrm{m}^{-3}$; $A = \sqrt{2h\nu Z/\varepsilon_0} = 6.3 \times 10^{-2}\,\mathrm{V\,m^{-1}}$; $u = 2 \times 10^{-14}\,\mathrm{J\,m^{-3}}$.
(c) $Z = 2 \times 10^{13}\,\mathrm{m}^{-3}$; $A = 180\,\mathrm{V\,m^{-1}}$; $u = 1\,\mu\mathrm{J\,m^{-3}}$.

2.3 Occupation number

(b) $kT = 4.14 \times 10^{-21}\,\mathrm{J} = 25.8\,\mathrm{meV}$; $f_1^{\mathrm{Boltz}} - f_2^{\mathrm{Boltz}} \sim 1.8 \times 10^{-35}$.
(c) $f_1^{\mathrm{Boltz}} - f_2^{\mathrm{Boltz}} \sim 0.54 - 0.46 = 0.08$.

2.4 Oscillation condition

(a) In one case, the condition is $G_1 G_1 V u/2 = u/2$ and in an other case, $V G_1 G_1 u/2 = u/2$. Show that both cases lead to $GV = 1$.
(b) For both directions, we obtain the product GV and the same sum of the phases.

2.5 Brewster angle
(a) 54.4°. (b) 56.3°. (c) 61.2°. (d) 60.4°.

Problems of Chap. 3

3.1 $\delta\nu/(c/2L) = 2 \times 10^5$.
3.2 $V = R_1 R_2$; $s_{\mathrm{eff}} = 10$; $\tau_{\mathrm{p}} = 6.7\,\mathrm{ns}$; $l_{\mathrm{p}} = 2\,\mathrm{m}$; $Q = 63$.

3.3 Resonator with air

(a) $\nu_1 = c/(2nL)$.
(b) $\delta\nu = c/(2L) - (c/n)/(2L) = c/(2L)(1 - 1/n) = 160\,\mathrm{kHz}$; $\delta\nu/\nu \sim 3 \times 10^{-9}$.

3.4 Energy $= \varepsilon_0 a_1 T^{-1} \int_0^L \int_0^T E^2(z,t)\mathrm{d}z\mathrm{d}t = \varepsilon_0 a_1 a_2 A^2 T^{-1} \int_0^L \int_0^T \sin^2 kz \sin^2 \omega t\, \mathrm{d}z\mathrm{d}t = \varepsilon_0 a_1 a_2 L A^2/4$; $u = \varepsilon_0 A^2/4$.
3.5 $V = R_1 R_2$; $\tau_{\mathrm{p}} = 1/(1 - V)$.
3.6 Photon density

(a) $Z_{\mathrm{FP}}/\tau_{\mathrm{p}} = Z/T$; $Z_{\mathrm{FP}} = Q/(2\pi)Z$; $Q = 2\pi l/(1 - R)$.
(b) We obtain the same result, but $Q = \pi l/(1 - R)$.

3.7 $R_{\mathrm{FP}} = 1 - T_{\mathrm{FP}} = 4R(1 - R)^{-2} \sin^2 \delta/2[1 + 4R(1 - R)^{-2} \sin^2 \delta/2]^{-1}$.
3.8 $T_{\mathrm{FP}} = 1/[1 + 4R(1 - R)^{-2} \sin^2 \delta/2]$, where $R = \sqrt{R_1 R_2}$.

3.9 Fabry–Perot interferometer with absorbing mirrors

(a) $T_{FP} = (1+(A_m^2/T_m^2)^{-1}(1+4R(1-R)^{-2}\sin^2\delta/2)^{-1}$; $T_{FP,max} = (1+A_m^2/T_m^2)^{-1}$.
(b) $1/(1+A_m^2/T_m^2) < 0.98$; $A_m/T_m < 0.1$.

3.10 Fabry–Perot interferometer for obliquely incident radiation

(a) $\delta = k \times 2L\cos\theta + 2\varphi$.
(b) $\varphi = 0$; $\delta = k \times 2L\cos\theta = zl \times 2\pi$; $2L\cos\theta = zl \times \lambda$.

Problems of Chap. 4

4.2 Absolute number of two-level systems

(a) $N_{tot} = 10^{15}$. (b) $N_{tot} = 10^{10}$. (c) $N_{tot} = 10^4$.

Problems of Chap. 5

5.1 $L + (n-1)L' = 57.6\,\text{cm}$.

5.2 Photon density

(a) The laser beam has only a slightly larger diameter at 10 m distance from the laser and the laser power is of the order of 1 W.
(b) Assuming that the luminescence radiation is emitted isotropically, the power reaching an area of 1 cm diameter is $P_{fluor} = P_0 \times 2\pi\sin^2(\alpha/2)$, where α is the angle corresponding to the area. It follows that $\alpha \sim 5 \times 10^{-4}$; $P_{fluor} \sim P_0 \times 2\pi \times \alpha^2/2 \sim 0.4\,\mu\text{W}$.

5.3 $g(\lambda)\mathrm{d}\lambda = g(\nu)\mathrm{d}\nu$; $g(\nu) = g(\lambda)/|\mathrm{d}\nu/\mathrm{d}\lambda|$; $\nu = c/\lambda$; $\mathrm{d}\nu/\mathrm{d}\lambda = -c/\lambda^2$; $g(\nu) = \lambda^2 g(\lambda)/c$.

5.4 Population of the upper laser level

(a) $r = N_2/\tau_{rel}^* = 3.3 \times 10^{29}\,\text{m}^{-3}\,\text{s}^{-1}$; volume $= \pi r^2 L' = 7.9 \times 10^{-10}\,\text{m}^3$; $P_{pump} = 1.5 \times 3.3 \times 10^{29} \times 7.9 \times 10^{-10} \times 2.4 \times 10^{-19}\,\text{W} = 9.4\,\text{W}$; the factor 1.5 takes account of the quantum efficiency.
(b) $N_{tot} = 10^{24} \times 7.9 \times 10^{-10} = 7.9 \times 10^{14}$.
(c) Energy $= 10^{24} \times 7.9 \times 10^{-10} \times 1.5 \times 1.6 \times 10^{-19}\,\text{J} = 190\,\mu\text{J}$; energydensity = energy/volume $= (N_{tot}/\text{volume}) \times h\nu = 240\,\text{kJ/m}^3 = 240\,\text{J}$ per liter.
(d) $P_{pump} = 94\,\text{W}$. [The reason is the stimulated emission (Sect. 8.8).]

Problems of Chap. 6

6.1 Photon density

(a) $Z = D(\nu)\mathrm{d}\nu\bar{n} = (8\pi\nu^2/c^3)kT/h\nu \sim 6 \times 10^7\,\text{m}^{-3}$.

(b) $Z = 6 \times 10^{10}\,\mathrm{m}^{-3}$.
(c) $Z = (8\pi\nu^2/c^3)\mathrm{d}\nu\exp(-h\nu/kT) = 4 \times 10^{-34}\,\mathrm{m}^{-3}$.

6.2 Number of thermal photons in a mode of a laser resonator

(a) $\bar{n} = \exp(-h\nu/kT) \sim 2 \times 10^{-29}$.
(b) $\bar{n} = 1/[\exp(h\nu/kT) - 1] \sim 0.25$.
(c) $\bar{n} = kT/h\nu \sim 6$.

Problems of Chap. 7

7.1 Amplification of radiation in titanium–sapphire

(a) $\alpha = 8\,\mathrm{m}^{-1}$. (b) $G_1 - 1 = 0.5$.
(c) $\alpha(1\,\mu\mathrm{m})/\alpha(\lambda_0) \sim 0.5$; $\alpha(1\,\mu\mathrm{m}) \sim 4\,\mathrm{m}^{-1}$; $G_1 - 1 = 0.25$.

7.2 $\sigma_{\mathrm{nat}} = (\lambda/n)^2/2\pi = 3.2 \times 10^{-14}\,\mathrm{m}^2$; $\tau_{\mathrm{sp}} = 3.8\,\mu\mathrm{s}$; $\Delta\nu_{\mathrm{nat}} = 1/2\pi\tau_{\mathrm{sp}} = 4.2 \times 10^4\,\mathrm{Hz}$; $\Delta\nu_0 = 1.1 \times 10^{14}\,\mathrm{Hz}$; $\sigma_{21}/\sigma_{\mathrm{nat}} = 1.5\Delta\nu_{\mathrm{nat}}/\Delta\nu_0 = 5.7 \times 10^{-10}$; $\sigma_{21} = 1.8 \times 10^{-23}\,\mathrm{m}^2$.

7.3 Two-dimensional gain medium

(a) $\alpha = 2{,}000\,\mathrm{m}^{-1} = 20\,\mathrm{cm}^{-1}$. (b) $G_1 - 1 = 1.5 \times 10^{-3}$.

Problems of Chap. 8

8.1 $\tau_{\mathrm{p}} = (2nL/c)(1 - R)^{-1} = 6 \times 10^{-8}\,\mathrm{s}$; $l_{\mathrm{p}} = (c/n)\tau_{\mathrm{p}} = 10\,\mathrm{m}$; $(N_2 - N_1)_{\mathrm{th}} = 1/l_{\mathrm{p}}\sigma_{21} = 3 \times 10^{21}\,\mathrm{m}^{-3}$.

8.2 $r_{\mathrm{th}} = (N_2 - N_1)_{\mathrm{th}}/\tau^*_{\mathrm{rel}} = 8 \times 10^{26}\,\mathrm{m}^{-3}\,\mathrm{s}^{-1}$;
$Z_\infty = (10r_{\mathrm{th}} - r_{\mathrm{th}})\tau_{\mathrm{p}} = 9r_{\mathrm{th}}\tau_{\mathrm{p}} = 4.3 \times 10^{19}\,\mathrm{m}^{-3}$; $P_{\mathrm{out}} = Z_\infty a_1 a_2 L h\nu/\tau_{\mathrm{p}} = 9\,\mathrm{W}$; $r_{\mathrm{out}}/r = Z_\infty/(\tau_{\mathrm{p}} r) = 9r_{\mathrm{th}}/10r_{\mathrm{th}} = 0.9$.

8.3 $(N_2 - N_1)_0 = 10 \times (N_2 - N_1)_{\mathrm{th}} = 3 \times 10^{22}\,\mathrm{m}^{-3}$; $\gamma_0 = b_{21}(N_2 - N_1)_0 = 1.3 \times 10^8\,\mathrm{s}^{-1}$; $\kappa = 1/\tau_{\mathrm{p}} = 1.6 \times 10^7\,\mathrm{s}^{-1}$; $Z_0 = 1/a_1 a_2 L = 2 \times 10^7\,\mathrm{m}^{-3}$; $t_{\mathrm{on}} = 18\,\mathrm{ns}$.

8.4 If the active medium has a smaller length than the resonator, the threshold condition is $(N_2 - N_1)_\infty = -\ln V/(2nL'\sigma_{21})$, where L' is the length of the active medium. It follows for that case that the gain coefficient $\alpha = (N_2 - N_1)\sigma_{21}$ has to be larger than the reciprocal of the effective photon path length in the crystal, $l'_{\mathrm{p}} = l_{\mathrm{p}}L'/L$, $\alpha \geq 1/l'_{\mathrm{p}} = -\ln V/(2nL')$ or $2\alpha L' \geq -\ln V$. We find, with $G = \exp(2\alpha L')$, that the condition of gain, $GV \geq 1$, is fulfilled.

Problems of Chap. 10

10.1 $\nu_{110} = \nu_{101} = \nu_{011} = c/(\sqrt{2}a) = 21.2\,\mathrm{GHz}$; $\nu_{111} = \sqrt{3}c/(2a) = 26\,\mathrm{GHz}$.

10.2 Degeneracy of modes of a rectangular cavity resonator
(a) 3. (b) 2. (c) No degeneracy. (d) 2.
10.3 Density of modes of a cavity resonator

(a) $D(\nu) = 8\pi n^3 \nu^2/c^3 = 1.7 \times 10^5\,\mathrm{m}^{-3}\,\mathrm{Hz}^{-1}$ for $\nu = 4.3 \times 10^{14}$ Hz; n = 1.
(b) $1.0 \times 10^6\,\mathrm{m}^{-3}\,\mathrm{Hz}^{-1}$.
(c) $8.3 \times 10^6\,\mathrm{m}^{-3}\,\mathrm{Hz}^{-1}$.

10.5 Mode density on different scales

(a) $D(\nu)\mathrm{d}\nu = D(h\nu)\mathrm{d}(h\nu)$; $D(h\nu) = D(\nu)\mathrm{d}\nu/\mathrm{d}(h\nu) = D(\nu)/h$.
(b) $D(\nu)\mathrm{d}\nu = D(\omega)\mathrm{d}\omega$; $D(\omega) = D(\nu)\mathrm{d}\nu/\mathrm{d}\omega = D(\nu)/(2\pi)$.
(c) $D(\nu)\mathrm{d}\nu = D(\lambda)\mathrm{d}\lambda$; $D(\lambda) = D(\nu) \times \mathrm{d}\nu/\mathrm{d}\lambda = cD(\nu)/\lambda^2$.

10.6 $\nu = (c/2)\sqrt{a_1^{-2} + L^{-2}} = c/(2a_1)\sqrt{1 + a_1^2/L^2} \sim c/(2a_1)[1 + a_1/(2L^2)]$; $\mathrm{d}\nu/\mathrm{d}L \sim -(ca_1/(2L^3)$; $\mathrm{d}\nu/\nu \sim (a_1^2/L^2)\mathrm{d}L/L$.
10.7 Density of modes in free space
We consider a propagating wave $E = A\exp[\mathrm{i}(\omega t - \boldsymbol{kr})]$. We apply periodic boundary conditions: $E(x+L, y+L, z+L) = E(x, y, z)$ for each value of t; L is the length of the periodicity interval assumed to be equal in all spatial directions. This leads to the conditions: $\exp(\mathrm{i}k_x L) = 1$; $\exp(\mathrm{i}k_y L) = 1$; $\exp(\mathrm{i}k_z L) = 1$. It follows that: $k_x = l \times 2\pi/L$; $k_y = m \times 2\pi/L$; $k_z = n \times 2\pi/L$; $k^2 = (2\pi/L)^2(l^2+m^2+n^2)$, with $l, m, n = 0, \pm1, \pm2, \ldots$. We find, with $\omega = ck$, that $\omega^2 = (2\pi c/L)^2(l^2+m^2+n^2)$. The mode density in k space is $D(k) = (L^3/\pi^2)k^2$ and in ω space $D^*(\omega) = \omega^2 L^3/(\pi^2 c^3)$. With $D^*(\omega)d\omega = D^*(\nu)\mathrm{d}\nu$, we obtain $D^*(\nu) = (8\pi\nu^2/c^3)L^3$.

Problems of Chap. 11

11.1 Gaussian beam

(b) The ratio of the intensity of the radiation within the beam radius r_0 to the total intensity is

$$\int_0^{r0} 2\pi r \exp(-r^2/r_0^2)\mathrm{d}r \Big/ \int_0^{\infty} 2\pi r \exp(-r^2/r_0^2)\mathrm{d}r = 1 - 1/\mathrm{e} = 0.63.$$

We used $\int 2xe^{-x^2}\mathrm{d}x = -\exp(-x^2)$.
(c) $r_p = I_p/I_{\mathrm{tot}} = 1 - \exp(-r_p^2/r_0^2)$; $r_p/r_0 = \sqrt{-\ln(1-p)}$.
(d) $r_p = 1.52\, r_0$.
(e) $r_p = 1.73\, r_0$.
(f) A Taylor expansion of $p(r_p)$ with respect to r_p yields $p \sim 1 - r_p^2/r_0^2$.

11.2 $\theta_{0,\mathrm{u}} = \sqrt{2}\lambda/(\pi r_0) = 1.3 \times 10^{-3}$ (=4.5 arc minutes)
11.3 The angle of divergence is $\Theta = 0.1(\sqrt{2}/\pi)\lambda/r_0 = 2 \times 10^{-6}$.

11.4 Density of photons in a Gaussian beam. The number of photons emitted per second by the laser is $P_{out}/(h\nu) = 6 \times 10^{14}\,s^{-1}$. A detector of diameter D monitors radiation within the angle $\vartheta = D/d$, where d is the distance from the laser. The portion of radiation within the angle ϑ is $\sin^2\vartheta/\sin\theta^2 \sim \vartheta^2/\theta^2$. It follows for the number of photons per second:

(a) $\vartheta = 10^{-7}$; $\vartheta^2/\theta^2 = 2 \times 10^{-3}$; $3 \times 10^{11}\,s^{-1}$.
(b) $\vartheta = 7 \times 10^{-11}$; $\vartheta^2/\theta^2 \sim 10^{-9}$; $\sim 10^6\,s^{-1}$.

Problems of Chap. 13

13.1 Ultrashort pulses
(a) $t_p \sim 1/\Delta\nu_0 \sim 0.3\,ps$. (b) $t_p \sim 10\,ps$. (c) $t_p = 30\,ps$.
13.2 Excited Ti^{3+} ions are collected during the round trip time $T = 10^{-8}\,s$. The density of excited Ti^{3+} is $rT = 3 \times 10^{20}\,m^{-3}$. Accordingly, the energy in a pulse is $rTa_1a_2L' \times h\nu = 19\,nJ$ and the pulse power $= 1.9\,MW$. The average power is 1.9 W.
13.3 $2 \times 10^7\,W$.

Problems of Chap. 14

14.1 Helium–neon laser: line broadening and gain cross section

(a) $\Delta\nu_D = 1.5 \times 10^9\,Hz$. (b) $\Delta\nu_c \sim 10^6\,Hz$. (c) $\Delta\nu_{nat} = 1.6 \times 10^6\,Hz$.
(d) $\Delta\nu_0 = 1.6 \times 10^7\,Hz$. (e) $\sigma_{21}(\nu_0) = (1/c)h\nu B_{21}g_G(\nu_0) = 1.0 \times 10^{-16}\,m^2$.

14.2 Helium–neon laser: threshold condition, output power and oscillation onset time

(a) $\tau_p = T/(1-R_1R_2) = 1.5 \times 10^{-7}\,s$; $l_p = c\tau_p = 45\,m$. $(N_2-N_1)_{th} = 1/(\sigma_{21}l_p) = 2 \times 10^{14}\,m^{-3}$.
(b) $(N_2 - N_1)_{th} \times a_1a_2L = 2 \times 10^8$; number of excited neon atoms in the laser.
(c) $r_{th} = (N_2 - N_1)_{th} \times a_1a_2L/\tau^*_{rel} = 2 \times 10^{15}\,s^{-1}$; $r_{out} \times a_1a_2L = 9r_{th}a_1a_2L = 2 \times 10^{16}\,s^{-1}$; $P_{out} = r_{out} \times a_1a_2Lh\nu = 13\,mW$.
(d) $Z_0 = (a_1a_2L)^{-1} = 5 \times 10^5\,m^{-3}$; $Z_\infty = r_{out}\tau_p = 2.7 \times 10^9\,m^{-3}$; $\alpha_{th} = (N_2 - N_1)_{th} \times \sigma_{21} = 0.02\,m^{-1}$; $\gamma_{th} = c\alpha_{th} = 6 \times 10^6\,s^{-1}$; $\kappa = 6 \times 10^6\,s^{-1}$ (because $\kappa = \gamma_{th}$ at threshold); $\gamma_0 = 10\,\gamma_{th} = 6 \times 10^7\,s^{-1}$; $t_{on} = \ln(Z_\infty/Z_0)/(\gamma_0 - \kappa) = 160\,ns$.

14.3 Doppler effect

(a) $\nu = \nu_0(1 \pm v/c)$; $\delta\nu = (2v/c)\nu_0 = 3 \times 10^9\,Hz$.
(b) The homogeneous width of the line due to $2 \rightarrow 1$ spontaneous transitions is $\Delta\nu = 1/(2\pi\tau_{rel}) = 1.6 \times 10^8\,Hz$, where τ_{rel} is the lifetime of the lower laser level. This corresponds to a velocity range $-v$, v or to $|v| = c\delta\nu/\nu_0 = 100\,m\,s^{-1}$.

(c) The gain curve has a minimum of a halfwidth of 160 kHz. In the line center, the gain is smaller than outside because outside (80 kHz away from the center) ions of the velocity +v contribute to gain in half a round trip and ions of the velocity −v contribute during the other half round trip. The Lamb dip can be used for frequency stabilization of a helium–neon laser.

14.4 CO_2 laser

(a) $\Delta\nu_D = 2\nu_0\sqrt{(2kT/mc^2)\ln 2} = 5.6\times10^7$ Hz; $m = m_C + 2m_O = 44\,m_p = 7.3\times10^{-26}$ kg; m_p = proton mass. $\sigma_{21}(\nu_0) = 0.94c^2A_{21}/(8\pi\nu^2\Delta\nu_D) = 1\times10^{-21}\,\mathrm{m}^2$. $G_{th}V = 1$; $V = 0.7$; $G_{th} = 1.43$; $G_{th} = \exp[\alpha_{th}\times 2L]$; $\alpha_{th} = \ln(G_{th})/2L = 0.18\,\mathrm{m}^{-1}$. $(N_2 - N_1)_{th} = \alpha_{th}/\sigma_{21} = 1.8\times10^{20}\,\mathrm{m}^{-3}$. $r_{th} = (N_2 - N_1)_{th}/\tau^*_{rel} = 4.5\times10^{19}\,\mathrm{m}^{-3}\,\mathrm{s}^{-1}$; $P = ra_1a_2Lh\nu$; $r = P/a_1a_2Lh\nu) = 3\times10^{25}\,\mathrm{m}^{-3}\,\mathrm{s}^{-1}$; $r \sim 10^6\,r_{th}$; the pump rate is about 10^6 times larger than the threshold pump rate.

(b) Because of the extremely long lifetime of the upper laser level with respect to spontaneous emission, the oscillation builds up as soon as the population difference exceeds $(N_2 - N_1)_{th}$. Stronger pumping then leads to generation of laser radiation. By collisions of the CO_2 molecules with each other, a quasithermal distribution of the populations of the different rotational levels of the excited state is maintained and the pump energy is converted to laser radiation (and energy of relaxation).

(c) We treat, for simplicity, the CO_2 gas as an ideal gas. At 273 K and normal pressure, an ideal gas (mole volume 22.4 l) contains 6×10^{23} molecules. This corresponds to about $3\times10^{25}\,\mathrm{m}^{-3}$. We use this number for CO_2 at room temperature and normal pressure (1 bar). At a pressure of 10 mbar, the density of available CO_2 molecules is $3\times10^{22}\,\mathrm{m}^{-3}$. At room temperature, excited CO_2 molecules are in different rotational states. About 1% of the molecules are in a particular rotational state. Thus, about 3×10^{21} molecules per m^3 are available for laser transitions. Assuming that half of the molecules are in an excited state we find that the density of molecules in a vibrational-rotational state is $1.5\times10^{21}\,\mathrm{m}^{-3}$. This leads to $\alpha \sim 8\times\alpha_{th} \sim 1.4\,\mathrm{m}^{-1}$ and to a single path gain of $G_1 = \exp(\alpha L) = 4$.

(d) For a collision-broadened line, the gain cross section is $\sigma_{21} = c^2A_{21}/(8\pi\nu^2)\,g(\nu)$. With increasing pressure $g(\nu)$ broadens and the cross section in the line center, $\sigma_{21}(\nu_0) = c^2A_{21}/(8\pi\nu^2)\times2/(\pi\Delta\nu_c)$, is inversely proportional to the gas pressure p. It follows that $\alpha(\nu_0)$ is independent of pressure above a pressure of about 10 mbar. At this pressure, $2\Delta\nu_c \sim \Delta\nu_D$, the gain coefficient we calculated is the maximum gain coefficient for the TEA and the high-pressure CO_2 laser.

In a TEA laser, the pulse duration of the radiation is about 200 ns. It is much larger than the duration (20 ns) of the electrical excitation pulse. During about 20 round trip transits of the radiation through the active medium, a fast redistribution occurs for the population of the levels involved in a laser oscillation. If we assume that about 1% of the excited molecules contribute to the laser oscillation, we find the pulse energy $E_{pulse} = 0.3$ J and the pulse power $E_{pulse}/t_{pulse} \sim 1$ MW.

(e) $Z_0 = (a_1 a_2 L)^{-1} = 10^4\,\text{m}^{-3}$; $\tau_p = (2L/c)/(1-R) = 2.2 \times 10^{-8}\,\text{s}$; $Z_\infty = P_{out}\tau_p/(a_1 a_2 L h\nu) = 1.1 \times 10^{21}\,\text{m}^{-3}$; $t_{on} = T \ln(Z_\infty/Z_0)/(GV) = 24\,\text{ns}$.

Problems of Chap. 15

15.2 $\sigma_{21}(\text{YAG})/\sigma_{21}(\text{TiS}) = \Delta\nu_0(\text{TiS})/\Delta\nu_0(\text{YAG}) \times \tau_{sp}(\text{TiS})/\tau_{sp}(\text{YAG}) \times \lambda^2(\text{YAG})/\lambda^2(\text{TiS}) \sim 10$.
15.3 An N_2 molecule consists of two atoms, a molecule has a single vibrational frequency (and all molecules in an N_2 gas have the same frequency) while the Ti^{3+} ions in Al_2O_3 belong to a system with a large number of atoms (ions), namely of $N \sim 10^{25}\,\text{m}^{-3}$, with $3N$ vibrational frequencies.
15.4 Laser tandem pumping

(a) $\eta = \eta_1 \times \eta_2 \times \eta_3 \times \eta_4 \sim 25\%$; $\eta_1 \sim 0.8$ (efficiency of a semiconductor laser); $\eta_2 \sim 0.8$ (Nd^{3+}:YVO_4 laser); $\eta_3 = 0.5$ (frequency doubling); $\eta_4 = 0.53\,\mu\text{m}/0.68\,\mu\text{m} = 0.78$.
(b) The Nd^{3+}:YVO_4 laser produces a laser beam with a small angle of aperture. Therefore, a column of a small diameter can be excited in titanium–sapphire allowing generation of a narrow laser beam. Direct pumping with a semiconductor laser beam, which has a large divergence, leads to excitation of the whole titanium–sapphire crystal. This results in strong heating.

Problems of Chap. 16

16.1 Dye laser

(a) GV; $V = 0.7$; $G_{th} = 1.43$; $G_{th} = \exp(\alpha_{th} L')$; $\alpha_{th} = (1/L') \ln G_{th} = 350\,\text{m}^{-1}$; $(N_2 - N_1)_{th} = \alpha_{th}/\alpha_{21} = 3.5 \times 10^{21}\,\text{m}^{-3}$.
(b) $r_{th} = (N_2 - N_1)_{th}/\tau^*_{rel} = 7 \times 10^{29}\,\text{m}^{-3}\,\text{s}^{-1}$; $r_{th} a_1 a_2 L' = 3 \times 10^{19}\,\text{s}^{-1}$; $P_{out} = 9 r_{th} a_1 a_2 L' h\nu = 0.8\,\text{W}$.

16.2 $P_{pulse} = 190\,\text{MW}$; $P = 1.9\,\text{mW}$.

Problems of Chap. 19

19.1 Acceleration energies

(a) $E = \sqrt{\lambda_w/\lambda}\, m_0 c^2 = 2.9\,\text{MeV}$. (b) $E = 500\,\text{MeV}$.

19.2 $E = \sqrt{\lambda_w(1+k^2)8m_0c^2)/2}/\sqrt{\lambda}$; $\text{d}E = \text{d}\lambda = \sqrt{\cdots}(-1)/(2\lambda^{3/2})$; $\text{d}\nu/\nu = -\text{d}\lambda/\lambda = 2\text{d}E/E$.

Problems of Chap. 20

20.1 $\lambda_{\rm deBroglie} = h/(mv)$. The ratio is $m_0 = m_{\rm e} = 1/0.07 = 14$.
20.2 Number of states

(a) $\epsilon = 26\,{\rm meV} = 4.2 \times 10^{-21}$ J; $d\epsilon = 1.6 \times 10^{-22}$ J; $D(\epsilon)d\epsilon = 1.5 \times 10^{23}\,{\rm m}^{-3}$.
(b) $D^{\rm 2D}(\epsilon)d\epsilon = 1.0 \times 10^{16}\,{\rm m}^{-2}$.
(c) $D^{\rm 1D}(\epsilon)d\epsilon = 7.6 \times 10^{7}\,{\rm m}^{-1}$.

20.3 $\delta\nu = c/(2nL) = 41.6\,{\rm GHz}$.

Problems of Chap. 21

21.1 Wave vector of nonequilibrium electrons in GaAs
$k = (1/\hbar)\sqrt{2m_{\rm e}\epsilon} = 4.3 \times 10^8\,{\rm m}^{-1}$; $1.4 \times 10^8\,{\rm m}^{-1}$; $4.3 \times 10^7\,{\rm m}^{-1}$; $\lambda_{\rm g} = hc/E_{\rm g} = 870\,{\rm nm}$; gap wavelength (vacuum wavelength); $q_{\rm p} = 2n\pi/\lambda_{\rm g} = 2.6\times 10^7\,{\rm m}^{-1}$. It follows that $q_{\rm p}$ is small compared to k for the first two values of k.
(b) $q_{\rm p} = k$ for $\epsilon = \hbar^2 q_{\rm p}^2/(2m_{\rm e}) = 0.4$ meV.
21.2 $q_{\rm p} \ll k_1 = (1/\hbar)\sqrt{2m_{\rm e}\epsilon_{\rm c}}$; $\epsilon_{\rm c} \gg \hbar^2 q_{\rm p}^2/(2m_{\rm e}) = 0.4\,{\rm meV}$; $q_{\rm p} \ll k_2 = (1/\hbar)\sqrt{2m_{\rm h}\epsilon_{\rm v}}$; $\epsilon_{\rm v} \gg \hbar^2 q_{\rm p}^2/(2m_{\rm h}) = 0.4/6\,{\rm meV}$.
21.3 Electron and holes in an undoped GaAs quantum film in thermal equilibrium

(a) $E_{\rm Fe} = E_{\rm Fh} = E_{\rm F}$.
(b) Since $E_{\rm Fe} = E_{\rm c} + \epsilon_{\rm Fe} = E_{\rm F}$ and $E_{\rm Fv} = E_{\rm V} - \epsilon_{\rm Fh} = E_{\rm F}$, it follows that $E_{\rm c} + \epsilon_{\rm Fe} = E_{\rm v} - \epsilon_{\rm Fh}$ and $-\epsilon_{\rm Fe} - \epsilon_{\rm Fh} = E_{\rm g}$. The gap energy is positive because $\epsilon_{\rm Fe}$ and $\epsilon_{\rm Fh}$ have negative signs for small electron and hole densities.
(c) $N_{\rm thermal}^{\rm 2D} = \sqrt{D_{\rm e}^{\rm 2D} D_{\rm h}^{\rm 2D} kT \exp[-E_{\rm g}^{\rm 2D}/(2kT)]} = 2 \times 10^4\,{\rm m}^{-2}$; with $kT = 26\,{\rm meV}$; $E_{\rm g}^{\rm 2D} = 1.4\,{\rm eV}$; $N_{\rm thermal}^{\rm 2D}$ is by many orders of magnitude smaller than $N_{\rm tr}^{\rm 2D} = 1.4 \times 10^{16}\,{\rm m}^{-2}$.

Problems of Chap. 22

22.1 $N^{\rm 2D} \sim 3.3 \times 10^{16}\,{\rm m}^{-2}$; $\epsilon_{\rm Fv} \sim 6\,{\rm meV}$; $\epsilon_{\rm Fc} \sim 84\,{\rm meV}$; $E_{\rm Fc} - E_{\rm Fv} \sim 78\,{\rm meV}$.
22.2 Quantum well laser
Average photon path length $l_{\rm p} = 0.9L = 0.45\,{\rm mm}$, $N_{\rm th}^{\rm 2D} - N_{\rm tr}^{\rm 2D} = (1/3) \times 1.3(\sigma_{\rm eff} l_{\rm p}/a_1)^{-1} = 1.3 \times 10^{15}\,{\rm m}^{-2}$: $j = 3N_{\rm th}^{\rm 2D} e/\tau_{\rm sp} = 3.1 \times 10^6\,{\rm A\,m}^{-2}$; $I = 0.3$ A.
22.3 Photons in a quantum well laser

(a) $h\nu = 1.42\,{\rm eV} = 2.3 \times 10^{-19}$ J; $\tau p = 5 \times 10^{-12}$ s; $Z a_1 a_2 L h\nu/\tau_{\rm p} = 2P_{\rm out}$; $Z = 2.0 \times 10^{18}\,{\rm m}^{-3}$.
(b) $Z_{\rm tot} = Z a_1 a_2 L = 10^5$.

(c) $N_{\rm tot} \sim N_{\rm tr}^{\rm 2D} a_2 L = 1.2 \times 10^9$; $Z_{\rm tot}$ is much smaller than $N_{\rm tot}$.

Problems of Chap. 23

23.1 $q_{\rm p} = 2\pi n/\lambda_{\rm g} = 2.6 \times 10^7\,{\rm m}^{-1}$; $\pi/(a/2) = 1.1 \times 10^{10}\,{\rm m}^{-1} \gg q_{\rm p}$.
23.2 Indirect gap semiconductor

(a) $h\nu = E_{\rm g}^{\rm ind} + \hbar\omega_{\rm phonon}$; $0 = 2\pi/a + q_{\rm phonon}$.
(b) $E_{\rm g}^{\rm ind} = h\nu + \hbar\omega_{\rm phonon}$; $2\pi/a = q_{\rm phonon}$.

Problems of Chap. 24

24.1 GaN quantum well

(a) $D^{\rm 2D}$(GaN) = 3 × $D^{\rm 2D}$ (GaAs).
(b) To obtain the same occupation number difference, the nonequilibrium electron density has to be larger by a factor of three. If the Einstein coefficient B_{21} has the same value, the gain is by a factor $\nu_2 = \nu_1$ larger for GaN (ν_2 = frequency of a laser with a GaN-based quantum well and ν_1 = frequency of a laser with a GaAs-based quantum well).

Problems of Chap. 26

26.1 We consider a two-dimensional plane wave,

$$\Psi = \Psi_0 {\rm e}^{{\rm i}[\boldsymbol{kr} - (E/\hbar)t]},$$

where $\boldsymbol{k}$ and $\boldsymbol{r}$ are two-dimensional vectors within the plane. We apply periodic boundary conditions for the x and y direction, $k_x L = m \times 2\pi$ and $k_y L = {\rm n} \times 2\pi$, where m and n are integers and L is the periodicity length (for the directions along x and y). In k space, the area of a ring of radius k and width dk is $2\pi k$ dk. The area containing one k point is $(2\pi/L)^2$. The density of k states is $\bar{D}^{\rm 2D}(k) = k{\rm d}k/(2\pi)L^2$. The density of states in the energy space follows from the relation $\bar{D}^{\rm 2D}(\epsilon){\rm d}\epsilon = 2\bar{D}^{\rm 2D}(k){\rm d}k$, where the factor 2 takes into account that there are two spin directions for an electron. Making use of the dispersion relation $\epsilon = \hbar^2k^2/(2m)$ and of ${\rm d}\epsilon/{\rm d}k = \hbar^2 k/m$ we find $\bar{D}^{\rm 2D} = 2\bar{D}^{\rm 2D}(k)\,{\rm d}k/{\rm d}\epsilon = m/(\pi\hbar^2)L^2$ and $D^{\rm 2D}(\epsilon) = m/(\pi\hbar^2)$ (=density of states per unit of energy and unit of area).

26.2 Subpicosecond quantum well laser

(a) Yes. It is in principle possible to have a gain profile of a halfwidth of about 50 meV. The necessary frequency width is $\Delta\nu_0 = \Delta E_0/h = 12\,\text{THz}$; $t_{\text{pulse}} \sim 1/\Delta\nu_0 \sim 10^{-13}\,\text{s} = 100\,\text{fs}$.
(b) The pulse separation is $T = 2nL/c = 2 \times 10^{-11}$ s. Most likely, the number of photons available in a pulse is not sufficient for the saturation of a semiconductor reflector.
(c) If an external reflector is used, an active Q-switching technique should be applicable and the generation of subpicosecond pulses should be possible.

Problems of Chap. 27

27.1 The periodic boundary condition for a one-dimensional system yields the k values $k = s \times 2\pi/L$. The density of states in k space is $\bar{D}^{1D}(k) = 2 \times L/(2\pi)$ because there are two states ($\pm k$) in an interval $2\pi/L$.
$\epsilon = \hbar^2k^2/2m$; $d\epsilon/dk = (\hbar/m)k = \hbar\sqrt{2\epsilon/m}$; $\bar{D}^{1D}(\epsilon)\, d\epsilon = 2\bar{D}^{1D}(k)\, dk$;
$\bar{D}^{1D}(\epsilon) = (2L/\pi)\, dk/d\epsilon = L/(\pi\hbar)\sqrt{2m/\epsilon}$ and $D^{1D} = (\pi/\hbar)\sqrt{2m/\epsilon}$.
27.2 $f_2 = 0$; $f_1 = 1$; $G - 1 = (n/a_2c)h^2 \text{v} B_{21} n^{1D}_{0,\text{nat}}/\Delta\epsilon_{\text{nat}} = 0.26$.
27.3 $I_{\text{th}} = 0.8 \times Ne/\tau_{\text{sp}} = 0.75$ nA.
27.4 Bipolar laser as two-level laser

(a) $D_r(E)$; $E = E_g + \epsilon$ = pair level energy; $\epsilon = \epsilon_2 + \epsilon_1$; ϵ_2 and ϵ_1 are the energies of the electron and the hole that constitute a radiative pair. $D_r(\epsilon)$ is the 3D, 2D, 1D or 0D density of states, depending on the dimensionality of the semiconductor.
(b) The gain characteristic H_{21} is proportional to $f_p - \bar{f}_p$, where f_p is the probability that the pair level is occupied and $\bar{f}_p$ is the probability that the pair level is empty; $f_p - \bar{f}_p = 2f_p$.
(c) $f_p - f_p > 1/2$, since the absorption coefficient is proportional to $\bar{f}$ and the stimulated emission to f.
The condition must correspond to the condition $f_2 - f_1 = 0$ or $f_2 - f_1 = f_e - (1 - f_h = 0$. It follows that $f_p - \bar{f}_p = f_2 - f_1 + 1/2 = f_e + f_h - 1/2$. f_2 and f_1 are the occupation numbers for the electrons in the conduction band and the valence band and f_e and f_h are, in the electron-hole picture, the occupation numbers for the electrons and holes, respectively (see Sect. 21.10).

27.5 Laser operated with a gain medium with a naturally broadened line

(a) To the knowledge of the author: no.

An electron-hole pair can be considered as an occupied single electron pair level (=occupied upper laser level). The lower laser level is the vacuum level. The lifetime of the vacuum level is infinitely large (or large compared to the spontaneous lifetime of the pair). Thus, we have no lifetime broadening of the

lower laser level, supposed that the population of the levels of the electron and the hole, which constitute a radiative electron-hole pair, occurs sufficiently fast.
(b) $\sigma_{21} = (\lambda/n)^2/2\pi = 9 \times 10^{-15}\,\mathrm{m}^2$ for $n = 3.6$.

Problems of Chap. 31

31.3

(a) $A_\infty = 0.18\,\mathrm{V}$.
(b) $\gamma_0 = a/C = 10 \times 10^9\,\mathrm{s}^{-1}$; $\kappa = G/C = 0.77 \times 10^9\,\mathrm{s}^{-1}$; $\gamma_0 - \kappa = 2.3 \times 10^8\,\mathrm{s}^{-1} \ll \omega_0 = 6 \times 10^{10}\,\mathrm{s}^{-1}$.

31.4 Van der Pol equation

(a) The equation follows from (31.47) by introducing $\epsilon = (\gamma_0 - k)/\omega_0$ and $y = U\sqrt{(3\omega_0 b/C)(\gamma_0 - k)^{-1}}$.
(b) The ansatz $y = A \cos \tau$ leads to $A = 2$.

References

1. A.E. Siegman, *Lasers* (University Science, 1986)
2. W.T. Silfvast, *Laser Fundamentals*, 2nd edn. (Cambridge University Press, 2003)
3. O. Svelto, *Principles of Lasers*, 5th edn. (Plenum Press, 2010)
4. S. Hooker, C. Webb, *Laser Physics* (Oxford University Press, 2010)
5. P.W. Milonni, J.H. Eberly, *Lasers*, 1st edn. (Wiley, 1998)
6. A. Yariv, *Optical Electronics*, 3rd edn. (Holt, Rinehort & Winston, 1985)
7. W. Koechner, *Solid-State Laser Engineering*, 6th edn. (Springerm, 2005)
8. J.T. Verdeyen, *Laser Electronics*, 3rd edn. (Prentice Hall, 1995)
9. C.C. Davis, *Lasers and Electro-Optics* (Cambridge University Press, 1996)
10. W. Demtröder, *Laser spectroscopy*, vol. 1, *Basic Principles*, 5th edn. (Springer, 2014); vol. 2, *Experimental Techniques* (Springer, 2015)
11. F. Kneubühl, M. Sigrist, *Laser*, 6th edn. (Teubner Studienbücher der Physik, 2005)
12. P. Meystre, M. Sargent III, *Elements of Quantum Optics*, 4th edn. (Springer, 2007)
13. M. Fox, *Quantum Optics* (Oxford University Press, 2006)
14. M.R. Spiegel, S. Lipschutz, J. Liu, *Mathematical Handbook of Formulas and Tables*, 3rd edn. (McGraw Hill, 2008)
15. I. Gradstein, I. Ryshik, *Tables of Series, Products and Integrals* (Harri Deutsch, 1981)
16. M. Abramowitz, I.A. Stegun, *Handbook of Mathematical Functions with Formulas, Graphs, and Mathematical Tables* (U.S. Govt. Printing Office, Washington, DC, 1984)
17. R.N. Bracewell, *The Fourier Transform and its Applications*, 2nd edn. (McGraw Hill, 1986)
18. J. Mathews, R.L. Walker, *Mathematical Methods of Physics*, 2nd edn. (Benjamin, 1970)
19. I.N. Bronstein, K.A. Semendjajew, *Taschenbuch der Mathematik*, 6th edn. (Harry Deutsch, 1961)
20. K. Jänich, *Mathematik 1* (Springer, 2001)
21. B.A. Lengyel, Evolution of masers and lasers 1966. A. J. Phys. **34**, 903 (1966)
22. C.H. Townes, *How the Laser Happened* (Oxford University Press, 1999)
23. N. Bloembergen, Physical review records the birth of the laser area. Phys. Today 28 (1993)
24. J. Hecht, *The Laser Guidebook* (McGraw Hill, 1986)
25. J. Hecht, *Laser Pioneers* (Academic, 1992)
26. M. Meschede, *Optics, Light and Lasers* (Wiley, 2004)
27. E. Hecht, *Optics*, 4th edn. (Addison Wesley, 2002)
28. M. Born, W. Wolf, *Principles of Optics*, 6th edn. (Pergamon Press, 1987)
29. M.V. Klein, *Optics* (Wiley, 1970)
30. H. Paul, *Photonen* (Teubner, 1995)

K.F. Renk, *Basics of Laser Physics*, Graduate Texts in Physics,
DOI 10.1007/978-3-319-50651-7

31. M. Fox, *Optical Properties of Solids* (Oxford University Press, 2007)
32. J.P. Gordon, H.J. Zeiger, C.H. Townes, The maser-new type of microwave amplification, frequency standard, and spectrometer. Phys. Rev. **99**, 1264 (1954)
33. P.F. Moulton, Spectroscopic and laser characteristics of Ti:Al_2O_3. J. Opt. Soc. Am. B **3**, 125 (1986)
34. H.S. Carslaw, J.C. Jaeger, *Conduction of Heat in Solids*, 2nd edn. (Clarendon, 1959)
35. C.J. Foot, *Atomic Physics* (Oxford University Press, 2008)
36. R. Loudon, *The Quantum Theory of Light*, 2nd edn. (Clarendon Press, 1983)
37. W. Brunner, W. Radloff, K. Junge, *Quantenelektronik* (VEB Deutscher Verlag der Wissenschaften, 1975)
38. A. Einstein, Zur Quanttheorie der Strahlung, Mitt. Phys. Ges. Zürich **16**, 47 (1916) and Physikalische Zeitschrift **18**, 121 (1917); English translations in B.L. van Werden (Ed.), Sources of Quantum Mechanics (North-Holland 1967) and in D. ter Haar (Ed.), The Old Quantum Theory (Pergamon, 1967)
39. A.L. Schawlow, C.H. Townes, Infrared and optical masers. Phys. Rev. **112**, 1940 (1958)
40. K. Shimoda, *Introduction to Laser Physics*, 2nd edn. (Springer, 1986)
41. H. Haken, *Laser Theory, Handbuch der Physik*, vol. XXV/2c (Springer, 1970)
42. M. Sargent, M.O. Scully, W.E. Lamb, Jr., *Laser Physics* (Addison-Wesley, 1974)
43. S. Strogatz, *Nonlinear Systems and Chaos* (Perseus Publishing, 1994)
44. E. Ott, *Chaos in Dynamical Systems* (Cambridge University Press, 2002)
45. J. Ohtsubo, *Semiconductor Lasers; Stability, Instability and Chaos*, 2nd edn. (Springer, 2008)
46. H.A. Kramers, La diffusion de la lumi'ere, Atti Cong. Inter. Fisica (Transactions of Volta Centenary Congress), Como, vol. 2, p. 545 (1927)
47. R. de L. Kronig, On the theory of the dispersion of X-rays. J. Opt. Soc. Am. **12**, 547 (1926)
48. C. Füchtbauer, Die Absorption in Spektrallinien im Lichte der Quantentheorie. Physik. Zeitschr. **21**, 322 (1920)
49. M. Czerny, Messungen im Rotationsspektrum des HCl im langwelligen Ultrarot. Zeitschr. Physik. **34**, 227 (1925)
50. R. Tolman, Duration of molecules in upper quantum states. Phys. Rev. **23**, 693 (1924)
51. H.A. Kramers, The quantum theory of dispersion. Nature **113**, 673 (1924)
52. R. Ladenburg, Untersuchungen über die anomale dispersion angeregter Gase. Z. Physik. **48**, 15 (1928)
53. H. Kopfermann, R. Ladenburg, Anomale dispersion in angeregtem Neon. Z. Physik. **48**, 26 (1928)
54. H. Kopfermann, R. Ladenburg, Experimental proof of negative dispersion. Nature **122**, 438 (1928)
55. R. Ladenburg, Negative dispersion in angeregtem Neon. Z. Physik. **65**, 167 (1930)
56. R. Ladenburg, Dispersion in electrically excited gases. Rev. Mod. Phys. **5**, 243 (1933)
57. L. Hoddeson, E. Braun, J. Teichmann, S. Weart, *Out of the Crystal Maze* (Oxford University Press, 1992)
58. L. Allen, J. H. Eberly, *Optical Resonance and Two-Level Atoms* (Dover Publications, 1987)
59. J.D. Jackson, *Classical Electrodynamics* (Wiley, 1962)
60. S. Ramo, J.R. Whinnery, T. Van Duzer, *Fields and Waves in Communication Electronics*, 3rd edn. (Wiley, 1993)
61. D.M. Pozar, *Microwave Engineering* (Addison Wesley Publishing Company, 1990)
62. A.D. Olver, *Microwave and Optical Transmission* (Wiley, 1992)
63. P.F. Goldsmith, *Quasioptical Systems* (IEEE PRESS, 1998)
64. W.J. Smith, *Modern Optical Engineering*, 2nd edn. (McGraw-Hill, 1990)
65. A. Vanderlugt, *Optical Signal Processing* (Wiley, 1992)
66. W.S. Chang, *Principles of Lasers and Optics* (Cambridge University Press, 2005)
67. A.G. Fox, T. Li, Resonant modes in a laser interferometer. Bell Syst. Tech. J. **40**, 453 (1961)
68. G.D. Boyd, J.P. Gordon, Confocal multimode resonator for millimeter through optical wavelengths. Bell Syst. Tech. J. **40**, 489 (1961)
69. H. Kogelnik, T. Li, Laser beams and resonators. Proc. IEEE **5**, 1550 (1966)

70. L.G. Gouy, Sur une propriet'e nouvelle des ondes lumineuses. C. R. Acad. Sci. Paris **110**, 1251 (1890)
71. L.G. Gouy, Sur la propagation anomale des ondes. C. R. Acad. Sci. Paris **111**, 33 (1890)
72. L.G. Gouy, Sur la propagation anomale des ondes, Annales de Chimie et de Physique. Ser. **6**, 145 (1891)
73. F. Reiche, Über die anomale Fortpflanzung von Kugelwellen beim Durchgang durch Brennpunkte. Ann. der Physik. **29**, 65 (1909)
74. C.R. Carpenter, Am. J. Phys. **27**, 98 (1958)
75. F. Lindner et al., Gouy phase shift for a few-cycle laser pulses. Phys. Rev. Lett. **92**, 113001 (2004)
76. A.B. Ruffin et al., Direct observation of the Gouy phase shift with single-cycle terahertz pulses. Phys. Rev. Lett. **83**, 3410 (1999)
77. H.H. Telle, A.G. Ureña, R. J. Donovan, *Laser Chemistry* (Wiley, 2007)
78. D. Bäuerle, *Laser Processing and Chemistry*, 3rd edn. (Springer, 2000)
79. M.W. Berns, K.O. Greulich (Eds.), *Laser Manipulation of Cells and Tissues* (Elsevier , 2007)
80. A. Katzir, *Lasers and Optical Fibers in Medicine* (Academic Press, 1993)
81. A.N. Chester, S. Martucelli, A.M. Scheggi, *Laser Systems for Photobiology and Photomedicine* (Plenum Press, 1991)
82. H. Niemz, *Laser-Tissue Interactions* (Springer, 1996)
83. J. Eichler, G. Ackermann, *Holographie* (Springer, 1993)
84. M. Françon, *Holographie* (Springer, 1972)
85. H.A. Bachor, *A Guide to Experiments in Quantum Optics* (Wiley, 1988)
86. J.C. Diels, W. Rudolph, *Ultrashort Laser Pulse Phenomena* (Elsevier, 2006)
87. M.E. Fermann, A. Galvanaushas, G. Sucha (Eds.), *Ultrafast Lasers* (Marcel Dekker, 2003)
88. A.M. Weiner, *Ultrafast Optics* (Wiley, 2009)
89. J. Ye, S.T. Cundiff, *Femtosecond Optical Frequency Comb Technology* (Springer, 2005)
90. C. RulliJe re (Ed.), *Femtosecond Laser Pulses* (Springer, 1988)
91. R. Ell et al., Generation of 5-fs pulses and octave-spanning spectra directly from a Ti:sapphire laser. Opt. Lett. **26**, 373 (2001)
92. D. Mittleman (Ed.), *Sensing with Terahertz Radiation* (Springer, 2003)
93. R.E.Miles, P. Harrison, D. Lippens (Eds.), *Terahertz Sources and Systems* (Kluver Academic Publishers, 2001)
94. K. Sakai (Ed.), *Terahertz Optoelectronics* (Springer, 2005)
95. C. Kübler, R. Huber, A. Leitenstorfer, Ultrabroad terahertz pulses: generation and fieldresolved detection. Semicond. Sci. Technol. **20**, 128 (2005)
96. A. Sell, A. Leitenstorfer, R. Huber, Phase-locked generation and field-resolved detection of widely tunable terahertz pulses with amplitudes exceeding 100 MV/cm. Opt. Lett. **33**, 2767 (2008)
97. D.H. Auston, Subpicosecond electro-optic shock waves. Appl. Phys. Lett. **43**, 713 (1983)
98. D.H. Auston, K.P. Cheung, P.R. Smith, Picosecond photoconducting Hertzian dipoles. Appl. Phys. Lett. **45**, 284 (1984)
99. D.H. Auston, K.P. Cheung, Coherent time-domain far-infrared spectroscopy. J. Opt. Soc. Am. B **2**, 606 (1985)
100. B.B. Hu, M.C. Nuss, Imaging with terahertz waves. Opt. Lett. **20**, 1716 (1985)
101. T. Brabec, F. Krausz, Intense few-cycle laser fields: frontiers of nonlinear optics. Rev. Mod. Phys. **72**, 545 (2000)
102. A. Baltuska et al., Attosecond control of electronic processes by intense light fields. Nature **421**, 611 (2003)
103. H.C. Kapteyn, M.M. Murnane, I.P. Christov, Extreme nonlinear optics: coherent X-rays from lasers. Phys. Today 39 (2005)
104. A.V. Oppenheim, A.S. Wilsky, *Signals and Systems* (Prentice Hall, 1983)
105. R.A. Alfano (Ed.), *The Supercontinuum Laser Source*, 2nd edn. (Springer, 2006)
106. T.P. Softley, *Atomic Spectra* (Oxford University Press, 1994)
107. B.H. Bransden, C.J. Joachain, *Physics of Atoms and Molecules*, 2nd edn. (Prentice Hall, 2003)

108. L.I. Schiff, *Quantum Mechanics* (McGraw-Hill, 1955)
109. C. Cohen-Tannoudji, B. Diu, F. Laloe, *Quantum Mechanics*, 1st edn. (Wiley, 1977)
110. A. Corney, *Atomic and Laser Spectroscopy* (Clarendon Press, 1977)
111. H.G. Kuhn, *Atomic Spectra*, 2nd edn. (Longmans, 1969)
112. G. Herzberg, *Atomic Spectra and Atomic Structure* (Dover, 1944)
113. G. Herzberg, *Molecular Spectra and Molecular Structure* (D. van Nostrand, 1967)
114. H. Haken, H. Wolf, *The Physics of Atoms and Quanta*, 6th edn. (Springer, 2000)
115. M. Karplus, R.N. Porter, *Atoms and Molecules* (Benjamin, 1970)
116. Javan et al., Population inversion and continuous optical maser oscillations in a gas discharge containing a He-Ne mixture. Phys. Rev. Lett. **6**, 106 (1961)
117. B.A. Lengyel, *Introduction to Laser Physics* (Wiley, 1966)
118. W.L. Faust, R.A. McFarlane, Line strengths for noble-gas maser transitions; calculations of gain/inversion at various wavelengths. J. Appl. Phys. **35**, 2010 (1964)
119. S.D. Ganichev, W. Prettl, *Intense Terahertz Excitation of Semiconductors* (Oxford University Press, 2006)
120. T.H. Maiman, Optical maser action in Ruby. Nature **187**, 493 (1960)
121. S. Hüfner, *Optical Spectra of Transparent Rare Earth Compounds* (Academic Press, 1978)
122. G.H. Dieke, H.M. Crosswhite, The spectra of the doubly and triply ionized rare earths. Appl. Opt. **2**, 675 (1963)
123. E.H. Carlson, G.H. Dieke, The state of the Nd^{3+} ion as derived from the absorption and fluorescence spectra of NdCl3 and their zeeman effects. J. Chem. Phys. **34**, 1602 (1961)
124. T. Kushida, H.M. Marcos, J.E. Geusic, Laser transition cross section and fluorescence branching ratio for Nd^{3+} in yttrium aluminum garnet. Phys. Rev. **716**, 289 (1968)
125. W.M. Yen, P.M. Selzer (Eds.), *Laser Spectroscopy of Solids* (Springer, 1981)
126. W.M. Yen, *Laser Spectroscopy of Solids II* (Springer, 1989)
127. A. Klein, W. Bäumler, M. Landthaler, P. Babilas, Laser and IPL treatment of port-wine stains: therapy, options, limitations, and practical aspects. Lasers Med. Sci. **26**, 845 (2011)
128. W. Bäumler, H. Ulrich, A. Hartl, M. Landthaler, G. Sharifstein, Optical parameters for the treatment of leg veins using Nd:YAG lasers at 1.064 nm. Dermatol. Surg. Lasers *155*, 363 (2006)
129. T. Maisch et al., The role of singlet oxygen and oxygen concentration in photodynamic inactivation of bacteria. PNAS **104** (2007)
130. F.P. Schäfer (Ed.), *Dye Lasers*, 3rd edn. (Springer, 1990)
131. B.R. Benware et al., Demonstration of a high average power tabletop soft X-ray laser. Phys. Rev. Lett. **81**, 5804 (1998)
132. J.J. Rocca et al., Energy extraction and achievement of the saturation limit in a discharge-pumped table-top soft X-ray amplifier. Phys. Rev. Lett. **77**, 1476 (1996)
133. J. Dunn et al., Gain saturation regime for laser-driven tabletop, transient Ni-like ion X-ray lasers. Phys. Rev. Lett. **84**, 4834 (2000)
134. M.A. Noginov, *Solid-State Random Lasers* (Springer, 2005)
135. L.V. Keldysh, Kinetic theory of impact ionization in semiconductors. Sov. Phys. JETP **21**, 509 (1960)
136. L.V. Keldysh, Concerning the theory of impact ionization in semiconductors. Sov. Phys. JETP **37**, 1135 (1965)
137. B. Henderson, G.F. *Imbusch, Optical Spectroscopy of Inorganic Solids* (Clarendon Press, 1989)
138. B.F. Gächter, J.A. Koningstein, J. Chem. Phys. **60**, 2003 (1974)
139. P. Albers, E. Stark, G. Huber, Continuous-wave laser operation and quantum efficiency of titanium-doped sapphire. J. Opt. Soc. Am. B **134**, 134 (1986)
140. M. J. F. Digonnet, *Rare-Earth-Doped Fiber Lasers and Amplifiers* (Marcel Dekker, 2001)
141. E. Desurvire, *Erbium-Doped Fiber Amplifiers* (Wiley, 2002)
142. A. Mendez, T.F. Morse, *Speciality Optical Fibers Handbook* (Elsevier, 2007)
143. J.P. Dakin, R.G.W. Brown (Eds.), *Handbook of Optoelectronics*, vols. I and II (Taylor and Francis, 2006)

144. G.P. Agrawal, *Nonlinear Fiber Optics* (Academic Press, 1989)
145. K.F. Renk, Role of excited-impurity quasiparticles for amplification of radiation in an erbium-doped glass fiber amplifier. Appl. Phys. Lett. **96**, 131104 (2010)
146. G.H. Dieke, S. Singh, Absorption and fluorescence spectra with magnetic properties of $ErCl_3$. J. Chem. Phys. **35**, 555 (1961)
147. F. Varsanyi, G.H. Dieke, Energy levels of hexagonal $ErCl_3$. J. Chem. Phys. **36**, 2951 (1962)
148. T. Holstein, S.K. Lyo, R. Orbach, Spectral-spatial diffusion in inhomogeneously broadened systems. Phys. Rev. Lett. **36**, 891 (1976)
149. T. Holstein, S.K. Lyo, R. Orbach, Phonon-assisted energy transport in inhomogeneously broadened systems. Phys. Rev. B **15**, 4693 (1977)
150. R.M. Macfarlane, R.M. Shelby, Homogeneous line broadening of optical transitions of ions and molecules in glasses. J. Lumin. **36**, 179 (1987)
151. W.M. Yen, R.T. Brundage, Fluorescence line narrowing in inorganic glasses: linewidth measurements. J. Lumin. **36**, 209 (1987)
152. T. Förster, Zwischenmolekulare Energiewanderung und Fluoreszenz. Ann. Phys. **2**, 55 (1948)
153. E. Desurvire, J.L. Zyskind, J.R. Simpson, Spectral gain hole-burning at 1.53 μm in erbiumdoped fiber amplifiers. IEEE Photon. Technol. Lett. **2**, 246 (1990)
154. J.N. Sandoe, P.H. Sarkies, S. Parke, Variation of the Er^{3+} cross section for stimulated emission with glass composition. J. Phys. D: Appl. Phys. **5**, 1788 (1972)
155. E. Desurvire, J.R. Simpson, P.C. Becker, High-gain erbium-doped traveling-wave fiber amplifier. Opt. Lett. **12**, 888 (1987)
156. Y. Sun, J.L. Zyskind, A. Srivastava, Average inversion level, modeling, and physics of erbium-doped fiber amplifiers. IEEE. J. Sel. Top. Quan. Electron. **3**, 991 (1997)
157. C.R. Giles, E. Desurvire, Modeling erbium-doped fiber amplifiers. IEEE J. Lightwave Technol. **9**, 271 (1991)
158. G.C. Valley, Modeling cladding-pumped Er/Yb fiber amplifiers. Opt. Fiber Technol. **7**, 21 (2001)
159. J. Limpert et al., High-power femtosecond Yb-doped fiber amplifier. Opt. Expr. **10**, 628 (2002)
160. P.F. Wysocki, N. Park, D. DiGiovanni, Erbium-ytterbium codoped fiber amplifier. Opt. Lett. **21**, 1744 (1996)
161. S.D. Jackson, A. Sabella, D.G. Lancaster, Application and development of high-power and highly efficient silica-based fiber lasers operating at 2 μm. IEEE. J. Sel. Top. in Quan. Elect. **34**, 991 (2009)
162. G.D. Goodno et al., Low-phase-noise, single-frequency, single-mode 608 W thulium fiber amplifier. Opt. Lett. **34**, 1204 (2009)
163. S.D. Jackson, E. Bugge, G. Ebert, Directly diode-pumped holmium fiber lasers. Opt. Lett. **32**, 2496 (2007)
164. S.D. Jackson, A. Sabella, D.G. Lancaster, Diode-pumped 1.7-W erbium 3-μ fiber laser. Opt. Lett. **24**, 1133 (1999)
165. X. Zhu, R. Jain, Watt-level Er-doped and Er-Pr-codoped ZBLAN fiber amplifiers at the 2.7-2.8 μm wavelength range. Opt. Lett. **33**, 208 (2008)
166. C.A. Brau, *Free-Electron Lasers* (Academic Press, 1990)
167. C.E. Webb, J.D.C. Jones (Eds.), *Handbook of Laser Technology and Applications* (Jones Publications, 2004) (Chapter B 5.1, free electron lasers and synchrotron light sources)
168. L.R. Elias, W.M. Fairbank, J.M. Madey, H.A. Schwettman, T.I. Smith, Observation of stimulated emission of radiation by relativistic electrons in a spatially periodic transverse magnetic field. Phys. Rev. Lett. **36**, 717 (1976)
169. D.A. Deacon, L.R. Elias, J.M. Madey, G.J. Ramian, H.A. Schwettman, T.I. Smith, First operation of a free-electron laser. Phys. Rev. Lett. **38**, 892 (1977)
170. S.V. Milton, Exponential gain and saturation of a self-amplified spontaneous emission freeelectron laser. Science **292**, 2037 (2001)
171. V. Ayvazyan et al., Generation of GW radiation pulses from a VUV free-electron laser operating in the femtosecond regime. Phys. Rev. Lett. **88**, 104802 (2002)

172. V. Ayvazyan et al., First operation of a free-electron laser generating GW power radiation at 32 nm wavelength. Eur. Phys. J. D **37**, 297 (2006)
173. Z. Huang, K. Kim, Review of X-ray free-electron laser theory. Phys. Rev. Spec. Top. Accel. Beams **10**, 034801 (2007)
174. B.W. McNeil, N.R. Thompson, X-ray free-electron lasers. Nature Photon. **4**, 814 (2010)
175. J.M. Madey, Stimulated emission of Bremsstrahlung in a periodic magnetic field. J. Appl. Phys. **42**, 1906 (1971)
176. F.A. Hopf, P. Meystre, M.O. Scully, Classical theory of a free-electron laser. Opt. Comm. **18**, 413 (1976)
177. C. Kittel, *Introduction to Solid State Physics*, 7th edn. (Wiley, 1996)
178. J. Singleton, *Band Theory and Electronic Properties of Solids* (Oxford University Press, 2001)
179. H. Ibach, H. Lüth, *Solid State Physics*, 3rd edn. (Springer, 2003)
180. J.M. Ziman, *Principle of the Theory of Solids*, 2nd edn. (Cambridge University Press)
181. L.D. Landau, E.M. Lifshitz, L.P. *Pitaevski, Statistical Physics: Theory of the Condensed State*, vol. 2; *Course of theoretical physics*, vol. 9 (Pergamon Press, 1980)
182. U. Rössler, *Solid State Theory: An Introduction*, 2nd edn. (Springer, 2009)
183. H. Haug, S.W. Koch, *Quantum Theory of Optical and Electronic Properties of Semiconductors*, 4th edn. (World Scientific, 2004)
184. C.F. Klingshirn, *Semiconductor Optics* (Springer, 1997)
185. K.F. Brennan, *The Physics of Semiconductors* (Cambridge University Press, 1999)
186. K. Seeger, *Semiconductor Physics: An Introduction*, 4th edn. (Springer, 1999)
187. H.C. Casey Jr., M.B. Panish, *Heterostructure Lasers (Fundamental Principles)* (Academic Press, Part A, 1987)
188. S. Zory, *Quantum Well Lasers* (Academic Press, 1993)
189. L.A. Coldren, S.W. Corzine, *Diode Lasers and Photonic Integrated Circuits* (Wiley, 1995)
190. J. Agrawal, *Semiconductor Lasers* (AIP Press, 1995)
191. T. Numai, *Fundamental of Semiconductor Lasers* (Springer, 2004)
192. W.W. Chow, S.W. Koch, *Semiconductor-Laser Fundamentals* (Springer, 1999)
193. C.Y. Chang, F. Kai, *GaAs High-Speed Devices* (Wiley, 1994)
194. J. Cheng, N.K. Dutta, *Vertical-Cavity Surface-Emitting Lasers: Technology and Applications* (Gordon and Breach Science Publisher, 2000)
195. S.F. Yu, *Analysis and Design of Vertical Cavity Surface Emitting Lasers* (Wiley, 2003)
196. S. Nakamura, G. Fasol, *The Blue Laser Diode* (Springer, 1997)
197. H.K. Choi, *Long-Wavelength Infrared Semiconductor Lasers* (Wiley, 2004)
198. S.L. Chuang, *Physics of Optoelectronic Devices* (Wiley, 1995)
199. N.K. Dutta, M. Wang, *Semiconductor Optical Amplifiers* (World Scientific, 2006)
200. W.B. Leigh, *Devices for Optoelectronics* (Marcel Dekker, 1996)
201. M.G. Bernard, G. Duraffourg, Laser conditions in semiconductors. Phys. Stat. Sol. **1**, 699 (1961)
202. H. Welker, Über neue halbleitende Verbindungen. Z. Naturforschung **7a**, 744 (1952)
203. S. Nakamura et al., Room temperature continuous-wave operation of InGaN multi-quantum-well- structure laser diodes with a long lifetime. Appl. Phys. Lett. **70**, 868 (1997)
204. N.M. Johnson, A.V. Nurmikko, S.P. DenBaars, Blue diode lasers. Phys. Today 31 (2000)
205. M.A. Haase, J. Qiu, J.M. DePuydt, H. Cheng, Appl. Phys. Lett. **59**, 1272 (1991)
206. G.F. Neumark, R.M. Park, M. DePuydt, Blue-green diode lasers. Phys. Today 26 (1994)
207. P. Yeh, *Optical Waves in Layered Media* (Wiley, 1988)
208. A. Yariv, P. Yeh, *Photonics*, 6th edn. (Oxford University Press, 2007)
209. E. Rosencher, B. Vinter, *Optoelectronics* (Cambridge University Press, 2002)
210. B.E.A. Saleh, M.C. Teich, *Photonics*, 2nd edn. (Wiley, 2007)
211. C. Yeh, F. Shimabukuro, *The Essence of dielectric Waveguides* (Springer, 2008)
212. J.D. Joannopoulos, R.D. Meade, J.N. Winn, *Photonic Crystals* (Princeton University Press, 1995)
213. H. Benestry, V. Berger, J.-M. G'erard, D.Maystre, A. Tchelnokov, *Photonic Crystals* (Springer, 2005)

214. K. Vahala, *Optical Microcavities* (World Scientific, 2004)
215. J. Carrol, J. Whiteaway, D. Plump, *Distributed Feedback Semiconductor Lasers* (SPIE Optical Engineering Press, 1998)
216. L. Novotny, B. Hecht, *Nano-Optics* (Cambridge University Press, 2006)
217. J.W.S. Rayleigh, On the remarkable phenomenon of crystalline reflection described by Prof. Stokes. Phil. Mag. **26**, 256 (1888)
218. E. Yablonovitch, Inhibited spontaneous emission in solid state physics and electronics. Phys. Rev. Lett. **58**, 2059 (1987)
219. W. Culshaw, High resolution millimeter wave Fabry–Perot interferometer. IRE Trans. Microw. Theory Tech. **MTT-8**, 182 (1960)
220. K.F. Renk, L. Genzel, Interference filters and Fabry-Perot interferometers for the far infrared. Appl. Opt. **1**, 642 (1962)
221. V.M. Ustinov, A.W. Zhukov, A. Yu, *Egorov*, N.A. Maleev, Quantum Dot Lasers (Oxford University Press, 2003)
222. P. Harrison, *Quantum Wells, Wires and Dots*, 3rd edn. (Wiley, 2009)
223. J. Faist, F. Capasso, D.L. Sivco, C. Sirtori, A.L. Hutchinson, A.Y. Cho, Quantum cascade laser Science **264**, 553 (1998)
224. R. Köhler, A. Tredicucci, F. Beltram et al., Terahertz heterostructure laser. Nature **417**, 156 (2002)
225. B.S. Williams, Terahertz quantum-cascade lasers. Nature Photon. **1**, 517 (2007)
226. R. Shankar, *Principles of Quantum Mechanics*, 2nd edn. (Springer, 2008)
227. E. Merzbacher, *Quantum Mechanics*, 2nd edn. (Wiley, 1970)
228. L. Liboff, *Introductory Quantum Mechanics* (Holden-Day, 1980)
229. F. S. Levin, *Quantum Theory* (Cambridge University Press, 2002)
230. K. Kopitzki, P. Herzog, *Einführung in die Festkörperphysik*, 4th edn. (Teubner, 2002)
231. R. de L. Kronig, W.G. Penney, Quantum mechanics of electrons in crystal lattices. Proc. Roy. Soc. (London) **A 130**, 499 (1931)
232. G. Bastard, *Wave Mechanics Applied to Semiconductor Heterostructures* (Les Editions de Physique, 1988)
233. G. Bastard, Superlattice band structure in the envelope-function approximation. Phys. Rev. B **24**, 5693 (1981)
234. S. Yngvesson, *Microwave Semiconductor Devices* (Kluwer Academic Publishers, 1991)
235. D.M. Pozar, *Microwave Engineering* (Addison Wesley Publishing Company, 1990)
236. S.M. Sze, *Physics of Semiconductor Devices*, 2nd edn. (Wiley, 1981)
237. S.M. Sze (Ed.), *Modern Semiconductor Device Physics*, 2nd edn. (Wiley, 1998)
238. M. Shur, *GaAs Devices and Circuits* (Plenum, 1987)
239. H. Kroemer, Gunn effect—bulk instabilities, in *Topics in Solid State and Quantum Electronics*, ed. by W.D. Hershberger (Wiley, 1972)
240. I. Bahl, P. Bhartia, *Microwave Solid State Circuit Design* (Wiley, 1988)
241. J.C. Slater, *Microwave Electronics* (Dover Publications, 1950)
242. M.J. Howes, D.V. Morgan (Eds.), *Microwave Devices* (Wiley, 1976)
243. B. van der Pol, Proc. IRE **22**, 1051 (1934)
244. K. Kurokawa, Some basic characteristics of broadband negative resistance oscillator circuits. Bell Syst. Tech. J. **48**, 1937 (1969)
245. K. Kurokawa, *An Introduction to the Theory of Microwave Circuits* (Academic, 1969)
246. K.F. Renk, B.I. Stahl, Operation of a semiconductor superlattice oscillator. Phys. Lett. A **375**, 2644 (2011)
247. T.C. Sollner, E.R. Brown, W.D. Goodhue, H.Q. Le, Microwave and millimeter-wave resonant-tunneling devices, in *Physics of Quantum Electron Devices*, ed. by F. Capasso (Springer, 1990)
248. E.R. Brown, J.R. Söderström, C.D. Parker, L.L. Mahoney, K.M. Molvar, McGill, Oscillations up to 712 GHz in InAs/AlSb resonant-tunneling diodes. Appl. Phys. Lett. **58**, 2291 (1991)
249. E.R. Brown, T.C. Sollner, C.D. Parker, W.D. Goodhue, C.L. Chen, Oscillations up to 420 GHz in GaAs/AlAs resonant-tunneling diodes. Appl. Phys. Lett. **55**, 1777 (1989)
250. V.I. Sugakov, *Lectures in Synergetics* (World Scientific, 1998)

251. F. Bloch, Über die Quantenmechanik der Elektronen in Kristallgittern. Zeitschrift für Physik. **52**, 555 (1928)
252. C. Zener, A theory of the electrical breakdown of solid dielectrics. Proc. Roy. Soc. London Ser. A **145**, 523 (1934)
253. L.V. Keldish, Effect of ultrasonics on the electron spectrum of crystals. Sov. Phys. Solid State **4**, 1658 (1962)
254. L. Esaki, R. Tsu, Superlattice and negative differential conductivity in semiconductors. IBM J. Res. Dev. **14**, 61 (1970)
255. S.A. Ktitorov, G.S. Simin, V. Ya, Sindalovskii, Bragg reflections and the high-frequency conductivity of an electronic solid-state plasma. Sov. Phys. Sol. State **13**, 1872 (1972)
256. A.A. Ignatov, K.F. Renk, E.P. Dodin, Esaki-Tsu superlattice oscillator: Josephson-like dynamics of carriers. Phys. Rev. Lett. **70**, 1996 (1993)
257. A.A. Ignatov, E. Schomburg, J. Grenzer, K.F. Renk, E.P. Dodin, THz-field induced nonlinear transport and dc voltage generation in a semiconductor superlattice due to Bloch oscillations. Z. Phys. B **98**, 187 (1995)
258. H. Kroemer, On the nature of the negative-conductivity resonance in a superlattice Bloch oscillator, cond-mat/0007428 (2000)
259. E. Schomburg, N.V. Demarina, K.F. Renk, Amplification of a terahertz field in a semiconductor superlattice via phase-locked k-space bunches of Bloch oscillating electrons. Phys. Rev. B **67**, 155302 (2003)
260. N.V. Demarina, K.F. Renk, Bloch gain for terahertz radiation in semiconductor superlattices of different miniband widths mediated by acoustic and optical phonons. Phys. Rev. B **67**, 155302 (2003)
261. A. Sibille, J.F. Palmier, H. Wang, F. Mollot, Observation of Esaki-Tsu negative differential velocity in GaAs/AlAs superlattices. Phys. Rev. Lett. **64**, 52 (1990)
262. A.A. Ignatov, E. Schomburg, K.F. Renk, W. Schatz, J.F. Palmier, F. Mollot, Response of a Bloch oscillator to a THz-field. Ann. Physik. **3**, 137 (1994)
263. K. Unterrainer, S.J. Allen et al., Inverse bloch oscillator: strong terahertz-photocurrent resonances at the Bloch frequency. Phys. Rev. Lett. **76**, 2973 (1996)
264. P.G. Savvidis, B. Kolasa, G. Lee, S.J. Allen, Resonant crossover of terahertz loss to gain in a Bloch oscillating InAs/AlSb super-superlattice. Phys. Rev. Lett. **92**, 196802 (2004)
265. A. Lisauskas, E. Mohler, H. Roskos, N. Demarina, Towards superlattice terahertz amplifiers and lasers, in *Terahertz Frequency Detection and Identification of Materials and Objects*, ed. by R.E. Miles et al. (Springer, 2007), p. 31
266. H. Kroemer, Large-amplitude oscillation dynamics and domain suppression in a superlattice Bloch oscillator, cond-mat/0009311 (2000)
267. J. Feldmann, K. Leo et al., Optical investigation of Bloch oscillations in a semiconductor superlattice. Phys. Rev. B **46**, 7252 (1992)
268. W. Waschke, H.G. Roskos, R. Schwegler, K. Leo, H. Kurz, K. Köhler, Coherent submillimeter wave emission from Bloch oscillations in a semiconductor superlattice. Phys. Rev. Lett. **70**, 3319 (1993)
269. K. Leo, *High-Field Transport in Semiconductor Superlattices* (Springer, 2003)
270. A. Wacker, Semiconductor superlattices; a model system for nonlinear transport. Phys. Rep. **357**, 1 (2002)
271. E.E. Mendez, G. Bastard, Wannier-stark ladders and Bloch oscillations in superlattices. Phys. Today 34 (1993)
272. H. Willenberg, G.H. Döhler, J. Faist, Intersubband gain in a Bloch oscillator and quantum cascade laser. Phys. Rev. B **67**, 08531 (2003)
273. J. Stark, Beobachtungen über den Effekt des elektrischen Feldes auf Spektrallinien. Annalen der Physik. **43**, 965 (1914)
274. G.H. Wannier, Wave functions and effective Hamiltonian for Bloch electrons in an electric field. Phys. Rev. **117**, 432 (1960)
275. G.H. Wannier, Dynamics of band electrons in electric and magnetic fields. Rev. Mod. Phys. **34**, 645 (1962)

276. R.G. Hunsperger, *Integrated Optics: Theory and Technology*, 6th edn. (Springer, 2009)
277. R. Menzel, *Photonics*, 2nd edn. (Springer, 2007)
278. F.G. Smith, T.A. King, *Optics and Photonics; An Introduction* (Wiley, 2000)
279. X. Li, *Optoelectronic Devices* (Cambridge University Press, 2009)
280. R.G. Hunsperger (Ed.), *Photonic Devices and Systems* (Marcel Dekker, 1994)
281. T. Schneider, *Nonlinear Optics in Telecommunications* (Springer, 2004)
282. J. Eberspächer, H.-J. Vögel, C. Bettstetter, C. Hartmann, *GSM—Global System for Communications* (Wiley, 2008)
283. E.F. Schubert, *Light-Emitting Diodes* (Cambrigde University Press, 2005)
284. K. Müller, U. Scherf, *Organic Light-Emitting Devices* (Wiley, 2006)
285. M. Reufer et al., Low-threshold polymeric distributed feedback lasers with metallic contacts. Appl. Phys. Lett. **84**, 3262 (2004)
286. D. Schneider et al., Deep blue widely tunable organic solid-state laser based on a spirobifluorene derivative. Appl. Phys. Lett. **84**, 4693 (2004)
287. E. Holzer, A. Penzkofer et al., Corrugated neat thin-film conjugated polymer distributedfeedback lasers. Appl. Phys. B **74**, 333 (2002)
288. P. Andrew, W.L. Barnes, Förster energy transfer in an optical microcavity. Science **290**, 785 (2000)
289. S.V. Frolov et al., Lasing and stimulated emission in π-conjugated polymers. IEEE J. Quant. Electr. **36**, 2 (2000)
290. V. Bulovic et al., Transform-limited. Narrow-linewidth lasing action in organic semiconductor microcavities. Science **279**, 553 (1998)
291. A. Haugeneder et al., Mechanism of gain narrowing in conjugated polymer thin films. Appl. Phys. B **66**, 389 (1998)
292. V.G. Kozlov et al., Study of lasing action based on Förster energy transfer in optically pumped organic semiconductor thin films. J. Appl. Phys. **84**, 4096 (1998)
293. V.G. Kozlov et al., Laser action in organic semiconductor waveguide and doubleheterostructure devices. Nature **389**, 362 (1997)
294. M. Berggren et al., Stimulated emission and lasing in dye-doped organic thin films with Förster transfer. Appl. Phys. Lett. **71**, 2230 (1997)
295. N. Tessler et al., Lasing from conjugated-polymer microcavities. Nature **382**, 695 (1996)
296. F. Hide et al., Semiconducting polymers: a new class of solid-state laser materials. Science **273**, 1833 (1996)
297. R.W. Boyd, *Nonlinear Optics*, 2nd edn. (Academic Press, 2002)
298. N. Bloembergen, *Nonlinear Optics* (Benjamin, 1965)
299. F. Zernike, J.E. Midwinter, *Applied Nonlinear Optics* (Wiley, 1973)
300. E. Hamamura, Y. Kawabe, A. Yamanaka, *Quantum Nonlinear Optics* (Academic Press, 1989)
301. G.P. Agrawal, *Nonlinear Fiber Optics* (Springer, 2007)
302. D.L. Mills, *Nonlinear Optics* (Springer, 1991)
303. T. Schneider, *Nonlinear Optics in Telecommunications* (Springer, 2004)
304. B.E.A. Saleh, M.C. Teich, *Photonics*, 2nd edn. (Wiley, 2007)
305. P.P. Banerjee, *Nonlinear Optics* (Marcel Dekker, 2004)
306. Y. Guo, C.K. Kao, E.H. Li, K.S. Chiang, *Nonlinear Photonics* (Springer, 2002)
307. E. Hanamura, Y. Kawave, A. Yomanaka, *Quantum Nonlinear Optics* (Springer, 2007)
308. M. Eichhorn, *Laser Physics* (Springer, 2014)
309. W. Nagournay, *Quantum Electronics for Atomic Physics and Telecommunication*, 2nd edn. (Oxford University Press, 2014)
310. G. Brooker, *Modern Classical Optics* (Oxford University Press, 2008)
311. B.P. Abbott et al., Observation of gravitational waves from a binary black hole merger. Phys. Rev. Lett. **116**, 061102 (2016)
312. T.C. Marshall, *Free-Electron Lasers* (Macmillan Publishing Company, 1985)
313. C. Pellegrini, A. Marinelli, S. Reiche, The physics of free-electron lasers. Rev. Mod. Phys. **88**, 015006 (2016)

314. C. Bostedt, S. Boutet, D.M. Fritz et al., Linac coherent light source: the first five years. Rev. Mod. Phys. **88**, 015007 (2016)
315. L. Young, E.P. Kanter, B. Krässig et al., Nature **476**, 09177 (2010)
316. G. Doumy, C. Roedig, S.-K. Son et al., Nonlinear atomic response to intense ultrashort X-rays. Phys. Rev. Lett. **106**, 083002 (2011)
317. E. Allaria et al., Highly coherent and stable pulses from the FERMI seeded free-electron laser in the extreme ultraviolet. Nature Photon. **6**, 699 (2012)
318. T. Sato, A. Iwasaki, S. Owada et al., Full-coherent free-electron laser seeded by 13th and 15th order harmonics of near-infrared femtosecond laser pulses. J. Phys. B: Mol. Opt. Phys. **46**, 164006 (2013)
319. S. Ackermann, A. Azima, S. Bajt et al., Generation of 19- and 38-nm radiation at a free-electron laser directly seeded at 38 nm. Phys. Rev. Lett. **106**, 114801 (2013)
320. T. Maltzezopoulos, M. Mittenzwey, A. Azima et al., A high-harmonic generation source for seeding a free-electron laser at 38 nm. Appl. Phys. B **115**, 45 (2014)
321. D.G. Gauthier, P.R. Ribic et al., Generation of phase-locked pulses from a seeded free-electron laser. Phys. Rev. Lett. **116**, 024801 (2016)
322. E.A. Schneidmiller, M.V. Yurkov, Physical review special topics—accelerators and beams. **15**, 080702 (2012)
323. G. Penn, Simple method to suppress the fundamental in a harmonic free-electron laser. Phys. Rev. Spec. Top. Accel. Beams **18**, 060703 (2015)
324. G.R.M. Robb, R. Bonifazio, Coherent and spontaneous emission in the quantum free-electron laser. Phys. Plasmas **19**, 073101 (2012)
325. P. Kling, E. Giese, R. Endrich et al., What defines the quantum regime of the free-electron laser? New J. Phys. **17**, 123019 (2015)
326. H.P. Freund, M. Shinn, S.V. Benson, Simulation of high-average power free-electron laser oscillator. Phys. Rev. Spec. Top. Accel. Beams **10**, 030702 (2007)
327. W.B. Colson, Classical free-electron laser theory, in *Laser Handbook*, vol. 6, ed. by W.B. Colson, C. Pellegrini, A. Renieri (Elsevier Science Publishers, 1990), p. 115
328. J.B. Murphy, C. Pellegrini, Introduction to the physics of the free-electron laser, in *Laser Handbook*, vol. 6, ed. by W.B. Colson, C. Pellegrini, A. Renieri (Elsevier Science Publishers, 1990), p. 9
329. W.B. Colson, S.K. Ride, The free-electron laser: Maxwell's equations driven by single-particle currents. Phys. Quan. Electron. **7**, 377 (1980)
330. N.M. Kroll, P.L. Morton, M.N. Rosenbluth, Free-electron lasers with variable parameter wigglers, IEEE J. Quan. Electron. **QE-17**, 1436 (1981)
331. K. Tiedke, A. Azima, N. von Bargen et al., The soft X-ray free-electron laser FLASH at DESY: beamlines, diagnostics, and end-stations. New J. Phys. **11**, 023029 (2009)
332. W. Steiner, *Transoral Laser Microsurgery for Cancer in the Upper Aerodigestive Tract* (Straub Druck+Medien, Schramberg, Germany, 2014)
333. M. Holthaus, Collapse of minibands in far infrared irradiated superlattices. Phys. Rev. Lett. **69**, 351 (1992)

Index

K.F. Renk, *Basics of Laser Physics*, Graduate Texts in Physics,
DOI 10.1007/978-3-319-50651-7

O

P

U

V

W

X

Y

Z

Printed in the United States
By Bookmasters

Printed in the United States
By Bookmasters